T0189796

Lecture Notes in Artificial Intelligence   10245

Subseries of Lecture Notes in Computer Science

LNAI Series Editors

Randy Goebel
University of Alberta, Edmonton, Canada
Yuzuru Tanaka
Hokkaido University, Sapporo, Japan
Wolfgang Wahlster
DFKI and Saarland University, Saarbrücken, Germany

LNAI Founding Series Editor

Joerg Siekmann
DFKI and Saarland University, Saarbrücken, Germany

More information about this series at http://www.springer.com/series/1244

Leszek Rutkowski · Marcin Korytkowski
Rafał Scherer · Ryszard Tadeusiewicz
Lotfi A. Zadeh · Jacek M. Zurada (Eds.)

Artificial Intelligence and Soft Computing

16th International Conference, ICAISC 2017
Zakopane, Poland, June 11–15, 2017
Proceedings, Part I

 Springer

Editors

Leszek Rutkowski
Częstochowa University of Technology
Częstochowa
Poland

Marcin Korytkowski
Częstochowa University of Technology
Częstochowa
Poland

Rafał Scherer
Częstochowa University of Technology
Częstochowa
Poland

Ryszard Tadeusiewicz
AGH University of Science and Technology
Kraków
Poland

Lotfi A. Zadeh
University of California
Berkeley, CA
USA

Jacek M. Zurada
University of Louisville
Louisville, KY
USA

ISSN 0302-9743   ISSN 1611-3349   (electronic)
Lecture Notes in Artificial Intelligence
ISBN 978-3-319-59062-2   ISBN 978-3-319-59063-9   (eBook)
DOI 10.1007/978-3-319-59063-9

Library of Congress Control Number: 2017941502

LNCS Sublibrary: SL7 – Artificial Intelligence

Printed on acid-free paper

This Springer imprint is published by Springer Nature
The registered company is Springer International Publishing AG
The registered company address is: Gewerbestrasse 11, 6330 Cham, Switzerland

Preface

This volume constitutes the proceedings of 16th International Conference on Artificial Intelligence and Soft Computing ICAISC 2017, held in Zakopane, Poland, during June 11–15, 2017. The conference was organized by the Polish Neural Network Society in cooperation with the University of Social Sciences in Łódź, the Institute of Computational Intelligence at the Częstochowa University of Technology, and the IEEE Computational Intelligence Society, Poland Chapter. Previous conferences took place in Kule (1994), Szczyrk (1996), Kule (1997), and Zakopane (1999, 2000, 2002, 2004, 2006, 2008, 2010, 2012, 2013, 2014, 2015, and 2016) and attracted a large number of papers and internationally recognized speakers: Lotfi A. Zadeh, Hojjat Adeli, Rafal Angryk, Igor Aizenberg, Shun-ichi Amari, Daniel Amit, Piero P. Bonissone, Jim Bezdek, Zdzisław Bubnicki, Andrzej Cichocki, Swagatam Das, Ewa Dudek-Dyduch, Włodzisław Duch, Pablo A. Estévez, Erol Gelenbe, Jerzy Grzymala-Busse, Martin Hagan, Yoichi Hayashi, Akira Hirose, Kaoru Hirota, Adrian Horzyk, Eyke Hüllermeier, Hisao Ishibuchi, Er Meng Joo, Janusz Kacprzyk, Jim Keller, Laszlo T. Koczy, Tomasz Kopacz, Zdzislaw Kowalczuk, Adam Krzyzak, James Tin-Yau Kwok, Soo-Young Lee, Derong Liu, Robert Marks, Evangelia Micheli-Tzanakou, Kaisa Miettinen, Krystian Mikołajczyk, Henning Müller, Ngoc Thanh Nguyen, Andrzej Obuchowicz, Erkki Oja, Witold Pedrycz, Marios M. Polycarpou, José C. Príncipe, Jagath C. Rajapakse, Šarunas Raudys, Enrique Ruspini, Jörg Siekmann, Roman Słowiński, Igor Spiridonov, Boris Stilman, Ponnuthurai Nagaratnam Suganthan, Ryszard Tadeusiewicz, Ah-Hwee Tan, Shiro Usui, Thomas Villmann, Fei-Yue Wang, Jun Wang, Bogdan M. Wilamowski, Ronald Y. Yager, Xin Yao, Syozo Yasui, Gary Yen, and Jacek Zurada. The aim of this conference is to build a bridge between traditional artificial intelligence techniques and so-called soft computing techniques. It was pointed out by Lotfi A. Zadeh that "soft computing (SC) is a coalition of methodologies which are oriented toward the conception and design of information/intelligent systems. The principal members of the coalition are: fuzzy logic (FL), neurocomputing (NC), evolutionary computing (EC), probabilistic computing (PC), chaotic computing (CC), and machine learning (ML). The constituent methodologies of SC are, for the most part, complementary and synergistic rather than competitive." These proceedings present both traditional artificial intelligence methods and soft computing techniques. Our goal is to bring together scientists representing both areas of research. This volume is divided into five parts:

- Neural Networks and Their Applications
- Fuzzy Systems and Their Applications
- Evolutionary Algorithms and Their Applications
- Computer Vision, Image, and Speech Analysis
- Bioinformatics, Biometrics, and Medical Applications

The conference attracted a total of 274 submissions from 37 countries and after the review process, 133 papers were accepted for publication.

The ICAISC 2017 scientific program included the following special sessions:

1. "Granular and Human-Centric Approaches in Data Mining and Analytics" – Special session devoted to the 70th anniversary of Prof. Janusz Kacprzyk, organized by:

   - Witold Pedrycz, University of Alberta, Edmonton, Canada and Systems Research Institute, Polish Academy of Sciences, Warsaw, Poland
   - Rudolf Kruse, Otto von Guericke University, Magdeburg, Germany
   - Leszek Rutkowski, Czestochowa University of Technology, Poland
   - Jacek Żurada, University of Louisville, USA
   - Ronald R. Yager, Hagan School of Business, Iona College, New Rochelle, NY, USA
   - Sławomir Zadrożny, Systems Research Institute, Polish Academy of Sciences, Warsaw, Poland

2. "Biology as a Source of Technical Inspirations" — Special session devoted to the 70th anniversary of Prof. Ryszard Tadeusiewicz, organized by:

   - Krzysztof Cios, Virginia Commonwealth University, USA
   - Leszek Rutkowski, Czestochowa University of Technology, Poland
   - Jacek Żurada, University of Louisville, USA
   - Bogdan M. Wilamowski, Auburn University, USA

3. "Advances in Single-Objective Continuous Parameter Optimization with Nature-inspired Algorithms" organized by:

   - Swagatam Das, Indian Statistical Institute, India
   - P.N. Suganthan, Nanyang Technological University, Singapore
   - Janez Brest, University of Maribor, Slovenia
   - Roman Senkerik, Tomas Bata University in Zlin, Czech Republic
   - Rammohan Mallipeddi, Kyungpook National University, Republic of Korea

4. "Stream Data Mining" organized by:

   - Piotr Duda, Czestochowa University of Technology, Poland
   - Maciej Jaworski, Czestochowa University of Technology, Poland

I would like to thank our participants, invited speakers, and reviewers of the papers for their scientific and personal contribution to the conference.

Finally, I thank my co-workers Łukasz Bartczuk, Piotr Dziwiński, Marcin Gabryel, Marcin Korytkowski, Agnieszka Piersiak-Puchała, and the conference secretary Rafał Scherer for their enormous efforts to make the conference a very successful event. Moreover, I would like to acknowledge the work of Marcin Korytkowski, who designed the Internet submission system.

June 2017   Leszek Rutkowski

Organization

ICAISC 2017 was organized by the Polish Neural Network Society in cooperation with the University of Social Sciences in Łódź, the Institute of Computational Intelligence at Częstochowa University of Technology.

ICAISC Chairs

Honorary Chairs

Lotfi Zadeh (USA)
Hojjat Adeli (USA)
Jacek Żurada (USA)

General Chairs

Leszek Rutkowski (Poland)

Co-chairs

Włodzisław Duch (Poland)
Janusz Kacprzyk (Poland)
Józef Korbicz (Poland)
Ryszard Tadeusiewicz (Poland)

ICAISC Program Committee

Rafał Adamczak, Poland
Cesare Alippi, Italy
Shun-ichi Amari, Japan
Rafal A. Angryk, USA
Jarosław Arabas, Poland
Robert Babuska, The Netherlands
Ildar Z. Batyrshin, Russia
James C. Bezdek, Australia
Marco Block-Berlitz, Germany
Leon Bobrowski, Poland
Piero P. Bonissone, USA
Bernadette Bouchon-Meunier, France
Tadeusz Burczynski, Poland
Andrzej Cader, Poland
Juan Luis Castro, Spain
Yen-Wei Chen, Japan

Wojciech Cholewa, Poland
Kazimierz Choroś, Poland
Fahmida N. Chowdhury, USA
Andrzej Cichocki, Japan
Paweł Cichosz, Poland
Krzysztof Cios, USA
Ian Cloete, Germany
Oscar Cordón, Spain
Bernard De Baets, Belgium
Nabil Derbel, Tunisia
Ewa Dudek-Dyduch, Poland
Ludmiła Dymowa, Poland
Andrzej Dzieliński, Poland
David Elizondo, UK
Meng Joo Er, Singapore
Pablo Estevez, Chile

János Fodor, Hungary
David B. Fogel, USA
Roman Galar, Poland
Adam Gaweda, USA
Joydeep Ghosh, USA
Juan Jose Gonzalez de la Rosa, Spain
Marian Bolesław Gorzałczany, Poland
Krzysztof Grąbczewski, Poland
Garrison Greenwood, USA
Jerzy W. Grzymala-Busse, USA
Hani Hagras, UK
Saman Halgamuge, Australia
Rainer Hampel, Germany
Zygmunt Hasiewicz, Poland
Yoichi Hayashi, Japan
Tim Hendtlass, Australia
Francisco Herrera, Spain
Kaoru Hirota, Japan
Adrian Horzyk, Poland
Tingwen Huang, USA
Hisao Ishibuchi, Japan
Mo Jamshidi, USA
Andrzej Janczak, Poland
Norbert Jankowski, Poland
Robert John, UK
Jerzy Józefczyk, Poland
Tadeusz Kaczorek, Poland
Władysław Kamiński, Poland
Nikola Kasabov, New Zealand
Okyay Kaynak, Turkey
Vojislav Kecman, New Zealand
James M. Keller, USA
Etienne Kerre, Belgium
Frank Klawonn, Germany
Jacek Kluska, Poland
Leonid Kompanets, Poland
Przemysław Korohoda, Poland
Jacek Koronacki, Poland
Jan M. Kościelny, Poland
Zdzisław Kowalczuk, Poland
Robert Kozma, USA
László Kóczy, Hungary
Rudolf Kruse, Germany
Boris V. Kryzhanovsky, Russia
Adam Krzyzak, Canada
Juliusz Kulikowski, Poland

Věra Kůrková, Czech Republic
Marek Kurzyński, Poland
Halina Kwaśnicka, Poland
Soo-Young Lee, Korea
George Lendaris, USA
Antoni Ligęza, Poland
Zhi-Qiang Liu, Hong Kong, SAR China
Simon M. Lucas, UK
Jacek Łęski, Poland
Bohdan Macukow, Poland
Kurosh Madani, France
Luis Magdalena, Spain
Witold Malina, Poland
Jacek Mańdziuk, Poland
Antonino Marvuglia, Luxembourg
Andrzej Materka, Poland
Jacek Mazurkiewicz, Poland
Jaroslaw Meller, Poland
Jerry M. Mendel, USA
Radko Mesiar, Slovakia
Zbigniew Michalewicz, Australia
Zbigniew Mikrut, Poland
Sudip Misra, USA
Wojciech Moczulski, Poland
Javier Montero, Spain
Eduard Montseny, Spain
Kazumi Nakamatsu, Japan
Detlef D. Nauck, Germany
Antoine Naud, Poland
Edward Nawarecki, Poland
Ngoc Thanh Nguyen, Poland
Antoni Niederliński, Poland
Robert Nowicki, Poland
Andrzej Obuchowicz, Poland
Marek Ogiela, Poland
Erkki Oja, Finland
Stanisław Osowski, Poland
Nikhil R. Pal, India
Maciej Patan, Poland
Witold Pedrycz, Canada
Leonid Perlovsky, USA
Andrzej Pieczyński, Poland
Andrzej Piegat, Poland
Vincenzo Piuri, Italy
Lech Polkowski, Poland
Marios M. Polycarpou, Cyprus

Danil Prokhorov, USA
Anna Radzikowska, Poland
Ewaryst Rafajłowicz, Poland
Sarunas Raudys, Lithuania
Olga Rebrova, Russia
Vladimir Red'ko, Russia
Raúl Rojas, Germany
Imre J. Rudas, Hungary
Enrique H. Ruspini, USA
Khalid Saeed, Poland
Dominik Sankowski, Poland
Norihide Sano, Japan
Robert Schaefer, Poland
Rudy Setiono, Singapore
Paweł Sewastianow, Poland
Jennie Si, USA
Peter Sincak, Slovakia
Andrzej Skowron, Poland
Ewa Skubalska-Rafajłowicz, Poland
Roman Słowiński, Poland
Tomasz G. Smolinski, USA
Czesław Smutnicki, Poland
Pilar Sobrevilla, Spain
Janusz Starzyk, USA
Jerzy Stefanowski, Poland
Vitomir Štruc, Slovenia
Pawel Strumillo, Poland
Ron Sun, USA
Johan Suykens, Belgium
Piotr Szczepaniak, Poland

Eulalia J. Szmidt, Poland
Przemysław Śliwiński, Poland
Adam Słowik, Poland
Jerzy Świątek, Poland
Hideyuki Takagi, Japan
Yury Tiumentsev, Russia
Vicenç Torra, Spain
Burhan Turksen, Canada
Shiro Usui, Japan
Michael Wagenknecht, Germany
Tomasz Walkowiak, Poland
Deliang Wang, USA
Jun Wang, Hong Kong, SAR China
Lipo Wang, Singapore
Zenon Waszczyszyn, Poland
Paul Werbos, USA
Slawo Wesolkowski, Canada
Sławomir Wiak, Poland
Bernard Widrow, USA
Kay C. Wiese, Canada
Bogdan M. Wilamowski, USA
Donald C. Wunsch, USA
Maciej Wygralak, Poland
Roman Wyrzykowski, Poland
Ronald R. Yager, USA
Xin-She Yang, UK
Gary Yen, USA
John Yen, USA
Sławomir Zadrożny, Poland
Ali M. S. Zalzala, United Arab Emirates

ICAISC Organizing Committee

Rafał Scherer, Secretary
Łukasz Bartczuk, Organizing Committee Member
Piotr Dziwiński, Organizing Committee Member
Marcin Gabryel, Finance Chair
Marcin Korytkowski, Databases and Internet Submissions

Additional Reviewers

R. Adamczak
M. Al-Dhelaan
T. Babczyński
M. Baczyński
M. Blachnik
L. Bobrowski
P. Boguś
G. Boracchi
B. Boskovic
J. Botzheim
J. Brest
T. Burczyński
R. Burduk
Y. Cheng
W. Cholewa
K. Choros
P. Cichosz
C. Coello Coello
B. Cyganek
J. Cytowski
R. Czabański
I. Czarnowski
F. Deravi
N. Derbel
A. Doi
W. Duch
L. Dymowa
A. Dzieliński
S. Ehteram
B. Filipic
I. Fister
D. Fogel
M. Fraś
A. Galuszka
E. Gelenbe
P. Głomb
F. Gomide
Z. Gomółka
M. Gorawski
M. Gorzałczany
D. Grabowski
E. Grabska
K. Grąbczewski

J. Grzymala-Busse
B. Hammer
Y. Hayashi
Z. Hendzel
F. Hermann
H. Hikawa
Z. Hippe
A. Horzyk
E. Hrynkiewicz
M. Hwang
J. Ishikawa
D. Jakóbczak
E. Jamro
A. Janczak
T. Jiralerspong
R. Jong-Hei
W. Kamiński
Y. Kaneda
W. Kazimierski
V. Kecman
E. Kerre
J. Kitazono
F. Klawonn
J. Kluska
F. Kobayashi
L. Koczy
Z. Kokosinski
A. Kołakowska
J. Konopacki
J. Korbicz
M. Kordos
P. Korohoda
J. Koronacki
M. Korytkowski
M. Korzeń
L. Kotulski
Z. Kowalczuk
M. Kraft
R. Kruse
A. Krzyzak
A. Kubiak
E. Kucharska
J. Kulikowski

O. Kurasova
V. Kurkova
M. Kurzyński
H. Kwaśnicka
J. Kwiecień
A. Ligęza
M. Ławryńczuk
J. Łęski
K. Madani
W. Malina
R. Mallipeddi
J. Mańdziuk
U. Markowska-Kaczmar
A. Martin
A. Materka
J. Mazurkiewicz
V. Medvedev
J. Mendel
D. Meyer
J. Michalkiewicz
Z. Mikrut
S. Misina
W. Mitkowski
M. Morzy
O. Mosalov
G. Nalepa
S. Nasuto
F. Neri
M. Nieniewski
R. Nowicki
A. Obuchowicz
S. Osowski
E. Ozcan
M. Pacholczyk
W. Palacz
K. Patan
A. Pieczyński
A. Piegat
V. Piuri
P. Prokopowicz
A. Przybył
R. Ptak
A. Radzikowska

Contents – Part I

Fuzzy Systems and Their Applications

Evolutionary Algorithms and Their Applications

Computer Vision, Image and Speech Analysis

Bioinformatics, Biometrics and Medical Applications

Contents – Part II

Various Problems of Artificial Intelligence

**Special Session: Advances in Single-Objective Continuous Parameter
Optimization with Nature-Inspired Algorithms**

Special Session: Stream Data Mining

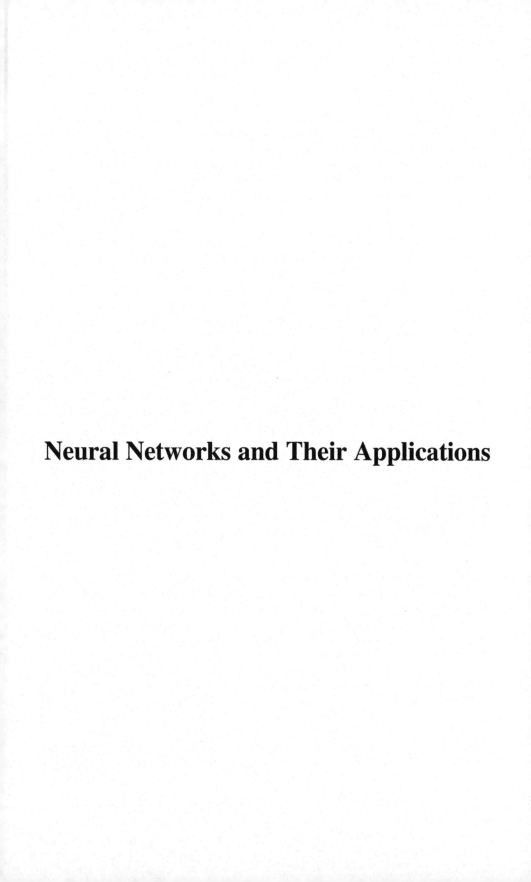

Neural Networks and Their Applications

Neural Networks and Their Applications

Author Profiling with Classification Restricted Boltzmann Machines

Mateusz Antkiewicz, Marcin Kuta[✉], and Jacek Kitowski

Department of Computer Science, Faculty of Computer Science, Electronics and
Telecommunications, AGH University of Science and Technology,
Al. Mickiewicza 30, 30-059 Krakow, Poland
mkuta@agh.edu.pl

Abstract. This paper discusses author profiling of English-language
mails and blogs using Classification Restricted Boltzmann Machines. We
propose an author profiling framework with no need for handcrafted
features and only minor use of text preprocessing and feature engineering. The classifier achieves competitive results when evaluated with the
PAN-AP-13 corpus: 36.59% joint accuracy, 57.83% gender accuracy and
59.17% age accuracy. We also examine the relations between discriminative, generative and hybrid training methods.

Keywords: Author profiling · Restricted Boltzmann Machines ·
Classification Restricted Boltzmann Machines · Discriminative training ·
Generative training · Hybrid training

1 Introduction

The author profiling problem is a classification task where, based on an input
document, the goal is to construct a profile describing the author's gender, age,
native language, personality traits, etc. Author profiling comes into play when
no set of candidate authors is available ($|\mathcal{C}| = 0$). Thus it may be perceived as a
variation of the authorship attribution problem. Author profiling finds applications in marketing (user segmentation), linguistic forensics (profiling of authors
of blackmail or offensive posts), terrorism prevention, suicide attempt detection
and analysis.

The quality of authorship attribution and author profiling systems depends
on a set of features applicable to the examined documents. Currently, such
features have to be handcrafted (which may require years of experience) and
extracted, selected or composed from core features. The number of applied features may run into the millions [6].

This paper presents an author profiling system which does not rely on handcrafted features or feature engineering. This represents the first application of
Restricted Boltzmann Machines to the author profiling problem. In this paper
we consider two dimensions of the author profile, i.e., gender and age. Two possible values for gender and three age groups are considered, yielding $2 \times 3 = 6$
possible profiles.

© Springer International Publishing AG 2017
L. Rutkowski et al. (Eds.): ICAISC 2017, Part I, LNAI 10245, pp. 3–13, 2017.
DOI: 10.1007/978-3-319-59063-9_1

The contribution of this paper is as follows:

- creation of author profiling application which does not require feature engineering or feature selection and performs only a tiny amount of text preprocessing
- comparison of generative and discriminative training in author profiling tasks
- evaluation of the performance of Restricted Boltzmann Machines compared to heavy feature-engineering approaches and classical machine learning classification algorithms applied to the PAN author profiling task.

2   Author Profile Dimensions

The author profiling problem is fully defined only when the applicable profile dimensions are specified. Author profile dimensions include gender, age, native language, personality traits, emotional state and health.

In languages such as Polish the difference between text authored by males and females may be easy to observe due to the syntactic peculiarities of inflective languages. In other languages, like English, it is much harder to infer the author's gender. Typical style features useful for gender discrimination are determiners (a, an, the) and prepositions, which are preferred by men. Women, on the other hand, tend to use pronouns more frequently. Content features characteristic of men include words related to technology, whereas for women words associated with personal life and relationships are more prevalent. Koppel [2] was able to classify a subset of documents from the British National Corpus with regard to gender, achieving the accuracy of approximately 80%.

Another important dimension in author profiling is age. Usually, age groups are defined as intervals with lower and upper age bounds provided. Age profiling exploits the observation that people tend to apply different vocabulary and style as they grow older. To-date attempts to discriminate documents with respect to age groups have yielded promising results. For instance, age classification performed with a corpus composed of blog notes and three age groups (13–17, 23–27, 33–47) resulted in an accuracy of 76% [3]. Considering the English language, more frequent presence of determiners and prepositions suggests that the author is older. Younger people are more likely to omit apostrophes in contractions. The most useful content features for identifying teenagers include words connected with mood and schoolwork. For people in their twenties the corresponding features include social life and professional career, and for people over 30 – family life [3].

3   Restricted Boltzmann Machines

Restricted Boltzmann Machines (RBMs) have been successfully applied to various problems including image classification, speech recognition and user rating of movies.

Restricted Boltzmann Machine [1,9] is an undirected graphical model, consisting of two layers: visible layer of inputs, $x \in \mathbb{R}^{X \times 1}$, and hidden layer, $h \in \mathbb{R}^{H \times 1}$, of variables, which are not observed directly (Fig. 1). There are X neurons in the visible layer and H units in the hidden layer. Matrix $W \in \mathbb{R}^{H \times X}$ is a weight matrix of connections between the visible and hidden layers, while element W_{ij} denotes the weight of the directionless edge connecting nodes h_i and v_j. Vector $b \in \mathbb{R}^{X \times 1}$ is the bias vector of the visible layer x, while vector $c \in \mathbb{R}^{H \times 1}$ is the bias vector of the hidden layer h. RBMs with binary inputs were used in this paper. A discussion of RBMs which omit this assumption can be found in [8].

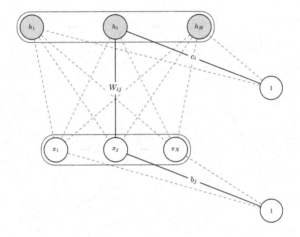

Fig. 1. Restricted Boltzmann Machine

The energy of RBM in configuration (x, h) is defined as:

$$E(x, h) = \exp(h^\top W x + b^\top x + c^\top h). \qquad (1)$$

Probability $p(x, h)$ of RBM configuration (x, h) is expressed in terms of energy $E(x, h)$ of this configuration, and is given by:

$$p(x, h; \theta) = \frac{1}{Z(\theta)} \exp(-E(x, h)). \qquad (2)$$

In the above equation $Z(\theta)$ is a normalization constant referred to as the partition function, $Z(\theta) = \sum_{x,h} \exp(-E(x, h))$.

RBMs have no lateral connections between nodes in the same layer. This lack of lateral connections makes inference easier as conditional probabilities factorize, e.g., $P(h|y, x) = \prod_i P(h_i|y, x)$.

Classification RBM (ClassRBM) is an extension of the RBM model which operates as a standalone classifier. It does not require an external classifier such as SVM, or to be a part of deeper multilayered architecture to perform classification.

The architecture of the Classification RBM is shown in Fig. 2. ClassRBM is equipped with an additional layer, e^y, encoding the true class of input x. In our case, layer e^y encodes the profile of input x. Vector $e^y \in \{0,1\}^{Y \times 1}$ represents a one-hot vector encoding of the target class, i.e.,

$$e^y = (\mathbb{1}[y=1], \ldots, \mathbb{1}[y=Y])^\top, \tag{3}$$

where $y \in \{1, \ldots, Y\}$ encodes the target class as a natural number, and $Y = |\mathcal{C}|$ is the number of classes (profiles). Weight matrix $U \in \mathbb{R}^{H \times Y}$ represents the strength of connections between nodes of the visible and target class layer, while $d \in \mathbb{R}^{Y \times 1}$ is the bias vector of the target class layer.

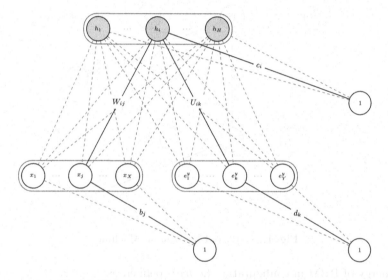

Fig. 2. Classification Restricted Boltzmann Machine

The energy of a Classification RBM [4,5] is defined as:

$$E(x, y, h) = \exp(h^\top W x + b^\top x + c^\top h + d^\top e^y + h^\top U e^y). \tag{4}$$

In this work we consider discriminative and generative RBM training protocols, along with their weighted sum referred to as hybrid training. Let us denote the learning corpus as dataset $\mathcal{D}_{\text{train}} = \{(x^{(i)}, y^{(i)}) \mid i \in \{1, .., |\mathcal{D}_{\text{train}}|\}\}$, where $x^{(i)}$ is the representation of the i-th conversation and $y^{(i)}$ is the author profile of the i-th conversation. In discriminative training the cost function is calculated on the basis of conditional probabilities $p(y^{(i)}|x^{(i)})$:

$$\mathcal{L}_{\text{discr}} = - \sum_{i|1}^{|\mathcal{D}_{\text{train}}|} \log p(y^{(i)}|x^{(i)}). \tag{5}$$

The cost function of generative training is determined using joint probabilities $p(y^{(i)}, x^{(i)})$:

$$\mathcal{L}_{\text{gen}} = - \sum_{i|1}^{|\mathcal{D}_{\text{train}}|} \log p(y^{(i)}, x^{(i)}). \tag{6}$$

Hybrid training integrates discriminative and generative terms in the cost function:

$$\mathcal{L} = \alpha \mathcal{L}_{\text{gen}} + (1 - \alpha) \mathcal{L}_{\text{discr}}, \tag{7}$$

where hyperparameter $\alpha \in [0,1]$ is the weight which determines the contribution of discriminative and generative parts to the overall cost. Hybrid training procedure applies contrastive divergence [1] for the generative term and stochastic gradient descent for the discriminative term.

4   Probabilities and Gradients

Predictions concerning the author profile y of text x (or, in general, concerning class y of point x) are made with the help of probability $P(y|x)$, i.e., the profile for which $P(y|x)$ achieves its peak value is selected:

$$\hat{y} = \underset{y \in \mathcal{C}}{\arg\max}\, P(y|x). \tag{8}$$

For the sake of notational simplicity, let $o_{yi}(x)$ stand for the total input from x [5]:

$$o_{yi}(x) \overset{\text{df}}{=} \sum_j W_{ij} x_j + c_i + U_{iy}. \tag{9}$$

Then probability $P(y|x)$ can be computed in terms of ClassRBM parameters $\theta = (W, U, c, d)$ in a computationally stable manner as follows:

$$P(y|x) = \frac{1}{\sum_{y'} \exp\left(d_{y'} - d_y + \sum_i \left[\text{softplus}(o_{y'i}(x)) - \text{softplus}(o_{yi}(x)) \right]\right)}. \tag{10}$$

Minimization of the loss function requires computing its gradient with respect to θ parameters. It is especially important that derivatives of the loss function be represented in matrix form, as it allows for efficient implementation in the Theano library.

4.1   Discriminative Training

For discriminative training we have to minimize loss $\mathcal{L}_{\text{discr}} = - \sum_i \log p(y^{(i)}|x^{(i)})$ (Eq. 5). For each data point (x, y) the gradient of $\log p(y|x)$ with respect to parameters $\theta = (W, U, c)$ is as follows:

$$\frac{\partial}{\partial \theta} \log p(y|x) = \sigma(o_{yi}(x)) \frac{\partial o_{yi}(x)}{\partial \theta} - \sum_{i,y'} p(y'|x) \cdot \sigma(o_{y'i}(x)) \frac{\partial o_{y'i}(x)}{\partial \theta} \tag{11}$$

For matrix W Eq. (11) assumes the following form:

$$\frac{\partial}{\partial W}\log P(y|x) = \left(\sigma\left(Wx+c+U_{\cdot y}\right) - \sigma\left((Wx+c)_{k=1}^{Y}+U\right)\begin{bmatrix}P(y=1|x)\\\vdots\\P(y=Y|x)\end{bmatrix}\right)x^{\mathsf{T}} \quad (12)$$

The gradient of Eq. (11) w.r.t. U equals ($\odot$ denotes the Hadamard product):

$$\frac{\partial}{\partial U}\log P(y|x) = \begin{bmatrix}\sigma(o_{11}(x)) & \cdots & \sigma(o_{Y1}(x))\\\vdots & \ddots & \vdots\\\sigma(o_{1H}(x)) & \cdots & \sigma(o_{YH}(x))\end{bmatrix} \odot \begin{bmatrix}[y=1]-P(y=1|x) & \cdots & [y=Y]-P(y=Y|x)\\\vdots & \ddots & \vdots\\[y=1]-P(y=1|x) & \cdots & [y=Y]-P(y=Y|x)\end{bmatrix}$$

$$(13)$$

The conditional log-likelihood gradient with respect to bias c is as follows:

$$\frac{\partial}{\partial c}\log P(y|x) = \sigma\left(Wx+c+U_{\cdot y}\right) - \sigma\left((Wx+c)_{k=1}^{Y}+U\right)\begin{bmatrix}P(y=1|x)\\\vdots\\P(y=Y|x)\end{bmatrix} \quad (14)$$

For $\theta = d$ the gradient of $\log p(y|x)$ assumes the following form:

$$\frac{\partial}{\partial d}\log P(y|x) = \begin{bmatrix}\mathbb{1}[y=1]-P(y=1|x)\\\mathbb{1}[y=2]-P(y=2|x)\\\vdots\\\mathbb{1}[y=Y]-P(y=Y|x)\end{bmatrix} \quad (15)$$

For $\theta = b$ we obtain $\frac{\partial}{\partial b}\log p(y|x) = 0$ as $p(y|x)$ does not depend on b.

4.2   Generative Training

In generative training, loss $\mathcal{L}_{\text{gen}} = -\sum_i \log p(y^{(i)}, x^{(i)})$ (Eq. 6) has to be minimized. For a data point (x, y) the gradient of $\log p(y, x)$ with respect to θ parameters assumes the following form:

$$\frac{\partial}{\partial\theta}\log p(y,x) = -\mathbb{E}_{h|y,x}\left[\frac{\partial}{\partial\theta}E(y,x,h)\right] + \mathbb{E}_{y',x',h'}\left[\frac{\partial}{\partial\theta}E(y',x',h')\right] \quad (16)$$

The resulting gradient is a sum of two expectations. The first expectation iterates over all possible states of the hidden layer, given visible input and target class. It is therefore tractable. The second expectation is intractable as it sums over all possible states of an RBM. Fortunately, the Contrastive Divergence (CD) algorithm can effectively approximate the second expectation [1].

5 Evaluation Datasets

Author profiling experiments have been conducted with the PAN-AP-13 corpus[1], which was released for the PAN-AP'13 competition. The PAN-AP-13 corpus contains conversations in English and Spanish, stored in XML files. There is a one-to-one correspondence between authors and XML files, with each XML file containing all conversations of an author. Each profiled author belongs to one of three age groups: 13–17, 23–27 and 33–47. In the English part the training set contains 413,564 conversations by 236,000 authors, and the test set contains 47,928 conversations by 21,200 authors. After preprocessing the training set contained 192,581,455 tokens and 893,926 word types.

Table 1. Breakdown of conversations into training, development and test sets and class balancing in the English part of the PAN-AP-13 corpus

Age group	Gender	Conversations			% of corpus
		Training	Development	Test	
13–17	Female	13,815	1,239	1,462	3.30
	Male	13,836	1,205	1,433	3.29
23–27	Female	79,057	7,702	9,095	19.13
	Male	76,797	6,718	8,657	18.40
33–47	Female	111,139	11,075	13,074	27.01
	Male	118,920	11,534	14,207	28.87
Σ		413,564	39,473	47,928	100.00

As shown in Table 1, the PAN-AP-13 dataset exhibits an age group imbalance (6:1 ratio between groups 23–27 and 13–17; 17:2 ratio between groups 33–47 and 13–17). A classifier trained on a skewed training set may be biased towards the overrepresented classes [10]. Moreover, the training set consists of over 400,000 conversations, which makes the learning of even one epoch slow. In order to address these issues, conversations from the most frequently represented classes (33–47 male, 33–47 female, 23–27 male and 23–27 female) were randomly under-sampled to 18,400 instances per class. Two underrepresented classes (13–17 male and 13–17 female) were randomly oversampled to reach 18,400 instances for each class. Following the resampling operation the training set contained 110,400 conversations, balanced with regard to age group as well as gender. This resampled set was used in further training.

---

[1] http://pan.webis.de/clef13/pan13-web/author-profiling.html.

Each conversation was represented as a vector of binary values, encoding the occurrence of words belonging to the list of X most frequent words from the training part of the PAN-AP-13 corpus, where X was subject to model selection and varied between 500 and 10000. No stopword removal was applied.

6 Experiments and Results

The PAN-AP-13 corpus is annotated with HTML tags (e.g., `<b>`, `</b>`) and BBCode tags (e.g., `[b]`, `[/b]`), which were filtered out. URL addresses were removed as well. Image tags were replaced with the values of their respective `alt` attributes because they contained descriptive and useful information in the context of profile classification. Fragments of text expressed as HTML entities were replaced with their unicode equivalents (e.g., the `&lt;` entity was replaced with the `<` character). Punctuation marks were treated as separate tokens. To avoid the overhead connected with handling large numbers of files, the corpus originally stored as a collection of XML files was converted into a single large file in the messagepack format.

Model Selection. To select the best model, the following list of hyperparameters was examined with the grid search algorithm:

- learning rate ε: $0.5, 0.1, 10^{-2}, 10^{-3}, 10^{-4}$
- discriminative and generative training ratio α: $0, 0.25, 0.5, 0.75, 1$
- size of visible layer X: $500, 1000, 2000, 5000, 10000$
- size of hidden layer H: $100, 500, 1000, 2000, 5000$.

The remaining parameters were selected in advance. Optimization of the cost function was performed in mini-batches of size 10, $k = 1$ in the CD-k algorithm, and the learning rate was annealed with iteration step t according to the formula $\frac{\varepsilon}{1+\mu t}$ with $\mu = 10^{-6}$.

Training Procedure. ClassRBM training was carried out as follows:

1. All ClassRBM instances are created, each with a different set of hyperparameters.
2. All bias vectors are initialized with zero values. Weights of matrices W and U are initialized to random numbers close to zero, sampled from the uniform distribution, $\mathcal{U}(-4\sqrt{6/(cols + rows)}, \sqrt{6/(cols + rows)})$, with *cols* and *rows* denoting the number of matrix columns and rows respectively.
3. All ClassRBM instances are trained using the same first 35% of the training corpus, i.e., approximately 40,000 samples.
4. All ClassRBM instances are evaluated using the development corpus. The ClassRBM instance with the highest accuracy for the development set is selected to continue the training process; remaining instances are abandoned.
5. The training process continues using the entire training corpus. Each iteration involves:

- taking a mini-batch sample from the training set,
- for each parameter θ computing the discriminative criterion gradient, and an approximation of the generative criterion gradient with CD-k,
- updating θ parameters of the ClassRBM with the gradient approximations.

6. Every 1000 iterations the ClassRBM instance is evaluated using the development set. If accuracy has not improved in comparison with the highest accuracy within the last 20 evaluations of the development set, the training process concludes (validation-based early stopping).
7. Parameters $\theta = (W, U, b, c, d)$ (weight matrices and biases) with the highest accuracy on the development set are returned as ClassRBM parameters.

6.1 Overall Results

The accuracy of author profiling is defined as the fraction of correctly identified author profiles (both age group and gender), measured using the test part of the corpus:

$$acc = \frac{\sum_{(x,(gender,age))\in\mathcal{D}_{\text{test}}} \mathbb{1}[\text{Age}(x) = age] \cdot \mathbb{1}[\text{Gender}(x) = gender]}{|\mathcal{D}_{\text{test}}|}, \quad (17)$$

where x is the concatenation of all conversations of a given author and $\text{Age}(x)$ and $\text{Gender}(x)$ are predictions made by the classifier. Age accuracy and gender accuracy are both defined in an analogous manner.

Note that while training is performed per mini-batch of conversations, evaluation on the test set is performed per author, to be consistent with PAN-AP'13 results reporting.

The highest accuracy was obtained with the hyperparameters $\varepsilon = 0.1$, $\alpha = 0.25$, $X = 10000$ and $H = 5000$, which were fixed on the development set. The training lasted 3 epochs, i.e., 33,000 iterations. The best model was found after 16,000 iterations. The ClassRBM evaluated using the development set achieved then 37.51% joint accuracy, 58.29% gender accuracy and 59.12% age accuracy.

When using the test set, the ClassRBM correctly recognized both age and gender with 36.59% accuracy, gender only with 57.83% accuracy and age only with 59.17% accuracy. The accuracy of the purely discriminatively trained model ($\alpha = 0$) with the other hyperparameters remaining unchanged was 35.13%. The corresponding result for the fully generative model ($\alpha = 1$) was 28.54%. This shows the advantage of hybrid learning over purely discriminative or generative training.

Results listed in Table 2 indicate that the proposed feature-learned classifier is only slightly less accurate than the winning approach and outperforms the majority of available solutions on the PAN-AP-13 corpus. This further suggests that the author profiling problem can be efficiently solved without using handcrafted features.

Table 2. Accuracy of the author profiling task (in %) obtained with Classification RBM, compared with results achieved during the PAN-AP'13 competition [7]

Team	Joint	Gender	Age
Meina	38.94	59.21	64.91
Pastor L	38.13	56.90	65.72
Seifeddine	36.77	58.16	58.97
Classification RBM	**36.59**	**57.83**	**59.17**
Santosh	35.08	56.52	64.08
Yong Lim	34.88	56.71	60.98
Ladra	34.20	56.08	61.18
Aleman	32.92	55.22	59.23
Gillam	32.68	54.10	60.31
Kern	31.15	52.67	56.90
Cruz	31.14	54.56	59.66
Pavan	28.43	50.00	60.55
Caurcel Diaz	28.40	50.00	56.79
H. Farias	28.16	56.71	50.61
Jankowska	28.14	53.81	47.38
Flekova	27.85	53.43	52.87
Weren	25.64	50.44	50.99
Sapkota	24.71	47.81	54.15
De-Arteaga	24.50	49.98	48.85
Moreau	23.95	49.41	48.24
Baseline	*16.50*	*50.00*	*33.33*
Gopal Patra	15.74	56.83	28.95
Cagnina	7.41	50.40	12.34

7   Conclusions

In this paper we constructed the ClassRBM classifier which performed author profiling, evaluated it on the PAN-AP-13 corpus, and compared its accuracy with the effectiveness of solutions submitted to the PAN-AP'13 competition. ClassRBM is a stand-alone classifier and does not rely on handcrafted features – instead, it automatically infers features from data, which distinguishes it from among other approaches to author profiling. The accuracy of feature-learning ClassRBM is competitive compared to classifiers which rely on sophisticated, handcrafted features. The proposed solution outperformed classifiers submitted by 18 different teams, and ranked fourth overall.

We also compared different ways of training ClassRBM, concluding that hybrid training is superior to purely discriminative or generative learning.

Acknowledgments. This research was partly supported by the PL-Grid Infrastructure. The research was also supported by the AGH University of Science and Technology (AGH-UST), grant no. 11.11.230.124 (statutory project).

References

1. Hinton, G.E.: Training products of experts by minimizing contrastive divergence. Neural Comput. **14**(8), 1771–1800 (2002)
2. Koppel, M., Argamon, S., Shimoni, A.R.: Automatically categorizing written texts by author gender. Lit. Linguist. Comput. **17**(4), 401–412 (2002)
3. Koppel, M., Schler, J., Argamon, S.: Computational methods in authorship attribution. J. Am. Soc. Inf. Sci. Technol. **60**(1), 9–26 (2009)
4. Larochelle, H., Bengio, Y.: Classification using discriminative restricted boltzmann machines. In: Proceedings of the 25th International Conference on Machine Learning, ICML 2008, pp. 536–543 (2008)
5. Larochelle, H., Mandel, M., Pascanu, R., Bengio, Y.: Learning algorithms for the classification restricted boltzmann machine. J. Mach. Learn. Res. **13**(1), 643–669 (2012)
6. Maharjan, S., Shrestha, P., Solorio, T., Hasan, R.: A straightforward author profiling approach in MapReduce. In: Bazzan, A.L.C., Pichara, K. (eds.) IBERAMIA 2014. LNCS (LNAI), vol. 8864, pp. 95–107. Springer, Cham (2014). doi:10.1007/978-3-319-12027-0_8
7. Rangel, F., Rosso, P., Koppel, M., Stamatatos, E., Inches, G.: Overview of the Author Profiling Task at PAN 2013. In: Forner, P., Navigli, R., Tufis, D., Ferro, N. (eds.) Working Notes for CLEF 2013 Conference, vol. 1179 (2013)
8. Salakhutdinov, R., Mnih, A., Hinton, G.: Restricted Boltzmann Machines for collaborative filtering. In: Ghahramani, Z. (ed.) Proceedings of the 24th International Conference on Machine Learning, ICML 2007, pp. 791–798 (2007)
9. Smolensky, P.: Information processing in dynamical systems: Foundations of harmony theory. In: Rumelhart, D.E., McClelland, J.L., PDP Research Group (eds.) Parallel Distributed Processing: Explorations in the Microstructure of Cognition, Volume 1: Foundations, pp. 194–281. MIT Press, Cambridge (1986)
10. Zheng, Z., Cai, Y., Li, Y.: Oversampling method for imbalanced classification. Comput. Inf. **34**(5), 1017–1037 (2016)

Parallel Implementation of the Givens Rotations in the Neural Network Learning Algorithm

Jarosław Bilski[1]([✉]), Bartosz Kowalczyk[1], and Jacek M. Żurada[2]

[1] Institute of Computational Intelligence,
Częstochowa University of Technology, Częstochowa, Poland
{Jaroslaw.Bilski,Bartosz.Kowalczyk}@iisi.pcz.pl
[2] Department Electrical and Computer Engineering,
University of Louisville, Louisville, KY 40292, USA
jacek.zurada@louisville.edu

Abstract. The paper describes a parallel feed-forward neural network training algorithm based on the QR decomposition with the use of the Givens rotation. The beginning brings a brief mathematical background on Givens rotation matrices and elimination step. Then the error criterion and its necessary transformations for the QR decomposition are presented. The paper's core holds an essential explanation to accomplish hardware-based parallel implementation. The paper concludes with a theoretical description of speed improvement gained by parallel implementation of the Givens reduction in the QR decomposition process.

Keywords: Feed-forward neural network · Neural network training algorithm · Optimization · QR decomposition · Givens rotation · Parallel Givens rotation · Parallel Givens flow

1 Introduction

Artificial neural networks are finding countless applications in both, the worlds of science and industry. This generates an everlasting demand for scientific research in those areas as in [2,3,15–17,19,23,24]. One of the most significant branches is the neural network teaching process. For that purpose many algorithms have been developed [7,10,22]. Some of them are broadly known, easy to implement but require a significant amount of time to establish convergence [12,21]. Other algorithms are much faster at the cost of more complex implementation and utilization of many more resources [8,14]. The next important branch in the area of artificial neural network research is optimization of already-known teaching algorithms. A well designed practice for that purpose is parallelization of the serial algorithm. Many research projects in this area have been attempted as in [4,6,9,11]. This paper focuses on delivering a hardware-based parallelization concept of feed-forward neural network teaching algorithm using the QR decomposition based on the Givens rotation. The principal goal of this idea is to complete the QR decomposition process in a lower iteration count than required by the serial variant of this algorithm.

© Springer International Publishing AG 2017
L. Rutkowski et al. (Eds.): ICAISC 2017, Part I, LNAI 10245, pp. 14–24, 2017.
DOI: 10.1007/978-3-319-59063-9_2

2   Givens Elimination Step

The Givens reduction is the process of applying orthogonal matrix $\mathbf{G}_{pq}$ (shown in Eq. 1) by left-sided multiplication of rotated vector $\mathbf{a}$.

$$
\mathbf{G}_{pq} =
\begin{bmatrix}
1 & & \cdots & & 0 \\
 & \ddots & & & \\
 & & c \cdots s & & \\
\vdots & & \vdots \ddots \vdots & & \vdots \\
 & & -s \cdots c & & \\
 & & & \ddots & \\
0 & & \cdots & & 1
\end{bmatrix}
\begin{matrix} \\ \\ p \\ \\ q \\ \\ \\ \end{matrix}
\tag{1}
$$
$$
\begin{matrix} & & p & & q & & \end{matrix}
$$

$$
\mathbf{a} \to \bar{\mathbf{a}} = \mathbf{G}_{pq}\mathbf{a}. \tag{2}
$$

Equation 2 performs the following operations

$$
\begin{aligned}
\bar{a}_p &= ca_p + sa_q \\
\bar{a}_q &= -sa_p + ca_q \\
\bar{a}_i &= a_i \quad (i \neq p, q; i = 1, \ldots, n).
\end{aligned}
\tag{3}
$$

In order to eliminate the value of a_q, parameters c and s should be calculated according to the following equations

$$
c = \frac{a_p}{\rho} \qquad s = \frac{a_q}{\rho}, \tag{4}
$$

where ρ due to numerical error minimization takes the form

$$
\rho = \begin{cases} a_p\sqrt{1 + (a_q/a_p)^2}, & for \quad |a_p| \geq |a_q| \\ a_q\sqrt{1 + (a_p/a_q)^2}, & for \quad |a_p| < |a_q| \end{cases}
\tag{5}
$$

When Eqs. 4 and 5 are applied to 3, we obtain the following

$$
\begin{aligned}
\bar{a}_p &= ca_p + sa_q = \rho \\
\bar{a}_q &= -sa_p + ca_q = 0.
\end{aligned}
\tag{6}
$$

3   Givens QR Decomposition

The QR decomposition is the process of transforming any non-singular matrix into the product of orthogonal $\mathbf{Q}$ and upper-triangle $\mathbf{R}$ matrices

$$
\mathbf{A} = \mathbf{QR}, \tag{7}
$$

where

$$Q^T Q = I, \tag{8}$$

$$Q^T = Q^{-1}, \tag{9}$$

$$r_{ij} = 0 \quad for \ i > j. \tag{10}$$

The orthogonal matrix $\mathbf{Q}$ does not need to be explicitly determined. It is enough to calculate c and s parameters of their respective rotations and apply them to input matrix $\mathbf{A}$. Acquisition of all c and s parameters for all $\mathbf{G}_{pq}$ matrices concludes the QR decomposition by transforming matrix $\mathbf{A}$ to an upper-triangle form

$$\mathbf{R} = \mathbf{G}_{m-1} \ldots \mathbf{G}_1 \mathbf{A}_1 = \mathbf{G}_{m-1,m} \ldots \mathbf{G}_{23} \ldots \mathbf{G}_{2m} \mathbf{G}_{12} \ldots \mathbf{G}_{1m} \mathbf{A}_1 = \mathbf{Q}^T \mathbf{A} \tag{11}$$

Since there is no need to explicitly create $\mathbf{G}_{pq}$ matrices, the orthogonal matrix $\mathbf{Q}$ can be calculated back from the respective rotations

$$\mathbf{Q} = \mathbf{G}_1^T \ldots \mathbf{G}_{m-1}^T = \mathbf{G}_{1m}^T \ldots \mathbf{G}_{12}^T \mathbf{G}_{2m}^T \ldots \mathbf{G}_{23}^T \ldots \mathbf{G}_{m-1,m}^T \tag{12}$$

4   QR Decomposition in Neural Network Weights Update

In order to calculate weight update the following error criterion is given

$$J(n) = \sum_{t=1}^{n} \lambda^{n-t} \sum_{j=1}^{N_L} \varepsilon_j^{(L)2}(t) = \sum_{t=1}^{n} \lambda^{n-t} \sum_{j=1}^{N_L} \left[d_j^{(L)}(t) - f\left(\mathbf{x}^{(L)T}(t) \mathbf{w}_j^{(L)}(n)\right) \right]^2 \tag{13}$$

By performing the derivation and linearization of Eq. 13 the normal equation is obtained

$$\sum_{t=1}^{n} \lambda^{n-t} f'^2 \left(s_i^{(l)}(t)\right) \left[b_i^{(l)}(t) - \mathbf{x}^{(l)T}(t) \mathbf{w}_i^{(l)}(n)\right] \mathbf{x}^{(l)T}(t) = \mathbf{0}. \tag{14}$$

Transforming Eq. 14 to the vector form reveals an entry point to the QR decomposition. The following equations are applicable for each layer, where i denotes the neuron index of layer l.

$$\mathbf{A}_i^{(l)}(n) \mathbf{w}_i^{(l)}(n) = \mathbf{h}_i^{(l)}(n), \tag{15}$$

where

$$\mathbf{A}_i^{(l)}(n) = \sum_{t=1}^{n} \lambda^{n-t} \mathbf{z}_i^{(l)}(t) \mathbf{z}_i^{(l)T}(t), \tag{16}$$

$$\mathbf{h}_i^{(l)}(n) = \sum_{t=1}^{n} \lambda^{n-t} f'\left(s_i^{(l)}(t)\right) b_i^{(l)}(t) \mathbf{z}_i^{(l)}(t), \tag{17}$$

where

$$\mathbf{z}_i^{(l)}(t) = f'\left(s_i^{(l)}(t)\right)\mathbf{x}^{(l)}(t). \tag{18}$$

$$b_i^{(l)}(n) = \begin{cases} b_i^{(L)}(n) = f^{-1}\left(d_i^{(L)}(n)\right) & for\ l = L \\ s_i^{(l)}(n) + e_i^{(l)}(n) & for\ l = 1\dots L-1, \end{cases} \tag{19}$$

$$e_i^{(k)}(n) = \sum_{j=1}^{N_{k+1}} f'\left(s_i^{(k)}(n)\right)w_{ji}^{(k+1)}(n)\,e_j^{(k+1)}(n) \qquad for\ k = 1\dots L-1. \tag{20}$$

Equation 15 is solved by the QR decomposition. The first step is to acquire orthogonal matrix $\mathbf{Q}^T$ and perform left-sided multiplication

$$\mathbf{Q}_i^{(l)T}(n)\,\mathbf{A}_i^{(l)}(n)\,\mathbf{w}_i^{(l)}(n) = \mathbf{Q}_i^{(l)T}(n)\,\mathbf{h}_i^{(l)}(n), \tag{21}$$

$$\mathbf{R}_i^{(l)}(n)\,\mathbf{w}_i^{(l)}(n) = \mathbf{Q}_i^{(l)T}(n)\,\mathbf{h}_i^{(l)}(n). \tag{22}$$

Finally, matrix $\mathbf{A}$ is transformed to $\mathbf{R}$, which is upper-triangle. The update of neuron weights is described by the following equations

$$\hat{\mathbf{w}}_i^{(l)}(n) = \mathbf{R}_i^{(l)-1}(n)\,\mathbf{Q}_i^{(l)T}(n)\,\mathbf{h}_i^{(l)}(n), \tag{23}$$

$$\mathbf{w}_i^{(l)}(n) = (1-\eta)\,\mathbf{w}_i^{(l)}(n-1) + \eta\,\hat{\mathbf{w}}_i^{(l)}(n). \tag{24}$$

5   Parallel Implementation

The parallel weight update system consists of three subsystems where each one is executed one by one. All the structures presented in this section are related to single neuron only. A complete layer parallel weight update system is of 3D cuboid structure whose size depends on the layer's neuron count. The first subsystem can be found in Fig. 1. It is responsible for formulating autocorrelation matrix $\mathbf{A}$ and vector $\mathbf{h}$ according to Eqs. 16 and 17. Block $\mathbf{S}$ executes Eq. 18. The definition of blocks $\mathbf{S}$, $\mathbf{A1}$ and $\mathbf{H1}$ can be found in Fig. 2. Matrix $\mathbf{A}$ and vector $\mathbf{h}$ determined after the first step cannot be changed until the current iteration of the teaching process is finished as they are used in oncoming iterations. The second system realizes the parallel QR decomposition based on Givens rotation according to Eq. 3, keeping in mind Eq. 6. The second system is supposed to calculate matrix $\mathbf{R}$ and vector $\mathbf{Q}^T\mathbf{h}$, which is done by the structure shown in Fig. 3. Blocks $\mathbf{A2}$ are holding respective copies of matrix $\mathbf{A}$. Similarly blocks $\mathbf{H2}$ are holding their respective copies from vector $\mathbf{h}$. From Eq. 16 we know that matrix $\mathbf{A}$ is a square matrix of size $N_{l-1} + 1$ x $N_{l-1} + 1$, where N_{l-1} denotes input layer vector size. To improve readability of oncoming equations the following substitution is performed $n = N_{l-1} + 1$. In most cases the complete QR decomposition cannot be done in a single step (except $N_{l-1} = 2$). An example parallel flow (for $N_{l-1} = 8$) of the QR decomposition based on the Givens rotation is shown in Fig. 5. Each parallel elimination step affects only two rows of

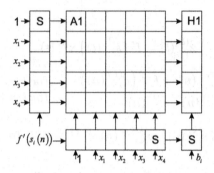

Fig. 1. Structure of the system responsible for formulating matrix **A** and vector **h**.

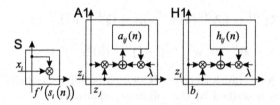

Fig. 2. Definitions of A1, H1 and S processing elements.

matrix **A** starting from column k, where $k \in \langle 1, 2, \ldots, N_{l-1} - 1 \rangle$. The starting point of a single rotation is represented by objects A2pk and A2qk shown in Fig. 4. Block A2pk calculates length (ρ) of vector $[a_{pk}, a_{qk}]^T$. Then, the value of A2pk block $(a_{pk}(n))$ is replaced by obtained ρ. Simultaneously, value $a_{qk}(n)$ in block A2qk is replaced by 0 according to Eq. 6. Block A2pk is also responsible for determining rotation c and s parameters which are applied for all right-hand objects starting from A2pk and A2qk blocks in matrix **A**. Also values p and q of vector **h** are directly affected by c and s parameters as shown in Fig. 4. As result of solving the QR decomposition in the second step matrix **R** and vector $\mathbf{Q}^T \mathbf{h}$ are obtained. The third-last system is responsible for updating weights of the respective neuron according to Eqs. 23 and 24. The parallel structure that is capable of doing that is presented in Fig. 6. Objects A3a and A3b determine the value of $\hat{w}_{ij}$ according to Eq. 25 while block W performs the weight update. The third system block definitions are shown in Fig. 7.

$$\hat{w}_{mj} = \frac{h_{mj} - \sum_{k=i+1}^{n} \hat{w}_{ik} a_{ik}}{a_{ii}} \tag{25}$$

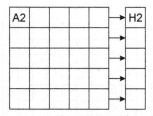

Fig. 3. Structure of the Givens rotation system.

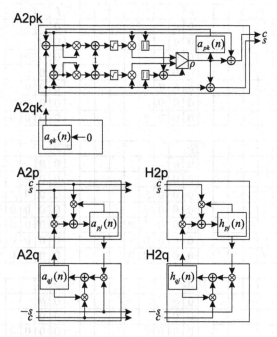

Fig. 4. Blocks of A2 and H2 structures.

6   Simulation Results

Using a parallel variant of the Givens QR decomposition it is possible to achieve the process performance given by Eq. 26

$$\xi_p \leq 2n - b,\qquad(26)$$

where $n = N_{l-1} + 1$ and b is given in Table 1. Performance of serial variant is given by Eq. 27

$$\xi_s = \sum_{k=1}^{n-1} n - k.\qquad(27)$$

The whole teaching process complexity of serial and parallel variants for single layer l which consists of N_l neurons is shown in Table 2. When a parallel structure

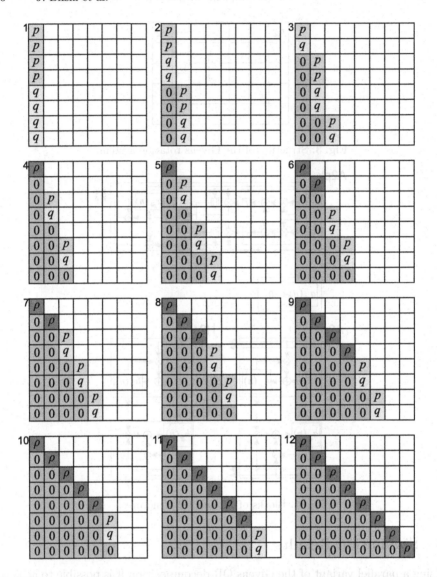

Fig. 5. The parallel QR decomposition based on the Givens rotation flow for 8×8 matrix. The first step is presented in the upper left corner of the figure. Labels A2 and column index k have been omitted to improve picture readability. Each block marked as p corresponds to A2pk object. Similarly, each q block reflects the A2qk object. The blocks marked as ρ indicate that the respective column is fully transformed

for the teaching process is used, all neurons are able to update their weights at the same time. Due to that, multiplication of complexity by N_l is omitted. During the simulation the parameters n (layer input size) and N (layer's neuron count) have been adjusted to obtain the results presented in Figs. 8 and 9.

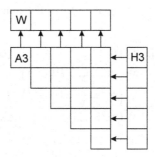

Fig. 6. The structure of the weight update system.

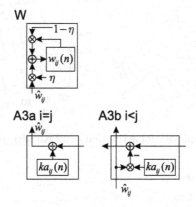

Fig. 7. Processing elements A3 and W structures. The only purpose of H3 objects is to provide values for further processing.

Table 1. Values of b depending on $x \in [x_{r_0}, x_{r_{end}}]$ for $y = 2x - b$.

x_{r_0}	$x_{r_{end}}$	b	x_{r_0}	$x_{r_{end}}$	b	x_{r_0}	$x_{r_{end}}$	b	x_{r_0}	$x_{r_{end}}$	b	x_{r_0}	$x_{r_{end}}$	b
2	3	3	202	229	16	776	835	29	1724	1815	42	3048	3165	55
4	7	4	230	266	17	836	895	30	1816	1905	43	3166	3287	56
8	14	5	267	298	18	896	959	31	1906	1998	44	3288	3405	57
15	23	6	299	338	19	960	1023	32	1999	2093	45	3406	3537	58
24	31	7	339	376	20	1024	1097	33	2094	2189	46	3538	3657	59
32	47	8	377	417	21	1098	1166	34	2190	2289	47	3658	3790	60
48	61	9	418	463	22	1167	1239	35	2290	2391	48	3791	3918	61
62	79	10	464	508	23	1240	1314	36	2392	2498	49	3919	4051	62
80	100	11	509	558	24	1315	1391	37	2499	2598	50	4052	4187	63
101	121	12	559	610	25	1392	1470	38	2599	2712	51	4188	4322	64
122	146	13	611	663	26	1471	1552	39	2713	2817	52	4323	4465	65
147	170	14	664	718	27	1553	1641	40	2818	2936	53	4466	4603	66
171	201	15	719	775	28	1642	1723	41	2937	3047	54	4604	4752	67

Table 2. Operation complexity of serial and parallel complete teaching processes for single layer l.

Operation	Serial	Parallel
$+/-$	$N_l \left(\frac{4n^3 + 15n^2 + 11n}{6} + 3 \right)$	$5n - 5$
$*/\div$	$N_l \left(\frac{8n^3 + 39n^2 + 25n}{6} + 5 \right)$	$12n - 13$
f/f'	$N_l \left(2n^2 - 2n + 1 \right)$	$2n - 2$

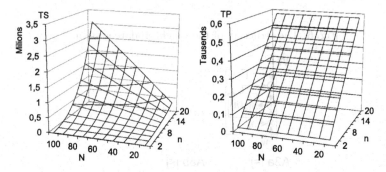

Fig. 8. Computation complexity for serial (left) and parallel (right) implementations.

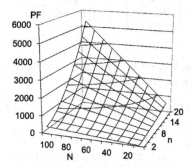

Fig. 9. Performance factor achieved by parallel implementation.

7   Conclusion

The paper presents a full parallel QR decomposition process. The use of described structures can bring satisfying acceleration even for small networks (up to 10 neurons) and very significant acceleration for big networks (over 100 neurons per layer). The serial complexity is of order $\mathcal{O}\left(n^3 N_l\right)$ while the parallel one is of $\mathcal{O}\left(n\right)$. During the research the performance factor established the highest value of 5335 for $n = 20$ and $N_l = 100$. The trend of growth can be seen in Fig. 9. Parallelization of teaching algorithms seems to be a good direction for further optimization research. A similar parallel approach might be applied for other algorithms [1,13,18,20]. The next step of the parallel Givens research is to acquire an equation to get the value of b. In the near future also a momentum variant of Givens QR decomposition will be attempted as proposed in [5].

References

1. Starczewski, A.: A clustering method based on the modified RS validity index. In: Rutkowski, L., Korytkowski, M., Scherer, R., Tadeusiewicz, R., Zadeh, L.A., Zurada, J.M. (eds.) ICAISC 2013. LNCS, vol. 7895, pp. 242–250. Springer, Heidelberg (2013). doi:10.1007/978-3-642-38610-7_23
2. Starczewski, A.: A new validity index for crisp clusters. Pattern Anal. Appl. (2015)
3. Starczewski, A., Krzyżak, A.: A modification of the silhouette index for the improvement of cluster validity assessment. In: Rutkowski, L., Korytkowski, M., Scherer, R., Tadeusiewicz, R., Zadeh, L.A., Zurada, J.M. (eds.) ICAISC 2016. LNCS, vol. 9693, pp. 114–124. Springer, Cham (2016). doi:10.1007/978-3-319-39384-1_10
4. Bilski, J., Smolag, J., Galushkin, A.I.: The parallel approach to the conjugate gradient learning algorithm for the feedforward neural networks. In: Rutkowski, L., Korytkowski, M., Scherer, R., Tadeusiewicz, R., Zadeh, L.A., Zurada, J.M. (eds.) ICAISC 2014. LNCS(LNAI), vol. 8467, pp. 12–21. Springer, Cham (2014). doi:10.1007/978-3-319-07173-2_2
5. Bilski, J.: Momentum modification of the RLS algorithms. In: Rutkowski, L., Siekmann, J.H., Tadeusiewicz, R., Zadeh, L.A. (eds.) ICAISC 2004. LNCS, vol. 3070, pp. 151–157. Springer, Heidelberg (2004). doi:10.1007/978-3-540-24844-6_18
6. Bilski, J.: Struktury rwnolege dla jednokierunkowych i dynamicznych sieci neuronowych. Akademicka Oficyna Wydawnicza EXIT, Warszawa (2013)
7. Bilski, J., Kowalczyk, B., Żurada, J.M.: Application of the givens rotations in the neural network learning algorithm. In: Rutkowski, L., Korytkowski, M., Scherer, R., Tadeusiewicz, R., Zadeh, L.A., Zurada, J.M. (eds.) ICAISC 2016. LNCS(LNAI), vol. 9692, pp. 46–56. Springer, Cham (2016). doi:10.1007/978-3-319-39378-0_5
8. Bilski, J., Wilamowski, B.M.: Parallel learning of feedforward neural networks without error backpropagation. In: Rutkowski, L., Korytkowski, M., Scherer, R., Tadeusiewicz, R., Zadeh, L.A., Zurada, J.M. (eds.) ICAISC 2016. LNCS(LNAI), vol. 9692, pp. 57–69. Springer, Cham (2016). doi:10.1007/978-3-319-39378-0_6
9. Bilski, J., Smolag, J., Żurada, J.M.: Parallel approach to the levenberg-marquardt learning algorithm for feedforward neural networks. In: Rutkowski, L., Korytkowski, M., Scherer, R., Tadeusiewicz, R., Zadeh, L.A., Zurada, J.M. (eds.) ICAISC 2015. LNCS(LNAI), vol. 9119, pp. 3–14. Springer, Cham (2015). doi:10.1007/978-3-319-19324-3_1
10. Bilski, J., Rutkowski, L.: Numerically robust learning algorithms for feed forward neural networks. In: Rutkowski, L., Kacprzyk, J. (eds.) Neural Networks and Soft Computing. Advances in Soft Computing, pp. 149–154. Physica-Verlag, Heidelberg (2003)
11. Bilski, J., Smolag, J.: Parallel approach to learning of the recurrent jordan neural network. In: Rutkowski, L., Korytkowski, M., Scherer, R., Tadeusiewicz, R., Zadeh, L.A., Zurada, J.M. (eds.) ICAISC 2013. LNCS(LNAI), vol. 7894, pp. 32–40. Springer, Heidelberg (2013). doi:10.1007/978-3-642-38658-9_3
12. Werbos, J.: Beyond Regression: New Tools for Prediction and Analysis in the Behavioral Sciences. Harvard University (1974)

13. Nowicki, R.K., Nowak, B.A., Starczewski, J.T., Cpalka, K.: The learning of neuro-fuzzy approximator with fuzzy rough sets in case of missing features. In: 2014 International Joint Conference on Neural Networks (IJCNN), pp. 3759–3766, July 2014

14. Bilski, J., Rutkowski, L.: A fast training algorithm for neural networks. IEEE Trans. Circ. Syst. Part II **45**, 749–753 (1998)

15. Zalasiński, M., Łapa, K., Cpałka, K.: New algorithm for evolutionary selection of the dynamic signature global features. In: Rutkowski, L., Korytkowski, M., Scherer, R., Tadeusiewicz, R., Zadeh, L.A., Zurada, J.M. (eds.) ICAISC 2013. LNCS (LNAI), vol. 7895, pp. 113–121. Springer, Heidelberg (2013). doi:10.1007/978-3-642-38610-7_11

16. Zalasiński, M., Cpałka, K., Rakus-Andersson, E.: An idea of the dynamic signature verification based on a hybrid approach. In: Rutkowski, L., Korytkowski, M., Scherer, R., Tadeusiewicz, R., Zadeh, L.A., Zurada, J.M. (eds.) ICAISC 2016. LNCS (LNAI), vol. 9693, pp. 232–246. Springer, Cham (2016). doi:10.1007/978-3-319-39384-1_21

17. Zalasiński, M., Cpałka, K., Hayashi, Y.: New method for dynamic signature verification based on global features. In: Rutkowski, L., Korytkowski, M., Scherer, R., Tadeusiewicz, R., Zadeh, L.A., Zurada, J.M. (eds.) ICAISC 2014. LNCS, vol. 8468, pp. 231–245. Springer, Cham (2014). doi:10.1007/978-3-319-07176-3_21

18. Zalasiński, M., Cpałka, K., Hayashi, Y.: A new approach to the dynamic signature verification aimed at minimizing the number of global features. In: Rutkowski, L., Korytkowski, M., Scherer, R., Tadeusiewicz, R., Zadeh, L.A., Zurada, J.M. (eds.) ICAISC 2016. LNCS (LNAI), vol. 9693, pp. 218–231. Springer, Cham (2016). doi:10.1007/978-3-319-39384-1_20

19. Mleczko, W.K., Kapuściński, T., Nowicki, R.K.: Rough deep belief network - application to incomplete handwritten digits pattern classification. In: Dregvaite, G., Damasevicius, R. (eds.) ICIST 2015. CCIS, vol. 538, pp. 400–411. Springer, Cham (2015). doi:10.1007/978-3-319-24770-0_35

20. Nowak, B.A., Nowicki, R.K., Starczewski, J.T., Marvuglia, A.: The learning of neuro-fuzzy classifier with fuzzy rough sets for imprecise datasets. In: Rutkowski, L., Korytkowski, M., Scherer, R., Tadeusiewicz, R., Zadeh, L.A., Zurada, J.M. (eds.) ICAISC 2014. LNCS (LNAI), vol. 8467, pp. 256–266. Springer, Cham (2014). doi:10.1007/978-3-319-07173-2_23

21. Tadeusiewicz, R.: Sieci Neuronowe. Akademicka Oficyna Wydawnicza, Warszawa (1993)

22. Nowak, B.A., Nowicki, R.K.: Learning in rough-neuro-fuzzy system for data with missing values. In: Wyrzykowski, R., Dongarra, J., Karczewski, K., Waśniewski, J. (eds.) PPAM 2011. LNCS, vol. 7203, pp. 501–510. Springer, Heidelberg (2012). doi:10.1007/978-3-642-31464-3_51

23. Knop, M., Kapuscinski, T., Mleczko, W.K.: Video key frame detection based on the restricted boltzmann machine. J. Appl. Math. Comput. Mech. **14**(3), 49–58 (2015)

24. Wozniak, M., Polap, D., Nowicki, R.K., Napoli, C., Pappalardo, G., Tramontana, E.: Novel approach toward medical signals classifier. In: 2015 International Joint Conference on Neural Networks (IJCNN), pp. 1–7, July 2015

Parallel Levenberg-Marquardt Algorithm Without Error Backpropagation

Jarosław Bilski[1]([✉]) and Bogdan M. Wilamowski[2]

[1] Institute of Computational Intelligence,
Częstochowa University of Technology, Częstochowa, Poland
`Jaroslaw.Bilski@iisi.pcz.pl`
[2] Auburn University, Auburn, AL 36849-5201, USA
`wilambm@auburn.edu`

Abstract. This paper presents a new parallel architecture of the Levenberg-Marquardt (LM) algorithm for training fully connected feedforward neural networks, which will also work for MLP but some cells will stay empty. This approach is based on a very interesting idea of learning neural networks without error backpropagation. The presented architecture is based on completely new parallel structures to significantly reduce a very high computational load of the LM algorithm. A full explanation of parallel three-dimensional neural network learning structures is provided.

Keywords: Forward-only computation · Neural network training · Parallel architectures

1 Introduction

Feedforward arificial neural networks have been studied by many scientists e.g. [20,24,30,32,33]. In a number of learning methods error backpropagation is used. This methodology is simple enough and has been often applied to learn feedforward networks, see e.g. [19,23,31]. In network learning there are two phases. In the first phase data are entered into network inputs and calculations are carried forward to network outputs. In the second phase errors calculated at network outputs are sent back to all neurons and a learning algorithm is applied to weights update. A new approach to calculate errors in hidden neurons is presented in [32]. In this new algorithm, all calculations are executed only forward in one phase. This eliminates the error backpropagation phase and allows to enter pipelined processing for learning neural networks. It should be noted that in the classical approach, neural networks learning algorithms, like other learning algorithms [14–18], are implemented on a serial computer. Some learning algorithms need a large number of computational operations and thus, serial implementation is time consuming and very slow. In this case, especially for large networks, computational load makes learning algorithms less useful. The Levenberg-Marquardt algorithm is one of the most computationally complex algorithms.

© Springer International Publishing AG 2017
L. Rutkowski et al. (Eds.): ICAISC 2017, Part I, LNAI 10245, pp. 25–39, 2017.
DOI: 10.1007/978-3-319-59063-9_3

An increasingly common solution to the computational load problem is the use of high-performance dedicated parallel structures, see e.g. [4–13, 25, 26]. This paper presents a new concept of parallel realization of the Levenberg-Marquardt learning algorithm without error bacpropagation. A single epoch of the parallel architecture needs much fewer computation cycles than a serial computer implementation. Efficiency of this new architecture seems to be very promising and is explained in the last part of this paper.

In this paper we introduce a parallel architecture for [32]. In Fig. 1 an example of a fully-connected feedforward network is shown. This network has nn (8) neurons and no (2) outputs. The input vector contains ni (3) input signals. The neuron model is shown in Fig. 2. In the fully-connected network each neuron is connected to all the inputs and all the previous neurons. It is easy to see that by removing some of the weight connections, a traditional multilayer neural network can be obtained.

The input vector to the $i - th$ neuron is given by:

$$[x_{-ni}, \ldots, x_0, \ldots, x_{i-1}]^T \tag{1}$$

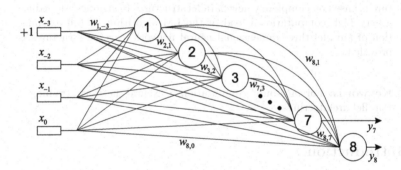

Fig. 1. A fully connected neural network with three inputs, eight neurons and two outputs.

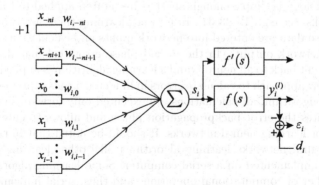

Fig. 2. A neuron model.

where:

$$x_j = \begin{cases} 1 & j = -ni \\ inp_{j+ni} & -ni+1 \le j \le 0. \\ y_j & 0 < j < i \end{cases} \tag{2}$$

The following two equations describe the recall phase of the network:

$$s_i = \sum_{j=-ni}^{i-1} w_{ij} x_j, \tag{3}$$

$$y_i = f(s_i), \tag{4}$$

where $f()$ is the neuron activation function. Correction of network weights is based on the goal function minimization, which is defined as the sum of squared errors of the network outputs for all the presented samples:

$$E = \frac{1}{2} \sum_{p=1}^{np} \sum_{m=1}^{no} \varepsilon_m^{(p)2} = \frac{1}{2} \sum_{p=1}^{np} \sum_{m=1}^{no} \left(d_m^{(p)} - y_m^{(p)} \right)^2. \tag{5}$$

where $\varepsilon_m^{(p)}$ is error and $d_m^{(p)}$ is the desired output in the $p-th$ probe on the $m-th$ output.

The Levenberg-Marquardt algorithm is a modification of the Newton method and is based on the first three elements of the Taylor series expansion of the goal function. In the classical case a change of weights is given by

$$\Delta(\mathbf{w}) = -\left[\nabla^2 \mathbf{E}(\mathbf{w}) \right]^{-1} \nabla \mathbf{E}(\mathbf{w}), \tag{6}$$

where the gradient vector is given by

$$\nabla \mathbf{E}(\mathbf{w}) = \mathbf{J}^T(\mathbf{w}) \varepsilon(\mathbf{w}), \tag{7}$$

the Hessian matrix equals

$$\nabla^2 \mathbf{E}(\mathbf{w}) = \mathbf{J}^T(\mathbf{w}) \mathbf{J}(\mathbf{w}) + \mathbf{S}(\mathbf{w}), \tag{8}$$

and the error vector $\varepsilon(\mathbf{w})$ is a concatenation of the errors for all the input patterns.

$$\varepsilon(\mathbf{w}) = \left[\varepsilon_1^{(1)}, \varepsilon_2^{(1)}, \cdots, \varepsilon_{no}^{(1)}, \cdots, \varepsilon_{no}^{(np)} \right]^T = [\varepsilon_1, \varepsilon_2, \cdots, \varepsilon_r, \cdots, \varepsilon_{no \times np}]^T. \tag{9}$$

The $\mathbf{J}(\mathbf{w})$ in (7) and (8) is the Jacobian matrix

$$\mathbf{J}(\mathbf{w}) = \begin{bmatrix} \frac{\partial \varepsilon_1^{(1)}}{\partial w_{10}} & \frac{\partial \varepsilon_1^{(1)}}{\partial w_{11}} & \cdots & \frac{\partial \varepsilon_1^{(1)}}{\partial w_{ij}} & \cdots & \frac{\partial \varepsilon_1^{(1)}}{\partial w_{nn,nn-1}} \\ \vdots & \vdots & \vdots & \vdots & \vdots & \vdots \\ \frac{\partial \varepsilon_{no}^{(1)}}{\partial w_{10}} & \frac{\partial \varepsilon_{no}^{(1)}}{\partial w_{11}} & \cdots & \frac{\partial \varepsilon_{no}^{(1)}}{\partial w_{ij}} & \cdots & \frac{\partial \varepsilon_{no}^{(1)}}{\partial w_{nn,nn-1}} \\ \vdots & \vdots & \vdots & \vdots & \vdots & \vdots \\ \frac{\partial \varepsilon_{no}^{(np)}}{\partial w_{10}} & \frac{\partial \varepsilon_{no}^{(np)}}{\partial w_{11}} & \cdots & \frac{\partial \varepsilon_{no}^{(np)}}{\partial w_{ij}} & \cdots & \frac{\partial \varepsilon_{no}^{(np)}}{\partial w_{nn,nn-1}} \end{bmatrix}. \tag{10}$$

The Levenberg-Marquardt algorithm is obtained assuming that $\mathbf{S}(\mathbf{w}) = \mu\mathbf{I}$. Now the Eq. (6) takes the form

$$\Delta(\mathbf{w}) = -\left[\mathbf{J}^T(\mathbf{w})\,\mathbf{J}(\mathbf{w}) + \mu\mathbf{I}\right]^{-1}\mathbf{J}^T(\mathbf{w})\,\varepsilon(\mathbf{w}). \tag{11}$$

The $\partial\varepsilon_m^{(p)}/\partial w_{ij}^{(p)}$ derivatives in the Jacobian matrix can be computed in the following way

$$
\begin{aligned}
\frac{\partial\varepsilon_m^{(p)}}{\partial w_{ij}} &= \frac{\partial\varepsilon_m^{(p)}}{\partial y_m^{(p)}}\frac{\partial y_m^{(p)}}{\partial w_{ij}} = -\frac{\partial y_m^{(p)}}{\partial y_i^{(p)}}\frac{\partial y_i^{(p)}}{\partial w_{ij}} \\
&= -\frac{\partial y_m^{(p)}}{\partial y_i^{(p)}}\frac{\partial y_i^{(p)}}{\partial s_i^{(p)}}\frac{\partial s_i^{(p)}}{\partial w_{ij}} = -\frac{\partial y_m^{(p)}}{\partial y_i^{(p)}}f'_i^{(p)}x_j^{(p)}.
\end{aligned}
\tag{12}
$$

By defining

$$\delta_{ki}^{(p)} \overset{\triangle}{=} \frac{\partial\,y_k^{(p)}}{\partial\,y_i^{(p)}}, \quad i \leq k, \tag{13}$$

a formula is obtained:

$$\frac{\partial\varepsilon_m^{(p)}}{\partial w_{ij}} = -\delta_{mi}^{(p)}f'_i^{(p)}x_j^{(p)} = c_{mi}^{(p)}x_j^{(p)}. \tag{14}$$

Of course, the $\delta_{ii}^{(p)}$ has the value

$$\delta_{ii}^{(p)} = \frac{\partial\,y_i^{(p)}}{\partial\,y_i^{(p)}} = 1. \tag{15}$$

The $\delta_{ij}^{(p)}(n)$ values are calculated as follows:

$$
\begin{aligned}
\delta_{ij}^{(p)} &= \frac{\partial\,y_i^{(p)}}{\partial\,y_j^{(p)}} = \frac{\partial\,y_i^{(p)}}{\partial\,s_i^{(p)}}\frac{\partial\,s_i^{(p)}}{\partial\,y_j^{(p)}} \\
&= f'_i^{(p)}\sum_{k=j}^{i-1}w_{ik}^{(p)}(n)\frac{\partial\,y_k^{(p)}}{\partial\,y_j^{(p)}}.
\end{aligned}
\tag{16}
$$

It results to:

$$\delta_{ij}^{(p)} = f'_i^{(p)}\sum_{k=j}^{i-1}w_{ik}^{(p)}\delta_{kj}^{(p)}. \tag{17}$$

The initial values of weights within the network are randomly selected (e.g. from the interval $[-0.5, 0.5]$).

The weights w_{ij} connecting the two neurons and the appropriate deltas δ_{ij} can be inserted into the table (see Table 1) to show the calculation sequence. The calculations of δ_{ij} can be made sequentially row by row from top to bottom of the table. The deltas δ_{ij} in the $i - th$ row can be calculated on the basis of the deltas and the weights from the previous rows (17).

By defining

$$
\begin{aligned}
\mathbf{A} &= -\left[\mathbf{J}^T(\mathbf{w})\,\mathbf{J}(\mathbf{w}) + \mu\mathbf{I}\right], \\
\mathbf{h} &= \mathbf{J}^T(\mathbf{w})\,\varepsilon(\mathbf{w}),
\end{aligned}
\tag{18}
$$

the Eq. (11) takes the following form

$$\Delta\mathbf{w} = \mathbf{A}^{-1}\mathbf{h}. \tag{19}$$

Table 1. The weights between pairs of neurons and deltas δ_{ij} Eqs. (15) and (17)

idx	1	2	$\cdots$	j	$\cdots$	i	$\cdots$	nn
1	δ_{11}	w_{21}	$\cdots$	w_{j1}	$\cdots$	w_{i1}	$\cdots$	w_{nn1}
2	δ_{21}	δ_{22}	$\cdots$	w_{j2}	$\cdots$	w_{i2}	$\cdots$	w_{nn2}
$\cdots$	$\cdots$	$\cdots$	$\cdots$	$\cdots$	$\cdots$	$\cdots$	$\cdots$	$\cdots$
j	δ_{j1}	δ_{j2}	$\cdots$	δ_{jj}	$\cdots$	w_{ij}	$\cdots$	w_{nnj}
$\cdots$	$\cdots$	$\cdots$	$\cdots$	$\cdots$	$\cdots$	$\cdots$	$\cdots$	$\cdots$
i	δ_{i1}	δ_{i2}	$\cdots$	δ_{ij}	$\cdots$	δ_{ii}	$\cdots$	w_{nni}
$\cdots$	$\cdots$	$\cdots$	$\cdots$	$\cdots$	$\cdots$	$\cdots$	$\cdots$	$\cdots$
nn	δ_{nn1}	δ_{nn2}	$\cdots$	δ_{nnj}	$\cdots$	δ_{nni}	$\cdots$	$\delta_{nn,nn}$

The elements of the gradient vector **h** are calculated as follows

$$h_k = \nabla E_k = \nabla E_{ij} = \frac{\partial E}{\partial w_{ij}} = \frac{\partial \frac{1}{2} \sum_{p=1}^{np} \sum_{m=1}^{no} \varepsilon_m^{(p)2}}{\partial w_{ij}}$$
$$= \sum_{p=1}^{np} \sum_{m=1}^{no} \varepsilon_m^{(p)} \frac{\partial \varepsilon_m^{(p)}}{\partial w_{ij}} = \sum_{r=1}^{np \times no} \varepsilon_r \frac{\partial \varepsilon_r}{\partial w_k} = \sum_{r=1}^{np \times no} \varepsilon_r j_{rk}, \tag{20}$$

where j_{rk} is an element of the Jacobian matrix placed in the $r - th$ row and the $k - th$ column and ε_r is $r - th$ element of the error vector. The elements of the **A** matrix are calculated as follows

$$a_{ij} = a_{ij}^0 + \sum_{r=1}^{np \times no} j_{ri} j_{rj}, \tag{21}$$

where

$$a_{ij}^0 = \begin{cases} 0, & i \neq j \\ \mu, & i = j \end{cases} \tag{22}$$

The Eq. (15) can be solved using the QR factorization

$$\mathbf{Q}^T \mathbf{A} \Delta \mathbf{w} = \mathbf{Q}^T \mathbf{h}, \tag{23}$$

$$\mathbf{R} \Delta \mathbf{w} = \mathbf{Q}^T \mathbf{h}. \tag{24}$$

This paper used the Householder reflection method for the QR factorization.

2 Parallel Realisation

The parallel structure of the presented algorithm uses the architecture which requires a number of simple processing elements. The presented parallel solution is realized in three main steps. In the first step the $\partial \varepsilon_m^{(p)} / \partial w_{ij}$ derivatives are calculated. In the next step the **A** matrix and the **h** vector (18) are computed. And in the last step the **QR** decomposition is applied to obtain the $\Delta \mathbf{w}$ and update all weights w_{ij}.

2.1 Calculating the Weight Derivatives Without Error Backpropagation

Fig. 3 shows the two-dimensional base layer of the parallel structure for learning a fully-connected network (Fig. 1) with three inputs, eight neurons and two outputs. It is based on Table 2. The processing elements for the base layer structure are depicted in Fig. 4. The parallel structure (Fig. 3) corresponds to the above Table 1. At the top the processing elements(PEs) connecting neurons with inputs x_i of the network have been added. The x_i input signals are entered into the structure by using pipelined processing (PEs Z). Then, the processing is

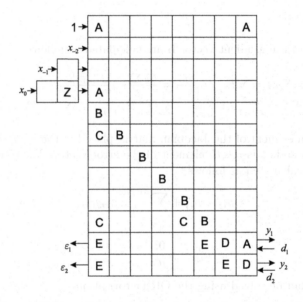

Fig. 3. The parallel two-dimensional base layer for learning neural networks.

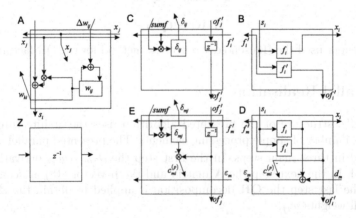

Fig. 4. The processing elements for the base layer for learning neural networks.

performed row by row. The A processing elements store the weights and calculate sums for the following neurons (3) in appropriate columns. The B and D processing elements match the elements of the main diagonal of Table 1. These elements calculate the value of the activation functions f_i and their derivatives f_i'. The network outputs and errors are calculated in the D PEs. Simultaneously, in the C PEs and E PEs the $\delta_{ij}^{(p)}$ deltas (3) are obtained. Additionally, the $c_{mi}^{(p)} = -f_i'^{(p)} x_j^{(p)}$ values are computed in the E PEs. On the presented base layer structure a three-dimensional structure is created (see Fig. 5) for calculating the $\partial \varepsilon_m^{(p)} / \partial w_{ij}$ derivatives (the Jacobian matrix elements). Its processing elements are shown in Fig. 6. The main goal is to provide a full pipelining during the network learning process. This has been fulfilled by supplementing four new processing elements S, T, Z2 and V. Sums from the Eq. (17) are calculated in a pipelined manner row by row through the T, S and finally, C/E processing elements to obtain the δ_{ij} values as a result. The Z2 PEs ensure appropriate previous values of the $\mathbf{x}$ vector to calculate the $c_{mi}^{(p)}$ values. In the V processing

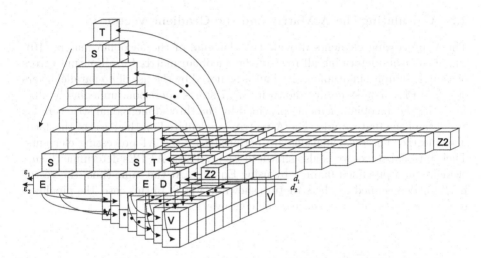

Fig. 5. The three-dimensional structure for calculating the Jacobian matrix elements.

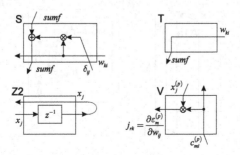

Fig. 6. Additional processing elements of the three-dimensional structure.

elements the $\partial \varepsilon_m^{(p)}/\partial w_{ij}$ derivatives are produced. Note that each layer of the V PEs creates its own vector of the $\partial \varepsilon_m^{(p)}/\partial w_{ij}$ derivatives for another error $\varepsilon_m^{(p)}$ on output m. The operation steps of the above three-dimensional structure are as follows:

- input data are delivered in a pipelined manner;
- successive values are calculated in rows;
- all weights are served simultaneously to the T and S processing elements;
- at the same time known values of δ_{ij} are transmitted from the C/E to the S processing elements that are above them;
- the T and S processing elements forming the "staircase" calculate in a pipelined manner the sums $sumf$ from the Eqs. (17);
- the Z2 PEs supply the previous values of the network inputs and neurons outputs;
- the V PEs calculate the $\partial \varepsilon_m^{(p)}/\partial w_{ij}$ derivatives.

2.2   Calculating the A Matrix and the Gradient Vector

The V processing elements provide the elements of the Jacobian matrix (10). These elements determine all rows of the Jacobian matrix for all output errors of a single sample simultaneously. This is achieved by the use of no parallel layers of the V PEs. The structure shown in Fig. 7 starts operation immediately after receiving the Jacobian elements provided by the three-dimensional structure for the data of the first sample. The **A** matrix is calculated on the basis of the Jacobian matrix. These calculations are performed by the A1 processing elements. Their internal structure is also shown in Fig. 7. These A1 PEs determine the a_{ij} elements in a pipelined manner by using Eq. (21). At the same time, the vector **h** (20) is determined by the structure shown in Fig. 8. In this case, the pipelined processing is also used to shorten the computation time.

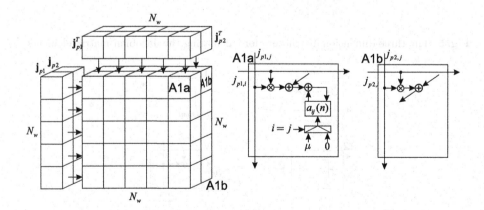

Fig. 7. The structure for computing the **A** matrix and its processing element.

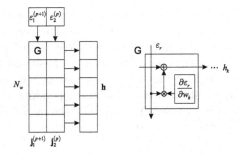

Fig. 8. The structure for computing the **h** vector and its processing element.

2.3   The QR Decomposition Based on the Householser Reflections

After calculating the **A** matrix and the **h** vector, the Eq. (19) can be solved using the QR factorization. This has been achieved by the application of the Hauseholder reflections. The parallel structure calculating matrices **R** and $\mathbf{Q}^T\mathbf{h}$ is presented in Fig. 9 and its processing elements in Fig. 10. Elements A2 and H2 transform the elements of the **A** matrix and the **h** vector. This step performs a sequence of the Householder reflections so as to reset the elements below the

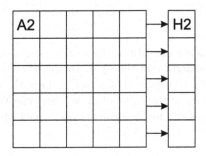

Fig. 9. The general structure for parallelization of the QR decomposition.

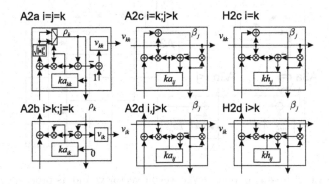

Fig. 10. The processing elements of the QR decomposition.

main diagonal of the **A** matrix. First, the elements in the first column are reset, then the second and so on, until the last but one. The vectors used to perform reflections are based on the columns of the **A** matrix, they include the elements from the main diagonal to the end of the column. It should be noted that the QR decomposition process requires the $w - 1$ matrix reflections. The A2 and H2 processing elements will operate differently depending on the phase (k) of the process. The A2a and A2b elements calculate the module of the **a** subvector and, on this basis, calculate the value

$$\rho_k = \begin{cases} \|\mathbf{a}_k\|_2 & for \ a_{kk} \leq 0 \\ -\|\mathbf{a}_k\|_2 & for \ a_{kk} > 0, \end{cases} \tag{25}$$

and the reflection vector

$$\mathbf{v}_k = \begin{bmatrix} \mathbf{0} \\ \bar{\mathbf{v}}_k \end{bmatrix}. \tag{26}$$

The elements of **v** vector are sent to the elements A2c and A2d which are located in the next columns. There, the values of the reflected vectors $\bar{\mathbf{a}}$ are calculated in the following way

$$\bar{\mathbf{a}}_j = \mathbf{a}_j - \mathbf{v_k}\beta, \tag{27}$$

where

$$\beta = \frac{\mathbf{v}_k^T \mathbf{a}_j}{\gamma}, \tag{28}$$

$$\gamma = v_1. \tag{29}$$

The H2c and H2d elements operate in the same manner on the **h** vector. After computation of the **R** matrix and the $\mathbf{Q^Th}$ vector, the equation (17) is solved. This is performed by the structure shown in Fig. 11 with its elements. The A3a and A3b elements determine the value of $\Delta(\mathbf{w})$, and the W elements update all the weights in the network. The W elements are a part of the A elements from the base layer structure (see Figs. 3 and 4)

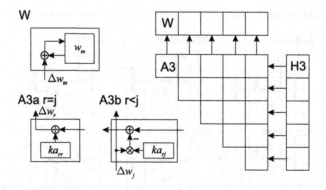

Fig. 11. The structure for computing the weight vector **w** and its processing elements.

3   Computational Results

In both cases, the number of computing cycles has been determined. Tables 2 and 3 show the numbers of computational cycles per one epoch for serial computing and for 3D parallel computing, respectively. The formulas for cycles of addition, multiplication and function computation are presented separately. Symbols i, n, o and p denote the numbers of inputs, neurons, outputs and patterns respectively. To considerable simplicity the formulas in Tables 2 and 3 symbol $w = in + \frac{1}{2}n^2 + \frac{1}{2}n$ has been introduced. The w symbol means number of weights in the network. Figure 12 shows the charts of the number of serial TS and parallel TP computing cycles and the charts of performance factors $PF = TS/TP$ for (a) the neural network with 2 inputs and 1 output, and (b) the network with 20 neurons and 100 samples. For variant (a) it was assumed that there are 10 to 100 learning patterns in each epoch and 2 to 20 neurons in a network. For variant b) it was assumed that there are 10 to 100 inputs and 1 to 10 outputs in a network.

Table 2. The number of cycles per one epoch for serial computing

Op.	Number of cycles per one epoch for serial computing
$+/-$	$\left(\frac{n^3}{6} + w^2 o + wo + ni + o - \frac{n}{6} \right) p + \frac{2w^3}{3} + \frac{3w^2}{2} + \frac{5w}{6} - 1$
$*$	$\left(\frac{n^3}{6} + w^2 o + 2wo + \frac{n^2}{2} + ni + no - \frac{1}{2}o^2 - \frac{2n}{3} - \frac{1}{2}o \right) p + \frac{2w^3}{3} + 2w^2 + \frac{14w}{6} - 3$
f/f'	$2np$

Table 3. The number of cycles per one epoch for 3D parallel computing

Operation	Number of cycles per one epoch for 3D parallel computing
$+/-$	$n + w^2 + 2w - 1$
$*$	$n + i + 8w - 3$
f/f'	$n + p - 1$

4   Conclusions

In this paper the parallel structures of the Levenberg-Marquardt learning algorithm without error backpropagation for a fully-connected feedforward neural network are described. The comparison of computational performance of the parallel structure with a sequential solution has been conducted. The number of computational cycles of the presented parallel architecture is of order $\mathcal{O}(n^4)$ only while in a serial solution this number is of order $\mathcal{O}(n^6)$. The performance factor $(PF = TS/TP)$ of parallel realization achieves about 680 and 6600 for presented examples (see Fig. 12) and it grows fast when the numbers of inputs, neurons, outputs and patterns grow. It can be clearly seen that the performance of the proposed architecture is very promising. A similar parallel approach could be used for other advanced learning algorithms of feedforward neural networks, see e.g. [1–3,11]. In later research it might be possible to make an attempt at designing a parallel realization of learning in other methods [27], and structures [28,29] and various fuzzy [21,22], and neuro-fuzzy structures [34,35].

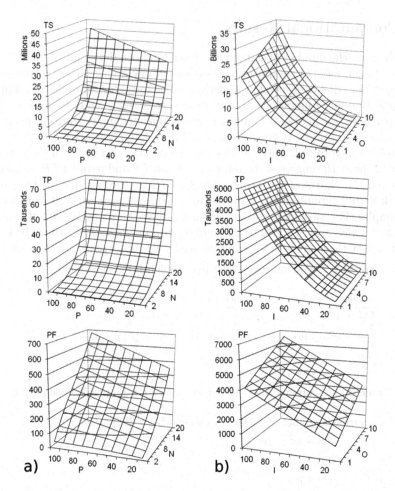

Fig. 12. The number of computing cycles for the serial TS and parallel TP version of the LM algorithm and the performance factors $PF = TS/TP$. In column a) the network with 2 inputs and 1 output is used, the learning patterns are changing from 10 to 100 and number of neurons from 2 to 20 neurons. In column b) the network with 20 neurons and 100 samples is used, the number of inputs was set from 10 to 100 and number of outputs from to 10.

References

1. Bilski, J.: The UD RLS algorithm for training the feedforward neural networks. Int. J. Appl. Math. Comput. Sci. **15**(1), 101–109 (2005)
2. Bilski, J.: Parallel Structures for Feedforward and Dynamical Neural Networks (in Polish). AOW EXIT, Warszawa (2013)
3. Bilski, J.: Momentum Modification of the RLS Algorithms. In: Rutkowski, L., Siekmann, J.H., Tadeusiewicz, R., Zadeh, L.A. (eds.) ICAISC 2004. LNCS (LNAI), vol. 3070, pp. 151–157. Springer, Heidelberg (2004). doi:10.1007/978-3-540-24844-6_18

4. Bilski, J., Litwiński, S., Smolag, J.: Parallel realisation of QR algorithm for neural networks learning. In: Rutkowski, L., Siekmann, J.H., Tadeusiewicz, R., Zadeh, L.A. (eds.) ICAISC 2004. LNCS(LNAI), vol. 3070, pp. 158–165. Springer, Heidelberg (2004). doi:10.1007/978-3-540-24844-6_19

5. Bilski, J., Rutkowski, L.: Numerically robust learning algorithms for feed forward neural networks. In: Rutkowski, L., Kacprzyk, J. (eds.) Neural Networks and Soft Computing. Advances in Soft Computing, vol. 19, pp. 3–14. Physica-Verlag, Heidelberg (2003)

6. Bilski, J., Smolag, J.: Parallel realisation of the recurrent RTRN neural network learning. In: Rutkowski, L., Tadeusiewicz, R., Zadeh, L.A., Zurada, J.M. (eds.) ICAISC 2008. LNCS(LNAI), vol. 5097, pp. 11–16. Springer, Heidelberg (2008). doi:10.1007/978-3-540-69731-2_2

7. Bilski, J., Smolag, J.: Parallel realisation of the recurrent Elman neural network learning. In: Rutkowski, L., Scherer, R., Tadeusiewicz, R., Zadeh, L.A., Zurada, J.M. (eds.) ICAISC 2010. LNCS(LNAI), vol. 6114, pp. 19–25. Springer, Heidelberg (2010). doi:10.1007/978-3-642-13232-2_3

8. Bilski, J., Smolag, J.: Parallel realisation of the recurrent multi layer perceptron learning. In: Rutkowski, L., Korytkowski, M., Scherer, R., Tadeusiewicz, R., Zadeh, L.A., Zurada, J.M. (eds.) ICAISC 2012. LNCS(LNAI), vol. 7267, pp. 12–20. Springer, Heidelberg (2012). doi:10.1007/978-3-642-29347-4_2

9. Bilski, J., Smolag, J.: Parallel approach to learning of the recurrent jordan neural network. In: Rutkowski, L., Korytkowski, M., Scherer, R., Tadeusiewicz, R., Zadeh, L.A., Zurada, J.M. (eds.) ICAISC 2013. LNCS(LNAI), vol. 7894, pp. 32–40. Springer, Heidelberg (2013). doi:10.1007/978-3-642-38658-9_3

10. Bilski, J., Smolag, J., Galushkin, A.I.: The parallel approach to the conjugate gradient learning algorithm for the feedforward neural networks. In: Rutkowski, L., Korytkowski, M., Scherer, R., Tadeusiewicz, R., Zadeh, L.A., Zurada, J.M. (eds.) ICAISC 2014. LNCS(LNAI), vol. 8467, pp. 12–21. Springer, Cham (2014). doi:10.1007/978-3-319-07173-2_2

11. Bilski, J., Smolag, J.: Parallel architectures for learning the RTRN and elman dynamic neural networks. IEEE Trans. Parallel Distrib. Syst. 26(9), 2561–2570 (2015)

12. Bilski, J., Smolag, J., Żurada, J.M.: Parallel approach to the levenberg-marquardt learning algorithm for feedforward neural networks. In: Rutkowski, L., Korytkowski, M., Scherer, R., Tadeusiewicz, R., Zadeh, L.A., Zurada, J.M. (eds.) ICAISC 2015. LNCS(LNAI), vol. 9119, pp. 3–14. Springer, Cham (2015). doi:10.1007/978-3-319-19324-3_1

13. Bilski, J., Wilamowski, B.M.: Parallel learning of feedforward neural networks without error backpropagation. In: Rutkowski, L., Korytkowski, M., Scherer, R., Tadeusiewicz, R., Zadeh, L.A., Zurada, J.M. (eds.) ICAISC 2016. LNCS(LNAI), vol. 9692, pp. 57–69. Springer, Cham (2016). doi:10.1007/978-3-319-39378-0_6

14. Cpalka, K.: A method for designing flexible neuro-fuzzy systems. In: Rutkowski, L., Tadeusiewicz, R., Zadeh, L.A., Żurada, J.M. (eds.) ICAISC 2006. LNCS, vol. 4029, pp. 212–219. Springer, Heidelberg (2006). doi:10.1007/11785231_23

15. Cpałka, K.: Design of Interpretable Fuzzy Systems. Springer, New York (2017)

16. Cpałka, K., Zalasiński, M., Rutkowski, L.: A new algorithm for identity verification based on the analysis of a handwritten dynamic signature. Appl. Soft Comput. 43, 47–56 (2016)

17. Gabryel, M.: The bag-of-features algorithm for practical applications using the MySQL database. In: Rutkowski, L., Korytkowski, M., Scherer, R., Tadeusiewicz, R., Zadeh, L.A., Zurada, J.M. (eds.) ICAISC 2016. LNCS, vol. 9693, pp. 635–646. Springer, Cham (2016). doi:10.1007/978-3-319-39384-1_56

18. Gabryel, M., Grycuk, R., Korytkowski, M., Holotyak, T.: Image indexing and retrieval using GSOM algorithm. In: Rutkowski, L., Korytkowski, M., Scherer, R., Tadeusiewicz, R., Zadeh, L.A., Zurada, J.M. (eds.) ICAISC 2015. LNCS(LNAI), vol. 9119, pp. 706–714. Springer, Cham (2015). doi:10.1007/978-3-319-19324-3_63

19. Fahlman, S.: Faster learning variations on backpropagation: an empirical study. In: Proceedings of Connectionist Models Summer School, Los Atos (1988)

20. Hagan, M.T., Menhaj, M.B.: Training feedforward networks with the Marquardt algorithm. IEEE Trans. Neuralnetw. 5(6), 989–993 (1994)

21. Lapa, K., Przybył, A., Cpałka, K.: A new approach to designing interpretable models of dynamic systems. In: Rutkowski, L., Korytkowski, M., Scherer, R., Tadeusiewicz, R., Zadeh, L.A., Zurada, J.M. (eds.) ICAISC 2013. LNCS(LNAI), vol. 7895, pp. 523–534. Springer, Heidelberg (2013). doi:10.1007/978-3-642-38610-7_48

22. Lapa, K., Cpałka, K., Wang, L.: New method for design of fuzzy systems for nonlinear modelling using different criteria of interpretability. In: Rutkowski, L., Korytkowski, M., Scherer, R., Tadeusiewicz, R., Zadeh, L.A., Zurada, J.M. (eds.) ICAISC 2014. LNCS, vol. 8467, pp. 217–232. Springer, Cham (2014). doi:10.1007/978-3-319-07173-2_20

23. Riedmiller, M., Braun, H.: A direct method for faster backpropagation learning: the RPROP algorithm. In: IEEE International Conference on Neural Networks, San Francisco (1993)

24. Rumelhart, D.E., Hinton, G.E., Williams, R.J.: Learning internal representations by error propagation. In: Rumelhart, D.E., McCelland, J. (eds.) Parallel Distributed Processing, vol. 1. The MIT Press, Cambridge (1986)

25. Smoląg, J., Bilski, J.: A systolic array for fast learning of neural networks. In: Proceedings of V Conference on Neural Networks and Soft Computing, Zakopane, pp. 754–758 (2000)

26. Smoląg, J., Rutkowski, L., Bilski, J.: Systolic array for neural networks. In: Proceedings of IV Conference on Neural Networks and Their Applications, Zakopane, pp. 487–497 (1999)

27. Starczewski, A.: A new validity index for crisp clusters. Pattern Anal. Appl. (2015). doi:10.1007/s10044-015-0525-8

28. Starczewski, J.: Advanced Concepts in Fuzzy Logic and Systems with Membership Uncertainty, vol. 284. Studies in Fuzziness and Soft Computing, pp. 1–304. Springer, New York (2013)

29. Starczewski, J.T., Pabiasz, S., Vladymyrska, N., Marvuglia, A., Napoli, C., Woźniak, M.: Self organizing maps for 3D face understanding. In: Rutkowski, L., Korytkowski, M., Scherer, R., Tadeusiewicz, R., Zadeh, L.A., Zurada, J.M. (eds.) ICAISC 2016. LNCS(LNAI), vol. 9693, pp. 210–217. Springer, Cham (2016). doi:10.1007/978-3-319-39384-1_19

30. Tadeusiewicz, R.: Neural Networks (in Polish). AOW RM, Warsaw (1993)

31. Werbos, J.: Backpropagation through time: What it does and how to do it. Proc. IEEE 78(10), 1550–1560 (1990)

32. Wilamowski, B.M., Yo, H.: Neural network learning without backpropagation. IEEE Trans. Neural Netw. 21(11), 1793–1803 (2010)

33. Wilamowski, B.M., Yo, H.: Improved computation for levenbeeg marquardt training. IEEE Trans. Neural Netw. 21(6), 930–937 (2010)

34. Zalasiński, M.: New algorithm for on-line signature verification using characteristic global features. In: Wilimowska, Z., Borzemski, L., Grzech, A., Świątek, J. (eds.) Information Systems Architecture and Technology: Proceedings of 36th International Conference on Information Systems Architecture and Technology – ISAT 2015 – Part IV. AISC, vol. 432, pp. 137–146. Springer, Cham (2016). doi:10.1007/978-3-319-28567-2_12

35. Zalasiński, M., Cpałka, K.: New algorithm for on-line signature verification using characteristic hybrid partitions. In: Wilimowska, Z., Borzemski, L., Grzech, A., Świątek, J. (eds.) Information Systems Architecture and Technology: Proceedings of 36th International Conference on Information Systems Architecture and Technology – ISAT 2015 – Part IV. AISC, vol. 432, pp. 147–157. Springer, Cham (2016). doi:10.1007/978-3-319-28567-2_13

Spectral Analysis of CNN for Tomato Disease Identification

Alvaro Fuentes[1(✉)], Dong Hyeok Im[2], Sook Yoon[3], and Dong Sun Park[4(✉)]

[1] Department of Electronics Engineering, Chonbuk National University, Jeonju, South Korea
afuentes@jbnu.ac.kr
[2] National Institute of Agricultural Sciences, Jeonju, South Korea
[3] Department of Multimedia Engineering, Mokpo National University, Mokpo, South Korea
syoon@mokpo.ac.kr
[4] IT Convergence Research Center, Chonbuk National University, Jeonju, South Korea
dspark@jbnu.ac.kr

Abstract. Although Deep Convolutional Neural Networks have been widely applied for object recognition, most of the works have often based their analysis on the results generated by a specific network without considering how the internal part of the network itself has generated those results. The visualization of the activations and features of the neurons generated by the network can help to determine the best network architecture for our proposed idea. By the application of deconvolutional networks and deep visualization, in this work, we propose an analysis to determine which kind of images with different color spectrum provide better information to generate a better accuracy of our CNN model. The focus of this study is mostly based on the identification of diseases and plagues on plants. Experimental results on images with different diseases from our Tomato disease dataset show that each disease contains valuable information in the infected part of the leaf that responds differently to other uninfected parts of the plant.

Keywords: Tomato disease · Deep Visualization · Neurons · Convolutional Neural Networks

1 Introduction

Diseases in plants cause considerable production and economic losses in the agriculture sector. Among the most common practices in pest and disease control in crops is still to spray pesticides uniformly over fields at different times during the cultivation cycle [1]. However, most diseases infestations might not be presented across the whole plant, rather in parts of the plant, such as leafs, fruits, stem, etc. Figure 1 shows some examples of how diseases and pest affect a tomato plant. The affection not only includes leafs, but also steam, fruit, root, etc. Therefore, excessive use of pesticides not only increase the costs of production but also in terms of healthier food it can increase the pesticide residual levels in the products.

© Springer International Publishing AG 2017
L. Rutkowski et al. (Eds.): ICAISC 2017, Part I, LNAI 10245, pp. 40–51, 2017.
DOI: 10.1007/978-3-319-59063-9_4

Fig. 1. A representation of the damage caused by diseases and pests in tomato plants.

Traditionally, in order to reduce the use of pesticides, several ways have been used such as targeting pesticides only in the infested part of the plant or plants in the field where they are needed. However, due to the variety of disease for any particular crop, it is still difficult even for a human to ensure that any particular disease is causing the damage, and also find the suitable technique to attack it on time before it is extended to the whole field and cause an extended damage.

Recent advances in computer technology have allowed extending their applications, for example, to plant protection and precision agriculture in general. Several image processing techniques have been used to identify some patterns in images that are treated as diseases in plants. Generally, object recognition and classification using images are some of the most challenging tasks in the area of Computer Vision. But through the recent advances in Deep Learning, the performance of systems aiming to detect or recognize an object in images have been widely improving compared to conventional image processing techniques.

As deep learning techniques have been probing outstanding performance on object recognition, large-scale datasets are needed to increase the accuracy. In the past few years, the PASCAL VOC Challenge 2007/12 [2], and the Large Scale Visual Recognition Challenge (ISLVR) [3] based on ImageNet Dataset have been widely used as benchmarks for image recognition and classification. Starting by the year 2012, Deep Convolutional Neural Networks such as AlexNet [4], followed by ZF [5], VGG [6], GoogleNet [7], ResNet [8], year by year several deep learning architectures are proposed to increase the accuracy and lower the error rate allowing the systems to become more intelligent. These applications of deep learning can be also suitable for different applications allowing users to use them in complex problems when fast and accurate information is needed.

In terms of plant disease detection, different approaches have been used, such as hyperspectral based techniques to identify infested parts of the leafs [1], and image-based techniques using Artificial Neural Networks [9, 10] to classify among diseases. These methods are combined with other image processing techniques to improve the accuracy.

Although Deep CNN have been widely applied for object recognition, most of the works often based their analysis only on the results generated by the network, but do not consider how the internal part of the network has generated those results. In other words, how each neuron in the network contributes to the training and testing stages to generate those results. To achieve that purpose, a technique called Deconvolution Neural Networks has been proposed in [11], which aims to visualize and understand the internal part of convolutional networks. This technique can be useful for discovering the performance contribution for different model layers.

Deep learning systems use images as a source of information, however, most of the images are RGB images, which means they contain color information from three channels taking at a fixed frequency in the visible spectrum. But for some applications, we believe that the results can be improved when the images are collected at a specific frequency with different wavelength values according to the characteristic of the problem itself. For instance, in plant disease identification some diseases show different patterns of the infested part of the plant that are more perceptible than others. In that way, color and spectrum information can be useful when collecting the images and using then to train a CNN.

In this work, we propose a strategy to analyze the influence of color and spectrum information of images to determine the performance of the internal layers and neurons of a CNN based on a deconvolutional method. Using the deep visualization toolbox [11], we aim to visualize the activations and feature maps of our trained CNN model and thus determine the strongest neurons and parts of the network which contribute to generate the expected results.

2   Related Works

Several techniques have been proposed to deal with plant disease identification. According to Sankaran et al. [12], these techniques can be classified into direct and indirect. Direct methods are closely related to a chemical analysis of the infected area of the plant itself. Indirect methods use physical techniques to detect diseases, such as:

- Plant properties/stress based disease detection
- Imaging techniques (Fluorescence and hyperspectral imaging).
- Spectroscopic techniques (Visible, infrared, fluorescence, multispectral bands).

For our analysis, we consider the use of images taken at a visible spectrum for the human eye or RGB images that show the symptoms that differentiate a disease from another.

As the advance of the technology era, recent works in the agricultural area have proposed the use of non-destructive methods to detect diseases in plants. Techniques based on images are used to detect diseases without causing any secondary impact in the plant. However, the accuracy of a real-time system is still a challenge in Computer Vision.

Recent advances in Deep learning have shown good performance especially in terms of accuracy. However, a trade-off of these systems is the need of a large number of data. Some approaches related to our study about plant disease identifications are mentioned below.

In [13], a method applies a neural network to distinguish diseases in wheat. This method is based on color features, shape features and texture features from disease images. Four types of neural networks are used to probe the best method Backpropagation, Radial Basis Function, Generalize Regression Network, Probabilistic Neural Network.

In [9], Convolutional Neural Networks are used to identify 13 different types of plant diseases out of healthy leaves. The images used in this work are download from the internet and correspond to different plants. Another approach in [10] also proposes the use of deep learning technique to identify 14 crop species and 26 diseases.

For our analysis, the problem becomes more challenge, since our work is focused in only one crop. Thus, diseases may be shown into different variations, such as color, shape, illumination, infection status, etc. Physically, some of them seem to be similar, however, they show some characteristics proper of each one. Therefore, in this work, we basically focus our attention in the study of how these characteristics can be used to recognize of the disease.

3   Spectral Analysis of CNN for Tomato Disease

Deconvolutional networks, as proposed in [11], are produced using an input image, which aims to highlight which pixels in that image contribute to a neuron firing. Therefore, interacting with each part of the Deep Neural Network can allow us to understand how they work internally.

3.1   Deep Visualization of CNN

The Deep Visualization allows us to plot the activation values of neurons in the layer of a convolutional network in response to an image or video, based on the model previously trained by the CNN. As deep as the architecture of the model is, the purpose is to visualize the weights and features generated by the network itself.

The main idea of deconvolutional networks is basically to map the activations at high layers back to the input pixel space and show what inputs patterns originally caused a given activation in the feature maps. This deconvolutional operation can be generated

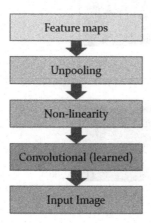

Fig. 2. Deconvolutional network for features visualization.

like a convolutional operation, but in reverse such as un-pooling feature maps and convolving un-pooled maps. In the Fig. 2, we show a representation of the process to visualize the feature maps from an input image. The visualization starts with the feature maps generated by the trained network followed by an un-pooling procedure and decon-volution to the original input image.

3.2   Color Sensitivity of RGB Images

The main purpose of this part is to evaluate the sensitivity of each disease for any specific color. As we have seen from the experience, although there are many inter- and extra-variations among each class, every disease present some specific characteristics, that make them differ from others, e.g. color, texture, shape. Each of those attributes shows different patterns. Thus, by this experiment we aim to visualize what kind of images perform better as input in the CNN Model.

To start with the analysis, we first visualize the distribution of colors in an RGB spectrum for each disease to evaluate the sensitivity to a specific color. Since the

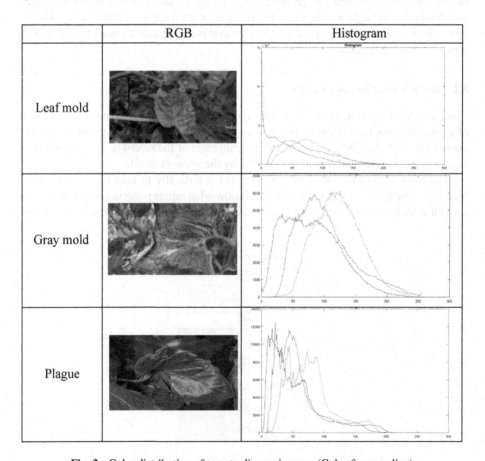

Fig. 3.   Color distribution of tomato disease images. (Color figure online)

common color of the plants is green, all diseases tend to get influenced by that color. But, since the infected part of the leaf shows a different color pattern, each disease seems to respond to that variation in terms of the distribution of colors in the image, as shown in the examples of Fig. 3.

To evaluate the sensitivity of each disease for any specific color, the proposed idea starts with the following procedure:

1. Divide the original image into each RGB channel.
2. Evaluate the response of the neurons in the Convolutional Neural Network of the previously trained CNN model to every image using the deep network visualization tool.

Figure 4, shows an overview of the procedure. In the representation, as mentioned above, RGB channels are extracted from the original image, which are posteriorly used as input to the deep visualization tool. Consequently, the visualization procedure illustrates the activation of each neuron in the network according to the patterns of the input image. As in the example for leaf mold, due to the patterns of the diseases, its activation mostly comes from the Red Color.

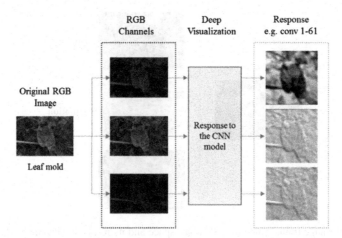

Fig. 4. Color sensitivity and response to the CNN model. (Color figure online)

3.3 Sensitivity to Color with Different Wavelength Values

This experiment consists on evaluating the sensitivity of tomato diseases to different wavelengths applied to the images in order to determine what colors show more influence on the infected part of the leaf.

3.3.1 Visible Spectrum of Images

Images are a representation of the visible light at a specific color spectrum sensible to the human eyes. In terms of frequency, a human eye is able to visualize frequencies in the length between 390 to 700 nm. Following this consideration, in order to determine the sensitivity of tomato diseases, we determine what color spectrum can generate better results or contribute to the activation of neurons more than others.

Having an RGB image as input, we first modify the frequency using different wavelengths, as shown in Fig. 5.

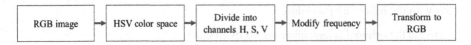

Fig. 5. Procedure to modify the frequency of images

Once the image has been transformed to HSV color space, we modify the channel H (HUE) as shown in Fig. 6. Following, we generate different images, similar to the original RGB image, but with the modified frequencies corresponding to each visible color of the spectrum. Those images are posteriorly used as the input of the deconvolutional network to visualize the activations and feature maps to our pre-trained CNN model.

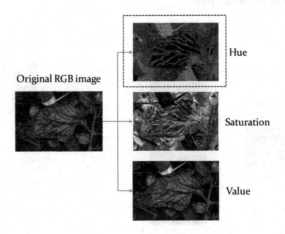

Fig. 6. RGB image transformed to HSV color space and its respective channels. (Color figure online)

4   Experimental Results

Our CNN model (above and below mentioned as our pre-trained CNN) has been previously trained and tested with an Intel Core I7 3.50 GHz Processor and a GPU GeForce GTX Titan X. The model is based on a VGG-16 deep network architecture using Faster R-CNN. For the purpose of this work, our final CNN trained model has been used to perform the experiments. However, since the goal of this work is to the study the internal visualization of network, we mostly focus on the analysis of the deep visualization of activations and features. Following, we present some details of our experiments.

4.1 Dataset Description

Our dataset consists of about 5000 images, which contain tomato diseases and pests in different amounts depending on the conditions when they were taken from. The classes were defined based on the experience, according to the most common diseases that affect the tomato production in South Korea. Therefore, we have identified the following classes, diseases such as Leaf mold, Gray Mold, Canker, Plague, Powdery mildew, and pests such as Leaf miners, Greenhouse whitefly, and some infested plants by secondary effects like due to Nutritional deficiency and Temperature.

Disease in tomato can be produced by different factors such as temperature, humidity excess of nitrogen, fertilizer, shade, light, etc. Based on them we consider the following characteristics for the selection of the samples to be presented in the following work. These are the stage of infection (early, last), type of symptom, Side of the leaf (front, back), type of fungus, color, and shape.

4.2 CNN Activations and Features Visualization

Images containing the infected part as the main part of the foreground provide better information that helps to distinguish some specific patterns of them. Some diseases show some proper features that usually differentiate them from other classes.

4.2.1 Activations of Neurons

Following the idea of the deep visualization, we use the trained model on tomato dataset and applied it to visualize the weights and features for different input images, as the sample shown in Fig. 7.

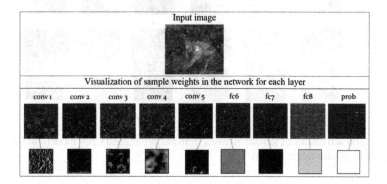

Fig. 7. Activations of our pre-trained CNN model to tomato diseases images

Starting in the first conv layer (conv1), this process allows us to visualize the weights and features of each neuron in the layers until the last one (prob) where the neurons with higher probability are used to classify the input image as its respective class.

48   A. Fuentes et al.

4.2.2   RGB Color Sensitivity

In Fig. 8, we show the responses of some sample images representing each class of tomato diseases and pest for different RGB color channels. Due to its own characteristics, each disease shows to be activated by different colors depending on the pattern of the infected area of the leaf. In the first column, the responses to the original RGB channels are represented. In the following columns, we have interchanged the RGB channels to generate new images. This experiment allows us to determine that some diseases tend to be activated by specific colors such as red in leaf mold or green in gray mold.

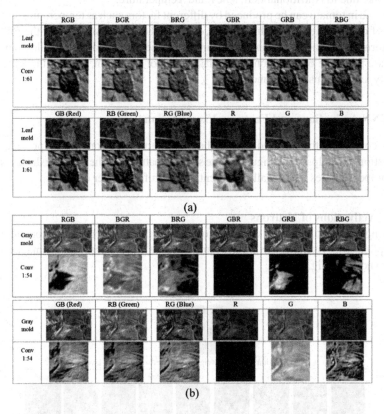

Fig. 8. Neuron activations of tomato diseases from our pre-trained CNN model. (a) Leaf mold, (b) Gray mold. (Color figure online)

4.2.3   Feature Maps

The propose of this experiment starts by generating images with different wavelengths considering RGB images as reference. We aim to visualize which wavelength contributes better to the performance for the network.

Starting from the original RGB image, each image has been generated following the procedure mentioned earlier in the Figs. 5 and 6. Each image used in this experiment corresponds to a different wavelength image in a range between 380 to 750 nm of the following HUE color palette. Each color represents different wavelength, as shown in the Table 1.

Table 1. Wavelengths used to generate the images.

Color	Violet	Blue	Green	Yellow	Orange	Red
Wavelength	380–450	450–495	495–570	570–590	590–620	620–570

In Fig. 9, we show a representation of images with different wavelengths and their response to the weights generated by our pre-trained network in the first convolutional layer. As visualize in Fig. 9, diseases contain features which make them distinctive from others. These characteristics are invariable to the wavelength and thus the information of the infested part of the leaf is preserved. To evaluate the performance in this experiment, we used the same neuron located in the first convolutional layer (conv1-54).

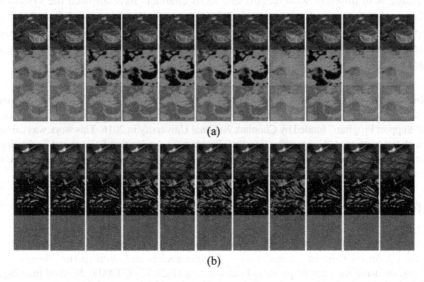

Fig. 9. Feature map responses of tomato disease from our pre-trained CNN model. Starting from the original RGB image (left), each image (to the right) corresponds to a different wavelength image in a range from 0–10 of the HUE color palette. (a) Plague, (b) Canker. (Color figure online)

Images with different wavelengths and RGB channels have allowed the system to perceive what type of color is more sensitive to any specific disease or infested part of the plant. Some diseases like leaf mold and plague shows better visualization at a wavelength which corresponds to the blue, green and violet based colors. Unlike, Canker which infected part of the leaf tend to react better to a frequency related to red. This information will eventually support the re-design of our Deep Learning Model in order to improve the performance and extend the solution to other diseases as well.

5 Conclusion and Future Work

During the last few years, the performance of convolutional neural networks has been drastically increased compared to traditional hand-craft feature methods of computer vision. However, their applications to different areas demand a deep study of the network itself. Our study proposed a strategy for an understanding of the performance of the internal part of the network such as to visualize the activation and feature maps from an input image and its response to the network. As an application for tomato diseases and pest identification.

Images with diseases and pests contain their own characteristics that differentiate them from others such as patterns, colors, status of infection, location in the plant, etc. Neurons in the CNN layers are activated differently based on the information of color or pattern that each image contains.

Image with different wavelengths and RGB channels have allowed the system to perceive what type of color is more sensitive to any specific disease or infested part of the plant. Therefore, by this work, we were able to determine some parameters that will posteriorly help us to re-design of our CNN model and improve the recognition rate as future work.

Acknowledgments. This work was supported by the Brain Korea 21 PLUS Project, National Research Foundation of Korea. This research was supported by the "Research Base Construction Fund Support Program" funded by Chonbuk National University in 2016. This work was carried out with the support of "Cooperative Research Program for Agriculture Science and Technology Development (Project No. PJ0120642016)" Rural Development Administration, Republic of Korea.

References

1. Moshou, D., Bravo, C., Oberti, R., West, J., Bodria, L., Mc Cartney, A., Ramon, H.: Plant disease detection based on data fusion of hyperspectral and multi-spectral fluorescence imaging using Kohonen maps. Real-Time Imaging **11**(2), 75–83 (2005). Spectral Imaging II
2. Everingham, M., Van Gool, L., Williams, C.K., Winn, J., Zissermann, A.: The pascal visual object classes (VOC) challenge. Int. J. Comput. Vision **88**(2), 303–338 (2010). Springer
3. Russakowsky, O., Deng, J., Su, H., Krause, J., Satheesh, S., Huang, Z., Karphathy, A., Khosla, A., Bernstein, M., Berg, A.: ImageNet large scale visual recognition challenge. Int. J. Comput. Vis. (IJCV) **115**(3), 211–252 (2015). Springer
4. Krizhevsky, A., Sutskever, I., Hinton. G.: ImageNet classification with deep convolutional neural networks. In: Advances in Neural Information Processing Systems, vol. 25, pp. 1097–1105 (2012)
5. Zeiler, M.D., Fergus, R.: Visualizing and understanding convolutional networks. In: Fleet, D., Pajdla, T., Schiele, B., Tuytelaars, T. (eds.) ECCV 2014. LNCS, vol. 8689, pp. 818–833. Springer, Cham (2014). doi:10.1007/978-3-319-10590-1_53
6. Simoyan, K., Zisserman, A.: Very deep convolutional networks for large-scale image recognition. In: International Conference on Learning Representations - ICLR 2015 (2015)

7. Szegedy, C., Liu, W., Jia, Y., Sermanet, P., Reed, S., Anguelov, D., Erhan, D., Vanhouke, D., Ravinovich, A.: Going deeper with convolutions. In: Computer Vision and Pattern Recognition - CVPR 2015 (2015)
8. He, K., Zhang, X., Ren, S., Sun, J.: Deep residual learning for image recognition. In: Computer Vision and Pattern Recognition - CVPR 2016 (2016)
9. Sladojevic, S., Arsenovic, M., Anderla, A., Culibr, D., Stefanovic, D.: Deep neural networks based recognition of plant diseases by leaf image classification. Comput. Intell. Neurosci. **6**, 1–11 (2016)
10. Mohanty, S., Hughes, D., Salathe, M.: Using deep learning for image-based plant disease detection. Front. Plant Sci. **7**, 1419 (2016)
11. Yosinski, J., Clune, J., Nguyen, A., Fuchs, T., Lipson, H.: Understanding neural networks through deep visualization. In: International Conference on Machine Learning, Deep Learning Workshop, Lille, France (2015)
12. Sankaran, S., Mishra, A., Ehsani, R.: A review of advanced techniques for detecting plant disease. Comput. Electron. Agric. **72**(1), 1–13 (2010)
13. Wang, H., Li, G., Ma, Z., Li, X.: Application of neural networks to image recognition of plant diseases. In: International Conference on Systems and Informatics (ICSAI) (2012)

From Homogeneous Network to Neural Nets with Fractional Derivative Mechanism

Zbigniew Gomolka[1]($\boxtimes$), Ewa Dudek-Dyduch[2], and Yuriy P. Kondratenko[3,4]

[1] Faculty of Mathematics and Natural Sciences Department of Computer Engineering, University of Rzeszow, ul. Pigonia 1, 35-959 Rzeszow, Poland
zgomolka@ur.edu.pl
[2] Department of Biomedical Engineering and Automation,
AGH University of Science and Technology Cracow, Kraków, Poland
edd@agh.edu.pl
[3] Intelligent Information Systems Department, Petro Mohyla Black Sea State University, 68th Desant-nykiv Street, Mykolaiv 54-003, Ukraine
y_kondrat2002@yahoo.com
[4] Cleveland State University, 2121 Euclid Avenue, Cleveland, OH 44115, USA

Abstract. The paper refers to ANNs of the feed-forward type, homogeneous within individual layers. It extends the idea of network modelling and design with the use of calculus of finite differences proposed by Dudek-Dyduch E. and then developed jointly with Tadeusiewicz R. and others. This kind of neural nets was applied mainly to different features extraction i.e. edges, ridges, maxima, extrema and many others that can be defined with the use of classic derivative of any order and their linear combinations. Authors extend this type ANNs modelling by using fractional derivative theory. The formulae for weight distribution functions expressed by means of fractional derivative and its discrete approximation are given. It is also shown that the application of discrete approximation of fractional derivative of some base functions allows for modelling the transfer function of a single neuron for various characteristics. In such an approach smooth control of a derivative order allows to model the neuron dynamics without direct modification of the source code in IT model. The new approach presented in the paper, universalizes the model of the considered type of ANNs.

1 Introduction

The paper refers to ANNs of the feed-forward type such that neurones of one layer have the same transfer functions and homogeneuos couplings. The paper is inspired by early research of Dudek-Dyduch [4,5] who has proposed modelling of ANNs with the use of calculus of finite differences. Then this way of modelling has been developed by Tadeusiewicz [8], Dyduch [6], Gomolka [10,12] and others. This kind of neural nets were applied mainly to different features extraction i.e. edges, ridges, maxima, extrema and many others that can be defined with the use of classic derivative of any order and their linear combinations. The paper extends the idea by introducing also fractional order derivatives and their

L. Rutkowski et al. (Eds.): ICAISC 2017, Part I, LNAI 10245, pp. 52–63, 2017.
DOI: 10.1007/978-3-319-59063-9_5

discrete approximation into modelling. Then the features that are extracted by the network can be also defined by fractional order derivatives. Moreover, both the weight distribution functions and transfer functions can be described using derivatives of fractional order. More potential tasks for neural networks and more areas of their application can be achieved by introducing into the mathematical model the weight distribution function with fractional derivative mechanism that would allow for smooth transitions between integer orders of selected derivatives. A homogenous network [4, 20] for analysis of one variable function $f(x)$ consists of the layer of receptors that measure the values of a signal in discrete points x_n and consecutive neuron layers. We assumed, for such kind of one-dimensional networks following relations. The input function of the jth layer $e_j(n)$ might be expressed by means of the weight distribution function $w_j(i)$:

$$e_j(n) = \sum_{i=-\infty}^{+\infty} w_j(i) f_{j-1}(n+i) = \sum_{i=-q_j}^{q_j} w_j(i) f_{j-1}(n+i) \qquad (1)$$

where f_j, q_j denote the output function and range of weight distribution function of the jth layer respectively, $j > 0$, n – number of neuron in the considered layer and f_0 denotes input signal at the receptor layer [4, 7, 8]. The flow equation becomes:

$$f_j(n) = h_j(e_j(n)) = h_j\left(\sum_{i=-q_j}^{q_j} w_j(i) f_{j-1}(n+i)\right) \qquad (2)$$

$$f_{j+1}(n) = h_{j+1}(e_{j+1}(n)) =$$
$$h_{j+1}\left(\sum_{i=-q_{j+1}}^{q_{j+1}} w_{j+1}(i) h_j\left(\sum_{l=-q_j}^{q_j} w_j(l) f_{j-1}(n+i+l)\right)\right) \qquad (3)$$

where h_j stands transfer function of the jth layer. For such a model the natural role of neuron couplings consists in some weighted averaging of the signals that reach the neuron. In the co-author earlier works [5, 6] attention was paid to the additional role of these couplings, namely the measurement of differences or a linear combination of differences of adequate signals. There were also given the formula enabling one to determine the weight distribution function in such a way that the input signal is proportional to the required difference (any order, simple or central) of the output function of the preceding layer and, what follows, to any linear combination of the differences. The algorithm was also proposed that enables one to present the input signal, as a linear combination of central differences. The function of weight distribution can be identified with the operator, realized by them. Let us recall that if a weight distribution function is given by the following formula:

$$w(i) = \begin{cases} (-1)^{r-i} \cdot \binom{r}{i} & for\ i \in [0, r] \\ 0 & for\ i \notin [0, r] \end{cases} \qquad (4)$$

then the input signal e is equal to the r-th difference of the given layer output function:

$$e(n) = \sum_{i=-\infty}^{+\infty} w(i) \cdot f(n+i) = \Delta^r f(n) \tag{5}$$

In [7] a synthesis of network that indicates distribution of relative maxims of one-dimensional input signal is given. The network consists of receptors layer (0-layer) and two layers of neurons. The homogeneous weight distribution function of the first layer $w_1(i)$ and the second layer $w_2(i)$ is of the form:

$$w_1(i) = \begin{cases} -\alpha \; for & i = 0 \\ \alpha \; for & i = 1 \\ 0 \; for \; i \neq 0, i \neq 1 \end{cases} \qquad w_2(i) = \begin{cases} \beta \; for & i = 0 \\ -\beta \; for & i = 1 \\ 0 \; for \; i \neq 0, i \neq 1 \end{cases} \tag{6}$$

where parameters $\alpha > 0$ and $\beta > 0$. Both weight functions were set using Eqs. (4) and (5) based on the conditions that define relative maxima. The transfer functions h_1, h_2 are of sigmoid form or its piece linear approximation, where threshold and saturation points has been appropriate adjusted. With this notation the network for maxima detection works well and earlier experiments has proved its functionality [4,9–12].

2   Weight Distribution with Fractional Calculus

The differential calculus of fractional order [18,19] derives its origin from the second half of the 17-th century, when Leibnitz in discussion with L'Hospital asked about existence of the 1/2 order derivatives. Leibniz replied "It leads to a paradox, from which one day useful consequences will be drawn". Looking for these "useful consequences" we took this sentence as the leitmotiv for the next part of this work. Actually there are a few mathematical models that enable to determine a discrete approximation of fractional order derivative [1,3,16–19,22]. Here and later we use known Grunwald-Letnikov definition:

$$_{x_0}D_x^v f(x) = \lim_{h \to 0} \frac{1}{h^v} \sum_{j=0}^{[(x-x_0)/h]} (-1)^j \binom{v}{j} f(x - jh) \tag{7}$$

where $\binom{v}{j}$ means Newton's binomial, while v is fractional order of derivative function $f(x)$, x_0 - states the range of differentiation, h - step of discretization. For a given discrete function $f(x)$ of a real variable x defined on the interval $\langle x_0, x \rangle$ where $0 \leq x_0 \leq x$, we assume the backward difference of v order $_{x_0}\Delta_x^{(v)}$ (fractional or integer) where $v \in \mathbb{R}^+$:

$$_{x_0}\Delta_x^{(v)} f(x) = \sum_{i=0}^{\lfloor (x-x_0)/h \rfloor} a_i^{(v)} f(x - ih) \tag{8}$$

or in the equivalent form:

$$
{}_{x_0}\Delta_x^{(v)} f(x) = \begin{bmatrix} a_0^{(v)} a_1^{(v)} \cdots a_{\lfloor(x-x_0)/h\rfloor}^{(v)} \end{bmatrix} \begin{bmatrix} f(x) \\ f(x-h) \\ \vdots \\ f(x_0) \end{bmatrix} \tag{9}
$$

where the consecutive coefficients $a_i^{(v)}$ are defined as:

$$
a_i^{(v)} = \begin{cases} 1 & for & i=0 \\ (-1)^i \frac{v(v-1)\,(v-2)...(v-i+1)}{i!} & for\ i = 1,2,3,...,N \end{cases} \tag{10}
$$

where N is a number of measurements. For practical reasons it is usually assumed $x_0 = 0$, i.e.:

$$
{}_0\Delta_x^{(v)} f(x) = \sum_{i=0}^{\lfloor x/h \rfloor} a_i^{(v)} f(x - ih) \tag{11}
$$

The ordinary progressive difference might be defined as:

$$
{}_x\Delta_\infty^{(v)} f(x) = \sum_{i=0}^{\infty} a_i^{(v)} f(x + ih) \tag{12}
$$

and similarly, in vector form:

$$
{}_x\Delta_\infty^{(v)} f(x) = \begin{bmatrix} a_0^{(v)} a_1^{(v)} \cdots a_\infty^{(v)} \end{bmatrix} \begin{bmatrix} f(x) \\ f(x+h) \\ \vdots \\ f(\infty) \end{bmatrix} \tag{13}
$$

Consecutive coefficients $a_i^{(v)}$ remain in the relationship:

$$
a_i^{(v)} = a_{i-1}^{(v)} \left(1 - \frac{v+1}{i} \right) \tag{14}
$$

$$
a_i^{(v)} - a_{i-1}^{(v)} = -a_{i-1}^{(v)} \left(\frac{v+1}{i} \right) \tag{15}
$$

$$
a_i^{(v-1)} - a_{i-1}^{(v-1)} = -a_{i-1}^{(v)} \left(\frac{v}{i} \right) \tag{16}
$$

Let us assume for certain N that $f(x) = 0$ for $x \leq x_N \leq \infty$ where $x_N = x + Nh$. Then we have:

$$
{}_x\Delta_{x_N}^{(v)} f(x) = \sum_{i=0}^{\infty} a_i^{(v)} f(x + ih) = \sum_{i=0}^{\lfloor (x_N-x)/h \rfloor} a_i^{(v)} f(x + ih) \tag{17}
$$

or as a vectors product of finite dimensions:

$$_x\Delta_{x_N}^{(v)}f(x) = \begin{bmatrix} a_0^{(v)} & a_1^{(v)} & \cdots & a_{\lfloor(x_N-x)/h\rfloor}^{(v)} \end{bmatrix} \begin{bmatrix} f(x) \\ f(x+h) \\ \vdots \\ f(x_N) \end{bmatrix} \tag{18}$$

At the Fig. 1 we presented exemplary non-integer coefficients $a_i^{(v)}$ for different values of v and assuming $N = 10$. In this paper we adopted such mechanism for calculating the difference of fractional order to achieve new model of the neuron weight distribution and transfer function. In words of image processing that family of weight distribution may acts also as a set of universal kernels for parallel signal processing, e.g. edges or ridges detection at wide spectrum frequencies [2,4,9,11,13–15,21].

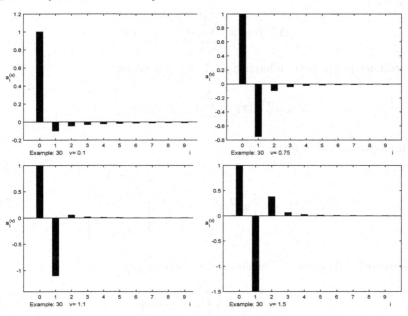

Fig. 1. The fractional weights of v order with N=10 non zero elements

3   Fractional Derivative Inside Neuron Transfer Function

Fractional order derivatives can smoothly change the form of a single neuron transition function, as it is shown below. Considering sigmoid and Gauss (19) as two exemplary transfer functions which most often appear in the ANN learning algorithms, two base functions have been assumed, "sigmoid like function" $h_S(x)$ and "Gauss like function" $h_G(x)$ respectively:

$$h_S^{v_S}(x) = \frac{1}{1+e^{-x}}, \quad h_G^{v_G}(x) = \frac{1}{e^{x^2}} \tag{19}$$

where v_S and v_G denote fractional derivative of appropriate order. The analytical determination of derivative and respectively, integral of total order of the above base functions resulted in obtaining a chosen part of series, presented in the Table 1. In the series presented in the Fig. 2 there was assumed $h_S^{v_S}(x)$, $h_G^{v_G}(x)$ such that $v_S \in \langle 0, \Delta v_S, 3 \rangle$, $v_G \in \langle -2, \Delta v_G, 1 \rangle$, $\Delta v_S = \Delta v_G = 1$. Vertical arrows marked with bold letters D and J show further integer derivatives and integrals respectively of the basic $h_S^{v_S}$ and $h_G^{v_G}$.

Table 1. Series of basis functions

D Sigmoid series	Gauss series
$h_S^{v_S(0)}(x) = \log(1+e^x)$	$h_G^{v_G(-2)}(x) = \iint \left(\frac{1}{e^{x^2}} dx + \frac{\sqrt{\pi}}{2} \right) dx$
$h_S^{v_S(1)}(x) = \frac{1}{1+e^{-x}}$	$h_G^{v_G(-1)}(x) = \frac{\sqrt{\pi}}{4} erf(x) + \frac{1}{2}$
$h_S^{v_S(2)}(x) = \frac{e^x}{(e^x+1)^2}$	$h_G^{v_G(0)}(x) = \frac{1}{e^{x^2}}$
$h_S^{v_S(3)}(x) = -\frac{e^x(e^x-1)}{(e^x+1)^3}$	$h_G^{v_G(1)}(x) = -2\frac{1}{e^{x^2}}x$
$h_S^{v_S(4)}(x) = \frac{e^x(-4e^x+e^{2x}+1)}{(e^x+1)^4}$	$h_G^{v_G(2)}(x) = e^{-x^2}(4x^2-2)$ J

Positive values v_S and v_G are the derivatives of non-integer orders, while negative values v_G are often referred to as antiderivatives. Exemplary transfer functions waveforms of the obtained series were presented below. Particular functions in the consecutive rows of Table 1, composing the series of base functions are shifted in relation to each other of the value $v_S(i) - v_G(i) = 2$. Such assumption allows to group transfer function in relation to their similar shape. In practice it will allow us to use both functions interchangeably controlling the series of derivative or antiderivative computed with respect to base function. Assuming Eq. (19) for sigmoid and gauss respectively, we obtained formulas:

$$_{x_0}\Delta_x^{(v)} h_S(x) = \begin{bmatrix} a_0^{(v)} a_1^{(v)} & \cdots & a_{x_0}^{(v)} \end{bmatrix} \begin{bmatrix} \log(1+e^x) \\ \log(1+e^{x-h}) \\ \vdots \\ \log(1+e^{x_0}) \end{bmatrix} \tag{20}$$

$$_{x_0}\Delta_x^{(v)} h_G(x) = \begin{bmatrix} a_0^{(v)} a_1^{(v)} & \cdots & a_{x_0}^{(v)} \end{bmatrix} \begin{bmatrix} \frac{1}{e^{x^2}} \\ \frac{1}{e^{(x-h)^2}} \\ \vdots \\ \frac{1}{e^{(x_0)^2}} \end{bmatrix} \tag{21}$$

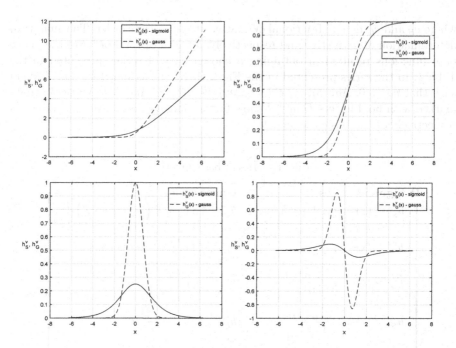

Fig. 2. Selected transfer functions for $v_s = \langle 0, \Delta v_s, 3\rangle$, $v_G = \langle -2, \Delta v_G, 1\rangle$, $h_S^{v_S^{(i)}}(x)$ and $h_G^{v_G^{(i)}}(x)$ respectively

A series of simulation experiments have been carried out for various vector lengths of coefficients of Grunwald-Letnikov non-integer order derivative. Figure 3 shows exemplary characteristics assuming the length of fractional coefficients vector $\boldsymbol{a}^{(v)}$, $n = 250$ and input base function in the form of $h_S^{v_S^{(0)}}(x) = \log(1 + e^x)$. Similar experiments were conducted with the use of Reimann-Liouville derivative-integral obtaining analogous fluency of hypothetical neuron transfer function shaping (Fig. 4).

4   The Fractional Mechanism Within 2D Homogeneous Network

By extending the network for one dimensional signal processing described in Sect. 1 to the two dimensional case we got architecture presented in the Fig. 5. It consists of two subsystems working on two perpendicular directions X and Y respectively. The extended form of the weight distribution given by the matrix W_1^x, W_1^y, W_2^x, W_2^y regarding fractional distribution becomes (for the first and the second layer respectively in the X and Y direction):

$$W_{1[M_x \times M_y \times K_x^1]}^x \text{ and } W_{2[M_x \times M_y \times K_x^2]}^x \tag{22}$$

$$W_{1[M_x \times M_y \times K_y^1]}^y \text{ and } W_{2[M_x \times M_y \times K_y^2]}^y \tag{23}$$

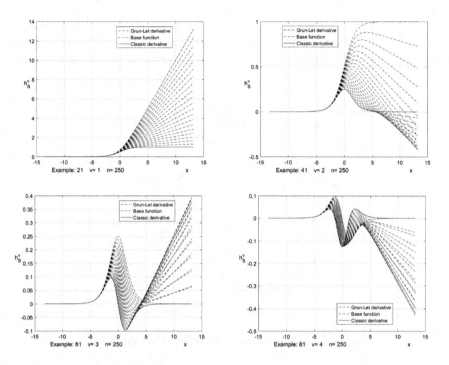

Fig. 3. Waveforms of discrete Grunwald-Letnikov derivatives for fluently shaping order v and used vector of N fractional coefficients

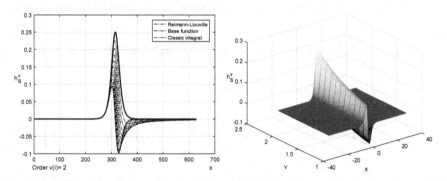

Fig. 4. Waveforms of discrete Reimanna-Liouville derivative-integral for fluently shaping order v within range $v_S = \langle 2, \Delta v_G, 3 \rangle$, $\Delta v_G = 0.1$

$$w_1^x(i,j,k) = \begin{cases} a_k^{(v_{x_1})} & for\ i,j \in [0, M_x - 1] \wedge k \in [0, K_x] \\ 0 & for\ i,j \notin [0, M_x - 1] \vee k \notin [0, K_x] \end{cases} \quad (24)$$

$$w_1^y(i,j,k) = \begin{cases} a_k^{(v_{y_1})} & for\ i,j \in [0, M_y - 1] \wedge k \in [0, K_y] \\ 0 & for\ i,j \notin [0, M_y - 1] \vee k \notin [0, K_y] \end{cases} \quad (25)$$

where (i,j) are coordinates of neurones, M_x, M_y stands the number of the neurons in particular layer of subsystem X and Y respectively, K_x and K_y denotes number of non zero coefficients of $a^{(v)}$ at the direction X and Y respectively.

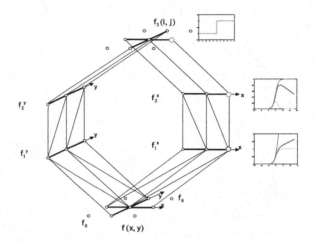

Fig. 5. Two dimensional network for ridges detection

Transfer functions in the same layers for the both directions are the same and can be written in the form (in general case they do not have to be the same):

$$f_1^x (i,j) = {}_{K_x} \Delta_i^{(v)} (e_1 (i,j)) = {}_{K_x} \Delta_{e_i}^{(v)} e_1^x \left(\sum_{k=0}^{k=K_x} w_1^x (i,j,k) f_0 (i-k,j) \right) \quad (26)$$

and consequently on perpendicular direction:

$$f_1^y (i,j) = {}_{K_y} \Delta_j^{(v)} (e_1 (i,j)) = {}_{K_y} \Delta_{e_j}^{(v)} e_1^y \left(\sum_{k=0}^{k=K_y} w_1^y (i,j,k) f_0 (i,j-k) \right) \quad (27)$$

respectively. The weights distribution functions in the second layer can be formulated as follow:

$$w_2^x(i,j,k) = \begin{cases} a_k^{(v_{x_2})} & for\ i,j \in [0, M_x - 1] \wedge k \in [0, K_x] \\ 0 & for\ i,j \notin [0, M_x - 1] \vee k \notin [0, K_x] \end{cases} \quad (28)$$

and

$$w_2^y(i,j,k) = \begin{cases} a_k^{(v_{y_2})} & for\ i,j \in [0, M_y - 1] \wedge k \in [0, K_y] \\ 0 & for\ i,j \notin [0, M_y - 1] \vee k \notin [0, K_y] \end{cases} \quad (29)$$

respectively. Above equations do not include boundary conditions like classical or circular convolution has. In consequence of the above notation the output signals in the second layer can be written in the form:

$$f_2^x(i,j) = \kappa_x \Delta_i^{(v)}(e_2(i,j)) = \kappa_x \Delta_i^{(v)} e_2^x \left(\sum_{k=0}^{k=K_x} w_2^x(i,j,k) f_1(i-k,j) \right) \quad (30)$$

and

$$f_2^y(i,j) = \kappa_y \Delta_j^{(v)}(e_2(i,j)) = \kappa_y \Delta_j^{(v)} e_2^y \left(\sum_{k=0}^{k=K_y} w_2^y(i,j,k) f_1(i,j-k) \right) \quad (31)$$

The output signals of the second layers $f_2^x(i,j)$ and $f_2^y(i,j)$ respectively, have to be accumulated by the third layer. Because both matrixes of the output signals F_2^x and F_2^y are of the same size, we can build matrix $F_2 = \left[F_2^x \ F_2^y \right]^T$. The last third layer accumulates signals from the both directions so its weights may be assumed also by an augmented eye weight matrix which makes summation of f_2^x and f_2^y possible. This notation refers in some details to the multi-diagonal matrix notation described in earlier works [4,12]. Thus, in the general case, the network output signal $f_k(i,j)$ which is simultaneously output signal of the last layer determines the transformation $\aleph$ made by the entire network (two dimensional case)4:

$$f_k(i,j) = \aleph F(i,j) \quad (32)$$

which can be additionally presented as a sequence of several superpositions of the given type. The range of convolution Eqs. (26) and (27) will depend on the value of v, and will determine the number of nonzero elements of weight vector. The number of non-zero vector of weight coefficients can be limited by quantitative or qualitative method, as in the case of data compression algorithms. That also has important impact on the number of necessary computations of the whole system.

5   Conclusion

The mechanism of fractional derivative for modelling of feed-forward neural network architectures has been presented in the paper. Fractional range presumed derivative can smoothly modify the types of tasks performed by the neural network. The step of discrete derivative approximation must take into account the class of the signal fed to the input of the network. In the same way, mechanism of fractional order derivative can be used for modelling transfer functions of single neuron. Using dynamics variable adjustment of such function will allow the construction of network learning algorithms without the necessity of source code modification that implements particular types of traditional functions, e.g. sigmoid, hyperbolic, tangent. For integer values of v, presented neural network behaves as classical net with architecture presented in earlier works. Further experiments with wide range fluctuation of v and its impact on the whole performance of the net will be the next stage of the presented work. Accuracy in approximation of derivative or derivative-integral values in calculations of step discretization, determines the accuracy of numerical approximation. Proposed base functions with adopted fractional derivative mechanism have been successfully tested with classic back-prop net architecture under the XOR problem.

Acknowledgments. The research were conducted in the scientific cooperation between Faculty of Mathematics and Natural Sciences Department of Computer Engineering at University of Rzeszow and AGH University of Science and Technology Cracow, Department of Biomedical Engineering and Automation. The studies were conducted in the laboratory of Computer Graphics and Digital Image Processing at Center for Innovation and Transfer of Natural Sciences and Engineering Knowledge of Rzeszow University. Grants: WMP/GD-11/2016 and AGH-UST 11.120.417.

References

1. Chi, C., Gao, F.: Simulating fractional derivatives using matlab. J. Softw. **8**(3), 572–578 (2013)
2. Chang, O., Constante, P., Gordon, A.: A novel deep neural network that uses space-time features for tracking and recognizing a moving object (2017). https://doi.org/10.1515/jaiscr-2017-0009. Accessed 23 Feb 2017
3. Vance, D.: Fractional derivatives and fractional mechanics, 2 June 2014
4. Dudek-Dyduch, E., Tadeusiewicz, R.: Neural networks indicating maxima and ridges in two-dimensional signal. In: Engineering Applications of Artificial Neural Networks, pp. 485–488. FAIS- Helsinki (1995)
5. Dudek-Dyduch, E.: Artificial neuron network indicating distribution of local maxima of input signals Ph.D. thesis, AGH-University of Science and Technology (1977)
6. Dudek-Dyduch, E., Dyduch, T.: Application of neural networks in 3D object recognition system. Int. J. Pattern Recog. Artif. Intell. **12**(4), 491–504 (1998)
7. Dudek-Dyduch, E.: Synthesis of feed forward neural network indicating extremes. Int. J. Syst. Anal. Model. Simul. **24**(1996), 135–151 (1996)
8. Dudek-dyduch, E., Tadeusiewicz, R., Horzyk, A.: Neural network adaptation process effectiveness dependent of constant training data availability. Neurocomputing **72**(13), 3138–3149 (2009). ISSN:0925–2312
9. Gomolka, Z., Twarog, B., Bartman, J.: Improvement of image processing by using Homogeneous neural networks with fractional derivatives theorem. Discret. Contin. Dyn. Syst. **31**(Supplement), 505–514 (2011)
10. Gomolka, Z., Twarog, B.: Artificial intelligence methods for image processing. In: Symbiosis of Engineering and Computer Science, pp. 93–124 (2010)
11. Gomolka, Z., Twarog, B., Kwiatkowski, B.: The fractional order operators applied in the image processing. In: Computing in Science and Technology, Monographs in AI, pp. 77–96 (2013)
12. Gomolka, Z.: Neural network for fringe image analysis, Academy of Mining and Metallurgy Cracow, Ph.D. thesis (2000)
13. Kondratenko, Y., Gordienko, E.: Implementation of the neural networks for adaptive control system on FPGA. In: DAAAM International, Vienna, Austria, EU, pp. 0389–0392 (2012)

14. Kondratenko, Y.P., Kozlov, O.V., Gerasin, O.S., Zaporozhets, Y.M.: Synthesis and research of neuro-fuzzy observer of clamping force for mobile robot automatic control system. In: IEEE First International Conference on Data Stream Mining & Processing (DSMP), pp. 90–95 (2016). doi:10.1109/DSMP.2016.7583514

15. Korytkowski, M., Rutkowski, L., Scherer, R.: Fast image classification by boosting fuzzy classifiers. Inf. Sci. **327**, 175–182 (2016)

16. Neel, M.C., Joelson, M.: Generalizing Grunwald-Letnikovs formulas for fractional derivatives, ENOC-2008, Saint Petersburg, Russia, June 30–July 4 (2008)

17. Chakraborty, M., Maiti, D., Konar, A., Janarthanan, R.: A study of the Grunwald-Letnikov definition for minimizing the effects of random noise on fractional order differential equations. In: 4th IEEE International Conference on Information and Automation for Sustainability (2008)

18. Ostalczyk, P.: On simplified forms of the fractional-order backward difference and related fractional-order linear discrete-time system description. Bull. Polish Acad. Sci. Techn. Sci. **63**(2), 423–433 (2015)

19. Ostalczyk, P.: Discrete Fractional Calculus Applications in Control and Image Processing. World Scientific Publishing Co., Pte. Ltd., Singapore (2016)

20. Tadeusiewicz, R.: Neural networks as a tool for modelling of biological systems. Bio Algorithms Med. Syst. **11**(3), 135–144 (2015). doi:10.1515/bams-2015-0021

21. Tadeusiewicz, R., Chaki, R., Chaki, N.: Exploring Neural Networks with C#. CRC Press, Taylor & Francis Group, Boca Raton (2014)

22. Garg, V., Singh, K.: An improved Grunwald-Letnikov fractional differential mask for image texture enhancement. IJACSA **3**(3), 130–135 (2012)

Neurons Can Sort Data Efficiently

Adrian Horzyk[(✉)]

Department of Automatics and Biomedical Engineering, AGH University
of Science and Technology, Mickiewicza Av. 30, 30-059 Cracow, Poland
horzyk@agh.edu.pl

Abstract. This paper introduces an efficient sorting algorithm that uses
new models of receptors and neurons which apply the time-conditional
approach characteristic for nervous systems. These models have been
successfully applied to automatically construct neural graphs that con-
solidate representation of all sorted objects and relations between them.
The introduced parallely working algorithm sorts objects simultaneously
for all attributes constructing an active associative neural graph representing all sorted objects in linear time. The sequential version works in
linear or sub-linearithmic time. The paper argues that neurons can be
used for efficient sorting of objects and the constructed network can be
further used to explore relationships between these objects.

Keywords: Brain-inspired computations · Neuron models · Sorting
algorithms · ASSORT · Associative representation · Computational
complexity

1 Introduction

Brains are powerful biological computational machines that allow us to quickly
and intelligently react to various situations, learn, represent knowledge, and
think [6,9,14]. They use neurons that can automatically adapt, specialize, and
cooperate to represent complex data and various combinations of input stimuli
coming from other neurons or various receptors [5,11]. The cooperation is possible due to the conditionally triggered plasticity processes which allow neurons
to arrange data in order and represent objects [8,10]. Spiking neurons more precisely reproduce biological processes of real neurons, but they are still difficult to
use to solve real engineering tasks and compete with other computational methods [7]. Modern deep learning strategies hierarchically combine various neural
networks and usually achieve better generalization results than other learning
strategies [3,12]. The contemporary neural networks and learning strategies do
not yet model knowledge or intelligence satisfactorily, so scientists conduct further research and investigations of unexplained features and functions of biological neurons to reveal useful mechanisms that can expand our knowledge and
the abilities of artificial neural networks [2,9,13].

Sorting algorithms are very useful and widely used in contemporary computer
science because they determine the efficiency of search operations and other algorithms based on the Turing machine computational model. The major part of

© Springer International Publishing AG 2017
L. Rutkowski et al. (Eds.): ICAISC 2017, Part I, LNAI 10245, pp. 64–74, 2017.
DOI: 10.1007/978-3-319-59063-9_6

sorting algorithms compares sorted data and uses the results of comparisons to change the order of elements, e.g. quicksort or heapsort. A few others count up data of the same values, e.g. counting sort or radix sort, in order to compute the destination positions of sorted elements in the defined order [1]. The introduced ASSORT-2 algorithm uses the local measures of similarities that are computed by sensors which differentiate moments of neural activities that determine in which positions new sorted elements will be placed. In this algorithm, no values are directly compared or counted up, but activated neurons stimulate other connected neurons in which a plasticity process can be conditionally started, which leads to an insertion of new sorted elements in order. This paper introduces an improved brain-inspired sorting algorithm based on new models of neurons and sensors with built-in plasticity rules that automatically constructs and develops an associative neural graph to achieve a neural structure representing sorted objects by all attributes simultaneously. The correctness of the described algorithm will be proven and its computational complexity will be estimated in the following sections.

In the previous research, the ASSORT algorithm was proposed. It operated only on the positive numbers and the neurons combined with the built-in sensors [5]. This paper extends this model with an ability to sort any numbers and even symbolic data that can be ordered and for which a measure of distance between them can be defined. It also enriches this algorithm with new models of sensors and neurons with predefined plastic built-in rules that automatically organize and connect neurons to achieve a neural structure of sorted objects simultaneously for all attributes using time as a computational factor. Thus, this paper argues that it is beneficial to expand artificial models of neurons with some predefined plasticity mechanisms to automatically develop an appropriate neural network structure for given data as it is working in nature thanks to a genetic code inside nuclei of biological neurons [8,10,11]. This paper reveals how plasticity rules can make neurons to work collectively using brain-inspired mechanisms [4,5] and sort objects without comparisons of sorted values or computing their positions in the sorted sequence. In the ASSORT-2 algorithm, sorted values are presented to sensors that stimulate neurons with the strength corresponding to the similarities of the represented values by the sensors to the value presented on the input sensory field where these sensors are located. A single rule will be used to conditionally start a plasticity process that can appropriately and automatically construct and configure a structure of neurons representing all objects sorted according to all their attributes simultaneously.

The introduced sorting strategy inspired by biological neural plasticity processes and the biological receptors and the senses [8,10,11] is very efficient and can be used for sorting data as well as other fast sorting algorithms as quick sort, heap sort, or counting sort [1]. The ASSORT-2 algorithm is parallel in its nature as well as biological neural processes in nervous systems. It works in linear time. It has been also adapted to work on a sequential Turing machine where it has sub-linearithmic or linear time complexity depending on the sorted data.

2 Models of Neurons, Receptors, and the Senses

Biological neurons are living cells that are preprogrammed by the genetic code to perform various functions in nervous systems. This code enables the diversity of neurons, their functions, and structures. Artificial neurons should also have an artificial genetic code that determines their behavior and conditional plasticity rules to optimize their structure for given data and attributes. Therefore, we have to create a neural structure from scratch, defining some local plasticity rules for neural network elements replacing the real genetic code. In the presented model, sensors, sensory neurons, and object neurons use such rules.

The sensory neurons are preprogrammed to create exactly two connections to other sensory or extreme neurons (described in Subsect. 2.2), which is necessary for sequential sorting. Each sensory neuron is also preprogrammed to connect to a single sensor. Sensors are highly sensitive to the represented unique values and less sensitive to all other values of the attribute represented by the sensory field associated with these sensors. Each sensor measures the similarity of the represented value to an input value presented on the input sensory field in which this sensor is placed. The plastic input sensory fields can represent the biological senses and other sensors that we cannot meet in nature, but they are useful from the practical point of view in order to optimize neural structures for various data and attributes.

We assume to develop an associative neural graph structure from scratch (Fig. 1) that will be used to store all objects $\{o_1, ..., o_N\}$ sorted simultaneously according to all attributes $\{a_1, ..., a_K\}$ with aggregated duplicates. The sorted objects are introduced to this neural graph subsequently. Each sorted object is defined as $o_n = (v_{n_1}^{a_1}, ..., v_{n_K}^{a_K})$ where $v_{n_k}^{a_k}$ is a value of its attribute a_k. The attribute values of the sorted objects are simultaneously presented on the adequate sensory inputs fields that distribute these values to all connected sensors.

2.1 Sensory Fields and Sensors

Receptors of the biological senses (sight, hearing, touch, taste, smell) are capable of reacting to a certain type and limited range of input stimuli with different strength according to its type, location, and the intensity, length, and frequency of an input stimulus. The biological senses, as well as other attributes, are here modeled using plastic input sensory fields that consist of sensors which are sensitive to a limited range of values defined by sorted objects. All input sensory fields are created from scratch, so it is necessary to automatically define and update ranges of sensed values after presented input data on them. Therefore we need to create a variable r^{a_k} in each sensory field F^{a_k} to store and update the experienced range of data values of the attribute a_k:

$$r^{a_k} = v_{max}^{a_k} - v_{min}^{a_k} \quad where\ v_{max}^{a_k} = max\{v_i^{a_k}\},\ v_{min}^{a_k} = min\{v_i^{a_k}\} \quad (1)$$

where $v_{min}^{a_k}$ and $v_{max}^{a_k}$ are the minimum and maximum already experienced values from all values $v_i^{a_k}$ of the attribute a_k on the sensory field F^{a_k} stored in

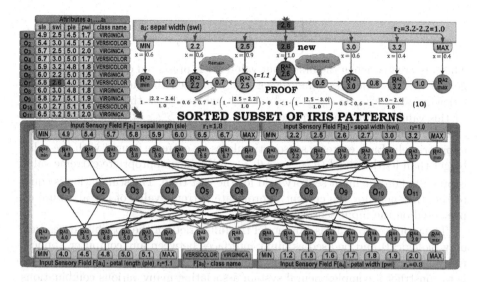

Fig. 1. The associative neural graph constructed using the associative sorting for all attributes simultaneously with an illustration of the proof (13)–(14) on the sample data.

the extreme sensors $S_{min}^{a_k}$ and $S_{max}^{a_k}$. Each plastic input sensory field F^{a_k} is preprogrammed to represent a new value v^{a_k} of the object attribute a_k and create new sensors $S_v^{a_k}$ for the presented value v^{a_k} that is sufficiently different $(\forall_i d(v^{a_k}, v_i^{a_k}) > \varepsilon^{a_k})$ from the other values $v_i^{a_k}$ already represented by the existing sensors $S_{v_i}^{a_k}$, where the difference $d(v^{a_k}, v_i^{a_k})$ between v^{a_k} and $v_i^{a_k}$ is defined as follows:

$$d(v^{a_k}, v_i^{a_k}) = |v^{a_k} - v_i^{a_k}| \qquad (2)$$

The information about the sufficiently different input value v^{a_k} from the values $v_i^{a_k}$ represented by the existing sensors $S_{v_i}^{a_k}$ can be driven from the lack of the strong enough reactions of these sensors to this value presented on the input sensory field F^{a_k}. If the minimum difference (2) of all existing sensors $S_{v_i}^{a_k}$ representing the values $v_i^{a_k}$ is bigger than the defined minimum difference $(\forall_i d(v^{a_k}, v_i^{a_k}) > \varepsilon^{a_k})$ then a new sensor $S_v^{a_k}$ is created by the input sensory field F^{a_k} to represent this new input value (Fig. 1). If necessary the range r^{a_k} and the minimum $v_{min}^{a_k}$ and maximum $v_{max}^{a_k}$ experienced values represented in the extreme sensors are automatically updated (1). Each parallelly working sensor $S_{v_i}^{a_k}$ reports back its input sensory field F^{a_k} if it differs from the presented input value v^{a_k} less than the defined minimum difference $(d(v^{a_k}, v_i^{a_k}) \le \varepsilon^{a_k})$ in order to prevent the creation of a new sensor for the value v^{a_k}. Sorting requires the maximum precision and no approximation, so we assume here that $\forall_{a_k} \varepsilon^{a_k} = 0$.

The extreme sensors $S_{min}^{a_k}$ and $S_{max}^{a_k}$ are highly sensitive to the existing and new extreme values presented on the input sensory field F^{a_k}:

$$x_{min}^{a_k} = \begin{cases} \frac{v_{max}^{a_k} - v^{a_k}}{r^{a_k}} & if\ r^{a_k} > 0 \\ v_{min}^{a_k} - v^{a_k} + 1 & if\ r^{a_k} = 0 \end{cases} \qquad (3)$$

$$x_{max}^{a_k} = \begin{cases} \frac{v^{a_k} - v_{min}^{a_k}}{r^{a_k}} & if \ r^{a_k} > 0 \\ v^{a_k} - v_{max}^{a_k} + 1 & if \ r^{a_k} = 0 \end{cases} \tag{4}$$

They are created when the first input value v^{a_k} is presented on the input sensory field F^{a_k}. After the first value sensor $S_{v_i}^{a_k}$ is created they are initialized in the following way: $v_{min}^{a_k} = v_{max}^{a_k} = v_i^{a_k} = v^{a_k}$, where $v_i^{a_k}$ is the value represented by this value sensor. The extreme sensors $S_{min}^{a_k}$ and $S_{max}^{a_k}$ stimulate appropriate extreme neurons $R_{min}^{a_k}$ and $R_{max}^{a_k}$ with the strength computed by (3) and (4) appropriately. Each extreme sensor stimulates its extreme neuron with the value bigger than its threshold ($\theta_{R_{v_{min}}^{a_k}}$ or $\theta_{R_{v_{max}}^{a_k}}$) if a new appropriate extreme value is presented on the input sensory field for the first time. The stimulation is equal to this threshold for the current appropriate extreme value of the attribute a_k presented on the input sensory field.

The input sensory fields can model the full variety of the possible senses (not only biological) and be used as input interfaces between any computer application and an internal neural representation of objects and their relations. The goal is to construct a complex neural system associating many various combinations of input stimuli [5]. The input sensory fields contain sensors. The same values of each attribute are aggregated by the same sensors, so we usually create much fewer sensors than the number of sorted objects. All value sensors $S_{v_i}^{a_k}$ located in the sensory field F^{a_k} simultaneously react to the presented input value v^{a_k} with different strength (5) according to the difference of the represented and sensed values:

$$x_{v_i}^{a_k} = \begin{cases} 1 - \frac{|v_i^{a_k} - v^{a_k}|}{r_i^{a_k}} & if \ r^{a_k} > 0 \\ \frac{|v_i^{a_k}|}{|v_i^{a_k}| + |v_i^{a_k} - v^{a_k}|} & if \ r^{a_k} = 0 \end{cases} \tag{5}$$

Each such sensor $S_{v_i}^{a_k}$ is highly sensitive to the value $v_i{}^{a_k}$ it represents and appropriately less sensitive to the other presented values from the already experienced range (1). The stimulated value sensors $S_{v_i}^{a_k}$ continuously stimulate the connected sensory neurons $R_{v_i}^{a_k}$ and charge them as long as the value v^{a_k} is presented on the input sensory field F^{a_k}. Thus, each sensory neuron can achieve its activation threshold in a different time (6) according to the strength it has been stimulated by the sensor and the period of time it has been stimulated:

$$t_{v_i^{a_k}} = \begin{cases} \frac{r^{a_k}}{\theta_{R_{v_i}^{a_k}} \left(r^{a_k} - |v_i^{a_k} - v^{a_k}| \right)} & if \ |v_i^{a_k} - v^{a_k}| < r^{a_k} \\ \infty & if \ |v_i^{a_k} - v^{a_k}| = r^{a_k} \\ 1 + \frac{|v_i^{a_k} - v^{a_k}|}{|v_i^{a_k}|} & if \ r^{a_k} = 0 \end{cases} \tag{6}$$

2.2   Extreme, Sensory and Object Neurons

In nature, actions elicit reactions, so immediately after the creation of the sensors $S_{v_i}^{a_k}$, $S_{min}^{a_k}$ and $S_{max}^{a_k}$ new appropriate sensory neurons $R_{v_i}^{a_k}$ and extreme neurons $R_{min}^{a_k}$ and $R_{max}^{a_k}$ are created, stimulated, and activated. Each sensory neuron is preprogrammed to create exactly two weighted connections to any sensory or extreme neurons. Each extreme neuron is preprogrammed to create exactly a single connection to any sensory neuron. In this computational model, sensory and extreme neurons use virtual noticeboards $B^{a_1}, ..., B^{a_K}$ built-in all sensory fields $F^{a_1}, ..., F^{a_K}$ to send advertisements about their will to connect. These noticeboards match advertisements and send back to advertisers information about the matched neurons to which advertisers should connect. Each noticeboard B^{a_k} can contain only two announcements, so they are processed in constant time. The connections weights between sensory and extreme neurons are equal the activation thresholds of the postsynaptic neurons:

$$
\begin{aligned}
w_{R^{a_k}_{min}, R^{a_k}_{v_{min}}} &= \theta_{R^{a_k}_{v_{min}}} & w_{R^{a_k}_{max}, R^{a_k}_{v_{max}}} &= \theta_{R^{a_k}_{v_{max}}} \\
w_{R^{a_k}_{v_{min}}, R^{a_k}_{min}} &= \theta_{R^{a_k}_{min}} & w_{R^{a_k}_{v_{max}}, R^{a_k}_{max}} &= \theta_{R^{a_k}_{max}}
\end{aligned}
\tag{7}
$$

The activation threshold of the sensory neuron $R^{a_k}_{v_i}$ is denoted as $\theta_{R^{a_k}_{v_i}}$ and of the extreme neurons as $\theta_{R^{a_k}_{min}}$ and $\theta_{R^{a_k}_{max}}$. The activation thresholds of all extreme and sensory neurons are fixed here: $\forall_{k,i}\ \theta_{R^{a_k}_{min}} = \theta_{R^{a_k}_{max}} = \theta_{R^{a_k}_{v_i}} = 1$ and they do not change. The activation functions of the extreme neurons (8) are defined in such a way to activate an appropriate extreme neuron when a new or existing extreme value is presented on the input sensory field:

$$
y^{a_k}_{min} = f(x^{a_k}_{min}) = \begin{cases} 1\ if\ x^{a_k}_{min} \geq \theta^{a_k}_{min} \\ 0\ if\ x^{a_k}_{min} < \theta^{a_k}_{min} \end{cases}
$$

$$
y^{a_k}_{max} = f(x^{a_k}_{max}) = \begin{cases} 1\ if\ x^{a_k}_{max} \geq \theta^{a_k}_{max} \\ 0\ if\ x^{a_k}_{max} < \theta^{a_k}_{max} \end{cases}
\tag{8}
$$

The extreme value presented on the input sensory field causes excitation of an appropriate extreme neuron exactly to its activation threshold $x^{a_k}_{min} = \theta_{R^{a_k}_{min}}$ or $x^{a_k}_{max} = \theta_{R^{a_k}_{max}}$. In turn, any new extreme value stimulates this neuron stronger $x^{a_k}_{min} > \theta_{R^{a_k}_{min}}$ or $x^{a_k}_{max} > \theta_{R^{a_k}_{max}}$. Such a strong, above-threshold stimulation starts the plasticity process of the extreme neuron, which is processed as follows:

1. Disconnect the extreme neuron from the currently connected sensory neuron.
2. Send a new advertisement to the noticeboard B^{a_k} in order to connect it to another sensory neuron.
3. Connect it to a new matched sensory neuron when the noticeboard answers.

The sensory neuron $R^{a_k}_{v_i}$, which represents a highly similar value to the one presented on the input sensory field F^{a_k}, will be activated the fastest and then it stimulates the connected neurons $R^{a_k}_{v_j}$ (sometimes also $R^{a_k}_{min}$ or $R^{a_k}_{max}$) throughout the synapses which weights are computed after:

$$
w_{R^{a_k}_{v_i}, R^{a_k}_{v_j}} = 1 - \frac{|v^{a_k}_i - v^{a_k}_j|}{r^{a_k}}
\tag{9}
$$

Objects are defined by combinations of attribute values, where each object o_n is represented by the object neuron O_n and each attribute value v_i is represented by the sensory neuron $R^{a_k}_{v_i}$. Therefore, each object neuron O_n is defined by a certain combination of connected sensory neurons (Fig. 1). The weighted connections between the object neurons and the sensory neurons are fixed as $w_{O_n, R_{v_i}^{a_k}} = \theta_{R_{v_i}^{a_k}}$. The excitation level of each sensory neuron is computed as the weighted sum of possible input stimuli coming from its connected sensor, sensory neurons and object neurons using the following formula:

$$X_{R_{v_i}^{a_k}} = t_{v_i^{a_k}} \cdot x_{v_i}^{a_k} + \sum_{j}^{R_{v_j}^{a_k} \rightsquigarrow R_{v_i}^{a_k}} y_{R_{v_j}^{a_k}} \cdot w_{R_{v_j}^{a_k}, R_{v_i}^{a_k}} + \sum_{n}^{O_n \rightsquigarrow R_{v_i}^{a_k}} y_{O_n} \cdot w_{O_n, R_{v_i}^{a_k}} \qquad (10)$$

where y_{O_n} denotes the output value of the object neuron O_n and $y_{R_{v_j}^{a_k}}$ is the output value of the sensory neuron $R_{v_j}^{a_k}$ computed as follows:

$$y_{R_{v_j}^{a_k}} = \begin{cases} 1 & if \ X_{R_{v_j}^{a_k}} \geq \theta_{R_{v_j}^{a_k}} \\ 0 & if \ X_{R_{v_j}^{a_k}} < \theta_{R_{v_j}^{a_k}} \end{cases} \qquad (11)$$

The connection plasticity condition for sensory neurons that is crucial for the presented ASSORT-2 algorithm is defined in the following way (Fig. 1): Disconnect the sensory neuron $R_{v_j}^{a_k}$ from the sensory neuron $R_{v_i}^{a_k}$ when it is less stimulated by this neighbor neuron than by its value sensor $S_{v_j}^{a_k}$ taking into account the defined sensory minimum difference ε^{a_k}:

$$0 < y_{R_{v_i}^{a_k}} \cdot w_{R_{v_i}^{a_k}, R_{v_j}^{a_k}} < x_{v_j^{a_k}} - \varepsilon^{a_k} \qquad (12)$$

The disconnected neurons $R_{v_j}^{a_k}$ and $R_{v_i}^{a_k}$ send advertisements to the noticeboard B^{a_k} and wait for answers. If possible the noticeboard automatically matches appropriate advertisements in constant time and sends advertisers back the information about the associated neurons. The matched advertisers simply connect to one another.

Only one of the connected sensory neurons $R_{v_j}^{a_k}$ can satisfy this plasticity condition (12) (Fig. 1). If so, the two neurons $R_{v_i}^{a_k}$ and $R_{v_j}^{a_k}$ are disconnected and send advertisements to the noticeboard B^{a_k} that already contains two advertisements sent previously by just created sensory neuron $R_v^{a_k}$. In result, these advertisements are matched and two new weighted connections are created (9).

Proof. Each sensory neuron is always connected in the right order because

$$\left| v_i^{a_k} - v_j^{a_k} \right| > \left| v_j^{a_k} - v^{a_k} \right| if \ and \ only \ if \ v_j^{a_k} < v^{a_k} < v_i^{a_k} \ or \ v_j^{a_k} > v^{a_k} > v_i^{a_k} \qquad (13)$$

Therefore, the plasticity condition (12) is true only in the above cases (Fig. 1):

$$0 < y_{R_{v_i}^{a_k}} \cdot w_{R_{v_i}^{a_k}, R_{v_j}^{a_k}} = 1 \cdot \left(1 - \frac{\left| v_i^{a_k} - v_j^{a_k} \right|}{r^{a_k}} \right) < 1 - \frac{\left| v_j^{a_k} - v^{a_k} \right|}{r^{a_k}} = x_{v_j^{a_k}} - \varepsilon^{a_k} \qquad (14)$$

In the parallel implementation, neurons always incept a neuron representing a new value in constant time, so the ASSORT-2 is always performed in linear time.

The object neurons $\{O_1, ..., O_{\widehat{N}}\}$ are automatically created for the presented objects $\{o_1, ..., o_N\}$ if the parallel stimulation of sensory neurons has not activated any existing object neurons, where $\widehat{N} \leq N$ because identical objects are represented by the same object neurons. Each neuron O_n is defined by a certain combination of the sensory neurons $\{R_{v_{n_1}}^{a_1}, ..., R_{v_{n_K}}^{a_K}\}$ connected to the sensors $\{S_{v_{n_1}}^{a_1}, ..., S_{v_{n_K}}^{a_K}\}$ representing the object values $\{v_{n_1}{}^{a_1}, ..., v_{n_K}{}^{a_K}\}$. The activation threshold of each object neuron is set to the number of the connected sensory neurons $\theta_{O_n} = \left\| R_{v_{n_k}}^{a_k} \rightsquigarrow O_n \right\| = K$. The object neuron excitation is computed as:

$$X_{O_n} = \sum_k^{R_{v_{n_k}}^{a_k} \rightsquigarrow O_n} y_{R_{v_{n_k}}^{a_k}} \cdot w_{R_{v_{n_k}}^{a_k}, O_n} \tag{15}$$

All connection weights are $w_{R_{v_{n_k}}^{a_k}, O_n} = 1$. Thus, the object neuron O_n is activated only if all sensory neurons defining it have been activated earlier according to:

$$y_{O_n} = \begin{cases} 1 & if \ X_{O_n} \geq \theta_{O_n} \\ 0 & if \ X_{O_n} < \theta_{O_n} \end{cases} \tag{16}$$

The ASSORT-2 algorithm develops an associative neural graph (Fig. 1) that makes data contextually available, replacing time-consuming search algorithms.

3   Simplistic Sequential Neural Associative Sorting

The simplistic sequential neural associative sorting SS-ASSORT-2 is intended to sequential Turing machines. It only uses the strongest stimulated sensors for each attribute to start the plasticity insertion process of new sensory neurons. The sensors in the input sensory fields $F^{a_1}, ..., F^{a_K}$ are organized using B-trees with duplicates aggregation (Fig. 2) which enable to find the most similar already represented values to the values presented on the input sensory fields in logarithmic time. The sensor numbers $J^{a_1}, ..., J^{a_K}$ for the attributes $a_1, ..., a_K$ are always less or equal the number of sorted objects ($\forall_k J^{a_k} \leq N$) because of the aggregations of the same attribute values in the same sensors. As a result, the strongest stimulated sensors representing the most similar values to the presented values are usually found in sublogarithmic or even constant time, according to the numbers of unique attribute values. If all attributes have constant numbers of unique values ($\forall_k J^{a_k} \ll N$) we can assume that the search complexity for these B-trees is $O(log_2 J^{a_k}) \approx O(1)$. Therefore, the simplistic sequential neural associative sorting SS-ASSORT-2 is processed in linear time if this condition is true. In the other cases, where for $\forall_k J^{a_k} \leq N$, it has usually sub-linearithmic complexity $O(N \cdot (log_2 J^{a_1} + ... + log_2 J^{a_K}))$. If $\forall_k J^{a_k} \approx N$ then the complexity is linearithmic. Summarizing, the best time complexity for the SS-ASSORT-2 algorithm on a sequential machine is linear, the worst one is linearithmic, e.g. when all or almost all sorted attribute values are different, and the expected one is sub-linearithmic dependently on the numbers of unique values defining object attributes.

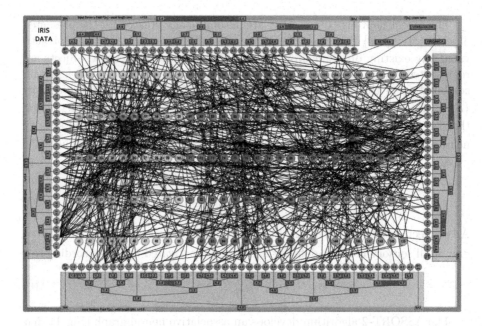

Fig. 2. The simplistic sequential associative neural graph constructed for all Iris data using B-trees to organize sensors in the sensory fields, which is able find out the most similar ones in $O(log_2 J^{a_k})$ time for each attribute a_k.

The presented algorithm has been sucessuffy tested on various data, e.g. Iris or Wine data. However it is very difficult to illustrate such structures, the presented one in Fig. 2 associates all similar attribute values and all similar objects of this data set. In this structure, all training objects are sorted after all attributes simultaneously. This great advantage can be appreciated only when it is used to find similar objects or specify different dependences between objects.

4   Conclusions and Remarks

This paper described the neural associative sort ASSORT-2 that uses the neural platform to automatically arrange data represented in sensors and neurons in order. The objects defined by various sets of parameters can be now sorted by neurons according to all attributes simultaneously in linear time. The presented brain-inspired computations work locally, use time as the computational factor and do not require loops, complex mathematical formulas, conditions, direct comparisons of attribute values or counting them up. On the basis of a few preprogrammed simple rules, each neuron can locally decide about starting its plasticity process that creates or reconfigures its connections and sets or updates weights. The presented models of neurons owe their opportunities threshold activation functions which enable to activate them in different moments. The sequence of activations of neurons is the key factor that allows for local plasticity decision making and sorting of values represented by neurons connected to sensors which charge them according to the different similarities to presented values on input sensory fields. Thus, this automatically self-developed and self-organized neural structure allows us to replace time-consuming search routines by more efficient ones working on appropriately connected neurons representing desired relations. Such neural networks can then be used to provide us with various answers about associated objects which can be received usually in constant time. Contemporary computer architectures based on the Turing machine are not the efficient platforms for execution of such neural computations. On the other hand, this paper also introduced the efficient simplistic sequential associative sorting algorithm SS-ASSORT-2 that can sort objects according to all attributes simultaneously in sub-linearithmic or even linear time on a sequential Turing machine according to the numbers of unique attribute values.

This paper proved that special neurons can be used to sort data and can do it efficiently in comparison to other sorting algorithms, taking into account their computational complexities. The additional outcome of this algorithm is the ability to create associative neural graph structures that allow us for fast automatic data mining and reasoning about similarities, differences, correlations, extremes, classes etc. These interesting features, abilities, and applications will be described in future papers. This research was supported by AGH 11.11.120.612.

References

1. Cormen, T.H., Leiserson, C.E., Rivest, R.L., Stein, C.: Introduction to Algorithms, 3rd edn. MIT Press, Cambridge (2009)
2. Duch, W.: Brain-inspired conscious computing architecture. J. Mind Behav. **26**, 1–22 (2005)
3. Graupe, D.: Deep Learning Neural Networks. World Scientific, Singapore (2016)
4. Horzyk, A.: How does generalization and creativity come into being in neural associative systems and how does it form human-like knowledge? Neurocomputing **144**, 238–257 (2014). Elsevier
5. Horzyk, A.: Artificial Associative Systems and Associative Artificial Intelligence, pp. 1–276. EXIT, Warsaw (2013)
6. Horzyk, A.: Human-like knowledge engineering, generalization, and creativity in artificial neural associative systems. In: Skulimowski, A.M.J., Kacprzyk, J. (eds.) Knowledge, Information and Creativity Support Systems: Recent Trends, Advances and Solutions. AISC, vol. 364, pp. 39–51. Springer, Cham (2016). doi:10. 1007/978-3-319-19090-7_4
7. Izhikevich, E.: Neural excitability, spiking, and bursting. Int. J. Bifurcat. Chaos **10**, 1171–1266 (2000)
8. Kalat, J.W.: Biological Grounds of Psychology. PWN, Warsaw (2006)
9. Larose, D.T.: Discovering Knowledge from Data. Introduction to Data Mining. PWN, Warsaw (2006)
10. Longstaff, A.: Neurobiology. PWN, Warsaw (2006)
11. Nowak, J.Z., Zawilska, J.B.: Receptors and Mechanisms of Signal Transfer. PWN, Warsaw (2004)
12. Rutkowski, L.: Techniques and Methods of Artificial Intelligence. PWN, Warsaw (2012)

13. Tadeusiewicz, R.: New trends in neurocybernetics. Comput. Method. Mater. Sci. **10**(1), 1–7 (2010)
14. Zongyao, S., Xiaolei, L.: Mining local association patterns from spatial dataset. In: 7th International Conference on Fuzzy Systems and Knowledge Discovery (2010)

Avoiding Over-Detection: Towards Combined Object Detection and Counting

Philip T.G. Jackson and Boguslaw Obara[✉]

School of Engineering and Computing Sciences,
Durham University, South Road, Durham DH1 3LE, UK
{p.t.g.jackson,boguslaw.obara}@durham.ac.uk

Abstract. Existing object detection frameworks in the deep learning field generally over-detect objects, and use non-maximum suppression (NMS) to filter out excess detections, leaving one bounding box per object. This works well so long as the ground-truth bounding boxes do not overlap heavily, as would be the case with objects that partially occlude each other, or are packed densely together. In these cases it would be beneficial, and more elegant, to have a fully end-to-end system that outputs the correct number of objects without requiring a separate NMS stage. In this paper we discuss the challenges involved in solving this problem, and demonstrate preliminary results from a prototype system.

Keywords: Object detection · Deep learning · Overlapping objects · Clustered objects · Non-max suppression

1 Introduction

Object detection is the task of localising and classifying all objects present in an image [18]. While the field of deep learning has produced many object detection networks with excellent true positive rate, they tend to suffer from low precision, i.e. high false positive rate. Usually the network outputs many bounding boxes per object, and these over-detections are filtered by non-max suppression (NMS) [17], leaving one box per object. NMS is a fixed post-processing step that is not learnt from the data, and typically relies on a user-chosen overlap threshold (0.7 used in [16]). Furthermore, NMS is unaware of the contents of the boxes it prunes, and so has no way to know if the ground-truth boxes really do overlap.

The question arises of how it may be possible to train a deep neural network to output exactly one box per object, without the need for a separate non-learned filtering step. Aside from being more elegant, this approach may have potential for greater accuracy, particularly in the case of detecting many small, densely clustered objects. In these cases, traditional NMS may struggle to tell if two boxes overlap because they are localising the same object or if they are localising different objects which are very close. This is especially true when objects of the same class are not only close but genuinely do overlap. With very high numbers

© Springer International Publishing AG 2017
L. Rutkowski et al. (Eds.): ICAISC 2017, Part I, LNAI 10245, pp. 75–85, 2017.
DOI: 10.1007/978-3-319-59063-9_7

of densely packed objects, another problem may also emerge: because detection networks emit a fixed number of boxes, it may become necessary to coordinate these boxes such that they are properly distributed among the many objects present. Over-detection in these cases may not only raise the false positive rate, but also lower the true positive rate; if there are only enough boxes to detect everything once then over-detecting one object may leave no boxes for another.

Close and overlapping objects occur in crowd footage, autonomous vehicle visual feeds, and histological images from biomedical microscopy, such as those in Fig. 1. In this paper we choose cell microscopy as a test case, and use the Simulating Microscopy Images with Cell Populations (SIMCEP) [10] system to generate large quantities of synthetic images with perfect ground-truth annotation for training and testing. The simplicity of this benchmark, which can be solved to reasonable accuracy without deep learning [19], allows us to focus solely on the over-detection problem. SIMCEP allows the user to generate artificial cell populations with varying degrees of clustering and overlap, and so makes an excellent testing ground for a dense object detection framework. Using simulated images allows us to generate essentially unlimited quantities of training data, bypassing the scarcity of labelled data that is normally the biggest constraint when training deep networks to solve bio-imaging problems. It is hoped that systems trained on SIMCEP images may still be applicable to real-world histological images via transfer learning. Fluorescence microscopy image analysis often requires objects to be counted as well as localised, so a one-box-per-cell system, which can be seen as combined localisation and counting, would be quite relevant in this field.

(a)   (b)

Fig. 1. (a) mouse embryo, an extreme case of overlapping objects consisting of a ball of around 20 cells. (b) Human HT29 colon cancer cells, packed very closely. Both images from the Broad Bioimage Benchmark Collection [12].

2   Related Work

2.1   Deep Learning Methods for Object Detection

Object detection in deep learning is largely dominated by the Region Convolutional Neural Network (R-CNN) family of models. The original R-CNN [5] uses a selective search based method [22] to propose interesting-looking regions, only using the CNN to generate feature vectors for each region and a support vector machine (SVM) approach to then score them for each class. Fast R-CNN [4] is an iteration on this work, speeding the process up mostly by generating all convolutional features for the image in a single pass and pooling sub-sets of them for different region proposals, rather than running each proposed region through the CNN separately. Faster R-CNN [16] improves further by using the same convolutional network for both proposing regions and classifying their contents. This saves computational time and results in slightly more accurate bounding boxes, as well as being a more elegant system. Almost the whole pipeline is performed by the network, only the NMS is done separately.

Faster R-CNN is a fully convolutional network (FCN), so images of arbitrary size can be passed and the feature maps will grow or shrink accordingly. The final convolutional layer outputs feature vectors describing overlapping square regions in the image; these are used by the region proposal network (RPN) to predict a fixed number of bounding boxes per region. The RPN's output tensor consists of multiple "detectors": groups of neurons representing bounding box parameters and confidence levels. Each output box is described relative to a different fixed "anchor" box. The anchors are Faster R-CNN's answer to the problem of expressing an unordered set of boxes with a fixed-size tensor. The loss function must decide at training time which boxes from the RPN are to match with which ground-truth boxes, and which boxes should have high class probability (i.e. the RPN's confidence that box contains an object). In practice, all boxes whose anchors overlap sufficiently with a ground-truth box are trained to have high class probability and incur regression loss on their deviation from the ground-truth box. Output boxes whose anchors do not overlap sufficiently with any ground-truth box only incur loss for having high class probability. This can be seen as giving each detector a different "jurisdiction", in which it is responsible for matching any ground-truth box with a certain position, aspect ratio and size.

Another relevant detection framework is YOLO [15], which differs from Faster R-CNN principally in that is not an FCN. Although this requires that images are resized to a fixed dimension before processing, it also means that the feature maps are of constant size. This allows the final layer to be converted to a fixed-size vector that describes the entire image, in a similar manner to AlexNet [9]. This allows the classifier to make use of global image context, resulting in higher accuracy compared to Faster R-CNN, whose classifier only pools convolutional features from within the proposed bounding boxes. YOLO assigns responsibility to its output boxes in a different way to Faster R-CNN. Unlike Faster R-CNN, the jurisdictions of the detectors are not pre-defined, rather, responsibility for detecting a given ground-truth box is assigned at training time to whichever detector outputs a box with the greatest intersection over union (IoU) with that box. The authors claim this leads to detectors learning to specialise in different sizes, aspect ratios and classes of object.

Although the above methods excell at detecting small numbers of large objects in datasets such as Pascal VOC [3], they are less well tested on large numbers of small objects. In particular, they all tend to over-detect objects, outputting many bounding boxes which must then be pruned by NMS to leave only one box per instance. The problem of learning to count has been explicitly investigated in [20], whose authors show that a network trained only on the multiplicity of a target object type will learn features that are also useful for classification and localisation of said objects. Although the results are encouraging, they do not tackle the problem of coordinating object detectors to output exactly one bounding box per ground-truth object.

2.2 Deep Learning Methods for Cell Detection

The greatest obstacle in applying deep learning approaches to biomedical image processing is the scarcity of labeled training data. Deep neural networks generally require many thousands of labeled images to train effectively, but individual problems in biomedicine tend to avail neither thousands of images nor enough trained experts to label them all. Many proposed methods [1,11,14] circumvent this problem by using CNNs to perform pixel-wise binary classification. These networks take small image patches as input and output the probability of the central pixel in the patch being part of a target object. Although this is a harder task than whole-image classification, it can yield thousands of training examples per image, since each pixel and its neighbourhood becomes an example in the training set. For example, [2] trains a CNN to identify the central voxels of zebrafish dopaminergic neurons in 3D images. This is part of a larger pipeline, which first uses an SVM to narrow down the set of potential voxels, so that the CNN need not be applied to every possible location in the image. The output probability map is then smoothed and individual cells are detected as local probability maxima. [14] uses a CNN to detect lipid deposits in retinal images, by classifying the central pixel of 65×65 image patches. Since these deposits are diffuse, amorphous objects, pixel-wise classification is appropriate here and there is no attempt to define the number of deposits present.

In [8], an FCN is trained to classify histological images at a whole-image level. Although it is only trained with whole-image labels, it is still able to localise individual cells by deriving class probability maps from the final convolutional layers, in a manner inspired by [21,23]. FCNs are particularly useful when processing histological images due to their ability to naturally scale to images of arbitrary size, without needing to downsample large images to a fixed size. [11] train a standard CNN to classify the central pixel of image patches, then convert it to an FCN to perform pixel-wise classification over a whole image in one pass. This has performance benefits over processing patches one-by-one, since computations can be shared among overlapping image patches.

A standard CNN based on the design of Krizhevsky [9] is used to count human embryonic cells in [6]. Since the cells in these images show very high overlap, the act of counting is treated as a classification task and the cells themselves are not localised.

3   Method

When attempting to design a network that produces output of variable length, one immediately hits two technical limitations:

- Existing deep learning frameworks process data in "tensors", N-dimensional arrays whose shape is always a hyperrectangle. This includes the output tensor. Outputting a different number of boxes for each image in a batch would be like outputting a matrix with variable length rows, which is not supported.
- In order for the network to learn the correct number of boxes, this number needs to be somehow differentiable. That means the number of boxes produced must vary smoothly with respect to the network parameters; a small parameter change should result in a small improvement in the number of boxes.

These constraints can be satisfied by outputting a fixed number of boxes with confidence scores attached - as is the case in existing detection frameworks. The problem now is how to assign confidence scores such that each object gets exactly one high confidence box that matches its corresponding ground truth box.

3.1   Loss Function

To train a network to behave in such a way, a loss function is required that is minimised if and only if the network outputs exactly one matching box with high confidence for each ground-truth box. This is difficult, because the order in which the boxes are emitted should not matter. Loss functions in supervised learning work by penalising deviation from some target output, but if a network emits N output boxes per image and an image has M objects, then there are $\frac{N!}{(N-M)!}$ possible correct outputs, corresponding to different orderings of the boxes. Faster R-CNN and YOLO solve this problem by establishing "jurisdictions" for their output boxes, whereby the loss function demands that a box should have high confidence if a ground-truth box falls into its jurisdiction.

Ideally, we would like the loss function to be minimised no matter which detectors are used to label the objects, so long as there is only one each. To this end, we define a loss function that assigns responsibility for ground-truth boxes based on both the output box parameters (centre coordinates, width and height) and confidence scores. We define a responsibility matrix R, where R_{ij} is the responsibility of detector i for object j, and

$$R_{ij} = \frac{C_i}{D_{ij}} \qquad (1)$$

$$D_{ij} = (x_i - x_j^*)^2 + (y_i - y_j^*)^2 + (w_i - w_j^*)^2 + (h_i - h_j^*)^2 \qquad (2)$$

where x, y, w, h are centre coordinates and width and height, normalised to $[0, 1]$ relative to the image dimensions and mean box size, respectively, $*$ denotes ground-truth, and C_i is the confidence of detector i.

At training time, each ground-truth box j selects the detector that is most responsible for it:

$$R_j^* = \arg\max_i R_{ij} \qquad (3)$$

This chosen detector incurs a regression loss $D_{R_j^*,j}$, causing it to better localise the object for which it was responsible. All detectors also incur regression loss on their confidence, where target confidence C_i^* is 1 if detector i is responsible for an object, and 0 otherwise. The total loss is then:

$$L = \frac{1}{N} \sum_i (C_i - C_i^*)^2 + \frac{1}{M} \sum_j D_{R_j^*,j} \qquad (4)$$

where M and N are the number of ground-truth objects and detectors, respectively. This responsibility scheme is similar to that used by [15], but differs in that ours takes into account box confidence, allowing it to penalise over-detection if too many high confidence boxes are emitted.

Using detector confidence to establish responsibility allows the network to choose for itself which detector will be responsible. If detectors $1 - 5$ localise object j, then their regression losses D_{ij} for $i = 1..5$ will be similar, and so the highest responsibility will go to detector k with highest confidence C_k. This chosen detector will get a target confidence C_k^* of 1 while the others get 0. This reinforces detector k as the detector responsible for that object; next time the same object is seen, C_k will be higher, while others will be lower. This can be seen as a kind of learnt NMS.

We found that if confidence is not used to determine responsibility ($R_{ij} = \frac{1}{D_{ij}}$), the network outputs many boxes per object which all have roughly equal confidence well below 0.5. This is because the network cannot predict which box will be closest to the ground-truth since they are all close, and so cannot predict which should have confidence 1 and which should have 0. Moving all but one box away from the object would be a solution, but this would only produce discontinuous, non-differentiable changes in loss as the responsibility assignment changes suddenly, so the network cannot learn to do this.

3.2   Model Architecture

A recurrent neural network (RNN) would be the obvious choice to minimise the loss function described above. If bounding boxes are emitted sequentially rather than simultaneously, then each one can be dependent on the ones that came before it. In this way, a detector can avoid outputting a high confidence box on an object that has already been detected. Despite this attractiveness though, our best results out of the many architectures trialled came not from an RNN but from an FCN. This architecture is specified in Table 1.

Everything from `conv1` to `conv7` is a relatively standard convolution/maxpooling stack, with some slightly unusual features (stride of 2 in `conv6`) which allow the stack to output feature maps whose effective receptive fields in the image overlap by half (effective receptive field size is 64×64 pixels, effective stride is 32×32). This overlap ensures that every object lies fully within at least one neuron's receptive field. `boxes` emits bounding box parameters and `boxes_global` is a custom layer that performs a simple transformation from local coordinate space global image space. `concat` joins the feature maps of `boxes_global` and `conv7`, allowing the remaining three layers to predict confidence scores based on both the boxes themselves and the image features they were predicted from. We observed a modest improvement in performance due to this addition. The final three layers, then, can be seen as a learnt filtering stage that replaces the traditional NMS postprocessing. A Theano/Lasagne implementation is available at https://github.com/philipjackson/avoiding-overdetection.

Table 1. A specification of our network architecture. Unless otherwise stated, each layer takes the previous layer's output as input. Nonlinearities are leaky rectified linear [13] with $\alpha = 0.1$ unless otherwise stated. B is a hyperparameter denoting the number of detectors per "window" (i.e. position in the final feature map, `conv7`). $B = 9$ in our experiments.

Network layers		
Name	Type	Parameters
conv1	Convolution	num_filters=32, filter_size=(5,5)
pool1	Maxpool	pool_size=(2,2)
conv2	Convolution	num_filters=48, filter_size=(3,3)
pool2	Maxpool	pool_size=(2,2)
conv3	Convolution	num_filters=64, filter_size=(3,3)
pool3	Maxpool	pool_size=(2,2)
conv4	Convolution	num_filters=86, filter_size=(3,3)
pool4	Maxpool	pool_size=(2,2)
conv5	Convolution	num_filters=128, filter_size=(1,1)
conv6	Convolution	num_filters=128, filter_size=(2,2), stride=(2,2)
conv7	Convolution	num_filters=128, filter_size=(1,1)
boxes	Convolution	num_filters=4*B, filter_size=(1,1), nonlinearity=identity
boxes_global	Coord Transform	
concat	Concatenation	inputs=boxes_global,conv7
filter1	Convolution	num_filters=16*B, filter_size=(3,3)
filter2	Convolution	num_filters=16*B, filter_size=(1,1)
confidence	Convolution	num_filters=B, filter_size=(1,1), nonlinearity=sigmoid

4   Results

We trained our model on a set of 17000 SIMCEP images using the Adam optimizer [7], and validated against a set of 3000. The images were of size 224×224

pixels and contained anywhere from 1 to 15 cells. The parameters of SIMCEP were adjusted to randomise obfuscating features such as blur, Gaussian noise and uneven lighting, and the cells show varying levels of clustering and overlap.

A selection of results is shown in Fig. 2. We interpret any detection with a confidence above 0.5 as a positive, and so the number of such detections is the network's estimate of the number of cells present. Across our validation set, the root mean square of the deviation of this estimate from the true count was 2.28. Further quantitative results are shown in Table 2.

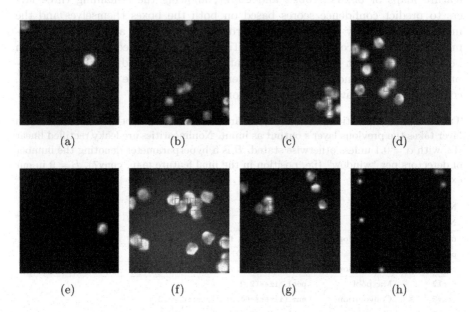

Fig. 2. A sample of detection results on SIMCEP images. Confidence is represented in the transparency of the boxes; all output boxes with confidence above 0.1 are shown. Instead of post-processing with NMS, we simply take boxes with confidence above 0.5 (shown in red) as positive detections. Boxes with confidence below 0.5 are shown in blue. (Color figure online)

Table 2. True and false positive rates on training and validation sets. A true positive is counted as any output box with an intersection over union (IoU) above 60% with a ground-truth box, but each ground-truth box can only be paired with a single output box. So if two output boxes cover the same object, then this counts as one true positive and one false positive. Output boxes with less than 60% IoU with any ground-truth box are always false positives.

	True positive rate	False positive rate	F_1-score
Training set	75.4%	19.2%	0.774
Validation set	75.3%	19.4%	0.773

5   Conclusion

For images containing objects whose bounding boxes overlap heavily due to occlusion or dense clustering, NMS cannot reliably remove excess bounding boxes emitted by the network, since ground-truth bounding boxes with identical classes may truly overlap significantly. An end-to-end system that outputs the correct number of boxes without the need for post-processing NMS is therefore preferable. In this paper, we discuss the problem and take some early steps towards solving it, demonstrating a system that can localise densely clustered objects and simultaneously approximate the correct number of boxes. Rather than performing regression directly on the number of objects, we encode this number implicitly in the number of high confidence boxes emitted by the network.

We propose that, unless an alternative output encoding can be found which shows a one-to-one mapping between output values and unordered sets of boxes, supervised learning itself is unsuitable for this task. There are many ways for a network to output the same set of boxes, depending on which box it places on which object (or for an RNN, which order it outputs them in), but supervised learning requires us to arbitrarily choose one of them, and penalise all the others. In this paper we partially solve this problem by choosing which box should have high confidence based partly on the confidence values themselves, however the results are far from perfect. We put forward three reasons for this, and suggest how they may be countered by applying reinforcement learning instead of supervised learning.

Firstly, because we assign responsibility for an object to the detector with the maximum responsibility for it, our loss function is discontinuous, due to the arg max operation. This is likely to cause problems for supervised learning, which is based on direct optimization of the loss function by gradient descent. Reinforcement learning trains a network to optimize a reward function which may be related to the network's output in a complex, non-differentiable or even unknown way. In particular, reward functions do not prescribe a target output for every input, and so they completely bypass the problem of choosing which detector should label which object. This makes reward functions a much more natural way to express the goal of one box per object.

The second reason is that in order to teach a network to output the right number of boxes, that number must somehow be made smooth and differentiable, despite the fact that we ultimately want an integral number. To derive this hard number, we currently threshold the confidence levels at 0.5. Not only is this threshold somewhat arbitrary, but worse still, it is effectively a post-processing step that the network itself is unaware of, and indeed cannot be trained to optimize because it is non-differentiable. This too can be solved with reinforcement learning by building the thresholding step into the reward function, because discontinuous reward functions are alllowed.

The third reason is that our FCN architecture outputs all the boxes in parallel. This means that each detector is unaware of what the others are doing, so it is difficult for them to coordinate themselves so as to avoid over-detection. Using an RNN that outputs boxes in series would solve this problem, as the output on one time step can be conditioned on that of previous ones. This also fits well with reinfrocement learning, since a reward signal can be administered on every time step; this would accelerate training compared to a single overall reward signal per image.

References

1. Ciresan, D., Giusti, A., Gambardella, L.M., Schmidhuber, J.: Deep neural networks segment neuronal membranes in electron microscopy images. In: Advances in Neural Information Processing Systems, vol. 25, pp. 2843–2851. Curran Associates, Inc. (2012)
2. Dong, B., Shao, L., Da Costa, M., Bandmann, O., Frangi, A.F.: Deep learning for automatic cell detection in wide-field microscopy zebrafish images. In: IEEE International Symposium on Biomedical Imaging, pp. 772–776 (2015)
3. Everingham, M., Van Gool, L., Williams, C.K.I., Winn, J., Zisserman, A.: The PASCAL visual object classes challenge 2012 (VOC2012) results. http://www.pascal-network.org/challenges/VOC/voc2012/workshop/index.html
4. Girshick, R.: Fast R-CNN. In: IEEE International Conference on Computer Vision, pp. 1440–1448 (2015)
5. Girshick, R., Donahue, J., Darrell, T., Malik, J.: Rich feature hierarchies for accurate object detection and semantic segmentation. In: IEEE Conference on Computer Vision and Pattern Recognition, pp. 580–587 (2014)
6. Khan, A., Gould, S., Salzmann, M.: Deep convolutional neural networks for human embryonic cell counting. In: Hua, G., Jégou, H. (eds.) ECCV 2016. LNCS, vol. 9913, pp. 339–348. Springer, Cham (2016). doi:10.1007/978-3-319-46604-0_25
7. Kingma, D., Ba, J.: Adam: a method for stochastic optimization. In: International Conference on Learning Representations (2015)
8. Kraus, O.Z., Ba, J.L., Frey, B.J.: Classifying and segmenting microscopy images with deep multiple instance learning. Bioinformatics 32(12), 52–59 (2016)
9. Krizhevsky, A., Sutskever, I., Hinton, G.E.: Imagenet classification with deep convolutional neural networks. In: Advances in Neural Information Processing Systems, pp. 1–9 (2012)
10. Lehmussola, A., Ruusuvuori, P., Selinummi, J., Huttunen, H., Yli-Harja, O.: Computational framework for simulating fluorescence microscope images with cell populations. IEEE Trans. Med. Imag. 26(7), 1010–1016 (2007)
11. Litjens, G., Sánchez, C.I., Timofeeva, N., Hermsen, M., Nagtegaal, I., Kovacs, I., Hulsbergen-van de Kaa, C., Bult, P., van Ginneken, B., van der Laak, J.: Deep learning as a tool for increased accuracy and efficiency of histopathological diagnosis. Sci. Rep. 6 (2016)
12. Ljosa, V., Sokolnicki, K.L., Carpenter, A.E.: Annotated high-throughput microscopy image sets for validation. Nat. Meth. 9(7), 637 (2012)
13. Maas, A.L., Hannun, A.Y., Ng, A.Y.: Rectifier nonlinearities improve neural network acoustic models. In: International Conference on Machine Learning, vol. 30 (2013)
14. Prentai, P., Lonari, S.: Detection of exudates in fundus photographs using convolutional neural networks. In: International Symposium on Image and Signal Processing and Analysis, pp. 188–192 (2015)
15. Redmon, J., Divvala, S., Girshick, R., Farhadi, A.: You only look once: unified, real-time object detection. In: Computing Research Repository (2015)

16. Ren, S., He, K., Girshick, R., Sun, J.: Faster R-CNN: towards real-time object detection with region proposal networks. In: Advances in Neural Information Processing Systems, pp. 91–99 (2015)
17. Rothe, R., Guillaumin, M., Gool, L.: Non-maximum suppression for object detection by passing messages between windows. In: Cremers, D., Reid, I., Saito, H., Yang, M.-H. (eds.) ACCV 2014. LNCS, vol. 9003, pp. 290–306. Springer, Cham (2015). doi:10.1007/978-3-319-16865-4_19
18. Russakovsky, O., Deng, J., Su, H., Krause, J., Satheesh, S., Ma, S., Huang, Z., Karpathy, A., Khosla, A., Bernstein, M.S., Berg, A.C., Li, F.: Imagenet large scale visual recognition challenge, vol. 1409 (2014)
19. Ruusuvuori, P., Manninen, T., Huttunen, H.: Image segmentation using sparse logistic regression with spatial prior. In: 2012 Proceedings of the 20th European Signal Processing Conference (EUSIPCO), pp. 2253–2257, August 2012
20. Seguí, S., Pujol, O., Vitria, J.: Learning to count with deep object features. In: IEEE Conference on Computer Vision and Pattern Recognition, pp. 90–96 (2015)
21. Simonyan, K., Vedaldi, A., Zisserman, A.: Deep inside convolutional networks: visualising image classification models and saliency maps (2013)
22. Uijlings, J.R., van de Sande, K.E., Gevers, T., Smeulders, A.W.: Selective search for object recognition. Int. J. Comput. Vis. 104(2), 154–171 (2013)
23. Zeiler, M.D., Fergus, R.: Visualizing and understanding convolutional networks. In: Fleet, D., Pajdla, T., Schiele, B., Tuytelaars, T. (eds.) ECCV 2014. LNCS, vol. 8689, pp. 818–833. Springer, Cham (2014). doi:10.1007/978-3-319-10590-1_53

Echo State Networks Simulation
of SIR Distributed Control

Tibor Kmet[1]($\boxtimes$) and Maria Kmetova[2]

[1] Department of Informatics, Constantine the Philosopher University,
Tr. A. Hlinku 1, 949 74 Nitra, Slovakia
`tkmet@ukf.sk`
[2] Department of Mathematics, Constantine the Philosopher University,
Tr. A. Hlinku 1, 949 74 Nitra, Slovakia
`mkmetova@ukf.sk`
`http://www.ukf.sk`

Abstract. Echo State Networks (ESNs) have been shown to be effective for a number of tasks, including motor control, dynamic time series prediction, and memorising musical sequences. In this paper, we propose a new task of ESNs in order to solve distributed optimal control problems for systems governed by parabolic differential equations with discrete time delay using an adaptive critic designs. The optimal control problems are discretised by using a finite element method in time and space, then transcribed into a nonlinear programming problems. To find optimal controls and optimal trajectories ESNs adaptive critic designs are used to approximate co-state equations. The efficiency of our approach is demonstrated for a SIR distributed system to control the spread of diseases.

Keywords: Echo State Networks · SIR distributed model · Distributed control problem with discrete time delay · Adaptive critic synthesis · Numerical examples

1 Introduction

ESNs are a type of three-layered recurrent network with sparse, random, and crucially, untrained connections within the recurrent hidden layer. Networks have been shown to be effective in a variety of domains, however, we could find no prior work in applying ESNs to distributed optimal control problems. The problem is motivated by better understanding of real world systems eventually for the purpose of being able to influence these systems in a desired way. The solution of a distributed control problem is characterised by the state (evolving forward in time) and co-state (evolving backward in time) equations with initial and terminal conditions, respectively. We pursue the one-shot multigrid strategy as proposed in [2]. A one-shot multigrid algorithm means solving the optimality system for the state, the co-state and the control variables in parallel in the multigrid process evolving forward in time. The finite element approximation plays an important

© Springer International Publishing AG 2017
L. Rutkowski et al. (Eds.): ICAISC 2017, Part I, LNAI 10245, pp. 86–96, 2017.
DOI: 10.1007/978-3-319-59063-9_8

role in the numerical treatment of optimal control problems. This approach has been extensively studied in the papers e.g. [3,4,12,13] for parabolic optimal control problems. Through discretisation the optimal control problem is transcribed into a finite-dimensional nonlinear programming problem (NLP-problem). Optimal control problems have thus been a stimulus to develop optimisation codes for large-scale NLP-problems [2,5,13]. Then neural networks are used as a universal function approximation to solve co-state variable forward in time with "adaptive critic designs" [16,18]. The paper presented extends adaptive critic neural network architecture proposed by [11] to solve optimal distributed control problem for systems governed by parabolic differential equations with control and state constraints and discrete time delay with ESNs. In this paper, we propose the application of ESNs for solving nonlinear optimal control problems with time delay. Section 2 describes the typical ESNs architecture, using for the proposed adaptive critic algorithm. In Sect. 3, we present a mathematical model describing the population dynamics of infectious disease and optimal control problem of vaccine coverage threshold needed for disease control and eradication. In Sect. 4, we discuss a space-time discretisation approach in which both control and state variables are discretised. We use augmented Lagrangian techniques and architecture of the proposed adaptive critic neural network synthesis for the optimal control problem with delay. We also present a new algorithm to solve distributed optimal control problems. Simulations and illustrative examples are presented. Finally, Sect. 5 concludes the paper.

2   Echo State Networks

In this section we describe the standard ESNs, which consist of a large, randomly connected neural network, the reservoir, which is driven by an input signal and projects to output units. During the training, only the connections from the reservoir to these output units are learned. A key requisite for output-only training is the echo state property (ESP), which means that the effect of initial conditions should vanish as time passes. The general layout of ESNs is illustrated in Fig. 1. It consists of K input nodes, N_r reservoir nodes, and L output nodes. The standard discrete-time ESNs are defined as follows:

$$y(t+1) = f(Wy(t) + W^{bf}y^{out}(t) + W^{in}v(t)) \qquad (1)$$
$$y^{out}(t) = g(W^{out}[y(t); v(t)]),$$

where $W \in R^{N_r \times N_r}$ is the internal connections weight matrix or the reservoir, $W^{in} \in R^{N_r \times K}$ is the input matrix, $W^{bf} \in R^{N_r \times L}$ is the feedback matrix, $W^{out} \in R^{L \times (N_r + K)}$ is the output matrix, and $y(t) \in R^{N_r \times 1}$, $v(t) \in R^{K \times 1}$, $y^{out}(t) \in R^{L \times 1}$ are the internal, input and output vectors at time t respectively. The state activation function $f = (f_1, \ldots, f_{N_r})^T$ is a sigmoid function (usually $f_i = tanh$) applied component-wise with $f(0) = 0$, and the output activation function is $g = (g_1, \ldots, g_L)^T$, where each g_i is usually the identity or a sigmoid function. $[;]$ denotes the vector concatenation and y^T denotes the transpose of the vector y. For every sample, $y(0)$ is initialised as 0. The key

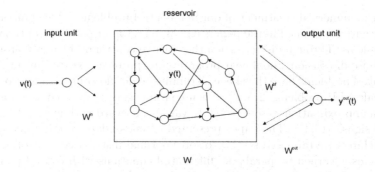

Fig. 1. The basic structure of ESNs. Solid arrows denote the fixed connections and dashed arrows denote the trainable connections.

idea in reservoir computing [15,17] is to feed time series to a reservoir by modelling the dynamics of the system which generates the time series. The reservoir is then read by a readout function in order to make predictions using the constructed model. When training the model, only the readout function is modified, the complex dynamic modelling behaviour of the reservoir is left unchanged. In ESNs [7], the reservoir consists of a recurrent artificial neural network with sigmoid activation functions and the echo state property which ensures good modelling abilities. A recurrent artificial neural network is said to have the *echo state property* when its state is uniquely determined by the input time series. This implies the state forgetting property: the initial state of the reservoir has no impact on the state after feeding a possibly infinite time series. If the largest absolute eigenvalue of the matrix W $|\omega|_{max} < 1$, the network states $y(t)$ become asymptotically independent of initial conditions $y(0)$ and depend only on the input history $v(t)$. The spectral radius ω is defined as the largest absolute eigenvalue of the matrix W. It has been shown that reservoirs, whose spectral radius is larger than one, i.e., $|\omega|_{max} > 1$, do not have the echo state property, but in practice the spectral radius is chosen close to one to achieve a suitable dynamic response [8]. ESNs [7,9] provide an architecture and supervised learning principle for recurrent neural networks. The main idea is (i) to drive a random, large, fixed recurrent neural network with the input signal, thereby inducing in each neuron within this "reservoir" network a nonlinear response signal, and (ii) combine a desired output signal by a trainable linear combination of all of these response signals. The internal weights of the underlying reservoir network are not changed by the learning; only the reservoir-to-output connections are trained. Supervised learning, or what statisticians know as nonparametric regression, is the problem of estimating a function, $Y(v) \in R^{L \times 1}$ given only a training set of pairs of input–output points, $\{(v(t), Y(t)\}_{t=1}^{P}$ sampled, usually with noise, from the function. For a training set with P patterns, the optimal weight vector can be found by minimising the sum of squared errors

$Err(W^{out}) = \sum_{t=1}^{P} \sum_{i=1}^{L} (W_i^{out}[y(t); v(t)] - Y_i(t))^2$ and is given by

$$W^{out} = (H^T H)^{-1} H^T Y \ or \ W^{out} = H^T (HH^T)^{-1} Y, \tag{2}$$

where $H = ([y(1); v(1)], \ldots, [y(P); v(P)])$.

3   SIR Model with Delay and Spatial Diffusions

Mathematical models describing the population dynamics of infectious disease have played an important role in better understanding epidemiological patterns and disease control. One of the most popular models of the infectious diseases is the classical SIR model [1,19,20]. In this model, the whole population is divided into three compartments which describe separated groups of individuals: susceptible which are able to contract the disease (denoted by S, x_1), infective which are capable of transmitting the disease (marked by I, x_2) and recovered which are permanently immune (denoted by R, x_3). The letters represent the number of individuals in each compartment at a particular time t and space p, and the whole population size N, x_4 is the sum of above fractional groups, i.e. $S + I + R = N$. In reality, the environment in which an individual lives is often heterogeneous making it necessary to distinguish the locations, and due to the large mobility of people within a country or even worldwide, spatially uniform models are not sufficient to give a realistic picture of a disease's transmission. For this reason, the effect of dispersion of the population in a bounded habitat has been taken into consideration, and in this situation the governing equations for the population densities become a system of reaction diffusion equations with time delay. The time evolutions of the populations compartments in the SIR model is described by four nonlinear partial differential equations:

$$\frac{\partial x_i(p,t)}{\partial t} = D \frac{\partial^2 x_i(p,t)}{\partial p^2} + F_i\left(x(p,t), x(p,t-\tau), u(p,t), t\right), \ i = 1, \ldots, 4, \tag{3}$$

where

$$F_1\left(x(p,t), x(p,t-\tau), u(p,t), t\right) = \left(b - \mu \frac{r x_4(p,t)}{K_c}\right) x_4(p,t) - \frac{\beta x_1(p,t) x_2(p,t-\tau)}{x_4(t-\tau)}$$

$$+ \omega u(p,t) \frac{x_2(p,t)}{x_4(p,t)} - \left(d + (1-\mu)\frac{r x_4(p,t)}{K_c}\right) x_1(p,t)$$

$$F_2\left(x(p,t), x(p,t-\tau), u(p,t), t\right) = \frac{\beta x_1(p,t) x_2(p,t-\tau)}{x_4(p,t-\tau)} - u(p,t) \frac{x_2(p,t)}{x_4(p,t)} -$$

$$\left(d + (1-\mu)\frac{r x_4(p,t)}{K_c}\right) x_2(p,t) - \alpha x_2(p,t)$$

$$F_3\left(x(p,t), x(p,t-\tau), u(p,t), t\right) = \alpha x_2(p,t) + (1-\omega)u(p,t) \frac{x_2(p,t)}{x_4(p,t)}$$

$$- \left(d + (1-\mu)\frac{r x_4(p,t)}{K_c}\right) x_3(p,t)$$

$$F_4\left(x(p,t), x(p,t-\tau), u(p,t), t\right) = r x_4(p,t) \left(1 - \frac{x_4(p,t)}{K_c}\right)$$

with Neumann boundary condition

$$\frac{\partial x_i}{\partial p}(0,t) = \frac{\partial x_i}{\partial p}(1,t) = 0 \qquad (4)$$

and initial conditions

$$x_i(p,t) = \phi_i(p,t) \geq 0, \ 0 \leq p \leq 1, \ t \in [-\tau, 0], \ i = 1, \ldots, 4. \qquad (5)$$

Here t denotes the time, p represents the spatial location, D is the diffusion coefficient, $b > 0$, $d > 0$, $\alpha > 0$ and $\beta > 0$ are the birth, death, recovery and contact rate, respectively. $r = b - d$ is the intrinsic growth rate, μ is the convex combination constant, K_c is the carrying capacity of the population, τ is a non-negative constant represented a time delay on the infected individuals I and the total individuals N during the spread of diseases, and u is the vaccination coverage of infected individuals. Let us consider the whole population size x_4 given by the following equation:

$$\frac{\partial x_4(p,t)}{\partial t} = D\frac{\partial^2 x_4(p,t)}{\partial p^2} + rx_4(p,t)\left(1 - \frac{x_4(p,t)}{K_c}\right). \qquad (6)$$

It can be shown [14] that there is no spatially non-homogenous stationary solution for a scalar reaction-diffusion equation in one dimension with Neumann boundary conditions. Thus, the only stable solutions are spatially homogeneous and the solutions have similar properties as in the case without diffusion. Equation (6) has two spatially constant equilibrium $\bar{N}_1 = 0$, $\bar{N}_2 = K_c$, where N_1 is unstable and N_2 is asymptotically stable. Moreover, if we assume that the initial condition $x_4(p,0)$ is separated from 0, then we can show using Theorem 1 [6] that $x_4(p,t)$ tends to K_c as $t \to \infty$.

Theorem 1. *If* $x_4(p,0) \geq \alpha$ *for* $p \in [0,1]$ *and* $\alpha \in (0, K_c)$ *then* $\lim_{t\to\infty} x_4(p,t) = K_c$.

3.1   Distributed Optimal Control Problem

We set an optimal control problem in the SIR model to control the spread of diseases. The main goal of this problem is to investigate the optimal vaccine coverage threshold needed for disease control and eradication [10]. From these facts, our optimal control problem is given by the following. Find a distributed control $u(p,t)$ to minimise the objective functional

$$J(u) = \int_a^b \int_{t_0}^{t_f} a_1 x_1(p,t) + a_2 x_2(p,t) + \frac{1}{2}u(p,t)^2 dt dp, \qquad (7)$$

subject to the state system (3) and (4), where a_1, a_2 are small positive constants to keep a balance in the size of S and I, respectively. The theory of necessary conditions for the optimal control problem of form (7) is well developed, see, e.g. [5,13]. The augmented Hamiltonian function for problem (7) is given by

$$\mathcal{H}(x, x_\tau, u, \lambda) = a_1 x_1(p, t) + a_2 x_2(p, t) + \frac{1}{2}u(p, t)^2 + \sum_{j=1}^{4} \lambda_j F_j(x, u) =$$

$$\frac{1}{2}u(p, t)^2 + u(p, t)\frac{x_2(p, t)}{x_4(p, t)}\left(\omega\lambda_1(p, t) - \lambda_2(p, t) + (1 - \omega)\lambda_3(p, t)\right) + G(p, t) \quad (8)$$

where $\lambda \in R^4$ is the adjoint variable. Let $(\hat{x}, \hat{u})$ be an optimal solution for (7). Then the necessary optimality condition for (7) implies [5] that there exists a piecewise continuous and piecewise continuously differentiable adjoint function $\lambda : Q \to R^4$ satisfying

$$\frac{\partial\lambda}{\partial t} = D\frac{\partial^2\lambda}{\partial p^2} - \frac{\partial\mathcal{H}}{\partial x}(\hat{x}, \hat{x}_\tau, \hat{u}, \lambda) - \chi_{[t_0, t_f - \tau_x]}\frac{\partial\mathcal{H}}{\partial x_\tau}(\hat{x}_{+\tau}, \hat{x}, \hat{u}_{+\tau}, \lambda_{+\tau}), \quad (9)$$

$$\lambda(p, t_f) = 0, \frac{\partial\lambda(a, t)}{\partial p} = \frac{\partial\lambda(b, t)}{\partial p} = 0, \quad (10)$$

$$0 = \frac{\partial\mathcal{H}}{\partial u}(\hat{x}, \hat{x}_\tau, \hat{u}, \lambda). \quad (11)$$

According to the optimality condition (11), we have

$$\frac{\partial\mathcal{H}(\hat{x}, \hat{x}_\tau, \hat{u}, \lambda)}{\partial u} = \hat{u}(p, t) + \frac{\hat{x}_2(p, t)}{\hat{x}_4(p, t)}\left(\omega\lambda_1(p, t) - \lambda_2(p, t) + (1 - \omega)\lambda_3(p, t)\right) = 0.$$

Now, using the property of the control space $u \in \langle 0, u_{max}\rangle$, we get

$$\hat{u}(p, t) = 0, \; if, \; \frac{x_2(p, t)}{x_4(p, t)}\left(-\omega\lambda_1(p, t) + \lambda_2(p, t) - (1 - \omega)\lambda_3(p, t)\right) \leq 0,$$

$$\hat{u}(p, t) = \frac{x_2(p, t)}{x_4(p, t)}\left(-\omega\lambda_1(p, t) + \lambda_2(p, t) - (1 - \omega)\lambda_3(p, t)\right), \quad (12)$$

$$if, \; 0 \leq \frac{x_2(p, t)}{x_4(p, t)}\left(-\omega\lambda_1(p, t) + \lambda_2(p, t) - (1 - \omega)\lambda_3(p, t)\right) < u_{max},$$

$$\hat{u}(p, t) = u_{max}, \; if, \; \frac{x_2(p, t)}{x_4(p, t)}\left(-\omega\lambda_1(p, t) + \lambda_2(p, t) - (1 - \omega)\lambda_3(p, t)\right) \geq u_{max}.$$

There are some numerical methods to solve the challenges of obtaining an optimal control $\hat{u}$. The first is based on solving the optimal systems (3) and (9) which consist of eight partial differential equations and boundary conditions in time also. The second method starting with an initial guess for the adjoint Eq. (9), then we solve the state Eq. (3) and the control Eq. (12) by a forward method in time. These state and control values are used to solve the adjoint equations by backward methods in time. We proposed a new method to solve distributed optimal control problem (7) using ESNs, where state x and co-state variables λ are solved forward in time.

4 Discretisation and Adaptive Critic Neural Networks Solution of the Distributed Optimal Control

In the optimal control problems the objective is to devise a strategy of action, or control law, that minimises the desired performance cost Eq. (7). In 1977, Werbos [18] introduced an approach for approximate dynamic programming, which later became known under the name of adaptive critic design (ACD). A typical design of ACD consists of three modules: action, model (plant), and critic. We need to determine three pieces of information: How to adapt the critic network; How to adapt the model network; and How to adapt the action network.

The action consists of a parametrised control law. The critic approximates the value-related function and captures the effect that the control law has on the future cost. At any given time the critic provides a guidance on how to improve the control law. In return, the action can be used to update the critic. An algorithm that successively iterates between these two operations converges to the optimal solution over time. The plant dynamics are discrete, time-invariant, and deterministic, and they can be modelled by a difference Eq. (10). The action and critic networks are chosen as echo state networks. Assume that the rectangle $Q = \{(p, t) : a \le p \le b, t_0 \le t \le t_f\}$ is subdivided into N by M rectangles with sides $h_t = \frac{t_f - t_0}{N}$, $K_\tau = \frac{\tau}{h_t}$ and $h_s = \frac{b-a}{M}$. Start at the bottom row, where $t = t_0$, and the solution is $x(p_i, t_0) = \phi_s(p_i, t_0)$. A method for computing the approximations to $x(p, t)$ at grid the points in successive rows $\{x(p_i, t_j) : i = 0, 1, \ldots, N, \ j = 0, 1, \ldots, M\}$ will be developed. Let us denote $x(p_i, t_j)$ by x_{ij}. The difference formulae for $x_t(p, t)$ and $x_{pp}(p, t)$ are

$$x_t(p, t) \approx \frac{x_{i,j+1} - x_{ij}}{h_t}, \quad x_p(p, t) \approx \frac{x_{i+1,j} - x_{ij}}{h_s}$$

and

$$x_{pp}(p, t) \approx \frac{x_{i-1,j} - 2x_{i,j}) + x_{i+1,j}}{h_s^2}.$$

Discretisation of Eqs. (3) and (4) is given by

$$x_{ij+1} = x_{i,j} + h_t D \frac{x_{i-1,j} - 2x_{ij} + x_{i+1,j}}{h_s^2} + h_t f(x_{ij}, x_{ij-K_\tau}, u_{ij}), \quad (13)$$

$$x_{i0} = \phi_i, x_{0j} = x_{1j}, x_{Nj} = x_{N-1,j}. \quad (14)$$

Discretisation of Eqs. (9 – 11) is given by

$$\lambda_{ij} = \lambda_{i,j+1} + h_t D \frac{\lambda_{i-1,j+1} - 2\lambda_{ij+1} + \lambda_{i+1,j+1}}{h_s^2} + h_t \lambda_{i,j+1} f_{x_{ij}}(x_{ij}, x_{ij-K_\tau}, u_{ij}) +$$

$$\lambda_{i,j+1+K_\tau} f_{x_{\tau ij}}(x_{ij+K_\tau}, x_{ij}, u_{ij+K_\tau}), \quad (15)$$

$$\lambda(iM) = 0, \lambda_{0j} = \lambda_{1j}, \lambda_{Nj} = \lambda_{N-1,j}, \quad (16)$$

$$0 = h_t(h_s \quad u_{ij} + \lambda_{ij+1} f_{u_{ij}}(x_{ij}, x_{ij-K_\tau}, u_{ij},)), \ j = 0, \ldots, M-1, \ i = l, \ldots, N-1. \quad (17)$$

Algorithm 1. Algorithm to solve the optimal control problem.

Input: Choose t_0, t_f, a, b, N, M, $K_\tau = \tau/h_t$ - number of steps, time and
space steps h_t, h_s, ϵ - stopping tolerance for critic neural networks, τ -
time delay $x_{p_i,t} = \phi_{(p_i,t)}$ -initial values, $t \in [-\tau, 0]$.

Output: Set of final approximate optimal control $\hat{u}(p_i, t_0 + jh_t) = \hat{u}_{ij}$ and
optimal trajectory $\hat{x}(p_i, t_0 + (j+1)h_t) = \hat{x}_{i,j+1}$, $j = 0, \ldots, M-1$,
respectively

1 Set randomly the initial weight of Eq. (1) $\mathbf{W} = (W^{in}, W, W^{bf})$, W with echo
state property

2 Set randomly the initial estimate of $\hat{u}_{i0}$, $\hat{\lambda}_{i0}$

3 **for** $j \leftarrow 0$ **to** $M - 1$ **do**

4   **for** $i \leftarrow 1$ **to** $N - 1$ **do**

5   Set randomly u_{ij} and λ_{ij}

6   **while** $err \geq \epsilon$ **do**

7   Compute W_a^{out} and W_c^{out} using Eq. (2) with training set
$v = (x_{ij}, x_{ij-K_\tau})$, $Y = u_{ij}$ and $v = (x_{ij}, x_{ij-K_\tau})$, $Y = \lambda_{ij}$,
respectively

8   **for** $s \leftarrow 0$ **to** K_τ **do**

9   Compute $u_{i,j+s}$ and $\lambda_{i,j+1+s}$ using action (W_a^{out}) and critic
(W_c^{out}) by Eq. (1), respectively with $v = (x_{ij+s}, x_{ij-K_\tau+s})$ and
$x_{i,j+1+s}$ by Eq. (13)

10   Compute $x_{i,j+1}$ using Eq. (13) with x_{ij}, x_{ij-K_τ} and u_{ij}

11   Compute $\lambda_{i,j+1}$ using Eq. (1) with x_{ij+1}, x_{ij+1-K_τ} and W_c^{out}

12   Compute u_{ij}, using Eq. (17) with x_{ij}, x_{ij-K_τ} and λ_{ij+1}

13   Compute λ_{ij}^t, using Eq. (15) with x_{ij}, x_{ij-K_τ}, x_{ij+K_τ}, u_{ij}, λ_{ij+1}
and λ_{ij+1+K_τ}

14   *Set* $err =\parallel \lambda_{ij}^t - \lambda_{ij} \parallel$

15           $\lambda_{ij} = \lambda_{ij}^t$

16   Set $\hat{\lambda}_{i,j} = \lambda_{i,j}^t$, $\hat{u}_{i,j} = u_{i,j}$

17   Compute $\hat{x}_{i,j+1}$ using Eq. (9) with x_{ij} and $\hat{u}_{i,j}$

18   Set $\lambda_{0j} = \lambda_{1j}$, $\lambda_{Nj} = \lambda_{N-1,j}$

19   **return** $\hat{\lambda}_{i,j}$, $\hat{u}_{i,j}$, $\hat{x}_{i,j+1}$

Equations (15–17) represent the discrete version of Eqs. (9 – 11) with
$\lambda = (-1, \lambda_1, \ldots, \lambda_4)$ and $f = (a_1 x_1 + a_2 x_2 + 1/2u^2, F_1, \ldots, F_4)$. The adaptive
critic neural network procedure of the optimal control problem is summarised in
Algorithm 1. In the adaptive critic synthesis, the action and critic network were
selected such that they consist of 8 input neurons, 1000 neurons in reservoir and
1 and 4 neurons in output, respectively.

4.1   Numerical Simulation

The solution of distributed optimal control problem (1) with state and control
constraints using adaptive critic neural network and NLP methods are displayed

in Figs. 2, 3, 4 and 5. We have plotted susceptible, infected and recovered individuals with and without control by considering values of parameters [20] as $b = 0.07$, $d = 0.0123$, $\alpha = 0.0476$, $\beta = 0.21$, $\mu = 0.014$, $\omega = 0.35$, $\epsilon = 1$, $\tau = 0.5, K_c = 140$, $a_1 = 0$, $a_2 = 1$. The numerical results show that the number of susceptible individuals (Fig. 3) increase after the optimal control treatment and small number of individuals are infected from population $x_4(p,t)$, which converges to spatially constant equilibrium K_c, see Fig. 2. Figure 4 represents the population of infected individuals with and without control $u(p,t)$. In Fig. 5, the number of recovered individuals increase.

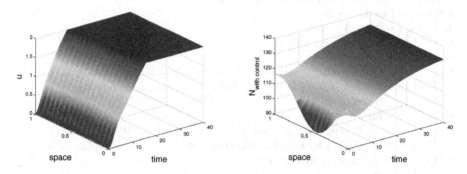

Fig. 2. Adaptive critic neural network simulation of optimal control $\hat{u}(p,t)$ and total population $N(p,t)$ with initial condition $\psi_s(t) = (8,3,2,13)(8 + cos(2\pi p))$ for $t \in [-0.5, 0]$.

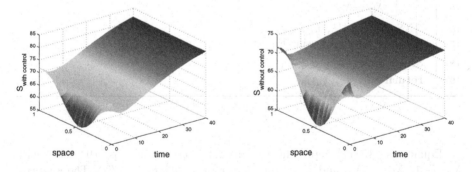

Fig. 3. The plot represents the population of susceptible individuals both with control and without control with initial condition $\psi_s(t) = (8,3,2,13)(8 + cos(2\pi p))$ for $t \in [-0.5, 0]$.

The proposed ESNs are able to meet the convergence tolerance values that we choose, which leads to satisfactory simulation results. Our results are quite similar to those obtained in [6] by using Pontriagin's maximum principle without diffusion.

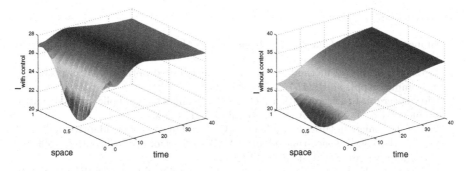

Fig. 4. The plot represents the population of infected individuals both with control and without control with initial condition $\psi_s(t) = (8, 3, 2, 13)(8 + cos(2\pi p))$ for $t \in [-0.5, 0]$.

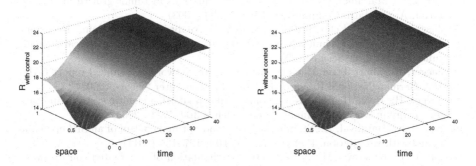

Fig. 5. The plot represents the population of recovered individuals both with control and without control with initial condition $\psi_s(t) = (8, 3, 2, 13)(8 + cos(2\pi p))$ for $t \in [-0.5, 0]$.

5 Conclusion

In the current work, we presented efficient ESNs adaptive critic design opti- misation algorithm for distributed optimal control. Using discretisation of time and space variable the optimal control problem is transcribed into a discrete- time high-dimensional nonlinear programming problem which is characterised by state and co-state equations involving forward by ESNs adaptive critic design in time, respectively. This approach is applicable to a wide class of nonlinear systems. The method was tested on the SIR model. Using MATLAB, we present a comparison between optimal control and without control. It is easy to see that the optimal control is much more effective for reducing the number of infected individuals. In order to illustrate the overall picture of the epidemic, the num- bers of infected, susceptible and recovered individuals under the optimal control and without control are shown in figures.

Acknowledgment. The paper was worked out as a part of the solution of the scientific project KEGA 010UJS-4/2014.

References

1. Brauer, F., Castillo-Chavez, C.: Mathematical Models in Population Biology and Epidemiology. Springer, New York (2001)
2. Borzi, A.: Multigrid methods for parabolic distributed optimal control problems. J. Comput. Appl. Math. **157**, 365–382 (2003)
3. Chryssoverghi, I.: Discretization methods for semilinear parabolic optimal control problems. Int. J. Numer. Anal. Model. **3**, 437–458 (2006)
4. Clever, D., Lang, J., Ulbrich, S., Ziems, J.C.: Combination of an adaptive multilevel SQP method and a space-time adaptive PDAE solver for optimal control problems. Procedia Comput. Sci. **1**, 1435–1443 (2012)
5. Gollman, L., Kern, D., Mauer, H.: Optimal control problem with delays in state and control variables subject to mixed control-state constraints. Optim. Control Appl. Meth. **30**, 341–365 (2009)
6. Forys, U., Marciniak-Czochra, A.: Logistic equations in tumor growth modelling. Int. J. Appl. Math. Comput. Sci. **13**, 317–325 (2003)
7. Jaeger, H.: The "Echo State" Approach to Analysing and Training Recurrent Neural Networks. Technical report GMD 148, German National Research Institute for Computer Science (2001)
8. Jaeger, H.: Short Term Memory in Echo State Networks. Technical report GMD 152, German National Research Institute for Computer Science (2002)
9. Jaeger, H., Haas, H.: Harnessing nonlinearity: predicting chaotic systems and saving energy in wireless communication. Science **304**, 7880 (2004)
10. Kar, T.K., Batabyal, A.: Stability analysis and optimal control of an SIR epidemic model with vaccination. BioSystems Sci. **104**, 127–135 (2011)
11. Kmet, T., Kmetova, M.: Neural networks simulation of distributed control problems with state and control constraints. In: Villa, A.E.P., Masulli, P., Pons Rivero, A.J. (eds.) ICANN 2016. LNCS, vol. 9886, pp. 468–477. Springer, Cham (2016). doi:10.1007/978-3-319-44778-0_55
12. Knowles, G.: Finite element approximation of parabolic time optimal control problems. SIAM J. Control Optim. **20**, 414–427 (1982)
13. Mittelmann, H.D.: Solving elliptic control problems with interior point and SQP methods: control and state constraints. J. Comput. Appl. Math. **120**, 175–195 (2000)
14. Murray, J.D.: Mathematical Biology. Springer, Berlin (1993)
15. Ongenae, F., Van Looy, S., Verstraeten, D., Verplancke, T., Decruyenaere, J.: Time series classification for the prediction of dialysis in critically ill patients using echo state networks. Eng. Appl. Artif. Intell. **26**, 984–996 (2013)
16. Padhi, R., Unnikrishnan, N., Wang, X., Balakrishnan, S.N.: Adaptive-critic based optimal control synthesis for distributed parameter systems. Automatica **37**, 1223–1234 (2001)
17. Verstraeten, D., Schrauwen, B., D'Haene, M., Stroobandt, D.: An experimental unification of reservoir computing methods. Neural Networks **20**(3), 414–423 (2007)
18. Werbos, P.J.: Approximate dynamic programming for real-time control and neural modelling. In: White, D.A., Sofge, D.A. (eds.) Handbook of Intelligent Control: Neural Fuzzy, and Adaptive Approaches, pp. 493–525 (1992)
19. Yoshida, N., Hara, T.: Global stability of a delayed SIR epidemic model with density dependent birth and death rates. J. Comput. Appl. Math. **201**, 339–347 (2007)
20. Zaman, G., Kang, Y.H., Jung, I.H.: Optimal treatment of an SIR epidemic model with time delay. BioSystems **98**, 43–50 (2009)

The Study of Architecture MLP with Linear Neurons in Order to Eliminate the "vanishing Gradient" Problem

Janusz Kolbusz[1], Pawel Rozycki[1(✉)], and Bogdan M. Wilamowski[2]

[1] University of Information Technology and Management in Rzeszow,
Sucharskiego 2, 35-225 Rzeszow, Poland
{jkolbusz,prozycki}@wsiz.rzeszow.pl
[2] Auburn University, Auburn, AL 36849-5201, USA
wilambm@auburn.edu
http://wsiz.rzeszow.pl

Abstract. Research in deep neural networks are becoming popular in artificial intelligence. Main reason for training difficulties is the problem of vanishing gradients while number of layers increases. While such networks are very powerful they are difficult in training. The paper discusses capabilities of different neural network architectures and presents the proposition of new multilayer architecture with additional linear neurons, that is much easier to train that traditional MLP network and reduces effect of vanishing gradients. Efficiency of suggested approach has been confirmed by several exeriments.

Keywords: Deep neural networks · Vanishing gradient · Nonlinearity

1 Introduction

The current approach of solving complex problems and processes usually includes the following steps: at first we are trying to understand the problem, and then we are trying to describe them in the form of mathematical formulas. The problem is that such complex problems are usually described by a huge amount of data, which are very difficult to understand and process by human's brain. Moreover, the use of mathematical or statistical methods often do not give a satisfactory solution due to the multidimensional and nonlinear nature of the problem. Methods based on artificial intelligence that use neural networks provide effective solutions to such complex problems. There are many neural networks technologies with different architectures such as SLP/MLP, BMLP, FCC, RBF networks, LVQ, PCA and SVM but currently especially interesting are those based on neural networks with deep architecture. Training deep multilayer neural networks is known to be very hard. The standard learning strategy

This work was supported by the National Science Centre, Cracow, Poland under Grant No. 2013/11/B/ST6/01337.

© Springer International Publishing AG 2017
L. Rutkowski et al. (Eds.): ICAISC 2017, Part I, LNAI 10245, pp. 97–106, 2017.
DOI: 10.1007/978-3-319-59063-9_9

consisting of randomly initializing the weights of the network and applying gradient descent using backpropagation is known empirically to find poor solutions for networks with three or more hidden layers. For that reason, artificial neural networks have been limited to one or two hidden layers [1]. Deep Learning has achieved significant success in the past few years on many challenging tasks [2–5], but there is still much work needed to understand why deep architectures are able to learn such effective representations.

Deep multilayer neural networks were not successfully trained, since then several methods have been developed to more successful train them [6–8] showing experimentally advantages of using deeper over less deep architectures. Most of these experimental results were obtained with new approaches in initialization or training mechanisms. Our objective here is to understand better why standard gradient descent from random initialization is doing so poorly in the case of deep neural networks. Some recent research shows that non-linear activations functions like sigmoid are not suited for deep networks with random initialization because of its mean value, which can drive especially the top hidden layer into saturation that explaining the plateaus sometimes seen when training neural networks. Once the deeper networks start converging, a degradation problem occurs. Due to this, the accuracy degrades rapidly after it is saturated. The training error increases as we add more layers to a deep model, as mentioned in [16].

These deep architectures are attractive because they provide a significantly larger computing power than shallow neural networks. It seems that deep neural networks are good candidates for modeling complex multidimensional nonlinear systems, but the training process of these networks is very difficult [9–11]. This is because with an increase of the network depth the network became less transparent for training that impedes the convergence of deeper networks [13]. This is usually called the vanishing gradient problem [11] that is a well-known nuisance in neural networks with many layers [18]. As the gradient information is backpropagated, repeated multiplication or convolution with small weights renders the gradient information ineffectively small in earlier layers. Several approaches exist to reduce this effect in practice, for example through careful initialization [12], hidden layer supervision [19]. Recent research propose to solve this problem by normalized initialization [12–14] and Batch Normalization [15] where instead of normalized initialization and keeping a lower learning rate, Batch Normalization makes normalization a part of the model and performs it for each mini-batch. To solve this problem several authors introduced additional connections to improve the information flow across several layers. Highway Networks [17] have parametrized connections, known as information highways, which allow information to flow unimpeded into deeper layers. During the training phase this connection are adjusted to control the amount of information allowed on these highways.

Our experiments confirming the last approach show that the vanishing gradient problem can be practically eliminated by introducing additional connections across layers in the BMLP architecture. In this paper it will be shown that the similar goal can be achieved using traditional MLP architecture with additional linear neurons in each layer.

2   Nonlinearity capabilities of deep neural networks

As mentioned in previous chapter the deep neural networks despite of problems with training are able to solve much complex and more nonlinear problems than shallow networks. The power of given architecture depends on nonlinearity that can be modeled by network configured in this architecture. The question is how the nonlinearity of network depends on its architecture. Answer could be key for proper dimensioning of network for given class of problem. This is a complex problem and it probably will not be possible to solve it in general way.

The deep neural networks have huge capabilities, that allow to solve complex nonlinear problems. The rapid development of computing machines led to renewed interest in deep neural networks. Research shows that non-linearity, which may be modeled by a string transfer function of neurons in the deep neural network increases exponentially with the depth of the network. This was confirmed by analysis using the trigonometric or exponential approach to function modeling activate neurons. [8, 20]

As already mentioned, the deep neural networks are difficult to learn. The problem of "vanishing gradient" is illustrated in Fig. 1 where with increasing depth the training success rate initially increases too, but then the success rate decreases reaching zero with about 6 hidden layers. The presented results were achieved using the algorithm NBN (Neuron-by-Neuron) [22] solving the well-known benchmark the two-spiral problem.

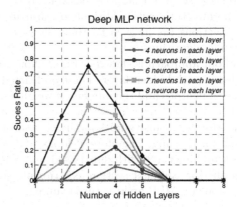

Fig. 1. Success rates for training the two-spiral problems using MLP networks with number of hidden layers (from 1 to 8) and different numbers of neurons in each layer. For each network, all the hidden layers consist of the same number of neurons.

3   Approach for Resolving Vanishing Gradient Problem

As mentioned in [22] our experiments show that the vanishing gradient problem
can be significantly reduced, if not eliminated when additional connections across
layers are introduced. The approach of deep neural network learning that we here
propose is based on the introduction to the MLP network architecture additional
linear neurons that allow to transmit the input signal through the various hidden
layers of the neural network.

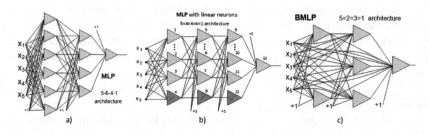

Fig. 2. Neural network architectures used in research: (a) MLP Multi-Layer Perceptron, (b) MLP Multi-Layer Perceptron with linear (green) neurons, (c) BMLP (Bridged
MLP).

In the other words we suggest that it is possible to train the deep neural
networks using MPL architectures with additional linear neurons instead of using
the cross layered connections (as in BMLP) that allows to use much simpler
gradient descent training algorithms.

Implementation of our approach is shown in Fig. 2. Note that only one linear
neuron has been added in each hidden layer. These additional neurons increases
both the total number of neurons and the number of weights in the network, but
in the contrary to MPL architecture it make the network trainable.

The reason is very simple because additional linear neurons became the network more transparent for training. Therefore, it looks that it would be possible
to train deep network architectures with supervised algorithms using gradient
based methods. Described in the next section experiments show the significant
influence of linear neurons on MLP architecture training.

4   Experimental Results

To confirm the validity of the proposed learning of deep neural network in solving complex problems four experiments with well-known benchmarks have been
prepared [23]: two-spiral classification problem, Schwefel and Peaks functions
approximation, and Parity-7 problem. All problems have been tried to resolve
with neural networks with different architectures from multi-layer MLP, multi-
layer MLP with linear neurons (MLPL), BMLP deep network with second-order
NBN algorithm. In all experiments NBN 2.08 software [24] been missed. Note

that MLP and BMLP architectures have been shown as the reference architectures that allow to evaluate efficiency of suggested approach.

Carried out experiments consisted on training of neural network with NBN algorithm for different network architectures solving the same problem. The process of learning network was repeated 100 times with a maximum number of iterations of 500. The resulting "success rate" in range between 0 and 1 percent reflects the cases when the SSE (square sum error) reached value of 0.01, or in other words the number of successful attempts learning network for all trials. The obtained results are presented in Tables 1, 2, 3 and 4 and Figs. 4, 5 and 6.

The obtained results show that adding linear neurons has a big impact on the success of the neural network training. In all cases, the network with new architecture allow to reach higher success rate then these reached for MPL

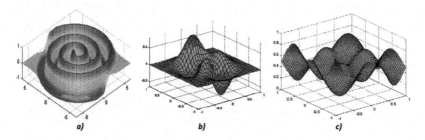

Fig. 3. Visualization of problems used in experiments: (a) two-spiral, (b) peaks, (c) Schwefel function.

Table 1. Success rate of two-spiral classification problem for different MLP, MLP with linear neurons and BMLP architectures with NBN algorithm.

Layers	1			2			3			4		
Neurons	MLP	MLPL	BMLP	MLP	MLPL	BMLP	MLP	MLPL	BMLP	MLP	MLPL	BMLP
1	0	0	0	0	0	0	0	0	0	0	0	0
2	0	0	0	0	0	0	0	0.06	0	0	0.16	0.30
3	0	0	0	0	0	0	0	0.10	0.18	0	0.20	0.70
4	0	0	0	0	0	0	0	0.14	0.82	0.09	0.22	0.91
5	0	0	0	0	0	0.25	0	0.30	0.91	0.05	0.47	0.98
6	0	0	0	0	0	0.37	0	0.32	0.93	0	0.50	1
7	0	0	0	0	0	0.53	0	0.38	0.99	0	0.54	1
8	0	0	0	0	0.15	0.79	0	0.40	1	0	0.62	1
Layers	5			6			7			8		
Neurons	MLP	MLPL	BMLP	MLP	MLPL	BMLP	MLP	MLPL	BMLP	MLP	MLPL	BMLP
1	0	0	0	0	0	0	0	0	0	0	0	0
2	0	0.28	0.24	0	0.28	0.38	0.12	0.28	0.53	0.41	0.36	0.78
3	0.11	0.57	0.85	0.30	0.64	0.92	0.50	0.73	0.94	0.74	0.78	0.97
4	0.22	0.65	0.93	0.35	0.71	0.94	0.43	0.76	0.98	0.51	0.81	0.99
5	0.08	0.81	0.99	0.09	0.87	1	0.12	0.89	1	0.18	0.90	1
6	0	0.83	1	0	0.90	1	0	0.91	1	0	0.94	1
7	0	0.85	1	0	0.93	1	0	0.95	1	0	0.96	1
8	0	0.86	1	0	0.94	1	0	0.98	1	0	1	1

Table 2. Success rate of Peaks function for different MLP, MLP with linear neurons and BMLP architectures with NBN algorithm.

Layers	1			2			3			4		
Neurons	MLP	MLPL	BMLP	MLP	MLPL	BMLP	MLP	MLPL	BMLP	MLP	MLPL	BMLP
1	0	0	0	0	0	0	0	0	0	0	0	0
2	0	0	0	0	0	0	0	0	0	0	0	0
3	0	0	0	0	0	0	0	0	0	0	0	0.07
4	0	0	0	0	0	0	0	0	0.20	0	0.52	0.57
5	0	0	0	0	0	0	0	0.58	0.25	0	0.98	1
6	0	0	0	0	0.01	0	0	0.98	1	0.01	1	1
7	0	0	0	0	0.12	0.04	0	1	1	0	1	1
8	0	0	0	0	0.24	0.07	0	1	1	0	1	1
Layers	5			6			7			8		
Neurons	MLP	MLPL	BMLP	MLP	MLPL	BMLP	MLP	MLPL	BMLP	MLP	MLPL	BMLP
1	0	0	0	0	0	0	0	0	0	0	0	0
2	0	0	0	0	0	0	0	0.02	0.04	0	0.54	0.91
3	0	0	0.78	0.04	0.34	1	0.87	0.79	1	0.98	0.83	1
4	0.02	0.70	0.83	0.52	0.86	1	0.77	0.85	1	0.90	0.90	1
5	0.12	0.90	1	0.35	0.93	1	0.50	0.94	1	0.65	0.98	1
6	0.11	0.92	1	0.16	0.95	1	0.35	0.97	1	0.48	1	1
7	0.04	0.96	1	0.14	0.98	1	0.20	1	1	0.41	1	1
8	0.03	0.98	1	0.04	1	1	0.16	1	1	0.20	1	1

Table 3. Success rates for Schwefel function approximation with MLP, MLP with linear neurons and BMLP architectures and NBN algorithm.

Layers	1			2			3			4		
Neurons	MLP	MLPL	BMLP	MLP	MLPL	BMLP	MLP	MLPL	BMLP	MLP	MLPL	BMLP
1	0	0	0	0	0	0	0	0	0	0	0	0
2	0	0	0	0	0	0	0	0	0	0	0	0
3	0	0	0	0	0	0	0	0.02	0	0	0	0.34
4	0	0	0	0	0	0	0	0.18	0	0.09	0.22	0.78
5	0	0	0	0	0.12	0	0	0.32	0.42	0.06	0.36	0.98
6	0	0	0	0	0.26	0	0	0.36	0.56	0.01	0.46	1
7	0	0	0	0	0.38	0.11	0	0.44	0.89	0	0.56	1
8	0	0	0	0	0.18	0.23	0	0.64	0.98	0	0.68	1
Layers	5			6			7			8		
Neurons	MLP	MLPL	BMLP	MLP	MLPL	BMLP	MLP	MLPL	BMLP	MLP	MLPL	BMLP
1	0	0	0	0	0	0	0	0	0	0	0	0
2	0	0.02	0	0.06	0.14	0	0.10	0.14	0	0.12	0.44	0
3	0	0.10	0.01	0.17	0.86	0	0.21	0.86	0	0.32	0.91	0.01
4	0.12	0.18	0.14	0.23	0.93	0	0.28	0.93	0	0.37	1	0.14
5	0.18	0.20	0.19	0.29	1	0.15	0.31	1	0	0.45	1	0.19
6	0.09	0.42	0.12	0.39	1	0.04	0.43	1	0	0.52	1	0.12
7	0.02	0.47	0.05	0.48	1	0	0.51	1	0	0.59	1	0.05
8	0	0.84	0	0.89	1	0	0.90	1	0	0.93	1	0

Table 4. Success rates for Parity-7 problem with MLP, MLP with linear neurons and BMLP architectures and NBN algorithm.

Layers	1			2			3			4		
Neurons	MLP	MLPL	BMLP	MLP	MLPL	BMLP	MLP	MLPL	BMLP	MLP	MLPL	BMLP
1	0	0	0	0	0	0	0	0	0	0	0	0.09
2	0	0	0	0	0.26	1	0	0	1	0	0.43	1
3	0	0	0.77	0.15	0.75	1	0.10	0.9	1	0.12	0.86	1
4	0	0.26	0.94	0.66	0.91	1	0.7	0.96	1	0.64	0.97	1
5	0	0.89	1	0.91	1	1	0.95	1	1	0.89	0.99	1
6	0.95	1	1	0.98	1	1	0.98	1	1	0.98	1	1
7	0.99	1	1	1	1	1	0.99	1	1	0.99	1	1
8	0.78	1	1	1	1	1	1	1	1	0	1	1
Layers	5			6			7			8		
Neurons	MLP	MLPL	BMLP	MLP	MLPL	BMLP	MLP	MLPL	BMLP	MLP	MLPL	BMLP
1	0	0	1	0	0	0	1	0	0	1	0	0.01
2	0	0.73	1	0.06	0	0.76	1	0	0.78	1	0	0.82
3	0.02	0.87	1	0.17	0.05	0.88	1	0.10	0.89	1	0.19	0.90
4	0.47	0.91	1	0.23	0.51	0.93	1	0.37	0.93	1	0.41	0.94
5	0.75	0.96	1	0.29	0.79	0.97	1	0.77	0.98	1	0.83	1
6	0.90	0.98	1	0.39	0.91	0.99	1	0.94	1	1	0.96	1
7	0.98	1	1	0.48	1	1	1	1	1	1	1	1
8	0.99	1	1	0.89	1	1	1	1	1	1	1	1

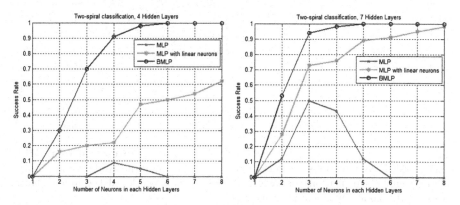

Fig. 4. Success rates for training the Two-spiral classification using MLP, MLPL, BMLP architectures networks with number of hidden layers (4 and 7) and different numbers of neurons in each layer. For each network, all the hidden layers consist of the same number of neurons.

architecture that in most cases was unable to train at all. Note that BMLP achieved much better results compared with other tested architectures but training this type of neural network requires the algorithm, which allows to connect of neurons across layers of the network. Presented results show that linear neurons allow to reduce or even eliminate the impact of "vanishing gradient" during training of network without connections through layers.

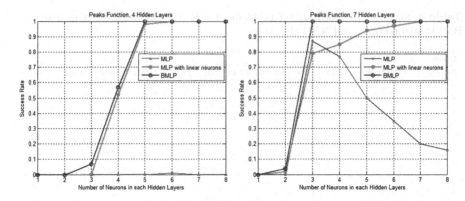

Fig. 5. Success rates for training the Peaks function using MLP, MLPL, BMLP architectures networks with number of hidden layers (4 and 7) and different numbers of neurons in each layer. For each network, all the hidden layers consist of the same number of neurons.

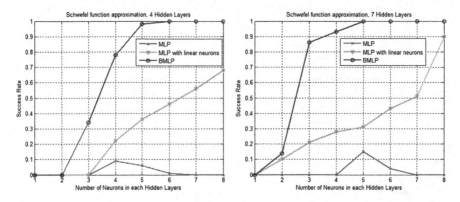

Fig. 6. Success rates for training the Schwefel function approximation using MLP, MLPL, BMLP architectures networks with number of hidden layers (4 and 7) and different numbers of neurons in each layer. For each network, all the hidden layers consist of the same number of neurons.

5   Conclusions

The deeper neural networks allow to solve more complex problems. This is a result of the ability to model non-linear functions more as the activation function. Also experimental results confirm advantage of deep over shallow neural networks in ability to solve high nonlinear problems. It was shown experimentally that power of neural network grows linearly with its width and exponentially with its depth. Also it is shown that commonly used MLP architecture is not suitable for deep learning because if number of hidden layer exceeds 3 than success rate for training decrease rapidly (see Figs. 1, 2, 3, 4, 5, 6 and 7 and Tables 1, 2, 3 and 4) and it is almost impossible to train MLP architecture with 6 or more hidden

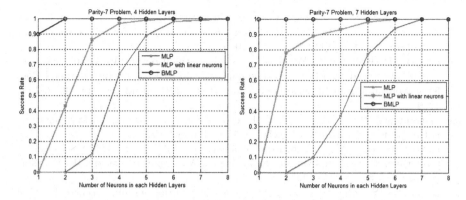

Fig. 7. Success rates for training the Parity-7 problem using MLP, MLPL, BMLP architectures networks with number of hidden layers (4 and 7) and different numbers of neurons in each layer. For each network, all the hidden layers consist of the same number of neurons.

layers. By introduction of linear neurons the vanishing gradient problem can be practically eliminated. Presented architecture should allow to use well-known algorithms gradient learning neural networks, which will be the basis for later studies. The advantage of the presented solution of the "vanishing gradient" problem by adding the linear neurons into the MLP architecture is possibility to train successfully similar and simple (much simpler than BMLP) multilayer network that allow to use well-known gradient method. As a disadvantages it may be included increase number of neurons and weight that affects the network training time.

References

1. Larochelle, H., et al.: Exploring strategies for training deep neural networks. J. Mach. Learn. Res. **10**, 1–40 (2009)
2. Krizhevsky, A., Sutskever, I., Hinton, G.E.: Imagenet classification with deep convolutional neural networks. In: Advances in Neural Information Processing Systems, pp. 1097–1105 (2012)
3. Simonyan, K., Zisserman, A.: Very deep convolutional networks for large-scale image recognition. arXiv preprint arXiv:1409.1556 (2014)
4. Mnih, V., Kavukcuoglu, K., Silver, D., Rusu, A.A., Veness, J., Bellemare, M.G., Graves, A., Riedmiller, M., Fidjeland, A.K., Ostrovski, G., et al.: Human-level control through deep reinforcement learning. Nature **518**(7540), 529–533 (2015)
5. Silver, D., Huang, A., Maddison, C.J., Guez, A., Sifre, L., van den Driessche, G., Schrittwieser, J., Antonoglou, I., Panneershelvam, V., Lanctot, M., et al.: Mastering the game of go with deep neural networks and tree search. Nature **529**(7587), 484–489 (2016)
6. Wilamowski, B.M., Bo, W., Korniak, J.: Big data and deep learning. In: 2016 IEEE 20th Jubilee International Conference on Intelligent Engineering Systems (INES). IEEE (2016)

7. Wilamowski, B.M., Korniak, J.: Learning architectures with enhanced capabilities and easier training. In: 2015 IEEE 19th International Conference on Intelligent Engineering Systems (INES). IEEE (2015)
8. Rozycki, P., Kolbusz, J., Wilamowski, B.M.: Estimation of deep neural networks capabilities based on a trigonometric approach. In: IEEE 20th International Conference on Intelligent Engineering Systems (INES 2016), Budapest, pp. 30–2, June 2016
9. Wilamowski, B.M., Yu, H.: Neural network learning without backpropagation. IEEE Trans. Neural Networks **21**(11), 1793–1803 (2010)
10. Hunter, D., Hao, Y., Pukish, M.S., Kolbusz, J., Wilamowski, B.M.: Selection of proper neural network sizes and architectures A comparative study. IEEE Trans. Industr. Inf. **8**, 228–240 (2012)
11. Hochreiter, S.: The vanishing gradient problem during learning recurrent neural nets and problem solutions. Int. J. Unc. Fuzz. Knowl. Based Syst. **06**, 107 (1998)
12. Glorot, X., Bengio, Y.: Understanding the difficulty of training deep feedforward neural networks. In: AISTATS (2010)
13. He, K., Zhang, X., Ren, S., Sun, J.: Delving deep into rectifiers: surpassing human-level performance on imagenet classification. In: ICCV (2015)
14. LeCun, Y., Bottou, L., Orr, G.B., Müller, K.-R.: Efficient backProp. In: Orr, G.B., Müller, K.-R. (eds.) Neural Networks: Tricks of the Trade. LNCS, vol. 1524, pp. 9–50. Springer, Heidelberg (1998). doi:10.1007/3-540-49430-8_2
15. Ioffe, S., Szegedy, C.: Batch normalization: accelerating deep network training by reducing internal covariate shift. In: ICML (2015)
16. He, K., J. Sun, J.: Convolutional neural networks at constrained time cost. In: CVPR (2015)
17. Srivastava, R.K., Greff, K., Schmidhuber, J.: Highway networks. arXiv preprint arxiv:1505.00387 (2015)
18. Bengio, Y., Simard, P., Frasconi, P.: Learning long-term dependencies with gradient descent is dificult. IEEE Trans. Neural Networks **5**(2), 157–166 (1994)
19. Lee, C.Y., Xie, S., Gallagher, P., Zhang, Z., Tu, Z.: Deeply-supervised nets. arXiv preprint arxiv:1409.5185 (2014)
20. Rozycki, P., Kolbusz, J., Korostenskyi, R., Wilamowski, B.M.: Estimation of deep neural networks capabilities using polynomial approach. In: Rutkowski, L., Korytkowski, M., Scherer, R., Tadeusiewicz, R., Zadeh, L.A., Zurada, J.M. (eds.) ICAISC 2016. LNCS (LNAI), vol. 9692, pp. 136–147. Springer, Cham (2016). doi:10.1007/978-3-319-39378-0_13
21. Wilamowski, B.M., Yu, H.: Improved computation for levenberg marquardt training. IEEE Trans. Neural Networks **21**(6), 930–937 (2010)
22. Rozycki, P., Kolbusz, J., Wilamowski, B.M.: Dedicated deep neural network architectures and methods for their training. In: IEEE 19th International Conference on Intelligent Engineering Systems (INES 2015), Bratislava, pp. 73–78, 3–5 September 2015
23. Hunter, D.: Utilizing Dual Neural Networks as a Tool for Training, Optimization, and Architecture Conversion. Ph.D. thesis, Auburn University (2013)
24. Wilamowski, B.M., Yu, H.: NNT - Neural Networks Trainer. http://nng.wsiz.rzeszow.pl/

Convergence and Rates of Convergence of Recursive Radial Basis Functions Networks in Function Learning and Classification

Adam Krzyżak[1]($\boxtimes$) and Marian Partyka[2]

[1] Department of Computer Science and Software Engineering,
Concordia University, 1455 de Maisonneuve Blvd. West,
Montreal H3G 1M8, Canada
krzyzak@cs.concordia.ca
[2] Department of Knowledge Engineering,
Faculty of Production Engineering and Logistics,
Opole University of Technology, ul. Ozimska 75, 45-370 Opole, Poland
m.partyka@po.opole.pl

Abstract. In this paper we consider convergence and rates of convergence of the normalized recursive radial basis function networks in function learning and classification when network parameters are learned by the empirical risk minimization.

Keywords: Nonlinear regression · Classification · Recursive radial basis function networks · MISE convergence · Strong convergence · Rates of convergence

1 Introduction

In artificial neural network literature three types of neural network architectures prevail: multilayer perceptrons (MLP), radial basis function (RBF) networks and normalized radial basis function (NRBF) networks. These neural network models have been applied in variety of tasks, including interpolation, classification, data smoothing and regression. Theoretical analysis of MLP can be found among others in, e.g., Cybenko [8], White [43], Hornik et al. [23], Barron [3], Anthony and Bartlett [1], Devroye et al. [11], Györfi et al. [20], Ripley [37], Haykin [22], Hastie et al. [21] and Kohler and Krzyżak [24] (the latter paper analyzed so called deep networks and dimensionality reduction) and analysis of RBF networks (both standard and normalized) in, e.g., Moody and Darken [34], Park and Sandberg [35,36], Girosi and Anzellotti [16], Xu et al. [44], Krzyżak et al. [28], Krzyżak and Linder [29], Krzyżak and Niemann [30], Györfi et al. [20] and Krzyżak and Schäfer [31].

Research of the first author was supported by the Natural Sciences and Engineering Research Council of Canada under Grant RGPIN-2015-06412.

L. Rutkowski et al. (Eds.): ICAISC 2017, Part I, LNAI 10245, pp. 107–117, 2017.
DOI: 10.1007/978-3-319-59063-9_10

In this article we consider normalized radial basis function (RBF) networks with one hidden layer of at most k nodes with a fixed kernel $\phi : \mathcal{R}_+ \to \mathcal{R}$:

$$f_k(x) = \frac{\sum_{i=1}^{k} w_i \phi \left(\|x - c_i\|_{A_i} \right)}{\sum_{i=1}^{k} \phi \left(\|x - c_i\|_{A_i} \right)} \tag{1}$$

where

$$\|x - c_i\|_{A_i}^2 = [x - c_i]^T A_i [x - c_i].$$

which are class of functions satisfying the following conditions:

(i) radial basis function condition: $\phi : \mathcal{R}_0^+ \to \mathcal{R}^+$ is a left-continuous, decreasing function, the so-called *kernel*.

(ii) centre condition: $c_1, ..., c_k \in \mathcal{R}^d$ are the so-called *centre vectors* with $\|c_i\| \leq R$ for all $i = 1, ..., k$.

(iii) receptive field condition: $A_1, ..., A_k$ are symmetric, positive definite, real $d \times d$-matrices each of which satisfies the eigenvalue inequalities $\ell \leq \lambda_{min}(A_i) \leq \lambda_{max}(A_i) \leq L$. Here, $\lambda_{min}(A_i)$ and $\lambda_{max}(A_i)$ are the minimal and the maximal eigenvalue of A_i, respectively. A_i specifies the *receptive field* about the centre c_i.

(iv) weight condition: $w_1, ..., w_k \in \mathcal{R}$ are the *weights* satisfying $|w_i| \leq B$ for all $i = 1, ..., k$.

Throughout the paper we use the convention $0/0 = 0$. Common choices for the kernel satisfying (i) are:

- **Window type kernels.** These are kernels for which some $\delta > 0$ exists such that $\phi(t) \notin (0, \delta)$ for all $t \in \mathcal{R}_0^+$. The classical naive kernel $\phi(t) = \mathbf{1}_{[0,1]}(t)$ is a member of this class.
- **Non-window type kernels with bounded support.** These comprise all kernels with support of the form $[0, s]$ which are right-continuous in s. For example, for $\phi(t) = \max\{1 - t, 0\}$, $\phi(x^T x)$ is the Epanechnikov kernel.
- **Kernels with unbounded support**, i.e. $\phi(t) > 0$ for all $t \in \mathcal{R}_0^+$. The most famous example of this class is $\phi(t) = \exp(-t)$. Then $\phi(x^T x)$ is the classical Gaussian kernel.

Let us denote the parameter vector $(w_0, ..., w_k, c_1, ..., c_k, A_1, ..., A_k)$ by θ. It is assumed that the kernel is fixed, while network parameters $w_i, c_i, A_i, i = 1, ..., k$ are learned from the data. Normalized RBF networks are generalizations of standard RBF networks defined by

$$f_k(x) = \sum_{i=1}^{k} w_i \phi \left(\|x - c_i\|_{A_i} \right) + w_0. \tag{2}$$

The most popular choices of radial function ϕ are:

- $\phi(x) = e^{-x^2}$ (Gaussian kernel)
- $\phi(x) = e^{-x}$ (exponential kernel)

- $\phi(x) = (1 - x^2)_+$ (truncated parabolic or Epanechnikov kernel)
- $\phi(x) = \frac{1}{\sqrt{x^2+c^2}}$ (inverse multiquadratic)

All these kernels are nonincreasing. In the literature on approximation by means of radial basis functions the following monotonically increasing kernels were considered

- $\phi(x) = \sqrt{x^2 + c^2}$ (multiquadratic)
- $\phi(x) = x^{2n} \log x$ (thin plate spline)

They play important role in interpolation and approximation with radial functions [17], but are not considered in the present paper.

Standard RBF networks have been introduced by Broomhead and Lowe [7] and Moody and Darken [34]. Their approximation error was studied by Park and Sandberg [35,36]. These result have been generalized by Krzyżak, Linder and Lugosi [28], who also showed weak and strong universal consistency of RBF networks for a large class of radial kernels in the least squares estimation problem and classification. The rate of approximation of RBF networks was investigated by Girosi and Anzellotti [16]. The rates of convergence of RBF networks trained by complexity regularization have been investigated in regression estimation problem by Krzyżak and Linder [29].

Normalized RBF networks (1) have been originally investigated by Moody and Darken [34] and Specht [40]. Further results were obtained by Shorten and Murray-Smith [39]. Some convergence results for the regression estimation problem have been discussed in [31].

Other nonparametric regression estimation techniques include Nadaraya-Watson kernel estimate and recursive kernel estimate (considered later in the paper) [19,20], nearest-neighbor estimate [12,20], partitioning estimate [4,20], orthogonal series estimate [18,20], tree estimate [6,21] and Breiman random forest [5,38].

This paper investigates mean integrated square error (MISE) convergence and strong convergence as well as rates of convergence of the normalized recursive RBF network estimation in nonlinear function learning and classification. The paper is organized as follows. In Sect. 2 the algorithm for nonlinear regression learning is presented. In Sect. 3 the normalized recursive RBF network classifier is discussed. In Sect. 4 convergence properties of the learning algorithms are investigated and Sect. 5 presents conclusions.

2   Nonlinear Function Learning

Let $(X, Y), (X_1, Y_1), (X_2, Y_2), \ldots, (X_n, Y_n)$ be independent, identically distributed, $\mathcal{R}^d \times \mathcal{R}$–valued random variables with $\mathbf{E}Y^2 < \infty$, and let $R(x) = \mathbf{E}(Y|X = x)$ be the corresponding nonlinear regression function. Let μ be the distribution of X. It is well-known that regression function R minimizes L_2 error:

$$\mathbf{E}|R(X) - Y|^2 = \min_{f:\mathcal{R}^d \to \mathcal{R}} \mathbf{E}|f(X) - Y|^2.$$

Our aim is to estimate R from the i.i.d. observations of random vector (X, Y)

$$D_n = \{(X_1, Y_1), \ldots, (X_n, Y_n)\}$$

using RBF network (1). We train the network by choosing its parameters that minimize the L_2 risk

$$\frac{1}{n} \sum_{j=1}^{n} |f(X_j) - Y_j|^2 \tag{3}$$

on the training data D_n, that is we choose RBF network m_n in the class

$$\mathcal{F}_n = \{f_k = f_\theta : \theta \in \Theta_n\} = \left\{ \frac{\sum_{i=1}^{k} w_i \phi \left(\|x - c_i\|_{A_i} \right)}{\sum_{i=1}^{k} \phi \left(\|x - c_i\|_{A_i} \right)} \right\}$$

where

$$\Theta_n = \{\theta = (w_1, \ldots, w_{k_n}, c_1, \ldots, c_{k_n}, A_1, \ldots, A_{k_n})\}.$$

so that

$$\frac{1}{n} \sum_{j=1}^{n} |m_n(X_j) - Y_j|^2 = \min_{f \in \mathcal{F}_n} \frac{1}{n} \sum_{j=1}^{n} |f_\theta(X_j) - Y_j|^2. \tag{4}$$

We measure the performance of the RBF network estimate by the MISE error

$$\mathbf{E}|m_n(X_1) - R(X_1)|^2 = \mathbf{E} \int |m_n(x) - m(x)|^2 \mu(dx).$$

Even though direct analysis of m_n has been carried out in [31] using Vapnik-Chervonenkis dimension and covering numbers [20, 41, 42] it is not fully satisfactory as it is pretty complex and learning of parameters by empirical risk minimization imposes heavy computational burden. In the reminder of the paper we will explore in the analysis of m_n its proximity to the recursive kernel regression function estimate

$$r_n(x) = \frac{\sum_{i=1}^{n} Y_i K \left(\frac{x - X_i}{h_i} \right)}{\sum_{i=1}^{n} K \left(\frac{x - X_i}{h_i} \right)} \tag{5}$$

where $K : \mathcal{R}^d \to \mathcal{R}$ is a kernel and $h_i, i = 1, 2, \ldots n$ is a smoothing sequence (bandwith) of positive real numbers. The estimate has been introduced by Devroye and Wagner [10] and investigated by Krzyżak and Pawlak [27], Greblicki and Pawlak [19] and Krzyżak [25]. It can be computed recursively as follows:

$$r_0(x) = g_0(x) = 0$$
$$g_n(x) = g_{n-1}(x) + K_{h_n}(x - X_n)$$

and

$$r_n(x) = r_{n-1}(x) + g_n^{-1}(x)(Y_n - r_{n-1}(x))K_{h_n}(x - X_n)$$

where $K_h(x) = K(\frac{x}{h})$.

Estimator (5) need not be recomputed entirely when additional observation is combined with the previous ones. In addition, implementing a nonrecursive kernel regression estimate based on n observations requires storing the observations $X_1, \ldots X_n$, whereas implementing m_n requires storing m_{n-1} and g_{n-1}.

The properties of the standard kernel regression estimate with applications to analysis of RBF networks have been discussed in detail in [30]. This approach leads to very simple and efficient training. Assume that K is spherically symmetric, i.e. $K(x) = K(||x||)$. Let parameters of (1) be trained as follows

$$k_n = n, A_i = \frac{1}{h_i^2}I, w_i = Y_i, \ c_i = X_i, \ i = 1, \cdots, n. \tag{6}$$

Thus we obtain the plug-in recursive RBF network

$$g_n(x) = \frac{\sum_{i=1}^{n} \phi(\frac{||x-X_i||}{h_i})Y_i}{\sum_{i=1}^{n} \phi(\frac{||x-X_i||}{h_i})} = \frac{\sum_{i=1}^{n} K(\frac{x-X_i}{h_i})Y_i}{\sum_{i=1}^{n} K(\frac{x-X_i}{h_i})}. \tag{7}$$

As a consequence of the simple bound

$$\mathbf{E}|m_n(X_1) - R(X_1)|^2 \leq \mathbf{E}|g_n(X_1) - R(X_1)|^2 \tag{8}$$

plug-in recursive RBF network provides an upper bound on performance of m_n on D_n.

3   Recursive Classification Rules

Let (Y, X) be a pair of random variables taking values in the set $\{1, ..., M\}$, whose elements are called classes, and in R^d, respectively. The problem is to classify X, i.e. to decide on Y. Let us define a posteriori class probabilities

$$p_i(x) = P\{Y = i|X = x\}, i = 1, \cdots, M, x \in R^d.$$

The Bayes classification rule

$$\Psi^*(X) = i \text{ if } p_i(X) > p_j(X), j < i, \text{ and } p_i(X) > p_j(X), j > i$$

minimizes the probability of error. The Bayes risk L^* is defined by

$$P\{\Psi^*(X) \neq Y\} = \inf_{\Psi:R^d \to \{1,...,M\}} P\{\Psi(X) \neq Y\}.$$

The local Bayes risk is equal to $P\{\Psi^*(X) \neq Y \mid X = x\}$. Observe that $p_i(x) = E\{I_{\{Y=i\}} \mid X = x\}$ may be viewed as a regression function of the indicator of the event $\{Y = i\}$. Given the learning sequence $V_n = \{(Y_1, X_1), ..., (Y_n, X_n)\}$ of independent observations of the pair (Y, X), we may learn $p_i(x)$ using recursive RBF nets mimicking (4), i.e.,

$$\frac{1}{n}\sum_{j=1}^{n}|\hat{p}_{in}(X_j) - I_{\{Y_j=i\}}|^2 = \min_{f \in \mathcal{F}_n} \frac{1}{n}\sum_{j=1}^{n}|f_Y(X_j) - Y_j|^2. \tag{9}$$

We construct an empirical recursive classification rule Ψ_n, which classifies every $x \in R^d$ to any class maximizing $\hat{p}_{in}$. In order to simplify learning process we will consider simple plug-in classification rules considered in Sect. 3. We propose plug-in recursive RBF classifier with parameters learned by (6) resulting in the recursive classification rule Ψ_n which classifies every $x \in R^d$ to any class maximizing

$$p_{in} = \frac{\sum_{j=1}^n I_{\{Y_j=i\}} K\left(\frac{x-X_j}{h_j}\right)}{\sum_{j=1}^n K\left(\frac{x-X_j}{h_j}\right)}. \tag{10}$$

The global performance of Ψ_n is measured by $L_n = P\{\Psi_n(X) \neq \theta \mid V_n\}$ and the local performance by $L_n(x) = P\{\Psi_n(x) \neq \theta \mid V_n\}$. A rule is said to be weakly, strongly, or completely Bayes risk consistent (BRC) if $L_n \to L^*$, in probability, almost surely, or completely, respectively, as $n \to \infty$. Thanks to relation (see [25])

$$|L_n(X) - L^*(X)| \leq \sum_{i=1}^M |p_{in}(X) - p_i(X)| \tag{11}$$

any convergence result obtained for regression estimate (7) is also valid for Ψ_n.

In the next section we will study convergence and rates of plug-in recursive RBF network regression estimate g_n and recursive RBF estimate m_n as well as classification rules induced by them.

4   Consistency and Rates of Convergence

In the first part of this section we present convergence results (consistency and the rates) for the recursive RBF learning function learning and classification algorithms. In the second part of the section we outline the proofs.

4.1   Convergence Results

We have the following convergence and rates of convergence results for the recursive plug-in RBF network g_n and classification rule Ψ_n. Inequality (8) enables us to apply convergence and rates results of g_n to the recursive RBF network m_n trained and evaluated on the sequence D_n. Likewise inequality (11) enables us to deduce convergence and rates of convergence for recursive RBF classification rules.

Theorem 1. *Let* $EY^2 < \infty$,

$$c_1 I_{S_{0,r}} \leq \phi(||x||) \leq c_2 I_{S_{0,R}}, \quad 0 < r < R < \infty, \quad c_1, c_2 > 0 \tag{12}$$

$$\sum_{i=1}^n h_i^d \to \infty \quad , \sum_{i=1}^n h_i^d I_{\{h_i > \epsilon\}} / \sum_{i=1}^n h_i^d \to 0, \, as \, n \to \infty,$$

$$\limsup_n \frac{1}{n} \sum_{i=1}^n \left(\frac{h_i}{\overline{h}}\right)^{2d} = \gamma < \infty, \tag{13}$$

where $\bar{h} = \min_{1 \leq i \leq n} h_i$.

Then

$$E(g_n(X) - R(X))^2 \to 0 \ as \ n \to \infty$$
$$E(m_n(X_1) - R(X_1))^2 \to 0 \ as \ n \to \infty$$
and
$$E(L_n(X_1) - L^*(X_1))^2 \to 0 \ as \ n \to \infty.$$

Theorem 1 provides MISE convergence of the recursive RBF plug-in estimates $g_n(x), m_n$ and L_n for all distributions of the data with the bounded second moment condition $EY^2 < \infty$. Radial functions satisfying condition (12) are functions with compact support separated away from zero at the origin. Such functions do not include Gaussian kernel. Using kernel trick introduced in [13] one can show that condition (12) can be relaxed to

$$\phi(||x||) \geq cI_{S_{0,r}}, \int \sup_{y \in S_{x,r}} \phi(||y||)dx < \infty, \quad c, r > 0. \tag{14}$$

However, enlarging the class of kernels results in necessity to impose stricter condition on the outputs, namely $|Y| \leq M < \infty$.

Condition (14) means that envelope of ϕ is bounded away from zero at the origin and is Riemann integrable and ϕ may have infinite support. It is satisfied by Gaussian and exponential kernels. Assumption (12) is satisfied for arbitrary finite ϕ with compact support and bounded away from zero at the origin.

Theorem 2 provides MISE convergence of the recursive RBF plug-in algorithms. Note that the rate of convergence is obtained for Lipschitz regression. As Devroye [9] points out there is no free-lunch, i.e., there are no distribution-free rates of convergence.

Theorem 2. *Let μ denote the probability measure of X with a compact support, and let (12) hold. Let smoothing bandwidth h_i satisfy (13). Also let*

$$\sup_x E(Y^2|X = x) \leq \sigma^2 < \infty$$
$$|R(x) - R(y)| \leq \beta ||x - y||^\alpha, \quad 0 < \alpha \leq 1, \quad \beta > 0. \tag{15}$$

Then

$$E(g_n(X) - R(X))^2 = O(n^{-\frac{2\alpha}{2\alpha+d}})$$
$$E(m_n(X_1) - R(X_1))^2 = O(n^{-\frac{2\alpha}{2\alpha+d}})$$
and
$$E(L_n(X_1) - L^*(X_1))^2 = O(n^{-\frac{2\alpha}{2\alpha+d}}). \tag{16}$$

The final result concerns an exponential bound from which almost sure convergence of the learning algorithms follows. For the sake of brevity we only present the result for g_n.

Theorem 3. *Let (12) and (13) hold. Then for every distribution of (X, Y) with $E|Y|^{2+\delta} < \infty$ with $\delta > 0$ and for every $\epsilon > 0$, there exist constants c and n_0 such that for all $n \geq n_0$,*

$$P\{\int |g_n(x) - g(x)|\mu(dx) > \epsilon\} \leq 2\exp(-c \cdot n). \tag{17}$$

4.2   Outlines of Proofs

The sketches of the proofs are given below. We only provide proofs for g_n as the proofs for remaining algorithms are similar.

PROOF of Theorem 1. Let $K_i = K((X_1 - X_i)/h_i), i = 1, \ldots, n$. We start by noticing that we have for any function g

$$E(R(X_1) - g_n(X_1))^2$$
$$\leq 4E(R(X_1) - g(X_1))^2 + 4E(\sum_{j=1}^{n}(Y_j - R(X_j))K_j / \sum_{j=1}^{n} K_j)^2$$
$$+4E\left(\frac{\sum_{j=1}^{n}(R(X_j) - g(X_j))K_j}{\sum_{j=1}^{n} K_j}\right)^2 + 4E\left(\frac{\sum_{j=1}^{n} g(X_j)K_j}{\sum_{j=1}^{n} K_j} - g(X_1)\right)^2$$
$$= 4(A + B + C + D).$$

For any $\epsilon > 0$ we can find continuous, compactly supported $g \in L_2(\mu)$ such that $\int(R(x) - g(x))^2 d\mu(x) < \epsilon/16$. Hence

$$A \leq \epsilon/16.$$

Using Jensen's inequality we can bound term C by $\epsilon/16$ and term D by

$$\sup_{x,y:||x-y||\leq\delta} |g(x) - g(y)|^2 < \epsilon/16.$$

Term B can be bounded by using truncation argument for Y, conditions (13) and Lemma 3 of [26].

PROOF of Theorem 2. Let's bound MISE as follows

$$E(R(X_1) - g_n(X_1))^2$$
$$\leq 2E(\sum_{i=1}^{n}(Y_i - R(X_i))K_i / \sum_{i=1}^{n} K_i)^2 + 2E(\sum_{i=1}^{n}(R(X_i) - R(X_1))K_i / \sum_{i=1}^{n} K_i)^2$$
$$= 2(A + B).$$

One can show that compactness of μ implies

$$A = O\left(\frac{1}{\sum_{i=1}^{n} h_i^d}\right). \tag{18}$$

Using Jensen's inequality and Lipschitz assumption for term B we have

$$B = O\left(\frac{\sum_{i=1}^{n} h_i^{2d+2\alpha}}{\sum_{i=1}^{n} h_i^d}\right). \tag{19}$$

The rate result (16) follows from (18) and (19).

PROOF of Theorem 3. The exponential bound (17) follows from the exponential inequality of Hoeffding type for martingale difference sequences called Azuma inequality (see Azuma [2]) or in a more straightforward way from the fundamental McDiarmid's inequality [33], which can be stated as follows:
    Let $X_1, \cdots, X_n$ be independent random variables and assume that

$$\sup_{x_i, x_i'} |f(x_1, ..., x_i, ...x_n) - f(x_1, ..., x_i', ...x_n)| \leq c_i, 1 \leq i \leq n. \tag{20}$$

Then

$$P\{|f(X_1, \cdots, X_n) - Ef(X_1, \cdots, X_n)| \geq \epsilon\} \leq 2\exp(-2\epsilon^2/\sum_{i=1}^{n} c_i^2). \tag{21}$$

Let $K((x - x_i)/h_i) = K_{h_i}(x - x_i)$ and take

$$f(X_1, Y_1, \cdots, X_n, Y_n) = \sum_{i=1}^{n} \int (Y_i - R(x))K_{h_i}(x - X_i)/\sum_{i=1}^{n} EK_{h_i}(x - X_i)\mu(dx).$$

Following [13] the left-hand side of (20) is bounded by

$$\sup_{x_i, x_i', y_i, y_i'} \int \frac{|y_i K_{h_i}(x-x_i) - y_i' K_{h_i}(x-x_i')|}{\sum_{i=1}^{n} EK_{h_i}(x-X_i)} \mu(dx)$$
$$\leq 4M \sup_y \int \frac{K_{h_i}(x-y)}{\sum_{i=1}^{n} EK_{h_i}(x-X_i)} \mu(dx) \leq 4M\frac{\rho}{n}\left(\frac{h_i}{h}\right)^d.$$

Mimicking the proof of Theorem 1 in [26] (we omit the details) the result follows from (13) and (21).

5   Conclusions

We have analyzed MISE and strong convergence and the rates of convergence of the recursive normalized radial basis function regression estimates and classification rules learned from data by the empirical risk minimization. The analysis has been simplified by taking advantage of the relationship between the empirical risk minimization and the recursive Nadaraya-Watson kernel regression estimation.

References

1. Anthony, M., Bartlett, P.L.: Neural Network Learning: Theoretical Foundations. Cambridge University Press, Cambridge, UK (1999)
2. Azuma, K.: Weighted sums of certain dependent random variables. Tohoku Math. J. **19**(3), 357–367 (1967)
3. Barron, A.R.: Universal approximation bounds for superpositions of a sigmoidal function. IEEE Trans. Inf. Theory **39**, 930–945 (1993)
4. Beirlant, J., Györfi, L.: On the asymptotic L_2-error in partitioning regression estimation. J. Stat. Plan. Infer. **71**, 93–107 (1998)

5. Breiman, L.: Random forests. Mach. Learn. **45**, 5–32 (2001)
6. Breiman, L., Friedman, J.H., Olshen, R.A., Stone, C.J.: Classification and regression trees. In: Wadsworth Advanced Books and Software, Belmont, CA (1984)
7. Broomhead, D.S., Lowe, D.: Multivariable functional interpolation and adaptive networks. Complex Syst. **2**, 321–323 (1988)
8. Cybenko, G.: Approximations by superpositions of sigmoidal functions. Math. Control Signals Syst. **2**, 303–314 (1989)
9. Devroye, L.: Any discrimination rule can have arbitrary bad probability of error for finite sample size. IEEE Trans. Pattern Anal. Mach. Intell. **PAMI–4**, 154–157 (1982)
10. Devroye, L.P., Wagner, T.J.: On the L1 convergence of the kernel estimators of regression functions with applications in discrimination. Zeitschrift Wahrscheinlichkeitstheorie verwandte Gebiete **51**(1), 15–25 (1980)
11. Devroye, L., Györfi, L., Lugosi, G.: Probabilistic Theory of Pattern Recognition. Springer, New York (1996)
12. Devroye, L., Györfi, L., Krzyżak, A., Lugosi, G.: On the strong universal consistency of nearest neighbor regression function estimates. Ann. Stat. **22**, 1371–1385 (1994)
13. Devroye, L., Krzyżak, A.: An equivalence theorem for L_1 convergence of the kernel regression estimate. J. Stat. Plann. Infer. **23**, 71–82 (1989)
14. Duchon, J.: Sur l'erreur d'interpolation des fonctions de plusieurs variables par les D^m-splines. RAIRO-Anal. Numèrique **12**(4), 325–334 (1978)
15. Faragó, A., Lugosi, G.: Strong universal consistency of neural network classifiers. IEEE Trans. Inf. Theory **39**, 1146–1151 (1993)
16. Girosi, F., Anzellotti, G.: Rates of convergence for radial basis functions and neural networks. In: Mammone, R.J. (ed.) Artificial Neural Networks for Speech and Vision, pp. 97–113. Chapman and Hall, London (1993)
17. Girosi, F., Jones, M., Poggio, T.: Regularization theory and neural network architectures. Neural Comput. **7**, 219–267 (1995)
18. Greblicki, W., Pawlak, M.: Fourier and Hermite series estimates of regression functions. Ann. Inst. Stat. Math. **37**, 443–454 (1985)
19. Greblicki, W., Pawlak, M.: Necessary and sufficient conditions for Bayes risk consistency of a recursive kernel classification rule. IEEE Trans. Inf. Theory **IT–33**, 408–412 (1987)
20. Györfi, L., Kohler, M., Krzyżak, A., Walk, H.: A Distribution-Free Theory of Nonparametric Regression. Springer, New York (2002)
21. Hastie, T., Tibshirani, R., Friedman, J.: The Elements of Statistical Learning; Data Mining, Inference and Prediction, 2nd edn. Springer, New York (2009)
22. Haykin, S.O.: Neural Networks and Learning Machines, 3rd edn. Prentice-Hall, New York (2008)
23. Hornik, K., Stinchocombe, S., White, H.: Multilayer feed-forward networks are universal approximators. Neural Netw. **2**, 359–366 (1989)
24. Kohler, M., Krzyżak, A.: Nonparametric regression based on hierarchical interaction models. IEEE Trans. Inf. Theory **63**, 1620–1630 (2017)
25. Krzyżak, A.: The rates of convergence of kernel regression estimates and classification rules. IEEE Trans. Inf. Theory **IT–32**, 668–679 (1986)
26. Krzyżak, A.: Global convergence of recursive kernel regression estimates with applications in classification and nonlinear system estimation. IEEE Trans. Inf. Theory **38**, 1323–1338 (1992)

27. Krzyżak, A., Pawlak, M.: Distribution-free consistency of a nonparametric kernel regression estimate and classification. IEEE Trans. Inf. Theory **IT–30**, 78–81 (1984)
28. Krzyżak, A., Linder, T., Lugosi, G.: Nonparametric estimation and classification using radial basis function nets and empirical risk minimization. IEEE Trans. Neural Netw. **7**(2), 475–487 (1996)
29. Krzyżak, A., Linder, T.: Radial basis function networks and complexity regularization in function learning. IEEE Trans. Neural Netw. **9**(2), 247–256 (1998)
30. Krzyżak, A., Niemann, H.: Convergence and rates of convergence of radial basis functions networks in function learning. Nonlinear Anal. **47**, 281–292 (2001)
31. Krzyżak, A., Schäfer, D.: Nonparametric regression estimation by normalized radial basis function networks. IEEE Trans. Inf. Theory **51**, 1003–1010 (2005)
32. Lugosi, G., Zeger, K.: Nonparametric estimation via empirical risk minimization. IEEE Trans. Inf. Theory **41**, 677–687 (1995)
33. McDiarmid, C.: On the method of bounded differences. Surv. Comb. **141**, 148–188 (1989)
34. Moody, J., Darken, J.: Fast learning in networks of locally-tuned processing units. Neural Comput. **1**, 281–294 (1989)
35. Park, J., Sandberg, I.W.: Universal approximation using Radial-Basis-Function networks. Neural Comput. **3**, 246–257 (1991)
36. Park, J., Sandberg, I.W.: Approximation and Radial-Basis-Function networks. Neural Comput. **5**, 305–316 (1993)
37. Ripley, B.D.: Pattern Recognition and Neural Networks. Cambridge University Press, Cambridge (2008)
38. Scornet, E., Biau, G., Vert, J.-P.: Consistency of random forest. Ann. Stat. **43**(4), 1716–1741 (2015)
39. Shorten, R., Murray-Smith, R.: Side effects of normalising radial basis function networks. Int. J. Neural Syst. **7**, 167–179 (1996)
40. Specht, D.F.: Probabilistic neural networks. Neural Netw. **3**, 109–118 (1990)
41. Vapnik, V.N., Chervonenkis, A.Y.: On the uniform convergence of relative frequencies of events to their probabilities. Theory Probab. Appl. **16**, 264–280 (1971)
42. Vapnik, V.N.: Estimation of Dependences Based on Empirical Data, 2nd edn. Springer, New York (1999)
43. White, H.: Connectionist nonparametric regression: multilayer feedforward networks that can learn arbitrary mappings. Neural Netw. **3**, 535–549 (1990)
44. Xu, L., Krzyżak, A., Yuille, A.L.: On radial basis function nets and kernel regression: approximation ability, convergence rate and receptive field size. Neural Netw. **7**, 609–628 (1994)

Solar Event Classification Using Deep Convolutional Neural Networks

Ahmet Kucuk[✉], Juan M. Banda, and Rafal A. Angryk

Department of Computer Science, Georgia State University, Atlanta, GA, USA
{akucuk1,jbanda,rangryk}@cs.gsu.edu

Abstract. The recent advances in the field of neural networks, more specifically deep convolutional neural networks (CNN), have considerably improved the performance of computer vision and image recognition systems in domains such as medical imaging, object recognition, and scene characterization. In this work, we present the first attempt into bringing CNNs to the field of Solar Astronomy, with the application of solar event recognition. With the objective of advancing the state-of-the-art in the field, we compare the performance of multiple well established CNN architectures against the current methods of multiple solar event classification. To evaluate the effectiveness of deep learning in the solar image domain, we experimented with well-known architectures such as LeNet-5, CifarNet, AlexNet, and GoogLenet. We investigated the recognition of four solar event types using image regions extracted from the high-resolution full disk images of the Sun from the NASA's Solar Dynamics Observatory (SDO) mission. This work demonstrates the feasibility of using CNNs by obtaining improved results over the conventional pattern recognition methods used in the field.

Keywords: Image classification · Solar event classification · Deep learning · Convolutional neural networks

1 Introduction

The effectiveness of deep neural networks in computer vision has been demonstrated over the years in several different works [20,24,37,38]. One of the first major successes of deep learning was handwritten character recognition in 1990s [24]. More recently, Krizhevsky et al. won the 2012 ImageNet Large Scale Visual Recognition Competition (ILSRVC) with his graphics processing unit (GPU) supported method. Nowadays, the usage of deep convolutional neural networks (CNN) in computer vision tasks has exploded [20]. Several variations and techniques for CNNs are developed to be used in wide range of applications including object detection [37], image recognition [18], and autonomous driving systems [10]. Today, we are witnessing the effectiveness of deep learning in industry-level applications of major companies like Google's PlaNet for image geolocation [41] or Facebook's DeepFace for face recognition [38], among other applications.

© Springer International Publishing AG 2017
L. Rutkowski et al. (Eds.): ICAISC 2017, Part I, LNAI 10245, pp. 118–130, 2017.
DOI: 10.1007/978-3-319-59063-9_11

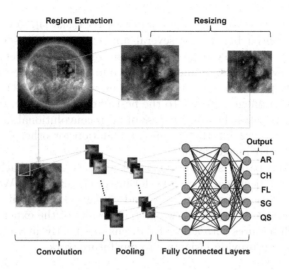

Fig. 1. Example of solar event classification using deep convolutional neural networks.

To eliminate the filtering effect of the atmosphere during solar observations, NASA launched a satellite named Solar Dynamic Observatory (SDO) into orbit [42]. Since 2010, SDO has been successfully taking high-resolution full-disk images of the Sun in 10 different wavelengths [25]. SDO generates 1.5 TB image data in a day, which makes it one of the largest solar dataset available. The data is used by solar physicists for a better understanding of the Sun, which is crucial because coronal mass ejection and solar flares on the Sun might endanger astronauts by changing the radiation levels or interrupt Global Positioning System (GPS) and intercontinental communications or damage power grids [22]. For this purpose, several computer vision tasks including object recognition [13], image classification [5], and image retrieval [7] have been utilized on this data for solar physics research. However, we believe that emerging deep learning techniques can further improve the effectiveness of previous studies.

Deep learning has several advantages compared to conventional techniques in computer vision. One of these advantages is that deep learning does not require extensive domain knowledge or expertise to select the features on the images because a general purpose learning algorithm can configure the parameters on its own; thus, the network can automatically extract the most useful features for its decision process [23]. This phenomenon is extremely compelling for the solar domain because of the fuzzy boundaries of solar events in the images. Additionally, as it is illustrated in Fig. 1, deep learning models can be fed with raw image data, which avoids daily feature extraction process required by conventional methods [7]. Although the training process can be time-consuming for large CNNs, response time in production is reasonable for real-time applications [17].

We believe that deep learning can be an alternative to conventional methods, which extract the image parameters and use well-known machine learning algorithms, to solve solar related computer vision tasks [4]. Our approach has a promising potential benefit for several applications such as Content-Based Image Retrieval (CBIR) systems for solar images [5], event detection, validation of existing event detection modules [39], and solar event prediction [30]. To introduce the deep learning in the solar domain, we selected the task of classifying solar events in image regions. To the best of our knowledge, this is the first work that demonstrates the effectiveness of deep convolutional neural networks on solar event classification that forms a foundation for other computer vision tasks related to solar image domain.

The rest of this paper is organized as follows: Sect. 2 gives background information about both deep learning and solar image classification. We then present the solar image data preparation steps and configuration details of used CNN architectures in Sect. 3. In Sect. 4, we present results of the experiments on different CNN architectures with a detailed discussion. Lastly, in Sect. 5, we present our conclusions and provide insights into our future work.

2   Background

2.1   Convolutional Neural Networks

LeCun et al. successfully applied CNN on a handwritten digit recognition task for the first time in the 1990s [24]. At the time, lack of large datasets and computational power needed for training the deeper models to solve more complicated tasks were an obstacle to the popularity of the technique. The last decade brought very large datasets and made higher computational power available at a very affordable prices. Along with increasing central processing unit (CPU) computing power, GPU computing has been adapted for deep learning for faster computational power in certain tasks. Thanks to these advancements and novel improvements in CNN architecture, deep CNNs gained major success in computer vision tasks such as ILSVRC competitions [18,20,37]. The success gave community an alternative to conventional methods like feature engineering.

A simple CNN differs from other deep learning architectures in its convolutional layers that are specifically designed for image data. Convolutional layers use a kernel to filter the image to extract local features from an image and it is usually followed by a pooling layer that reduces the number of features extracted while increasing the robustness of output to shifts and distortions [24]. In fully connected layers, extracted features are collected to detect higher level features [24]. To eliminate non-significant neurons and emphasize significant ones, an activation function is used. CNNs can be trained using back propagation that allows the network to adjust weights of the network to reduce the output of loss function. In other words, this process simulates hand-picking of features that is usually performed by domain experts in most of the traditional computer vision approaches.

Several techniques based on simple CNN architecture are proposed to address different problems of the learning process. The most relevant techniques include Rectified Linear Unit (ReLU), data augmentation, and dropout. Using ReLU activation increases the training speed significantly when it is compared to *tanh* activation function [20, 28]. Data augmentation is artificially populating the existing data to reduce overfitting [20]. Dropout is randomly picking some neurons to be silenced temporarily. Theoretically, a network with n neurons behaves like an assembly of up to 2^n model while making decision [36]. This technique is also very effective against overfitting. Besides the architectural advancements, the efficiency of deep learning has significantly improved thanks to hardware support. GPU usage during training process increase the training speed significantly so that larger models can be trained to solve more complex models in reasonable times.

Network in Network (NIN) structure is another technique introduced recently that has proved its effectiveness [26]. NIN structure consists of the multilayer perceptron (MLP) and global averaging pooling layer. Multilayer perceptrons, which consist of several layers to increase the effectiveness of local feature extraction, are used instead of convolution filters. Fully connected layers, followed by convolutional layers, were replaced with global average pooling that introduce feature map for each class. Unlike fully connected layer, global average pooling is not prone to overfitting. Inception model introduced in [37] is inspired by NIN structure. The main idea in their work is to increase the depth of a network by adding 1×1 convolutional layers and enhancing the feature re-use. While increasing the network depth, this strategy decreases the number of parameters used in the network.

Increasing the depth and width of a deep learning model is the simplest way to increase the performance of the network. Although increment in the depth size brings problems with it such as overfitting, computational complexity, and vanishing gradient, deep models (with 22 layers) have been trained in these works [35, 37]. Even deeper models with 152 layers have been trained using residual learning technique [18].

While advancements on deep learning architectures are continuing at a fast pace, CNN applications produced significant results in real life problems such as face verification by Facebook [38], vehicle type classification [15], species recognition for wild animal monitoring [11], land use classification in remote sensing images [9]. CNNs also started to give good results on medical imaging, which is interesting for our work because of the similarity between medical images and solar images [6]. Usages of CNNs on medical image includes, but are not limited to, glaucoma detection [12], segmentation of neuronal membranes [14], and identification of breast cancer [40].

2.2   Solar Event Classification

With the start of the SDO mission in 2010 the Feature Finding Team (FFT) has been in charge of identifying solar events present in SDO images via two different approaches: (a) traditional computer vision via specialized solar event detection modules, and (b) the trainable module, as described in [27]. The specialized solar event detection modules are in charge of detecting individual events like Flares [13], Coronal Holes and Active Regions [8] and Sigmoids [31]. The trainable module utilizes statistical classification algorithms to identify solar phenomena and has been detailed in [1].

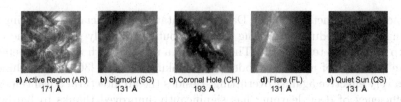

a) Active Region (AR)   b) Sigmoid (SG)   c) Coronal Hole (CH)   d) Flare (FL)   e) Quiet Sun (QS)
171 Å   131 Å   193 Å   131 Å   131 Å

Fig. 2. Example image regions for each event type with corresponding wavelengths

The SDO images are processed by extracting a set of ten image parameters: entropy, fractal dimension, the mean intensity, the third and fourth moments, relative smoothness, the standard deviation of the intensity, Tamura contrast, Tamura directionality and uniformity. These parameters have been tested to perform the best based on computational expense and classification accuracy for SDO images. More details as well as their formulas are beyond the scope of this work and can be found in [2, 3, 7].

Once the image parameters have been extracted, multiple classification algorithms have been tested in [4] allowing researchers to identify which ones perform the best depending on individual combinations of image parameters and types of solar events [1]. With the advantage of building a general purpose classifier that is able to identify more than one different type of solar event at a time, the current state of the art in terms of solar event classification has been achieved in [5]. The average classification accurracy for Active Regions (AR), Coronal Holes (CH), Flares (FL), Sigmoids (SG) has reach 70% while the individual accurracies are 79%, 84%, 72%, and 68%. These values will be considered our baseline results to compare against in this work.

In order to provide researchers with training and testing data for classification and retrieval analyses, the FFT team has released multiple datasets which include solar image parameters corresponding to labeled events and can be found in [33, 34]. It was not until [21] where researchers had also have the access to the corresponding high-resolution image files for the solar events in datasets [33, 34]. This allows for analyses like ones presented in this work.

3 Methods

3.1 Data

We used the Large-Scale Solar Dynamic Observatory Dataset (LSDO) that can be found in [21]. The data used in this work consists of solar image regions that are labeled as a particular type of solar event. We conducted several preprocessing steps to prepare an appropriate dataset for our task. The first preprocessing step was to ensure that we feed our models with unique records that does not contain overlaps of multiple event types. Event records in LSDO are reported from several different detector modules, which means that the same solar activity can be reported multiple times, if different types of events co-occur or different module can independently report the same events multiple times. To discard the possible duplicate reports, we only used reports from Spatial Possibilistic Clustering Algorithm (SPoCA) for AR and CH [39], Flare Detective for FL [16], and Sigmoid Sniffer for SG [31]. In addition, we found out that some events occur spatiotemporally close to each other which may lead to ambiguity in labeling. To remove this ambiguity, we eliminated the events that are spatiotemporally overlapping (i.e. co-occurring).

Before we extract the regions from 4 K full disk images of the Sun using bounding box information, we conducted an additional step. The network models that we use for experiments are designed for square regions. However, bounding boxes of solar events have different sizes and aspect ratios. We shortened the list of regions by selecting the ones with the aspect ratio $min(w, h)/max(w, h)$, ratio between width (w) and height (h) greater than ratio threshold 0.90, which we picked according to the distribution of ratio for each event type to get desired number of regions. We introduced ratio threshold because event regions with low aspect ratio may change the characteristic of the event pattern on the region significantly when it is rescaled to be fit into the models. Finally, we extracted regions and labeled them accordingly. Samples of each class is shown in Fig. 2.

Moreover, to evaluate our models under more realistic conditions, we created a fifth class, Quiet Sun (QS), which consists of unlabeled regions of the Sun. QS represents regions of the Sun without any specific activity. To create QS regions, in each image we randomly picked a region in size of 32×32 to 256×256 pixels and checked if the selected region is spatiotemporally overlapping with any event records in all dataset. We avoided to pick regions close to the edge of the image in order not to include the background region behind the Sun. We repeated this process until we found a non-overlapping region or gave up on that image. In Fig. 2, region e is an example QS region.

Instead of having a single dataset, we created three datasets to use in different CNN models and test different aspects of the models. For each event type, we used the optimal reported wavelength, which is listed as follows: 171 Å for AR, 193 Å for CH, 131 Å for FL and SG. We chose 1,500 regions with the highest resolution for each event type on the specified wavelength. For the fifth class (i.e. QS), we used 500 regions from each wavelength (131 Å, 171 Å, 193 Å) to get 1,500 regions so that this class contains the wavelength of all other event types. We used 32×32, 128×128, 256×256 pixels region sizes to evaluate and understand the behavior of models against different image sizes. To resize images, we used Lanczos algorithm to get better quality pacthes [29]. For convenience, we named them as follows: "R32" for 32×32, "R128" for 128×128, and "R256" for 256×256 regions.

3.2 CNN Architectures

Several different CNN architectures are presented to solve computer vision problems in different domains and in Sect. 2.1, we mention the most significant techniques employed by these models. However, for solar image domain, there is no previously proposed CNN architecture and selection of a CNN model for a particular dataset is not trivial. To see the performance of CNN architectures in

Table 1. Highest average testing accuracy of models with class-based classification accuracy

Model	Dataset	Epochs	Accuracy in %					
			AR	CH	FL	SG	QS	Average
LeNet-5	R32	65	97.7	97.0	87.0	92.0	94.0	93.5
CifarNet	R32	80	99.7	97.3	91.3	85.0	96.3	93.9
CifarNet	R128	80	78.3	88.7	86.3	92.0	82.3	85.5
CifarNet	R256	75	99.0	99.3	75.0	82.0	85.0	88.1
AlexNet	R128	35	97.3	95.0	85.0	88.3	92.7	91.7
AlexNet	R256	40	98.0	95.7	85.0	84.0	91.0	90.7
GoogLenet	R256	40	98.0	99.3	84.0	91.0	98.3	94.1

solar image classification task, we used several different models that have success in other domains. These models consist of different number of layers and employ different techniques, which allow us to measure the effectiveness of deep learning on the solar domain under different configurations.

We used four different CNN architectures: LeNet-5, CifarNet, AlexNet, and GoogLenet. LeNet-5 is trained by using stochastic gradient descent (SGD) with learning rate of 0.01, momentum of 0.9, batch size of 100, and weight decay of 0.0005. CifarNet model is trained by using SGD with learning rate of 0.001, momentum of 0.9, batch size of 100, and decay of 0.004. In AlexNet model, we used data augmentation. Random subregions of the input image are given to the network as an input in order to reduce overfitting. For 256×256 regions, 227×227 sub-images are used as it is used in [20]. For 128×128 regions, we selected 96×96 sub-images for data augmentation. We trained this mode using SGD with learning rate of 0.01, momentum of 0.9, batch size of 100, and decay of 0.0005. GoogLenet is trained by using SGD with learning rate of 0.01, momentum of 0.9, batch size of 50, and decay of 0.0002.

4   Experiments

4.1   Experimental Setup

Experiments are conducted on a dedicated server with 128 cores (Intel Xeon CPU E7-8860 with Ubuntu 14.04). We used the CPU-only version of popular deep learning framework, *Caffe* [19]. OpenBLAS is a dependency of *Caffe* and we configured OpenBLAS library to compile for 128 cores. All the results are derived from random split of 80% training data and 20% test data for each dataset.

4.2   Evaluation of the Models

Our goal is not only to accurate classification of the regions, but also to evaluate the general behavior of deep learning on solar imagery. Table 1 depicts best

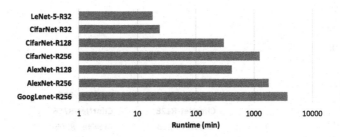

Fig. 3. Runtime of models for 10 epochs in log scale

average accuracy of each model, class-based accuracy, and number of epochs. We experimented on different models and different datasets (different image sizes). Some model-dataset combination were not listed in Table 1 because some of the models can not be feed with all image sizes and some of the experiments never give interpretable results after several parameter search.

In Fig. 3, we present 10 epochs runtime of each model in log scale. There is a considerable performance gap between fastest model, LeNet-5, and slowest model, GoogLeNet. CifarNet is close to AlexNet when it is trained on same dataset. Another implication of Fig. 3 is runtime by image size. As size of images increases, runtime changes significantly which can be clearly observed in CifarNet experiments.

Region size is important aspect of our experiments that needed to be investigated, because solar events are reported on a wide range of sizes (e.g. 16×16 to 512×512) and most of the time, regions are small part in a high resolution full disk image (4096×4096). Aspect ratio of region is also important, but it is beyond the scope of the paper because datasets are prepared from square like regions as it is described in Sect. 3. We experimented with three different region sizes: 32×32, 128×128, 256×256. Smaller regions tend to give better results, which may imply that for classification of 5 classes, CNN models do not need high-resolution detail.

For a better comparison among experiments, we provide accuracy of models calculated every epochs in Fig. 4. We had to stop training of GoogLenet model at 40 epochs because of the very long runtime but it does not influence our analysis since the model already converged at 40 epochs. We can share following observations considering the overall picture. LeNet-5 is the fastest model with reasonable accuracy but as the complexity of task increases, performance of LeNet-5 might not sustain because of its simplicity. CifarNet classify accurately in R32 dataset but it converges later than LeNet-5. GoogLenet gives the best result. However, when we consider the size of GoogLenet and its training time, GoogLenet is a complex model for our classification task. In R128 and R256 experiments, CifarNet does not give close distribution of accuracies among classes. On the other hand, AlexNet is performing well on those datasets. In Fig. 4, we observe that AlexNet converges after 15 epochs, while it takes more than 30 epochs for CifarNet and GoogLenet. Accuracy of AlexNet is less than

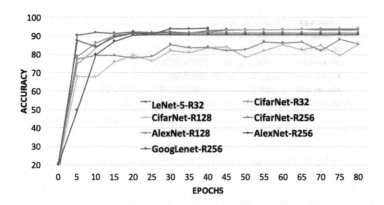

Fig. 4. Classification accuracy of models in epochs

GoogLenet but the model is significantly faster. As a result, LeNet-5 performs well within limited time in our classification task while AlexNet might be a good candidate model for more complex classification tasks.

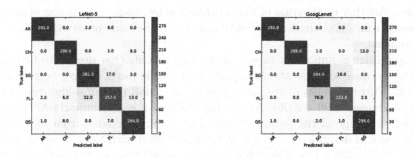

Fig. 5. Confusion matrix for LeNet-5 and GoogLenet

Figure 5 illustrates the confusion matrix for LeNet-5 and GoogLenet. Both models have similar distribution of error. One of the important observations is very minor transitivity between FL and SG classes, which may stems from the wavelength used for FL and SG is same, 131 Å. We believe that having QS regions in three different wavelengths significantly reduces this kind of transitivity. CH events are generally not occurring close to other events like AR. As expected, there is less confusion between CH and other events. However, having confusion between QS and FL is unexpected and may show that the model is having difficulty to differentiation FL and QS regions that are generated from 131 Å images.

Table 2. Accuracy comparison between traditional and deep learning approaches

Experiment/Method	Epochs	Accuracy in %					
		AR	CH	FL	SG	QS	Average
Image parameters	N/A	79.0	85.0	72.0	68.0	N/A	76.0
LeNet-5-R32	65	97.7	97.0	87.0	92.0	94.0	93.5
CifarNet-R32	80	99.7	97.3	91.3	85.0	96.3	93.9
AlexNet-R128	35	97.3	95.0	85.0	88.3	92.7	91.7
GoogLenet-R256	40	98.0	99.3	84.0	91.0	98.3	94.1

4.3 Comparison with Conventional Methods

Comparison between traditional methods and a deep learning approach is important to understand the effectiveness of our efforts. For comparison, we found the best results previously introduced for solar image classification tasks in [5] and the baseline results are shown in Table 2 at "Image Parameters" row. These results are produced using the dataset in [32]. Banda et al. are using ten different image parameters that are well performing on solar data [5]. Then, they present the best performing combinations of multiple metrics and classifiers, from which we picked the best results for each class. Our deep learning models are giving better accuracy on each class and overall accuracy. Without creating a domain specific model, achieving better accuracy for each class shows that deep learning approach is valuable and encouraging for using deep learning on solar image domain.

5 Conclusions

We presented results for four well-known deep convolutional neural networks models on a solar image classification task. We tested these models with three datasets with different region sizes. Our experiments show that for basic solar image classification task, well-known algorithms performs well while LeNet-5 and AlexNet are slightly better option in terms of time and complexity. Also, accuracy of our models is better than traditional approaches, based on extracting statistical image features and using machine learning classification algorithms. For future work, we believe that spending more time on a personalized CNN architecture for solar domain can lead to an optimal model. In addition, ensemble models have shown to produce better results in other image classification/object detection tasks [14,18,37]. Based on the fact that a model is better on classifying particular event types while its overall accuracy is low, ensemble of multiple models can increase the performance of classification. Finally, the scope of this work was limited to four event types, and adding more types to the classification task (e.g. Filaments, Sunspots) may help us to built better CNN models and lead to new and valuable observations.

References

1. Banda, J., Angryk, R., Martens, P.: Steps toward a large-scale solar image data analysis to differentiate solar phenomena. Sol. Phys. **288**, 435–462 (2013)
2. Banda, J.M., Angryk, R.A.: On the effectiveness of fuzzy clustering as a data discretization technique for large-scale classification of solar images. In: IEEE International Conference on Fuzzy Systems, FUZZ-IEEE 2009, pp. 2019–2024. IEEE (2009)
3. Banda, J.M., Angryk, R.A.: An experimental evaluation of popular image parameters for monochromatic solar image categorization. In: FLAIRS Conference, pp. 380–385 (2010)
4. Banda, J.M., Angryk, R.A.: Selection of image parameters as the first step towards creating a CBIR system for the solar dynamics observatory. In: International Conference on Digital Image Computing: Techniques and Applications, DICTA 2010, Sydney, Australia, December 1–3, 2010, pp. 528–534 (2010)
5. Banda, J.M., Angryk, R.A.: Large-scale region-based multimedia retrieval for solar images. In: Rutkowski, L., Korytkowski, M., Scherer, R., Tadeusiewicz, R., Zadeh, L.A., Zurada, J.M. (eds.) ICAISC 2014. LNCS (LNAI), vol. 8467, pp. 649–661. Springer, Cham (2014). doi:10.1007/978-3-319-07173-2_55
6. Banda, J.M., Angryk, R.A., Martens, P.C.: On the surprisingly accurate transfer of image parameters between medical and solar images. In: 18th IEEE International Conference on Image Processing, ICIP 2011, Brussels, Belgium, September 11–14, 2011, pp. 3669–3672 (2011)
7. Banda, J.M., Angryk, R.A., Martens, P.C.: Imagefarmer: introducing a data mining framework for the creation of large-scale content-based image retrieval systems. Int. J. Comput. Appl. **79**(13), 8–13 (2013)
8. Barra, V., Delouille, V., Kretzschmar, M., Hochedez, J.F.: Fast and robust segmentation of solar euv images: algorithm and results for solar cycle 23. Astron. Astrophys. **505**(1), 361–371 (2009)
9. Castelluccio, M., Poggi, G., Sansone, C., Verdoliva, L.: Land use classification in remote sensing images by convolutional neural networks. CoRR abs/1508.00092 (2015)
10. Chen, C., Seff, A., Kornhauser, A.L., Xiao, J.: Deepdriving: learning affordance for direct perception in autonomous driving. In: 2015 IEEE International Conference on Computer Vision, ICCV 2015, Santiago, Chile, December 7–13, 2015, pp. 2722–2730 (2015)
11. Chen, G., Han, T.X., He, Z., Kays, R., Forrester, T.: Deep convolutional neural network based species recognition for wild animal monitoring. In: 2014 IEEE International Conference on Image Processing, ICIP 2014, Paris, France, October 27–30, 2014, pp. 858–862 (2014)
12. Chen, X., Xu, Y., Wong, D.W.K., Wong, T.Y., Liu, J.: Glaucoma detection based on deep convolutional neural network. In: 37th Annual International Conference of the IEEE Engineering in Medicine and Biology Society, EMBC 2015, Milan, Italy, August 25–29, 2015, pp. 715–718 (2015)
13. Christe, S., Hannah, I.G., Krucker, S., McTiernan, J., Lin, R.P.: RHESSI microflare statistics. I. Flare-finding and frequency distributions. Astrophys. J. **677**(2), 1385 (2008)
14. Ciresan, D.C., Giusti, A., Gambardella, L.M., Schmidhuber, J.: Deep neural networks segment neuronal membranes in electron microscopy images. In: Advances in Neural Information Processing Systems 25: 26th Annual Conference on Neural Information Processing Systems 2012, pp. 2852–2860 (2012)

15. Dong, Z., Pei, M., He, Y., Liu, T., Dong, Y., Jia, Y.: Vehicle type classification using unsupervised convolutional neural network. In: 22nd International Conference on Pattern Recognition, ICPR 2014, Stockholm, Sweden, August 24–28, 2014, pp. 172–177 (2014)

16. Grigis, P., Davey, A., Martens, P., Testa, P., Timmons, R., Su, Y.: SDO feature finding team: the SDO flare detective, 402.08, vol. 41, May 2010

17. Gurghian, A., Koduri, T., Bailur, S.V., Carey, K.J., Murali, V.N.: Deeplanes: end-to-end lane position estimation using deep neural networks. In: Proceedings of the IEEE Conference on Computer Vision and Pattern Recognition Workshops, pp. 38–45 (2016)

18. He, K., Zhang, X., Ren, S., Sun, J.: Deep residual learning for image recognition. In: 2016 IEEE Conference on Computer Vision and Pattern Recognition, CVPR 2016, Las Vegas, NV, USA, June 27–30, 2016, pp. 770–778 (2016)

19. Jia, Y., Shelhamer, E., Donahue, J., Karayev, S., Long, J., Girshick, R.B., Guadarrama, S., Darrell, T.: Caffe: convolutional architecture for fast feature embedding. In: Proceedings of the ACM International Conference on Multimedia, MM 2014, Orlando, FL, USA, November 03–07, 2014, pp. 675–678 (2014)

20. Krizhevsky, A., Sutskever, I., Hinton, G.E.: Imagenet classification with deep convolutional neural networks. In: Advances in Neural Information Processing Systems 25: 26th Annual Conference on Neural Information Processing Systems 2012, pp. 1106–1114 (2012)

21. Kucuk, A., Banda, J.M., Angryk, R.A.: A large scale solar dynamic observatory image dataset for computer vision applications. Nature Scientific Data (2017, to be Submitted). http://lsdo.dmlab.cs.gsu.edu/

22. Langhoff, S.R., Straume, T.: Highlights of the space weather risks and society workshop. Space Weather 10(6) (2012)

23. LeCun, Y., Bengio, Y., Hinton, G.: Deep learning. Nature **521**(7553), 436–444 (2015)

24. LeCun, Y., Bottou, L., Bengio, Y., Haffner, P.: Gradient-based learning applied to document recognition. Proc. IEEE **86**(11), 2278–2324 (1998)

25. Lemen, J.R., Akin, D.J., Boerner, P.F., Chou, C., Drake, J.F., Duncan, D.W., Edwards, C.G., Friedlaender, F.M., Heyman, G.F., Hurlburt, N.E., et al.: The atmospheric imaging assembly (AIA) on the solar dynamics observatory (SDO). Solar Dyn. Obs. **275**, 17–40 (2011). Springer

26. Lin, M., Chen, Q., Yan, S.: Network in network. CoRR abs/1312.4400 (2013)

27. Martens, P., Attrill, G., Davey, A., Engell, A., Farid, S., Grigis, P., Kasper, J., Korreck, K., Saar, S., Savcheva, A., et al.: Computer vision for the solar dynamics observatory (SDO). Sol. Phys. **275**(1–2), 79–113 (2012)

28. Nair, V., Hinton, G.E.: Rectified linear units improve restricted boltzmann machines. In: Proceedings of the 27th International Conference on Machine Learning (ICML 2010), June 21–24, 2010, Haifa, Israel, pp. 807–814 (2010)

29. Parlett, B.N., Scott, D.S.: The lanczos algorithm with selective orthogonalization. Math. Comput. **33**(145), 217–238 (1979)

30. Qahwaji, R., Colak, T.: Automatic short-term solar flare prediction using machine learning and sunspot associations. Sol. Phys. **241**(1), 195–211 (2007)

31. Raouafi, N., Bernasconi, P.N., Georgoulis, M.K.: The sigmoid sniffer and the advanced automated solar filament detection and characterization code modules, 402.32, vol. 216, May 2010

32. Schuh, M., Angryk, R., Ganesan Pillai, K., Banda, J., Martens, P.: A large scale solar image dataset with labeled event regions. In: Proceedings of International Conference on Image Processing (ICIP), pp. 4349–4353 (2013)

33. Schuh, M.A., Angryk, R.A., Martens, P.C.: A large-scale dataset of solar event reports from automated feature recognition modules. J. Space Weather Space Clim. **6**, A22 (2016)
34. Schuh, M.A., Angryk, R.A., Pillai, K.G., Banda, J.M., Martens, P.C., et al.: A large-scale solar image dataset with labeled event regions. In: ICIP, pp. 4349–4353 (2013)
35. Simonyan, K., Zisserman, A.: Very deep convolutional networks for large-scale image recognition. CoRR abs/1409.1556 (2014)
36. Srivastava, N., Hinton, G.E., Krizhevsky, A., Sutskever, I., Salakhutdinov, R.: Dropout: a simple way to prevent neural networks from overfitting. J. Mach. Learn. Res. **15**(1), 1929–1958 (2014)
37. Szegedy, C., Liu, W., Jia, Y., Sermanet, P., Reed, S.E., Anguelov, D., Erhan, D., Vanhoucke, V., Rabinovich, A.: Going deeper with convolutions. In: IEEE Conference on Computer Vision and Pattern Recognition, CVPR 2015, Boston, MA, USA, June 7–12, 2015, pp. 1–9 (2015)
38. Taigman, Y., Yang, M., Ranzato, M., Wolf, L.: Deepface: closing the gap to human-level performance in face verification. In: 2014 IEEE Conference on Computer Vision and Pattern Recognition, CVPR 2014, Columbus, OH, USA, June 23–28, 2014, pp. 1701–1708 (2014)
39. Verbeeck, C., Delouille, V., Mampaey, B., De Visscher, R.: The spoca-suite: software for extraction, characterization, and tracking of active regions and coronal holes on euv images. Astron. Astrophys. **561**, A29 (2014)
40. Wang, D., Khosla, A., Gargeya, R., Irshad, H., Beck, A.H.: Deep learning for identifying metastatic breast cancer. CoRR abs/1606.05718 (2016)
41. Weyand, T., Kostrikov, I., Philbin, J.: Planet - photo geolocation with convolutional neural networks. CoRR abs/1602.05314 (2016)
42. Withbroe, G.: Living with a star. Bull. Am. Astron. Soc. **32**, 839 (2000)

Sequence Learning in Unsupervised and Supervised Vector Quantization Using Hankel Matrices

Mohammad Mohammadi[1], Michael Biehl[2], Andrea Villmann[3],
and Thomas Villmann[1(✉)]

[1] Computational Intelligence Group,
University of Applied Sciences Mittweida, Mittweida, Germany
`thomas.villmann@hs-mittweida.de`
[2] Intelligent System Group, University Groningen, Groningen, The Netherlands
[3] Berufliches Schulzentrum Döbeln-Mittweida, Mittweida, Germany

Abstract. In the present contribution we consider sequence learning by means of unsupervised and supervised vector quantization, which should be invariant regarding to shifts in the sequences. A mathematical tool to achieve a respective invariant representation and comparison of sequences are Hankel matrices with an appropriate dissimilarity measure based on subspace angles. We discuss their mathematical properties and show how they can be incorporated in prototype based vector quantization schemes like neural gas and self-organizing maps for clustering and data compression in case of unsupervised learning. For classification learning we refer to the closely related supervised learning vector quantization scheme. Particularly, median variants of these vector quantizers allow an easy application of Hankel matrices. A possible application of the Hankel matrix approach could be the analysis of DNA sequences, as it does not require the alignment of sequences due to its invariance properties.

1 Introduction

Learning of sequences or time series is a challenging task in machine learning. Time series may describe temporal behavior in technical or natural systems. Texts like articles in newspapers or books are sequences of letters. Nucleotide sequences in DNA constitute another important example of sequence data with the need for automated analysis.

Frequently, data processing tasks deal with the analysis of a set of sequences [1]. The alignment of sequences is frequently considered an essential preprocessing step, aiming at comparability. For example, if a sequence of video frames should be classified with respect to showing a certain event, but with unknown starting point, the video sequences first have to be standardized such that those events, if available in a sequence, are located at a defined position in the sequence. Another example are nucleotide sequences to be classified with and without a

© Springer International Publishing AG 2017
L. Rutkowski et al. (Eds.): ICAISC 2017, Part I, LNAI 10245, pp. 131–142, 2017.
DOI: 10.1007/978-3-319-59063-9_12

certain structural motif determining their functional behavior. Yet, data alignment is often time consuming and may destroy other information. If we want to avoid this pre-processing step, the task is to achieve invariance with respect to the position in the sequence.

A suitable tool for this purpose is the so-call Hankel matrix formalism [2]. Due to their algebraic properties Hankel matrices, generated from sequence data, show invariance properties regarding shifts in data. Therefore, proximity measures for sequences based on Hankel matrices provide interesting properties beneficial for adequate processing.

Supervised and unsupervised vector quantization are standard methodologies for analysis of high-dimensional data and are, therefore, also appropriate for sequence data analysis. Prominent algorithms for clustering and data compression are the affinity propagation clustering (AP, [3]), the neural gas algorithm (NG, [4]) and the self-organizing map (SOM, [5]). The latter is also frequently used in visual data analysis [6]. These methods belong to the class of prototype based algorithms realizing the representation principle, i.e. the prototypes play the role of codebooks representing the data [7]. Supervised vector quantizers for classification learning comprise the classical learning vector quantization (LVQ, [8]) as well as statistical variants like soft learning vector quantization or soft nearest prototype classification [9,10]. In LVQ, frequently the prototypes are class representative [11,12]. However, also border-sensitive variants are known [13].

In this paper we investigate whether dissimilarity measures based on Hankel matrices can be integrated also in vector quantization approaches like NG/SOM and LVQ. We will rather focus our considerations and discussion on the mathematical properties of respective approaches than provide numerical results for applications and simulations. One important aspect is the possibility to use Hankel matrices also in bioinformatics for DNA-sequence analysis. We will show, how the concept of Hankel matrices can be transferred also to such non-numerical sequences. Thus, the paper can be seen as a mathematical justification to apply Hankel matrices in generalized vector quantization algorithms.

2   Hankel Matrices - Mathematical Description and Properties

2.1   General Definition of Hankel Matrices

Hankel matrices are matrices $\mathbf{A} = (a_{ij})$ with the restriction $a_{ij} = a_{i+1,j-1}$ for their elements [2]. If a sequence of elements $s(k) = a_s \in \mathbb{S}$ is given, where $\mathbb{S}$ is a set of objects, the respective Hankel matrix restriction corresponds to the condition $a_{ij} = s(i + j - 2)$ [14]. Hankel matrices are closely related to Toeplitz matrices, which are characterized by the relation $a_{ij} = a_{i+1,j+1} = s(i - j)$. Particularly, both matrix types belong the set of persymmetric matrices for which $a_{ij} = s(f(i + j))$ is valid with a unique discrete function. For persymmetric matrices the equality $\mathbf{AJ} = \mathbf{JA}^T$ holds, where

$$\mathbf{J} = \begin{pmatrix} 0\,0\,0\,\ldots\,0\,0\,1 \\ 0\,0\,0\,\ldots\,0\,1\,0 \\ 0\,0\,0\,\ldots\,1\,0\,0 \\ \vdots\,\vdots\,\vdots\,\ddots\,\vdots\,\vdots\,\vdots \\ 0\,0\,1\,\ldots\,0\,0\,0 \\ 0\,1\,0\,\ldots\,0\,0\,0 \\ 1\,0\,0\,\ldots\,0\,0\,0 \end{pmatrix}$$

is the exchange matrix. For exchange matrices the relation $\mathbf{J} = \mathbf{J}^{-1}$ holds.

Frequently, Hankel and Toeplitz matrices are applied to cyclic sequences, i.e. $s_i = s_{i+L}$ holds with L being the periodicity. Corresponding persymmetric matrices are denoted as circulant matrices [15].

A (block) Hankel matrix $\mathbf{H}(\mathbf{y}) \in \mathbb{C}^{N \times D}$ generated from a sequence $\mathbf{y} = (y_1, \ldots, y_{D+N})^T$ is given by

$$\mathbf{H}(\mathbf{y}) = \begin{pmatrix} y_1 & y_2 & y_3 & \cdots & y_D \\ y_2 & y_3 & y_4 & \cdots & y_{D+1} \\ y_3 & y_4 & y_5 & \cdots & y_{D+2} \\ \vdots & \vdots & \vdots & \ddots & \vdots \\ y_N & y_{N+1} & y_{N+2} & \cdots & y_{D+N} \end{pmatrix} \tag{1}$$

with $y_k \in \mathbb{C}^m$. Hankel matrices are frequently considered in control theory [16]. Suppose a linear time invariant dynamical system (LTIS)

$$y_k = \mathbf{C}x_k + w_k \text{ with } x_k = \mathbf{A}x_{k-1} \tag{2}$$

where $x_k \in \mathbb{R}^D$ and w_k is a (Gaussian) noise with zero mean and $\mathbf{C}$ and $\mathbf{A}$ are constant matrices and $m < D$ is usually valid. The vector x_k determines the system state whereas y_k is the corresponding external measurement. For those systems Hankel matrices are suitable tools to identify LTIS's. More specifically, if two sequences $\mathbf{y}$ and $\mathbf{y}^\circ$ of length N are generated by the same LTIS and $\mathbf{H}(\mathbf{y})$ and $\mathbf{H}(\mathbf{y})$ are the corresponding Hankel matrices of the same order D, then their column vectors $\mathbf{h}_k(\mathbf{y})$ and $\mathbf{h}_k(\mathbf{y}^\circ)$ span the same linear subspace. In this sense, Hankel matrix representations are invariant with respect to time shifts in time series generated by the same LTIS. More generally, the subspace spanned by the column vectors of the Hankel matrix is invariant with respect to affine transformations of the states x_k. This invariance property makes them interesting as tools for system identification or, more generally, sequence data analysis. Note explicitly at this point that the set of Hankel matrices $\mathbf{H} \in \mathbb{C}^{N \times D}$ forms a mathematical vector space $\mathcal{H}$ with the sum of Hankel matrices as vector space operation.

2.2 Dissimilarity Measures for Hankel Matrices

One-Dimensional Sequences $\{y_k\}$. First, we consider one-dimensional sequences $\mathbf{y} = \{y_k\}$ of length L. Hence, the entries of the corresponding Hankel matrix $\mathbf{H}(\mathbf{y}) \in \mathbb{C}^{N \times D}$ with $N + D \leq L$ are real or complex numbers, i.e. $h_{ij} \in \mathbb{R}$ or $h_{ij} \in \mathbb{C}$. An appropriate norm of Hankel matrices $\mathbf{H}$ is the Frobenius-norm

$$\|\mathbf{H}\|_F = \sqrt{\text{trace}(\mathbf{P})} \tag{3}$$

with the product $\mathbf{P} = \mathbf{H}^*\mathbf{H}$ is the determined by the Hermitian inner product of the column vectors of $\mathbf{H}$, i.e. $P_{ik} = \langle \mathbf{h}_i, \mathbf{h}_k \rangle = \sum_{j=1}^{n} \overline{h}_{ij} h_{kj}$ where $\mathbf{h}_j$ denotes a column vector of $\mathbf{H}$. Then, the Frobenius distance for two Hankel matrices $\mathbf{H}_1 = \mathbf{H}(\mathbf{y}_1)$ and $\mathbf{H}_2 = \mathbf{H}(\mathbf{y}_2)$ generated from two sequences $\mathbf{y}_1$ and $\mathbf{y}_2$ becomes

$$d_F(\mathbf{H}_1, \mathbf{H}_2) = \|\mathbf{H}_1 - \mathbf{H}_2\|_F \tag{4}$$

determined by the Frobenius norm (3). Note at this point that the inner product generating the Frobenius norm is

$$\langle \mathbf{H}_1, \mathbf{H}_2 \rangle_F = trace(\mathbf{H}_1^*\mathbf{H}_2) \tag{5}$$

which can be easily calculated by the trace operator [17]. It is reffered to the matrix Frobenius inner product (MFIP).

Yet, comparing Hankel matrices by the Frobenius distance would not reflect their invariance property of the column vectors regarding affine transformations in the states x_k of a LTIS. For this purpose the dynamic subspace angle (DSA) is an appropriate quantity [18]. In the context considered here, the DSA is defined as the canonical correlation between the subspaces spanned by the Hankel matrices. However, this DSA method requires a precise estimation of the noiseless Hankel matrices [19], which can be achieved for LTIS's only for $\|w_k\| \ll 1$. Therefore, a surrogate dissimilarity measure (score)

$$\sigma_F(\mathbf{H}_1, \mathbf{H}_2) = 4 - \left\|\hat{\mathbf{H}}_1 + \hat{\mathbf{H}}_2\right\|_F^2 \tag{6}$$

was suggested in [20], where $\hat{\mathbf{H}}_1 = \frac{\mathbf{H}_1\mathbf{H}_1^T}{\|\mathbf{H}_1\mathbf{H}_1^T\|_F}$ and $\hat{\mathbf{H}}_2 = \frac{\mathbf{H}_2\mathbf{H}_2^T}{\|\mathbf{H}_2\mathbf{H}_2^T\|_F}$ are the normalized squared variants of $\mathbf{H}_1$ and $\mathbf{H}_2$, respectively. We denote $\sigma_F(\mathbf{H}_1, \mathbf{H}_2)$ as the Hankel-subspace-angle-dissimilarity (HSD). It is not longer a mathematical distance [21], but it remains to be a semi-metric [22,23]. Yet, in [24,25] it is argued that the similarity measure

$$s_F(\mathbf{H}_1, \mathbf{H}_2) = \left\|\hat{\mathbf{H}}_1^*\hat{\mathbf{H}}_2\right\|_F^2 \tag{7}$$

is more robust than the HSD (6). It is denoted as Hankel product similarity (HPS). For the HPS we have the inequality

$$0 \leq s_F(\mathbf{H}_1, \mathbf{H}_2) = \left|\left\langle \hat{\mathbf{H}}_1, \hat{\mathbf{H}}_2 \right\rangle_F\right| \leq \left\|\hat{\mathbf{H}}_1\right\|_F \cdot \left\|\hat{\mathbf{H}}_2\right\|_F \tag{8}$$

according to the Cauchy-Schwarz-inequality. Because both $\hat{\mathbf{H}}_1$ and $\hat{\mathbf{H}}_2$ are normalized matrices, we get $s_F(\mathbf{H}_1, \mathbf{H}_2) \leq 1$. Hence, the quantity

$$\delta_F(\mathbf{H}_1, \mathbf{H}_2) = 1 - \left|\left\langle \hat{\mathbf{H}}_1, \hat{\mathbf{H}}_2 \right\rangle_F\right| \tag{9}$$

is a dissimilarity measure denoted as Hankel-Product-Dissimilarity (HPD). Obviously, it is not a mathematical distance but a semi-metric as it is the case for the HSD $\sigma_F\left(\mathbf{H}_1, \mathbf{H}_2\right)$.

Multi-dimensional Sequences $\{y_k\}$. In the following we assume multi-dimensional sequences $\left\{y_k = \left(\mathfrak{y}_k^1, \ldots, \mathfrak{y}_k^m\right)^T\right\}$, i.e. $y_k \in \mathbb{R}^m$ or $y_k \in \mathbb{C}^m$. The length of sequence is assumed to be $L \geq N + D$ such that the corresponding Hankel $(N \times D)$-matrix $\mathbf{H}(\mathbf{y})$ contains m-dimensional vectors as entries $\mathfrak{h}_{ij}$, i.e.

$$
\mathbf{H}(\mathbf{y}) = \begin{pmatrix}
\begin{pmatrix} \mathfrak{y}_1^1 \\ \vdots \\ \mathfrak{y}_1^m \end{pmatrix} & \begin{pmatrix} \mathfrak{y}_2^1 \\ \vdots \\ \mathfrak{y}_2^m \end{pmatrix} & \begin{pmatrix} \mathfrak{y}_3^1 \\ \vdots \\ \mathfrak{y}_3^m \end{pmatrix} & \cdots & \begin{pmatrix} \mathfrak{y}_D^1 \\ \vdots \\ \mathfrak{y}_D^m \end{pmatrix} \\
\begin{pmatrix} \mathfrak{y}_2^1 \\ \vdots \\ \mathfrak{y}_2^m \end{pmatrix} & \begin{pmatrix} \mathfrak{y}_3^1 \\ \vdots \\ \mathfrak{y}_3^m \end{pmatrix} & \begin{pmatrix} \mathfrak{y}_4^1 \\ \vdots \\ \mathfrak{y}_4^m \end{pmatrix} & \cdots & \begin{pmatrix} \mathfrak{y}_{D+1}^1 \\ \vdots \\ \mathfrak{y}_{D+1}^m \end{pmatrix} \\
\begin{pmatrix} \mathfrak{y}_3^1 \\ \vdots \\ \mathfrak{y}_3^m \end{pmatrix} & \begin{pmatrix} \mathfrak{y}_4^1 \\ \vdots \\ \mathfrak{y}_4^m \end{pmatrix} & \begin{pmatrix} \mathfrak{y}_5^1 \\ \vdots \\ \mathfrak{y}_5^m \end{pmatrix} & \cdots & \begin{pmatrix} \mathfrak{y}_{D+2}^1 \\ \vdots \\ \mathfrak{y}_{D+2}^m \end{pmatrix} \\
\vdots & \vdots & \vdots & \ddots & \vdots \\
\begin{pmatrix} \mathfrak{y}_N^1 \\ \vdots \\ \mathfrak{y}_N^m \end{pmatrix} & \begin{pmatrix} \mathfrak{y}_N^1 \\ \vdots \\ \mathfrak{y}_N^m \end{pmatrix} & \begin{pmatrix} \mathfrak{y}_N^1 \\ \vdots \\ \mathfrak{y}_N^m \end{pmatrix} & \cdots & \begin{pmatrix} \mathfrak{y}_{D+N}^1 \\ \vdots \\ \mathfrak{y}_{D+N}^m \end{pmatrix}
\end{pmatrix}
\tag{10}
$$

which can be written formally as a column vector

$$
\mathbf{H}^\circ(\mathbf{y}) = \begin{pmatrix} \mathbf{H}_{\mathbf{y}}^1 \\ \vdots \\ \mathbf{H}_{\mathbf{y}}^N \end{pmatrix}
\tag{11}
$$

of matrices

$$
\mathbf{H}_{\mathbf{y}}^j = \begin{pmatrix} \mathfrak{y}_1^1 & \cdots & \mathfrak{y}_D^1 \\ \vdots & \ddots & \vdots \\ \mathfrak{y}_j^m & \cdots & \mathfrak{y}_{D+j}^m \end{pmatrix} \in \mathbb{R}^{m \times D}
\tag{12}
$$

and

$$
\left(\mathbf{H}^\circ(\mathbf{y})\right)^T = \left(\left(\mathbf{H}_{\mathbf{y}}^1\right)^T | \ldots | \left(\mathbf{H}_{\mathbf{y}}^N\right)^T\right)
\tag{13}
$$

is the transpose of $\mathbf{H}^\circ(\mathbf{y})$.

Therefore, we can write the product $\mathbf{T}(\mathbf{y}) = \mathbf{H}(\mathbf{y})\mathbf{H}(\mathbf{y})^T$ as a dyadic product, i.e. $\mathbf{T}(\mathbf{y}) = \mathbf{H}^\circ(\mathbf{y}) \otimes \mathbf{H}^\circ(\mathbf{y})$ is a tensor of second order[1].

The Frobenius inner product $\langle \mathbf{T}_1, \mathbf{T}_2 \rangle_{TF}$ of tensors $\mathbf{T}_k = \mathbf{A}_k \otimes \mathbf{B}_k$ generated by the dyadic product can be written as

$$\langle \mathbf{T}_1, \mathbf{T}_2 \rangle_{TF} = trace\left((\mathbf{A}_1 \otimes \mathbf{B}_1)^T \cdot (\mathbf{A}_2 \otimes \mathbf{B}_2) \right) \tag{14}$$

with

$$trace\left((\mathbf{A}_1 \otimes \mathbf{B}_1)^T \cdot (\mathbf{A}_2 \otimes \mathbf{B}_2) \right) = \langle \mathbf{A}_1, \mathbf{A}_2 \rangle_F \cdot \langle \mathbf{B}_1, \mathbf{B}_2 \rangle_F \tag{15}$$

by means of the dyadic structure of the tensors.

Therefore, we get for the tensor Frobenius inner product $\langle \mathbf{T}(\mathbf{y}), \mathbf{T}(\mathbf{z}) \rangle_{TF}$ the expression

$$\langle \mathbf{T}(\mathbf{y}), \mathbf{T}(\mathbf{z}) \rangle_{TF} = \langle \mathbf{H}^\circ(\mathbf{y}), \mathbf{H}^\circ(\mathbf{z}) \rangle_F \cdot \langle \mathbf{H}^\circ(\mathbf{y}), \mathbf{H}^\circ(\mathbf{z}) \rangle_F \tag{16}$$

with

$$\langle \mathbf{H}^\circ(\mathbf{y}), \mathbf{H}^\circ(\mathbf{z}) \rangle_F = \sum_k \langle \mathbf{H}_\mathbf{y}^k, \mathbf{H}_\mathbf{z}^k \rangle_F = \sum_k trace\left(\left(\mathbf{H}_\mathbf{y}^k\right)^* \mathbf{H}_\mathbf{z}^k \right) \tag{17}$$

as the underlying Frobenius inner product for matrices [17]. Hence, the Frobenius tensor norm reads in our case as

$$\|\mathbf{T}(\mathbf{y})\|_{TF}^2 = \left(\sum_k trace\left(\left(\mathbf{H}_\mathbf{y}^k\right)^* \mathbf{H}_\mathbf{y}^k \right) \right)^2 \tag{18}$$

as the desired tensor norm and $\hat{\mathbf{T}}(\mathbf{y}) = \frac{\mathbf{T}(\mathbf{y})}{\|\mathbf{T}(\mathbf{y})\|_F}$ becomes the normalized tensor.

We now re-consider the HSD $\sigma_F(\mathbf{H}_1, \mathbf{H}_2)$ from (6) for the m-dimensional data sequence case and obtain

$$\sigma_{TF}(\mathbf{H}(\mathbf{y}), \mathbf{H}(\tilde{\mathbf{y}})) = 4 - \left\| \hat{\mathbf{T}}(\mathbf{y}) + \hat{\mathbf{T}}(\tilde{\mathbf{y}}) \right\|_{TF}^2 \tag{19}$$

as the respective quantity. Adequately, we obtain for the HPS

$$s_{TF}(\mathbf{H}(\mathbf{y}), \mathbf{H}(\tilde{\mathbf{y}})) = \left\| \hat{\mathbf{T}}(\mathbf{y})^* \cdot \hat{\mathbf{T}}(\tilde{\mathbf{y}}) \right\|_{TF} \tag{20}$$

and the HPD rewrites as

$$\delta_{TF}(\mathbf{H}(\mathbf{y}), \mathbf{H}(\tilde{\mathbf{y}})) = 1 - \left\langle \hat{\mathbf{T}}(\mathbf{y}), \hat{\mathbf{T}}(\tilde{\mathbf{y}}) \right\rangle_{TF} \tag{21}$$

in case of multi-dimensional sequences.

---

[1] The dyadic product here is

$$\mathbf{H}^\circ(\mathbf{y}) \otimes \mathbf{H}^\circ(\mathbf{y}) = \begin{pmatrix} \mathbf{H}_\mathbf{y}^1 \cdot \left(\mathbf{H}_\mathbf{y}^1\right)^T & \cdots & \mathbf{H}_\mathbf{y}^1 \cdot \left(\mathbf{H}_\mathbf{y}^N\right)^T \\ \vdots & \ddots & \vdots \\ \mathbf{H}_\mathbf{y}^N \cdot \left(\mathbf{H}_\mathbf{y}^1\right)^T & \cdots & \mathbf{H}_\mathbf{y}^N \cdot \left(\mathbf{H}_\mathbf{y}^N\right)^T \end{pmatrix}$$

i.e. the tensor is a matrix of matrices.

3 Unsupervised and Supervised Neural Vector Quantization for Hankel matrices

Vector quantization is to distribute a set W of codebook vectors $W = \{\mathbf{w}_k\}_{k=1...K} \subseteq V$, where V is a vector space frequently identified with the Euclidean space $\mathbb{R}^n$. The codebook vectors should represent the data and are, therefore, often denoted as prototypes. The training data $\mathbf{v}_j$ are supposed also to be from V. In case of supervised learning, additionally, the class labels $c_j = c(\mathbf{v}_j) \in \mathscr{C}$ are available as well as the prototypes are equipped with prototype labels ζ_k. For a short illustrative overview of prototype based vector quantization we refer to [7].

3.1 Unsupervised Neural Vector Quantization

The task in unsupervised vector quantization is to generate a codebook distribution such that the data are represented as good as possible, which implies the designation prototypes. Therefore, the codebook vector are also denoted as prototypes. The representation of the data $\mathbf{v}$ takes place as a winner take all rule (WTAR)

$$s(\mathbf{v}) = \operatorname{argmin}_k [d(v, \mathbf{w}_k)] \tag{22}$$

such that $\mathbf{w}_s$ is the respective representative or best matching prototype with respect to the dissimilarity measure d frequently the squared Euclidean distance.

The most prominent unsupervised vector quantization algorithm is the K-means algorithm [26], although it frequently does not show superior performance due to its sensitivity with respect to the initialization. A more robust and stable variant with better convergence properties is the neural gas algorithm (NG, [4]). The online learning of NG is realizing a stochastic gradient descent learning on the cost function

$$E_{NG}(W) = \sum_{j=1}^{K} \int_V h_\lambda(\kappa_j(\mathbf{v}, W)) d_j(\mathbf{v}) P(\mathbf{v}) d\mathbf{v} \tag{23}$$

with $P(\mathbf{v})$ being the data density, $h_\lambda(x) = \exp(-x/\lambda)$ is the so-called neighborhood function with neighborhood range scale λ and $d_j(\mathbf{v}) = (\mathbf{v} - \mathbf{w}_j)^2$. The function $\kappa_j(\mathbf{v}, W) = \sum_{i=1}^{K} H(d_j(\mathbf{v}) - d_i(\mathbf{v}))$ determines the winning rank where $H(x)$ is the Heaviside function being zero for non-positive arguments and 1 for positive values. The online update for a given $\mathbf{v}$ is obtained as

$$\Delta\mathbf{w}_j \propto -\varepsilon h_\lambda(\kappa_j(\mathbf{v}, W)) \frac{\partial d_j(\mathbf{v})}{\partial \mathbf{w}_j} \tag{24}$$

with $0 < \varepsilon \ll 1$ being the learning rate. Thus, the derivative $\frac{\partial d_j(\mathbf{v})}{\partial \mathbf{w}_j}$ is essentially needed for this approach.

An alternative is the median variant M-NG [27]. It performs an expectation maximization strategy to optimize the prototypes, which are restricted to be data points. For this purpose, the quantities $\omega_{jk} = \kappa_j(\mathbf{v}_k, W)$ are considered as hidden variables which are optimized for given prototypes alternating with the optimization of the prototypes while fixing the ω_{jk}. For further properties of the algorithm we refer to [27]. For the M-NG, more general dissimilarity measures than the Euclidean distance or other differentiable dissimilarities can be used which may violate properties like the triangle inequality [23]. The M-NG only assumes very basic dissimilarity properties. Moreover, the algorithm requires only the knowledge of the pairwise dissimilarity values for the data due to the restriction for the prototypes to remain data points.

Another frequently applied vector quantizer is the self-organizing map (SOM, [5]). In SOM an external grid is supposed denoted as neuron lattice and frequently assumed to be a two-dimensional rectangular or hexagonal grid. The prototypes are here assigned to the nodes of the grid and the neighborhood cooperativeness during prototype learning is realized via the neighborhood structure in the neuron lattice. The original SOM does not follow a gradient descent scheme [28]. Yet, an SOM with an WTAR slightly different from (22) was proposed in [29]. The performance of SOM is frequently inferior to NG, however, SOM offers excellent visualization abilities [6], provided it realizes a topographic mapping [30]. A batch/median SOM variant was suggested in [31].

3.2 Supervised Neural Vector Quantization

Supervised neural vector quantization is mainly guided by the family of learning vector quantizers (LVQ) [8]. The idea for LVQ was to approximate a Bayes classifier based on vector quantization. The resulting heuristic adaptation rule for the prototypes is a combination of prototype attraction and repulsion where the best matching prototype $\mathbf{w}_s$ is shifted towards the training vector $\mathbf{v}_k$ if $c_k = \zeta_s$ and pushed away if $c_k \neq \zeta_s$. An LVQ variant based on the cost function

$$E_{GLVQ}(W) = \sum_{k=1}^{N_V} f(\mu(\mathbf{v}_k, W)) \tag{25}$$

was introduced in [32]. Here f is a sigmoid or identity function and

$$\mu(\mathbf{v}_k, W) = \frac{d(\mathbf{v}_k, \mathbf{w}^+) - d(\mathbf{v}_k, \mathbf{w}^-)}{d(\mathbf{v}_k, \mathbf{w}^+) + d(\mathbf{v}_k, \mathbf{w}^-)} \tag{26}$$

is the classifier function with $\mathbf{w}^+$ the best matching prototype $\mathbf{w}_s$ with the correct class label $c_k = \zeta_s$ and $\mathbf{w}^-$ the best matching prototype $\mathbf{w}_{s^\circ}$ with an incorrect class label $c_k \neq \zeta_s$. Thus, (25) approximates the classification error [13]. Learning takes place as stochastic gradient descent learning according to

$$\Delta \mathbf{w}^\pm \propto \mp \varepsilon \frac{\partial f}{\partial \mu} \frac{\partial \mu}{\partial d(\mathbf{v}_k, \mathbf{w}^\pm)} \frac{\partial d(\mathbf{v}_k, \mathbf{w}^\pm)}{\partial \mathbf{w}^\pm} \tag{27}$$

realizing the attraction-repulsion-scheme. Many variants have been proposed in the recent years. We refer to [11,12] for recent overviews. The most interesting variant regarding complex data with general dissimilarities is the median variant M-GLVQ [33], which only requires dissimilarity values between data generated by some general dissimilarity measure and the prototypes are restricted to be data points as for M-NG.

3.3   Hankel Matrices and Neural Vector Quantization

As we have seen in the previous subsections, unsupervised and supervised vector quantization rely on data dissimilarities or similarities but not necessarily require the Euclidean distance. Prominent alternatives already in use are divergences or kernel distances [34, 35].

Obviously, the application of Hankel matrices dissimilarities in sequence analysis by neural vector quantizers is not a problem when applying median variants. Thus M-NG/M-SOM can be immediately applied for unsupervised learning and clustering whereas M-GMLVQ would be applicable for classification learning. The advantage of this strategy would be that the computationally expensive calculations for the dissimilarity determination between the data is only needed in the beginning, because the prototypes are restricted to be data points in median learning. Thus, all four provided dissimilarities $d_{F,m}$, $\sigma_{F,m}$, $s_{F,m}$ and $\delta_{F,m}$ are applicable.

The online versions of the previously discussed vector quantizers require the calculation of the derivatives of the underlying dissimilarity or similarity measure. The introduced quantities $d_{F,m}$, $\sigma_{F,m}$, $s_{F,m}$ and $\delta_{F,m}$ are all differentiable functions in dependence on the Hankel-matrix entries. However, the respective formulas become very complex because of the special internal structure of the Hankel matrices such that numerical instabilities may occur. Thus an application has to be done carefully.

Note that in both cases the prototypes are Hankel matrices themselves. For the online learning this is due to the fact that the set $\mathcal{H}$ of Hankel matrices is a vector space as mentioned above.

4   Existing Application Scenarios and New Perspectives for DNA Sequence Analysis

So far, Hankel matrices were applied in machine learning for face emotion and 3d-activity recognition. For the emotion recognition task, from image sequences were first extracted features using Haar-wavelets. These feature sequences then serve as measurements of a dynamical face representation [24, 25]. In activity recognition systems short video sequences with only a few frames are considered, which can be seen also as image sequence. From these sequences features are extracted which again serve as time series to be encoded by Hankel matrices [20, 21].

In engineering, the main focus for application of Hankel matrices as data descriptors is system identification by means of measured time series. As pointed out, the invariant behavior regarding affine transformations, which includes shifts, is the key aspect of Hankel matrices to make them favorable compared to other techniques. However, although the use of Hankel matrices is well-known in engineering and control, those system identification scenarios were not investigated by machine learning approaches so far, at least to our best knowledge.

Now we want to draw the attention to another possible application area. The analysis and classification of DNA-sequences with respect to their suspected behavioral properties in biological systems frequently depends on short structural motifs contained within the sequences [36]. However, the localization of these motifs may vary from sequence to sequence. Therefore, a costly alignment procedure is usually applied in preprocessing. An approach avoiding this alignment is the so-called Frequency Chaos Game Representation (FCGR) [37]. This technique was used to classify sequences by deep networks [38]. A Markov chain based similarity measure for DNA sequences was proposed in [39] preventing the alignment likewise. Yet, both methods require other computational expensive techniques for compensation. Here we favor another perspective. Consider a nucleotide sequence $s_1 s_2 \ldots s_n$ with $s_k \in \{A, C, T, G\}$ coding the nucleotides. A common vector representation of the nucleotides is in terms of the canonical unit vectors in $\mathbb{R}^4$, i.e. each nucleotide sequence is represented by the corresponding sequence of canonical unit vectors forming the sequence $y_1, \ldots, y_n$ with $y_k \in \mathbb{R}^4$. Using this coding scheme, we are able to apply the Hankel-matrix representation of the sequence for DNA sequence analysis. The invariance property of the Hankel matrices immediately provides the independence of the representation regarding the localization of the structural motif within the nucleotide sequence. In this way we are able to apply the unsupervised and supervised vector quantization schemes for clustering and classification of DNA sequences without any precise knowledge about the localization of the structural motifs.

5   Conclusion

In this paper we discussed the Hankel-matrix representation of sequences for use in supervised and unsupervised prototype based vector quantization. For this purpose the mathematical properties of Hankel matrices and respective dissimilarity measures are considered. Usually, similarities are computationally expensive such that online learning schemes for vector quantization become infeasible. However, recently proposed median variants of vector quantization algorithms for classification and clustering offer an attractive alternative, because they do not require re-calculations of dissimilarities between the prototypes and the data.

As one new perspective benefiting from this strategy, we suggest to investigate DNA-sequences depending on structural motifs. Vector representation of the nucleotides leads to vector sequences of DNA sequences, which could serve as input for Hankel-matrix based sequence analysis using vector quantization approaches.

References

1. Mokbel, B., Paaßen, B., Schleif, F.-M., Hammer, B.: Metric learning for sequences in relational LVQ. Neurocomputing **169**, 306–322 (2015)
2. Golub, G.H., Van Loan, C.F.: Matrix Computations. Johns Hopkins Studies in the Mathematical Sciences, 4th edn. John Hopkins University Press, Baltimore (2013)
3. Frey, B.J., Dueck, D.: Clustering by message passing between data points. Science **315**, 972–976 (2007)
4. Martinetz, T.M., Berkovich, S.G., Schulten, K.J.: 'Neural-gas' network for vector quantization and its application to time-series prediction. IEEE Trans. Neural Netw. **4**(4), 558–569 (1993)
5. Kohonen, T.: Self-Organizing Maps. Springer Series in Information Sciences, vol. 30. Springer, Heidelberg (1995)
6. Vesanto, J.: SOM-based data visualization methods. Intell. Data Anal. **3**(7), 123–456 (1999)
7. Biehl, M., Hammer, B., Villmann, T.: Prototype-based models in machine learning. Wiley Interdisc. Rev. Cogn. Sci. **7**(2), 92–111 (2016)
8. Kohonen, T.: Learning vector quantization. Neural Netw. **1**(Supplement 1), 303 (1988)
9. Seo, S., Obermayer, K.: Soft learning vector quantization. Neural Comput. **15**, 1589–1604 (2003)
10. Seo, S., Bode, M., Obermayer, K.: Soft nearest prototype classification. IEEE Trans. Neural Netw. **14**, 390–398 (2003)
11. Kaden, M., Lange, M., Nebel, D., Riedel, M., Geweniger, T., Villmann, T.: Aspects in classification learning - review of recent developments in learning vector quantization. Found. Comput. Decis. Sci. **39**(2), 79–105 (2014)
12. Villmann, T., Bohnsack, A., Kaden, M.: Can learning vector quantization be an alternative to SVM and deep learning? J. Artif. Intell. Soft Comput. Res. **7**(1), 65–81 (2017)
13. Kaden, M., Riedel, M., Hermann, W., Villmann, T.: Border-sensitive learning in generalized learning vector quantization: an alternative to support vector machines. Soft Comput. **19**(9), 2423–2434 (2015)
14. Gray, R.M.: Toeplitz and circulant matrices: a review. Found. Trends Commun. Inf. Theor. **2**(3), 155–239 (2006)
15. Davis, P.J.: Circulant Matrices. Wiley, New York, Chichester, Brisbane (1979)
16. Lai, D., Chen, G.: Dynamical systems identification from time-series data: a Hankel matrix approach. Math. Comput. Model. **24**(3), 1–10 (1996)
17. Lange, M., Nebel, D., Villmann, T.: Non-euclidean principal component analysis for matrices by hebbian learning. In: Rutkowski, L., Korytkowski, M., Scherer, R., Tadeusiewicz, R., Zadeh, L.A., Zurada, J.M. (eds.) ICAISC 2014. LNCS, vol. 8467, pp. 77–88. Springer, Cham (2014). doi:10.1007/978-3-319-07173-2_8
18. Viberg, M.: Subspace-based methods for the identification of linear time-invariant systems. Automatica **31**(12), 1835–1851 (1995)
19. Li, B., Camps, O.I., Sznaier, M.: Activity recognition using dynamic subspace angles. In: Proceedings of the IEEE Conference on Computer Vision and Pattern Recognition, CVPR 2011, Providence, USA, pp. 3193–3200 (2012)
20. Li, B., Camps, O.I., Sznaier, M.: Cross-view activity recognition using Hankelets. In: Proceedings of the IEEE Conference on Computer Vision and Pattern Recognition, CVPR 2012, Providence, USA, pp. 1362–1369 (2012)

21. Presti, L.L., LaCascia, M., Sclaroff, S., Camps, O.: Hankelet-based dynamical systems modeling for 3D action recognition. Image Vis. Comput. **44**, 29–43 (2015)
22. Pekalska, E., Duin, R.P.W.: The Dissimilarity Representation for Pattern Recognition: Foundations and Applications. World Scientific, Singapore (2006)
23. Nebel, D., Kaden, M., Bohnsack, A., Villmann, T.: Types of (dis-)similarities and adaptive mixtures thereof for improved classification learning. Neurocomputing (2017, in press)
24. Presti, L.L., La Cascia, M.: Ensemble of Hankel matrices for face emotion recognition. In: Murino, V., Puppo, E. (eds.) ICIAP 2015. LNCS, vol. 9280, pp. 586–597. Springer, Cham (2015). doi:10.1007/978-3-319-23234-8_54
25. Presti, L.L., LaCascia, M.: Boosting Hankel matrices for face emotion recognition and pain detection. Compu. Vis. Image Underst. **156**, 19–33 (2017)
26. Linde, Y., Buzo, A., Gray, R.M.: An algorithm for vector quantizer design. IEEE Trans. Commun. **28**, 84–95 (1980)
27. Cottrell, M., Hammer, B., Hasenfuß, A., Villmann, T.: Batch and median neural gas. Neural Netw. **19**, 762–771 (2006)
28. Erwin, E., Obermayer, K., Schulten, K.: Self-organizing maps: ordering, convergence properties and energy functions. Biol. Cybern. **67**(1), 47–55 (1992)
29. Heskes, T.: Energy functions for self-organizing maps. In: Oja, E., Kaski, S. (eds.) Kohonen Maps, pp. 303–315. Elsevier, Amsterdam (1999)
30. Villmann, T., Der, R., Herrmann, M., Martinetz, T.: Topology preservation in self-organizing feature maps: exact definition and measurement. IEEE Trans. Neural Netw. **8**(2), 256–266 (1997)
31. Kohonen, T., Somervuo, P.: How to make large self-organizing maps for nonvectorial data. Neural Netw. **15**(8–9), 945–952 (2002)
32. Sato, A., Yamada, K.: Generalized learning vector quantization. In: Touretzky, D.S., Mozer, M.C., Hasselmo, M.E. (eds.) Proceedings of the 1995 Conference Advances in Neural Information Processing Systems, vol. 8, pp. 423–429. MIT Press, Cambridge (1996)
33. Nebel, D., Hammer, B., Frohberg, K., Villmann, T.: Median variants of learning vector quantization for learning of dissimilarity data. Neurocomputing **169**, 295–305 (2015)
34. Mwebaze, E., Schneider, P., Schleif, F.-M., Aduwo, J.R., Quinn, J.A., Haase, S., Villmann, T., Biehl, M.: Divergence based classification in learning vector quantization. Neurocomputing **74**(9), 1429–1435 (2011)
35. Villmann, T., Haase, S., Kaden, M.: Kernelized vector quantization in gradient-descent learning. Neurocomputing **147**, 83–95 (2015)
36. D'haeseleer, P.: What are DNA sequence motifs? Nat. Biotechnol. **24**, 423–425 (2006)
37. Almeida, J.S., Carricio, J.A., Maretzek, A., Noble, P.A., Fletcher, M.: Analysis of genomic sequences by chaos game representation. Bioinformatics **17**(5), 429–437 (2001)
38. Rizzo, R., Fiannaca, A., LaRosa, M., Urso, A.: Classification experiments of DNA sequences by using a deep neural network and chaos game representation. In: Proceedings of the International Conference on Computer Systems and Technologies - CompSysTech 2016, Palermo, Italy, pp. 222–228 (2016)
39. Blaisdell, B.E.: A measure of the similarity of sets of sequences not requiring sequence alignment. Proc. Nat. Acad. Sci. U.S.A. **83**, 5155–5159 (1986)

Discrete Cosine Transformation as Alternative to Other Methods of Computational Intelligence for Function Approximation

Angelika Olejczak[1], Janusz Korniak[1(✉)], and Bogdan M. Wilamowski[2]

[1] University of IT and Management, Rzeszow, Poland
angelika.olejczak@gmail.com, jkorniak@wsiz.pl
[2] Auburn University, Auburn, AL, USA
wilam@ieee.org

Abstract. The discrete cosine transform (DCT) is commonly known in signal processing. In this paper DCT is used in computational intelligence to show its usefulness. Proposed DCT method is used to reduce the size of system which results in faster processing with limited and controlled precision lost. Proposed method is compared to other ones like Fuzzy Systems, Neural Networks, Support Vector Machines, etc. to investigate the ability to solve sample problem. The results show that the method can be successfully used and the results are comparable or better to those achieved by other methods considered as powerful ones.

Keywords: Approximation · Neural networks · Computational intelligence

1 Introduction

The computational intelligence is a growing field of science. Wide range of possible applications, ability to solve complex problems make these methods very attractive. But, on the other hand these various potential applications are challenges for scientists. There is not any one universal method, good for all possible problems. Therefore, there are a lot of different methods of computational intelligence.

Fuzzy systems, artificial neural networks and evolutionary computation become popular methods for problems difficult to solve by human due to large amount of data or problems which are not well understood. Especially, many fields require the function approximation. Neural networks, fuzzy systems as well as evolutionary algorithms can be used for this purpose. On the other hand, functions estimation can be solved using orthogonal systems functions [1–3]. Fuzzy systems found many applications for modeling problems difficult to define using mathematical models [4] and in the industry for control algorithms [5]. The main drawbacks of that system are the lack of smoothness in the approximation.

This work was supported by the National Science Centre, Krakow, Poland, under grant No.2015/17/B/ST6/01880.

© Springer International Publishing AG 2017
L. Rutkowski et al. (Eds.): ICAISC 2017, Part I, LNAI 10245, pp. 143–153, 2017.
DOI: 10.1007/978-3-319-59063-9_13

Today the most popular neural network architecture is MLP (Multi-Layer Perceptron) [6–8], but to achieve satisfying results it has to consist of many processing units (neurons). From the learning process perspective this approach is very time-consuming as well as finding a proper number of hidden units and neurons in them to build a successful architecture. Moreover networks with too many neurons may lead to poor generalization ability, resulting in not very good testing error, despite small training error.

However the size of traditional ANN can be significantly reduced (up to 100 times), if another topology like FCC (Fully Connected Cascade) is used instead of MLP architecture [9]. Unfortunately for such special topology (FCC) a dedicated learning algorithm has to be used [10,11]. For MLP and especially SLP architecture (MLP with single hidden layer) a first-order EBP (Error Back Propagation) is commonly used. But in this case for FCC topology with arbitrarily connected neurons there has been another recently developed second-order NBN algorithm [10], which can handle the same problem much faster than EBP.

These powerful architectures mentioned above are not organized like layer-by-layer, so one cannot use the most of presently existing learning software. Actually Lang and Witbrock have done a simple modification of Error Back Propagation (EBP) algorithm in 1988 [12], for their networks [13] which was then fulfilled in the Stuttgart Neural Network System (SNNS). That was implemented only for a first order EBP method. Another LM algorithm, adopted for training of neural networks [14] and used in MATLAB Neural Network Toolbox [15], cannot handle architecture with arbitrarily connected neurons. That is the reason why NBN algorithm was needed.

Meanwhile many researchers due to frustration with traditional neural networks [16] moved to other alternative approaches e.g. ELM (Extreme Learning Machine) [17] or SVM (Support Vector Machines) [18]. Even though SVM and ELM are not only fast, but also efficient systems with shallow type of architecture, solving problems demands using an excessive number of RBF units or neurons. It is also worth to mention that such big number (of neurons) is used very inefficiently, because weights for neurons are selected randomly (ELM) and support vectors are chosen from a training set (SVM), thus only weights of output linear neurons are adjusted by learning algorithm in both of them.

More recently developed ErrCor algorithm [16,19,20] is capable to solve the problems performed by SVM or ELM with 10 to 100 times smaller architecture and similar or even smaller error. Another advantage of ErrCor algorithm is that the final solution can be always reached in a single training run.

The purpose of this paper is to show the ability to use Discrete Cosine Transform (DCT) as an alternative to other methods for function approximation. The Sect. 2 reviews basics of DCT for one and multidimensional data. Following section shows that DCT can be used to decrease training points without significant lost of training precision.

2   Discrete Cosine Transform

Discrete Cosine Transform (DCT) is related to the Discrete Fourier Transform but it operates only on real numbers. The most widely recognized application of this transform is a lossy image compression in *jpeg* standard. There are several variants of DCT [21] however, the type-II is the most common:

$$y_k = \omega_k \sum_{n=1}^{N} x_n cos \left(\frac{\pi}{2N}(2n-1)(k-1) \right), \quad k = 1, \ldots, N, \tag{1}$$

where

$$\omega_k = \begin{cases} \frac{1}{\sqrt{N}} & k = 1, \\ \sqrt{\frac{2}{N}} & 2 \le k \le N, \end{cases} \tag{2}$$

The Eq. 1 defines one dimensional transform however, extension for two or more dimensions are known:

$$y_{k_1,k_2} = \omega_{k_1}\omega_{k_2} \sum_{n_1=1}^{N_1} \sum_{n_2=1}^{N_2} x_{n_1,n_2} cos \left(\frac{\pi}{2N_1}(2n_1-1)(k_1-1) \right)$$

$$cos \left(\frac{\pi}{2N_2}(2n_2-1)(k_2-1) \right) \quad k_1, k_2 = 1, \ldots, N_{1,2} \tag{3}$$

where both ω_{k_1} and ω_{k_2} can be expressed by Eq. 2. N-DCT transform can by extended in the same way. For example 3-DCT is as follow:

$$y_{k_1,k_2,k_3} = \omega_{k_1}\omega_{k_2}\omega_{k_3} \sum_{n_1=1}^{N_1} \sum_{n_2=1}^{N_2} \sum_{n_3=1}^{N_3} x_{n_1,n_2,n_3} cos \left(\frac{\pi}{2N_1}(2n_1-1)(k_1-1) \right)$$

$$cos \left(\frac{\pi}{2N_2}(2n_2-1)(k_2-1) \right) cos \left(\frac{\pi}{2N_3}(2n_3-1)(k_3-1) \right)$$

$$k_1, k_2, k_3 = 1, \ldots, N_{1,2,3}, \tag{4}$$

The discrete cosine transform (DCT) is frequently used together with Inverse Cosine Transform (IDCT). Since DCT allows to compute coefficients representing different frequencies of function (signal). IDCT is used to compute discrete function values form DCT coefficients. In this way DCT is invertible transform. Transforming vector X with DCT and then transforming the result coefficients with IDCT gives the original X vector. However, if some coefficients, responsible for high frequencies are omitted IDCT produce discrete data which looses details. This feature of DCT and IDCT pair is used for decreasing number of training points of data. Ability to use DCT to training data reduction is verified in next section.

IDCT is defined as follow:

$$x_k = \sum_{n=1}^{N} \omega_k y_n cos \left(\frac{\pi}{2N}(2n-1)(k-1) \right), \quad k = 1, \ldots, N, \tag{5}$$

where ω_k is defined by Eq. 2. Similarly to DCT, IDCT can be easily extended to N-IDCT.

3   Reduction of the System Size with DCT

Many systems allow to gather high number of data which can be used for training process. However, only part of this data is necessary for correct training. The DCT can be used for reduction amount of training point.

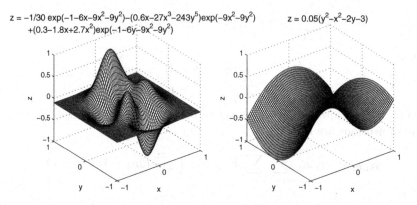

Fig. 1. Peaks function (left) used to generate training patterns and Ribbon function (right) used to generate training patterns.

In the experiment the peek and the ribbon functions have been used. Both ones are illustrated in Fig. 1. Input data contains 900 training points distributed as regular grid and 400 testing points normalized in the range of $\langle -1, 1 \rangle$. Training patterns are transformed by the DCT algorithm to compute frequency coefficients. Then all frequency components smaller than a threshold set in the range $\langle 0, 0.2 \rangle$ were eliminated, where 0 represents lack of reduction and higher values increase coefficients reduction. After that IDCT transformation was applied to reduced set of coefficients. These compressed patterns have been compared to the original functions to compute errors in original training points and other testing points. The results are illustrated in Fig. 2.

The training error is significantly small for wide range of threshold compression of learning data for 'peaks' and 'ribbon' (Fig. 1). Good results are achieved up to 90% reduction. Testing errors for both these functions show that having even roughly 10% of training data allows DCT algorithm to approximate unknown data with satisfying error.

The DCT input data must be provided as regular grid. This is a significant drawback of the method. Many systems allows to gather learning data as irregular sets. To remove this limitation a method of non-regular to regular data conversion should be implemented.

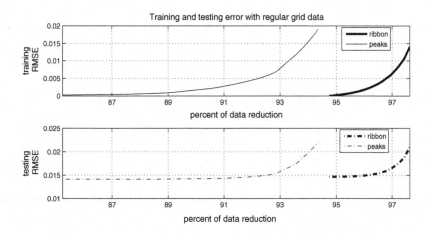

Fig. 2. Training and testing error as function of data compression threshold.

4 Converting Randomly Distributed Patterns into a Regular Grid

The problem of scattered data interpolation is known. Data interpolation algorithms can use linear, cubic, spline and other methods. Example of the implementation of these algorithms is Matlab's *griddata* function. The *griddata* function uses several different methods for regular grid points interpolation from irregular data points. Most of them are triangulation-based, what results in generating not a number (in Matlab called NaN) values especially at the edges. This means that there are some values, which cannot be computed by *griddata* what is the main drawback of this routine.

To illustrate this observation the following experiment has been carried out. 100 randomly distributed data points were generated for the peaks and ribbon functions. Inputs and outputs were normalized to $\langle -1, 1 \rangle$ before applying them to *griddata* function. The grid size for interpolation was chosen from 5 to 80 and the number of not defined values in relation to all grid points was presented on Fig. 3 as the not defined values rate, where 0 represents zero not defined values in the *griddata* output. As one can see for both tested functions, the not defined values rate is the highest when the grid size N is small and although computing interpolated values on larger grid can decrease this rate a lot, it is not still completely eliminated. This dependency is actual even for more data points as it is presented in Table 1.

The problem of not defined points can be eliminated in several ways:

- Training points can be recorded on wider range then needed. Thus, the not defined values, which are mostly located on the edges or extended area, are outside needed domain. However, internal not defined values can occur. Moreover, extending the range of domain may be not possible in all real cases.

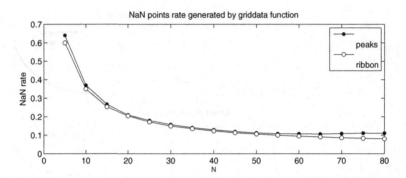

Fig. 3. The number of not defined points in relation to grid size for linear approximation of two functions.

Table 1. The rate of not defined points in output of linear approximation with *griddata* of two functions. Grid size N is equal 80, 40 and 20.

$N = 80$			$N = 40$			$N = 20$		
NaN rate		Data points	NaN rate		Data points	NaN rate		Data points
Peak	Ribbon		Peak	Ribbon		Peak	Ribbon	
0.2148	0.1464	50	0.2350	0.1762	50	0.2875	0.2425	50
0.1105	0.0803	100	0.1288	0.1225	100	0.2100	0.2050	100
0.0838	0.0728	200	0.1169	0.1100	200	0.1950	0.1900	200
0.0591	0.0566	500	0.1013	0.1006	500	0.1900	0.1900	500
0.0544	0.0541	1000	0.1000	0.0994	1000	0.1900	0.1900	1000
0.0514	0.0512	2000	0.0975	0.0975	2000	0.1900	0.1900	2000

- Lacking points can be approximated based on neighboring points as an additional data processing step.
- More advance algorithm like Matlab's 'v4' which is based on bi-harmonic spline interpolation. However, the method is limited only to 2 dimensions.

   To solve the problem of lacking values in the output of griddata function the second approach has been applied. For each occurrence of not defined point proposed algorithm searches for k (in our case k = 4) nearest neighbors, where nearest means lowest Euclidean distance to the not defined point. Having all nearest neighbors positions and their values known from the *griddata* result, the value of lacking point is computed as a weighted average from all nearest neighbors. The *weights* are inversely proportional to the distance to the searched not defined point.

   The proposed solution has been verified with the numerical experiment. In this experiment 100 random patterns have been generated as input data for griddata function (with default linear method of approximation). The resulted

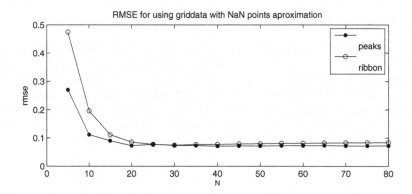

Fig. 4. The error of approximation NaN points for two functions.

approximated data includes some number of not defined values. Then, knowing the position of not defined points it was possible to compute their values by taking weighted average from a few nearest neighbors. In this way all needed grid values has been achieved. Finally the complete grid data can be used to compare with original function values. As it is presented in Fig. 4 the mean square error decreases to approximately 0.1 for appropriately big grid size.

To illustrate the difference between *griddata* results and developed modification another approximation has been performed using both methods. The Fig. 5 clearly shows how initially lacking values in the corners were successfully computed.

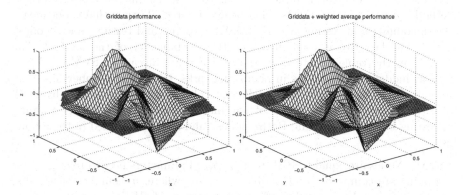

Fig. 5. The illustration of the griddata results with NaN points and the results after processing with implemented method.

5 Comparison of Selected Methods

In this section DCT approximation with compression is compared with other methods. For this purpose the peak function problem has been solved with DCT (compression threshold is 0.01), TSK Fuzzy system, ANN with two different network topologies, SVN, ELM and ErrCor. ANN uses network with single hidden layer with linear output unit (ANN-SLP) and fully connected cascade (ANN-FCC) with additional connections between hidden layers with single neuron per each hidden layer (Fig. 6).

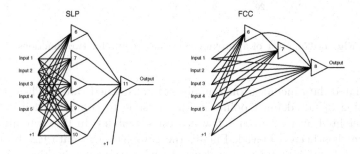

Fig. 6. The MPL (multi-layer perception) with single hidden layer, called SLP and Fully connected cascade FCC used for experiment.

For the experiment training data grid with size 40×40 and testing data grid with size 70×70 have been used. Training inputs and outputs were normalized to range $\langle -1, 1 \rangle$ and testing inputs as well. To show the difference between these methods several measurements have been gathered. They include root-mean-square training and testing errors (RMSE), training and testing time, "nodes" and P. The last two is used to demonstrate complexity of the method. The "nodes" in the table denote number of RBF units in case of SVM,ELM and ErrCor, hidden layer neurons in neural networks, triangular membership functions in TSK fuzzy system and coefficients obtained after reduction by DCT algorithm. The P is the number of parameters needed to store after training for testing by each algorithm. Both "nodes" and P are calculated in the following way:

- SVM - Nodes are equal to RBF units n_{RBF}. Parameters needed to store after training are computed as: $P = n_w + n_c + n_d$, where: n_w number of weights, $n_w = n_{RBF} + 1$; n_c number of center locations of RBFs, $n_c = 2n_{RBF}$; n_d number of RBF diameters values, $n_d = 1$.
- ELM and ErrCore - Nodes are equal to RBF units n_{RBF}. Parameters needed to store after training are computed as: $P = n_w + n_c + n_d$, where: $n_w = n_{RBF} + 1, n_c = 2n_{RBF}, n_d = n_{RBF}$.
- ANN-SLP - Nodes are equal to number of neurons in hidden layer n_h. Parameters are described as: $P = n_w$, where: $n_w = (n_i + 1)n_h + n_h + 1$, n_i number of inputs, $n_i = 2$ in this case.

- ANN-FCC - Nodes are equal to all neurons n_n. Parameters are described as: $P = n_w$, where $n_w = (n_i+1)n_n+(n_n-1)n_n/2$, $n_i = 2$; n_n number of neurons, $n_n = n_h + 1$.
- TSK - Nodes are equal to overall number of triangular membership functions $(2n_{md})$ used in all dimensions dim (dim=2). Parameters are computed as: $P = n_{md}^2 + 2dim$, where: n_{md} number of membership functions in each of dimensions, $n_{md} = 10$ in this case.
- DCT - Nodes are equal to number of frequency coefficients retained after process of its reduction n_coeff and number of stored parameters is equal to number of nodes.

Table 2. The results of different methods comparison.

Algorithm	Training RMSE	Testing RMSE	Training time (s)	Testing time (s)	Nodes	P
TSK fuzzy	0.0548	0.0551	N/A	1.2940	20	104
ANN-SLP	0.0061	0.0065	1539.092	1.2796	29	117
ANN-FCC	0.0051	0.0058	123.9989	0.2095	9	63
SVM	0.0630	0.0629	0.0066	0.0112	94	284
ELM	0.0695	0.0695	0.1432	0.4276	100	401
ErrCor	0.0023	0.0035	48.4040	0.7970	10	41
DCT	8.21e-04	0.0076	6.59e-05	4.70e-06	100	100

Obtained results are presented in Table 2 and provides following observations. DCT testing error is comparable to errors obtained for ANN-SLP and ANN-FCC except ErrCor which gives lowest testing error. Other methods TSK fuzzy, SVM and ELM give higher errors. Notice that DCT gives much lower training errors then all other methods what undeniably confirms ability to perform approximation with only 10 percent of initial number of coefficients. DCT undoubtedly outperforms the rest of algorithms when considering training and testing time because approximation results can be obtained within a single run without training process with large number of iterations as in neural networks. However ANN-FCC and ErrCor are the most efficient in terms of number of nodes needed to store. But training time is nearly million times longer than obtained with proposed DCT method. Moreover in case of DCT or ErrCor the training values are obtained in a single run, unlike in ANNs where many attempts have to be made before proper results are achieved. Here 50 trials had been conducted before the best results were obtained, thus the training process is actually much longer. TSK fuzzy system reaches smaller error than SVM or ELM with the same number of nodes but testing time is much longer. On the other hand in case of SVM and ELM training time is significantly shorter than in ANNs (still not as good as in DCT method) but the final network is definitely too large which results in up to four times higher number of stored parameters than in proposed DCT.

6 Conclusions

Thanks to the compression feature of DCT the system size can be significantly reduced. There are numerous systems which allow to gather large amount of training points in the observed domain. Therefore, for such systems DCT can be applied to significantly reduce input data with controlled lost of precision. That reduced data can be used for further manipulation like approximation or data conversion to obtain more suitable data form. Important example of application is data conversion from scattered to regular grid.

Last years have brought several powerful methods in computational intelligence. Although, they have some drawbacks and therefore other alternative methods should be considered. As shown in the previous section DCT can be successfully used to solve problem.

Because, DCT algorithm is not complex, it is easy to apply to multidimensional problems. Presented results are two or three dimensional problems for easy visualization, but the method works fine also for any other dimensions.

References

1. Greblicki, W., Pawlak, M.: Fourier and hermite series estimates of regression functions. Ann. Inst. Stat. Math. **37**(1), 443–454 (1985)
2. Rutkowski, L., Rafajlowicz, E.: On optimal global rate of convergence of some non-parametric identification procedures. IEEE Trans. Autom. Control **34**(10), 1089–1091 (1989)
3. Rafajlowicz, E., Pawlak, M.: On function recovery by neural networks based on orthogonal expansions. Nonlinear Anal. Theory Methods Appl. **30**(3), 1343–1354 (1997)
4. Kayacan, E., Kayacan, E., Khanesar, M.A.: Identification of nonlinear dynamic systems using type-2 fuzzy neural networks a novel learning algorithm and a comparative study. IEEE Trans. Ind. Electron. **62**(3), 1716–1724 (2015)
5. Takagi, T., Sugeno, M.: Fuzzy identification of systems and its applications to modeling and control. IEEE Trans. Syst. Man Cybern. **SMC-15**(1), 116–132 (1985)
6. Bengio, Y.: Learning deep architectures for AI. Found. Trends Mach. Learn. **2**(1), 1127 (2009). Also published as a book. Now Publishers
7. Dahl, G.E., Ranzato, M., Mohamed, A., Hinton, G.E.: Phone recognition with the mean-covariance restricted boltzmann machine. In: NIPS2010
8. Wu, X., Rozycki, P., Wilamowski, B.: Hybride constructive algorithm for single layer feeforward network learning. IEEE Trans. Neural Netw. Learn. Syst. **26**(8), 1659–1668 (2015)
9. Wilamowski, B.M.: Neural network architectures and learning algorithms- how not to be frustrated with neural networks. IEEE Ind. Electron. Mag. **3**(4), 56–63 (2009)
10. Wilamowski, B.M., Yu, H.: Improved computation for levenberg marquardt training. IEEE Trans. Neural Netw. **21**(6), 930–937 (2010)
11. Tiantian, X., Hao, Y., Hewlett, J., Rozycki, P., Wilamowski, B.: Fast and efficient second-order method for training radial basis function networks. IEEE Trans. Neural Netw. Learn. Syst. **23**(4), 609–619 (2012)
12. Lang, K.L., Witbrock, M.J.: Learning to tell two spirals apart. In: Proceedings of the 1988 Connectionists Models Summer School. Morgan Kaufman (1988)

13. Fahlman, S.E., Lebiere, C.: The cascade-correlation learning architecture. In: Touretzky, D.S. (ed.) Advances in Neural Information Processing Systems 2, pp. 524–532. Morgan Kaufmann, San Mateo (1990)
14. Hagan, M.T., Menhaj, M.B.: Training feedforward networks with the Marquardt algorithm. IEEE Trans Neural Netw. **5**(6), 989–993 (1994)
15. Demuth, H.B., Beale, M.: Neural Network Toolbox: For Use with MATLAB. Mathworks, Natick (2000)
16. Różycki, P., Kolbusz, J., Bartczak, T., Wilamowski, B.M.: Using parity-N problems as a way to compare abilities of shallow, very shallow and very deep architectures. In: Rutkowski, L., Korytkowski, M., Scherer, R., Tadeusiewicz, R., Zadeh, L.A., Zurada, J.M. (eds.) ICAISC 2015. LNCS, vol. 9119, pp. 112–122. Springer, Cham (2015). doi:10.1007/978-3-319-19324-3_11
17. Hunter, D., Hao, Y., Pukish, M.S., Kolbusz, J., Wilamowski, B.M.: Selection of proper neural network sizes and architecturesa comparative study. IEEE Trans. Ind. Inform. **8**, 228–240 (2012)
18. Smola, A.J., Schlkopf, B.: A tutorial on support vector regression. NeuroCOLT2 Technical report NC2-TR-1998-030 (1998)
19. Yu, H., Xie, T., Hewlett, J., Rozycki, P., Wilamowski, B.M.: Fast and efficient second order method for training radial basis function networks. IEEE Trans. Neural Netw. **24**(4), 609–619 (2012)
20. Cecati, C., Kolbusz, J., Siano, P., Rozycki, P., Wilamowski, B.: A novel RBF training algorithm for short-term electric load forecasting: comparative studies. IEEE Trans. Ind. Electron. **62**(10), 6519–6529 (2015)
21. Rao, K., Yip, P.: Discrete Cosine Transform: Algorithms, Advantages, Applications. Academic Press, Boston (1990). ISBN0-12-580203-X

Improvement of RBF Training by Removing of Selected Pattern

Pawel Rozycki[1]([⊠]), Janusz Kolbusz[1], Oleksandr Lysenko[1],
and Bogdan M. Wilamowski[2]

[1] University of Information Technology and Management in Rzeszow,
Sucharskiego 2, 35-225 Rzeszow, Poland
{prozycki,jkolbusz}@wsiz.rzeszow.pl, a11ias@hotmail.com
[2] Auburn University, Auburn, AL 36849-5201, USA
wilambm@auburn.edu

Abstract. Number of training patterns has a huge impact on artificial
neural networks training process, not only because of time-consuming
aspects but also on network capacities. During training process the error
for the most patterns reaches low error very fast and is hold to the end
of training so can be safely removed without prejudice to further training process. Skilful removal of patterns during training allow to achieve
better training results decreasing both training time and training error.
The paper presents some implementations of this approach for Error
Correction algorithm and RBF networks. The effectiveness of proposed
methods has been confirmed by several experiments.

Keywords: ANN training improvement · Error correction · Training
set reduction

1 Introduction

The rapid development of intelligent computational systems allowed to solve
thousands of practical problems using neural networks. We can build intelligent
systems, setting weights with random values initially, and then use an algorithm
that will teach this system adjusting these weights in order to solve complex
problems. It is interesting that such a system can achieve a higher level of competence than teachers. Such systems can be very useful wherever decisions are
taken, even if the man is not able to understand the details of their actions.
Most scientists use the MLP (*Multi-Layer Perceptron*) architecture and the EBP
(*Error Back Propagation*) algorithm [1,6]. However, since the EBP algorithm is
not efficient, usually using inflated the number of neurons which mean that the
network with a high degree of freedom consume their capabilities to learn the
noise. Consequently, after the step of learning system was score responsive to
the patterns that are not used during the learning, and it resulted in frustration.

This work was supported by the National Science Centre, Cracow, Poland under
Grant No. 2013/11/B/ST6/01337.

L. Rutkowski et al. (Eds.): ICAISC 2017, Part I, LNAI 10245, pp. 154–164, 2017.
DOI: 10.1007/978-3-319-59063-9_14

A new breakthrough in intelligent systems is possible due to new, better architectures and better, more effective learning algorithms. Although EBP caused a real breakthrough, it turned out to be a very slow algorithm, not capable of learning other than MLP, compact network architectures. Most visible progress in this field was develop the LM (*Levenberg-Marquardt*) algorithm to train the neural network. This algorithm is able to teach the network by 100 to 1000 times less iterations, but its usage to more complex problems is significantly limited, since the size of the Jacobian matrix is proportional to the number of patterns. Modification of existing algorithms and development of new algorithms for network learning will allow for faster and more effective network teaching. While MLP architecture is still very popular it has a very limited capabilities [1]. New deeper neural network architectures like BMLP (*Bridged MLP*) [1–3] or DNN (*Dual Neutral Networks*) [2] with the same number of neurons can solve problems up to 100 times more complex [2,4]. Therefore, it can be concluded that the way neurons interconnections in the network is fundamental and the use of appropriate architecture has a significant impact on the solution of given problem. Maintaining the same number of neurons can increase network capacity, even a hundred times depending on architecture [2,4,5]. However, a problem arises in that the currently known network learning algorithms, such as EBP, or LM do not deal with such network architectures. Moreover, the LM algorithm is not able to train datasets with a high number of patterns that restricts LM algorithm for solving a relatively small problems. Good alternative that can learn these new architectures is NBN (*Neuron-by-Neuron*) algorithm [7–9]. It is faster than LM and can be used for all architectures, including BMLP, FCC, DNN and MLP, and gives good training results. However, recently published *ErrCor* (*Error Correction*) algorithm [10] allow to get even better results showing that the shallow architecture can be still in the game.

The mentioned limitation on the number of training patterns during training with many popular algorithms make them practically unusable for huge problems. Many possible solutions for this issue can be found in [11–13] where training set reduction is done in few general ways (a) by random selection of subset patterns, (b) selection most important patterns according given cost function and (c) by clustering. The aim of this paper is proposition of resolve this problem during the training process by removing trained patterns in the similar way as outliers elimination described in [14].

2   Reduction of the Number of Training Patterns

The number of patterns used in training has a significant impact on training time. In the most training algorithms almost all calculations strongly depend on the number of patterns. Reduction of patterns that are used during training can, therefore, make this process faster. The challenge is how to reduce the training time removing training patterns without affect the accuracy.

In many cases the training error for most patterns reaches a low level very fast and is hold until the end of training process. These patterns are usually not important in the next process of training if they reach a given low level, so they can be then removed. It allow to increase impact of patterns with larger error into training process.

Considering removing of training patterns some questions should be answered:

- What should be the criterion for removing patterns or which patterns should be removed?
- When the patterns should be removed?
- Is patterns should be removed permanently or can be restored during training process?

As mentioned the criterion for pattern reduction should be a low level of error achieved for given pattern. This value, assigned in paper as CT (Cut-off Threshold), should be below expected value of maximal training error. In most cases especially for real world problems there is not possible to determine this value a priori because this value depends on complexity of given problem and cannot be easy specified so some trails are needed to be done. The pattern reduction can be done during training process and the exact moment of reduction can be dependent on used training algorithm and can be done after given number of training epochs or after some other specific step of training like changing network architecture in the case of constructive ANN methods. While this approach can be applied to any kind of training algorithm and many different ANN, both shallow and deep architectures like SLP, MLP, FCC or BMLP, in the next section we propose some implementations for the SLP architecture and $ErrCor$ algorithm.

2.1 Error Correction Algorithm

Error Correction algorithm ($ErrCor$) is constructive training method that is used to train neural networks that are build using RBF units with Gaussian activation function denied by (1).

$$\varphi_h\left(x_p\right) = exp\left(-\frac{\|x_p - c_h\|^2}{\sigma_h}\right) \tag{1}$$

where: c_h and σ_h are the center and width of RBF unit h, respectively. $\|\cdot\|$ represents the computation of Euclidean Norm. The output of such network is given by:

$$O_p = \sum_{h=1}^{H} w_h\varphi_h\left(x_p\right) + w_o \tag{2}$$

where: w_h presents the weight on the connection between RBF unit h and network output. w_0 is the bias weight of output unit. Note that the RBF networks can be implemented using neurons with sigmoid activation function in MLP architecture [17,18].

The main idea of the $ErrCor$ algorithm is to increase the number of RBF units one by one adjusting all RBF units in network by training them after adding of each unit. The new unit is initially set to compensate largest error in the current error surface and after that all units are trained changing both centers and widths as well as output weights, so $ErrCor$ algorithm is not only constructive that allow to achieve a networks with proper number of units but also deterministic. Details of algorithm can be found in [10,15].

Pseudo code of the Error Correction algorithm is shown in Fig. 1. More important training parameters are:

- N_{MAX} – maximal number of RBFs used in network; $ErrCor$ as constructive algorithm starts with one RBF unit and in each main loop next RBF unit is added. Number of units that can be added is limited by this parameter;
- $max\_iterations$ – maximal number of iterations determine what is the max number of execute the loop for training of network after adding each RBF unit;
- $desired\_R\_error$ – the level of relative change of training error for which the network in given configuration is treated as trained as good as possible; if these value is reached the training process for given number of RBFs is broken and next RBFs is added;
- $desired\_error$ – the level of desired error that determine successful training; if the network reaches this level of error the training process is broken even if number of RBFs is below N_{MAX}.

2.2   Proposed Methodologies

For the $ErrCor$ training method the five variants of training patterns reduction had been studied.

Method 1. First variant of proposed technique permanently remove patterns that match removing conditions after training of each N_{RBF} units. In this case removed pattern is not restored to training data set until the end of the training process (removing procedure is implemented in line 25 of $ErrCor$ pseudo code in Fig. 1).

Method 2. In the second variant all training patterns are checked every N_{RBF} trained units and only patterns that match removing conditions are removed from training process. In this case removed pattern can be restored to training process in the next step if the error for this pattern increased in the meantime (removing procedure is implemented in the same place as for method 1).

Method 3. In the third variant patterns are removed permanently every given number of iterations N_{ITER} (implemented in line 23 on listing of $ErrCor$ pseudo code)

Method 4. In the forth variant patterns are removed every given number of iterations N_{ITER} but all patterns are restored when new RBF unit is added to network (removing procedure is implemented in the same place as for method 3).

```
 1: set error vector err as desired outputs
 2: for n ← 1 to N_MAX do                                    ▷ for all new RBF units
 3:     find err_max as maximum of absolute value of err
 4:     create a new RBF unit with the center c_h at the location of pattern with
        err_max by setting width σ_h and corresponding weight w_h to 1
 5:     calculate MSE(iter = 1)
 6:     for iter ← 2 to max_iteration do
 7:         for all patterns do
 8:             calculate Jacobian vector
 9:             calculate sub quasi Hessian matrix
10:             calculate sub gradient vector
11:         end for
12:         calculate quasi-Hessian matrix Q and gradient vector g
13:         update network parameters
14:         calculate new output value actual
15:         find error vector err as a difference of desired and actual outputs
16:         calculate MSE(iter)
17:         while error is not reduced do
18:             adjust the parameter using the LM scheme
19:         end while
20:         if abs( MSE(iter−1)−MSE(iter)) / MSE(iter−1) ) < desired_R_error then
21:             break
22:         end if
23:         patterns removing for methods 3–5
24:     end for
25:     patterns removing for methods 1–2
26:     if MSE(n) < desired_error then
27:         break
28:     end if
29: end for
```

Fig. 1. Pseudocode of ErrCor algorithm

Method 5. In the last variant all patterns are checked every given number of iterations N_{ITER} and can be removed or restored depending on current value of error err (removing procedure is implemented in the same place as for method 3).

The next section shows experimental results of applying all five methods for some benchmark datasets.

3   Experimental Results

Experiments have been done for the following benchmark approximations [15,16]: Peaks function, Rastrigin function and Schaffer function. All experiments have been done in the MATLAB software on Intel Core i7 8 GB environment. To compare proposed methods the following parameters have been measured: training time and Root Mean Square Error for training ($RMSET$) and validation ($RMSEV$) data. As the reference the same parameters have been measured for original $ErrCor$ algorithm [10].

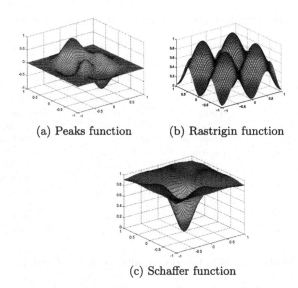

(a) Peaks function   (b) Rastrigin function

(c) Schaffer function

Fig. 2. Surfaces of functions used in experiments

3.1 Peaks Function

The first approximation experiment with described methods has been prepared for Peaks function given by Eq. (3) shown in Fig. 2a.

$$
\begin{aligned}
z(x, y) = & -\tfrac{1}{30}\, e^{\left(-1-6x-9x^2-9y^2\right)} \\
& - \left(0.6x - 27x^3 - 243y^5\right)\, e^{\left(-9x^2-9y^2\right)} \\
& + \left(0.3 - 1.8x + 2.7x^2\right)\, e^{\left(-1-6y-9x^2-9x^2\right)}
\end{aligned}
\tag{3}
$$

The training dataset contains 720 randomly generated points while validation dataset contains 180 points. Experiments been done for three cut-off threshold CT values from 0.00001 to 0.00003 and for N_{RBF} in range $< 1, 4 >$ (for methods 1, 2) and N_{ITER} in range $< 5, 30 >$ (for methods 3–5).

Achieved results are shown in Tables 1 and 2. As can be observed in most cases both $RMSET$ and $RMSEV$ reaches better result than pure $ErrCor$. Only method 3 proved to be ineffective where the training time indeed dropped significantly but the training error were 2–3 times higher. Others methods gave similar results to each other so that, unfortunately, the best method can not be identified clearly, but it can be said that suggested methods allows to reduce both training error and training time. Best result, bolded in Table 1, have been achieved using method 2 for $CT = 0.00002$ and $N_{RBF} = 1$.

Table 1. Peaks function results: method 1 and method 2

CT	N_{RBF}	Training time [s]	RMSET	RMSEV	Training time [s]	RMSET	RMSEV
Original ErrCor		198.28	0.00006514	0.00007145			
		Method 1			Method 2		
0.00001	1	175.62	0.00005132	0.00005422	160.18	0.00003602	0.00004069
	2	169.24	0.00003764	0.00004131	163.76	0.00003594	0.00003935
	3	177.93	0.00004270	0.00005042	167.83	0.00004660	0.00004888
	4	207.92	0.00004408	0.00016918	165.71	0.00003100	0.00003427
0.00002	1	105.90	0.00016330	0.00016248	179.75	**0.00002906**	**0.00003015**
	2	162.36	0.00004320	0.00004555	162.37	0.00004155	0.00004580
	3	167.11	0.00004447	0.00004529	159.86	0.00005233	0.00005920
	4	180.02	0.00004067	0.00004329	165.81	0.00004974	0.00004974
0.00003	1	137.29	0.00010990	0.00012450	173.34	0.00004589	0.00004996
	2	148.45	0.00005027	0.00005345	160.88	0.00005119	0.00016614
	3	148.75	0.00009407	0.00009776	160.23	0.00003993	0.00004247
	4	134.08	0.00007005	0.00007238	165.12	0.00004638	0.00005397

Table 2. Peaks function results: method 3,4 and 5

CT	N_{ITER}	Training time [s]	RMSET	RMSEV	Training time [s]	RMSET	RMSEV	Training time [s]	RMSET	RMSEV
Original ErrCor		198.28	0.00006514	0.00007145						
		Method 3			Method 4			Method 5		
0.00001	5	83.81	0.00016312	0.00017736	189.71	0.00005267	0.00005763	174.66	0.00009168	0.00010314
	10	109.05	0.00015312	0.00015400	180.45	0.00005122	0.00006021	143.98	0.00004925	0.00006836
	15	125.62	0.00012623	0.00014639	180.44	0.00004489	0.00005147	164.06	0.00004164	0.00004274
	20	122.71	0.00013321	0.00012397	192.03	0.00006416	0.00007006	164.19	0.00006422	0.00007784
	25	123.58	0.00008055	0.00008741	193.61	0.00004039	0.00004486	166.42	0.00005237	0.00005441
	30	139.69	0.00006311	0.00006658	201.79	0.00004193	0.00004618	172.54	0.00005314	0.00005683
0.00002	5	70.13	0.00025263	0.00030500	143.23	0.00009292	0.00010029	154.55	0.00004153	0.00004441
	10	87.51	0.00036390	0.00041616	139.05	0.00004287	0.00004505	171.59	0.00005479	0.00005952
	15	76.62	0.00019746	0.00020533	141.84	0.00003681	0.00004113	166.06	0.00004150	0.00004706
	20	87.74	0.00022901	0.00022911	173.48	0.00005490	0.00005816	136.31	0.00003652	0.00003570
	25	105.89	0.00014402	0.00015402	176.98	0.00004435	0.00004765	170.50	0.00005431	0.00006230
	30	88.63	0.00017820	0.00019562	191.53	0.00005108	0.00005288	158.27	0.00003449	0.00003616
0.00003	5	43.06	0.00029388	0.00032568	127.94	0.00010110	0.00010998	120.11	0.00005920	0.00006184
	10	74.33	0.00040997	0.00044360	134.62	0.00006126	0.00006912	129.20	0.00004772	0.00005161
	15	70.65	0.00032367	0.00032367	196.91	0.00006010	0.00006270	180.62	0.00006924	0.00015879
	20	82.48	0.00019794	0.00019942	172.30	0.00006082	0.00007071	156.92	0.00007356	0.00007608
	25	82.41	0.00010142	0.00012024	194.00	0.00008494	0.00009371	141.84	0.00004261	0.00007282
	30	89.15	0.00017587	0.00018446	285.94	0.00005418	0.00005521	143.96	0.00004963	0.00005224

3.2 Rastrigin Function

The second experiment has been prepared for Rastrigin function given by function (4) is shown in Fig. 2b.

$$f(x, y) = 20 + (x^2 - 10cos(2\pi x)) + (y^2 - 10cos(2\pi y)) \tag{4}$$

The training and verification data sets contain 2100 and 900 randomly selected points, respectively. Values of CT, N_{RBF} and N_{ITER} parameters had been set the same like in first experiment.

Results are shown in Tables 3 and 4. Results confirm observations from first experiment prepared for Peaks function. In the most cases errors and training

Table 3. Rastrigin function results for method 1 and method 2

CT	N_{RBF}	Training time [s]	RMSET	RMSEV	Training time [s]	RMSET	RMSEV
Original ErrCor		600.69	0.0003835	0.0003906			
		Method 1			Method 2		
0.00001	1	539.29	0.0002935	0.0003051	524.51	0.0002542	0.0002594
	2	504.73	0.0002343	0.0002788	557.27	0.0003235	0.0003215
	3	691.01	0.0003250	0.0003295	532.70	0.0003330	0.0003328
	4	546.54	0.0003210	0.0003198	544.59	0.0003342	0.0003343
0.00002	1	559.49	0.0003192	0.0003298	576.07	0.0003185	0.0003201
	2	557.02	0.0002571	0.0002626	581.20	**0.0001477**	**0.0001532**
	3	590.74	0.0002386	0.0002434	554.03	0.0002068	0.0002013
	4	550.17	0.0002548	0.0002602	637.23	0.0004466	0.0004872
0.00003	1	518.89	0.0002282	0.0002364	594.02	0.0002722	0.0002656
	2	595.10	0.0001629	0.0001715	590.66	0.0002548	0.0002636
	3	672.27	0.0003247	0.0003259	447.53	0.0002613	0.0002661
	4	512.83	0.0004335	0.0004799	646.34	0.0002964	0.0003138

Table 4. Rastrigin function results for methods 3–5

CT	N_{ITER}	Training time [s]	RMSET	RMSEV	Training time [s]	RMSET	RMSEV	Training time [s]	RMSET	RMSEV
Original ErrCor		600.69	0.0003835	0.0003906						
		Method 3			Method 4			Method 5		
0.00001	5	429.95	0.0003663	0.0003762	578.90	0.0002201	0.0002202	545.36	0.0003873	0.0003809
	10	482.46	0.0002470	0.0002542	546.53	0.0003048	0.0003259	529.49	0.0002434	0.0002664
	15	484.74	0.0004035	0.0004338	552.16	0.0001558	0.0001628	528.62	0.0002938	0.0002938
	20	485.19	0.0002866	0.0002876	513.48	0.0003350	0.0003478	530.28	0.0002752	0.0002992
	25	499.52	0.0003153	0.0003120	544.27	0.0002586	0.0002685	530.01	0.0003364	0.0003400
	30	504.62	0.0003474	0.0003483	502.98	0.0003290	0.0003290	494.42	0.0003144	0.0003170
0.00002	5	323.20	0.0006120	0.0007603	537.33	0.0003062	0.0003243	533.09	0.0003813	0.0003816
	10	352.55	0.0003288	0.0003243	532.57	0.0003610	0.0003735	560.15	0.0002743	0.0002981
	15	447.41	0.0003806	0.0003832	543.95	0.0001804	0.0001978	541.26	0.0004274	0.0004749
	20	410.53	0.0004548	0.0004826	534.28	0.0002541	0.0002614	552.63	0.0002456	0.0002685
	25	494.89	0.0003309	0.0003281	538.67	0.0002602	0.0002606	541.93	0.0002325	0.0002442
	30	470.28	0.0002636	0.0003768	535.22	0.0002627	0.0002687	525.52	0.0002479	0.0002763
0.00003	5	278.72	0.0005414	0.0005324	565.79	0.0003250	0.0003273	564.65	0.0002737	0.0002973
	10	314.37	0.0005926	0.0006259	514.45	0.0003090	0.0003092	506.72	0.0005028	0.0005480
	15	324.50	0.0004706	0.0004708	562.60	0.0002028	0.0002084	516.08	0.0003033	0.0003045
	20	414.03	0.0003886	0.0003984	507.22	0.0005283	0.0005621	556.75	0.0002720	0.0002958
	25	431.03	0.0003167	0.0003246	524.39	0.0004550	0.0004910	513.33	0.0002204	0.0002185
	30	444.51	0.0003816	0.0004118	536.74	0.0002736	0.0002968	515.21	0.0002833	0.0002989

time are lower than these achieved for pure ErrCor. In this case, method 3 gave not such bad results but clearly worse than other methods. Again, best result have been achieved using method 2, in this case for $CT = 0.00002$ and $N_{RBF} = 2$.

Table 5. Schaffer function results for method 1 and method 2

CT	N_{RBF}	Training time [s]	$RMSET$	$RMSEV$	Training time [s]	$RMSET$	$RMSEV$
Original ErrCor		17.6142		0.0070983	0.010035		
		Method 1			Method 2		
0.0001	1	24.4539	0.007075	0.009540	34.1358	0.006775	0.009238
	2	14.5325	0.006843	0.009868	15.0072	0.010386	0.012526
	3	11.9950	0.010392	0.012503	59.5133	NaN	NaN
	4	13.6592	0.010228	0.013150	20.2732	0.006769	0.008705
0.0002	1	36.1252	0.006588	0.009743	36.0065	0.006345	0.008086
	2	13.3991	0.019159	0.083296	21.0788	0.006766	0.008705
	3	13.9280	0.014630	0.016132	14.9345	0.010386	0.012517
	4	21.0090	0.005970	0.008579	13.4906	0.010387	0.012525
0.0003	1	23.1846	0.006680	0.008448	21.4202	0.006574	0.007953
	2	10.9314	0.009938	0.011148	14.2794	0.010386	0.012524
	3	23.6867	0.007142	0.010069	16.2057	0.009383	0.011854
	4	17.6142	0.007098	0.010035	13.0395	0.010386	0.012524

Table 6. Schaffer function results for methods 3–5

CT	N_{ITER}	Training time [s]	$RMSET$	$RMSEV$	Training time [s]	$RMSET$	$RMSEV$	Training time [s]	$RMSET$	$RMSEV$
Original ErrCor		17.6142	0.0070983	0.010035						
		Method 3			Method 4			Method 5		
0.0001	5	19.3955	0.007104	0.009098	16.5347	0.007502	0.016821	40.3093	0.006398	0.009185
	10	17.4771	0.013326	0.015874	38.3611	0.006094	0.007722	40.8045	**0.005326**	**0.007205**
	15	19.9957	0.010260	0.013047	36.8225	0.005697	0.008126	29.4327	0.006540	0.008225
	20	24.5160	0.006496	0.008395	21.6583	0.006774	0.008721	39.0197	0.006243	0.008234
	25	23.2449	0.007759	0.029973	20.1826	0.006925	0.008874	23.8640	0.007891	0.010062
	30	36.1263	0.007146	0.010104	34.9781	0.008849	0.011577	35.9352	0.005676	0.007986
0.0002	5	11.8116	0.013826	0.030933	27.7139	0.006041	0.008469	31.1944	0.006550	0.009567
	10	14.7851	0.010070	0.011648	27.1981	0.007844	0.030065	38.6671	0.006206	0.008651
	15	22.1337	0.011197	0.013147	37.5649	0.006327	0.008118	34.2823	0.005722	0.008152
	20	22.0903	0.007120	0.009077	19.6996	0.006952	0.008902	34.5039	0.005415	0.007878
	25	27.8958	0.006171	0.009015	14.2324	0.010398	0.012465	28.5269	0.008189	0.010541
	30	23.3772	0.007183	0.009623	35.4444	0.007341	0.009191	34.9908	0.006282	0.009208
0.0003	5	9.4365	0.025701	0.028074	16.9347	0.007261	0.034311	31.5590	0.007582	0.009262
	10	15.4384	0.011218	0.013381	40.9584	0.006270	0.008951	31.4714	0.006987	0.009185
	15	13.5180	0.011433	0.014721	34.4163	0.006557	0.007982	34.1413	0.006524	0.009682
	20	14.2874	0.012782	0.014230	15.1858	0.010386	0.012526	33.2569	0.006372	0.008155
	25	21.8395	0.007167	0.009267	30.0586	0.008759	0.011390	31.2345	0.006307	0.007916
	30	20.4763	0.008293	0.009039	35.8281	0.006508	0.008723	35.2985	0.006207	0.032363

3.3   Schaffer Function

The last experiment have been done for Second Schaffer function. This function is given by Eq. (5) shown in Fig. 2c.

$$z(x,y) = 0.5 + \frac{sin^2\left(x^2 - y^2\right) - 0.5}{\left[1 + 0.001\left(x^2 - y^2\right)\right]^2} \tag{5}$$

The training and verification data sets contain 630 and 270 randomly selected points, respectively, and values of N_{RBF} and N_{ITER} parameters had been set the same like in previous experiments while CT had been set in range from 0.0001 to 0.0003. Achieved results have been shown in Tables 5 and 6. As can be observed in this case results are not so good as in previous experiments. In most cases training time is higher than for pure ErrCor and training errors also in many cases are much higher. Such a case is caused by relatively high level of error achieved in short training time. It means that network was not able to train more successfully and calculations was broken very early without training error decreasing. As a consequence the number of iterations was very small and even presented methods were not very helpful, and additional processing time for removal procedure only significant increased the training time. Best result has been achieved by method 5 for $CT = 0.0001$ and $N_{ITER} = 10$, and again it was a worse method 3.

4   Conclusions

This is obvious that training patterns are important for neural network training. However, appropriate selection of these patterns can play an important role in obtaining good results. The approach proposed in this paper suggests that patterns that are already trained can be removed from training dataset during training process to improve training decreasing both training time and training error. The five proposed methods for ErrCor algorithms have been implemented and applied to resolve some benchmark problems. The results confirm the validity of the proposals. The further work in this area will be focused to extend this approach on soft pattern deletion and to apply presented approach to other training algorithms and methods for different architectures.

References

1. Rumelhart, D.E., Hinton, G.E., Williams, R.J.: Learning representations by back-propagating errors. Nature **323**, 533–6 (1986)
2. Fahlman, S.E., Lebiere, C.: The cascade-correlation learning architecture. In: Advances in Neural Information Processing Systems 2. pp. 524–532. Morgan Kaufmann, San Mateo (1990)
3. Wilamowski, B.M., Bo, W., Korniak, J.: Big data and deep learning. In: 20th Jubilee IEEE International Conference on Intelligent Engineering Systems (INES 2016), 30 June–July 2, pp. 11–16 (2016)
4. Lang, K.L., Witbrock, M.J.: Learning to tell two spirals apart. In: Proceedings of the 1988 Connectionists Models Summer School. Morgan Kaufman (1998)
5. Wilamowski, B.M., Korniak, J.: Learning architectures with enhanced capabilities and easier training. In: 19th IEEE International Conference on Intelligent Engineering Systems (INES 2015), 03–05 September, pp. 21–29 (2015)
6. Bengio, Y.: Learning deep architectures for AI. Found. Trends in Mach. Learn. **2**(1), 1127 (2009)

7. Ciresan, D.C., Meier, U., Gambardella, L.M., Schmidhuber, J.: Deep big simple neural nets excel on handwritten digit recognition. CoRR (2010)
8. Wilamowski, B.M., Yu, H.: Neural network learning without backpropagation. IEEE Trans. Neural Netw. **21**(11), 1793–80 (2010)
9. Werbo, P.J.: Back-propagation: past and future. In: Proceeding of International Conference on Neural Networks, San Diego, CA, vol. 1, pp. 343–354 (1988)
10. Yu, H., Reiner, P., Xie, T., Bartczak, T., Wilamowski, B.M.: An incremental design of radial basis function networks. IEEE Trans. Neural Netw. Learn. Syst. **25**(10), 1793–80 (2014)
11. Nguyen, G.H., Bouzerdoum, A., Phung, S.L.: Efficient supervised learning with reduced training exemplars. In: 2008 IEEE International Joint Conference on Neural Networks (IEEE World Congress on Computational Intelligence), Hong Kong, pp. 2981–2987 (2008)
12. Lozano, M.T.: Data reduction techniques in classification processes. Ph.D. Dissertation, Universitat Jaume I, Spain (2007)
13. Chouvatut, V., Jindaluang, W., Boonchieng, E.: Training set size reduction in large dataset problems. In: 2015 International Computer Science and Engineering Conference (ICSEC), Chiang Mai, pp. 1–5 (2015)
14. Kolbusz, J., Rozycki, P.: Outliers elimination for error correction algorithm improvement. In: CS&P Proceedings 24th International Workshop Concurrency, Specification & Programming, (CS & P 2015), vol. 2, pp. 120–129 (2015)
15. Xie, T.: Growing and learning algorithms of radial basis function networks. Ph.D. Dissertation, Auburn University, USA (2013)
16. Dieterich, J., Hartke, B.: Empirical review of standard benchmark functions using evolutionary global optimization. Appl. Math. **3**(10A), 1552–64 (2012)
17. Wilamowski, B.M., Jaeger, R.C.: Implementation of RBF type networks by MLP networks. In: IEEE International Conference on Neural Networks (ICNN 96), pp. 1670–1675 (1996)
18. Wu, X., Wilamowski, B.M.: Advantage analysis of sigmoid based RBF networks. In: Proceedings of the 17th IEEE International Conference on Intelligent Engineering Systems (INES 13), pp. 243–248 (2013)

Exploring the Solution Space of the Euclidean Traveling Salesman Problem Using a Kohonen SOM Neural Network

Ewa Skubalska-Rafajłowicz[✉]

Department of Computer Engineering, Faculty of Electronics,
Wrocław University of Science and Technology, Wrocław, Poland
ewa.rafajlowicz@pwr.edu.pl

Abstract. In this paper we present a new approach to solving the Euclidean traveling salesman problem (ETSP) using SOM Kohonen maps with chain topology. The Kohonen learning rule is used with random parameters providing different neuron locations. Any new neuron configuration allows us to obtain a new ETSP solution. This new approach to exploring the solution space of the ETSP is easy to implement and suitable for relatively large ETSP problems. Furthermore, the approach could be combined both with other global optimization methods as genetic algorithms and with simple TSP solving heuristic procedures. The method is illustrated by simulations used for solving some TSPLIB problems.

Keywords: SOM Kohonen map · The Euclidean traveling salesman problem

1 Introduction

The Traveling Salesman Problem (TSP) is a famous combinatorial optimization problem [3,17,20] started in the 1930s.

Given a list of places (cities) and the costs of travel between these cities, the TSP problem is to find the order in which the cities are visited forming a closed tour in such a way that the costs of traveling are minimized. Each city should be visited exactly once.

The special case of TSP is the Euclidean traveling salesman problem (ETSP), where cities are given as points in the Euclidean space (usually R^2) and cost of traveling between each pair of cities is defined as the Euclidean distance (Euclidean metric).

It is known that the Euclidean TSP, similarly as a more general TSP, is NP-hard [24]. There are known many exact and approximate methods of solving TSP and ETSP (see for example [2,28] and the literature cited therein).

Bartholdi and Platzman [7,25] have used the Sierpiński space-filling curve for a good, very quickly obtained approximate solution of ETSP. Such a solution is at most $O(log(N))$ times longer than the optimal tour [29].

© Springer International Publishing AG 2017
L. Rutkowski et al. (Eds.): ICAISC 2017, Part I, LNAI 10245, pp. 165–174, 2017.
DOI: 10.1007/978-3-319-59063-9_15

In this paper we concentrate on using a Kohonen SOM network [19] with closed chain topology for solving ETSP [1,2,4–6,9,11,14,15,21–23,30,31]. Closely related to SOM netwoks are the adaptive ring algorithm [16] and the elastic net method of Durbin and Willshow [12,13].

The next Section provides a precise formulation of the ETSP problem. The Kohonen SOM algorithm for ETSP is briefly described in Sect. 3. Section 4 presents the main results of the paper. In this Section the new algorithm for exploring the solution space of the ETSP problem is formulated and discussed. Finally, results of computational experiments are shown in Sect. 5.

2   Short Formulation of the ETSP Problem

In the ETSP problem every city is represented by a point on a map. Usually it is a point in 2-D space or in general a point in some compact subset of R^d, $d \geq 2$. We will assume that this set is a unit cube $I^d = [0,1] \times \ldots \times [0,1]$.

Let $X = \{x_1, \ldots x_N\}$ be a given set of points in I^d. Our goal is to find a sequence of points X, i.e., a permutation of point indexes $\Pi = (\pi(1), \pi(2) \ldots, \pi(N))$, such that

$$C_{TSP}(\Pi, X) = \sum_{i=2}^{N} ||x_{\pi(i)} - x_{\pi(i-1)}|| + ||x_{\pi(N)} - x_{\pi(1)}|| \qquad (1)$$

is minimized over all possible cyclic permutations. $||.||$ denotes the Euclidean norm.

Due to a symmetry property of the Euclidean distance between a pair of points and since it can be assumed that $\pi(1) = 1$, it is possible to find $0.5(N-1)!$ different closed paths between points from set X.

Testing all $0.5(N-1)!$ possible sequences one can always find the shortest closed path connecting all N points, but the computational effort for this exhaustive search strategy, rises exponentially with N and rapidly becomes unmanageable. For $N = 10$ it is 181440 different solutions, for $N = 20$ the number of solutions rises to 60822600000000000, and for $N = 30$ it is definitely unmanageable to perform the brute force search.

It is known that such SOM networks minimize the sum of squared Euclidean distances (SED) rather than the sum of Euclidean distances between city-points [10]. Nevertheless, in practice the Euclidean metric and SED produce the same or very similar optimal (or sub-optimal) solutions.

3   The Kohonen SOM Algorithm for Solving ETSP

Let M denote the number of neurons. It is assumed that $M \geq N$ [1,6]. Recall that N is the number of data points in set X. Each neuron is represented by a point in R^d labeled by a number $j = 1, 2, \ldots, M$. Coordinates of such a point, let's say labeled by j, form a vector $W_j \in I^d$, i.e., W_j is the j-th neuron weight. Each data point $x_i \in X$ can be uniquely represented by neuron j^* such that

$$i^{\star}(x_i) = arg \min_{j=1,\ldots,M} ||x_i - W_j||^2. \tag{2}$$

Such a neuron is often called "a winner". In the case of ties we will randomly pick exactly one winning neuron.

Position (weights) of the winning neuron is then corrected according to the Kohonen learning rule.

3.1 SOM Learning

The Kohonen learning rule is applied for updating neuron weights (their positions in the data space). Denote by $W_j(n)$ the weight of j-th neuron in n-iteration of the learning process.

The process of learning is some kind of stochastic approximation process [27]. In every iteration only one data point from X is chosen at random and used for neuron position updating.

Kohonen's method of learning is very efficient since not only a winning neuron selected according to (2) is updated but its neighbors in the neuron's chain topology are also updated according to:

$$W_j(n+1) = W_j(n) + \alpha(n)h(x_i, j, n)(x_i - W_j(n)), \; j = 1, \ldots, M \tag{3}$$

where $h(x_i, j, n)$ is a neighborhood function taking values in $[0, 1]$ and $\alpha(n)$ is a learning rate. $\alpha(n)$ is usually very small and decreases to zero with n. The common choice of the neighborhood function is the Gaussian function

$$h(x_i, j, n) = \exp{(-\frac{|i^{\star}(x_i) - j|_{\star}^2}{2\sigma(n)^2})}, \tag{4}$$

where symbol $|r - s|_{\star}$ denotes $min\{|r - s|, M - r + s, M - s + r\}$ and $\sigma(n) \to 0$ when $n \to \infty$. The are many different methods of changing σ and α [5,9,18,31].

4   Exploring ETSP Solution Space

Recall that we have assumed that all ETSP's data points are taken from the unit square, i.e., each $x_i \in [0, 1] \times [0, 1]$. Appropriate scaling of ETSP points was performed in the following test examples. Adequate rescaling of subsequent paths forming the tour under consideration allows us to retain the true goal function value. Although, both problems: an original and its rescaled version are not identical from the point of the Kohonen algorithm, we did not observe any important differences in the results (tested on 1000 random point examples).

4.1   Starting Solutions

There are known different methods of obtaining a starting solution (starting positions of SOM neurons) for SOM based algorithms solving ETSP [5,6,19]. It is possible to start with random values (neurons coordinates). One can also construct a polygon of neurons, which approximates a circle or a rhombic frame [5].

We have used randomly (uniformly) generated neuron positions along the Sierpiński space filling curve. The Sierpiński curve could be applied for ordering ETSP city-points [7, 29].

4.2 Computing the Tour Corresponding to the Actual Neuron's Locations

As a consequence of finding a winner (2) for each data point, every neuron $j \in \{1, \ldots, M\}$ may be closest to the exactly one data point from X, may be "empty", or it may happen that two (or even more than two) data points are connected to the same neuron [9, 14]. Nevertheless, it is relatively easy to obtain the permutation of cities adequate to the neuron weights.

Let $A(j)$ denote the set of data points connected to neuron j. It is clear that

$$\bigcup_{j=1}^{M} A(j) = \{1, 2, \ldots, N\}.$$

Furthermore, $A(j)$, $j = 1, \ldots, M$ define at least partial order in the set of data points. In the case when there exist $A(j)$ such that $|A(j)| > 1$, strict order can be obtained by random choice or using the lexicographical order defined in X.

We represent A_j by lists rather than sets, which allows us to retain ordering. So, we obtain a linear order of data points from X uniquely providing sequence Π. The computational complexity of that ordering is $O(NM \log M)$.

The similar approach for finding a ETSP solution was proposed in [9] in a final state of the Kohonen algorithm when positions of the neurons are slightly different than true point locations. We propose to use the method as a tool for new solution generation.

Thus, our goal is to obtain on the one hand an adequate level of randomness in the location of the chain of neurons, but on the other hand the neurons positions should be located at some level of agreement with neuron labels. The last condition is a reason that we decided to use the Kohonen learning rule for generating the subsequent neuron chain locations.

Furthermore, it should be noted that in every iteration of the SOM algorithm only a few neuron weights are effectively updated, namely the winning neuron and its closest neighbors. When $|i^\star(x_i) - j|$ is larger than (for example) $3\sigma(n)$ then $h(x_i, j, n)$ is very small and the weights W_j should not be changed.

The new permutation can be obtained in $O(N)$ time instead of $O(NM \log M)$.

We denote the value of already found best tour according to (1) by Q_{opt} and the corresponding best (already found) tour (list of M elements) by Π_{opt}. This means that Q_{opt} and Π_{opt} are used for storing the best solution and the best permutation found so far, respectively.

4.3 Algorithm

Step1. Initialization: Let n be a discrete learning time starting from 0. The algorithm inputs are points $X = \{x_1, \ldots x_N\} \subset I^2$. Choose $M \geq N$, (for example, let $M = 2N$). The neuron's weights W are initialized at random using the uniform distribution along the Sierpiński space-filling curve. Find the actual tour (permutation of cities) corresponding to neuron weights W.

Set $A_j = \{\}$ $j = 1, \ldots, M$. For $i = 1, 2, \ldots, N$ find $j = i^\star(x_i)$ and add index i to the list A_j. Obtain a starting sequence Π by concatenating lists A_j, $j = 1, \ldots, M$ (empty lists are rejected). The main loop of the algorithm consists of the following steps:

Step 2. If $n > n_{max}$ go to Step 6. Otherwise, randomly select parameter $\alpha(n)$ from interval $[0.1, 1.0]$ and parameter $\sigma(n)$ from the interval $[1., 3.0]$. Draw at random a city-point x from set X. Let say $x = x_i$.

Step 3. Find neuron $i^\star(x_i)$ closest to x_i (according to 2). Update weights according to the Kohonen rule (3). Only the neurons $j \in J_a$ such that

$$J_a = \{j : |i^\star(x_i) - j| < 3\,\sigma(n)\}$$

are updated.

Step 4. For $i = 1, 2, \ldots, N$ find

$$i^\star_{new}(x_i) = arg \min_{j \in J_a} ||x_i - W_j||^2. \tag{5}$$

If

$$||x_i - W_{i^\star_{new}(x_i)}||^2 < ||x_i - W_{i^\star(x_i)}||^2$$

remove index i from $A_{i^\star}$ and add i to the list $A_{i^\star_{new}}$.

Step 5. Obtain new sequence Π by concatenating lists A_j, $j = 1, \ldots, M$ (empty lists are rejected).

Step 6. Compute $Q(\Pi, X)$ according to (1). If $Q(\Pi, X) \leq Q_{opt}$ set: $Q_{opt} = Q(\Pi, X)$, $\Pi_{opt} = \Pi$ and $W_{opt} = W$. Set $n = n + 1$ and go to Step 2.

Step 7. Return the best results Π_{opt} and Q_{opt}.

In the presented Algorithm Kohonen's learning rule parameters are chosen at random and since exploring the solution space is assumed to be a never-ending process, these parameters are changing in certain regions (value's intervals separated from zero).

5   Simulation Study Using TSPLIB Examples

The method proposed in the paper is illustrated by simulations used for solving some TSPLIB [26] problems. The results of simulations using other, often very sophisticated methods, can be found in [2].

5.1   *Pbc3038* Problem (TSPLIB)

The optimal solution of *pbc3038* problem from TSPLIB [26] (at http://www.iwr. uni-heidelberg.de/groups/comopt/software/TSPLIB95) is known and length of the optimal tour equals 137694. The problem consists of 3038 data points. We have finished our search after 260000 simple iterations obtaining the problem solution with 155069 (i.e., 1.126 times optimal solution). It should be emphasized that one iteration means corrections of weights using only one data point (one city). Figure 1 shows in which way the length of the TSP tour is changing in the process of exploration of the solution space. In this search we have reduced the range of α parameter to $[0.1; 0.3]$ after 200000 iterations.

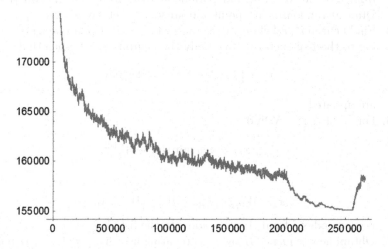

Fig. 1. The length of the TSP tour (versus a number of simple iterations) for *pbc3038* problem in the process of exploration of the solution space

The best tour obtained after 140000 simple iterations is depicted in Fig. 2.

5.2   *pbc1173* Problem (TSPLIB)

The optimal solution of *pbc1173* problem from TSPLIB is 56892. The problem consists of 1173 data points. Figure 3 shows in which way the length of the TSP tour is changing in the process of exploration of the solution space. We have finished the search after 200000 simple iterations obtaining the length of the best tour equal to 64786.1. In the other, longer trial we have obtained the best value of the goal function after 800000 equal to 62901.2, i.e., 1.127 times longer than the optimal value. Figure 4 shows the results of these simulations.

5.3   *bier127* Problem TSPLIB

The *bier127* problem from TSPLIB [26]) appears in some papers [5,6] and even the optimal solution is obtained by many simulations (repetitions). Optimal

Fig. 2. The best tour obtained after 140000 simple iterations (*pbc3038*). Dots indicate the neurons positions

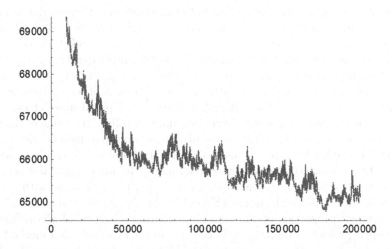

Fig. 3. The length of the TSP tour (versus a number of simple iterations) for *pbc1173* problem in the process of exploration of the solution space

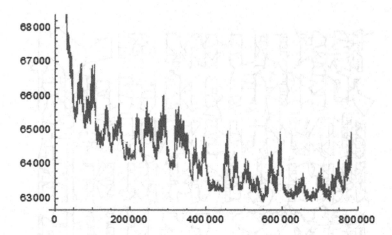

Fig. 4. The length of the TSP tour for *pbc1173* problem in the process of exploration of the solution space during 800000 iterations

solution value of *bier127* problem is 118282. It occurred that after ~130000 simple iterations the relative error of the best obtained tour with respect to the optimal tour was 4.7% and after ~260000 next iterations the error was 4.45%. In the next trial, the corresponding error after ~130000 iterations was equal to 3.37%. The smallest problem (*bier127*) occurred relatively hard from the point of view the exploration of the solutions space.

6   Comments and Conclusions

In every our test problem the first part of simulations was very similar in every repetition. We do not provide details, but this observation is a result of many short-time repetitions of the exploring process. Thus, one could start computing the goal function values after, for example, 100000 simple iterations.

In this paper we concentrate on the behavior of the proposed exploring algorithm. First, the subsequent solutions are close to each other and changes in the length of the tour are slow but rather consistent. Up and down changes are observed but the general trend is down. Second, controlling the value of the goal function allows us to adjust the range of the learning parameters. Especially, when the algorithm is trapped close to the local optimum one can expand the range of the parameters variability. Furthermore, the approach could be combined with other global optimization methods as genetic algorithms, with decomposition methods, and with simple TSP solving heuristic procedures [2,8,17,20].

Theoretical properties of the exploring solution space SOM algorithm are still an open problem. Similarly, as in the case of the classical SOM method it is not proved that the optimal solution of the ETSP is obtainable (or under which conditions it is obtainable).

References

1. Angeniol, B., Vaubois, C., Texier, J.-Y.L.: Self-organizing feature maps and the traveling salesman problem. Neural Netw. **1**, 289–293 (1988)
2. Avsar, B., Aliabadi, D.E.: Parallelized neural network system for solving Euclidean traveling salesman problem. Appl. Soft Comput. **34**, 862–873 (2015)
3. Applegate, D.L., Bixby, R.E., Chvatal, V., Cook, W.J.: The Traveling Salesman Problem: a Computational Study. Princeton University Press, Princeton (2006)
4. Aras, N., Oommen, B.J., Altinel, I.K.: Kohonen Network incorporating explicit statistics and its application to the traveling salesman problem. Neural Netw. **12**, 1273–1284 (1999)
5. Bai, Y., Zhang, W., Hu, H.: An efficient growing ring SOM and its application to TSP. In: Proceedings of the 9th WSEAS International Conference on Applied Mathematics, Istanbul, Turkey, May 27–29, 2006, pp. 351–355 (2006)
6. Bai, Y., Zhang, W., Jin, Z.: An new self-organizing maps strategy for solving the traveling salesman problem. Chaos Solitons Fractals **28**, 1082–1089 (2006)
7. Bartholdi, J.J., Platzman, L.K.: Heuristics based on spacefilling curves for combinatorial problems in Euclidean space. Manage. Sci. **34**, 291–305 (1988)
8. Bianchi, L., Knowles, J., Bowler, J.: Local search for the probabilistic traveling salesman problem: correction to the 2-p-opt and 1-shift algorithms. Eur. J. Oper. Res. **162**(1), 206–219 (2005)
9. Budinich, M.: A self-organising neural network for the travelling salesman problem that is competitive with simulated annealing. Neural Comput. **8**(2), 416–424 (1996)
10. Budinich, M., Rosario, B.: A neural network for the travelling salesman problem with a well behaved energy function. In: Ellacott, S.W., Mason, J.C., Anderson, I.J. (eds.) Mathematics of Neural Networks: Models, Algorthms and Applications, pp. 134–139. Kluwer Academic Publishers, Boston (1997)
11. Cochrane, E.M., Beasley, J.E.: The co-adaptive neural network approach to the euclidean travelling salesman problem. Neural Netw. **16**(10), 1499–1525 (2003)
12. Durbin, R., Szeliski, R., Yuille, A.: An analysis of the elastic net approach to the traveling salesman problem. Neural Comput. **1**, 348–358 (1989)
13. Durbin, R., Willshaw, D.: An analogue approach to the travelling salesman problem using an elastic net method. Nature **326**, 689–691 (1987)
14. Favata, S., Walker, R.: A study of the application of Kohonen-type neural networks to the traveling salesman problem. Biol. Cybern. **64**, 463–468 (1991)
15. Fort, J.C.: Solving combinatorial problem via self-organizing process: an application of the Kohonen algorithm to the traveling salesman problem. Biol. Cybern. **59**, 33–40 (1988)
16. Hueter, G.J.: Solution of the traveling salesman problem with an adaptive ring. In: Proceedings of the IEEE International Conference on Neural Networks, vol. I, pp. 85–92 (1988)
17. Karp, R.M., Steele, J.M.: Probabilistic analysis of heuristics. In: Lawlwr, E.L., Lenstra, J.K., Rinnooy Kan, A.G.H., Shmoys, D.B. (eds.) The Traveling Salesmen Problem: A Guided Tour of Combinatorial Optimization. Wiley, Chichister, New York, Brisbane, Toronto, Singapore (1985)
18. Kohonen, T.: The self-organizing map. Proc. IEEE **78**(9), 1464–1480 (1990)
19. Kohonen, T.: Self-organizing Map. Springer, New York (2001)
20. Lawler, E., Lenstra, J., Rinnooy, K.A.: The Traveling Salesman Problem: A Guided Tour of Combinatorial Optimization. Wiley, New York (1985)

21. Leung, K.S., Jin, H.D., Xu, Z.B.: An expanding self-organizing neural network for the traveling salesman problem. Neurocomputing **62**, 267–292 (2004)
22. La Maire, B.F.J., Mladenov, V.M.: Comparison of neural networks for solving the travelling salesman problem. In: IEEE Proceedings of the NEUREL 2012 (2012). doi:10.1109/NEUREL.2012.6419953
23. Matsuyama, Y.: Self-organizing neural networks and various euclidean traveling salesman problems. Syst. Comput. Jpn. **2**, 101–112 (1992)
24. Papadimitriou, C.H.: The Euclidean traveling salesman problem is NP-complete. Theor. Comput. Sci. **4**, 237–244 (1978)
25. Platzman, L.K., Bartholdi, J.J.: Spacefilling curves and the planar traveling salesman problem. J. ACM **36**, 719–737 (1989)
26. Reinelt, G.: TSPLIB - a traveling salesman problem library. ORSA J. Comput. **3**(4), 376–384 (1991)
27. Ritter, H., Martinetz, T., Schulten, K.: Neural Computation and Self Organizing Maps: An Introduction. Addison-Wesley, New York (1992)
28. Saadatmand-Tarzjan, M., Khademi, M., Akbarzadeh-T, M.R., Moghaddam, H.A.: A novel constructive-optimizer neural network for the traveling salesman problem. IEEE Trans. Syst. Man Cybern. Part B Cybern. **37**(4), 754–770 (2007)
29. Steele, J.M.: Efficacy of spacefilling heuristics in euclidean combinatorial optimization. Oper. Res. Lett. **8**, 237–239 (1989)
30. Vieira, F.C., Dria Neto, A.D., Costa, J.A.: An efficient approach to the travelling salesman problem using self-organizing maps. Int. J. Neural Syst. **13**(2), 59–66 (2003)
31. Xu, X., Jia, Z., Ma, J., Wang, J.: A self-organizing map algorithm for the traveling salesman problem. In: IEEE Proceedings of the Fourth International Conference on Natural Computation, pp. 431–435 (2008). doi:10.1109/ICNC.2008.569

Resolution Invariant Neural Classifiers for Dermoscopy Images of Melanoma

Grzegorz Surówka$^{(\boxtimes)}$ and Maciej Ogorzałek

Faculty of Physics, Astronomy and Applied Computer Science,
Jagiellonian University, 30-151 Kraków, Poland
{grzegorz.surowka,maciej.ogorzalek}@uj.edu.pl

Abstract. This article contributes to the Computer Aided Diagnosis (CAD) of melanoma pigmented skin cancer. We test back-propagated Artificial Neural Network (ANN) classifiers for discrimination in benign and malignant skin lesions. Features used for the classification are derived from wavelet decomposition coefficients of the dermoscopy image. We show the most efficient ANN setups as a function of the structure of hidden layers and the network learning algorithms. Our network topologies are limited for the sake of restrictions in memory and processing power of smartphones which are more and more popular as hand-held 'mobile' CAD devices for melanoma. We claim resolution invariance of the selected wavelet features.

Keywords: Melanoma · CAD · Wavelets · ANN

1 Introduction

1.1 Medical Background

Computer aided classification of malignant melanoma plays a crucial role in prevention of this tumor. Melanoma, as contrasted with the other forms of skin cancer, is extremely dangerous due to early metastases. Early diagnosis of malignancy is a life-saving factor. Lack of specialists, too late detections and an increased melanoma morbidity rate have become a medical problem lately and, on that account, a challange for computer assisted diagnosis (CAD) [1–3]. The standard therapy, biopsy, is not feasible for all the patients due to treatment costs and some health reasons. Even if affordable, excision must be done at an early stage, when standard clinical diagnosis based on the ABCD criteria may fail [4]. For that reason, analysis of skin lesions images have become a useful diagnostic tool. The most popular and cheapest form of lesion screening is dermoscopy (ELM-Epiluminescence Microscopy) [5]. Dermoscopy illuminates and enlarges the skin and when the image is stored on a computer allows for comparing past and current lesion advances. Clinical diagnosis of Melanoma is based on presence of classic dermoscopic features based on shape, color and structure content with help of the semi-quantitative descriptive measure ABCD(E), the 7-Point Checklist or Menzies [6].

© Springer International Publishing AG 2017
L. Rutkowski et al. (Eds.): ICAISC 2017, Part I, LNAI 10245, pp. 175–186, 2017.
DOI: 10.1007/978-3-319-59063-9_16

1.2   Wavelets

Melanoma at an early stage or the so called featureless melanoma [7] cannot be diagnosed with the aforementioned measures due to low recognition sensitivity of the geometric/coloristic features [8]. For that reason methods for wavelet based decomposition of the skin lesion images have been proposed [9].

The discrete wavelet transform (DWT) is a signal decomposition with a basis of real orthonormal functions obtained through translation and dilation of a kernel function (the so called mother wavelet). DWT is known to be a multiresolution approximation of the signal but when the translation is performed with arbitrarily small shift the algorithm is too computational-intensive (too redundant). As a workaround the Discrete Dyadic Wavelet Transform (DDWT) is defined. In DDWT the latter basic operations are performed at the pace of the powers of two. In DDWT a signal is represented as:

$$s(n) = \sum_r c_{0,r}\phi_{0,r} + \sum_{p=0}^{u}\sum_r d_{p,r}\psi_{p,r}(n)$$

where $p, r, n \in \mathcal{Z}$ and:

$$c_{p,r} = \sum_n s(n)\phi^*(n2^{-p} - r)$$

$$d_{p,r} = \sum_n s(n)\psi^*(n2^{-p} - r)$$

The mother wavelet ψ fulfils the 'two-scale' difference equation and the scaling function ϕ is subject to special relations with the kernel function. Mallat [10] demonstrated that for the coefficients it holds true:

$$c_{p-1,r} = \sum_n l(n)c_p(n - r)$$

$$d_{p-1,r} = \sum_n h(n)c_p(n - r)$$

i.e. one iteration of the transform filters out the input sequence with the low-pass $l(n)$ and high-pass filter $h(n)$ and decimates the output signals by two.

The derived filters should provide certain properties such as: transformation reversibility, orthogonality of the scaling function to its wavelet (and its translations) and finite response.

Details of the Mallat algorithm implemented for the 2D signals (images) can be found in Sect. 2.2.

Analyzis of the skin frequency and scale information with the wavelet features to distinguish between benign and malignant skin progression can be found in [11, 12].

1.3   Melanoma CAD with ANN

First ANN classification of melanoma and benign lesions comes from [13] where about 80% correct classifications were reported. The network topology (14-X-1, X-hidden neurons) accepted two asymmetry- and twelve color-based features.

In [14] authors analyzed basic, shape and color features with different normalization conditions and concluded that on both dichotomous and trichotomous tasks the ANNs performed similarly as logistic regression and SVMs, and better than the k-nearest neighbors and decision trees (sensitivity: 91%, specificity: 94% for 1619 lesion images).

In [15] authors analyzed 48 parameters belonging to four categories: geometry, color, texture, and color clusters inside the lesion and performed step-wise feature selection to identify an optimal subset of 10 variables (starting from the most significant: red multicomponent, decile of red, border homogeneity, mean value of red, grey-blue areas, contrast, interruptions of the border, mean skin-lesion gradient, background regions imbalance, variance of the border gradient). The clinical/dermoscopic equivalents of those variables are: multicomponent pattern/homogeneity, lesion darkness, border cleanliness, mean color of the lesion, grey-blue areas, network analysis, variation in the border cleanliness, grading of the border, color asymmetry, intensity in the border interruptions. Distinguishing melanoma from benign lesions with these optimal features gave the maximum sensitivity of 93% and specificity 92.75%.

Rajab et al. [16] investigated the neural network edge detection in the iterative thresholding segmentation. They drew conclusions on the method performance over a range of different border irregularity properties and signal-to-noise ratio.

An automatic neural skin cancer classification system was developed in [17] for dermoscopy imagess and optimized for different types of neural network topologies and different preprocessing modes. The authors reported best recognition accuracy of the 3-layers back-propagation neural network classifier as 89.9% and of the auto-associative neural network as 80.8%. In the system some features were extracted through 2-D wavelet packet decomposition under performance tests of seven different wavelet bases (the best wavelet base was Bior5.5, and the most stable experimentally Db1 or Db10). Some opimization was done to the number and structure of the hidden layers. The best topology was reported for the three hidden layers with 40-25-10 neurons.

In [18] a back-propagation neural network was used for segmentation. The results were compared with the ground truth images which showed that the method had slightly worse segmentation accuracy and slower training period than Extreme Learning Machine (ELM).

Group [19] segmented dermoscopic images using Maximum Entropy Threshold and extracted features using Gray Level Co-occurrence Matrix (GLCM) for classification into cancerous or non-cancerous cases using back-propagated neural networks. The reported accuracy is 88%.

Paper [20] proposed the flow: feature extraction, dimensionality reduction and classification to discriminate skin lesions into 'normal' and 'abnormal' skin cancer classes. In the first stage authors used discrete wavelet transforms, in the second stage PCA. In the classification stage a feed forward back-propagated artificial neural network and k-nearest neighbor paradigmes were applied. Accuracy of those experiments was: 95% (ANN) and 97.5% (kNN).

Discrimination between the six Menzies color classes in the calibrated RGB dermoscopy images were studied in [21]. The Jeffries–Matusita and transformed divergence separability distances were used to evaluate the color class separability. A nonlinear cluster transformation allowed for almost the total separation of each color class in the feature space. Several neural networks in competition were used as classifiers. Classification achieved 93% of sensitivity, 62% of specificity and 74% of accuracy (average). Authors claim that it might be possible to evaluate a lesion based on the presence of Menzies colors in the dermoscopic image, mimicking the human diagnosis.

In [22] statistical features and dermoscopic features (ABCD: Asymmetry, Border, Color and Diameter) for detection and diagnosis of melanoma were used. Segmentation-based thresholding plus statistical feature extraction using GLCM were used to calculate the Total Dermoscopy Score (TDS). The combined result of the TDS parameter and a neural network classification yielded accuracy of 88% which was claimed to be efficient for the skin cancer detection and diagnosis.

In [23] several methods of melanoma classification were proposed: a multilayered perceptron, a Bayesian classifier and the K nearest neighbors algorithm. These methods worked independently and also in combination making a collaborative decision support system. The performance factors obtained for seven neurons in a single hidden layer were: classification rate: 86.73%, sensitivity: 78.43%, specificity: 95.74%, slightly less than in the collaborative method: classification rate: 87.76%, sensitivity: 78.43%, specificity: 97.87%.

The state-of-the-art of melanoma skin lesion classification by ANN and other learning paradigms can be found in the reviews [1–3,24].

1.4   Motivation

According to the consumers' and doctors' demand for personal decision-supporting tools there have been developed for years various self-examination/self-diagnosing optical extensions and grip panels for mobile phones. Nowadays there is a growing market for handy dermatoscope-like devices with optics and ARM-based processors for a 'mobile' CAD and melanoma diagnosis [25–27]. Since the developed machine learning and image recognition and interpretation algorithms may demonstrate high complexity and small handheld devices have limited processing power and memory, it is of great importance to search for methods that analyze optimally both full-size dermoscopic images and preserve high efficiency also in downgraded image resolutions.

The latest research presented in [28], showed that the reverse bi-orthogonal (RBio) and bi-orthogonal (Bior) wavelets prove to be the most efficient, robust, and resolution invariant wavelet families for the machine learning of dermoscopic images. This was analyzed in the ensemble of different model types and optimized for various quality measures. In this work we want to check efficiency and classification performance of a homogeneous model for the wavelet base that follows conclusions of [28]. We pick the RBio3.1 wavelet base and study back-propagated artificial neural network (ANN) classifiers of the melanoma (RBio3.1)-based features with limited topologies for the hidden layers to take into account execution platforms with limited processing resources.

Within these bounds our objectives are:

- to select the best NN topology and learning method in terms of absolute melanoma classification performance upon condition of efficient cost of ANN backpropagation learning for three different resolutions of the dermoscopic images,
- to search for a resolution-invariant model of ANN for the original clinical data and the descendant downgraded image resolutions.

Below we show methodology of our machine learning experiments and present and discuss the results.

2   Data Analysis

2.1   Signal Processing

The database collected for this study included anonymous images of the moles from 185 patients of one medical clinic in Poland. The examinations were performed with plain digital camera with an extra dermoscopy extension and immersion liquid to remove light reflections. The primordial JPEG pixel resolution was 2272×1704 and the RGB color depth 24-bit. The resection and hist-pat examination of the moles allowed to assign labels to 102 malignant melanoma and 83 dysplastic nevus cases. The 83 non-melanoma images were picked up randomly from the predominant majority of about 2000 displastic lesions. In our analysis there were no 'unknown' or 'don't care' labels.

Melanoma incidence rate may fluctuate over countries, but clinical statistics shows an average of about 5% melanoma images as a fraction of all the dermoscopic images of the pigmented nevi. This means that the melanoma class is under-represented compared to the benign class. Learning classifiers from such cases would require special rules to properly treat the imbalanced class i.e. to draw equal attention to the minority class [29,30]. In this experiment we build classifiers for almost equal classes (102 malignant melanoma versus 83 dysplastic nevus cases) and we do not take into account the (clinical) class imbalance problem.

In the analysis there were three sets of images: the original set 2272×1704 (A) and the two downscaled sets (by averaging neighbor values in 2×2 elements) of 1136×852 (B) and 568×426 (C) pixels respectively.

2.2   Wavelet Features

There were no apparent artefacts on the analyzed dermoscopy images (black borders, hairs, droplets of immersion fluid, etc.) or there were few negligible distortions so no preprocessing tasks to the images took place.

To support wavelet transformations the dermoscopy images of all three sets (A, B, C) were transformed to indexed images with linear, monotonic color maps of double precision numbers. Each iteration of the wavelet decomposition downscales the input image by 2 both in rows and columns and three such iterations were done.

Images are two-dimensional signals and the wavelet transform to the images are done according to the Mallat algorithm [31]. One iteration of this algorithm produces 4 downscaled sub-images which can be considered as LL, LH, HL and HH filters (L-low-pass, H-high-pass filter) after one-dimensional wavelet transform on the rows and then on the columns.

DDWT can be applied recursively to the low-frequency sub-band only, but in our analysis we used the wavelet packets so each of the four filters was subject to further wavelet decompositions (not only LL).

Altogether in three iterations $1+4+16 = 21$ different transformation branches were produced. In one branch the following 12 simple features were calculated: $(e_i,\ i = 1,2,3,4)$ - energies of the sub-images, $(e_i/e_{max},\ i = 1,2,3,4)$ - maximum energy ratios and $(e_i/\Sigma e_k, k \neq i,\ i = 1,2,3,4)$ - fractional energy ratios, after [32, 33]. Energy was defined as a sum of absolute values of the pixels. This procedure was repeated for the three sets A, B, C of different image resolutions yielding $21 \times 12 = 252$ attributes in each single set.

For reasons presented in the Introduction for the skin texture analysis we took RBio3.1 wavelet base. Reverse bi-orthogonal wavelets (wavelet pairs) have the property of perfect reconstruction i.e. if X-image, A-reconstructed image of approximations and D-reconstructed image of details, then X = A+D. This property is possible due to two separate filter sets, one for decomposition and another one for image reconstruction. This wavelet is symmetric function and is not orthogonal ($X^2 \neq A^2 + D^2$).

Search for the best subset of features [34] can be used to (i) reduce bias (overtraining), (ii) reduce computational burden and (iii) enhance classification performance. This is usually done because the simplest approach, an exhaustive or random search to evaluate the best feature set, is infeasible or even computationally prohibitive. In this work we do not take advantage of any feature selection or extraction algorithms. This follows the results presented in [35] where widely known feature selection mechanisms (CFS, PCA, GSFS) were applied to melanoma classification problem. It was concluded that although application of feature selection algorithms may reduce the complexity of the classification, the performance is highly dependent upon the classifier. Therefore, it was opted to use all the features and preserve them for some late-selections. It is also the objective of this work - to search for efficient ANN classifiers in terms of their topologies and/or error minimization approaches and not by the feature selection of the data base.

2.3   Machine Learning

In our study we used a static feedforward back-propagation artificial neural network (ANN) to classify the dermoscopy images based on the calculated 252

Table 1. Training algorithms used for ANN classification of the dataset.

Mark	Method	Parameters
L1	Levenberg-Marquardt	$\mu = 0.001$
L2	Bayesian Regularization	$\mu = 0.005$
L3	Broyden-Fletcher-Goldfarb-Shanno	-
L4	Conjugate Gradient with Powell-Beale restarts	-
L5	Fletcher-Powell Conjugate Gradient	-
L6	Polak-Ribiére Conjugate Gradient	-
L7	Gradient Descent	$learn\_rate = 0.01$
L8	Gradient Descent with Adaptive Learning	$learn\_rate = 0.01$
L9	Gradient Descent with Momentum	$learn\_rate = 0.01$
		$momentum = 0.9$
LA	Variable Learning rate Gradient Descent	$learn\_rate = 0.01$
		$momentum = 0.9$
LB	One Step Secant	-
LC	Resilient Backpropagation	$learn\_rate = 0.01$
		$\Delta = 0.07$
LD	Scaled Conjugate Gradient	-

wavelet features. As a preprocessing phase normalization of the input was done and the labels were fixed to '1' (malanoma) and '0' (dysplastic nevus).

The ANN structure was:

- 252 input nodes that represent the wavelet features,
- a number of hidden nodes grouped into one or two hidden layers: [10], [20], [10-10], [10-20], [20-20],
- 2 output neurons, each one activated on the vectors belonging to one class only (so in the binary classification in a mutually exclusive way).

Two setups were analyzed within the scope of the activation functions:

- NN1: hyperbolic tangent sigmoid transfer function (a1) for the hidden layers [37] and linear activation function (a2) for the output layer [37]. This is a generic ANN to model any kind of input to output mapping.
- NN2: hyperbolic tangent sigmoid transfer function (a1) for the hidden layers and also hyperbolic tangent sigmoid transfer function (a1) for the output layer. This ANN should perform more efficiently while classifying inputs according to target classes.

The mean square error (mse) and cross-entropy (ent) were used as performance functions.

For the sake of cross-validation (CV) we randomly divided our data into training (70/100), validation (15/100) and testing (15/100) set. Every epoch all the training samples were presented simultaneously to the network to train it. The validation data was used to evaluate the prediction errors hence to optimize and update the weights in the backpropagation phase. The testing set was used to calculate all the performance coefficients. As stopping conditions the maximum number of epochs reached, maximum time exceeded and performance gradient fallen below 10^{-6} were fixed.

Several training algorithms for ANNs have been proposed in the literature to enhance the convergence speed and reduce the generalization error of the network [36–38]. In this work we do not discuss mathematical properties of those algorithms, rather focus on the classification interest when they are run with 'standard' base parameters. The analyzed backpropagation training algorithms are shown (with initial parameters where applicable) in Table 1.

Although in our ANN learning we meet the CV paradigm and validate during training, to promote better generalization for the performance function 'mse' (mean squared error) we applied also the performance regularization ratio (0.01) which takes into account not only minimizing the error but also the weights and biases (for L2 set to 0).

Due to performance aspects the Matlab library was used for calls to the ANN training algorithms. The code was run on the CUDA-based NVidia GTX 1070 GPU.

In the literature referenced in Introduction there are no methodical studies how the ANN structure (hidden layers) affect the melanoma classification performance. Arbitrary values for both the number of the hidden layers and the number of neurons in the hidden layers are published. Usually 2- and 3- hidden layers are presented. Taking into account the computational burden reported and own attempts on both CPU- and GPU-based parallel computing platforms and, last but not least, performance analyzes and benchmarks for the ARM-based mobile devices, we limited ourselves to up to 2 hidden layers with pretty small (up to 20) neurons on each layer.

3   Results and Discussion

We analyzed two networks: NN1 and NN2 (see Sect. 2.3) each tought according to two performance measures: (mse) and (ent). The best CV performance for the discrimination of melanoma from dysplastic nevi is reached with NN2(mse) which is full sigmoid-like network. Performance of NN1(mse/ent) and NN2(ent) is slightly worse (by about 8-3%) and more sparse in terms of the learning algorithms and topologies. Below we present the results for NN2(mse).

Our objective was to find the best performing ANN for the classification of melanoma dermoscopy images under the assumption that some ARM-based mobile devices may use it for the computer assisted diagnosis of melanoma and (as a next step) even be trained on a local database of dermoscopy images. For that reason only limited topologies were taken into account (starting with 10 hidden neuron on one hidden layer up to 20×20 hidden neurons on two layers). Although we take into account only about 10, 20, 100, 200, and 400 weights this

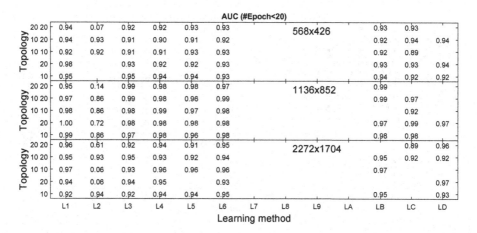

Fig. 1. AUC values for resolutions A, B, C as a function of hidden layers and learning algorithm (#Epoch<20).

is not a bad result compared to the literature. Close by performance our experiments show coarsely how complex in terms of epochs the back-propagation algorithms can perform. Statistics for the $5 \times 13 = 65$ different setups shows the following grouping of results in terms of pairs (*number_of_epochs*, *setups_finished*): $(10, 17)$, $(15, 31)$, $(20, 38)$, $(30, 44)$, $(50, 45)$, $(100, 51)$. As a necessary condition for further analysis we took the threshold of 20 epochs (median).

In Fig. 1 we plot numerical results of AUC for resolutions A, B and C for the five setups of the hidden layers and for the thirteen different back-propagation algorithms assumed that the number of epochs is below 20. Absence of L7-LA proves that methods based on (variations of) gradient decent converges very slowly (maximum number of epochs even above 1000) and are out of range of mobile hand-held devices and are not feasible as CAD applications.

AUC is a good overall measure for how a classifier performs. When the classifier is used however one should pick up the threshold thereby the tradeoff between the sensitivity and specificity. This is called the ROC operating point and its 'manual' optimal selection does depend on the subject of interest. In this study we determined the optimal operating point from probabilistic considerations published in [39]. In Fig. 2 we present those optimal points from trainings with different learning- and topology- setups. First we remark that the above numerical procedure fail for few points (middle field and right upper corner). All the other points are optimal operating points. Strings 'Lx' in Fig. 2 stand for the best-performing back-propagation algorithms and the colors represent the structure of the hidden layers. There are no single winners and we should mind that all the points come from the best-converging networks. It seems that different combinations of learning algorithms and topologies reach upper-left ROC region. Taking into account possible system limitations of the developed applications (e.g. image resolution limits, the number of the weights, memory limits of

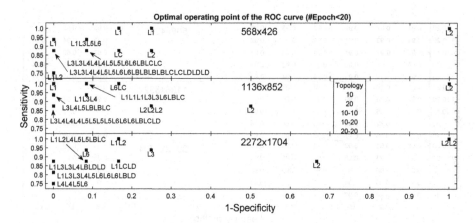

Fig. 2. Optimal points (not components of one ROC) from trainings with different learning functions and different hidden layers setups. Resolutions A, B, C are presented.

the training cycle) one can optimally select a pair ($learning\_alg, hidden\_layers$). L1 (Levenberg-Marquardt) shows extreme robustness for almost all topologies for resolutions A, B and C. L6 (Polak-Ribiére Conjugate Gradient) seems to be the second best for A,B,C - it is less represented but reaches high AUC values in the upper-left corner.

Figure 2 confirms the resolution invariance of the wavelet features for the original clinical data and the descendant downgraded image resolutions. The magnitude of the the operating points is preserved for 1136×852 and 568×426 and, unexpectedly, outperforms the initial 2272×1704 conditions. In the ensemble learning [28] the downgraded resolutions showed still high but slightly worse results. In our experiments with ANN, which is a homogeneous (not complex) model, better sensitivity and specificity is reached for dermoscopy images with smaller resolutions. This is an interesting finding that should be related to how the RBio3.1 wavelet 'analyzes' the skin texture. Our working hypothesis is that the structure of this wavelet base coincides either with the optimal 'frequency' of the human skin or with the texture 'signature' of the melanoma. This problem should be addressed in some future experiments.

References

1. Korotkov, K., Garcia, R.: Computerized analysis of pigmented skin lesions: a review. Artif. Intell. Med. **56**(2), 69 (2012)
2. Masood, A., Al-Jumaily, A.: Computer aided diagnostic support system for skin cancer: a review of techniques and algorithms. Int. J. Biomed. Imaging **2013**(7), 323268 (2013)
3. Oliveira, R.B., Papa, J.P., Pereira, A.S., Tavares, J.: Computational methods for pigmented skin lesion classification in images: review and future trends. Neural Comput. Appl. 1–24 (2016)

4. Skvara, H., Teban, L., Fiebiger, M., Binder, M., Kittler, H.: Limitations of dermoscopy in the recognition of Melanoma. Arch Dermatol. **141**, 155–160 (2005)
5. Stolz, W., Semmelmayer, U., Johow, K., Burgdorf, W.H.C.: Principles of dermatoscopy of pigmented skin lesions. Semin. Cutan. Med. Surg. **22**(1), 9–20 (2003)
6. Johr, R.H.: Dermatoscopy: alternative melanocytic algorithms - the ABCD rule of dermatoscopy, menzies scoring method, and 7-point checklist. Clin. Dermatol. **20**, 240247 (2002)
7. Kittler, H., Pehamberger, H., Wolff, K., Binder, M.: Follow-up of melanocytic skin lesions with digital epiluminescence microscopy: patterns of modifications observed in early melanoma, atypical nevi, and common nevi. J. Am. Acad. Dermatol. **43**(3), 467–476 (2000)
8. Goodson, A.G., Grossman, D.: Strategies for early melanoma detection: approaches to the patient with nevi. J. Am. Acad. Dermatol. **60**(5), 719–735 (2009)
9. Chang, T., Kuo, C.C.J.: Texture analysis and classification with tree-structured wavelet transform. IEEE Trans. Image Process. **2**(4), 429–441 (1993)
10. Mallat, S., Zhong, S.: Characterization of signals from multiscale edges. IEEE Trans. Pattern Anal. Mach. Intell. **14**, 710–732 (1992)
11. Mahmoud, K.A., Al-Jumaily, A., Takruri, M.: The automatic identification od melanoma by wavelet and curvelet analysis: study based on neural network classification. In: 11th International Conference on Hybrid Intelligent Systems 2011, pp. 680–685 (2011)
12. Aswin, R.B., Jaleel, J.A., Salim, S.: Implementation of ANN classifier using MATLAB for skin cancer detection. In: ICMiC13, pp. 87–94 (2013)
13. Ercal, F., Chawla, A., Stoecker, W.V., Lee, H., Moss, R.H.: Neural network diagnosis of malignant melanoma from color images. IEEE Trans. Biomed. Eng. **41**(9), 837–845 (1994)
14. Dreiseitl, S., Ohno-Machado, L., Kittler, H., Vinterbo, S., Billhardt, H., Binder, M.: A comparison of machine learning methods for the diagnosis of pigmented skin lesions. J. Biomed. Inform. **34**, 2836 (2001)
15. Rubegni, P., Burroni, M., Cevenini, G., Perotti, R., Dell'Eva, G., Barbini, P., Fimiani, M., Andreassi, L.: Digital dermoscopy analysis and artificial neural network for the differentiation of clinically atypical pigmented skin lesions: a retrospective study. J. Invest. Dermatol. **119**, 471–474 (2002)
16. Rajab, M.I., Woolfson, M.S., Morgan, S.P.: Application of region-based segmentation and neural network edge detection to skin lesions. Comput. Med. Imaging Graph. **28**, 61–68 (2004)
17. Lau, H.T., Al-Jumaily, A.: automatically early detection of skin cancer: study based on neural network classification. In: International Conference of Soft Computing and Pattern Recognition. IEEE, pp. 375–380 (2009)
18. Vennila, G.S., Suresh, L.P., Shunmuganathan, K.L.: Dermoscopic image segmentation and classification using machine learning algorithms. Am. J. Appl. Sci. **8**(11), 1159 (2012)
19. Jaleel, J.A., Salim, S., Aswin, R.B.: Computer aided detection of skin cancer, circuits. In: International Conference onPower and Computing Technologies (ICCPCT) (2013)
20. Elgamal, M.: Automatic skin cancer images classification. Int. J. Adv. Comput. Sci. Appl. **4**(3), 287–294 (2013)
21. Silva, C.S., Marcal, A.R.S.: Colour-based dermoscopy classification of cutaneous lesions: an alternative approach (2013). doi:10.1080/21681163.2013.803683

22. Achakanalli, S., Sadashivappa, G.: Skin cancer detection and diagnosis using image processing and implementation using neural networks and ABCD parameters (2014)
23. Ruiz, D., Berenguer, V., Soriano, A., Sanchez, B.: A decision support system for the diagnosis of melanoma: a comparative approach. Expert Syst. Appl. **38**, 15217–15223 (2011)
24. Maglogiannis, I., Kosmopoulos, D.: Computational vision systems for the detection of malignant melanoma. Oncol. Rep. **15**(Spec no. 4), 1027–1032 (2006)
25. Doukas, C., Stagkopoulos, P., Maglogiannis, I.: Skin lesions image analysis utilizing smartphones and cloud platforms. Methods Mol. Biol. **1256**, 435–458 (2015)
26. Filho, M.E., Ma, Z., Tavares, J.: A review of the quantification and classification of pigmented skin lesions: from dedicated to hand-held devices. J. Med. Syst. **39**(177), 1–12 (2015)
27. Kassianos, A.P., Emery, J.D., Murchie, P., Walter, F.: Smartphone applications for melanoma detection by 55community, patient and generalist clinician users: A review. Br. J. Dermatol. **172**(6), 1507–1518 (2015)
28. Surówka, G., Ogorzałek, M.: On optimal wavelet bases for classification of melanoma images through ensemble learning. In: Rutkowski, L., Korytkowski, M., Scherer, R., Tadeusiewicz, R., Zadeh, L.A., Zurada, J.M. (eds.) ICAISC 2016. LNCS, vol. 9692, pp. 655–666. Springer, Cham (2016). doi:10.1007/978-3-319-39378-0_56
29. He, H., Garcia, E.A.: Learning from imbalanced data. IEEE Trans. Knowl. Data Eng. **21**, 1263 (2009)
30. Wang, S., Minku, L.L., Yao, X.: Resampling-based ensemble methods for online class imbalance learning. IEEE Trans. Knowl. Data Eng. **26**, 1356 (2014)
31. Mallat, S.G.: A theory for multiresolution signal decomposition: the wavelet representation. IEEE Trans. Pattern Anal. Mach. Intell. **11**(7), 674 (1989)
32. Patwardhan, S.V., Dai, S., Dhawan, A.P.: Multi-spectral image analysis and classification of melanoma using fuzzy membership based partitions. Comput. Med. Imaging Graph. **29**, 287296 (2005)
33. Surówka, G., Merkwirth, C., Żabińska-Płazak, E., Graca, A.: Wavelet based classification of skin lesion images. Bio Alg. Med. Syst. **2**(4), 43–49 (2006)
34. Tang, J., Alelyani, S., Liu, H.: Feature selection for classification: a review. In: Aggarwal, C.C. (ed.) Data Classification: Algorithms and Applications, pp. 37–64. CRC Press, Boca Raton (2014)
35. Maglogiannis, I., Doukas, C.N.: Overview of advanced computer vision systems for skin lesions characterization. IEEE Trans. Inf. Techn. Biomed. **13**(5), 721–733 (2009)
36. Haykin, S.: Neural Networks: A Comprehensive Foundation, 2nd edn. Prentice Hall, Englewood Cliffs (1998). ISBN 0-13-273350-1
37. Hagan, M.T., Demuth, H.B., Beale, M.H., De Jesus, O.: Neural Network Design, 2nd edn., ISBN-10: 0–9717321-1-6, ISBN-13: 978-0-9717321-1-7
38. Battiti, R.: First- and second-order methods for learning: between steepest descent and newton's method. Neural Comput. **4**(2), 141 (1992)
39. Hajian-Tilaki, K.: Receiver operating characteristic (ROC) curve analysis for medical diagnostic test evaluation. Caspian J. Intern. Med. **4**(2), 627–635 (2013)

Application of Stacked Autoencoders to P300 Experimental Data

Lukáš Vařeka$^{(\boxtimes)}$, Tomáš Prokop, Roman Mouček, Pavel Mautner,
and Jan Štěbeták

Department of Computer Science and Engineering, University of West Bohemia,
Univerzitní 8, 306 14 Pilsen, Czech Republic
lvareka@kiv.zcu.cz

Abstract. Deep learning has emerged as a new branch of machine learning in recent years. Some of the related algorithms have been reported to beat state-of-the-art approaches in many applications. The main aim of this paper is to verify one of the deep learning algorithms, specifically a stacked autoencoder, to detect the P300 component. This component, as a specific brain response, is widely used in the systems based on brain-computer interface. A simple brain-computer interface experiment more than 200 school-age participants was performed to obtain large datasets containing the P300 component. After feature extraction the collected data were split into the training and testing sets. State-of-the art BCI classifiers (such as LDA, SVM, or Bayesian LDA) were applied to the data and then compared with the results of stacked autoencoders.

Keywords: Electroencephalography · Event-related potentials · P300 · Brain-computer interface · Stacked autoencoders · Deep learning

1 Introduction

Brain-computer interface (BCI) is a communication system in which users' messages or commands are recognized directly from the brain activity without using muscles [1]. Instead of using normal neuromuscular pathways, users explicitly try to manipulate their brain activity to produce signals that can be used to control devices. Although brain activity can be measured by various methods, electroencephalography (EEG) is commonly used in BCIs because only EEG and related methods can function in most environments, and require relatively simple and inexpensive equipment [2].

To be comfortable to use, BCIs require high reliability and bit rate that, besides other factors, depend on the classification algorithm chosen.

1.1 P300 Brain-Computer Interfaces

EEG-based BCIs rely on different phenomena of the brain, such as motor imagery, the P300 response, or steady-state visually evoked potentials. This paper focuses

© Springer International Publishing AG 2017
L. Rutkowski et al. (Eds.): ICAISC 2017, Part I, LNAI 10245, pp. 187–198, 2017.
DOI: 10.1007/978-3-319-59063-9_17

on the P300-based BCIs that are based on the P300 event-related potential (ERP) component. The P300 component is a positive deflection in the EEG signal occurring from 200 ms to 500 ms after the presentation of rare visual or auditory stimuli and in most subjects, it is a reliable and easy to detect event-related potential. An example of the P300 component is depicted in Fig. 1.

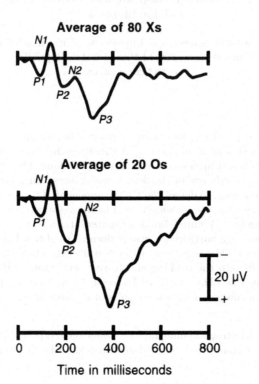

Fig. 1. Comparison of averaged EEG responses to common (non-target) stimuli (Xs) and rare (target) stimuli (Os). There is a clear P300 component following the Os stimuli. Negative is plotted upward [3].

The P300 speller [4] was the first P300-based BCI that allowed the user to type letters he or she concentrated on. The letters were arranged in rows and columns. In the past few years, P300 BCIs have emerged as one of the main BCI categories. P300 BCIs have consistently exhibited several appealing features – they are relatively fast, effective for most users, straightforward, and require practically no training of the user. Recent work has shown that P300 BCIs can be used for a wide range of different functions and can also work with disabled users in home settings [5].

The implementation of P300-based BCI systems is essentially a machine learning task. The P300 waveform that is expected to be most pronounced after target stimuli can be detected using well-chosen preprocessing, feature extraction, and classification algorithms. The objective of preprocessing is to increase signal to noise ratio. Frequencies below 0.5 Hz and above 30 Hz usually do not carry much information useful for the detection of the P300 component [3]. Therefore, bandpass filtering is a common preprocessing method. Since the P300 component is stimulus-locked and the background activity is randomly distributed, the P300 waveform can be extracted using averaging [3]. Various methods have been used for feature extraction, e.g. discrete wavelet transform, independent component analysis, or principal component analysis. The final step of the P300 component processing is the decision about the presence of the P300 component based on classification. Farwell and Donchin used step-wise discriminant analysis (SWDA) followed by peak picking and covariance evaluation [4]. Other methods have also been used for the P300 detection such as, support vector machine (SVM) [6], linear discriminant analysis (LDA) [7], and Bayesian LDA (BLDA) [8]. Although different features and classifiers have been compared [9], the comparison of all different features extraction and classification methods applied to the same dataset has not been performed yet. For comparison purposes, the benchmark P300 speller dataset from the BCI Competition 2003 [10] together with the papers reporting results achieved on this dataset have been used. Several approaches were able to reach 100% accuracy using only 4–8 averaged trials on the BCI Competition 2003 data [11].

1.2 Aims of this Paper

Deep learning classification models have been gaining attention in recent years. Although many different classification models have been used in P300 BCIs, deep learning models have not been widely explored so far regarding this application. Only deep belief networks from deep learning category have once been successfully applied to P300 detection [12]. To the authors' best knowledge, stacked autoencoders have not been used for the P300 detection so far. These models could outperform traditionally used classifiers because they are known to deal well with high-dimensional complex inputs.

The objective of this paper is to evaluate the benefits of using deep learning algorithms for simple P300 BCIs and verify if deep learning models can outperform other state-of-the-art classification methods used for P300 detection. The 'Guess the number' experiment was used as an example of the P300 BCI system. In this experiment, the subject secretly chooses a number between 1 and 9 and the BCI system tries to reveal the number.

This paper is organized as follows. Machine learning techniques used are theoretically described in Sect. 2. Subsequently, the proposed experiment is described. In Sect. 3, the details related to the obtained data and experimental conditions are given. Section 4.1 explains feature extraction and Sect. 4.2 describes the procedure used to train stacked autoencoders and classification models that were used for comparison. The results are given in Sect. 5 and discussed in Sect. 6.

2   Theoretical Background

2.1   Deep Learning Models

Deep learning models have emerged as a new area of machine learning since 2006 [13]. For many complex problems, deep learning models have proven to outperform traditional classification approaches (e.g. SVM) that are affected by the curse of dimensionality [14]. These problems cannot be efficiently solved by using neural networks with many layers (commonly referred to as deep neural networks) trained using backpropagation. The more layers the neural network contains, the lesser the impact of backpropagation on the first layers is. The gradient descent then tends to get stuck in local minima or plateaus which is why no more than two layer networks were used in most practical applications [13,14].

Typically, each layer is trained in a greedy way: once the previous layers are trained, a new layer is trained by encoding the input data provided by the preceding layers. Finally, a supervised fine-tuning stage of the whole network can be performed [14]. Deep networks models generally fall into several categories [14], including deep belief networks, stacked autoencoders, deep kernel machines, and deep convolutional networks.

2.2   Stacked Autoencoders

A single autoencoder (AA) is a neural network (see Fig. 2) that consists of an input layer, encoding layer and decoding layer. The encoding layer encodes the inputs of the network and the decoding layer decodes (reconstructs) the inputs. Consequently, the number of neurons in the decoding layer is equal to the input dimensionality. The goal of a single autoencoder is to compute a code h of an input instance x from which x can be recovered with high accuracy. This can be formalized as follows [14]:

$$f_{dec}(f_{enc}(x)) = f_{dec}(h) = \hat{x} \approx x \qquad (1)$$

with f_{enc} being the function computed by the encoding layer and f_{dec} being the function computed by the decoding layer.

The number of neurons in the encoding layer is lower than the input dimensionality. Therefore, in this layer, the network is forced to remove redundancy from the input by reducing dimensionality. The single autoencoder (being a shallow neural network) can easily be trained using the standard backpropagation algorithm with random weight initialization [15].

Stacking of autoencoders in order to boost performance of deep networks was originally proposed in [16]. A key function of stacked autoencoders (SAE) is unsupervised pre-training, layer by layer, as input is fed through. Once the first layer is pre-trained (neurons $h_1^{(1)}$, $h_2^{(1)}$, .., $h_4^{(1)}$ in Fig. 2), it can be used as an input of the next autoencoder. The final layer can deal with traditional supervised classification and the pre-trained neural network can be fine-tuned using backpropagation. An example of a stacked autoencoder is depicted in Fig. 3 [15].

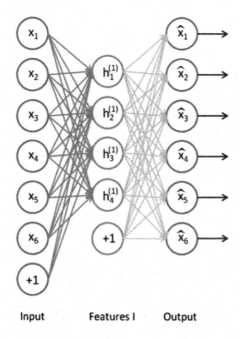

Input   Features I   Output

Fig. 2. Autoencoder. The input layer $(x_1, x_2, .., x_6)$ has the same dimensionality as the output (decoding layer). The encoding layer $(h_1^{(1)}, .., h_4^{(1)})$ has a lower dimensionality and performs the encoding [15].

3 Experimental Design

The 'Guess the number' experiment, based on visual stimulation, was originally developed to demonstrate the benefits of using BCI to public. The participant in the experiment is asked to choose a number between 1 and 9 and concentrate on it (i.e. this number is the target stimulus). Then, the subject is exposed to visual stimuli that include numbers between 1 and 9 randomly appearing on the monitor. During the experiment, both EEG signal and stimuli markers are recorded. Concurrently, experimenters observe average event-related potential (ERP) waveforms for each number and try to guess the number thought. Their guess is finally verified when the participant is asked to reveal the thought number.

3.1 Measurement

The experimental conditions and the procedure of data collection were described in detail in [17]. In this section, the most important points are mentioned.

The experiments were carried out in elementary and secondary schools, mainly located in the Pilsen region, the Czech Republic, between autumn 2014 and spring 2015. The measurements were taken at the time of regular school hours, typically in the morning. Most subjects were school-age children (average

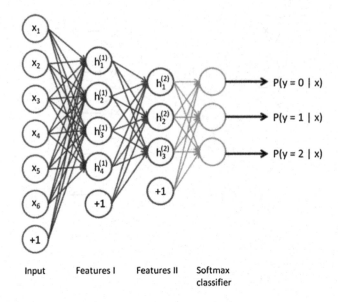

Fig. 3. Stacked autoencoder [15].

age 13.2): 135 males and 104 females. Unfortunately, the environment was usually quite noisy since many children and also many electrical devices were present in the room at the same time. However, in any case there was no movement of people beyond the monitor or in the close proximity of the participant.

The participants were stimulated with numbers flashing on the monitor in random order. The inter-stimulus interval was 1500 ms. The flashing numbers were white on the black background as shown in Fig. 4. The participants were sitting approximately 1 m from the monitor for as long as needed (approximately 10 min on average, stopped when the experimenters were convinced that they were able to guess the number thought).

3.2   Guess the Number Application for On-line and Off-line BCI Classification

A desktop Java application [18] using Swing for its graphical user interface was developed for the analysis of the experiments previously described. The purpose of this application is to enable off-line (experimental data are first collected and analyzed later) and on-line (data are streamed into the application during an experiment) classification. Off-line classification allows users to test preprocessing, feature extraction, and classification algorithms. Subsequently, suitable combinations can be selected for on-line classification.

Fig. 4. Numbers 1–9 were randomly shown on the monitor.

4   P300 Detection

4.1   Preprocessing and Feature Extraction

The following preprocessing steps were applied:

- Referencing. The signal from the root of the nose was used for referencing.
- Channel selection. For further processing, Fz, Cz, and Pz channels were selected. These EEG channels are related to the occurrence of the P300 component [3].
- Epoch extraction. The raw EEG signal was split into fixed-size segments, commonly referred to as epochs or trials. The length of epochs was set to 1000 ms.
- Baseline correction. To compensate for possible shifts in the baseline, the average of 100 ms before each epoch was subtracted from the whole epoch.
- Interval selection. In each epoch, only the most relevant time interval for the detection of the P300 component and its surroundings was chosen. 512 ms intervals starting 175 ms after the beginning of each epoch were used. These parameters were found experimentally [17].
- Discrete wavelet transform (DWT). Discrete wavelet transform was proposed for single trial event-related potential analysis [19]. For each EEG channel, 5-level DWT using Daubechies 8 mother wavelet was performed. Based on the results presented in [19], 16 approximation coefficients of level 5 were used from each EEG channel. Finally, all three results of DWT were concatenated to form a feature vector of dimension 48. This process is illustrated in Fig. 5.
- Vector normalizing. Finally, the feature vectors were normalized to contain only samples between −1 and 1.

Discrete wavelet transform coefficients

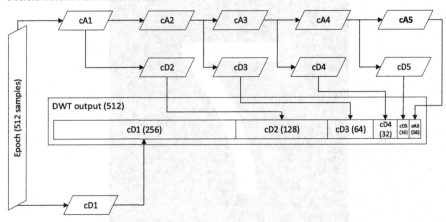

Fig. 5. Discrete wavelet transform coefficients. Input EEG signal has 512 samples. The number of coefficients obtained by DWT is in brackets. 5-level DWT was performed. cA1–cA5 represent approximation coefficients of different levels, cD1–cD5 represent detail coefficients. The feature vector was formed by cA5 coefficients.

Band-pass filtering was not performed because our previous results suggested that it did not improve classification accuracy [17].

4.2   Classification

The preprocessed data were split into training and testing datasets. The training dataset contained data from 13 subjects. These subjects were selected manually based on their P300 response to target stimuli. From each subject, all target trials were extracted. Additionally, the corresponding number of non-target was randomly selected. The number of target stimuli varied for each subject. However, in total, 262 target and 262 non-target randomly shuffled trials were used for training.

Supervised classifiers were used for classification. Several state-of-the-art classifiers were used to allow comparison with stacked autoencoders:

- Linear discriminant analysis (LDA). Fisher's LDA implementation from BCILAB [20] was used. The regularization was set to shrinkage, as recommended in [21].
- Bayesian LDA (BLDA). In [8], BLDA was proposed as a method superior to Fisher's LDA. The authors also attached an implementation that was used in our experiments.
- Support vector machines. The SVM implementation from WEKA Data Mining Software [22] was used. The C-SVC type of SVM with linear kernel function, cost 425, weight 1, and seed 1 was used.
- Multi-layer perceptron (MLP). The configuration was described in detail in [17].

Stacked autoencoders were implemented using Deeplearning4j library [23]. Their parameters were tuned empirically. Finally, the network was configured in the following way:

- the number of iterations was set to 1500,
- the optimization algorithm was set to Stochastic Gradient Descent,
- the learning rate was set to 0.005,
- the Nestorovs updater with momentum 0.9 was used.

The network contained three layers and its architecture was 48 - 24 - 12 - 2. The weights of all layers were initialized using the Xavier method. The first two layers used the Root-mean-square error cross entropy loss function and relu activation. The last layer used the Negative likelihood loss function and softmax activation.

For each classifier, the decision about the number thought was made as follows:

1. The classifier receives a feature vector and its expected classification class (target, non-target).
2. The classifier returns a score, typically in range between 0 and 1. The higher the score is, more likely the epoch belongs to the target number.
3. Scores to each number are summed. At the end of the classification of each number, all scores are averaged. The number with the highest averaged score is the winner.

5   Results

For each tested algorithm maximum and average classification success rate (the percentage of correctly revealed numbers) was computed. The feature extraction method was the same for all tested algorithms to ensure the same classification inputs. Because training algorithms were indeterministic, each of them was run for at least 400 iterations to obtain the average classification success rate. The results of classification available in Table 1 are sorted by the maximum classification success rate.

Table 1. Classification results

Method	Average success rate [%]	Maximum success rate [%]
SAE	74.00%	79.38%
BLDA	73.65%	77.16%
MLP	68.94%	76.7%
LDA	68.77%	75.63%
SVM	65.43%	73.71%
Human Expert [17]	—	64.43%

The highest average (74.00%) and maximum (79.38%) classification success rate was achieved by SAE followed by BLDA with 73.65% average and 77.16% maximum classification success rate. These two methods outperformed other algorithms. It is clearly visible that all algorithms were more successful in detecting the P300 component than the human expert (64.43%) was.

6   Discussion and Future Work

In our previous work, we described the 'Guess the number' experiment and discussed the feature extraction settings [17]. In this paper, the feature extraction method was unchanged and various classification algorithms were tuned and compared. The main attention was payed to deep learning neural network algorithms. These algorithms have rarely been used in electroencephalography and remain to be evaluated.

Based on our results, stacked autoencoders can be recommended for the P300 detection. Only BLDA was able to yield similar results which is consistent with findings of other studies, such as [8]. Apparently, the unsupervised pre-training prior to supervised fine-tuning helped to adapt weights better than the backpropagation alone [17]. This can be seen in Table 1, since the SAE network performed better than the MLP network.

On the other hand, deep belief networks that were also applied to the detection problem had only limited success. It turned out that it was difficult to find the appropriate parameters for them. In the future, more algorithms from the deep learning category can be explored and tuned to see if they can perform well in this machine learning task.

Based on the algorithms tested, the development of the BCI assistance system for disabled people is planned in the near future. The SAE network is easy and fast to train and the results showed that it could handle the single trial P300 classification problem. Therefore SAE could be a good choice for BCI assistance systems based on the P300 component detection.

Acknowledgements. This publication was supported by the UWB grant SGS-2016-018 Data and Software Engineering for Advanced Applications.

References

1. Wolpaw, J., Birbaumer, N., Heetderks, W., McFarland, D., Peckham, P., Schalk, G., Donchin, E., Quatrano, L., Robinson, C., Vaughan, T.: Brain-computer interface technology: a review of the first international meeting. IEEE Trans. Rehabil. Eng. **8**(2), 164–173 (2000)
2. McFarland, D.J., Wolpaw, J.R.: Brain-computer interfaces for communication and control. Commun. ACM **54**(5), 60–66 (2011)
3. Luck, S.: An Introduction to the Event-Related Potential Technique: Cognitive Neuroscience. MIT Press, Cambridge (2005)

4. Farwell, L.A., Donchin, E.: Talking off the top of your head: toward a mental prosthesis utilizing event-related brain potentials. Electroencephalogr. Clin. Neurophysiol. **70**(6), 510–523 (1988)

5. Fazel-Rezai, R., Allison, B.Z., Guger, C., Sellers, E.W., Kleih, S.C., Kübler, A.: P300 brain computer interface: current challenges and emerging trends. Front. Neuroeng. **5**(14), 14 (2012)

6. Thulasidas, M., Guan, C., Wu, J.: Robust classification of EEG signal for brain-computer interface. IEEE Trans. Neural Syst. Rehabil. Eng. **14**(1), 24–29 (2006)

7. Guger, C., Daban, S., Sellers, E., Holzner, C., Krausz, G., Carabalona, R., Gramatica, F., Edlinger, G.: How many people are able to control a P300-based brain-computer interface (BCI)? Neurosci. Lett. **462**(1), 94–98 (2009)

8. Hoffmann, U., Vesin, J.M., Ebrahimi, T., Diserens, K.: An efficient P300-based brain-computer interface for disabled subjects. J. Neurosci. Methods **167**(1), 115–125 (2008)

9. Mirghasemi, H., Fazel-Rezai, R., Shamsollahi, M.B.: Analysis of p300 classifiers in brain computer interface speller. Conf. Proc. IEEE Eng. Med. Biol. Soc. **1**, 6205–6208 (2006)

10. Blankertz, B., Muller, K., Curio, G., Vaughan, T., Schalk, G., Wolpaw, J., Schlogl, A., Neuper, C., Pfurtscheller, G., Hinterberger, T., Schroder, M., Birbaumer, N.: The BCI competition 2003: progress and perspectives in detection and discrimination of EEG single trials. IEEE Trans. Biomed. Eng. **51**(6), 1044–1051 (2004)

11. Cashero, Z.: Comparison of EEG Preprocessing Methods to Improve the Performance of the P300 Speller. Proquest, Umi Dissertation Publishing (2012)

12. Sobhani, A.: P300 classification using deep belief nets. Master's thesis, Colorado State University (2014)

13. Deng, L., Yu, D.: Deep learning: methods and applications. Found. Trends Sig. Process. **7**(34), 197–387 (2013)

14. Arnold, L., Rebecchi, S., Chevallier, S., Paugam-Moisy, H.: An introduction to deep-learning. In: Advances in Computational Intelligence and Machine Learning, ESANN 2011, pp. 477–488, April 2011

15. Ng, A., Ngiam, J., Foo, C.Y., Mai, Y., Suen, C.: UFLDL Tutorial (2010). http://ufldl.stanford.edu/wiki/index.php/UFLDL_Tutorial

16. Bengio, Y., Lamblin, P., Popovici, D., Larochelle, H.: Greedy layer-wise training of deep networks. In: Schölkopf, B., Platt, J., Hoffman, T. (eds.) Advances in Neural Information Processing Systems 19, pp. 153–160. MIT Press, Cambridge (2007)

17. Vareka, L., Prokop, T., Stebetak, J., Moucek, R.: Guess the number - applying a simple brain-computer interface to school-age children. In: Proceedings of the 9th International Joint Conference on Biomedical Engineering Systems and Technologies, vol. 4, BIOSIGNALS, pp. 263–270 (2016)

18. Vareka, L., Prokop, T., Stebetak, J., Moucek, R.: Guess the number (github repository) (2015). https://github.com/NEUROINFORMATICS-GROUP-FAV-KIV-ZCU/guess_the_number

19. Quiroga, R., Garcia, H.: Single-trial event-related potentials with wavelet denoising. Clin. Neurophysiol. **114**(2), 376–390 (2003)

20. Delorme, A., Mullen, T., Kothe, C., Acar, Z.A., Bigdely-Shamlo, N., Vankov, A., Makeig, S.: EEGLAB, SIFT, NFT, BCILAB, and ERICA: new tools for advanced EEG processing. Intell. Neurosci. **2011**, 10:10 (2011)

21. Blankertz, B., Lemm, S., Treder, M.S., Haufe, S., Müller, K.R.: Single-trial analysis and classification of ERP components - a tutorial. NeuroImage **56**(2), 814–825 (2011)

22. Frank, E., Hall, M., Holmes, G., Kirkby, R., Pfahringer, B., Witten, I.H., Trigg, L.: Weka-a machine learning workbench for data mining. In: Maimon, O., Rokach, L. (eds.) Data Mining and Knowledge Discovery Handbook, pp. 1269–1277. Springer, New York (2009)
23. Gibson, A., Nicholson, C., Patterson, J., Warrick, M., Black, A.D., Kokorin, V., Audet, S., Eraly, S.: Deeplearning4j: Distributed, open-source deep learning for Java and Scala on Hadoop and Spark, May 2016

NARX Neural Network for Prediction of Refresh Timeout in PIM–DM Multicast Routing

Nataliia Vladymyrska[1], Michał Wróbel[1], Janusz T. Starczewski[1,2(✉)], and Viktoriia Hnatushenko[3]

[1] Institute of Computational Intelligence,
Częstochowa University of Technology, Częstochowa, Poland
{natalia.vladymyrska,janusz.starczewski}@iisi.pcz.pl
[2] Institute of Information Technology,
Radom Academy of Economics, Radom, Poland
[3] National Metallurgical Academy of Ukraine, Dnipro, Ukraine

Abstract. In this paper, we propose a novel method for optimization of multicast routing. With the use of a NARX neural network, we predict a refresh timeout in PIM–DM algorithm.

Keywords: Neural network prediction · Multicast routing · Refresh timeout

1 Introduction

IP multicasting transmits information streams to a large number of users and achieves a balance between overloading a source and receivers, providing optimal network bandwidth [12]. This has beneficial effect the computer network throughput [15]. All supported multicasting protocol routers and routers with independent multicasting create a several copies of packets only at the points of divergence of the routes to ensure the most efficient data delivery. Most often, this type of communication is used in video conferencing, enterprise networks, distance learning, as well as for the software distribution, stock quotes, news, etc.

IP multicast is an emerging wide area network technology that provides the capability for efficient information dissemination from senders to potentially large sets of receivers. IP multicast distribution trees [11] or pathways are enabled via a combination of the LAN-based Internet Group Management Protocol working in concert with multicast routing protocols. The financial industry is beginning to utilize IP multicast as a delivery mechanism for near real-time information dissemination. However, financial services typically have stringent availability requirements. Therefore, the failure recovery characteristics of IP multicast are of interest to investigate.

2 PIM–DM

PIM–DM is a multicast routing protocol that uses an underlying unicast routing information base to flood multicast datagrams to all multicast routers.

© Springer International Publishing AG 2017
L. Rutkowski et al. (Eds.): ICAISC 2017, Part I, LNAI 10245, pp. 199–205, 2017.
DOI: 10.1007/978-3-319-59063-9_18

Prune messages are used to prevent future messages from propagating to routers without group membership information [1]. PIM–DM is a broadcast and prune protocol and is best suited for networks where most receivers are "densely" populated and bandwidth is plentiful. When a source commences sending UDP traffic with an IP destination group address, the first hop router distributes data through its interfaces except the interface at which data arrive. Subsequent routers do the same during this broadcast phase of the protocol. When the multicast channel reaches a leaf router (one that is LAN connected to potential receivers), the group information maintained by the router is examined and either the multicast data is forwarded onto the LAN or the router prunes back the channel. This restricts the group to only those receivers that have expressed interest to have the channel extended to them. In the standard configuration, the broadcast and prune processes repeat every three minutes. There is also the ability for a new receiver to asynchronously attach to the multicast channel via a grafting mechanism.

3   PIM–DM Protocol Overview

PIM–DM assumes that when a source starts sending, all downstream systems want to receive multicast datagrams. Initially, multicast datagrams are flooded to all areas of the network. PIM–DM uses RPF to prevent looping of multicast datagrams while flooding. If some areas of the network do not have group members, PIM–DM will prune off the forwarding branch by instantiating prune state. Prune state has a finite lifetime. When that lifetime expires, data will again be forwarded down the previously pruned branch. Prune state is associated with an (S,G) pair. When a new member for a group G appears in a pruned area, a router can "graft" toward the source S for the group, thereby turning the pruned branch back into a forwarding branch.

4   PIM–DM Protocol State

A Tree Information Base (TIB) holds the state of all the multicast distribution trees at this router. In this specification, we define PIM–DM mechanisms in terms of the TIB. However, only a very simple implementation would actually implement packet forwarding operations in terms of this state. Most implementations will use this state to build a multicast forwarding table, which would then be updated when the relevant state in the TIB changes.

PIM–DM does not maintain a keepalive timer associated with each (S,G) route. Within PIM–DM, route and state information associated with an (S,G) entry must be maintained as long as any timer associated with that (S,G) entry is active. When no timer associated with an (S,G) entry is active, all information concerning that (S,G) route may be discarded.

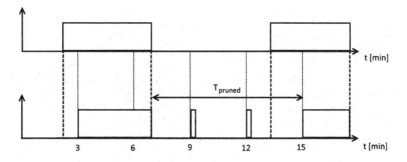

Fig. 1. Classic PIM–DM solution

5 Refresh Timeout

In this section, we present a new approach to make a re-flood time, called a refresh timeout, to be adaptive. Typically, PIM–DM would re-flood all the multicast traffic every 3 min. This value seems to be optimal for low volume multicast, but not higher bandwidth multicast packet streams. In our approach we estimate this time basing on the historical behaviour of the branch.

Empirical research indicate that sometimes transmission is really necessary, nevertheless, quite often a router receives PRUNE signal after a very short period time. The situation is presented in Fig. 1. The upper chart presents an activity of the end–user, while the lower chart presents, when router is sending data.

Our objective is to:

– minimize a number of unnecessary floods,
– minimize a time of waiting for the transmission

To do this, we propose a local adaptive mechanism that takes into account only standard signals transmitted to a router in PIM–DM. Namely, we apply a neural network to estimate a next time T_{pruned}, when the router should likely start its transmission.

In the situation under consideration, the router should re–flood a computer network after half of the standard time. Re–flooding after all this time is risky — transmission request may be earlier, hence the total time may be too long. We denote the value of time to the next re–flood as T_{n+1}.

6 Nonlinear Autoregressive Network for Prediction of Refresh Timeout

Various neural networks have been used for time series prediction [2,14]. A special kind of feedback neural networks called a nonlinear autoregressive network with exogenous inputs (NARX) has been developed to deal with filtering and prediction of time series [10,13]. The NARX model is a recurrent dynamic network (see for other types [3–9]), with feedback connections from arbitrary several

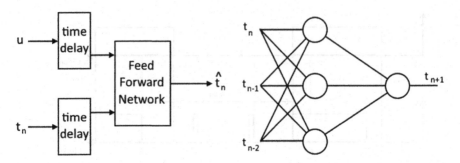

Fig. 2. Series-Parallel architecture of NARX

Fig. 3. Proposed NARX implementation

layers of the network. It is usually used as a nonlinear predictor, to predict the next value of the input signal, as well as a nonlinear filter, in which the target output is a noise-free version of the input signal. In this paper, it is applied in the modeling of nonlinear dynamic systems, which is the multicast router.

The NARX is defined with the following equation:

$$\widehat{t}(k) = f\left(t\left(k-1\right), t\left(k-2\right), \dots, t\left(k-n_t\right), u\left(k-1\right), u\left(k-2\right), \dots, u\left(k-n_u\right)\right) \tag{1}$$

where the next value of the output signal $t(k)$ depends on previous values of the output signal and previous values of an exogenous input signal.

Note that the output of the network is an estimate of the output of a modeled nonlinear dynamic system. In the standard architecture of NARX, the output, which is only an estimate, is fed back to the input of the feedforward neural network. However in our implementation, the actual output is available during the training process of the network. Therefore, we have made use of a series-parallel architecture, in which the true and more accurate output is used instead of the estimated output, as shown in Fig. 2. It is an important feature that we are able to use the standard backpropagation to train the architecture transformed into feedforward.

We consider a local approach in which a router sends datagrams to a single branch of group members. To provide an optimal time between receiving PRUNE signal and restarting transmission to this router we use a neural model within a NARX architecture presented in Fig. 3. The minimal neural network architecture (as we wish to obtain a good generalization ability) includes three neurons in the input layer and one neuron in the output layer (there is no hidden layer). Activation function in each neuron is expressed by $tanh(s)$.

We decided to represent the restart time by a relative value bounded in range $[-1, 1]$, denoted as t. Given the time T expressed in milliseconds, t is calculated as follows:

$$t = \begin{cases} -1 & \text{for } T \le T_{min} \\ 2\frac{T-T_{min}}{T_{max}-T_{min}} - 1 & \text{for } T_{min} < T < T_{max} \\ 1 & \text{for } T \ge T_{max} \end{cases} \tag{2}$$

The predicted time is calculated from the neural network output result as:

$$T = \frac{(t+1)(T_{max} - T_{min})}{2} + T_{min} \tag{3}$$

7 Simulation Results

As an assumption, we set T_{min} as 1 min (60000 ms) and T_{max} as 11 min (660000 ms). The NARX network returned T_{n+1} value predicted basing on previous three values T_n, T_{n_1} and T_{n_2}. We assumed, that times T_{-3}, T_{-2}, T_{-1} and T_0 equaled 3 min (180000 ms). We used the backpropagation algorithm to train the NARX. At beginning, we made 100 iterations with times t_{-1}, t_{-2}, t_{-3} presented on the inputs and t_0 on the output. Each time T_n was computed, we made 10 iterations of backpropagation with t_{n-1}, t_{n-2}, t_{n-3} on the inputs and t_n on the desired output. Next, we activated network with times t_n, t_{n-1}, t_{n-2} and t_{n+1}.

We carried out comparative tests of the proposed NARX–based PIM–DM protocol and the standard PIM–DM formulation on our implementation of a simulator. We simulated local part of the computer network with one sending router, one receiving router and one end–user. The end–user changed its state (wants transmission or not) after random time. Times of activity and inactivity have got uniform distribution with mean value T_a for activity and T_i for inactivity. We have been analysing number of re-floods and the time, when the end user was waiting for transmission. As a results of simulations, we get the results in Tables 1 and 2.

Table 1. Workload and waiting time using the standard PIM–DM protocol

T_a	T_i	Simulation time [h]	Number of floods	Waiting time for transmission [sec]
360	720	10	157	2154
360	720	10	162	2267
360	720	10	163	1927
180	90	10	165	12309
180	90	10	167	9633
180	90	10	155	9210
360	720	24	401	6283
360	720	24	381	6177
360	720	24	389	6173
180	90	24	407	24616
180	90	24	389	23019
180	90	24	394	22489
600	3600	24	440	1747
600	3600	24	417	1308
600	3600	24	441	2034

Table 2. Workload and waiting time using the NARX-based PIM–DM protocol

T_a	T_i	Simulation time [h]	Number of floods	Waiting time for transmission [sec]
360	720	10	64	5454
360	720	10	64	6709
360	720	10	63	5783
180	90	10	195	5634
180	90	10	217	5652
180	90	10	183	6230
360	720	24	154	17551
360	720	24	168	16135
360	720	24	178	16491
180	90	24	469	13080
180	90	24	505	12513
180	90	24	489	12429
600	3600	24	155	4282
600	3600	24	143	3712
600	3600	24	150	3572

8   Conclusions

In this paper, we have presented a preliminary research on the prediction of refresh timeouts. We have made the use of the NARX neural network simulated on our dedicated software for the PIM–DM protocol. We have simulated and tested the process of transferring data comparing our NARX-based version of the PIM–DM algorithm with the standard version of PIM–DM. As a result, we have obtained better abilities of the proposed network to adapt according an end-user behaviour.

If the user rarely is turned on, the NARX version of the multicasting protocol will be flooding rarely, so it will reduce the transmission network workload, as fewer superfluous floods will occur. However this will increase the waiting time for the transfer. If the end-users are turned on quite often, the proposed modification of the multicasting protocol will frequently flood, thus the workload will be significantly higher. Nevertheless, the user will wait for the transfer shorter. We have obtained a good compromise between the load on the computer network and the convenience of the end-user.

As we observe the simulation details, there are still some disadvantages of the predicted time, hence in our future work, we will try to decrease the workload of the network not worsening the comfort of the end-user by analyzing the impact of other possible inputs for the NARX model, especially, making use of control input $u(k)$ in the series–parallel NARX architecture presented in Fig. 2.

References

1. Adams, A., Nicholas, J., Siadak, W.: RFC 3973 protocol independent multicast – dense mode (PIM-DM): protocol specification (revised). Technical report (2005)
2. Bas, E.: The training of multiplicative neuron model based artificial neural networks with differential evolution algorithm for forecasting. J. Artif. Intell. Soft Comput. Res. **6**(1), 5–11 (2016)
3. Bilski, J., Smoląg, J.: Parallel realisation of the recurrent RTRN neural network learning. In: Rutkowski, L., Tadeusiewicz, R., Zadeh, L.A., Zurada, J.M. (eds.) ICAISC 2008. LNCS, vol. 5097, pp. 11–16. Springer, Heidelberg (2008). doi:10. 1007/978-3-540-69731-2_2
4. Bilski, J., Smoląg, J.: Parallel realisation of the recurrent Elman neural network learning. In: Rutkowski, L., Scherer, R., Tadeusiewicz, R., Zadeh, L.A., Zurada, J.M. (eds.) ICAISC 2010. LNCS, vol. 6114, pp. 19–25. Springer, Heidelberg (2010). doi:10.1007/978-3-642-13232-2_3
5. Bilski, J., Smoląg, J.: Parallel realisation of the recurrent multi layer perceptron learning. In: Rutkowski, L., Korytkowski, M., Scherer, R., Tadeusiewicz, R., Zadeh, L.A., Zurada, J.M. (eds.) ICAISC 2012. LNCS, vol. 7267, pp. 12–20. Springer, Heidelberg (2012). doi:10.1007/978-3-642-29347-4_2
6. Bilski, J., Smoląg, J.: Parallel approach to learning of the recurrent Jordan neural network. In: Rutkowski, L., Korytkowski, M., Scherer, R., Tadeusiewicz, R., Zadeh, L.A., Zurada, J.M. (eds.) ICAISC 2013. LNCS, vol. 7894, pp. 32–40. Springer, Heidelberg (2013). doi:10.1007/978-3-642-38658-9_3
7. Cierniak, R.: A new approach to image reconstruction from projections using a recurrent neural network. Int. J. Appl. Math. Comput. Sci. **18**(2), 147–157 (2008)
8. Cierniak, R.: A novel approach to image reconstruction problem from fan-beam projections using recurrent neural network. In: Rutkowski, L., Tadeusiewicz, R., Zadeh, L.A., Zurada, J.M. (eds.) ICAISC 2008. LNCS, vol. 5097, pp. 752–761. Springer, Heidelberg (2008). doi:10.1007/978-3-540-69731-2_72
9. Cierniak, R.: A statistical appraoch to image reconstruction from projections problem using recurrent neural network. In: Diamantaras, K., Duch, W., Iliadis, L.S. (eds.) ICANN 2010. LNCS, vol. 6353, pp. 138–141. Springer, Heidelberg (2010). doi:10.1007/978-3-642-15822-3_17
10. Diaconescu, E.: The use of NARX neural networks to predict chaotic time series. WSEAS Trans. Comp. Res. **3**(3), 182–191 (2008)
11. Dijkstra, E.W.: A note on two problems in connexion with graphs. Numer. Math. **1**(1), 269–271 (1959)
12. Harte, L.: Introduction to Data Multicasting, IP Multicast Streaming for Audio and Video Media Distribution. Althos, Slough (2008)
13. Lin, T., Horne, B.G., Tino, P., Giles, C.L.: Learning long-term dependencies in NARX recurrent neural networks. IEEE Trans. Neural Networks **7**(6), 1329–1338 (1996)
14. Nobukawa, S., Nishimura, H., Yamanishi, T., Liu, J.Q.: Chaotic states induced by resetting process in Izhikevich neuron model. J. Artif. Intell. Soft Comput. Res. **5**(2), 109–119 (2015)
15. Williamson, B.: Developing IP Multicast Networks. Cisco Press, Indianapolis (2000)

Evolving Node Transfer Functions in Deep Neural Networks for Pattern Recognition

Dmytro Vodianyk$^{(\boxtimes)}$ and Przemysław Rokita

Institute of Computer Science, Warsaw University of Technology,
Nowowiejska 15/19, 00-665 Warsaw, Poland
vodianyk@gmail.com, pro@ii.pw.edu.pl

Abstract. Theoretical results suggest that in order to learn complicated functions that can represent high-level features in the computer vision field, one may need to use deep architectures. The popular choice among scientists and engineers for modeling deep architectures are feed-forward Deep Artificial Neural Networks. One of the latest research areas in this field is the evolution of Artificial Neural Networks: NeuroEvolution. This paper explores the effect of evolving a Node Transfer Function and its parameters, along with the evolution of connection weights and an architecture in Deep Neural Networks for Pattern Recognition problems. The results strongly indicate the importance of evolving Node Transfer Functions for shortening the time of training Deep Artificial Neural Networks using NeuroEvolution.

1 Introduction

Deep systems are believed to play an important role in information processing of intelligent agents. A popular hypothesis which supports this belief is that deep systems can be exponentially more efficient at representing complicated functions than their shallow counterparts [1] and for certain problems and network architectures it is proven [2]. These systems include learning methods for a wide array of deep architectures, including neural networks with many hidden layers [3]. Much attention has recently been devoted to deep neural networks, because of their theoretical appeal, inspiration from biology, and because of their empirical success in the computer vision field [4,5].

When using a backpropagation algorithm for training a feed-forward deep neural network for pattern recognition problems, one may observe that the gradient tends to get smaller as it moves backwards through the hidden layers. This means that neurons in the earlier layers learn much more slowly than neurons in the later layers. There are fundamental reasons why this happens in many real-world neural networks, and this phenomenon is known as the vanishing gradient problem [6].

Much ongoing research aims to better understand challenges that can occur when training deep networks. In 2010 Glorot and Bengio [7] found evidence that using a sigmoid transfer function can potentially cause problems in training deep neural networks. In another paper [8] Sutskever and others studied an impact on

© Springer International Publishing AG 2017
L. Rutkowski et al. (Eds.): ICAISC 2017, Part I, LNAI 10245, pp. 206–215, 2017.
DOI: 10.1007/978-3-319-59063-9_19

deep learning of the random weight initialization and the momentum schedule in the momentum-based stochastic gradient descent.

The purpose of this research is to explore the effect of evolving a Node Transfer Function and its parameters, along with the evolution of connection weights and an architecture in Deep Neural Networks for pattern recognition problems. The paper concludes that for certain pattern recognition problems the choice of a node transfer function for each neuron is much more important than the choice of connection weights in deep neural networks. It provides a way to shorten the time of training Deep Artificial Neural Networks using the NeuroEvolutionary approach.

2   Evolution of Deep Neural Networks

Evolutionary Artificial Neural Networks (EANNs) refer to a special class of ANNs in which evolution is another fundamental form of adaptation in addition to learning [9]. One of the important features of EANNs is their adaptability to a dynamic environment. In other words, EANNs can adapt to an environment as well as changes in the environment. Evolution and learning of EANNs make adaptation to dynamic environments much more effective [10].

Evolutionary algorithms (EA) have been successfully applied for training deep neural networks [11]. Several NeuroEvolutionary systems have been successfully developed to solve various challenging tasks with remarkably better performance than traditional learning techniques [12].

The most common way to train an EANN is to evolve connection weights. It is possible to evolve a topology along with connection weights. Topology evolving methods include: GNARL [13], NEAT [14] and CGPANN [15]. EAs can be used to optimize a Node Transfer Function and its parameters of each neuron within a heterogeneous ANN. Indeed, a transfer function has been shown to be an important part of a neural network with one hidden layer [16]. Heterogeneous shallow EANNs perform much better on MNIST data when the connection weights and the transfer functions evolve simultaneously along with their parameters for each node [17]. It also was indicated that further research is required to investigate the impact of evolving a NTF along with connection weights and architecture of EANNs [17]. In this research we are taking an additional step forward – exploring the effect of evolving a NTF and its parameters, along with the evolution of connection weights and an architecture in Deep Neural Networks for Pattern Recognition problems.

The evolutionary approach to an ANN training process consists of two major phases. The first phase is to decide on the representation of connection weights and an architecture: in a form of binary strings or not. In the experiments performed in this paper, the connection weights are represented as real-number matrices. The second phase is to develop an evolutionary process, in which search operators, such as crossover and mutation, have to be defined in conjunction with the representation scheme. In this paper we use crossover and mutation operators for evolving connection weights, transfer functions and an architecture of a deep neural network. In the scope of this paper we developed an evolutionary deep system to perform tests of the proposed method on different pattern recognition problems. The evolutionary cycle of the system is illustrated on Fig. 1, where:

Decode – decoding of each individual (genotype) in the current generation into a set of connection weights, transfer functions and an architecture, constructing a corresponding Deep ANNs.

Evaluate – evaluating effectiveness of each deep neural network by computing its total mean square error between actual and target outputs. The fitness of an individual is determined by the error. The higher the error, the lower the fitness.

Select – parents selection based on their fitness for the further reproduction.

Generate – applying search operators, such as crossover and mutation for the connection weights, an architecture and transfer functions to selected parents in order to generate offspring, which form the next generation.

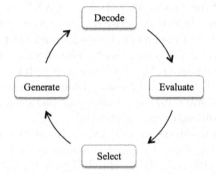

Fig. 1. An evolutionary cycle of the system.

3   Pattern Recognition Benchmarks

In order to test the proposed system and make sure that it performs better on pattern recognition problems with different training and test datasets sizes, three benchmarks were employed.

3.1   Brodatz Textures

The Brodatz Textures [18] are scans of a set of glossy black and white prints of the corresponding textures in the Brodatz book. While these prints are pictures of the same textures as in the book, in most cases they are not the same image as the one in the book. An example of the texture, which represents "Grass", can be seen on Fig. 2.

Since this is a small dataset, it was decided to use only training data for the performance validation of the deep evolutionary system. The first 26 images from the dataset were used for the experiment. Original images were resized to 30 by 30 pixels.

Fig. 2. Example of the Brodatz Texture.

3.2 Handwritten Digits

For this benchmark problem MNIST [19] data set was used. It contains tens of thousands of scanned images of handwritten digits, together with their correct classifications. MNIST's name comes from the fact that it is a modified subset of two data sets collected by NIST, the United States' National Institute of Standards and Technology. Figure 3 shows a few images from the data set.

The dataset was divided into two parts: the first part contains 1,000 handwritten digits to be used as a training data and the second part contains 2,000 images to be used as a test data. These images are 28 by 28 pixels in size.

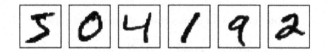

Fig. 3. Examples of images from MNIST.

3.3 COIL-100

The COIL-100 database [20] contains a set of 7,200 colored images of 100 objects on a black background. Each object in the database is represented by 72 images. Figure 4 shows an example of the image from the original dataset.

Images were resized to 20 by 20 pixels and then grayscaled using the following formula for calculating luminance [21]:

$$E_Y = 0.299 * E_R + 0.587 * E_G + 0.114 * E_B,$$

where:

E_Y – luminance of a pixel;
E_R, E_G, E_B – are red, green and blue components of a pixel.

For the training dataset it was chosen to use 40 images for each object and the other 32 images were used for the test dataset. In total 4,000 images were used for the training dataset and 3,200 for the test dataset.

Fig. 4. A random image from COIL-100 database.

4   Experimental Setting

4.1   Initial Architecture of the Deep Neural Network

The architecture of an artificial neural network in the experiment is not constant and can change during the evolution process. However, we still have to define an initial architecture of the neural network. It was decided to initialize an architecture of the network with one hidden layer only and see if it evolves into a deep neural network. Table 1 describes a shape of the networks used at the initial stage for each benchmark.

Table 1. Initial architectures of neural networks.

Benchmark problem	ANN shape
Brodatz Textures	$900 \rightarrow 60 \rightarrow 13$
Handwritten digits	$784 \rightarrow 60 \rightarrow 10$
COIL-100	$400 \rightarrow 60 \rightarrow 100$

It is important to notice that the evolved networks are truly deep: the number of hidden layers was varying between 4 and 5. For example, one of the evolutionary processes for COIL-100 benchmark produced a deep neural network with 4 hidden layers: $400 \rightarrow 64 \rightarrow 63 \rightarrow 69 \rightarrow 85 \rightarrow 100$.

4.2   Node Transfer Functions

The list of transfer functions was used to evolve a DNN. Table 2 shows which functions were used.

The default values of μ and σ for Unipolar Sigmoid, Sigmoid Prime, Hyperbolic and Bipolar functions are: $\mu = 0.0$, $\sigma = 1.0$ The default values of μ and σ for Gaussian function are: $\mu = 4.5$, $\sigma = 4.5$.

Table 2. List of node transfer functions.

Function name	Equation
Step	$g(x) = \begin{cases} 1, & \text{if } x \geq 0 \\ 0, & \text{if } x < 0 \end{cases}$
Unipolar Sigmoid	$g(x) = \dfrac{1}{1+e^{-\frac{x-\mu}{\sigma}}}$
Gaussian	$g(x) = e^{-\frac{(x-\mu)^2}{2\sigma^2}}$
Sigmoid Prime	$g(x) = \dfrac{e^{-\frac{x-\mu}{\sigma}}}{(1+e^{-\frac{x-\mu}{\sigma}})^2}$
Hyperbolic Sigmoid	$g(x) = \dfrac{e^{\frac{x-\mu}{\sigma}}-e^{-\frac{x-\mu}{\sigma}}}{e^{\frac{x-\mu}{\sigma}}+e^{-\frac{x-\mu}{\sigma}}}$
Bipolar Sigmoid	$g(x) = \dfrac{1-e^{-\frac{x-\mu}{\sigma}}}{e^{\frac{x-\mu}{\sigma}}+e^{-\frac{x-\mu}{\sigma}}}$

4.3 Representation of the Deep Evolutionary Neural Network

We are going to use a NeuroEvolutionary approach to train a deep neural network in our system. Therefore, we have to decide how the network should be represented and which search operators should be used for the evolutionary process.

A chromosome contains:

1. An array of matrices with connection weights which are represented as real numbers.
2. A matrix of NTFs for hidden layers and the output layer of the deep neural network.
3. A matrix with parameters which are represented as objects for node transfer functions.

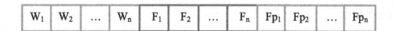

Fig. 5. Graphical representation of a chromosome.

Figure 5 shows a graphical representation of a chromosome in a DNN, where: n is a number of layers, $W_1, W_2, \ldots, W_n$ are matrices of connection weights between input-hidden, hidden-hidden and hidden-output nodes; $F_1, F_2, \ldots, F_n$ are arrays of transfer functions associated with each layer; $Fp_1, Fp_2, \ldots, Fp_n$ are arrays of parameters for each node transfer function.

4.4 Search Operators

In the experiment we decided to use both *crossover* and *mutation* search operators to evolve a deep neural network. A population size in one generation is 100 chromosomes. The population is generated with a random number of hidden layers, random connection weights and transfer functions.

We applied a randomized version of the crossover operator. The system takes two neighboring chromosomes and makes a copy of the first one, which forms a child chromosome. Then it iterates through all layers in the second chromosome and chooses at random a connection weight and a node transfer function to be copied into the child chromosome. The number of randomly chosen connection weights and transfer functions equals to $\frac{n}{2}$, where n is the number of neurons in the layer of the current iteration.

The mutation operator applies for both current chromosomes and for chromosomes generated as the result of the crossover procedure. A probability of the mutation in the connection weights and node transfer functions is 5%.

For connection weights, a mutation operator creates a random index for each layer and changes a connection weight value between $w_k - r$ and $w_k + r$, where w_k is the current value of connection weight and r is a randomly generated values between 0 and 1.

For node transfer functions, a mutation operator also creates a random index for each layer and replaces a function located at that position with a random transfer function from the predefined list. For evolving parameters in node transfer functions, the mutation operator changes a value of μ between -5.0 and 5.0. A value of σ changes between -5.0 and 5.0 for Gaussian function and between 1.0 and 5.0 for Sigmoid, Sigmoid Prime, Hyperbolic and Bipolar functions.

The generated population contains deep neural networks of different depth ranging from 1 to 5 hidden layers and it is not a subject to change during the evolutionary process. For an architecture, a mutation operator at first chooses a layer of the neural network, then it randomly chooses which operator to apply: *add* or *remove*. *Add* operator adds a new neuron to the previously chosen layer. *Remove* operators removes the last node from the layer. A probability of mutating an architecture of a deep neural network is 1%.

5   Results

For each benchmark problem 20 experiments were performed: 10 experiments for evolving a deep neural network with the sigmoid transfer function in each neuron and 10 experiments where node transfer functions were evolved for each neuron. Connection weights and architecture were simultaneously evolved during the NeuroEvolutionary process in all experiments. Each experiment contains 100 iterations.

The result of each experiment is a number, which represents the root-mean-square error (RMSE) associated with running a network on the test data set. RMSE is defined by means of the following formula:

$$C = \sqrt{\frac{\sum_{i=0}^{n}(y_i - t_i)^2}{n}}, \tag{1}$$

where:

n – a number of samples in the test data set;
C – RMSE;
y_i – the expected response of a neural network for the i_{th} sample from the test data set;
t_i – the actual response of a neural network for the i_{th} sample from the test data set.

Tables 3, 4 and 5 show results for each benchmark problem, where:

N – experiment number;
C_1 – RMSE when evolving connection weights and an architecture in a deep neural network;
C_2 – RMSE when evolving simultaneously connection weights and an architecture along with node transfer functions and its parameters in a deep neural network.

Table 6 shows an average RMSE associated with each evolutionary method.

Table 3. Results for Brodatz Textures.

N	1	2	3	4	5	6	7	8	9	10
C_1	0.69	0.68	0.68	0.7	0.68	0.71	0.69	0.69	0.69	0.67
C_2	0.63	0.66	0.65	0.65	0.6	0.66	0.64	0.63	0.64	0.63

Table 4. Results for handwritten digits.

N	1	2	3	4	5	6	7	8	9	10
C_1	0.66	0.68	0.68	0.67	0.68	0.71	0.66	0.67	0.69	0.67
C_2	0.65	0.66	0.66	0.65	0.65	0.64	0.66	0.65	0.65	0.65

Table 5. Results for COIL-100.

N	1	2	3	4	5	6	7	8	9	10
C_1	3.24	2.97	3.17	3.34	3.22	3.53	3.47	3.37	3.16	2.96
C_2	1.44	1.08	1.36	0.99	1.6	1.01	0.89	0.99	1.17	1.24

Table 6. Average RMSE for each evolutionary method and benchmark problem.

Cost\Benchmark	Brodatz Textures	Digits Recognition	COIL-100
$\langle C_1 \rangle$	0.688	0.677	3.243
$\langle C_2 \rangle$	0.639	0.652	1.177

6   Conclusions

In this paper we explored the effect of evolving a Node Transfer Function and its parameters, along with the evolution of connection weights and architecture in Deep Neural Networks for Pattern Recognition problems.

The results demonstrate that evolving both node transfer functions and the parameters for each node increases the performance of a deep neural network in pattern recognition problems. The performance boost is observed for all benchmark problems described in this paper. A **7%** boost was achieved for Brodatz Textures pattern recognition problem, **4%** for Digits Recognition and a dramatic **64%** boost for COIL-100. The improvement in performance for COIL-100 demonstrates that evolving a node transfer function can be used for an initial fine-tuning of a global minimum, which is an important discovery, because evolutionary algorithms are considered to be inefficient for this task.

This suggests the method of evolving node transfer functions and its parameters along with connection weights and architecture in deep neural networks should be considered to improve performance results for pattern recognition problems and should be included in research software for deep learning.

Further research can be conducted to extend a list of node transfer functions used for the described evolutionary process in deep neural networks.

References

1. Bengio, Y.: Learning deep architectures for AI. Found. Trends Mach. Learn. **2**, 1–127 (2009). Now Publishers
2. Pascanu, R., Montufar, G., Bengio, Y.: On the number of response regions of deep feedforward networks with piecewise linear activations. In: NIPS 2014, pp. 2924–2932 (2015)
3. Vincent, P., Larochelle, H., Bengio, Y., Manzagol, P.-A.: Extracting and composing robust features with denoising autoencoders. In: ICML, pp. 1096–1103 (2008)
4. Ranzato, M., Poultney, C., Chopra, S., LeCun, Y.: Efficient learning of sparse representations with an energy-based model. In: NIPS (2007)
5. Larochelle, H., Erhan, D., Courville, A., Bergstra, J., Bengio, Y.: An empirical evaluation of deep architectures on problems with many factors of variation. In: ICML, pp. 473–480 (2007)
6. Hochreiter, S., Bengio, Y., Franconi, P., Schmidhuber, J.: Gradient Flow in Recurrent Nets: The Difficulty of Learning Long-Term Dependencies. IEE Press, New York (2001)
7. Glorot, X., Bengio, Y.: Understanding the difficulty of training deep feedforward neural networks. In: AISTATS, pp. 249–256 (2010)
8. Sutskever, I., Martens, J., Dahl, G., Hilton, G.: On the importance of initialization and momentum in deep learning. In: ICML (3), vol. 28, pp. 1139–1147 (2013)
9. Kent, A., Williams, J.G. (eds.): Evolutionary Artificial Neural Networks. Encyclopedia of Computer Science and Technology, vol. 33, pp. 137–170. Marcel Dekker, New York (1995)
10. Yao, X.: Evolving artificial neural networks. Proc. IEEE **87**, 1423–1447 (2002)
11. David, O.E., Greental, I.: Genetic algorithms for evolving deep neural networks. In: GECCO, pp. 1451–1452 (2014)

12. Tirumala, S.S.: Implementation of evolutionary algorithms for deep architectures. In: AIC (2014)
13. Angeline, P.J., Saunders, G.M., Pollack, J.B.: An evolutionary algorithm that constructs recurrent neural networks. Neural Netw. **5**, 54–65 (1994)
14. Stanley, K.O., Miikkulainen, R.: Evolving neural networks through augmenting topologies. Evol. Comput. **10**, 99–127 (2002)
15. Mahsal, K.M., Masood, A.A., Khan, M., Miller, J.F.: Fastlearning neural networks using Cartesian genetic programming. Neurocomputing **121**, 274–289 (2013)
16. James, A.T., Miller, J.F.: NeuroEvolution: The Importance of Transfer Function Evolution (2013)
17. Vodianyk, D., Rokita, P.: Evolving node transfer functions in artificial neural networks for handwritten digits recognition. In: Chmielewski, L.J., Datta, A., Kozera, R., Wojciechowski, K. (eds.) ICCVG 2016. LNCS, vol. 9972, pp. 604–613. Springer, Cham (2016). doi:10.1007/978-3-319-46418-3_54
18. USC University of Southern California: Signal, Image Processing Institute, Ming Hsieh Department of Electrical Engineering. Textures, vol. 1. http://sipi.usc.edu/database/?volume=textures
19. The MNIST database of handwritten digits. http://yann.lecun.com/exdb/mnist/
20. Nene, S.A., Nayar, S.K., Murase, H.: Columbia Object Image Library (COIL-100). Technical report CUCS-006-96 (1996). http://www.cs.columbia.edu/CAVE/software/softlib/coil-100.php
21. Recommendation ITU-R BT.601-7 (2011)

A Neural Network Circuit Development via Software-Based Learning and Circuit-Based Fine Tuning

Changju Yang[1], Shyam Prasad Adhikari[1], Michal Strzelecki[2], and Hyongsuk Kim[1(✉)]

[1] Division of Electronics and Information Engineering,
Chonbuk National University, Jeonju 561-756, Korea
hskim@jbnu.ac.kr
[2] Institute of Electronics, Technical University of Lodz,
Wolczanska 211/215, 90-924 Lodz, Poland

Abstract. A development method of neural network with software-based learning and circuit-based fine tuning is proposed. The backpropagation is known as one of the most efficient learning algorithms. A weakness is that the hardware implementation is extremely difficult. The RWC algorithm which is very easy to implement its hardware circuits takes too many iterations for learning. In the proposed approach, the main learning is performed with a software version of the BP algorithm and then, learned weights are transplanted on a hardware version of a neural circuit. At the time of the weight transplantation, significant amount of output error would occur due to the characteristic difference between the software and the hardware. In the proposed method, such error is reduced via a complementary learning of RWC algorithm which is implemented in a simple hardware.

Keywords: Chip-in-the-loop · Backpropagation · RWC

1 Introduction

Backpropagation (BP) algorithm is regarded as the most powerful learning algorithm of neural networks [1, 2]. However, its algorithm is involved with huge amount of multiplications, sums and nonlinear functions. Also, the fact that measurement of weights and states of all nodes are required at every iteration is a very big burden. The difficulties are escalated when imperfections and mismatches are involved in the fabrication of circuit components.

Some researchers avoided the implementation difficulty of the backpropagation with the help of software called chip-in-the-loop learning algorithm (CIL) [3, 4], where complicated arithmetic required for the backpropagation algorithm is performed in software by a host computer and updating values are downloaded on the hardware version of neural networks at every iteration. Though the implementation burden of the complicated circuitry of the BP algorithm is reduced in this approach, the communication load for reading out the states and weights of neural network circuit at every iteration is very heavy.

© Springer International Publishing AG 2017
L. Rutkowski et al. (Eds.): ICAISC 2017, Part I, LNAI 10245, pp. 216–228, 2017.
DOI: 10.1007/978-3-319-59063-9_20

A little different approach from above is proposed also by [5], where learning is completed in software. After that, learned parameters (weights) are downloaded and programmed on the hardware version of neural network. However, there is no consideration for the error caused from the difference between the software version and its hardware version of neural networks.

Other group of researchers proposed a different kind of learning rule which is easier for hardware implementation [6–8]. Instead of using gradient descent directly, these algorithms utilize an approximation of the gradient, which is much easier for hardware implementation. Random Weight Change (RWC) algorithm [9] is one of the representative learning algorithms belonging to this category. Weight of each synapse is changed randomly by a fixed amount at each iteration. Only the weight changes with which error is reduced are taken for updating the weights. Therefore, learning procedure is simple and easy to be implemented with off-the-shelf circuit components [10, 11]. However, system implementation study had not been performed fully due to the lack of devices for neural synapses at that time.

The proposed algorithm is a hybrid learning of software-based backpropagation algorithm and circuit-based RWC learning. The software-based backpropagation is performed on the host computer firstly. Then, learned parameters (weights) are transplanted to the physical neural circuit via programming. Error created due to difference between the software and hardware versions of the neural network is eliminated via a complementary learning with the simple circuit of RWC algorithm. It is a practically useful method by taking only the advantages of these two approaches.

2 Memristor-Based Neural Network

Two major functions of biological neural synapses are analog multiplication and information storage. Building an analog multiplier artificially requires more than 10 transistors, which is a heavy burden for the implementation of artificial synapses per node. Though analog multiplication can be achieved with a single resistor via Ohms' law, namely, $v = i \times R$, resistor cannot be utilized for the artificial synapse since it is not programmable. Recently, a resistor-like but programmable element called Memristor has been fabricated successfully and opens the horizon in this field. One weakness of the memristor to be an artificial synapse is lack of negative value expression. Memristor bridge synapse is developed to overcome such weakness of the memristor [12]. It is composed of 4 memristors which can provide a signed weighting and regarded as a promising architecture for implementing synaptic weights in artificial neural networks.

2.1 The Memristor Bridge Synapse

Memristance (resistance of memristor) variation is nonlinear function of the input voltage [12]. When two identical memristors are connected in opposite polarity, total memristance becomes constant dramatically due to their complementary actions. This connection is called back-to-back connection or anti-serial connection. There are two types of back-to-back memristor (anti-serial memristor) pairs depending upon the directions of polarities as shown in Fig. 1. When these two memristor pairs are connected in parallel, a bridge type synapse is built as shown in Fig. 2.

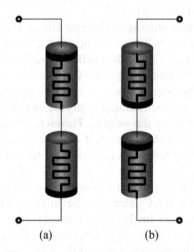

                    (a)   (b)

Fig. 1. Two types of anti-serial memristor pairs. Total resistances of both cases are constant while voltage variations at the middle points of both cases are different.

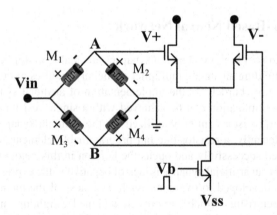

Fig. 2. Memristor bridge synaptic circuit. Weighting operation is performed by the memristor bridge circuit and voltage-to-current conversion is performed by the differential amplifier.

The memristor-based bridge circuit, shown in Fig. 2, can be used as a synapse in the proposed NN architecture. When a positive or a negative pulse V_{in} is applied at the input terminal, the memristance of each memristor is altered depending on its polarity [12].

By using the voltage divider formula, the output voltage between nodes A and B is given as,

$$V_{out} = V_A - V_B = \left(\frac{M_2}{M_1 + M_2} - \frac{M_4}{M_3 + M_4} \right) V_{in} \tag{1}$$

Equation (1) can be rewritten as a relationship between a synaptic weight ψ and a synaptic input signal V_{in} as follows:

$$V_{out} = \psi \times V_{in} \tag{2}$$

where,

$$\psi = \left(\frac{M_2}{M_1 + M_2} - \frac{M_4}{M_3 + M_4} \right) \tag{3}$$

where the range of ψ is $[-1.0, +1.0]$. This voltage is converted to corresponding current with the transconductance parameter g_m. The currents at the positive and the negative output terminals of the differential amplifier associated with the kth synapse are

$$\left. \begin{aligned} i_k^+ &= -\frac{1}{2} g_m \psi^k V_{in}^k \\ i_k^- &= \frac{1}{2} g_m \psi^k V_{in}^k \end{aligned} \right\} \tag{4}$$

where i_k^+ and i_k^- are the currents at the positive and the negative terminals, respectively.

2.2 Memristor-Based Neural Networks

Figure 3(a) shows a typical neural network where each neuron is composed of multiple synapses and one activation unit. The structure of a bigger neural network is simply the repeated connection of such synapses and neurons. The schematic of a memristor bridge synapse-based neuron in Fig. 3(a) is shown in Fig. 3(b). In Fig. 3(b), voltage inputs weighted by memristor bridge synapses are converted to currents by differential amplifiers.

In the proposed circuit, all positive terminals of the input synapses are connected together, as are the negative terminals and the sum of each signed currents is computed separately. The sum of each signed currents are

$$\left. \begin{aligned} i_{SUM}^+ &= -\frac{1}{2} \sum_k g_m \psi^K V_{in}^K \\ i_{SUM}^- &= \frac{1}{2} \sum_k g_m \psi^K V_{in}^K \end{aligned} \right\} \tag{5}$$

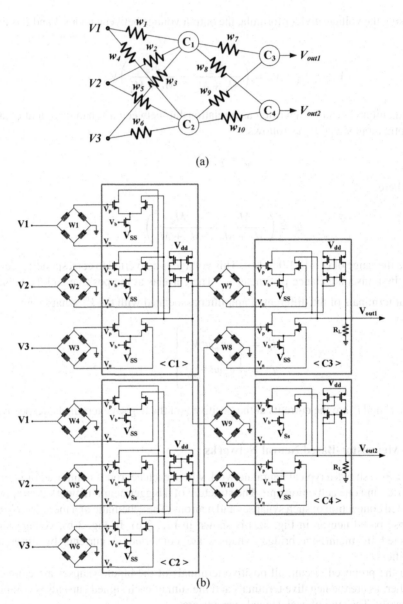

Fig. 3. An illustration of a memristor synapse-based multilayer neural network (a) a multilayer neural network (b) the schematic of memristor synapse-based multilayer neural circuit corresponding to the neural network (a).

where i^+_{SUM} and i^-_{SUM} are sum of currents at the positive and the negative terminals, respectively. The output current of the active load circuit is the difference between these two current components. It follows that,

$$i_{OUT} = \sum_k g_m \psi^K V_{in}^K \tag{6}$$

Assuming that a constant resistance R_L is connected at the output terminal, the output voltage of the neuron is

$$V_{OUT} = R_L \sum_k g_m \psi^K V_{in}^K \tag{7}$$

The voltage at the output is not linearly proportional to the current and is soon saturated when the output voltages exceeds $V_{DD} - V_{th}$ or $V_{SS} + V_{th}$, where V_{th} is the threshold voltage of the two transistors of the active load, or synapses. Thus, the range of V_{OUT} is restricted as follows:

$$V_{SS} + V_{th} \leq V_{OUT} \leq V_{DD} - V_{th} \tag{8}$$

Let the minimum voltage $V_{SS} + V_{th}$ be V_{MIN} and the maximum voltage $V_{DD} - V_{th}$ be V_{MAX}.

$$V_{out} = \begin{cases} R_L I_{OUT} & \text{if } \dfrac{V_{SS} + V_{th}}{R_{OUT}} \leq I_{out} \leq \dfrac{V_{DD} - V_{th}}{R_{OUT}} \\ V_{MAX} & \text{if } \dfrac{V_{DD} - V_{th}}{R_{OUT}} \leq I_{out} \\ V_{MIN} & \text{if } I_{out} \leq \dfrac{V_{SS} + V_{th}}{R_{OUT}} \end{cases} \tag{9}$$

Then, the circuit for a neural node is as in Fig. 4(a). The activation function is as shown in Fig. 4(b).

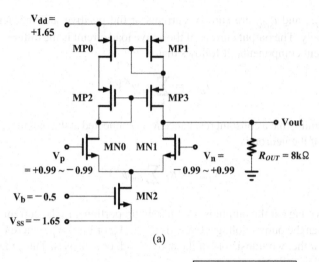

(a)

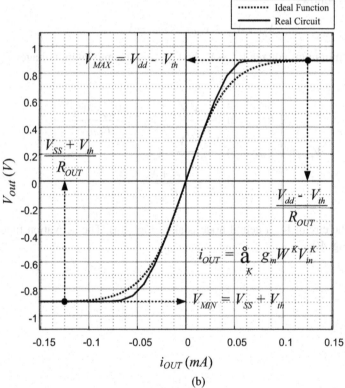

(b)

Fig. 4. A circuit for a neural node (a) and its activation function (b)

3   Proposed Hybrid Learning: Hardware Friendly Error-Backpropagation and Circuit-Based Complementary Learning with RWC

3.1   Random Weight Change Algorithm for Circuit-Based Learning

The learning algorithm for multilayer neural network should be as simple as possible for an easy implementation with hardware as described before. In this paper, Random Weight Change algorithm [9] is chosen as the most adequate candidate of the neural network learning. It requires only the simple circuitry as it does not involve complex derivative calculation of the activation function, or complex circuitry for back-propagation of error as shown in Fig. 5. It can be built by using circuit blocks for different elementary operations like summation, square, integration, comparison and random number generation.

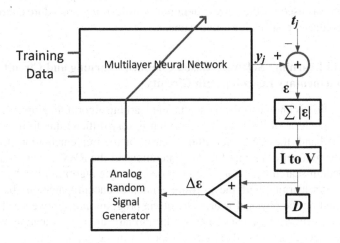

Fig. 5. A hardware architecture of the RWC learning algorithm

Learning processing starts with a neural network programmed with random initial weights and two capacitors set with big values which are used for the accumulation of errors in the learning loop. Then, an analog weight-updating vector (Δw) with a small magnitude is generated by an analog random weight generator and programmed additionally on the neural networks. Note that the analog random weight Δw is generated with an analog chaos generator [12] and is further simplified to $\pm\delta$ which has the same magnitude but random signs.

After updating with the random weighting vector (Δw), an input data x_j is presented at the memristor-based multilayer neural networks as shown in Fig. 5 and the output y_j of the neural network is obtained. The difference between the output of the neural network and target data t_j is computed. Then, its absolute value is taken by using an analog absolute circuit. The input signal of the analog absolute circuit is represented in current mode for the simplicity of current summation so that the error signals in current mode are summed with a simple wired connection in the case of neural networks with multiple output terminals.

To convert the current signal back to a voltage mode, a short time current charging technique of a capacitor is adopted in this paper, where the voltage of a capacitor charging with a current I during time T can be computed via the relationship $V = \dfrac{IT}{C}$.

The stored error voltage of the present iteration is compared with that of the previous iteration which was stored in another capacitor. Note that the previous error of the first iteration is set with the biggest possible value.

The next procedure is the updating of the weights depending upon the output value of the comparator. If the error of the current iteration is bigger than that of the previous iteration, the updated weight vector (ΔW) which is added for the current iteration is subtracted from the weight W. Then, a new small weight vector (ΔW) is generated and added to the weight W. However, if the error of the current iteration is smaller than that of the previous iteration, the weight W is updated with the previous learning vector (ΔW) which was used for the current iteration. This learning procedure continues until the error is reduced to a small enough value.

3.2   Hybrid Learning: Software-Based Confined Learning and Circuit-Based Complementary Learning with Circuits

Backpropagation is known as the most efficient learning algorithm. However, the hardware implementation of the learning algorithm is very difficult due to its complexity. On the other hand, the RWC algorithm is easy for the implementation in hardware. However, the number of required learning iterations of the RWC is very big due to its inherent random search behavior. The proposed learning algorithm is a hybrid one of these two. After learning with the software version of backpropagation, the parameter is transplanted on the physical neural circuit via programming. However, the weights obtained with software-based neural networks normally are not compatible with that of circuit-based neural networks due to the un-ideality in the implementation of hardware circuits. Programming on hardware weight is inaccurate due to the nonlinear characteristics of physical weights. For instance, weight implemented with memristor is nonlinear function of the programming voltage. Also, the un-ideality in the fabrication of the circuit makes the circuit-based neural network to be further deviation from the theoretical model. If the hardware circuit is not identical to the software version, the error will occur significantly when the weights learned with a software are transplanted on the hardware version of neural networks. Our proposed idea is the readjustment of the weight after the learned weights with the software are transplanted on the hardware-based neural network. The readjustment is performed with the hardware circuit of RWC which is easy to implement.

Since the location of the altered state after the learned weights are transplanted on the neural network circuit will not be deviated far away from that of the software version of the neural network, the error will decrease quickly during re-learning with RWC algorithm.

Figure 6 shows 3 steps of procedure of the hybrid learning of BP and RWC.

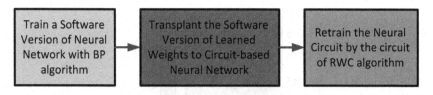

Fig. 6. Three steps of the learning procedure of the hybrid learning of the BP and the RWC algorithms. When weights learned with the software version of a neural network are downloaded on the hardware version of neural networks, learning error increases due to characteristics difference between software and hardware versions. The error decreases with an additional learning with the RWC circuit.

4 Simulation Results

The proposed research is on the hardware implementation of the neural networks and its learning system with the help of software. The hardware part as well as the software part is implemented with software and its simulation has been performed in a hardware described software, namely HSPICE. The synapses and nodes of neural networks are designed with memristor bridge and differential amplifier circuits for both software version of BP and hardware-based RWC, respectively. For the activation function of neural node, bipolar sigmoid function is employed.

The learning system is designed with hardware circuit of RWC as described in Sect. 3. For the simulations of hardware part, all the possible characteristics of circuits are included.

Parameters of memristors which are employed for memristor bridge synapse are $R_{ON} = 400\ \Omega$, $R_{OFF} = 40\ k\Omega$, $D = 10\ nm$, and $\mu_v = 10^{-14}\ m^2V^{-1}S^{-1}$ [9]. δ which is used for the RWC learning is a pulse of 1 V amplitude and width of 5 ms.

A problem considered for simulating the proposed learning method is to learn the workspace of a robot. If the map of the workspace is learned by an NN, faster operations of robots without collision with obstacles are possible. This is an important problem in robotics. After learning the map of the workspace, the robot can apply appropriate path planning algorithm to find its way to reach the goal safely in the workspace.

The workspace was considered as a grid and the inputs to NN are the co-ordinates of the grid. The network size used for this problem was 10 input $\times$ 20 hidden $\times$ 1 output nodes. The network was trained to learn the grid of size 21×21 as shown in Fig. 7. The co-ordinates in the grid without obstacle are labeled 1 (yellow) and those with obstacle were labeled -1 (dark blue). Each 2-dimensional co-ordinate position in the grid was converted to a 10-dimensional binary number. i.e., the co-ordinate position $(3, 2)$ in the workspace was converted to $(-1-1-111,-1-1-11-1)$ to allow more degree of freedom for learning.

The error vs. epoch curve of the training using BP $(\alpha = 0.001)$, and RWC $(\delta = 0.00025)$ is shown as in Fig. 8. Learning of this workspace is very difficult since it is a highly nonlinear function.

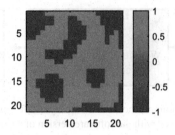

Fig. 7. A workspace for a robot navigation (Color figure online)

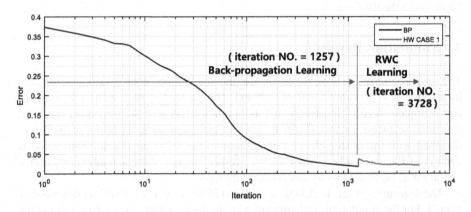

Fig. 8. A learning curve of a robot workspace problem with the proposed hybrid learning two stages of learning curve. After learning with the software version of BP algorithm, hardware-based learning was performed with the RWC algorithm. Though the error increased significantly temporally at the time of weight transplanting, it decreased by the complemental learning of RWC.

When the error reaches a threshold (0.01), the learned weights were transplanted to a neural network circuit. Then, hardware-based learning was performed with the RWC algorithm. Though the error increased significantly temporally at the time of weight transplanting, it decreased by the complemental learning of RWC.

Figure 9(a) shows the learned results of the original robot workspace in Fig. 7 with the software-based BP algorithm, where the left and the right ones are the results before and after a threshold, respectively. The learned result after the threshold is identical to the original one in Fig. 7. However, Fig. 9(b) is the output of the neural network circuit immediately after the transplant of the learned weights. Observe that learning errors appear at several places (red circled area) of the right one of Fig. 9(b). Those errors disappear when the complementary learning with the hardware circuit of RWC is conducted as shown in the right one of Fig. 9(c). This result shows the fact that the proposed hybrid learning system is a good solution for the hardware implementation of neural networks.

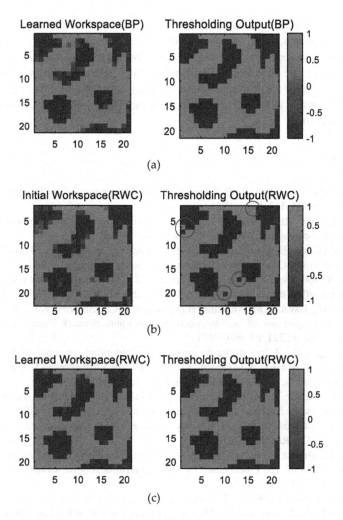

Fig. 9. Learned results of robot workspace with the proposed hybrid learning system. (a) Results with the software-based BP algorithm. Left and right ones are ones before and after a thresholding, respectively. (b) Outputs of the neural network circuit immediately after the learned weights are transplanted on it. Erroneous areas are marked in red circles. (c) Results after the complementary learning with the hardware circuit of the RWC. (Color figure online)

5   Conclusions

A hybrid learning method of software-based BP and hardware-based RWC is addressed in this paper. In the learning method, the software-based BP learning is conducted firstly and its learned weights are transplanted on the neural hardware circuits. Then, the hardware-based RWC learning is continued as a complementary learning.

The proposed learning method has been examined with a robot workspace learning problem. As expected, the error reduced with BP learning grows abruptly when weights are transplanted on the neural hardware circuit. However, after some amount of complementary learning, the error was reduced successfully below the threshold. Comparing the learning curves of the proposed hybrid learnings with those of RWC-only learnings, the required learning iterations of the proposed method are much less than those with RWC-only learning.

References

1. Rumelhart, D.E., Hinton, G.E., Williams, R.J.: Learning representations by back-propagating errors. Nature **323**, 533–536 (1986)
2. Werbos, P.J.: The Roots of Backpropagation: From Ordered Derivatives to Neural Networks and Political Forecasting. Wiley, New York (1994)
3. Tam, S.M., Gupta, B., Castro, H.A., Holler, M.: Learning on analog VLSI network chip. In: IEEE International Conference on Systems, Man, and Cybernetics, pp. 701–703 (1990)
4. Erkmen, B., Kahraman, N., Vural, R.A., Yildirim, T.: Conic section function neural network circuitry for offline signature recognition. IEEE Trans. Neural Netw. **21**, 667–672 (2010)
5. Prezioso, M., Bayat, F.M., Hoskins, B.D., Adam, G.C., Likharev, K.K., Strukov, D.B.: Training and operation of an integrated neuromorphic network based on metal-oxide memristors. Nature **521**, 61–64 (2015)
6. Cauwenberghs, G.: A fast stochastic error-descent algorithm for supervised learning and optimization. In: Hanson, S.J., Cowan, J.D., Lee, C. (eds.) Advances in Neural Information Processing Systems, p. 5. Morgan Kaufman Publishers, San Mateo (1993)
7. Maeda, Y., Hirano, H., Kanata, Y.: A Learning rule of neural networks via simultaneous perturbation and its hardware implementation. Neural Netw. **8**, 251–259 (1995)
8. Soudry, D., Di Castro, D., Gal, A., Kolodny, A., Kvatinsky, S.: Hebbian Learning Rules with memristors. CCIT Report # 840 (2013)
9. Hirotsu, K., Brooke, M.A.: An analog neural network chip with random weight change learning algorithm. In: Proceedings of 1993 International Joint Conference on Neural Networks, pp. 3031–3034 (1993)
10. Burton, B., Kamran, F., Harley, R.G., Habetler, T.G., Brooke, M.A., Poddar, R.: Identification and control of induction motor stator currents using fast online random training of neural network. IEEE Trans. Ind. Appl. **33**, 697–704 (1997)
11. Liu, J., Brooke, M., Hirotsu, K.: A CMOS feedforward neural-network chip with on-chip parallel learning for oscillation cancellation. IEEE Trans. Neural Networks **13**, 1178–1186 (2002)
12. Kim, H., Sah, M.P., Yang, C., Roska, T., Chua, L.O.: Memristor bridge synapses. Proc. IEEE **100**, 2061–2070 (2012)

Fuzzy Systems and Their Applications

Fuzzy Systems and Their Applications

A Comparative Study of Two Novel Approaches to the Rule-Base Evidential Reasoning

Ludmila Dymova, Krzysztof Kaczmarek, and Pavel Sevastjanov$^{(\boxtimes)}$

Institute of Computer and Information Science, Czestochowa University
of Technology, Dabrowskiego 73, 42-200 Czestochowa, Poland
`sevast@icis.pcz.pl`

Abstract. Recently two new approaches to the rule-base evidential reasoning were proposed in the literature. The first of them is based on the Atannasov's intuitionistic fuzzy sets theory $(A - IFS)$ and Dempster-Shafer theory (DST). The second one is based directly on the synthesis of fuzzy logic and DST. In this paper, using the simple, but real-world example, it is shown that the first approach in the critical case of high conflict provides more reasonable and intuitively obvious results.

Keywords: Rule-base evidential reasoning · Atannasov's intuitionistic fuzzy sets · Dempster-Shafer Theory

1 Introduction

The methodology of rule-base evidential reasoning $(RBER)$ is based on the tools of Fuzzy Sets theory (FST) and the Dempster-Shafer theory (DST). The synthesis of FST and DST was used for solving control and classification problems [3,4,17,21,25] when outputs are real values. Nevertheless, when we use $RBER$ in decision support systems, their outputs should be only the names or labels of actions or decisions. Obviously, in such cases, the methods, developed for the controlling can not be used. A more appropriate for the construction of decision support systems is the so-called $RIMER$ method [23,24], which is based on the Evidential Reasoning methodology [22]. On the other hand, there are two drawbacks of $RIMER$ method that substantially restrict its ability to deal with uncertainties that we can often meet in practice.

The first drawback is that in the $RIMER$ method, a degree of belief can be assigned only to a particular hypothesis, not to a group of them. The second drawback is that the $RIMER$ method does not provide a possibility for the combination of evidence from different sources using methodology of DST.

In [10,12,18], a new approach free of described above drawbacks was proposed and used for the solution of real-world problems.

Nevertheless, in all above mentioned approaches to the $RBER$ the conventional fuzzy logic was used. For example, consider the following rule:

$If\ x\ is\ High\ Then\ D$, where $High$ is the fuzzy class described by the membership function $\mu_{High}(x)$, D is a name of corresponding decision. On the other

© Springer International Publishing AG 2017
L. Rutkowski et al. (Eds.): ICAISC 2017, Part I, LNAI 10245, pp. 231–240, 2017.
DOI: 10.1007/978-3-319-59063-9_21

hand, in practice we often meet intersecting fuzzy classes, e.g., *High* and *Middle* and as a consequence of intersection we have $\mu_{High}(x) > 0$ and $\mu_{Middle}(x) > 0$. If $\mu_{High}(x) > \mu_{Middle}(x)$ then we assume that x *is High*. We can see that information of nonzero $\mu_{Middle}(x)$ is lost, although the difference between $\mu_{Middle}(x)$ and $\mu_{High}(x)$ may be negligible.

In the papers [13–15], we showed that this loss of information may provide undesirable results in the *RBER*. To avoid this problem, a new method for *RBER* based on the synthesis of Atanassov's intuitionistic fuzzy sets (*A-IFS*) [1] and *DST* was proposed in [13,14].

Another new method for *RBER* based of the treatment the values such as $\mu_{High}(x) > 0$ and $\mu_{Middle}(x) > 0$ as focal elements of basic probability assignment in the context of *DST* was developed in [15].

Since these two approaches are conceptually different, in this paper, using simple, but real-world example we compare them and show that the first approach in the critical case of high conflict provides more reasonable and intuitively obvious results. This paper is organised as follows. Section 2 presents the basic definition of *DST* and *A-IFS* needed for the subsequent analysis. In Sect. 3, we present the results of comparison of two above mentioned new approaches to the *RBER*. Concluding section summarises the paper.

2   Preliminaries

2.1   The Basics of *DST*

The basic definitions of Dempster-Shafer theory (*DST*) were introduced by A.P. Dempster [7]. Later G. Shafer [19] provided a more thorough definition of belief functions. Let A are subsets of X and X may be treated as a set of propositions or mutually exclusive hypotheses or answers. A *DS* belief structure is based on a mapping m, called basic probability assignment (*bpa*), from subsets of X into a unit interval, $m : 2^X \rightarrow [0,1]$ such that $m(\emptyset) = 0$, $\sum\limits_{A \subseteq X} m(A) = 1$. The subsets of X for which the mapping does not assume a zero value are called focal elements. Shafer [19] introduced a number of measures associated with *DS* belief structure.

The measure of belief is a mapping $Bel : 2^X \rightarrow [0,1]$ such that for any subset B of X it can be presented as $Bel(B) = \sum\limits_{\emptyset \neq A \subseteq B} m(A)$.

A second measure introduced by Shafer [19] is a measure of plausibility. The measure of plausibility associated with m is a mapping $Pl : 2^X \rightarrow [0,1]$ such that for any subset B of X it can be presented as $Pl(B) = \sum\limits_{A \cap B \neq \emptyset} m(A)$. It is easy to see that $Bel(B) \leq Pl(B)$. *DS* provides an explicit measure of ignorance about an event B and its complementary $\overline{B}$ as a length of an interval $[Bel(B), Pl(B)]$ called the belief interval (*BI*). It can also be interpreted as imprecision of the "true probability" of B [19].

The core of the evidence theory is the Dempster's rule of combination of evidence from independent different sources. With two belief structures m_1, m_2, the Dempster's rule of combination is defined as follows:

$$m_{12}(A) = \frac{\sum\limits_{B \cap C = A} m_1(B)m_2(C)}{1 - K}, A \neq \emptyset, m_{12}(\emptyset) = 0, \tag{1}$$

where $K = \sum\limits_{B \cap C = \emptyset} m_1(B)m_2(C)$ is called the degree of conflict which measures the conflict between the pieces of evidence. Zadeh [26] underlined that this rule involves counter-intuitive behaviors in the case of considerable conflict.

2.2   The Basics of A-IFS

The concept of A-IFS (the reasons for such notation are presented in [8]) is based on the simultaneous consideration of membership μ and non-membership ν of an element of a set to the set itself (see formal definition in [1]). It is postulated that $0 \leq \mu + \nu \leq 1$. Following to [1], we call $\pi_A(x) = 1 - \mu_A(x) - \nu_A(x)$ the hesitation degree of the element x in the set A. Hereinafter, we will call an object $\tilde{A} = \langle \mu_A(x), \nu_A(x) \rangle$ intuitiniostic fuzzy value (IFV).

The operations of addition $\oplus$ and multiplication $\otimes$ on $IFVs$ were defined by Atanassov [2] as follows. Let $A = \langle \mu_A, \nu_A \rangle$ and $B = \langle \mu_B, \nu_B \rangle$ be $IFVs$. Then

$$A \oplus B = \langle \mu_A + \mu_B - \mu_A\mu_B, \nu_A\nu_B \rangle, \tag{2}$$

$$A \otimes B = \langle \mu_A\mu_B, \nu_A + \nu_B - \nu_A\nu_B \rangle. \tag{3}$$

These operations were constructed in such a way that they produce $IFVs$. Using operations (2) and (3), in [6] the following expressions were obtained for any integer $n=1,2,..$:
$$nA = A \oplus ... \oplus A = \langle 1 - (1 - \mu_A)^n, \nu_A^n \rangle, \quad A^n = A \otimes ... \otimes A = \langle \mu_A^n, 1 - (1 - \nu_A)^n \rangle.$$

It was proved later that these operations produce $IFVs$ not only for integer n, but also for all real values $\lambda > 0$, i.e.,

$$\lambda A = \langle 1 - (1 - \mu_A)^\lambda, \nu_A^\lambda \rangle, \tag{4}$$

$$A^\lambda = \langle \mu_A^\lambda, 1 - (1 - \nu_A)^\lambda \rangle. \tag{5}$$

The operations (2)–(5) have good algebraic properties [20]:

An important problem is the comparison of $IFVs$. Therefore, the specific methods which are rather of heuristic nature were developed to compare $IFVs$. For this purpose, Chen and Tan [5] proposed to use the so-called score function (or net membership) $S(x) = \mu(x) - \nu(x)$. Let a and b be $IFVs$. It is intuitively appealing that if $S(a) > S(b)$ then a should be greater (better) than b, but if $S(a) = S(b)$ this does not always mean that a is equal to b. Therefore, Hong and Choi [16] in addition to the above score function introduced the so-called accuracy function $H(x) = \mu(x) + \nu(x)$ and showed that the relation between functions S and H is similar to the relation between mean and variance in statistics. Xu [20] used the functions S and H to construct order relations between any pair of intuitionistic fuzzy values a and b as follows:

$$If \text{ (S(a)} > \text{S(b))}, \text{ then } b \text{ is smaller than } a;$$
$$If \text{ (S(a)} = \text{S(b))}, \text{ then}$$
$$(1) If \text{ (H(a)=H(b))}, \text{ then } a=b; \qquad (6)$$
$$(2) If \text{ (H(a)} < \text{H(b))}, \text{ then } a \text{ is smaller than } b.$$

In [11], we have shown that operation (2)–(5) and (6) have some undesirable properties which may lead to the non-acceptable results in applications:

1. The addition (2) is not an addition invariant operation. Let a, b and c be $IFVs$. Then $a < b$ (according to (6)) does not always lead to $(a \oplus c) < (b \oplus c)$.

2. The operation (4) is not preserved under multiplication by a real-valued $\lambda > 0$, i.e., inequality $a < b$ (in sense of (6)) does not necessarily imply $\lambda a < \lambda b$.

2.3   Interpretation of A-IFS in the Framework of DST

It was shown in [9] that DST may serve as a good methodological base for interpretation of A-IFS. It was proved in [9] that IFV $\tilde{A} = \langle \mu_{\tilde{A}}(x), \nu_{\tilde{A}}(x) \rangle$ may be represented by the belief interval $BI_{\tilde{A}}(x) = [Bel_{\tilde{A}}(x), Pl_{\tilde{A}}(x)]$, where $Bel_{\tilde{A}}(x) = \mu_{\tilde{A}}(x)$ and $Pl_{\tilde{A}}(x) = 1 - \nu_{\tilde{A}}(x)$ (see [9] for formal definitions and more detail). This interpretation makes it possible to represent mathematical operations on $IFVs$ as operations on belief intervals. The use of the semantics of DST makes it possible to enhance the performance of A-IFS when dealing with the operations on $IFVs$. In [11], two sets of operations on $IFVs$ based on the interpretation of intuitionistic fuzzy sets in the framework of DST are proposed and analysed. The first set of operations is based on the treatment of belief interval as an interval enclosing a true probability. The second set of operations is based on the treatment of belief interval as an interval enclosing a true power of some statement (argument, hypothesis, ets). It was shown in [11], that the non-probabilistic treatment of belief intervals representing $IFVs$ performs better than the probabilistic one and operations based on the probabilistic and non-probabilistic treatments of belief intervals representing $IFVs$ perform better than operations on $IFVs$ defined in the framework of conventional A-IFS. Therefore, here we will use only the treatment of belief interval as an interval enclosing a true power of some statement.

Let $X = \{x_1, x_2, ..., x_n\}$ be a finite universal set. Assume A are subsets of X. It is important to note that in the framework of DST a subset A may be treated also as a question or proposition and X as a set of propositions or mutually exclusive hypotheses or answers. In such a context, a belief interval $BI(A) = [Bel(A), Pl(A)]$ may be treated as an interval enclosing a true power of statement (argument, proposition, hypothesis, etc) that $x_j \in X$ belongs to the set $A \subseteq X$. Obviously, the value of such a power lies in interval $[0, 1]$.

Therefore, a belief interval $BI(A) = [Bel(A), Pl(A)]$ as a whole may be treated as an imprecise (interval-valued) statement (argument, proposition, hypothesis, etc) that $x_j \in X$ belongs to the set $A \subseteq X$.

Based on this reasoning, we can say that if we pronounce this statement, we can obtain some result, e.g., as a reaction on this statement or as an answer to some question, and if we repeat this statement twice, the result does not change.

Such a reasoning implies the following property of addition operator:

$BI(A) = BI(A) \oplus_B BI(A) \oplus_B ... \oplus_B BI(A)$.

This is possible only if we define the addition $\oplus_B$ of belief intervals as follows: $BI(A) \oplus_B BI(A) = \left[\frac{Bel(A)+Bel(A)}{2}, \frac{Pl(A)+Pl(A)}{2} \right]$. So the addition of belief intervals is represented by their averaging.

Therefore, if we have n different statements represented by belief intervals $BI(A_i)$ then their sum $\oplus_B$ can be defined as follows:

$$BI(A_1) \oplus_B BI(A_2) \oplus_B .... \oplus_B BI(A_n) = \left[\frac{1}{n} \sum_{i=1}^{n} Bel(A_i), \frac{1}{n} \sum_{i=1}^{n} Pl(A_i) \right]. \quad (7)$$

The other operations on belief intervals are presented in [11]. It is justified in [11] that to compare belief intervals it is enough to compare their centres.

It is proved in [11] that introduced operations on belief intervals are free of undesirable properties (1) and (2) of conventional operations on $IFVs$.

3   Two New Approaches to the Rule-Based Evidential Reasoning: Comparative Study

To present our approach in a more transparent form, in this section we will use a simplified and relatively simple example of decision-making in the forex trading. Let us consider the currency pair EUR/USD (Euro/U.S. Dollar). The currency pair tells the reader how many U.S. dollars (the quote currency) are needed to purchase one euro (the base currency).

A two-way price quotation that indicates the best price at which a security can be sold and bought at a given point in time. The *Bid* price represents the maximum price that a buyer or buyers are willing to pay for a security. The *Ask* price represents the minimum price that a seller or sellers are willing to receive for the security. A trade or transaction occurs when the buyer and seller agree on a price for the security.

To simplify our subsequent analysis hereinafter we will use the averaged price $p = (Bid + Ask)/2$.

Suppose a trader makes transactions on the currency pair EUR/USD based on the opinions of two independent experts $(E1, E2)$. The experts on the base of analysis of changes of price propose the possible transactions: *Buy*, *Sell* and *Hold* (the lack of transactions). Here we will treat the decision *Hold* as an intermediate one when an expert hesitates in his/her choice between *Buy* and *Sell*. Therefore, the decision *Hold* in the spirit of *DST* will be treated as the compound decision $(Buy, Sell)$ In practice, experts intuitively and based on their experience transform the numerical information of prices into linguistic terms such as *Low*, *Medium*, *High* and use them as preconditions for possible decisions. In such a case, in our example (the pair EUR/USD) the expert's opinions concerned with the choice of transaction and based on the actual or predicted prises may be presented in the form of membership functions $\mu_{E1}^{Buy}(p)$, $\mu_{E1}^{Hold}(p)$,

$\mu_{E1}^{Sell}(p)$ (based on the opinion of expert $E1$) and $\mu_{E2}^{Buy}(p)$, $\mu_{E2}^{Hold}(p)$, $\mu_{E2}^{Sell}(p)$ (based on the opinion of expert $E2$). These membership functions are presented in Fig. 1, The values of them represent degrees of expert's belief in the reasonableness of transactions. Therefore, in the spirit of DST and [15] we can represent our decision system as follows:

$$IF \ (p = p_1) \ then$$
$$m_{E1}(Buy) = \mu_{E1}^{Buy}(p1), \ m_{E1}(Hold) = \mu_{E1}^{Hold}(p1), \ m_{E1}(Sell) = \mu_{E1}^{Sell}(p1),$$
$$IF \ (p = p_2) \ then$$
$$m_{E2}(Buy) = \mu_{E2}^{Buy}(p2), \ m_{E2}(Hold) = \mu_{E2}^{Hold}(p2), \ m_{E2}(Sell) = \mu_{E2}^{Sell}(p2),$$
$$(8)$$

where *bpas* $m_{E1}(Buy)$, $m_{E1}(Hold)$, $m_{E1}(Sell)$ and $m_{E2}(Buy)$, $m_{E2}(Hold)$, $m_{E2}(Sell)$ should be normalised [15]. It was shown in [15] that simple averaging of *bpas* seems to be a more reliable combination rule than the Dempster rule (1).

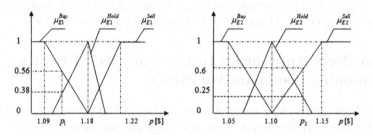

Fig. 1. The membership functions of transaction decision based on the opinion of two experts $E1$ and $E2$.

Therefore, using averaging rule, from (8) we get: $m(Buy) \approx 0.31$, $m(Hold) \approx 0.34$ and $m(Sell) \approx 0.35$. Hence from (8) we obtained the decision *Sell* which cannot be accepted as an appropriate one because in our case we have more arguments in favour of $= Hold$ than in favour of *Sell*. Moreover $m(Hold) \approx m(Sell)$ and the experts are not so cocksure in their estimations: ($\pi_{E1}^{Buy} = \pi_{E1}^{Hold}=0.06$ and $\pi_{E2}^{Hold} = \pi_{E2}^{Sell}=0.15$). It is important that there are no parameter π in the model (8).

Therefore, let us consider approach based on the synthesis of $A - IFS$ and DST [13,14]. From Fig. 1 we can see that p_1 belongs to the fuzzy class *Buy* with the value of membership function $\mu_{E1}^{Buy}(p_1) = 0.6$, but simultaneously it belongs to the competitive fuzzy class *Hold*, $\mu_{E1}^{Hold}(p_1) = 0.2$. Then following to [13], we can say that the non-membership function $\nu_{E1}^{Buy}(p_1)$ for the fuzzy class *Buy* is equal to $\mu_{E1}^{Hold}(p_1)$ for the competitive fuzzy class *Hold*. Based on such a reasoning, we can represent the final result using two IFVs: $\left\langle \mu_{E1}^{Buy}(p_1), \nu_{E1}^{Buy}(p_1) \right\rangle$, $\langle \mu_{E1}^{Hold}(p_1), \nu_{E1}^{Hold}(p_1) \rangle$, where $\nu_{E1}^{Buy}(p_1) = \mu_{E1}^{Hold}(p_1)$ i $\nu_{E1}^{Hold}(p_1) = \mu_{E1}^{Buy}(p_1)$.

Similarly ni the case of $p = p_2$ we get: $\langle \mu_{E2}^{Hold}(p_2), \nu_{E2}^{Hold}(p_2) \rangle$, $\langle \mu_{E2}^{Sell}(p_2), \nu_{E2}^{Sell}(p_1) \rangle$, where $\nu_{E2}^{Hold}(p_2) = \mu_{E2}^{Sell}(p_2)$ i $\nu_{E2}^{Sell}(p_2) = \mu_{E2}^{Hold}(p_2)$.

Then in the spirit of approach developed in our previous works we can present our decision system as follows:

$$IF\ (p = p_1)\ Then\ m_{E1}(Buy) = \left\langle \mu_{E1}^{Buy}(p_1), \nu_{E1}^{Buy}(p_1) \right\rangle,$$
$$m_{E1}(Hold) = \left\langle \mu_{E1}^{Hold}(p_1), \nu_{E1}^{Hold}(p_1) \right\rangle,\ m_{E1}(Sell) = \left\langle \mu_{E1}^{Sell}(p_1), \nu_{E1}^{Sell}(p_1) \right\rangle,\quad (9)$$
$$IF\ (p = p_2)\ Then\ m_{E2}(Buy) = \left\langle \mu_{E2}^{Buy}(p_2), \nu_{E2}^{Buy}(p_2) \right\rangle,$$
$$m_{E2}(Hold) = \left\langle \mu_{E2}^{Hold}(p_2), \nu_{E2}^{Hold}(p_2) \right\rangle,\ m_{E2}(Sell) = \left\langle \mu_{E2}^{Sell}(p_2), \nu_{E2}^{Sell}(p_2) \right\rangle.$$

The next step is the combination of obtained *bpas*. This may be done using the direct intuitionistic extension of the Dempster rule (1). If *bpas* are usual real values then after the use of Dempster rule (1) we obtain $\sum_{A \subseteq X} m(A) = 1$ since the normalisation (division by $1 - K$) is inherent part of (1). Obviously, if *bpas* are $IFVs$, the formal normalisation is impossible. On the other hand, all classical operations on $IFVs$ provide $IFVs$ as results. Therefore, using only the denominator of direct intuitionistic extension of (1) we get

$$m_{12}(A) = \oplus_{B \cap C = A}(m_1(B) \otimes m_2(C)). \quad (10)$$

On the base of (10) for our example we obtain

$$m_{E1E2}(Buy) = m_{E1}(Buy) \otimes m_{E2}(Buy) \oplus m_{E1}(Buy) \otimes m_{E2}(Hold) \oplus$$
$$m_{E1}(Hold) \otimes m_{E2}(Buy),$$
$$m_{E1E2}(Sell) = m_{E1}(Sell) \otimes m_{E2}(Sell) \oplus m_{E1}(Sell) \otimes m_{E2}(Hold) \oplus \quad (11)$$
$$m_{E1}(Hold) \otimes m_{E2}(Sell),$$
$$m_{E1E2}(Hold) = m_{E1}(Hold) \otimes m_{E2}(Hold),$$

Since the sum $m_{E1E2}(Buy) \oplus m_{E1E2}(Hold) \oplus m_{E1E2}(Sell)$ is IFV, we can say that combination rule (10) is normalised in sense of A–IFS.

Then from (9) and (11) we get $m_{E1E2}(Buy) = \langle 0.140, 0.211 \rangle$, $m_{E1E2}(Hold) = \langle 0.095, 0.664 \rangle$ and $m_{E1E2}(Sell) = \langle 0.228, 0.258 \rangle$.

To compare obtained $IFVs$ we use the score function S and from $S(x) = \mu(x) - \nu(x)$ we finally obtained $S(m_{E1E2}(Buy)) = -0.071$, $S(m_{E1E2}(Hold)) = -0.569$ and $S(m_{E1E2}(Sell)) = -0.03$. Since $S(m_{E1E2}(Sell)) > S(m_{E1E2}(Buy)) > S(m_{E1E2}(Hold))$ we can see that in this case the controversial decision *Sell* should be accepted.

Therefore, to obtain a more convincing result, in the spirit of [13,14] we will use the transformation of $IFVs$ into belief intervals and averaging rule for combination evidence from different sources. Then from (9) we get:

$$IF(p = p_1)\ Then\ m_{E1}(Buy) = BI_{E1}^{Buy}(p_1),\ m_{E1}(Hold) = BI_{E1}^{Hold}(p_1),$$
$$m_{E1}(Sell) = BI_{E1}^{Sell}(p_1),$$
$$IF(p = p_2)\ Then\ m_{E2}(Buy) = BI_{E2}^{Buy}(p_2),\ m_{E2}(Hold) = BI_{E2}^{Hold}(p_2),\quad (12)$$
$$m_{E2}(Sell) = BI_{E2}^{Sell}(p_2),$$

where $BI_{E1}^{Buy} = \left[\mu_{E1}^{Buy}(p_1), 1 - \nu_{E1}^{Buy}(p_1) \right]$, $BI_{E1}^{Hold} = \left[\mu_{E1}^{Hold}(p_1), 1 - \nu_{E1}^{Hold}(p_1) \right]$, $BI_{E1}^{Sell} = \left[\mu_{E1}^{Sell}(p_1), 1 - \nu_{E1}^{Sell}(p_1) \right]$, $BI_{E2}^{Buy} = \left[\mu_{E2}^{Buy}(p_2), 1 - \nu_{E2}^{Buy}(p_2) \right]$, $BI_{E2}^{Hold} = \left[\mu_{E2}^{Hold}(p_2), 1 - \nu_{E2}^{Hold}(p_2) \right]$, $BI_{E1}^{Sell} = \left[\mu_{E2}^{Sell}(p_2), 1 - \nu_{E2}^{Sell}(p_2) \right]$.

In this case, the averaging rule may be presented as follows:

$$m(D_i) = \frac{1}{n} \sum_{j=1}^{n} BI_j^{D_i} \tag{13}$$

where n is the number of sources of evidence, D_i is the ith decision (alternative), BI is a belief interval, $BI_j^{D_i} = [\mu_j^{D_i}, 1 - \nu_j^{D_i}]$.

For the simplicity, let us denote $\eta_j^{D_i} = 1 - \nu_j^{D_i}$ and $BI_j^{D_i} = [\mu_j^{D_i}, \eta_j^{D_i}]$. Then:

$$BI(D_i) = [\mu^{D_i}, \eta^{D_i}] = BI_1^{D_i} + BI_2^{D_i} + \ldots + BI_n^{D_i}$$
$$= \left[\mu_1^{D_i} + \mu_2^{D_i} + \ldots + \mu_n^{D_i}, \eta_1^{D_i} + \eta_2^{D_i} + \ldots + \eta_n^{D_i} \right], \tag{14}$$

$$m(D_i) = \frac{\mu(D_i) + \eta(D_i)}{2}. \tag{15}$$

The final decision is obtained using the following expression:

$$Decision = max\{m(D_i)\}. \tag{16}$$

For our example (see Fig. 1) from above expressions we get:

$$m(Buy) = \left[\frac{\mu_{E1}(Buy) + \mu_{E2}(Buy)}{2}, \frac{\eta_{E1}(Buy) + \eta_{E2}(Buy)}{2} \right] = [0.56, 0.62],$$

$$m(Hold) = \left[\frac{\mu_{E1}(Hold) + \mu_{E2}(Hold)}{2}, \frac{\eta_{E1}(Hold) + \eta_{E2}(Hold)}{2} \right] = [0.63, 0.84],$$

$$m(Sell) = \left[\frac{\mu_{E1}(Sell) + \mu_{E2}(Sell)}{2}, \frac{\eta_{E1}(Sell) + \eta_{E2}(Sell)}{2} \right] = [0.60, 0.75],$$

$$m(Buy) = 0.59,$$
$$m(Hold) = 0.735,$$
$$m(Sell) = 0.675,$$
$$Decision = max\{0.59, 0.735, 0.675\} \Rightarrow Hold.$$

We can see that using last approach for our critical example we obtain the decision *Hold* which seems to more convincing result than decision *Sell* we get with the use of other considered approaches.

4   Conclusion

In this paper, three different approaches to the rule-base evidential reasoning based on the synthesis of DST and fuzzy logic, $A - IFS$ and fuzzy logic and the DST with direct intuitionistic extension of Dempster's combination rule, $A - IFS$ and fuzzy logic and the DST with averaging rule of combination are compared using the critical real-world example. It is shown that synthesis of DST and fuzzy logic as well as the direct use of intuitionistic fuzzy values and classical operations on them may lead to the counter-intuitive results. This may be a consequence of bad properties of classical operation on intuitionistic fuzzy values. It is shown that the interpretation of $A - IFS$ in the framework of DST and the use of averaging rule make it possible to use more information in the evidential- reasoning and as a consequence to obtain reasonable results when the synthesis of classical fuzzy logic and DST is failed.

Acknowledgements. The research has been supported by the grant financed by National Science Centre (Poland) on the basis of decision number DEC-2013/11/B/ST6/00960.

References

1. Atanassov, K.T.: Intuitionistic fuzzy sets. Fuzzy Sets Syst. **20**, 87–96 (1986)
2. Atanassov, K.: New operations defined over the intuitionistic fuzzy sets. Fuzzy Sets Syst. **61**, 137–142 (1994)
3. Binaghi, E., Madella, P.: Fuzzy Dempster-Shafer reasoning for rule-based classifiers. Intell. Syst. **14**, 559–583 (1999)
4. Binaghi, E., Gallo, I., Madella, P.: A neural model for fuzzy Dempster-Shafer classifiers. Int. J. Approx. Reason. **25**, 89–121 (2000)
5. Chen, S.M., Tan, J.M.: Handling multicriteria fuzzy decision-making problems based on vague set theory. Fuzzy Sets Syst. **67**, 163–172 (1994)
6. Dey, S.K., Biswas, R., Roy, A.R.: Some operations on intuitionistic fuzzy sets. Fuzzy Sets Syst. **114**, 477–484 (2000)
7. Dempster, A.P.: Upper and lower probabilities induced by a muilti-valued mapping. Ann. Math. Stat. **38**, 325–339 (1967)
8. Dubois, D., Gottwald, S., Hajek, P., Kacprzyk, J., Prade, H.: Terminological difficulties in fuzzy set theory-the case of "Intuitionistic Fuzzy Sets". Fuzzy Sets Syst. **156**, 485–491 (2005)
9. Dymova, L., Sevastjanov, P.: An interpretation of intuitionistic fuzzy sets in terms of evidence theory: decision making aspect. Knowl. Based Syst. **23**, 772–782 (2010)
10. Dymova, L., Sevastianov, P., Bartosiewicz, P.: A new approach to the rule-base evidential reasoning: stock trading expert system application. Expert Syst. Appl. **37**, 5564–5576 (2010)
11. Dymova, L., Sevastjanov, P.: The operations on intuitionistic fuzzy values in the framework of Dempster-Shafer theory. Knowl. Based Syst. **35**, 132–143 (2012)
12. Dymova, L., Sevastianov, P., Kaczmarek, K.: A stock trading expert system based on the rule-base evidential reasoning using level 2 quotes. Expert Syst. Appl. **39**, 7150–7157 (2012)
13. Dymova, L., Sevastjanov, P., Tkacz, K.: The Use of Intuitionistic Fuzzy Values in Rule-Base Evidential Reasoning. In: Rutkowski, L., Korytkowski, M., Scherer, R., Tadeusiewicz, R., Zadeh, L.A., Zurada, J.M. (eds.) ICAISC 2013. LNCS (LNAI), vol. 7894, pp. 247–258. Springer, Heidelberg (2013). doi:10.1007/978-3-642-38658-9_23
14. Dymova, L., Sevastjanov, P.: A new approach to the rule-base evidential reasoning in the intuitionistic fuzzy setting. Knowl. Based Syst. **61**, 109–117 (2014)
15. Sevastjanov, P., Dymova, L., Kaczmarek, K.: A New Approach to the Rule-Base Evidential Reasoning with Application. In: Rutkowski, L., Korytkowski, M., Scherer, R., Tadeusiewicz, R., Zadeh, L.A., Zurada, J.M. (eds.) ICAISC 2015. LNCS (LNAI), vol. 9119, pp. 271–282. Springer, Cham (2015). doi:10.1007/978-3-319-19324-3_25
16. Hong, D.H., Choi, C.-H.: Multicriteria fuzzy decision-making problems based on vague set theory. Fuzzy Sets Syst. **114**, 103–113 (2000)
17. Ishizuka, M., Fu, K.S., Yao, J.T.P.: Inference procedure and uncertainty for the problem reduction method. Inform. Sci. **28**, 179–206 (1982)

18. Sevastianov, P., Dymova, L., Bartosiewicz, P.: A framework for rule-base evidential reasoning in the interval setting applied to diagnosing type 2 diabetes. Expert Syst. Appl. **39**, 4190–4200 (2012)
19. Shafer, G.: A Mathematical Theory of Evidence. Princeton University Press, Princeton (1976)
20. Xu, Z.: Intuitionistic preference relations and their application in group decision making. Inf. Sci. **177**, 2363–2379 (2007)
21. Yager, R.R.: Generalized probabilities of fuzzy events from belief structures. Inf. Sci. **28**, 45–62 (1982)
22. Yang, J.B.: Rule and utility based evidential reasoning approach for multi-attribute decision analysis under uncertainties. Eur. J. Oper. Res. **131**, 31–61 (2001)
23. Yang, J.B., Liu, J., Wang, J., Sii, H.S., Wang, H.: Belief rule-base inference methodology using the evidential reasoning approach - RIMER. IEEE Trans. Syst. Man Cybern. Part A-Syst. Hum. 36(2), 266–285 (2006)
24. Yang, J.B., Liu, J., Xu, D.L., Wang, J., Wang, H.: Optimization models for training belief-rule-based systems. IEEE Trans. Syst. Man Cybern. Part A Syst. Hum. 37(4), 569–585 (2007)
25. Yen, J.: Generalizing the Dempster-Shafer theory to fuzzy sets. IEEE Trans. Syst. Man Cybern. **20**, 559–570 (1990)
26. Zadeh, L.: A simple view of the Dempster-Shafer theory of evidence and its application for the rule of combination. AI Mag. **7**, 85–90 (1986)

STRIPS in Some Temporal-Preferential Extension

Krystian Jobczyk[1,2]([⊠]) and Antoni Ligeza[2]

[1] University of Basse-Normandie of Caen, Caen, France
krystian_jobczyk@op.pl
[2] AGH University of Science and Technology of Kraków, Kraków, Poland

Abstract. In 1971 N. Nilson introduced a very smart improvement of forward search methodology in classical planning. It is commonly known as STRIPS. However, the original STRIPS is not sensitive to temporal and preferential aspects of reasoning. Unfortunately, neither temporal, nor preferential extension of STRIPS is known. This paper is just aimed at proposing such an extension, called later TP-STRIPS. In addition, some of its meta-logical properties are proved. It is also shown how TP-STRIPS may be exploited in more practical contexts.

1 Introduction

The biggest difficulty with the main forward-search algorithm is a problem how to improve efficiency reducing the search space (even for a cost of losing of the algorithm completeness). The STRIPS method was chronologically one of the first attempts to overcome this difficulty. It was initially introduced and described in detail by R. Fikes and N. Nilson in [1] in 1971. It's computational features were discussed in detail two decades later in [13] in 1990. Since this time STRIPS has been exploited as the most useful method in the framework of the search-based planning paradigm. It is often associated to different planning languages such as PDDL and its different extensions – collectively denoted as PDDL+ – see: [2–4].

1.1 The Paper Motivation and Objectives

Independently of a broad applicability of STRIPS to planning languages – [2–4], planners and solvers – based on these languages – no temporal extension of this method is known. Indeed, some known restrictions of STRIPS – mentioned by N. Nilsson alone in [1] – refer to the original non-temporal depiction of STRIPS only. In addition, no extension of STRIPS that additionally covers some preferential aspects of planning is known. In a consequence, we have no knowledge about possible restrictions of these two extensions. These shortcomings form the main motivation factor of this paper.

According to it, this paper is aimed at proposing:

– some fuzzy temporal extension of STRIPS and
– some preferential extension of STRIPS.

© Springer International Publishing AG 2017
L. Rutkowski et al. (Eds.): ICAISC 2017, Part I, LNAI 10245, pp. 241–252, 2017.
DOI: 10.1007/978-3-319-59063-9_22

This paper is also focused on checking how these extended STRIPS's (collectively denoted later by TP-STRIPS) may be exploited in practice. In particular, we exploit this method making use of the convolution-based representation of fuzzy temporal constraints of Allen's sort – as introduced in [15]. It allows us to observe how TP-STRIPS might be naturally incorporated to considerations based on real-analysis. Some ideas of this paper stem from such author's papers focused on fuzziness and temporal reasoning as: [6–12].

1.2 The Paper Organization

Rest of the paper is organized as follows. Section 2 presents the original Nilsson's formulation of STRIPS. Section 3 presents fuzzy temporal and preferential extensions of STRIPS denoted by TP-STRIPS. It also contains two meta-logical properties of STRIPS. Section 4 describes TP-STRIPS in use with respect to some (sub)problems of Multi-Agent Schedule-Planning Problem. Section 5 contains concluding remarks.

2 Terminological Background of the Paper Analysis

We preface the main paper considerations by introducing a terminological background of a current analysis. We begin with the original Nilsson's formulation of STRIPS. For an introducing of TP-STRIPS we also formulate some remarks on preferences defined on a base of fuzzy temporal constraints of Allen's sort. We generally assume that all basis concepts of classical planning – such as notions of action, goals, relevance, plan etc. – are the reader known. All of them may be found in [16].

2.1 STRIPS – in the Original Nilsson's Depiction

STRIPS method was conceived by N. Nilsson as a unique forward search procedure. Beginning from the initial state s_0 to achieve a goal g it works as follows.

1. This algorithm works if a set of *states* is not empty,
2. Then we choose a state $s \in states$.
   - If $g \subseteq s$, then we take $\pi(s)$ as a desired plan[1].
   - Otherwise, we take a set $E(s)$ of actions applicable to s.
     (a) if $E(s)$ is empty – we remove s from *states*,
     (b) if does not – we choose an action $a \in E(s)$ and exchange s for $s^{'}$ by removing effects of a from s. Then a current plan $\pi(s^{'}) = \pi(s).a$ (The action a is added at the end.)
     (c) The same procedure is repeated for a set $E(s^{'})$, etc. until g will be achieved.

---

[1] We can see π as a functions from a set of states S and $E(S)$ — as set of actions applicable to S.

Elements of this procedure may be represented in the STRIPS-algorithm as follows (Effect$^-$ and Effect$^+$ denote negative and positive effects of actions a. For a detailed definition – see: [16]).

```
STRIPS-algorithm(A, s₀, g)
begin
  states= {s₀}
  π(s₀) = ⟨⟩
      E(s₀) = {a| a is a ground instance of an operator in A and precond(a)⊆ s₀}
      while true do
      if states = ∅ then return failure end if
      choose a state s ∈ state
      if g ⊆ s then return π(s) end if
      if E(s) = ∅ then
          remove s from states
      else
          choose and remove an action a ∈ E(s)
          s' ← s/ Effect⁻(a)∪Effect⁺(a)
          if s' ∉ states then
              π(s') = π(s).a
              E(s') = {a|a is a ground instance of an operation ∈ A and precond(a)⊆ s'}
          end if
      end if
  end
```

2.2   Fuzzy Temporal Constraints and Preferences Based on Them

Multi-Agent Schedule-Planning Problem are often temporally rendered in terms of time period (days or shifts) associated somehow to some agents. Let us, therefore, assume that: $d \in D = \{d_1, d_2 \ldots, d_k\}$, $z \in Z = \{z_1, z_2 \ldots, z_l\}$, $n \in N, a \in A)$, where:

– the pair $(d, z)^n_a$ represents a shift z of a day d indexed by an action a and an agent (nurse) n associated to $(d, z)^2$

are given. For each pair $(d, z)^n_a$ (a and n are parameters here) one can define a new (not-necessary continuous) function $f((d, z)^n_a)$, which, somehow, maps the initial and last interval, i.e. $(d_1, z_1)^a_n$ and $(d_k, z_l)^a_n$, to some linear functions defined on them and it maps other intervals to a function $\sum X_{n,d,z,a}{}^3$. We consider the function f as normalized, i.e. $0 \le \left| f(d, z)^n_a \right| \le 1$. An exemplary diagram of f-function is shown in Fig. 2.

_____

[2] We assume that $(d_i, z_j) \ne (d_k, z_l)$ are disjoint for $i \ne k$ and $j \ne l$.

[3] Because of generality of consideration, we omit possible ways of defining such a function. Anyhow, it may be defined, for example, as follows (Fig. 1):

$$f((d, z)^a_n) = \begin{cases} A(d_1, z_1) & \text{for } (d_1, z_1)^a_n \\ -B(d_k, z_l) & \text{for } (d_k, z_l)^a_n \\ \sum X_{n,d,z,a} & \text{otherwise} \end{cases} \qquad (1)$$

for $A(d, z)$ and $-B(d, z)$ being linear functions of arguments (d_1, z_1) and (d_k, z_l) (_resp._) for parameters $A, B > 0$.

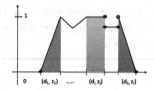

Fig. 1. Intervals (d_i, z_j) and function $f((d_i, z_j)$ (dark line on a picture) determining a fuzzy interval. This picture illustrates the fact that $f((d_i, z_j)$ should not necessary be a continuous function.

3 Fuzzy Temporal Constraints and Global Preferences in STRIPS

As earlier mentioned – STRIPS method itself refers to a case of classical planning. The "core" of this method consists in a plan construction by successive adding new (nondeterministically chosen) actions from a given set of admissible actions. The main choice criterion of them is their relevance to a task goal.

Our intention is to preserve the same idea in the proposed STRIPS-extension, denoted later as TP-STRIPS (*Temporal-Preferential STRIPS*). Independently of this similarity between the original version of STRIPS-algorithm and its TP-STRIPS, there is a couple of differences between these algorithms. It has already been said that STRIPS input contains: a set of actions $\mathcal{A}$, an initial state s and a goal state g. TP-STRIPS is intended to contain all these elements, but also:

– a fixed pair (d,z), where d denotes a fixed day and z is a fixed shift,
– a set of temporal constraints C and
– a unique function[4] called the preferential Pref(x): $(d, z) \mapsto [0, 1]$ and
– function $f(d, z)_n^a$ – described as earlier

An examplary preferential Pref(x) is marked in Fig. 3 by the red line.

The role of the preferential Pref(x) is to determine the preferred values of temporal constraints (more precisely, a preferred areas on a diagram of fuzzy intervals determined by function $f(d, z)_n^a$ – as in Fig. 2). Let us underline that we think about preferences only in terms of the preferential function Pref(x). We exploit these new components to propose a desired TP-STRIPS now. Because of a comparative nature of preferences – considering pairs of actions seems to be a more reasonable solution than considering a single action from $\mathcal{A}$ in this extension. This postulate approximates a nature of the temporal-preferential extension of STRIPS as follows:

1. After a nondeterministic choosing an action $a \in A$, we also nondeterministically chose other action a_1 from $A - \{a\}$ relevant to a given goal g,
2. We check whether a and a_1 respect temporal constraints from a given set C,

---

[4] We do not specify this function, we only assume a general condition of Lebesque integrability of it. It seems to be important because of the integral-based representation of preferences later. We omit, however, a detailed explanation.

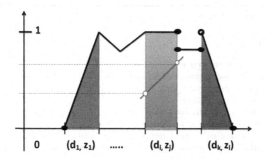

Fig. 2. Intervals (d_i, z_j) and function $f((d_i, z_j)$ (dark line on a picture) determining a fuzzy interval with the preferential function (red line) – introduced for (d_i, z_j). (Color figure online)

3. If chosen actions a_1 and a_2 respect temporal constraints from C, we compute $f(d, z)_n^{a_1}$ and $f(d, z)_n^{a_2}$ (for a fixed agent n) and compare preferences associated to each of them with values of the preferential function Pref(x).

An idea of TP-STRIPS. Let us specify TP-STRIPS in a more detailed way. The initial part of TS-STRIPS is the same as the original STRIPS-algorithm. The first difference consists in a requirement to check whether the chosen actions from action set $\mathcal{A}$ respect the imposed temporal constraints from a given set C – as already mentioned.

The second difference consists in a new 'preferential requirement' to check whether the actions – respecting constraints from C – are such that their temporal constraints (measured by a function f) take values greater or smaller than values of the preferential Pref(x). The procedure is now the following one: Taking two actions, say $\{a_1, a\}$ for a given pair (d, z) and agent set $\mathcal{N}$ – we compute $f(d, z)_{\mathcal{N}}^{a_1}$ and $f(d, z)_{\mathcal{N}}^{a}$.

- if $\forall x \in (d, z) \left(f(d, z)_{\mathcal{N}}^{a_1} > P(x) > f(d, z)_{N}^{a} \right)$, then we take a_1 and we take $\pi.a_1$ as a current plan.
- if $\forall x \in (d, z) \left(f(d, z)_{N}^{a} > \text{P(x)} > f(d, z)_{N}^{a_1} \right)$, then take a and we take $\pi.a$ as a current plan.

The TP-STRIPS algorithm is presented in the table.

```
TP-STRIPS (𝒜, 𝒩, C, Pref(x), s, goals, (d,z), f(d, z)ᴬₙ)
begin
    π ← the empty plan
    loop
    if s satisfies g then return π
    A ← {a| a is a ground instance of an opertor in O,
             and a is relevant for g}
    if A = ∅ then return failure
    choosenondeterministically an action a ∈ A
    choose nondeterministically an action a₁ ∈ A − {a}
    check whether a and a₁ respect temporal constraints from C
              if a respects C and a₁ does not then take a
                        s ← γ(s, a)
                        π ← π.a.
              if a₁ respects C and a does not then take a₁
                        s ← γ(s, a₁)
                        π ← π.a₁
              otherwise take {a, a₁}
              compare f(d, z)ₙ^{a₁} and f(d, z)ₙ^{a} with Pref(x)
                        if ∀x ∈ (d, z)(f(d, z)ₙ^{a₁} >Pref(x)> f(d, z)ₙ^{a}) then take a₁
                                  s ← γ(s, a₁)
                                  π ← π.π.a₁.
                        if ∀x ∈ (d, z)(f(d, z)ₙ^{a} >Pref(x)> f(d, z)ₙ^{a₁}) then take a
                                  s ← γ(s, a)
                                  π ← π.π.a.
                        if ∀x ∈ (d, z)(f(d, z)ₙ^{a₁} > f(d, z)ₙ^{a} > Pref(x)) then take a₁
                                  s ← γ(s, a₁)
                                  π ← π.π.a₁.
                        if ∀x ∈ (d, z)(f(d, z)ₙ^{a} > f(d, z)ₙ^{a₁} > Pref(x)) then take a
                                  s ← γ(s, a)
                                  π ← π.π.a.
              otherwise take ∅
end
```

It still remains a natural doubt whether this extended TP-STRIPS forms a decidable procedure and how its output size depends on the input size. The following lemma delivers an answer to these questions.

Lemma 1. *Assume that a set of actions $\mathcal{A}$ with $card(\mathcal{A}) = n$ and a set of temporal constraints C imposed on actions from $\mathcal{A}$ with $card(C) = m$ are given. Then TP-STRIPS is decidable and it polynomially depends on the input size.*

Proof. Let $\mathcal{A}$ and C are such as described in the lemma above, i.e. let $card(\mathcal{A}) = n$ and $card(C) = m$. Our output is N_{Choice} – a number of possible choices of actions from $\mathcal{A}$ in the whole TP-STRIPS procedure. Since in both the 'temporal' and the 'preferential' step of TP-STRIPS we always choose 2 actions from n-elemental set $\mathcal{A}$ of them, we can do it in $\binom{n}{2}$ ways. Each of such pairs of actions, say (a_i, a_j), for $1 \leq i, j \leq n$ should be now checked m-times to check which constraints from C are satisfied by it. Thus, we need $m\binom{n}{2}$ moves in the temporal step of TP-STRIPS. For each such a choice in the temporal step, we can choose 2 actions from maximally n-elemental set $\mathcal{A}$ in the 'preferential' step[5] Thus,

---

[5] This possibility holds if all actions remain 'good' as respecting temporal constraints from C.

the number of possible choices in these two steps N_{Choice} is constrained by $m\binom{n}{2}\binom{n}{2} = m\binom{n}{2}^2$. Since

$$m\binom{n}{2}\binom{n}{2} = m\binom{n}{2}^2 = m\left(\frac{n!}{2!(n-2)!}\right)^2 = \frac{m}{2}\left((n-1)n\right)^2 = \frac{m}{2}n^2(n-1)^2, \quad (2)$$

thus $N_{Choice} \leq O(n^4)$, what justifies its polynomial dependence on the input size.   □

It is noteworthy that the same property of under exponential complexity of TP-STRIPS is preserved by a more general situation, when temporal constraints are imposed not only on single actions, but also on pairs of actions, on their triples, ..., on k-tuples of actions from $\mathcal{A}$. This feature is formulated in the following lemma.

Lemma 2. *Assume that TP-STRIPS is modified such that in its 'temporal' step one consider single actions, their pairs, 3-tuples up to $n-1$-tuples from $\mathcal{A}$ – according to different types of temporal constraints from C (imposed on single actions, pairs, etc.). Then number of possible moves in 'temporal' step of TP-STRIPS is under exponentially dependent on the input size.*

Proof. Assume, as earlier, that $card(\mathcal{A}) = n$ and $card(C) = m$. Suppose that C_1 denotes constraints from C imposed on single actions, C_2 – denotes constraints imposed on pairs of actions, C_3 – imposed on 3-tuples,...,C_{n-1} – on $n-1$-tuples. Assume that $card(C_1) = m_1, card(C_2) = m_2, \ldots, card(C_{n-1}) = m_{n-1}$.

Then, obviously, $\sum_{k=0}^{n} m_{k-1} = m$ and we must successively choose 2 actions, the next – 3 actions, ..., $n-1$ actions from n-elemental $\mathcal{A}$ (Note that we can choose the same actions in new steps). Thus, all possible choices in the temporal step N_{choice}^{Temp} (our output):

$$N_{choice}^{Temp} = m_1\binom{n}{1} + m_2\binom{n}{2} + m_3\binom{n}{3} + m_4\binom{n}{4} + \ldots + m_{n-1}\binom{n}{n-1} \leq m\sum_{k=0}^{n}\binom{n}{k}. \quad (3)$$

Because of the known equality:

$$\sum_{k=0}^{n}\binom{n}{k} = 2^n \quad (4)$$

we get:

$$N_{choice}^{Temp} = m_1\binom{n}{1} + m_2\binom{n}{2} + m_3\binom{n}{3} + m_4\binom{n}{4} + \ldots + m_{n-1}\binom{n}{n-1} < m2^n. \quad (5)$$

Hence, $N_{choice}^{Temp} < O(exp(n))$, what already shows the thesis.   □

4   TP-STRIPS in Use

In this section we focus our attention on presenting how TP-STRIPS might be used in some problems of Multi-Agent Schedule-Planning Problem (M-A-S-PP) class. In order to illustrate this, we introduce the following problem. We make use of the convolution-based representation of Allen's relations as proposed by H-J. Ohlbach in [14, 15][6].

Problem. Let us consider an exemplary (simple) subproblem of M-AS-PP with two agents n_1 and n_2 performing some task with a goal $g = \{g_1, g_2\}$ up to day d_k from the initial state s_0[7] during a day d_1. They have an admissible set of actions $A = \{a_1, a_2, a_3, a_4, a_5, a_6, a_7\}$ such that:

- $\mathtt{precond}(a_4) = g_1$ and
- $\mathtt{precond}(a_1) = g_2$,
- $\mathtt{precond}(a_2) \subseteq s_0$ (only),
- a subgoal g_1 is quicker executable than g_2 (symbolically: $g_1 < g_2$) and
- temporal constraints imposed on actions a_5 and a_6 – having the same preconditions – are as presented in the picture and
- temporal constraints imposed on a_6 take values $(\frac{3}{2}, 4\frac{1}{2})$ over a time intervals $(1,2)$ (time units) and for a_5 take a constant value 0,1 in the same time intervals $(1,2)$.

Assume also that temporal constraints imposed on performing actions is approximated by the convolution[8].

$$h(x) = f(x - t) * g(t) = \int e^{x-t} \sin t \, dt. \tag{6}$$

Solution. Assume that – due to the original STRIPS procedure the following two plans are indicated as the appropriate ones.

- $\pi_1 = \langle a_2, a_7, a_3, a_4, a_5, a_1 \rangle$ and
- $\pi_2 = \langle a_2, a_7, a_3, a_4, a_6, a_1 \rangle$.

The choice between π_1 and π_2 will depend on temporal constraints and preferences imposed on a_5 and a_6. We will exploit the extended parts of TP-STRIPS-algorithm now.

Temporal and preferential component. If we do not impose any further restrictions on a choice of agents performing these two actions: a_5, a_6, both of them are appropriate from the perspective of temporal constraints. In fact, both actions respect temporal constraints represented graphically by the trapezium in Fig. 3 – as they completely belong to its interior.

Due to the appropriate step of TS-STRIPS-algorithm one should compare both of them from the point of view of a preference satisfaction. For that reason:

---

[6] Because of a very broad nature of the Ohlbach's approach, we omit his explications. They might be easily found in these works.

[7] In order to preserve generality of considerations, we omit a detailed specification of the initial state s_0 and a goal g. As such a pair of agents one can take, for example a pair:(crane, robot) etc.

[8] We can assume that $h(x)$ represents $meet(i, j)(x)$ Allen relation. Anyhow, this identification is redundant from the point of view of the current analysis.

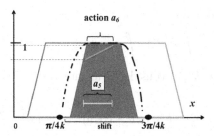

Fig. 3. Actions a_5 and a_6 performed during the shift $(\frac{1\pi}{4}k, \frac{3k\pi}{4})$ (of time units) with temporal constraints imposed on them (represented by values on y-axis normalized to [0,1]). To underline that a_5 and a_6 belong to skills of two different agents, they are marked by the green and red colors (resp.). (Color figure online)

– one needs compute the Preferential `Pref(x)`,
– secondly, one needs to check whether/which action a_5 or a_6 respects the preferences. It requires to check whether $f(d,z)_N^{a_5} > \texttt{Pref(x)}$ or $f(d,z)_N^a > \texttt{Pref(x)}$.

Computing convolution. It remains to consider the temporal constraints approximated by the given convolution.

$$h(x) = f(x-t) * g(t) = \int e^{x-t} \sin t \, dt. \tag{7}$$

Let us also define $\texttt{Pref(x)} = \varphi(f)(x)$, which we intend to consider $\varphi(f)(x)$ as the 'preferential line'. Without losing of generality one can assume that the actions a_6 and a_5 are represented by appropriate points of $\mathbb{R}^2$. It allows us to write

$$a_5 = \left\{ \left(x, \phi(x) \right) : 0 \le x \le \frac{\pi}{4} \text{ and } \phi(x) \le 1 \right\}, \tag{8}$$

$$a_6 = \left\{ \left(x, \psi(x) \right) : 0 \le x \le \frac{\pi}{4} \text{ and } \psi(x) \le 1 \right\}, \tag{9}$$

for some (Lebesque integrable) functions ϕ and ψ defined over $[0, \frac{\pi}{4}]$. Due to the modified STRIPS-algorithm – it is enough to check whether:

$$\forall x \in [0, \frac{\pi}{4}] \left(\phi(x) \le \texttt{Pref}(x) \text{ or } \texttt{Pref}(x) \le \phi(x) \right) \text{ and} \tag{10}$$

$$\forall x \in [0, \frac{\pi}{4}] \left(\psi(x) \le \texttt{Pref}(x) \text{ or } \texttt{Pref}(x) \le \phi(x) \right). \tag{11}$$

If we return to our convolution

$$h(x) = f(x-t) * g(t) = \int e^{x-t} \sin t \, dt. \tag{12}$$

then – due to well-known Convolution Theorem we obtain[9]:

$$F(h(x)) = \int e^{x-t} \, dz \int \sin t \, dt, \tag{13}$$

---

[9] We omit its formulation. It may be easily found in each handbook of real and abstract analysis. See, for example: [5].

where F denotes Fourier transform of $h(x)$. Since

$$\int e^{x-t}\,dz \int \sin t\,dt = e^x \cos t,\qquad (14)$$

thus $F * (h)(x)$ – as $F(h)(x)$ considered in integration limits $a, b \in [0, \frac{\pi}{4}]$ is as follows:

$$F * (h)(x) = e^x [\cos t]_b^a = e^x (\cos a - \cos b) = -2 \sin \frac{a+b}{2} \sin \frac{a-b}{2} e^x. \qquad (15)$$

Putting $A = -2 \sin \frac{a+b}{2} \sin \frac{a-b}{2}$, we obtain:

$$F * (h)(x) = Ae^x, \text{ so } h(x) = \frac{d}{dx} F * (h)(x) = Ae^x. \qquad (16)$$

Assume now two points $P_0 = (x_0, h(x_0))$ and $P_1 = (x_1, h(x_1))$ belonging to the diagram of $h(x)$. Because $P_0 = (x_0, \varphi(A, f(x_0)))$ and $P_1 = (x_1, \varphi(A, f(x_1)))$ a line passing through these points satisfies the following equation.

$$\varphi(f)(x) = \frac{\varphi(f)(x_1) - \varphi(f)(x_0)}{x_1 - x_0}(x - x_0) + \varphi(f)(x_0) \qquad (17)$$

Hence $\texttt{Pref}(x) := \varphi\big((f)(x), A\big)$ is as follows in our case:

$$\varphi((f(x), A) = \frac{A\big(e^{x_1} - e^{x_0}\big)}{x_1 - x_0}(x - x_0) + e^{x_0}. \qquad (18)$$

For $x_0 = 0$ and $x_1 = \frac{\pi}{4}$:

$$\varphi(f(x), A) = \frac{4A\big(e^{\frac{\pi}{4}} - e^0\big)}{\pi} x + e^{\frac{\pi}{4}} = \frac{4A\big(e^{\frac{\pi}{4}} - 1\big)}{\pi} x + e^{\frac{\pi}{4}}, \qquad (19)$$

for $A = -2 \sin \frac{a+b}{2} \sin \frac{a-b}{2} e^x$. It is not difficult to check that for $x \in [0, \frac{\pi}{4}]$:

$$e^{\frac{\pi}{4}} \leq \varphi(f(x), A) \leq 2(e^{\frac{\pi}{4}} - 1). \qquad (20)$$

After dividing by the normalization factor $N = 10$ we get:

$$\frac{e^{\frac{\pi}{4}}}{10} \leq \frac{\varphi(f(x), A)}{10} \leq \frac{2(e^{\frac{\pi}{4}} - 1)}{10}. \qquad (21)$$

Recall that $\phi(x)_{a_5} = \frac{1}{10}$ for all $x \in [0, \frac{\pi}{4}]$, so

$$\phi(x)_{a_5} \leq \frac{\varphi(f(x), A)}{10}. \qquad (22)$$

Simultaneuosly, $\frac{2(e^{\frac{\pi}{4}} - 1)}{10} \leq \frac{e}{5} \leq \phi(x)_{a_6}$, thus

$$\frac{\varphi(f(x), A)}{10} \leq \phi(x)_{a_6}. \qquad (23)$$

Therefore, we accept the action a_6 and reject a_5 (Fig. 4).

Finally, we choose a_6 instead of a_5 and the sequence $\pi = \{a_1, a_2, a_3, a_6, a_7\}$ as the required plan.

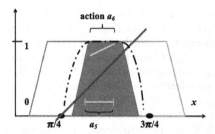

Fig. 4. Preferences (determined by the line running through $\frac{\pi}{4}$) imposed on temporal constraints (the trapesium) with fuzzy temporal constraints (the broken line).

5   Towards a Generalization and Closing Remarks

It has just been shown how the STRIPS method may be temporally and preferentially extended in order to support some more advanced types of reasoning. We also proved some meta-logical features of this extension. Anyhow, this approach referred to some rather ideal situation. Meanwhile, not always the situation is so ideal as the presented above. In fact, it is possible that either both of the following conditions

1. If $\forall x \in (d,z)(f(d,z)_N^{a_1} > \text{Pref}(\texttt{x}) > f(d,z)_N^{a})$ and
2. If $\forall x \in (d,z)(f(d,z)_N^{a} > \text{Pref}(\texttt{x}) > f(d,z)_N^{a_1})$

or (at least) one of them do not completely hold.

Unfortunately, solving this problem does not exhaust a list of all possible doubts. In fact, the situation, earlier described, may be not so ideal from another perspective, when functions representing temporal constraints imposed on actions are not linear or even continuous. Anyhow, an answer to this question slightly exceeds a terminological framework of this paper – as it requires more advanced and new portion of formal tools of real analysis and approximation theory. Nevertheless, it forms an interesting subject of further research.

References

1. Fikes, P., Nilsson, N.: Strips: a new appraoch to the application of theorem proving to problem solving. Artif. Intell. **2**(3–4), 189–208 (1971)
2. Fox, M., Long, D.: Pddl+: Planning with time and metric sources. Technical report, University of Durham (2001a)
3. Fox, M., Long, D.: Pddl2.1: an extension to pddl for expressing temporal planning domains. Technical report, University of Durham (2001b)
4. Fox, M., Long, D.: An extension to pddl for expressing temporal planning domains. J. Artif. Intell. Res. **20**, 61–124 (2003)
5. Hewitt, E., Stromberg, K.: Real and Abstract Analysis. Springer, New York (1965)
6. Jobczyk, K., Ligeza, A.: Fuzzy-temporal approach to the handling of temporal interval relations and preferences. In: Proceeding of INISTA, pp. 1–8 (2015)

7. Jobczyk, K., Ligeza, A.: Multi-valued halpern-shoham logic for temporal allen's relations and preferences. In: Proceedings of the Annual International Conference of Fuzzy Systems (FuzzIEEE) (2016, page to appear)
8. Jobczyk, K., Ligeza, A.: Systems of temporal logic for a use of engineering. Toward a more practical approach. In: Stýskala, V., Kolosov, D., Snášel, V., Karakeyev, T., Abraham, A. (eds.) Intelligent Systems for Computer Modelling. AISC, vol. 423, pp. 147–157. Springer, Cham (2016). doi:10.1007/978-3-319-27644-1_14
9. Jobczyk, K., Ligeza, A., Kluza, K.: Selected temporal logic systems: an attempt at engineering evaluation. In: Rutkowski, L., Korytkowski, M., Scherer, R., Tadeusiewicz, R., Zadeh, L.A., Zurada, J.M. (eds.) ICAISC 2016. LNCS (LNAI), vol. 9692, pp. 219–229. Springer, Cham (2016). doi:10.1007/978-3-319-39378-0_20
10. Jobczyk, K., Bouzid M., Ligeza, A., Karczmarczuk, J.: Fuzzy integral logic expressible by convolutions. In: Proceeding of ECAI 2014, pp. 1042–1043 (2014)
11. Jobczyk, K., Bouzid M., Ligeza, A., Karczmarczuk, J.: Fuzzy logic for representation of temporal verbs and adverbs 'often' and 'many times'. In: Proceeding of LENSL 2011 Tokyo (2014)
12. Jobczyk, K., Ligeza, A., Bouzid, M., Karczmarczuk, J.: Comparative approach to the multi-valued logic construction for preferences. In: Rutkowski, L., Korytkowski, M., Scherer, R., Tadeusiewicz, R., Zadeh, L.A., Zurada, J.M. (eds.) ICAISC 2015. LNCS (LNAI), vol. 9119, pp. 172–183. Springer, Cham (2015). doi:10.1007/978-3-319-19324-3_16
13. Lifschitz, V.: On the semantics of strips. In: Georgeff, M.P., Lansky, A.L., (eds.) Reasoning about Actions and Plans, pp. 1–9, 523–530 (1990)
14. Ohlbach, H.: Fuzzy time intervals and relations-the futire library. Research Report PMS-04/04, Inst. f. Informatik, LMU Munich (2004)
15. Ohlbach, H.: Relations between time intervals. In: 11th Internal Symposium on Temporal Representation And Reasoning, vol. 7, pp. 47–50 (2004)
16. Traverso, P., Ghallab, M., Nau, D.: Automated Planning: Theory and Practice. Elsevier, Amsterdam (2004). 1997

Geometrical Interpretation of Impact of One Set on Another Set

Maciej Krawczak$^{(\boxtimes)}$ and Grażyna Szkatuła

Systems Research Institute, Polish Academy of Sciences,
Newelska 6, 01–447 Warsaw, Poland
{krawczak,szkatulg}@ibspan.waw.pl

Abstract. In the paper, we describe new tacit problems during the process of comparison of objects. The direction of objects' comparison seems to have essential role because such comparison may not be symmetric. Thus, we can say that two objects may be viewed as an attempt to determine the degree to which they are similar or different. In this paper, we consider objects described by a set of nominal attributes which values are not precisely known or can be repeated in the object description. Two kinds of objects' descriptions are considered, the first the fuzzy and the second the multiset description. Asymmetric phenomena of comparing such descriptions of objects is emphasized and discussed.

Keywords: Fuzzy description of objects · Multiset description of objects · Nominal-valued attributes · Directional comparison of objects

1 Introduction

The role of similarity or dissimilarity of two objects is fundamental in many theories of cognitive as well as behavior knowledge, and therefore for comparison of objects there are commonly used different measures of objects' similarity.

It is important to notice that e.g., in psychological literature, similarity between objects can be asymmetric (e.g., Tversky [9]). The idea of asymmetries appearing in comparison of objects comes directly from the Tversky and Kahneman prospect theory (e.g., Tversky and Kahneman [10]). In short, this theory describes people rationality in decisions involving risk, and states, that people make decisions based on the potential value of losses and gains. The hypothetical value function is in general asymmetrical, see Fig. 1. The most evident characteristics of the prospect theory is that the same loss creates greater feeling of pain compared to the joy created by an equivalent gain. In the next considerations, Tversky considered objects represented by a sets of features, and proposed measuring of similarity via comparison of their common and distinctive features (e.g., Tversky [9]). Such assumptions generate different approaches to comparison of objects. Namely, comparing two objects A and B there are three following fundamental questions: "how similar are A and B?", "how similar is A to B?" and "how similar is B to A?". The first question does not distinguishes the

© Springer International Publishing AG 2017
L. Rutkowski et al. (Eds.): ICAISC 2017, Part I, LNAI 10245, pp. 253–262, 2017.
DOI: 10.1007/978-3-319-59063-9_23

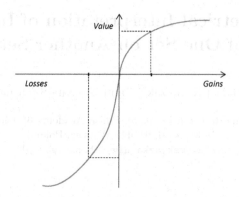

Fig. 1. A hypothetical value function [10]

directions of comparison and corresponds to symmetric similarity. In the next two questions the direction of comparison cannot be negligible and the similarity of the objects should not be symmetric.

Our present work is motivated by the need to develop a new look at comparison of objects described by either a set of incomplete and inconsistent nominal attributes. In some sense, the present paper is both a continuation as well as extension of the authors' previous papers on comparison of objects (cf. Krawczak and Szkatuła [1,3]).

Here, the concept of *the mutual impact of one set by another* is considered, and the concept can be considered as some extension of Levenshtein's distance (cf. Levenshtein [7]), however the presented concept is much more general.

It seems that the assumption of symmetry should not be established in advance, and therefore asymmetry of data should not be neglected. Such phenomena of asymmetry of the comparisons of objects mean, that one object has different impact on another object. It is obvious that objects are describe by attributes, and each object's description is represented by a respective set of the attributes' values. This way, the task of comparison of objects is converted into a task of comparison of sets of values of attributes.

Let us illustrate the idea of asymmetry of comparison of two crisp sets describing the objects. Thus, we consider a finite set V of nominal values, and two ordinary subsets A_i and A_j, where $A_i, A_j \subseteq V$. We can introduce a concept of the impact of set A_j by another set A_i, denoted by $(A_i \mapsto A_j)$, which is interpreted as a difference between these two sets. Analogically, the concept of the impact of set A_i by set A_j is denoted by $(A_j \mapsto A_i)$. Graphical illustration of the impact phenomena between two fixed subsets of the set V is depicted in Fig. 2.

Applying more complex description of objects we can discover several interesting phenomena involved in the process of comparison of objects.

In this paper, we consider a finite, non-empty set of objects (e.g., concepts, patterns, references, etc.), each object is described by a set of nominal attributes, it means the values of the attributes are not precisely known or can be repeated in the object description. On the one hand, the methodology of the fuzzy set

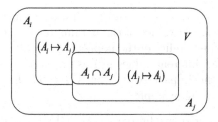

Fig. 2. A graphical illustration of the impact between two ordinary sets

theory allows us to model such imperfect, incomplete and inconsistent data, and on the other hand the multisets theory gives a very convenient mathematical methodology to describe and analyze qualitative data with repeated values of objects' attributes. Therefore, we considered two different cases of objects' description:

– in the first case, the objects are described by fuzzy sets,
– in the second case, by multisets.

For a better understanding of the considerations describing asymmetric results during comparisons of objects we use the geometrical interpretations presented in the forthcoming sections.

2   Matching of Fuzzy Sets

Let us consider a non-empty and finite set V of nominal elements, denoted by $V = \{v_1, v_2, \ldots, v_L\}$. In 1965, L. A. Zadeh introduced the concept of fuzzy set theory as an extension of the classical set theory (cf. Zadeh [11]). *A fuzzy set A in the set V* may be represented by a collection of ordered pairs written in the following form

$$A = \{(\nu,\ \mu_A(v)) \mid v \in V\} \tag{1}$$

where $\mu_A : V \to [0,1]$ is the membership function. Zadeh also introduced the fundamental operations for fuzzy sets, namely union, intersection and complementation.

Let us assume that there is a collection of fuzzy sets $FS(V)$ in the set V. Let us consider two fuzzy sets $A_i, A_j \in FS(V)$. We defined a novel concept of *the impact on one fuzzy set A_j by another fuzzy set A_i*, which is denoted by $(A_i \mapsto A_j)$, and interpreted as a difference between fuzzy sets

$$(A_i \mapsto A_j) = A_i \ominus A_j = \{(\nu,\ \mu_{A_i \mapsto A_j}(v)) \mid v \in V\} \tag{2}$$

where $\mu_{A_i \mapsto A_j}(v) := max\{\mu_{A_i}(v) - \alpha \cdot \mu_{A_j}(v)\}$ for a parameter $\alpha \geq 0$. In the opposite case, the impact on the fuzzy set A_i by the fuzzy set A_j is defined in a similar way

$$(A_j \mapsto A_i) = A_j \ominus A_i = \{(\nu,\ \mu_{A_j \mapsto A_i}(v)) \mid v \in V\} \tag{3}$$

where $\mu_{A_j \mapsto A_i}(v) := max\{\mu_{A_j}(v) - \beta \cdot \mu_{A_i}(v)\}$ for a parameter $\beta \geq 0$. This model is characterized by different values of the parameters α and β. In this sense one could have some extra flexibility of the impact. Obviously, the parameters α and β could be assumed as different values in Eqs. (2) and (3).

The meaning of the impact of one fuzzy set on the another fuzzy set is illustrated in the following example.

Example 1. Let us consider two fuzzy sets A_1 and A_2 in the set $V = \{v_1, v_2, v_3\}$, where $A_1 = \{(v_1, 0.4), (v_2, 0), (v_3, 0)\}$ and $A_2 = \{(v_1, 1), (v_2, 0.1), (v_3, 0.4)\}$. The impact on the fuzzy set A_2 by the fuzzy set A_1 is the empty fuzzy set because the following condition $(A_1 \mapsto A_2) = A_1 \ominus A_2 = \{(v_1, 0), (v_2, 0), (v_3, 0)\} = \varnothing$ is satisfied. On the other hand, the impact on the fuzzy set A_1 by the fuzzy set A_2 is the following fuzzy set $(A_2 \mapsto A_1) = A_2 \ominus A_1 = \{(v_1, 0.6), (v_2, 0.1), (v_3, 0.4)\}$.

The geometrical interpretation of the proposed concept of the fuzzy sets impact in 3D space is presented below.

Geometrical interpretation of fuzzy sets impacts

Therefore, we will consider a case characterized by $card(V) = 3$, i.e., $V = \{v_1, v_2, v_3\}$, so that we will consider the fuzzy set $A_j \in FS(V)$, denoted by $A_j = \{(v_1, \mu_{A_j}(v_1)), (v_2, \mu_{A_j}(v_2)), (v_3, \mu_{A_j}(v_3))\}$, where $\mu_A : V \rightarrow [0, 1]$. The fuzzy set A_j can be represented as a point in the three dimensional coordinates space, with coordinates $(\mu_{A_j}(v_1), \mu_{A_j}(v_2), \mu_{A_j}(v_3))$. For simplicity, the elements v_1, v_2, v_3 can be omitted, i.e., the point has coordinates $(\mu_{A_j}^1, \mu_{A_j}^2, \mu_{A_j}^3)$, see Fig. 3.

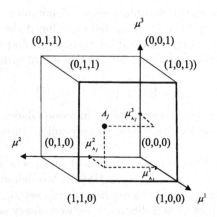

Fig. 3. The fuzzy set A_j represented as a point $(\mu_{A_j}^1, \mu_{A_j}^2, \mu_{A_j}^3)$, marked by •

The point (0,0,0) represents the fuzzy set with the elements v_1, v_2 and v_3 fully not belonging to this fuzzy set; the point (1,1,1) represents the elements v_1, v_2 and v_3 fully belonging to the fuzzy set; the point (1,0,0) represents the element v_1 fully belonging to the fuzzy set and the elements v_2 and v_3 fully not belonging to the fuzzy set; the point $(0,1,1)$ represents the element v_1 fully not belonging to the fuzzy set and the elements v_2 and v_3 fully belonging to the fuzzy set, etc. Any other combination of the values belonging to the fuzzy set with some degree can be represented inside the cube, as shown in Fig. 3.

For the fixed fuzzy set A_j there are eight areas (denoted by the area I, II, …VIII), as shown in Fig. 4.

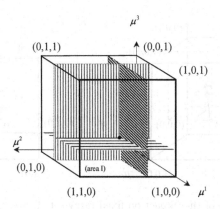

Fig. 4. Eight areas which designates a fixed fuzzy set A_j, marked by ●

Let us consider the fixed fuzzy set A_j and the area VII, i.e., the cubic marked by dotted line, as shown in Fig. 5.

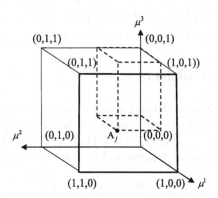

Fig. 5. Area of possible location of arbitrary fuzzy set A_i

Let us consider the any arbitrary fuzzy set $A_i \in FS(V)$ represented inside the cube, denoted by $A_i = \{(v_1, \mu_{A_i}(v_1), (v_2, \mu_{A_i}(v_2)), (v_3, \mu_{A_i}(v_3))\}$, with coordinates $(\mu^1_{A_i}, \mu^2_{A_i}, \mu^3_{A_i})$. Each point related to the fuzzy set A_i lying within specified area and the fuzzy set A_j satisfies some conditions. For the fixed fuzzy set

A_j and the arbitrary fuzzy sets A_i belonging to the area VII, the conditions $\mu^1_{A_j} \geq \mu^1_{A_i}$, $\mu^2_{A_j} \geq \mu^2_{A_i}$ and $\mu^3_{A_j} \leq \mu^3_{A_i}$ are satisfied. A three-dimensional interpretation of the impact on the fixed fuzzy set A_j and the arbitrary fuzzy set A_i is presented in Fig. 6. In figure, there are two impacts, i.e., $(A_j \mapsto A_i)$ and $(A_i \mapsto A_j)$, $\alpha = \beta = 1$. The arrow indicates direction of the impact.

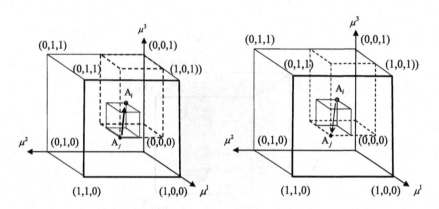

Fig. 6. Interpretation of the impact on fixed fuzzy set A_i and arbitrary fuzzy set A_j

Analyzing Fig. 6, one may notice that for the arbitrary fuzzy set A_i belonging to the area VII, the following conditions $\mu^1_{A_j \mapsto A_i} = \mu^1_{A_j} - \mu^1_{A_i}$, $\mu^2_{A_j \mapsto A_i} = \mu^2_{A_j} - \mu^2_{A_i}$ and $\mu^3_{A_j \mapsto A_i} = 0$ are satisfied, i.e., the segments marked in red indicate positive values $\mu^1_{A_j \mapsto A_i}$ and $\mu^2_{A_j \mapsto A_i}$, respectively. In the opposite case, the conditions $\mu^1_{A_i \mapsto A_j} = 0$, $\mu^2_{A_i \mapsto A_j} = 0$ and $\mu^3_{A_i \mapsto A_j} = \mu^3_{A_i} - \mu^3_{A_j}$ are satisfied, and the segment marked in red indicate positive value $\mu^3_{A_i \mapsto A_j}$.

3   Matching of Multisets

Let us consider the multisets defined in so-called multiplicative form (e.g., Petrovsky [8]), drawn from a non-empty and finite ordinary set V of nominal-valued elements, $V = \{v_1, v_2, \ldots, v_L\}$, v_{i+1}, $\forall i \in \{1, 2, \ldots, L-1\}$. The multiset S drawn from the ordinary set can be represented by a set of ordered pairs:

$$S = \{(k_S(v), v) \mid v \in V\} \tag{4}$$

where $k_S : V \rightarrow \{0, 1, 2, \ldots\}$. In (4) the function $k_S(.)$ is called *a counting function* or *the multiplicity function*, and the value of $k_S(v)$ specifies the number of occurrences of the element $v \in V$ in the multiset S. The element which is not included in the multiset S has its counting function equal zero. *The multiset space* is the set of all multisets with elements of V, such that no element occurs more than m times, and is denoted by $[V]^m$.

In the authors' previous papers (e.g., Krawczak and Szkatuła [4–6]) there was developed the definition of a novel *concept of impact* on one multiset S_2 by another multiset S_1, denoted by $(S_1 \mapsto S_2)$, which is interpreted as a difference between multisets, in the following way:

$$(S_1 \mapsto S_2) = S_1 \ominus S_2 = \{(k_{S_1 \mapsto S_2}(v), v) \mid v \in V\} \tag{5}$$

where $k_{S_1 \mapsto S_2}(v) := max\{k_{S_1}(v) - k_{S_2}(v), 0\}$. It is important to notice that the direction of the impact has significant meaning. In other words, one multiset can impact on another multiset with some degree. The counterpart definition is similar

$$(S_2 \mapsto S_1) = S_2 \ominus S_1 = \{(k_{S_2 \mapsto S_1}(v), v) \mid v \in V\} \tag{6}$$

where $k_{S_2 \mapsto S_1}(v) := max\{k_{S_2}(v) - k_{S_1}(v), 0\}$.

The meaning of the impact of one multiset on another multiset is illustrated in the following example.

Example 2. There is considered the following set $V = \{a, b, c, d, e\}$, and two exemplary multisets $S_1 = \{(1, a), (1, e)\}$ and $S_2 = \{(1, a), (1, d), (3, e)\}$, where $S_1, S_2 \in [V]^3$. The impact on the multiset S_2 by the multiset S_1 is the empty multiset, because $(S_1 \mapsto S_2) = S_1 \ominus S_2 = \varnothing$. The impact of the multiset S_1 by the multiset S_2 is the following multiset $(S_2 \mapsto S_1) = S_2 \ominus S_1 = \{(1, d), (2, e)\}$.

The geometrical interpretation of the proposed concept of the multisets' impact in 2D space is presented below.

Geometrical interpretation of multisets' impacts

Let us assume that $card(V) = 2$, i.e., $V = \{v_1, v_2\}$, and then consider two multisets $S_1 = \{(k_{S_1}(v_1), v_1), (k_{S_1}(v_2), v_2)\}$ and $S_2 = \{(k_{S_2}(v_1), v_1), (k_{S_2}(v_2), v_2)\}$, $S_1, S_2 \in [V]^m$. Each considered multiset can be represented as a point in 2-dimensional space, see in Fig. 7, and these two points have the following coordinates $(k_{S_1}(v_1), k_{S_1}(v_2))$ and $(k_{S_2}(v_1), k_{S_2}(v_2))$, respectively. According to Eq. (5), the impact on the multiset S_2 by the multiset S_1 is interpreted as a new multiset described as follows (cf. Krawczak and Szkatuła [4–6]):

$$(S_1 \mapsto S_2) = \{(k_{S_1 \mapsto S_2}(v_1), v_1), (k_{S_1 \mapsto S_2}(v_2), v_2)$$
$$= \{(max\{k_{S_1}(v_1) - k_{S_2}(v_1), 0\}, v_1), (max\{k_{S_1}(v_2) - k_{S_2}(v_2), 0\}, v_2)\}$$

And, in the opposite case, the impact on the multiset S_1 by the multiset S_2 has the similar definition (cf. Krawczak and Szkatuła [4–6]):

$$(S_2 \mapsto S_1) = \{(k_{S_2 \mapsto S_1}(v_1), v_1), (k_{S_2 \mapsto S_1}(v_2), v_2)$$
$$= \{(max\{k_{S_2}(v_1) - k_{S_1}(v_1), 0\}, v_1), (max\{k_{S_2}(v_2) - k_{S_1}(v_2), 0\}, v_2)\}$$

The two-dimensional geometrical interpretations of the impact on the exemplary multisets S_1 and S_2 are presented in Fig. 7.

Within the figure, there are indicated two impacts, i.e., the impact $(S_1 \mapsto S_2)$ in the left figure, and $(S_2 \mapsto S_1)$ in the right figure. The arrows indicate the directions of the impact.

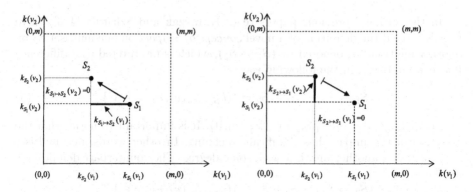

Fig. 7. The graphical interpretations of the impact on the multisets S_1 and S_2

Analyzing Fig. 7, one may notice that for the exemplary multisets $S_1, S_2 \in [V]^m$, the impact of one multiset by another creates a new multiset, obtained as the subtraction of these two multisets. Thus, the multisets' impact describes difference between multisets, and therefore the direction of the impact cannot be neglected. The following conditions $k_{S_1 \mapsto S_2}(v_1) = k_{S_1}(v_1) - k_{S_2}(v_1)$ and $k_{S_1 \mapsto S_2}(v_2) = 0$, as well as $k_{S_2 \mapsto S_1}(v_1) = 0$ and $k_{S_2 \mapsto S_1}(v_2) = k_{S_2}(v_2) - k_{S_1}(v_2)$, are satisfied. The segments marked by the thick lines indicate positive values of the counting functions $k_{S_1 \mapsto S_2}(v_1)$ and $k_{S_2 \mapsto S_1}(v_2)$, respectively. In the case of the impact $(S_1 \mapsto S_2)$, the beginning of the segment is the point $(k_{S_2}(v_1), k_{S_1}(v_2))$, and the end of the segment is the point $(k_{S_1}(v_1), k_{S_1}(v_2))$. While, for the opposite impact $(S_2 \mapsto S_1)$, the beginning of the segment is the point $(k_{S_2}(v_1), k_{S_1}(v_2))$, and the end is the point $(k_{S_2}(v_1), k_{S_2}(v_2))$.

The cases shown in Fig. 7 have been especially selected in order to obtain the impacts as single-element multisets, just indicated by the thick lines. Thus, the first impact, depictured at left side of Fig. 7, can be rewritten in the following multiset form

$$(S_1 \mapsto S_2) = \{(k_{S_1 \mapsto S_2}(v_1), v_1), (k_{S_1 \mapsto S_2}(v_2), v_2)\}$$
$$= \{(k_{S_1}(v_1) - k_{S_2}(v_1), v_1), (0, v_2)\}$$

while the second impact, depictured at right side of Fig. 7, can be rewritten as

$$(S_2 \mapsto S_1) = \{(k_{S_2 \mapsto S_1}(v_1), v_1), (k_{S_2 \mapsto S_1}(v_2), v_2)\}$$
$$= \{(0, v_1), (k_{S_2}(v_2) - k_{S_1}(v_2), v_2)\}$$

Thus, even such simple illustration made aware that the considered directional comparisons of multisets, in result the directional comparisons of objects, cannot be ignored.

4   Conclusions

In this paper we studied a new look at problems appearing during the process of comparison of objects. The comparison of two object should be considered

with regard to order of compared objects. Therefore, we called such phenomena
as the directional comparisons. For illustration of the considered problem, we
examined two kinds of objects' descriptions, namely the objects are described
by fuzzy sets, and the objects described by multisets. In results, we have noticed
the directional phenomena of comparisons of objects. In the paper we made use
the graphical illustrations in order to clarify the geometric interpretations of the
arisen problem. This way, we considered a finite, non-empty set of objects, and
each object was described by a set of attributes described by nominal values.
Within the first case the attributes' values are not precisely known while in
the second case repetitions of the objects' values were allowed. The direction of
objects' comparison seems to have essential role because such comparison may
not be symmetric.

In the authors' opinion, the considered asymmetric phenomena of comparing objects can constitute to developed the new measures objects' comparisons
(cf. Krawczak and Szkatuła [4–6]). However, it is impossible to indicate which
measure is better in general and the choice depends on the nature of data under
consideration. Therefore, for example, there does not exist the best measure for
evaluation of proximity between two arbitrary multisets. Here, we present an
example of comparison of a few measures for the multisets.

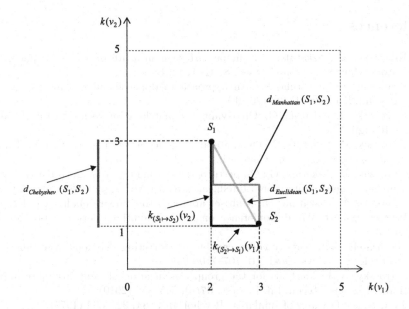

Fig. 8. A graphical illustration of few selected measures for fixed multisets S_1 and S_2

Let us consider the multisets drawn from an ordinary set V of nominal-
valued elements. Let us assume, that $card(V) = 2$, i.e., $V = \{v_1, v_2\}$, and then
consider two fixed multisets $S_1 = \{(2, v_1), (3, v_2)\}$ and $S_2 = \{(3, v_1), (1, v_2)\}$,
where $S_1, S_2 \in [V]^5$. The problem is to calculate degrees of proximity between

these multisets. We compare the counting functions of proposed impacts and the selected measures of distances, namely $d_{Chebyshev}(S_1, S_2) = \max\limits_{i \in \{1,2\}} |\ k_{S_1}(v_i) -$

$k_{S_2}(v_i)\ |$, $d_{Manhattan}(S_1, S_2) = \sum\limits_{i=1}^{2} |\ k_{S_1}(v_i) - k_{S_2}(v_i)\ |$, and $d_{Eyclidean}(S_1, S_2) =$

$\sqrt{\sum\limits_{i=1}^{2} |\ k_{S_1}(v_i) - k_{S_2}(v_i)\ |^2}$. The graphic illustration is shown in Fig. 8. It is easy to confirm that the different criteria of evaluation of the distances between multisets will lead to different results. Obviously, the Chebyshev measure $d_{Chebyshev} = 2$ (the purple segment) as well as Manhattan $d_{Manhattan} = 3$ (the red path shows one of possible realization) and Euclidean $d_{Eyclidean} = \sqrt{5}$ (the green segment) are symmetric. However, if the direction of comparison of multisets cannot be neglected, then the counting functions $k_{S_1 \mapsto S_2}(v_2) = 2$ and $k_{S_2 \mapsto S_1}(v_1) = 1$ of the impacts (two black segments) may be used.

The direction of the objects' comparison (or groups of objects' comparison) may have significant meaning, namely, to generate the classification rules, which distinguish no commutativity of the comparisons of the classes. Additionally, the methodology can be used to evaluate groups' distances in order to solve clustering tasks (cf. Krawczak and Szkatuła [2]).

References

1. Krawczak, M., Szkatuła, G.: On perturbation measure of sets - Properties. J. Autom. Mob. Robot. Intell. Syst. **8**, 41–44 (2014a)
2. Krawczak, M., Szkatuła, G.: An approach to dimensionality reduction in time series. Inf. Sci. **260**, 15–36 (2014b)
3. Krawczak, M., Szkatuła, G.: On asymmetric matching between sets. Inf. Sci. **312**, 89–103 (2015a)
4. Krawczak, M., Szkatuła, G.: On bilateral matching between multisets. In: Advances in Intelligent Systems and Computing, pp. 161–174 (2015b)
5. Krawczak, M., Szkatuła, G.: On perturbations of multisets. In: 2015 IEEE Symposium Series on Computational Intelligence, South Africa, pp. 1583–1589 (2015c)
6. Krawczak, M., Szkatuła, G.: Multiset approach to compare qualitative data. In: Proceedings 6th World Conference on Soft Computing, Berkeley, pp. 264–269 (2016)
7. Levenshtein, V.I.: Binary codes capable of correcting deletions, insertions, and rever-sals. Sov. Phys. Dokl. **10**, 707–710 (1966)
8. Petrovsky, A.B.: Methods for the group classification of multi-attribute objects (Part 1). Scient. Techn. Inf. Process. **37**(5), 346–356 (2010)
9. Tversky, A.: Features of similarity. Psychol. Rev. **84**, 327–352 (1977)
10. Tversky, A., Kahneman, D.: The framing of decisions and the psychology of choice. Science **211**, 453–458 (1981)
11. Zadeh, L.A.: Fuzzy sets. Inf. Control **8**, 338–353 (1965)

A Method for Nonlinear Fuzzy Modelling Using Population Based Algorithm with Flexibly Selectable Operators

Krystian Łapa[1(✉)], Krzysztof Cpałka[1], and Lipo Wang[2]

[1] Institute of Computational Intelligence,
Częstochowa University of Technology, Częstochowa, Poland
{krystian.lapa,krzysztof.cpalka}@iisi.pcz.pl
[2] Nanyang Technological University, Singapore
elpwang@ntu.edu.sg

Abstract. In this paper a new method based on a population-based algorithm with flexible selectable operators for nonlinear modeling is proposed. This method enables usage of any types of exploration and exploitation operators, typical for population-based algorithms. Moreover, in proposed approach each solution from population encodes activity and parameters of these operators. Due to this, they can be selected dynamically in the evolution process. Such approach eliminates the need for determining detailed mechanism of the population-based algorithm. For the simulations typical nonlinear modeling benchmarks were used.

Keywords: Population-based algorithms · Selection of evolutionary operators · Neuro-fuzzy systems · Nonlinear modeling · Exploration and exploitation

1 Introduction

Population-based algorithms are an artificial intelligence (AI) methods (see e.g. [3,6,8,15,23,49,71,75–77]) that belong to evolutionary computation subset (see e.g. [9,17,37,41,43,67,72,74]). The field of AI can be defined as a study of intelligent agents, machines that mimics functions associated with human minds functions (such as learning [30,42,56] and problem solving [65,66]). The goals of AI include, among others, data mining and processing (see e.g. [22,44,47,61,63,70]), language and image processing (see e.g. [1,28,36,64]), systems identification (see e.g. [27,50–55]), optimization (see e.g. [26,31]), etc. In this paper population-based algorithms are used for parameter optimization of neuro-fuzzy systems (see e.g. [20,21,48,59,60,62]) for nonlinear modelling (see e.g. [5,38,39,57]), however proposed method can be used in any optimization problem.

The population-based algorithm differ from traditional optimization methods: (a) they do not process parameters directly, but their encoded form, (b) they search solutions based not on a single point, but on a population of the points (the population contains individuals, each individual encodes single solution),

© Springer International Publishing AG 2017
L. Rutkowski et al. (Eds.): ICAISC 2017, Part I, LNAI 10245, pp. 263–278, 2017.
DOI: 10.1007/978-3-319-59063-9_24

(c) they use objective function directly, not its derivatives (this function determines the quality of the solutions in the population), (d) they use probabilistic mechanisms, not deterministic. As a result, they have advantage over other optimization techniques such as: analytical methods, random methods, etc. [12].

The idea of population-based algorithms relies on iterative processing of individuals. In each iteration a new generation of individuals is generated (characterized, by definition, by improved values of the objective function). Currently, many varieties of population-based algorithms exist and they can be divided to: (a) single-population algorithms (see e.g. [68]) and multi-population algorithms (see e.g. [40]), (b) algorithms generating single solutions (see e.g. [2]) and fronts of solutions (see e.g. [16]), (c) algorithms using single-objective function (see e.g. [73]), and multi-objective function (see e.g. [10,25]). The efficiency of these algorithm strongly depends on evolutionary operators used for exploration and exploitation of the search space (search space determines acceptable boundaries for parameters of solutions). These operators are used for creation of next generation of individuals. Operators from different algorithms have some similarities, for example the equivalent of mutation operator of the genetic algorithm is the revolution operator of the imperialist competitive algorithm [4], the equivalent of the crossover operator is assimilation operator. Evolutionary operators usually have a set of parameters that must be selected before starting of evolution process (this is done usually by trial and error method) or modified in this process (usually by specified equation or due to specified rules).

In this paper a new method based on population-based algorithm with flexible selectable operators for nonlinear modeling is proposed. This method enables usage any types of exploration and exploitation operators, typical for population-based algorithms. Moreover, in proposed approach each solution from population encodes activity and parameters of these operators. Due to this, they can be selected dynamically in the evolution process (choosing of them is performed automatically for the problem under consideration). Such approach eliminates the need for determining detailed mechanism of the population-based algorithm. For the simulations typical nonlinear modeling benchmarks were used.

The structure of this paper is as follows: in Sect. 2 a description of proposed method is placed, Sect. 3 contains simulation results and in Sect. 4 the conclusions are drawn.

2   Description of Proposed Method

Characteristics of the proposed method can be summarized as follows:

- The method is characterized by the ability of an automatic creation of model during the process of evolution. The most important feature of the method is the fact that during the process of evolution the selection and configuration of the evolution operators used for exploration and exploitation takes place simultaneously. This eliminates the need for selection of the type of operators and their parameters by trial and error method. This approach also enables to achieve an appropriate balance between exploration and exploitation of the search space.

- The method is based on the possibilities of fuzzy systems, which are a convenient tool for nonlinear modeling. Modeling can be done in different ways: (a) neuro-fuzzy system can model input-output dependences (see e.g. [18]), (b) neuro-fuzzy system can model elements of state variables matrix (see e.g. [7]), (c) neuro-fuzzy system can provide appropriate cooperation of partial models representing various operating states of the modeled object (see e.g. [24]), etc. In this paper the first type of modeling will be considered, and the description of system used in the modeling is given in Sect. 2.1.

- The method uses hybrid approach proposed by us earlier, enabling the simultaneous selection of real parameters and binary parameters (see e.g. [40]). Due to this, the structure and parameters used for modeling of fuzzy system and used in the process of evolution operators can be automatically selected.

The proposed method is based on the approach analogical to particle swarm optimization (PSO, [33]). PSO algorithm uses population of individuals, in which each individual encodes parameters of potential solution to the problem under consideration (marked as $\mathbf{X}_{ch}^{\mathrm{par}} = \{X_{ch,1}^{\mathrm{par}}, ..., X_{ch,L^{\mathrm{par}}}^{\mathrm{par}}\}$), velocity vector, used for modification of potential solution parameters ($\mathbf{X}_{ch}^{\mathrm{vel}} = \{X_{ch,1}^{\mathrm{vel}}, ..., X_{ch,L^{\mathrm{par}}}^{\mathrm{vel}}\}$), and best found so far parameters of potential solution $\mathbf{p}$ (marked as $\mathbf{X}_{ch}^{\mathrm{bst}} = \{X_{ch,1}^{\mathrm{bst}}, ..., X_{ch,L^{\mathrm{par}}}^{\mathrm{bst}}\}$). In the algorithm the best solution found in population is additionally remembered (marked as $\mathbf{X}^{\mathrm{glb}} = \{X_1^{\mathrm{glb}}, ..., X_{L^{\mathrm{par}}}^{\mathrm{glb}}\}$). In the evolution process, the parameters $\mathbf{X}_{ch}^{\mathrm{vel}}$ and $\mathbf{X}_{ch}^{\mathrm{par}}$ are subject to modification according to the following equation:

$$\begin{cases} X_{ch,g}^{\mathrm{vel}} := \begin{pmatrix} w \cdot X_{ch,g}^{\mathrm{vel}} + c_1 \cdot \mathrm{U}(0,1) \cdot \left(X_{ch,g}^{\mathrm{bst}} - X_{ch,g}^{\mathrm{par}} \right) + \\ + c_2 \cdot \mathrm{U}(0,1) \cdot \left(X_g^{\mathrm{glb}} - X_{ch,g}^{\mathrm{par}} \right) \end{pmatrix}, \\ X_{ch,g}^{\mathrm{par}} := X_{ch,g}^{\mathrm{par}} + X_{ch,g}^{\mathrm{vel}} \end{cases} \tag{1}$$

where $ch = 1, ..., Npop$ is index of individual in population, $Npop$ stands for number of individuals in population, $g = 1, ..., L^{\mathrm{par}}$ is index of real parameter (gene), L^{par} stands for number of real parameters (actual number of parameters is different, which is explained in Sect. 2.2), w is inertia weight (usually $w \in [0.8, 1.0]$), c_1 and c_2 are cognitive and social parameters (see [33]), $\mathrm{U}(a,b)$ is function returning random number from range $[a, b]$.

In this paper a generalization of (1) is proposed. The generalization was designed in a way any evolutionary operator of exploration and exploitation can be used. Moreover, operators and their parameters can be selected dynamically in evolution process. This is consistent with two facts on the population algorithms: (a) in some algorithms operators of exploration and exploitation are intermingled, (b) in modification of parameters multiple operators can be used simultaneously. The generalized form of Eq. (1) takes the following form:

$$\begin{cases} X_{ch,g}^{\mathrm{vel}} := w \cdot X_{ch,g}^{\mathrm{vel}} + \sum_{o=1}^{L^{\mathrm{op}}} X_{ch,o}^{\mathrm{op}} \cdot \mathrm{op}_o \left(\begin{matrix} X_{ch,g}^{\mathrm{par}}, X_g^{\mathrm{glb}}, X_{ch,g}^{\mathrm{bst}}, \\ X_{ch1,g}^{\mathrm{par}}, X_{ch2,g}^{\mathrm{par}}, h_{ch} \end{matrix} \right), \\ X_{ch,g}^{\mathrm{par}} := X_{ch,g}^{\mathrm{par}} + X_{ch,g}^{\mathrm{vel}}, \end{cases} \tag{2}$$

where vector $\mathbf{X}_{ch}^{\mathrm{op}} = \{X_{ch,1}^{\mathrm{op}}, ..., X_{ch,L^{\mathrm{op}}}^{\mathrm{op}}\}$ contains information on binary keys that stand for activation state of connected with them operators (in this paper an assumption that 0 value stands for excluded operator and vice versa), L^{op} stands for number of considered operators (see Table 1), $\mathrm{op}_o(\cdot)$ stand for functions representing operator with index o, $\mathbf{X}_{ch1}^{\mathrm{par}}$ and $\mathbf{X}_{ch2}^{\mathrm{par}}$ stand for parent individuals chosen by selection method (for example roulette wheel method, [58]), h_{ch} stands for difference between averaged location of individuals from current and previous generation (see e.g. [14]).

2.1 Description of Neuro-Fuzzy System Used for Nonlinear Modeling

As previously mentioned in this paper, for nonlinear modeling a multi-input, multi-output fuzzy system that maps $\mathbf{X} \to \mathbf{Y}$ is used, where $\mathbf{X} \subset \mathbf{R}^n$ and $\mathbf{Y} \subset \mathbf{R}^m$. The fuzzy rule base of the system consists of a collection of N fuzzy if-then rules that takes the following form:

$$R^k : \begin{bmatrix} \mathrm{IF}\ (x_1\ \mathrm{is}\ A_1^k)\ \mathrm{AND}\ ...\ \mathrm{AND}\ (x_n\ \mathrm{is}\ A_n^k) \\ \mathrm{THEN}\ (y_1\ \mathrm{is}\ B_1^k), ..., (y_m\ \mathrm{is}\ B_m^k) \end{bmatrix}, \tag{3}$$

where $\mathbf{x} = [x_1, ..., x_n] \in \mathbf{X}$, $\mathbf{y} = [y_1, ..., y_m] \in \mathbf{Y}$, $A_1^k, ..., A_n^k$ are fuzzy sets characterized by membership functions $\mu_{A_i^k}(x_i)$, $i = 1, ..., n$, $k = 1, ..., N$, n stands for number of inputs, $B_1^k, ..., B_m^k$ are fuzzy sets characterized by membership functions $\mu_{B_j^k}(y_j)$, $j = 1, ..., m$, $k = 1, ..., N$, m stands for number of outputs. In this paper a Gaussian membership type functions are considered [58].

In the Mamdani approach (see e.g. [58]) output signal $\bar{y}_j$, $j = 1, ..., m$, of the fuzzy system is described by the following formula (for more details see our previous papers, e.g. [18]):

$$\bar{y}_j = \frac{\sum_{r=1}^{R} \bar{y}_{j,r}^{\mathrm{def}} \cdot \overset{N}{\underset{k=1}{S}} \left\{ T \left\{ \overset{n}{\underset{i=1}{T}} \left\{ \mu_{A_i^k}(\bar{x}_i) \right\}, \mu_{B_j^k}\left(\bar{y}_{j,r}^{\mathrm{def}}\right) \right\} \right\}}{\sum_{r=1}^{R} \overset{N}{\underset{k=1}{S}} \left\{ T \left\{ \overset{n}{\underset{i=1}{T}} \left\{ \mu_{A_i^k}(\bar{x}_i) \right\}, \mu_{B_j^k}\left(\bar{y}_{j,r}^{\mathrm{def}}\right) \right\} \right\}}, \tag{4}$$

where $\bar{y}_{j,r}^{\mathrm{def}}$, $j = 1, ..., m$, $r = 1, ..., R$, are discretization points, R is a number of discretization points, $T\{\cdot\}$ is a t-norm, and $S\{\cdot\}$ is a t-conorm (see e.g. [35]).

2.2 Encoding of the Individuals

Used in this paper encoding refers to Pittsburgh approach [29]. Thus, the single individual $\mathbf{X}_{ch}$ encodes the information on:

- Keys of operators (binary parameters). They are encoded as $\mathbf{X}_{ch}^{\mathrm{op}}$ and they are used in Eq. (2) (see Table 1).
- Parameters of neuro-fuzzy system (4) and parameters of operators (for the list of operators parameters see Table 1):

$$
\mathbf{X}_{ch}^{\mathrm{par}} = \left\{
\begin{array}{l}
\bar{x}_{ch,1,1}^{A}, \sigma_{ch,1,1}^{A}, \ldots, \bar{x}_{ch,n,1}^{A}, \sigma_{ch,n,1}^{A}, \ldots \\
\bar{x}_{ch,1,N}^{A}, \sigma_{ch,1,N}^{A}, \ldots, \bar{x}_{ch,n,N}^{A}, \sigma_{ch,n,N}^{A}, \\
\bar{y}_{ch,1,1}^{B}, \sigma_{ch,1,1}^{B}, \ldots, \bar{y}_{ch,m,1}^{B}, \sigma_{ch,m,1}^{B}, \ldots \\
\bar{y}_{ch,1,N}^{B}, \sigma_{ch,1,N}^{B}, \ldots, \bar{y}_{ch,m,N}^{B}, \sigma_{ch,m,N}^{B}, \\
\bar{y}_{ch,1,1}^{\mathrm{def}}, \ldots, \bar{y}_{ch,1,R}^{\mathrm{def}}, \ldots, \bar{y}_{ch,m,1}^{\mathrm{def}}, \ldots, \bar{y}_{ch,m,R}^{\mathrm{def}}, \\
w, c_1, c_2, p_{c1}, p_{c2}, m_r, p_m, F^{\mathrm{DE}}, \\
CR, f_{\min}, f_{\max}, A^t, \alpha^t, \hat{A}, F^{\mathrm{BTO}}
\end{array}
\right\}
\tag{5}
$$
$$
= \left\{ X_{ch,1}^{\mathrm{par}}, \ldots, X_{ch,L^{\mathrm{par}}}^{\mathrm{par}} \right\},
$$

where $\{\bar{x}_{i,k}^{A}, \sigma_{i,k}^{A}\}$ stand for parameters of membership functions of input Gaussian fuzzy sets $A_1^k, \ldots, A_n^k$, $\{\bar{y}_{j,k}^{B}, \sigma_{j,k}^{B}\}$ stand for parameters of membership functions of output Gaussian fuzzy sets $B_1^k, \ldots, B_m^k$, L^{par} is the number of genes of part $\mathbf{X}_{ch}^{\mathrm{par}}$.
- Velocity vector of parameters $\mathbf{X}_{ch}^{\mathrm{par}}$ (see (5)). They are encoded as set $\mathbf{X}_{ch}^{\mathrm{vel}}$.
- Best location of individual $\mathbf{X}_{ch}^{\mathrm{bst}}$ (they are copy of best parameters $\mathbf{X}_{ch}^{\mathrm{par}}$).

A individual is thus a collection of components: $\mathbf{X}_{ch} = \{\mathbf{X}_{ch}^{\mathrm{op}}, \mathbf{X}_{ch}^{\mathrm{par}}, \mathbf{X}_{ch}^{\mathrm{vel}}, \mathbf{X}_{ch}^{\mathrm{bst}}\} = \{X_{ch,1}, \ldots, X_{ch,L}\}$ with a total length of genes equal to $L = 3 \cdot L^{\mathrm{par}} + L^{\mathrm{op}}$.

2.3   Evaluation of the Individuals

In this article the aim of the objective function (fitness function) is to minimize standardized error as follows:

$$
\epsilon = \frac{1}{m} \cdot \sum_{j=1}^{m} \frac{\frac{1}{Z} \cdot \sum_{z=1}^{Z} |d_{z,j} - \bar{y}_{z,j}|}{\max\limits_{z=1,\ldots,Z} \{d_{z,j}\} - \min\limits_{z=1,\ldots,Z} \{d_{z,j}\}},
\tag{6}
$$

where $d_{z,j}$ is expected value of j-th output for z-th data sample ($z = 1, \ldots, Z$), Z stands for number of data samples, $\bar{y}_j$ is neuro-fuzzy system (4) output value calculated for data sample $\bar{x}_z$. The purpose of normalization is to eliminate the differences between the errors of different outputs of the system (4) in case where $m > 1$. While evolutionary processing a function in a form (6) is used to evaluate accuracy of the neuro-fuzzy system encoded by component $\mathbf{X}_{ch}^{par}$ of individuals $\mathbf{X}_{ch}$, $ch = 1, 2, \ldots, Npop$ and it is marked as ff $(\mathbf{X}_{ch})$.

2.4   Processing of the Individuals

The proposed method for processing the population taking into account the Eq. (2) that allows use of any combination of evolutionary operators. This process is executed according to the following steps:

Table 1. Chosen functions that represent evolutionary operators of exploration and exploitation considered in this paper. In the table the following marks are additionally used: $\alpha = \mathrm{U}(0,1)$ stands for a random number generated individually for each parameter under modification, $\beta = \mathrm{U}(0,1)$ stands for a random number generated individually for each individual under modification, $RInd$ is randomly chosen index of parameter, $RSet$ contains a set of randomly chosen indexes of parameters, $\mathrm{ff}(\cdot)$ stand for objective function of individual, ff_{min} is a smallest value of objective function for current population, $\mathrm{U}_G(1,1)$ is random number according to Gaussian distribution with mean 1 and variance 1, A^t is parameter additionally multiplied by coefficient α^t each time when individual modification improves solution (see e.g. [73])

o	Base method	$\mathrm{op}_o\left(X^{\mathrm{par}}_{ch,g}, X^{\mathrm{glb}}_g, X^{\mathrm{bst}}_{ch,g}, X^{\mathrm{par}}_{ch1,g}, X^{\mathrm{par}}_{ch2,g}, h_g\right) =$	Parameters
0	PSO-best [33]	$c_1 \cdot \mathrm{U}(0,1) \cdot \left(X^{\mathrm{bst}}_{ch,g} - X^{\mathrm{par}}_{ch,g}\right)$	$c_1 \in [1.5, 2.5]$
1	PSO-global [33]	$c_2 \cdot \mathrm{U}(0,1) \cdot \left(X^{\mathrm{glb}}_g - X^{\mathrm{par}}_{ch,g}\right)$	$c_2 \in [1.5, 2.5]$
2	GA-cross1 [58]	$\begin{cases} \mathrm{U}(0,1) \cdot \left(X^{\mathrm{par}}_{ch1,g} - X^{\mathrm{par}}_{ch,g}\right) & \text{for } \beta < p_{c1} \\ 0 & \text{for otherwise} \end{cases}$	$p_{c1} \in [0.7, 1.0]$
3	GA-cross2 [58]	$\begin{cases} \left(\begin{array}{c} X^{\mathrm{par}}_{ch2,g} - X^{\mathrm{par}}_{ch,g} + \\ + \mathrm{U}(0,1) \cdot \left(X^{\mathrm{par}}_{ch1,g} - X^{\mathrm{par}}_{ch2,g}\right) \end{array}\right) \\ \qquad \text{for } \beta < p_{c2} \\ 0 \text{ for otherwise} \end{cases}$	$p_{c2} \in [0.7, 1.0]$
4	GA-mutation [58]	$\begin{cases} \mathrm{U}(-1,1) \cdot m_r & \text{for } \alpha < p_m \\ 0 & \text{for otherwise} \end{cases}$	$M_r \in [0.01, 0.20]$ $p_m \in [0.05, 0.50]$
5	DE-crossover [2]	$\begin{cases} F^{\mathrm{DE}} \cdot \left(X^{\mathrm{par}}_{ch1,g} - X^{\mathrm{par}}_{ch2,g}\right) \\ \qquad \text{for } (\alpha < CR) \text{ or } (ch = RInd) \\ 0 \text{ for otherwise} \end{cases}$	$F^{\mathrm{DE}} \in [0, 2]$ $CR \in [0, 1]$
6	BAT-movement [73]	$\mathrm{U}(f_{\min}, f_{\min} + f_{\max}) \cdot \left(X^{\mathrm{glb}}_g - X^{\mathrm{par}}_{ch,g}\right)$	$f_{min} \in [0.0, 0.5]$ $f_{max} \in [0.0, 1.0]$
7	BAT-walk [73]	$\mathrm{U}(-1,1) \cdot \sum\limits_{ch3=1}^{Npop} \dfrac{A^t_{ch3}}{Npop}$	$A^t \in [0.0, 0.5]$ $\alpha^t \in [0.9, 1.0]$
8	ABC-candidate [32]	$\begin{cases} \mathrm{U}(-1,1) \cdot \left(X^{\mathrm{par}}_{ch1,g} - X^{\mathrm{par}}_{ch,g}\right) & \text{for } ch = RInd \\ 0 & \text{for otherwise} \end{cases}$	–
9	FWA-explosion [68]	$\begin{cases} \dfrac{\mathrm{U}(-1,1) \cdot \hat{A} \cdot (\mathrm{ff}(\mathbf{X}_{ch}) - \mathrm{ff}_{\min})}{\sum\limits_{ch3=1}^{Npop}(\mathrm{ff}(\mathbf{X}_{ch3}) - \mathrm{ff}_{\min})} & \text{for } ch \in RSet \\ 0 & \text{for otherwise} \end{cases}$	$\hat{A} \in [0.1, 2.0]$
10	FWA-mutation [68]	$\begin{cases} X^{\mathrm{par}}_{ch,g} \cdot \mathrm{U}_G(1,1) - X^{\mathrm{par}}_{ch,g} & \text{for } ch \in RSet \\ 0 & \text{for otherwise} \end{cases}$	–
11	BTO-history [14]	$F^{\mathrm{BTO}} \cdot \mathrm{U}(0,1) \cdot h_i$	$F^{\mathrm{BTO}} \in [3, 4]$

– Step 1. Initialization of population. In this step all genes of all $Npop$ individuals are set randomly. It takes into account the specification of the problem under consideration and operators parameters ranges stated in Table 1. In case of genes $\mathbf{X}^{\mathrm{vel}}_{ch}$ the assumptions that the genes do not exceed 20% ranges of the corresponding genes from $\mathbf{X}^{\mathrm{par}}_{ch}$ are additionally taken.

- Step 2. Evaluation of population. This step aims to evaluate each individual by using fitness function (6).
- Step 3. Reproduction. It is based on mechanisms analogical to algorithm PSO (with improvements that makes generalization possible):
  - For each individual $\mathbf{X}_{ch}$ a copy is created (which allows to use of any selection method-see Step 4).
  - Binary genes of the copy $\mathbf{X}_{ch}^{op}$ are modified using crossover and mutation operators from genetic algorithm [58]. Parameters of these operators (p_m and p_{c1}) are encoded in set $\mathbf{X}_{ch}^{par}$.
  - Real genes of the copy $\mathbf{X}_{ch}^{vel}$ and $\mathbf{X}_{ch}^{par}$ are updated according to (2). Genes $\mathbf{X}_{ch}^{par}$ are also repaired (narrowed down to specified boundaries).
  - The copies of $\mathbf{X}_{ch}$ are evaluated by fitness function (6). If the fitness function value of the copy is better than the fitness function value of the individual, the genes $\mathbf{X}_{ch}^{bst}$ of the copy are set as genes $\mathbf{X}_{ch}^{par}$ of the copy.
- Step 4. New generation selection. Proposed approach allows us to use different strategy for selection new population individuals. In this paper we assumed that the following strategies will be used:
  - Strategy S1. In this strategy, the copy of an individual replaces the individual (as in the algorithm PSO).
  - Strategy S2. In this strategy, the copy of an individual replaces the individual only if the fitness function value of the copy is better than the fitness function value of the individual (similar to BAT algorithm [73]).
  - Strategy S3. In this strategy individuals and their modified copies are placed in temporary population, and then the best (according to fitness function) $Npop$ individuals are selected for next generation.
- Step 5. Stop condition. In this step a stop condition is checked (for example if the specified number of algorithm iterations is achieved). If this condition is not met, the algorithm goes back to Step 3, otherwise the best solution is presented and algorithm stops.

3   Simulations

The goals of the simulations were to test:

- Classic algorithm PSO ($X_{ch,g}^{op} = 1$, $g = 0, 1$, and $X_{ch,g}^{op} = 0$, $g = 2, 3, \ldots 11$), algorithm PSO enriched by using other evolutionary operators (PSO-OP) ($X_{ch,g}^{op} = 1$, $g = 0, 1$, and $X_{ch,g}^{op} \in \{0, 1\}$, $g = 2, 3, \ldots 11$) and proposed in this paper algorithm with flexible selectable evolutionary operators (OP) ($X_{ch,g}^{op} \in \{0, 1\}$, $g = 0, 1, \ldots 11$).
- Different approaches for selection of operators' parameters. Two approaches were considered: (a) static (operators' parameters were set as average values of ranges from literature-see Table 1) and (b) dynamic (operators parameters were selected in evolutionary process).
- Different strategies for selecting next generation of individuals (see Sect. 2.4-strategies S1, S2, and S3).

Table 2. Nonlinear benchmarks used in simulations

No.	Problem name and reference	Short name	Number of inputs (n)	Number of outputs (m)	Number of data rows (Z)
1.	Concrete slump test [13]	Slump	7	3	103
2.	Computer hardware [34]	Mcpu	9	1	209
3.	Yacht hydrodynamics [45]	Yacht	6	1	308
4.	Auto MPG [46]	Ampg	7	1	398
5.	Energy efficiency [69]	Energy	8	2	768
6.	Airfoil self-noise [11]	Airfoil	5	1	1503

Table 3. Averaged values of ff ($\cdot$) for considered benchmarks. Best results for each strategy are shown in bold

Strat.	Method	Param.	Slump	Mcpu	Yacht	Ampg	Energy	Airfoil	Avg.
S1	PSO	Static	0.1228	0.0327	0.0920	0.1053	0.2597	0.1373	0.1250
		Dynamic	0.1448	0.0469	0.1202	0.1277	0.3298	0.1477	0.1528
	PSO+OP	Static	0.1223	0.0327	**0.0805**	0.1048	0.2383	0.1339	0.1188
		Dynamic	0.1448	0.0452	0.1232	0.1249	0.3190	0.1385	0.1493
	OP	Static	0.1248	0.0339	0.0823	**0.1013**	0.2286	**0.1317**	0.1171
		Dynamic	**0.1201**	**0.0312**	0.0857	0.1045	**0.2195**	0.1346	**0.1159**
S2	PSO	Static	0.1051	0.0285	0.0719	0.0850	0.1986	0.1096	0.0998
		Dynamic	0.1025	0.0329	0.0746	0.0970	0.2124	0.1275	0.1078
	PSO+OP	Static	0.0994	0.0244	0.0581	0.0936	0.1890	0.1208	0.0976
		Dynamic	0.1082	0.0261	0.0660	0.0974	0.1802	0.1221	0.1000
	OP	Static	0.0874	0.0159	0.0379	0.0636	0.1683	0.0950	0.0780
		Dynamic	**0.0749**	**0.0152**	**0.0321**	**0.0609**	**0.1624**	**0.0811**	**0.0711**
S3	PSO	Static	0.0956	0.0260	0.0572	0.0818	0.1775	0.1052	0.0905
		Dynamic	0.1028	0.0298	0.0713	0.0938	0.1933	0.1092	0.1000
	PSO+OP	Static	0.0941	0.0248	0.0572	0.0820	0.1744	0.0993	0.0887
		Dynamic	0.0985	0.0244	0.0622	0.0871	0.1860	0.1174	0.0959
	OP	Static	0.0755	0.0093	0.0493	0.0575	0.1646	0.0775	0.0706
		Dynamic	**0.0636**	**0.0086**	**0.0431**	**0.0547**	**0.1627**	**0.0751**	*0.0663*

The simulations were performed for typical benchmarks from nonlinear modeling field (see Table 2). The parameters of simulations were set as follows: number of fuzzy rules $N = 3$, number of discretization points $R = 3$, triangular norms: algebraic, number of individuals $Npop = 100$, number of iterations: 1000, number of repeats of simulations for each problem and case: 20, selection method for choosing parents of the individual: roulette wheel. s The average simulation results are presented in Table 3, Figs. 1 and 2. The average number of used operators in population for best results (strategy S3 and algorithm OP) is shown

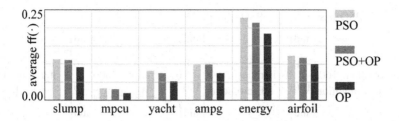

Fig. 1. Averaged results in relation to specified algorithms for considered benchmarks.

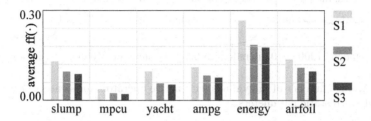

Fig. 2. Averaged results in relation to specified strategy for considered benchmarks.

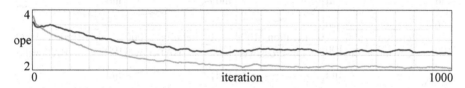

Fig. 3. Average number of used operators (ope) by single individual for strategy S3 and algorithm OP with static (gray line) and dynamic parameters (dark gray line).

in Figs. 4 and 5. The average number of used operators per individual for best results is shown in Fig. 3.

The simulations conclusions are as follows:

– Algorithm PSO+OP performs better than classic PSO algorithm, regardless of the strategy used for the selection of a new population (see Table 3 and Fig. 1). At the same time algorithm OP gives best results, regardless of the strategy used for the selection of a new population (see Table 3 and Fig. 1). Notwithstanding, the best results were obtained for algorithm OP and strategy S3 (see Table 3).

– Strategy S2 performs better then strategy S1 regardless of the used algorithm (see Table 3 and Fig. 2) and strategy S3 performs better then strategy S2 regardless of the used algorithm (see Table 3 and Fig. 2).

– Using dynamic parameters of operators mainly benefits cases that use of algorithm OP (see Table 3). At the same time the average number of active operators for a single individual decreased with the iteration of evolutionary algorithm (see Fig. 3).

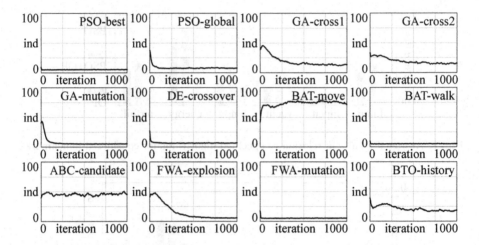

Fig. 4. Average number of individuals with active operators for strategy S3 and algorithm OP with static parameters.

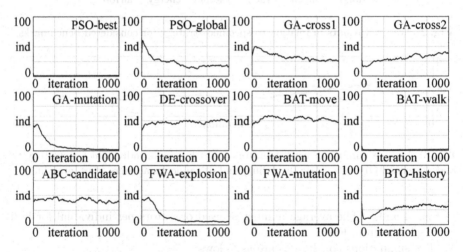

Fig. 5. Average number of individuals with active operators for strategy S3 and algorithm OP with dynamic parameters.

- In cases that use strategy S3 and algorithm OP the certain domination of specified evolutionary operators can be seen. It is presented in Fig. 4 (operators BAT-move and ABC-candidate dominates population) and Fig. 5 (operators DE-crossover and BAT-move dominates population).
- In cases that use strategy S3 and algorithm OP the number of used operators changes with the iteration of evolutionary algorithm. The examples are shown in Fig. 4 (activation of FWA-explosion and GA-cross1 decrease) and Fig. 5 (activation of GA-cross2 and BTO-history increase).

4 Conclusions

In this paper a new method based on population-based algorithm with flexible selectable evolutionary operators for nonlinear modeling was proposed. This method: (a) adjusts the behavior of evolutionary operators (both the exploration and exploitation operators) to problem under consideration and (b) provides a reasonable balance between exploration and exploitation of search space. Performed simulations shown, among others, that: (a) using flexible selectable operators ensures higher accuracy for nonlinear modeling benchmarks simultaneously with faster convergence of algorithm, (b) automatic selection of operators parameters is beneficial when is used together with flexible selectable operators (OP+S3).

In further studies in the field of evolutionary operators selection it is planned, among others, to develop a mechanism for flexible selectable operators in multi population-based algorithm that uses advantages of populations cooperation.

Acknowledgments. The project was financed by the National Science Centre (Poland) on the basis of the decision number DEC-2012/05/B/ST7/02138.

References

1. Aghdam, M.H., Heidari, S.: Feature selection using particle swarm optimization in text categorization. J. Artif. Intell. Soft Comput. Res. **5**(4), 231–238 (2015)
2. Ali, M., Pant, M., Abraham, A.: Unconventional initialization methods for differential evolution. Appl. Math. Comput. **219**(9), 4474–4494 (2013)
3. Almohammadi, K., Hagras, H., Alghazzawi, D., Aldabbagh, G.: Users-centric adaptive learning system based on interval type-2 fuzzy logic for massively crowded E-learning platforms. J. Artif. Intell. Soft Comput. Res. **6**(2), 81–101 (2016)
4. Atashpaz-Gargari, E., Lucas, C.: Imperialist competitive algorithm: an algorithm for optimization inspired by imperialistic competition. IEEE Congr. Evol. Comput. **7**, 4661–4666 (2007)
5. Bartczuk, Ł.: Gene expression programming in correction modelling of nonlinear dynamic objects. In: Borzemski, L., Grzech, A., Świątek, J., Wilimowska, Z. (eds.) Information Systems Architecture and Technology: Proceedings of 36th International Conference on Information Systems Architecture and Technology – ISAT 2015 – Part I. AISC, vol. 429, pp. 125–134. Springer, Cham (2016). doi:10.1007/978-3-319-28555-9_11
6. Bartczuk, Ł., Dziwiński, P., Starczewski, J.T.: A new method for dealing with unbalanced linguistic term set. In: Rutkowski, L., Korytkowski, M., Scherer, R., Tadeusiewicz, R., Zadeh, L.A., Zurada, J.M. (eds.) ICAISC 2012. LNCS, vol. 7267, pp. 207–212. Springer, Heidelberg (2012). doi:10.1007/978-3-642-29347-4_24
7. Bartczuk, Ł., Przybył, A., Koprinkova-Hristova, P.: New method for nonlinear fuzzy correction modelling of dynamic objects. In: Rutkowski, L., Korytkowski, M., Scherer, R., Tadeusiewicz, R., Zadeh, L.A., Zurada, J.M. (eds.) ICAISC 2014. LNCS, vol. 8467, pp. 169–180. Springer, Cham (2014). doi:10.1007/978-3-319-07173-2_16
8. Bello, O., Holzmann, J., Yaqoob, T., Teodoriu, C.: Application of artificial intelligence methods in drilling system design and operations: a review of the state of the art. J. Artif. Intell. Soft Comput. Res. **5**(2), 121–139 (2015)

9. Brasileiro, Í., Santos, I., Soares, A., Rabelo, R., Mazullo, F.: Ant colony optimization applied to the problem of choosing the best combination among M combinations of shortest paths in transparent optical networks. J. Artif. Intell. Soft Comput. Res. **6**(4), 231–242 (2016)

10. Brester, C., Semenkin, E., Sidorov, M.: Multi-objective heuristic feature selection for speech-based multilingual emotion recognition. J. Artif. Intell. Soft Comput. Res. **6**(4), 243–253 (2016)

11. Brooks, T.F., Pope, D.S., Marcolini, A.M.: Airfoil self-noise and prediction, Technical report, NASA RP-1218, July 1989

12. Che, J., Wang, J., Li, K.: A monte carlo based robustness optimization method in new product design process: a case study. Am. J. Ind. Bus. Manage. **4**(7), 360–369 (2014)

13. Cheng, Y.I.: Modeling slump flow of concrete using second-order regressions and artificial neural networks. Cem. Concr. Compos. **29**(6), 474–480 (2007)

14. Civicioglu, P.: Backtracking search optimization algorithm for numerical optimization problems. Appl. Math. Comput. **219**(15), 8121–8144 (2013)

15. Colchester, K., Hagras, H., Alghazzawi, D.: A survey of artificial intelligence techniques employed for adaptive educational systems within E-learning platforms. J. Artif. Intell. Soft Comput. Res. **7**(1), 47–64 (2017)

16. Corne, D.W., Knowles, J.D., Oates, M.J.: The pareto envelope-based selection algorithm for multiobjective optimization. In: Schoenauer, M., Deb, K., Rudolph, G., Yao, X., Lutton, E., Merelo, J.J., Schwefel, H.-P. (eds.) PPSN 2000. LNCS, vol. 1917, pp. 839–848. Springer, Heidelberg (2000). doi:10.1007/3-540-45356-3_82

17. Cpałka, K.: Design of Interpretable Fuzzy Systems. Springer, Heidelberg (2017)

18. Cpałka, K., Łapa, K., Przybył, A., Zalasiński, M.: A new method for designing neuro-fuzzy systems for nonlinear modelling with interpretability aspects. Neurocomputing 135, 203–217

19. Cpałka, K., Łapa, K., Przybył, A.: A new approach to design of control systems using genetic programming. Inf. Technol. Control **44**(4), 433–442 (2015)

20. Cpałka, K., Rebrova, O., Nowicki, R., Rutkowski, L.: On design of flexible neuro-fuzzy systems for nonlinear modelling. Int. J. Gen. Syst. **42**(6), 706–720 (2013)

21. Cpałka, K., Rutkowski, L.: Flexible takagi-sugeno fuzzy systems, neural networks. In: Proceedings of the 2005 IEEE International Joint Conference on IJCNN 2005, vol. 3, pp. 1764–1769 (2005)

22. Duda, P., Jaworski, M., Pietruczuk, L.: On pre-processing algorithms for data stream. In: Rutkowski, L., Korytkowski, M., Scherer, R., Tadeusiewicz, R., Zadeh, L.A., Zurada, J.M. (eds.) ICAISC 2012. LNCS, vol. 7268, pp. 56–63. Springer, Heidelberg (2012). doi:10.1007/978-3-642-29350-4_7

23. Dziwiński, P., Avedyan, E.D.: A new method of the intelligent modeling of the nonlinear dynamic objects with fuzzy detection of the operating points. In: Rutkowski, L., Korytkowski, M., Scherer, R., Tadeusiewicz, R., Zadeh, L.A., Zurada, J.M. (eds.) ICAISC 2016. LNCS (LNAI), vol. 9693, pp. 293–305. Springer, Cham (2016). doi:10.1007/978-3-319-39384-1_25

24. Dziwiński, P., Bartczuk, Ł., Przybył, A., Avedyan, E.D.: A new algorithm for identification of significant operating points using swarm intelligence. In: Rutkowski, L., Korytkowski, M., Scherer, R., Tadeusiewicz, R., Zadeh, L.A., Zurada, J.M. (eds.) ICAISC 2014. LNCS, vol. 8468, pp. 349–362. Springer, Cham (2014). doi:10.1007/978-3-319-07176-3_31

25. Ehrgott, M.: Multicriteria Optimization. Springer, Heidelberg (2000)

26. El-Samak, A.F., Ashour, W.: Optimization of traveling salesman problem using affinity propagation clustering and genetic algorithm. J. Artif. Intell. Soft Comput. Res. **5**(4), 239–245 (2015)
27. Gałkowski, T., Rutkowski, L.: Nonparametric recovery of multivariate functions with applications to system identification. Proc. IEEE **73**(5), 942–943 (1985)
28. Grycuk, R., Gabryel, M., Korytkowski, M., Scherer, R.: Content-based image indexing by data clustering and inverse document frequency. In: Kozielski, S., Mrozek, D., Kasprowski, P., Małysiak-Mrozek, B., Kostrzewa, D. (eds.) BDAS 2014. CCIS, vol. 424, pp. 374–383. Springer, Cham (2014). doi:10.1007/978-3-319-06932-6_36
29. Ishibuchi, H., Nakashima, T., Murata, T.: Comparison of the michigan and pittsburgh approaches to the design of fuzzy classification systems. In: Electronics and Communications in Japan (Part III Fundamental Electronic Science), vol. 80, pp. 10–19 (1997)
30. Jaworski, M., Er, M.J., Pietruczuk, L.: On the application of the parzen-type kernel regression neural network and order statistics for learning in a non-stationary environment. In: Rutkowski, L., Korytkowski, M., Scherer, R., Tadeusiewicz, R., Zadeh, L.A., Zurada, J.M. (eds.) ICAISC 2012. LNCS, vol. 7267, pp. 90–98. Springer, Heidelberg (2012). doi:10.1007/978-3-642-29347-4_11
31. Jaworski, M., Pietruczuk, L., Duda, P.: On resources optimization in fuzzy clustering of data streams. In: Rutkowski, L., Korytkowski, M., Scherer, R., Tadeusiewicz, R., Zadeh, L.A., Zurada, J.M. (eds.) ICAISC 2012. LNCS, vol. 7268, pp. 92–99. Springer, Heidelberg (2012). doi:10.1007/978-3-642-29350-4_11
32. Karaboga, D., Basturk, B.: Artificial bee colony (ABC) optimization algorithm for solving constrained optimization problems. In: Melin, P., Castillo, O., Aguilar, L.T., Kacprzyk, J., Pedrycz, W. (eds.) IFSA 2007. LNCS, vol. 4529, pp. 789–798. Springer, Heidelberg (2007). doi:10.1007/978-3-540-72950-1_77
33. Kennedy, J., Eberhart, R.: Particle swarm optimization. In: Proceedings of IEEE International Conference on Neural Networks, pp. 1942–1948 (1995)
34. Kibler, D., Aha, D.: Instance-based prediction of real-valued attributes. In: Proceedings of the CSCSI (Canadian AI) Conference, vol. 5, pp. 51–57 (1988)
35. Klement, E.P., Mesiar, R., Pap, E.: Triangular Norms. Springer, Heidelberg (2000)
36. Korytkowski, M., Rutkowski, L., Scherer, R.: Fast image classification, by boosting fuzzy classifiers. Inf. Sci. **327**, 175–182 (2016)
37. Leon, M., Xiong, N.: Adapting differential evolution algorithms for continuous optimization via greedy adjustment of control parameters. J. Artif. Intell. Soft Comput. Res. **6**(2), 103–118 (2016)
38. Li, X., Er, M.J., Lim, B.S., Zhou, J.H., Gan, O.P., Rutkowski, L.: Fuzzy regression modeling for tool performance prediction and degradation detection. Int. J. Neural Syst. **2005**, 405–419 (2010)
39. Łapa, K., Cpałka, K., Wang, L.: New method for design of fuzzy systems for nonlinear modelling using different criteria of interpretability. In: Rutkowski, L., Korytkowski, M., Scherer, R., Tadeusiewicz, R., Zadeh, L.A., Zurada, J.M. (eds.) ICAISC 2014. LNCS, vol. 8467, pp. 217–232. Springer, Cham (2014). doi:10.1007/978-3-319-07173-2_20
40. Łapa, K., Szczypta, J., Venkatesan, R.: Aspects of structure and parameters selection of control systems using selected multi-population algorithms. Artif. Intell. Soft Comput. Res. **9120**, 247–260 (2015)
41. Miyajima, H., Shigei, N., Miyajima, H.: Performance comparison of hybrid electromagnetism-like mechanism algorithms with descent method. J. Artif. Intell. Soft Comput. Res. **5**(4), 271–282 (2015)

42. Murata, M., Ito, S., Tokuhisa, M., Ma, Q.: Order estimation of Japanese paragraphs by supervised machine learning and various textual features. J. Artif. Intell. Soft Comput. Res. **5**(4), 247–255 (2015)
43. Nguyen, K.P., Fujita, G., Dieu, V.N.: Cuckoo search algorithm for optimal placement and sizing of static var compensator in large-scale power systems. J. Artif. Intell. Soft Comput. Res. **6**(2), 59–68 (2016)
44. Nikulin, V.: Prediction of the shoppers loyalty with aggregated data streams. J. Artif. Intell. Soft Comput. Res. **6**(2), 69–79 (2016)
45. Ortigosa, I., Lopez, R., Garcia, J.: A neural networks approach to residuary resistance of sailing yachts prediction. In: Proceedings of the International Conference on Marine Engineering MARINE (2007)
46. Quinlan, R.: Combining instance-based and model-based learning. In: Proceedings on the Tenth International Conference of Machine Learning, pp. 236–243 (1993)
47. Pietruczuk, L., Rutkowski, L., Jaworski, M., Duda, P.: How to adjust an ensemble size in stream data mining? Inf. Sci. **381**, 46–54 (2017)
48. Przybył, A., Er, M.J.: The idea for the integration of neuro-fuzzy hardware emulators with real-time network. In: Rutkowski, L., Korytkowski, M., Scherer, R., Tadeusiewicz, R., Zadeh, L.A., Zurada, J.M. (eds.) ICAISC 2014. LNCS (LNAI), vol. 8467, pp. 279–294. Springer, Cham (2014). doi:10.1007/978-3-319-07173-2_25
49. Przybył, A., Er, M.J.: A new approach to designing of intelligent emulators working in a distributed environment. In: Rutkowski, L., Korytkowski, M., Scherer, R., Tadeusiewicz, R., Zadeh, L.A., Zurada, J.M. (eds.) ICAISC 2016. LNCS, vol. 9693, pp. 546–558. Springer, Cham (2016). doi:10.1007/978-3-319-39384-1_48
50. Rutkowski, L.: On-line identification of time-varying systems by nonparametric techniques. IEEE Trans. Autom. Control **27**(1), 228–230 (1982)
51. Rutkowski, L.: On nonparametric identification with prediction of time-varying systems. IEEE Trans. Autom. Control **29**(1), 58–60 (1984)
52. Rutkowski, L.: Nonparametric identification of quasi-stationary systems. Syst. Control Lett. **6**(1), 33–35 (1985)
53. Rutkowski, L.: Real-time identification of time-varying systems by non-parametric algorithms based on parzen kernels. Int. J.Syst.Sci. **16**(9), 1123–1130 (1985)
54. Rutkowski, L.: A general approach for nonparametric fitting of functions and their derivatives with applications to linear circuits identification. IEEE Trans. Circuits and Syst. **33**(8), 812–818 (1986)
55. Rutkowski, L.: Application of multiple Fourier-series to identification of multivariable non-stationary systems. Int. J. Syst. Sci. **20**(10), 1993–2002 (1989)
56. Rutkowski, L.: Non-parametric learning algorithms in time-varying environments. Signal Process. **182**, 129–137 (1989)
57. Rutkowski, L.: Multiple Fourier series procedures for extraction of nonlinear regressions from noisy data. IEEE Trans. Signal Process. **41**(10), 3062–3065 (1993)
58. Rutkowski, L.: Computational Intelligence. Methods and Techniques. Springer, Heidelberg (2008)
59. Rutkowski, L., Cpałka, K.: A neuro-fuzzy controller with a compromise fuzzy reasoning. Control Cybern. **31**(2), 297–308 (2002)
60. Rutkowski, L., Cpałka, K.: Compromise approach to neuro-fuzzy systems. In: Proceedings of the 2nd Euro-International Symposium on Computation Intelligence. Frontiers in Artificial Intelligence and Applications, vol. 76, pp. 85–90 (2002)
61. Rutkowski, L., Jaworski, M., Pietruczuk, L., Duda, P.: A new method for data stream mining based on the misclassification error. IEEE Trans. Neural Netw. Learn. Syst. **26**, 1048–1059 (2015)

62. Rutkowski, L., Przybył, A., Cpałka, K.: Novel online speed profile generation for industrial machine tool based on flexible neuro-fuzzy approximation. IEEE Trans. Ind. Electron. **59**(2), 1238–1247 (2012)

63. Serdah, A.M., Ashour, W.M., Soft, C.R.: Clustering large-scale data based on modified affinity propagation algorithm. J. Artif. Intell. Soft Comput. Res. **6**(1), 23–33 (2016)

64. Staszewski, P., Woldan, P., Korytkowski, M., Scherer, R., Wang, L.: Query-by-example image retrieval in microsoft sql server. In: Rutkowski, L., Korytkowski, M., Scherer, R., Tadeusiewicz, R., Zadeh, L.A., Zurada, J.M. (eds.) ICAISC 2016. LNCS (LNAI), vol. 9693, pp. 746–754. Springer, Cham (2016). doi:10.1007/978-3-319-39384-1_66

65. Sugiyama, H.: Pulsed power network based on decentralized intelligence for reliable and lowloss electrical power distribution. J. Artif. Intell. Soft Comput. Res. **5**(2), 97–108 (2015)

66. Szarek, A., Korytkowski, M., Rutkowski, L., Scherer, R., Szyprowski, J.: Forecasting wear of head and acetabulum in hip joint implant. In: Rutkowski, L., Korytkowski, M., Scherer, R., Tadeusiewicz, R., Zadeh, L.A., Zurada, J.M. (eds.) ICAISC 2012. LNCS, vol. 7268, pp. 341–346. Springer, Heidelberg (2012). doi:10.1007/978-3-642-29350-4_41

67. Szczypta, J., Łapa, K., Shao, Z.: Aspects of the selection of the structure and parameters of controllers using selected population based algorithms. In: Rutkowski, L., Korytkowski, M., Scherer, R., Tadeusiewicz, R., Zadeh, L.A., Zurada, J.M. (eds.) ICAISC 2014. LNCS, vol. 8467, pp. 440–454. Springer, Cham (2014). doi:10.1007/978-3-319-07173-2_38

68. Tan, Y., Zhu, Y.: Fireworks algorithm for optimization. In: Tan, Y., Shi, Y., Tan, K.C. (eds.) ICSI 2010. LNCS, vol. 6145, pp. 355–364. Springer, Heidelberg (2010). doi:10.1007/978-3-642-13495-1_44

69. Tsanas, A., Xifara, A.: Accurate quantitative estimation of energy performance of residential buildings using statistical machine learning tools. Energy Buildings **49**, 560–567 (2012)

70. Wang, G., Zhang, S.: ABM with behavioral bias and applications in simulating china stock market. J. Artif. Intell. Soft Comput. Res. **5**(4), 257–270 (2015)

71. Woźniak, M., Gabryel, M., Nowicki, R.K., Nowak, B.A.: An application of firefly algorithm to position traffic in noSQL database systems. In: Kunifuji, S., Papadopoulos, G.A., Skulimowski, A.M.J., Kacprzyk, J. (eds.) Knowledge, Information and Creativity Support Systems. AISC, vol. 416, pp. 259–272. Springer, Cham (2016). doi:10.1007/978-3-319-27478-2_18

72. Yang, C., Moi, S., Lin, Y., Chuang, L.: Genetic algorithm combined with a local search method for identifying susceptibility genes. J. Artif. Intell. Soft Comput. Res. **6**(3), 203–212 (2016)

73. Yang, X.S.: A new metaheuristic bat-inspired algorithm. In: González, J.R., Pelta, D.A., Cruz, C., Terrazas, G., Krasnogor, N. (eds.) Nature Inspired Cooperative Strategies for Optimization, NICSO 2010, vol. 284, pp. 65–74. Springer, Heidelberg (2010)

74. Yin, Z., O'Sullivan, C., Brabazon, A.: An analysis of the performance of genetic programming for realised volatility forecasting. J. Artif. Intell. Soft Comput. Res. **6**(3), 155–172 (2016)

75. Zalasiński, M.: New algorithm for on-line signature verification using characteristic global features. Adv. Intell. Syst. Comput. **432**, 137–146 (2016)

76. Zalasiński, M., Cpałka, K.: New algorithm for on-line signature verification using characteristic hybrid partitions. In: Wilimowska, Z., Borzemski, L., Grzech, A., Świątek, J. (eds.) Information Systems Architecture and Technology: Proceedings of 36th International Conference on Information Systems Architecture and Technology – ISAT 2015 – Part IV. AISC, vol. 432, pp. 147–157. Springer, Cham (2016). doi:10.1007/978-3-319-28567-2_13
77. Zalasiński, M., Cpałka, K., Hayashi, Y.: A new approach to the dynamic signature verification aimed at minimizing the number of global features. In: Rutkowski, L., Korytkowski, M., Scherer, R., Tadeusiewicz, R., Zadeh, L.A., Zurada, J.M. (eds.) ICAISC 2016. LNCS, vol. 9693, pp. 218–231. Springer, Cham (2016). doi:10.1007/978-3-319-39384-1_20

Fuzzy Portfolio Diversification with Ordered Fuzzy Numbers

Adam Marszałek[1] and Tadeusz Burczyński[1,2(✉)]

[1] Computational Intelligence Department, Institute of Computer Science, Cracow University of Technology, Warszawska 24, 31-155 Cracow, Poland
amarszalek@pk.edu.pl
[2] Institute of Fundamental Technological Research, Polish Academy of Sciences, Pawinskiego 5B, 02-106 Warsaw, Poland
tburczynski@ippt.pan.pl

Abstract. In this paper, we consider a multi-objective portfolio diversification problem under real constraints in fuzzy environment, where the objective is to minimize the variance of portfolio and maximize expected return rate of portfolio. The return rates of assets are modeled using concept of Ordered Fuzzy Candlesticks, which are Ordered Fuzzy Numbers. The use of them allows modeling uncertainty associated with financial data based on high-frequency data. Thanks to well-defined arithmetic of Ordered Fuzzy Numbers, the estimators of fuzzy-valued expected value and covariance can be computed in the same way as for real random variables. In an empirical study, 20 assets included in the Warsaw Stock Exchange Top 20 Index are used to compare considered fuzzy model with crisp mean-variance model.

Keywords: Ordered fuzzy number · Kosinski's fuzzy number · Ordered fuzzy candlestick · Fuzzy portfolio diversification · Fuzzy returns · Multi-objective optimization · Financial high-frequency data

1 Introduction

Portfolio selection is the problem how to allocate investor's capital among different assets such that the investment goal can be achieved with minimal risk. The traditional portfolio optimization method initiated by Markowitz [17] employs variance as risk measure and have become a cornerstone of modern portfolio theory. However, several authors tried to explore alternative risk measures to replace variance [7,8]. The return rates of financial assets often are characterized as a random variable with a probability distribution. But there are many non-probabilistic factors that affect the financial markets such that the return of risky asset is fuzzy uncertainty. With the extensive application of fuzzy set theory, some researchers have investigated the portfolio selection problem in fuzzy environment [1,4,24,27].

In the above mentioned literatures, the authors used basic concept of fuzzy numbers [2,26]. In this paper, we presents an approach where the return rates

© Springer International Publishing AG 2017
L. Rutkowski et al. (Eds.): ICAISC 2017, Part I, LNAI 10245, pp. 279–291, 2017.
DOI: 10.1007/978-3-319-59063-9_25

of assets are modeled using concept of Ordered Fuzzy Candlesticks, which are Ordered Fuzzy Numbers. The use of them allows modeling uncertainty associated with financial data based on high-frequency data. The high-frequency financial data are observations which containing the most complete knowledge about quotations of the financial instrument.

2   Portfolio Diversification Problem

The classical Mean-Variance (MV) portfolio optimization model introduced by Markowitz [17–19] aims at determining the fractions x_i of a given capital to be invested in each asset i belonging to a predetermined set or market so as minimize the risk of the return of the whole portfolio, identified with its variance, while restricting the expected return of the portfolio to attain a specified value ρ. The classical MV problem can be written as follows

$$
\begin{aligned}
&\text{Minimize } \sum_{i=1}^{n}\sum_{j=1}^{n}\sigma_{ij}x_ix_j \\
&\text{Subject to} \\
&\sum_{i=1}^{n}\mu_ix_i = \rho \quad \text{(required expected return)} \\
&\sum_{i=1}^{n}x_i = 1 \quad \text{(budget constraint)}
\end{aligned}
\tag{1}
$$

where n assets are available, μ_i is the expected return of asset i and σ_{ij} is the covariance of returns of asset i and asset j.

The problem (1) is a convex quadratic programming problem which can be solved by a number of efficient algorithms with a moderate computational effort even for large instances. In our considerations we add to the MV model the realistic constraint that no more than K assets should be held in the portfolio and the quantity x_i of each asset should be limited within a given interval $[l_i, u_i]$. Furthermore, the transaction cost c_i of the asset i is also included. Under this assumptions the MV model is no longer a convex optimization problem because of the non-convexity of its feasible regions. This model can be defined and solved as following multi-objective optimization problem

$$
\begin{aligned}
&\text{Minimize } \sum_{i=1}^{n}\sum_{j=1}^{n}\sigma_{ij}x_ix_j \\
&\text{Maximize } \sum_{i=1}^{n}\mu_ix_i - \sum_{i=1}^{n}c_i|x_i - x_i^0| \\
&\text{Subject to} \\
&\sum_{i=1}^{n}x_i = 1 \quad \text{(budget constraint)} \\
&x_i = 0 \text{ or } l_i \le x_i \le u_i \quad \text{(quantity constraint)} \\
&\#\{i\colon x_i > 0\} \le K \quad \text{(cardinality constraint)} \\
&i = 1,\ldots,n
\end{aligned}
\tag{2}
$$

where x_i^0 are a current fractions of a given capital.

The solution of the multi-objective optimization problem is the set of Pareto optimal solutions (efficient frontier). One of the possibility to choose the final solution is to choose a portfolio with the highest Sharpe ratio (tangency portfolio). The Sharpe ratio, developed by W. F. Sharpe [25], is the ratio of a portfolio's total return minus the risk-free rate divided by the standard deviation of the portfolio, which is a measure of its risk. The Sharpe ratio is simply the risk premium per unit of risk, which is quantified by the standard deviation of the portfolio and it is defined as

$$SR = \frac{\mu_p - r_f}{\sigma_p} \tag{3}$$

where μ_p is the expected return of portfolio, σ_p is the standard deviation of portfolio and r_f is the risk-free rate.

3 Fuzzy Background Concepts

3.1 Ordered Fuzzy Numbers (OFN)

Ordered Fuzzy Numbers (called also the Kosiński's Fuzzy Numbers) introduced by Kosiński et al. in series of papers [9–13] are defined by ordered pairs of continuous real functions defined on the interval $[0, 1]$ i.e. $A = (f, g)$ with $f, g : [0, 1] \rightarrow \mathbb{R}$ as continuous functions.

Functions f and g are called the *up* and *down*-parts of the fuzzy number A, respectively. The continuity of both parts implies their images are bounded intervals, say UP and $DOWN$, respectively. In general, the functions f and g need not be invertible, and only continuity is required. If we assume, however, that these functions are monotonous, i.e., invertible, and add the constant function of x on the interval $[1_A^-, 1_A^+]$ with the value equal to 1, we might define the membership function

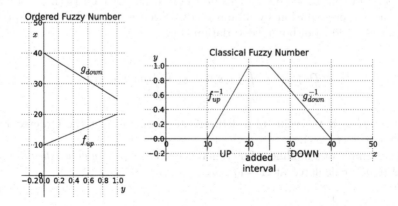

Fig. 1. Graphical interpretation of OFN and a OFN presented as fuzzy number in classical meaning

$$\mu(x) = \begin{cases} f^{-1}(x) & \text{if } x \in [f(0), f(1)], \\ g^{-1}(x) & \text{if } x \in [g(1), g(0)], \\ 1 & \text{if } x \in [1_A^-, 1_A^+], \end{cases} \tag{4}$$

if f is increasing and g is decreasing, and such that $f \leq g$ (pointwise). In this way, the obtained membership function $\mu(x)$, $x \in \mathbb{R}$ represents a mathematical object which resembles a convex fuzzy number in the classical sense. The Ordered Fuzzy Number and Ordered Fuzzy Number as a fuzzy number in classical meaning are presented in Fig. 1.

Furthermore, the basic arithmetic operations on Ordered Fuzzy Numbers are defined as the pairwise operations of their elements.

Let $A = (f_A, g_A)$, $B = (f_B, g_B)$ and $C = (f_C, g_C)$ are Ordered Fuzzy Numbers. The sum $C = A + B$, subtraction $C = A - B$, product $C = A \cdot B$, and division $C = A \div B$ are defined by formula

$$f_C(y) = f_A(y) * f_B(y), \qquad g_C(y) = g_A(y) * g_B(y) \tag{5}$$

where $*$ works for $+$, $-$, $\cdot$ and $\div$, respectively, and where $C = A \div B$ is defined, if the functions $|f_B|$ and $|g_B|$ are bigger than zero. In a similar way, multiply an Ordered Fuzzy Number A by a scalar $\lambda \in \mathbb{R}$, i.e. $C = \lambda \cdot A$ is defined by formula

$$f_C(y) = \lambda \cdot f_A(y), \qquad g_C(y) = \lambda \cdot g_A(y) \tag{6}$$

This definition leads to some useful properties. The one of them is existence of neutral elements of addition and multiplication. This fact causes that not always the result of an arithmetic operation is a fuzzy number with a larger support. This allows to build fuzzy models based on Ordered Fuzzy Numbers in the form of the classical equations without losing the accuracy.

Moreover, a universe $\mathcal{O}$ of all Ordered Fuzzy Numbers can be identified with $C^0([0,1]) \times C^0([0,1])$, hence the space $\mathcal{O}$ is topologically a Banach space [12]. A class of defuzzification operators of Ordered Fuzzy Numbers can be defined, as linear and continuous functionals on the Banach space $\mathcal{O}$. Each of them, say Def has a representation by a sum of two Stieltjes integrals with respect to functions ν_1 and ν_2 of bounded variation [14],

$$Def(A) = \int_0^1 f_A(s)d\nu_1(s) + \int_0^1 g_A(s)d\nu_2(s) \tag{7}$$

where $Def(A)$ is the value of a defuzzification operator at the Ordered Fuzzy Number $A = (f_A, g_A)$.

An example of a nonlinear functional is *center of gravity defuzzification* functional (CoG) calculated at $A = (f_A, g_A)$

$$CoG(A) = \int_0^1 \frac{f_A(s) + g_A(s)}{2}|f_A(s) - g_A(s)|ds\{\int_0^1 |f_A(s) - g_A(s)|ds\}^{-1} \tag{8}$$

provided $\int_0^1 |f_A(s) - g_A(s)| ds \neq 0$. Center of gravity operator defined above is equivalent to the center of gravity operator in classical fuzzy logic.

In addition, note that a pair of continuous functions (f, g) determines different Ordered Fuzzy Number than the pair (g, f). In this way, an extra feature to this object, named the orientation is appointed. Depending on the orientation, the Ordered Fuzzy Numbers can be divided into two types: a positive orientation, if the direction of Ordered Fuzzy Number is consistent with the direction of the axis Ox and a negative orientation, if the direction of the Ordered Fuzzy Number is opposite to the direction of the axis Ox.

3.2 Ordered Fuzzy Candlesticks (OFC)

Concept of Ordered Fuzzy Candlesticks was proposed by the authors in [20–22]. Generally, in this approach, a fixed time interval of financial high frequency data is identified with Ordered Fuzzy Number and it is called Ordered Fuzzy Candlestick. The general idea is presented in Fig. 2. Notice, that the orientation of the Ordered Fuzzy Number shows whether the Ordered Fuzzy Candlestick is long or short. While the information about movements in the price are contained in the shape of the f and g functions.

In our previous works listed two cases of construction of Ordered Fuzzy Candlesticks. The first assumes that the functions f and g are functions of predetermined type, moreover, the shapes of these functions should depend on two parameters (e.g. linear, etc.). Then the Ordered Fuzzy Candlestick for given time series can be defined as follows.

Let $\{X_t : t \in T\}$ be a given time series and $T = \{1, 2, \ldots, n\}$. The Ordered Fuzzy Candlestick is defined as an Ordered Fuzzy Number $C = (f, g)$ which satisfies the following conditions 1–4 (for long candlestick) or $1'$–$4'$ (for short candlestick).

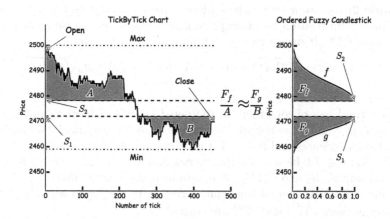

Fig. 2. Draft of general concept of ordered fuzzy candlestick

1. $X_1 \leq X_n$.
2. $f\colon [0,1] \to \mathbb{R}$ is continuous and increasing on $[0,1]$.
3. $g\colon [0,1] \to \mathbb{R}$ is continuous and decreasing on $[0,1]$.
4. $S_1 < S_2$, $f(1) = S_1$, $f(0) = \min\limits_{t\in T} X_t - C_1$, $g(1) = S_2$ and $g(0)$ is such that

   the ratios $\dfrac{F_g}{A}$ and $\dfrac{F_f}{B}$ are equal.

1'. $X_1 > X_n$.
2'. $f\colon [0,1] \to \mathbb{R}$ is continuous and decreasing on $[0,1]$.
3'. $g\colon [0,1] \to \mathbb{R}$ is continuous and increasing on $[0,1]$.
4'. $S_1 < S_2$, $f(1) = S_2$, $f(0) = \max\limits_{t\in T} X_t + C_2$, $g(1) = S_1$ and $g(0)$ is such that

   the ratios $\dfrac{F_f}{A}$ and $\dfrac{F_g}{B}$ are equal.

In the above conditions the center of Ordered Fuzzy Candlestick (i.e. added interval) is designated by parameters $S_1, S_2 \in [\min_{t\in T} X_t, \max_{t\in T} X_t]$ and can be compute as different kinds of averages (e.g. arithmetic, weighted or exponential). While C_1 and C_2 are arbitrary nonnegative real numbers, which further extend the support of fuzzy numbers and can be compute e.g. as standard deviation or volatility of X_t. The parameters A and B are positive real numbers, which determine the relationship between the functions f and g. They can be calculated as the mass of the desired area with the assumed density (see Fig. 2). Numbers F_f and F_g are the fields under the graph of functions f^{-1} and g^{-1}, respectively. The examples of realizations of Trapezoid and Gaussian Ordered Fuzzy Candlesticks are defined below and presented in Fig. 3.

Example 1: *Trapezoid OFC.* Suppose that f and g are linear functions in form

$$f(x) = (b_f - a_f)\, x + a_f \quad \text{and} \quad g(x) = (b_g - a_g)\, x + a_g \qquad (9)$$

then the Ordered Fuzzy Candlestick $C = (f,g)$ is called *a Trapezoid OFC*, especially if $S_1 = S_2$ then also can be called *a Triangular OFC*.

Example 2: *Gaussian OFC.* The Ordered Fuzzy Candlestick $C = (f,g)$ where the membership relation has a shape similar to the Gaussian function is called *a Gaussian OFC*. It means that f and g are given by functions

$$f(x) = f(z) = \sigma_f \sqrt{-2\ln(z)} + m_f \quad \text{and} \quad g(x) = g(z) = \sigma_g \sqrt{-2\ln(z)} + m_g$$
$$(10)$$

where e.g. $z = (1 - \alpha)x + \alpha$, α close to zero.

The second case of construction of Ordered Fuzzy Candlesticks assumes that the functions f and g are defined in similar way as the empirical distribution in the statistical sciences and it is called *an Empirical OFC*.

Let $\{X_t\colon t \in T\}$ be a given time series and $T = \{1, 2, \ldots, n\}$. The values of parameters S_1, S_2 and C_1, C_2 are determined based on a time series X_t. The new time series Y_t is created from time series X_t by sorting in ascending. Next, the two time series $Y_t^{(1)}$ and $Y_t^{(2)}$ are created as

$$Y_t^{(1)} = \{Y_i\colon i \in T \wedge Y_i \leq S_1\} \qquad t \in \{0, 1, \ldots, K_1\}$$

$$Y_t^{(2)} = \{Y_i : i \in T \wedge S_2 \leq Y_i\} \qquad t \in \{0, 1, \ldots, K_2\}$$

Now, based on these time series we define the two discrete functions on interval $[0, 1]$ with step $dx = \frac{1}{M}$ (i.e. $M + 1$ points) as

$$\Psi_1(k \cdot dx) = \begin{cases} Y^{(1)}_{\lfloor \frac{k}{dx} \rfloor} - \frac{M-k}{M} C_1 & \text{if } k \in \{0, 1, \ldots, M-1\} \\ S_1 & \text{if } k = M \end{cases}$$

$$\Psi_2(k \cdot dx) = \begin{cases} Y^{(2)}_{K_2 - \lfloor \frac{k}{dx} \rfloor} + \frac{M-k}{M} C_2 & \text{if } k \in \{0, 1, \ldots, M-1\} \\ S_2 & \text{if } k = M \end{cases}$$

Then the *empirical OFC* is an Ordered Fuzzy Number $C = (f, g)$ where the functions f and g are continous approximation of functions Ψ_1 and Ψ_2, respectively for long candlestick, whilst for short candlestick Ψ_2 and Ψ_1, respectively. The example of realization of the Empirical OFC is presented in Fig. 4.

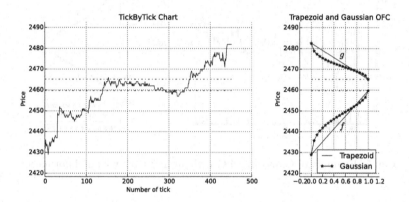

Fig. 3. Examples of trapezoid and Gaussian OFC

In technical analysis the prices are by far the most important. However, another piece of important information about price movement is volume. Volume is the number of entities traded during the time period under study. It is used to confirm trends and chart patterns. Any price movement up or down with relatively high volume is seen as a stronger, more relevant move than a similar move with weak volume [23].

In case of Ordered Fuzzy Candlestick, add extra information about volume is very easy. Enough to calculate the parameters S_1, S_2, A, B, C_1, C_2 using the density associated with volume or in case of Empirical OFC, enough to calculate functions Ψ_1 and Ψ_2 using prices repeated volume times. The example of Ordered Fuzzy Candlesticks without and with volume information are presented in Fig. 5.

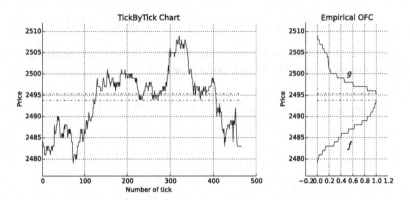

Fig. 4. Example of Empirical OFC

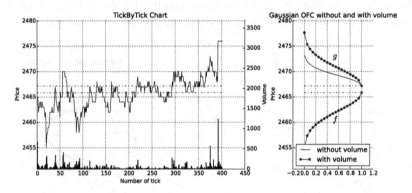

Fig. 5. Example of Gaussian OFC without and with volume information

3.3 Fuzzy Returns and Their Excepted Value and Covariance

In this approach, the fuzzy simply return of asset are modeled based on high-frequency data (tick-by-tick data) using concept of Ordered Fuzzy Candlestick. Let X_t, $t = 1, \ldots, T$ be a time series of quotations of given asset for given fixed time interval (e.g. day, week, month). The new time series R_t is created from time series X_t as follows

$$R_t = \frac{X_t - X_1}{X_1} \qquad t = 1, \ldots, T \tag{11}$$

Now, based on these time series the Ordered Fuzzy Candlestick $\overline{R} = (f_R, g_R)$ is defined and it is called *fuzzy simply return* for given time interval.

Let $\overline{R}_t^{(1)}$ and $\overline{R}_t^{(2)}$, $t = 1, \ldots, T$ be a sequences of fuzzy simple returns. Notice, that $\overline{R}_t$ are Ordered Fuzzy Numbers. Thanks to well-defined arithmetic of OFN, the estimators of the fuzzy-valued expected value and fuzzy-valued covariance of Ordered Fuzzy Numbers (as random variables) can be defined as follows

- Estimator of the fuzzy-valued expected value of $\overline{R}^{(1)}$:

$$\overline{\mu}_{\overline{R}^{(1)}} = \frac{1}{T} \sum_{t=1}^{T} \overline{R}_t^{(1)} \tag{12}$$

- Estimator of the fuzzy-valued covariance of $\overline{R}^{(1)}$ and $\overline{R}^{(2)}$:

$$\overline{\sigma}_{\overline{R}^{(1)}\overline{R}^{(2)}} = \overline{\mathrm{cov}}\left(\overline{R}^{(1)}, \overline{R}^{(2)}\right) = \frac{1}{T} \sum_{t=1}^{T} \left(\left(\overline{R}_t^{(1)} - \overline{\mu}_{\overline{R}^{(1)}}\right)\left(\overline{R}_t^{(2)} - \overline{\mu}_{\overline{R}^{(2)}}\right)\right) \tag{13}$$

Sometimes in practice, such as in uncertain programming theory, a scalar expected value is often required as a surrogate for a fuzzy random variables so that a decision-maker may employ the value to make decision. For this purpose, a defuzzification operator called *expected value* of Ordered Fuzzy Number $A = (f_A, g_A)$ is defined as follows

$$\mathcal{E}(A) = \frac{1}{2} \int_0^1 [f_A(s) + g_A(s)]ds \tag{14}$$

Expected value operator defined above is equivalent[1] to the definition of expected value operator proposed by Liu and Liu in [15, 16].

Now, the estimators of the scalar-valued expected value and scalar-valued covariance of Ordered Fuzzy Numbers can be defined as follows

- Estimator of the scalar-valued expected value of $\overline{R}^{(1)}$:

$$\mu_{\overline{R}^{(1)}} = \mathcal{E}\left(\overline{\mu}_{\overline{R}^{(1)}}\right) \tag{15}$$

- Estimator of the scalar-valued covariance of $\overline{R}^{(1)}$ and $\overline{R}^{(2)}$:

$$\sigma_{\overline{R}^{(1)}\overline{R}^{(2)}} = \mathcal{E}\left(\overline{\sigma}_{\overline{R}^{(1)}\overline{R}^{(2)}}\right) \tag{16}$$

4   Fuzzy Portfolio Diversification Problem

The fuzzy portfolio diversification problem is formulated as multi-objective optimization problem as follows

---

[1] For Ordered Fuzzy Number which can be represented as classical convex fuzzy number.

$$\text{Minimize } \mathcal{E}\left(\sum_{i=1}^{n}\sum_{j=1}^{n}\overline{\sigma}_{\overline{R}^{(i)}\overline{R}^{(j)}}x_i x_j\right)$$

$$\text{Maximize } \mathcal{E}\left(\sum_{i=1}^{n}\overline{\mu}_{\overline{R}^{(i)}}x_i\right) - \sum_{i=1}^{n}c_i|x_i - x_i^0|$$

Subject to   (17)

$$\sum_{i=1}^{n}x_i = 1 \quad \text{(budget constraint)}$$

$x_i = 0$ or $l_i \le x_i \le u_i$ (quantity constraint)

$\#\{i\colon x_i > 0\} \le K$ (cardinality constraint)

$i = 1, \ldots, n$

where n assets are available, fuzzy simple returns $\overline{R}^{(i)}$ of asset i are defined as Ordered Fuzzy Candlestisks, $\overline{\mu}_{\overline{R}^{(i)}}$ is the fuzzy-valued expected return of asset i and $\overline{\sigma}_{\overline{R}^{(i)}\overline{R}^{(j)}}$ is the fuzzy-valued covariance of returns of asset i and asset j. Other notations are the same as in the definition of the problem (2).

The Sharpe ratio for fuzzy model is defined as

$$SR = \frac{\mu_p - r_f}{\sigma_p} \tag{18}$$

where μ_p is the scalar-valued expected return of portfolio, σ_p is the scalar-valued standard deviation of portfolio and r_f is the risk-free rate.

5   Numerical Example

This section presents a numerical experiments of crisp portfolio diversification problem (2) and fuzzy portfolio diversification problem (17) by using securities from the Polish Stock Market. In particular, we have considered the monthly returns of 20 assets included in the stock index WIG20. The returns of the assets are characterized by monthly simple returns in case of crisp model. While in the case of fuzzy model, the monthly fuzzy simple returns are modeled by empirical Ordered Fuzzy Candlesticks with volume information and parameters S_1, S_2 as minimum and maximum of different kinds of averages (arithmetic, weighted and exponential), $C_1 = C_2 = 0$ and $M = 21$.

The analysis was conducted over three-month sample period using historical data from the past two years (24 monthly returns). The optimal portfolio weights were calculated, and then, these were applied to a portfolio held for a subsequent period of one month. The transaction cost of each assets is set to 0.38%, i.e., $c_i = 0.0038$ and the risk-free rate is approximated by the one-month Warsaw Interbank Bid Rate (WIBID). The lower l_i and upper u_i bounds are set to 0.01 and 0.15, respectively. The maximal count of diffrent asset K is set to 15. The IMmune GAme theory MultiObjective (IMGAMO) algorithm proposed by P. Jarosz et al. in [3,5,6] was used for solving both multi-objective problems.

The initial fractions of assets are set to $x_i^0 = 0.0$, $i = 1, \ldots, 20$ and risk-free rate r_f is set to $\frac{1}{12}\{1.36\%, 1.37\%, 1.40\%\}$, respectively. Figure 6 illustrate the

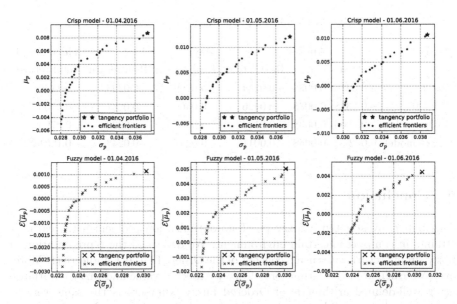

Fig. 6. The efficient frontiers and tangency portfolios

efficient frontiers and optimal tangency portfolios found by IMGAMO algorithm for the diversification at the beginning of each of the considered month (April, May, June 2016). The results show that for each month optimal portfolio is the same as for the first month but different for each of the models. The optimal weights of assets are presented in Table 1.

Table 1. Optimal tangency portfolios

Asset no	2, 4, 5, 8, 16, 17	10	19	1, 3, 6, 7, 9, 11 … 15, 18, 20
Crisp model	0.15	0.0	0.1	0.0
Fuzzy model	0.15	0.1	0.0	0.0

Out-of-sample experiments allow evaluation of the effectiveness of portfolio models by evaluate their ex-post performance at the end of each month. The performance are compared also with the official market index (WIG20) and naive diversification ($x_i = \frac{1}{n}$). The results are presented in Table 2. As can be seen the out-of-sample performance at the end of three-month period is the best for proposed fuzzy model.

Table 2. Out-of-sample performance of portfolios

| Model | | April | | May | June | Three-month |
	Return	Cost	Real return	Real return	Real return	Total return
Naive	−0.0216	0.0038	−0.0254	−0,0709	−0,0223	−0.1147
WIG20	−0.0388	0.0038	−0.0426	−0,0466	−0,0287	−0.1135
Crisp model	−0.0066	0.0038	−0.0104	−0,0387	−0,0322	−0.0793
Fuzzy model	0.0027	0.0038	−0.0011	−0,0247	−0,0243	**−0.0494**

6 Conclusion

This paper investigated a fuzzy portfolio optimization problem with real constraints in complex uncertain environment. We described the representation of fuzzy simply return rates using concept of the Ordered Fuzzy Candlesticks. Moreover, we calculated the estimators of fuzzy expected value and fuzzy covariance using well-defined arithmetic of Ordered Fuzzy Numbers in the same way as for real random variables. A numerical example was given to demonstrate the application and the effectiveness of our model. It should be mentioned that the approach proposed in this paper can be applied to other financial problems such as forecasting and pricing of financial instruments.

References

1. Carlsson, C., Fuller, R., Majlender, P.: A possibilistic approach to selecting portfolios with highest utility score. Fuzzy Sets Syst. **131**, 13–21 (2002)
2. Dubois, D., Prade, H.: Fuzzy Sets and Systems: Theory and Application. Academic Press, New York (1980)
3. Długosz, A., Jarosz, P.: Multiobjective optimization of electrothermal microactuators by means of Immune Game Theory Multiobjective Algorithm. In: Kleiber, M., et al. (eds.) Advances in Mechanics: Theoretical, Computational and Interdisciplinary Issues, pp. 141–145. Taylor & Francis Group, London (2016)
4. Huang, X.X.: Mean-semivariance models for fuzzy portfolio selection. J. Comput. Appl. Math. **217**(1), 1–8 (2008)
5. Jarosz, P., Burczyski, T.: Coupling of immune algorithms and game theory in multiobjective optimization. In: Rutkowski, L., Scherer, R., Tadeusiewicz, R., Zadeh, L.A., Zurada, J.M. (eds.) ICAISC 2010. LNCS (LNAI), vol. 6114, pp. 500–507. Springer, Heidelberg (2010). doi:10.1007/978-3-642-13232-2_61
6. Jarosz, P., Burczyński, T.: Biologically-inspired Methods and Game Theory in Multi-criterion Decision Processes. In: Bouvry, P., et al. (eds.) Intelligent Decision Systems in Large-Scale Distributed Environments, Studies in Computational Intelligence 362, pp. 101–124. Springer, Heidelberg (2011)
7. Konno, H., Yamazaki, H.: Mean-absolute deviation portfolio optimization model and its applications to Tokyo stock market. Manage. Sci. **37**, 519–531 (1991)
8. Konno, H., Shirakawa, H., Yamazaki, H.: A mean-absolute deviation-skewness portfolio optimization model. Ann. Oper. Res. **45**, 205–220 (1993)

9. Kosiński, W., Prokopowicz, P., Ślęzak, D.: Drawback of fuzzy arithmetic - new intuitions and propositions. In: Burczyński, T., Cholewa, W., Moczulski, W. (eds.) Proceedings PACM, Methods of Artificial Intelligence, Gliwice, pp. 231–237 (2002)
10. Kosiński, W., Prokopowicz, P., Ślęzak, D.: On algebraic operations on fuzzy numbers. In: Klopotek, M., Wierzchoń, S.T., Trojanowski, K. (eds.) Intelligent Information Processing and Web Mining, pp. 353–362. Springer, Heidelberg (2003)
11. Kosiński, W., Prokopowicz, P., Ślęzak, D.: Ordered fuzzy numbers. Bull. Polish Acad. Sci. Ser. Sci. Math. 51(3), 327–338 (2003)
12. Kosiński, W., Prokopowicz, P.: Algebra of fuzzy numbers mathematica applicanda. J. Polish Math. Soc. 32(46/05), 37–63 (2004)
13. Kosiński, W.: On soft computing and modelling. Image Process. Commun. 11(1), 71–82 (2006)
14. Kosiński, W., Frischmuth, K., Wilczyńska-Sztyma, D.: A new fuzzy approach to ordinary differential equations. In: Rutkowski, L., Scherer, R., Tadeusiewicz, R., Zadeh, L.A., Zurada, J.M. (eds.) ICAISC 2010. LNCS (LNAI), vol. 6113, pp. 120–127. Springer, Heidelberg (2010). doi:10.1007/978-3-642-13208-7_16
15. Liu, B., Liu, Y.-K.: Expected value of fuzzy variable and fuzzy expected value models. IEEE Trans. Fuzzy Syst. 10, 445–450 (2002)
16. Liu, Y.-K., Liu, B.: Fuzzy random variables: a scalar expected value operator. Fuzzy Optim. Decis. Making 2, 143–160 (2003). Kluwer Academic Publishers, Printed in The Netherlands
17. Markowitz, H.M.: Portfolio selection. J. Financ. 7(1), 77–91 (1952)
18. Markowitz, H.M.: Portfolio selection: efficient diversification of investments. Cowles Fundation for Research in Economics at Yal University. Monograph 16. Wiley, New York (1959)
19. Markowitz, H.M.: Mean-Variance Analysis in Portfolio Choice and Capital Markets. Basil Blackwell, Oxford (1987)
20. Marszałek, A., Burczyński, T.: Financial fuzzy time series models based on ordered fuzzy numbers. In: Pedrycz, W., Chen, S.-M. (eds.) Time Series Analysis, Model & Applications, ISRL 47, pp. 77–95. Springer, Heidelberg (2013)
21. Marszałek, A., Burczyński, T.: Modelling financial high frequency data using ordered fuzzy numbers. In: Rutkowski, L., Korytkowski, M., Scherer, R., Tadeusiewicz, R., Zadeh, L.A., Zurada, J.M. (eds.) ICAISC 2013. LNCS (LNAI), vol. 7894, pp. 345–352. Springer, Heidelberg (2013). doi:10.1007/978-3-642-38658-9_31
22. Marszałek, A., Burczyński, T.: Modeling and forecasting financial time series with ordered fuzzy candlesticks. Inf. Sci. 273, 144–155 (2014)
23. Murphy, J.J.: Technical Analysis of the Financial Markets. New York Institute of Finance, New York (1999)
24. Tanaka, H., Guo, P.: Portfolio selection based on upper and lower exponential possibility distributions. Eur. J. Oper. Res. 114, 115–126 (1999)
25. Sharpe, W.F.: The sharpe ratio. J. Portfolio Manage. 21(1), 49–58 (1994)
26. Zadeh, L.A.: Fuzzy sets. Inf. Control 8, 338–353 (1965)
27. Zhang, W.G., Zhang, X.L., Xiao, W.L.: Portfolio selection under possibistic mean-variance utility and SMO algorithm. Eur. J. Oper. Res. 197(2), 693–700 (2009)

Using a Hierarchical Fuzzy System for Traffic Lights Control Process

Bartosz Poletajew$^{(\boxtimes)}$ and Adam Slowik

Department of Electronics and Computer Science, Koszalin University of Technology,
Sniadeckich 2 Street, 75-453 Koszalin, Poland
bartoszpoletajew@o2.pl, aslowik@ie.tu.koszalin.pl

Abstract. The present study presents the applications of a hierarchical fuzzy system in the traffic light system control process. In this paper a hierarchical fuzzy system was designed. This system is based on the fuzzy system for traffic lights control FS-TLC. The main advantage of the solution presented herein is a significant reduction of the number of defined rules. Additionally, owing to the use of hierarchical fuzzy systems, it is possible to greatly reduce the time required for computations and to reduce the use of memory in the systems designed. Also that systems can be significantly more effective in operation, and they allow a creation of faster controllers that may be implemented in the hardware architecture.

1 Introduction

Fuzzy systems FLS [1] possess many applications [3,4], and one of them is the traffic light system control process [5,6]. The database of rules based on the IF ... THEN ... structure is the main component of every fuzzy system. Depending on the designed system, the rule base may have small dimensions yet it may also be very complex. If a traditional fuzzy system with a large number of input variables were used, then a rule database will be growing exponentially [7]. This is the main reason for the prolonged time of computations performed by a specific controller. A desirable feature of any fuzzy logic system is the shortest possible time for decision making. For this reason, optimization methods of the operation of control systems are sought, also those that operate based on FLS (see Fig. 1a) fuzzy logic principles. One solution includes the use of the modelling method of hierarchical fuzzy systems HFS [8,9]. Owing to this, it is possible to substantially reduce the number of fuzzy rules that were contained in the rule database. Traditional fuzzy systems are based on the assumptions proposed by Mamdani or Takagi-Sugeno-Kang (TSK). The difference between these two systems consists in a delicate modification of fuzzy rules. Hierarchical fuzzy systems are developed on the basis of traditional fuzzy systems. Figure 1 includes an example of the structure of fuzzy systems. As it is evident from examples 1b and 1c, outputs from the individual traditional systems serve at the same time as inputs for the FLS traditional system.

It should be emphasized that the number of the defined rules, which take up a certain portion of the memory, also has an impact on the total data processing time. Most frequently, systems are created whose operation is based on a

© Springer International Publishing AG 2017
L. Rutkowski et al. (Eds.): ICAISC 2017, Part I, LNAI 10245, pp. 292–301, 2017.
DOI: 10.1007/978-3-319-59063-9_26

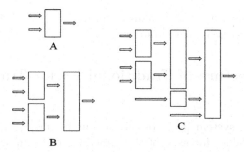

Fig. 1. Examples of fuzzy logic systems: simple FLS (A) and hierarchical fuzzy logic systems HFS (B, C)

traditional and invariable database of fuzzy rules [5,8]; however, there are also the so-called "adaptive" fuzzy systems, which dynamically create not merely a database of rules but rather a scope of data used during the performance of computations [6]. One should bear it in mind that the method of the creation and use of the so-called "database of rules" is dependent solely on the designer of a given system. It is also the designer who, based on observations and research, defines data and its ranges, which are responsible for the correct and desirable operation of the entire system. This paper is an extension of the idea of FS-TLC system that was presented in paper [5] but this time it is focused on optimization method by using hierarchical fuzzy logic systems. This paper consists of the following sections: Sect. 2 covers a brief description of the problem of rules and the issues related to the times of operations performed; Sect. 3 contains a comparison of the traditional fuzzy system for the traffic light control and a version created based on a hierarchical model; In Sect. 4 the assumptions of the taken experiment and the test results are shown; and finally in Sect. 5 conclusions are presented.

2 Problem of Rules

Usually the designer of the system is responsible for the criteria of an adequate selection of fuzzy rules. The system's designer, based on the research and expected results, decides about the number of rules and to what extent they will be complex. This also translates onto the process of the adequate computations. In many cases, the decision making process should be as fast as possible. This paper, is not focus merely on obtaining as short as possible time for a reduction of traffic at a given road intersection. An optimization of the computation process constitutes an important element of the operation of every system. By using a hierarchical fuzzy system, a six-fold reduction of the number of fuzzy rules in the rule database were obtained, and compared to the fuzzy logic system presented in paper [5] the computation time was reduced four times. An optimization of the operation of the system is one of the most desirable properties in currently designed controllers; hence, this paper is oriented onto an indication of the use of fuzzy hierarchical systems as an initial method of an optimization of the operation of systems based on fuzzy logic assumptions.

3   An Architecture of Traditional FS for Traffic Light Control

A traditional fuzzy system [5] is based on four linguistic input variables and on two output variables for each of the directions of the movement of vehicles (cf. Fig. 2). The linguistic input variables are described as follows: the number of vehicles in the north-south direction (VNS) and in the east-west direction (VEW), the total weight of vehicles in the north-south direction (WNS) and in the east-west direction (WEW). The linguistic output variables represent the activity time of the green light for the north-south (GNS) and the east-west (GEW) directions. The VNS and VEW linguistic variables are composed of three linguistic values: small (SM), medium (ME) and large (LA). The WNS and WEW linguistic variables are described with two linguistic values: light (LI) and heavy (HE). The GNS and GEW linguistic variables comprise five linguistic values: very short (XS), short (SH), average (AV), long (L) and very long (XL). Referring to the system [5], linguistic variables were modified so that linguistic values could total to 1 in any range. The linguistic input and output variables are presented in Fig. 3. In this system [5], PROD-MAX fuzzy operators were used. The FS-TLC [5] fuzzy system is composed of 72 fuzzy rules: 36 rules are assigned to the GNS output and further 36 rules are assigned to the GEW output. The rules were constructed like in FS-TLC [5]. The fuzzy rules coding scheme is as follows: for example the encoded fuzzy rule SM, SM, LI, LI, XS which is connected with the output GNS represents the decoded fuzzy rule:

IF VEW=SM AND VNS=SM AND WEW=LI AND WNS=LI THEN GNS=XS

All the fuzzy rules that are used in the FS-TLC [5] system are shown in Table 1. The height method was used in the defuzzification process.

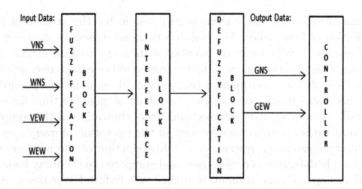

Fig. 2. The structure of FS-TLC

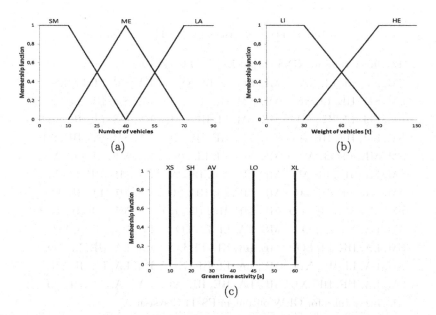

Fig. 3. Fuzzy system linguistic variables representation: the input VNS and VEW (a), WNS and WEW (b) and the output GNS and GEW (c)

4   Proposed Hierarchical Fuzzy Logic System for Traffic Lights Control

The hierarchical HFS-TLC fuzzy system was created based on the FS-TLC fuzzy system [5]. The linguistic input variables and the linguistic output variables were described in the same identical manner as in the FS-TLC [5] system (cf. Fig. 3). All the remaining processes of the fuzzy system, i.e. fuzzification, inferring and defuzzification remain unchanged, too. Considering the creation method of hierarchical fuzzy systems, the input data was distributed onto two separate fuzzy systems (cf. Fig. 4), where the output data is obtained from. In connection with this, the database of fuzzy rules was distributed onto these two fuzzy subsystems. Owing to this, it was possible to substantially reduce the number of rules in the database. As compared to the system of FS-TLC [5], where 72 fuzzy rules were constructed (36 rules for the GNS and GEW outputs respectively), in our HFS-TLC system, we obtained merely 12 fuzzy rules: 6 rules for the GNS output and 6 rules for the GEW output. This substantial reduction of the number of rules is also confirmed in the literature [7–10]. The fuzzy rules coding scheme for HFS-TLC is as follows: for example the encoded fuzzy rule SM, LI, XS which is connected with the output GNS represents the decoded fuzzy rule:

IF VEW=SM AND WEW=LI THEN GEW=XS

The set of rules for the HFS-TLC system is presented in Table 2.

Table 1. The fuzzy rules for FS-TLC system

The fuzzy rules for GNS output in FS-TLC system		
SM, SM, LI, LI, XS	ME, SM, LI LI, XS	LA, SM, LI, LI, XS
SM, SM, HE, LI, XS	ME, SM, HE, LI, XS	LA, SM, HE, LI, XS
SM, SM, LI, HE, SH	ME, SM, LI, HE, SH	LA, MS, LI, HE, SH
SM, SM, HE, HE, SH	ME, SM, HE, HE, SH	LA, SM, HE, HE, SH
SM, ME, LI, LI, AV	ME, ME, LI, LI, AV	LA, ME, LI, LI, AV
SM, ME, HE, LI, AV	ME, ME, HE, LI, AV	LA, ME, HE, LI, AV
SM, ME, LI, HE, LO	ME, ME, LI, HE, LO	LA, ME, LI, HE, LO
SM, ME, HE, HE, LO	ME, ME, HE, HE, LO	LA, ME, HE, HE, LO
SM, LA, LI, LI, LO	ME, LA, LI, LI, LO	LA, LA, LI, LI, LO
SM, LA, HE, LI, LO	ME, LA, HE, LI, LO	LA, LA, HE, LI, LO
SM, LA, LI, HE, XL	ME, LA, LI, HE, XL	LA, LA, LI, HE, XL
SM, LA, HE, HE, XL	ME, LA, HE, HE, XL	LA, LA, HE, HE, XL
The fuzzy rules for GEW output in FS-TLC system		
SM, SM, LI, LI, XS	ME, SM, LI LI, AV	LA, SM, LI, LI, LO
SM, SM, HE, LI, SH	ME, SM, HE, LI, LO	LA, SM, HE, LI, XL
SM, SM, LI, HE, XS	ME, SM, LI, HE, AV	LA, MS, LI, HE, LO
SM, SM, HE, HE, SH	ME, SM, HE, HE, LO	LA, SM, HE, HE, XL
SM, ME, LI, LI, XS	ME, ME, LI, LI, AV	LA, ME, LI, LI, LO
SM, ME, HE, LI, SH	ME, ME, HE, LI, LO	LA, ME, HE, LI, XL
SM, ME, LI, HE, XS	ME, ME, LI, HE, AV	LA, ME, LI, HE, LO
SM, ME, HE, HE, SH	ME, ME, HE, HE, LO	LA, ME, HE, HE, XL
SM, LA, LI, LI, XS	ME, LA, LI, LI, AV	LA, LA, LI, LI, LO
SM, LA, HE LI, SH	ME, LA, HE, LI, LO	LA, LA, HE, LI, XL
SM, LA, LI, HE, XS	ME, LA, LI, HE, AV	LA, LA, LI, HE, LO
SM, LA, HE, HE, SH	ME, LA, HE, HE, LO	LA, LA, HE, HE, XL

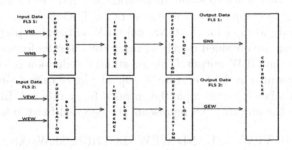

Fig. 4. The structure of HFS-TLC

Table 2. The fuzzy rules for outputs in HFS-TLC system.

The fuzzy rules for outputs in HFS-TLC system	
GNS	GEW
SM, LI, XS	SM, LI, XS
SM, HE, S	SM, HE, S
ME, LI, M	ME, LI, M
ME, HE, L	ME, HE, XL
LA, LI, L	LA, LI, L
LA, HE, XL	LA, HE, XL

5 Description of Experiments

In this paper, an experiment that is described in the literature [5] was used. All the conditions required for the experiment were fulfilled. This experiment consisted in a comparison of the operation of two different fuzzy systems: FS-TLC [5] and an adaptive system that is referred to as AEO [5,6]. The operation of both systems is related to the traffic light system control process and were tested on the situation when the traffic is rapidly growing in the given intersection. The simulation process was divided into eight stages as it was presented in paper [5]. The present study furthermore contains a comparison of the operation of the following systems: the base FS-TLC system [5], the adaptation system [5,6] and the hierarchical HFS-TLC fuzzy system. All these systems use the same random data generated, yet in the computation process, they use only the data that are required for computation. According to the assumptions, the results obtained for the FS-TLC [5] and HFS-TLC systems should be identical. Apart from this, time measurements will be performed in relation to the calculations made in order to verify the efficiency of the operation of the systems under comparison. These measurements were conducted on a computer based on the Intel Core2 Duo 3.18 GHz processor with Microsoft Windows 7 operating system. The application for the tests was created in the Microsoft Visual Studio 2012 C# environment.

During the experiment taken from literature [5], 25 simulation processes were conducted. The average simulation time (out of 25 trials performed) is: 2496[s] for the FS-TLC [5] systems and for HFS-TLC system and 2536[s] for the AEO system [6]. It needs to be added that for all simulations performed, the adaptational system [6] again won only once and time difference is equal to 8 s. It is evident based on the results obtained that the FS-TLC and HFS-TLC systems acquire the same results and they manage definitely better the traffic light system control problem under examination as compared to the AEO system. The times obtained by the individual fuzzy systems are presented in Table 3. The differences are much more noticeable in the results obtained during the measurements performed of computational times for input to the output for full simulation time. The minimum time for the each systems were: 12597[ns] for the HFS-TLC system, 41990[ns] for the FS-TLC [5] system and as much as

Table 3. The particular simulation times for fuzzy systems: FS-TLC, HFS-TLC and AEO

Simulation	FS-TLC Simulation time [s]	HFS-TLC Simulation time [s]	AEO [10] [s]	Difference [s]
1	2481	2481	2483	2
2	2489	2489	2575	86
3	2331	2331	2351	20
4	2505	2505	2535	30
5	2455	2455	2483	28
6	2443	2443	2485	42
7	2467	2467	2517	50
8	2453	2453	2517	64
9	2391	2391	2447	56
10	2527	2527	2559	32
11	2593	2593	2679	86
12	2473	2473	2475	2
13	2505	2505	2497	-8
14	2661	2661	2683	22
15	2527	2527	2599	72
16	2513	2513	2519	6
17	2507	2507	2561	54
18	2531	2531	2597	66
19	2489	2489	2511	22
20	2451	2451	2489	38
21	2577	2577	2621	44
22	2531	2531	2571	40
23	2467	2467	2503	36
24	2471	2471	2565	94
25	2551	2551	2571	20

160208[ns] for the adaptive system AEO. As it can be seen, the fuzzy system built in the hierarchical model obtained threefold faster time as compared to that of the FS-TLC system and almost thirteen times better time than that of the AEO system. Figure 5 contains complete characteristics of the minimum computational times; it can be seen that the FS-TLC [5] and HFS-TLC demonstrate more balanced results. The greatest differences can be observed while comparing the maximum times obtained during the computational processes performed. When comparing the results obtained in all of the simulation processes, an abrupt prolongation is noted of the performance time of calculations by the adaptive system. This is mainly connected with the algorithm [6] accepted by the authors,

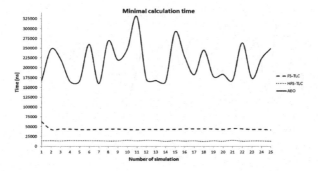

Fig. 5. The minimal calculation times for compared systems

which performs a search in its own database in order to determine the value that comes closest to the conditions on the road. The longest computational times obtained by the individual systems are as follows: 76228[ns] was obtained by the HFS-TLC system, 767125[ns] was obtained by the FS-TLC [5] system, while the adaptive system obtained 4998102[ns]. It can observed while analysing yet another time the characteristics of the maximum times for all the systems under examination that the adaptive system possesses a non-linear structure as compared to typically fuzzy systems (cf. Fig. 6). From the 25 simulations conducted, the average values were derived of times required to perform computational operations. It is to be noted that the HFS-TLC system obtained the average time that was not much higher (i.e. 19068[ns]) from its shortest time allocated to calculations. The traditional fuzzy system FS-TLC [5] obtained an average time on the level of 88625[ns] and this time is nearly twice as long as the minimum time. The AEO adaptive system gained a result of 1423578[ns] and this time is nearly 9 times as long as the minimum time. The characteristics of the average times obtained does do deviate from the remaining ones (cf. Fig. 7). The FS-TLC [5] and HFS-TLC fuzzy systems obtain the most linear course.

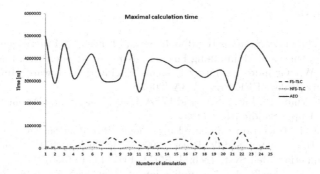

Fig. 6. The maximal calculation times for compared systems

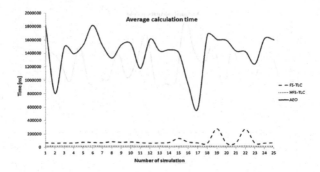

Fig. 7. The average calculation times for compared systems

6   Conclusions

It is clearly evident when analysing the results obtained during the experiment that the FS-TLC [5] and HFS-TLC fuzzy systems are much more efficient during the traffic light system control process. Additionally, it needs to be mentioned that it is the hierarchic HFS-TLC system that is the most effective one based on all the results obtained. Apart from an efficient operation, it is characterized by two principal advantages: a sixfold reduction of the fuzzy rules used comparing to the FS-TLC [5] and the shortest time required for the performance of all the computational operations. These are unquestionable arguments in favour of the need to create fuzzy systems in a hierarchical manner. In the future, the HFS-TLC hierarchical system under examination may be implemented in platforms that use the multithreading of the operations performed. Hierarchical fuzzy systems should be used in those systems that require as short as possible times of reactions and computational operations performed. Additionally, it needs to be emphasized that, when using hierarchical modelling of fuzzy systems, the demand of the systems designed for data memory can substantially be reduced.

References

1. Cox, E.: The Fuzzy Systems Handbook, Second Edition: A Practitioner's Guide to Building, Using, and Maintaining Fuzzy Systems. Academic Press, London (1999)
2. Pedrycz, W., Gomide, F.: Fuzzy Systems Engineering: Toward Human-Centric Computing. Wiley-IEEE Press, USA (2007)
3. Klir, G.J., Yuan, B.: Fuzzy Sets and Fuzzy Logic: Theory and Applications. Prentice Hall, Upper Saddle River (1995)
4. Mendel, J.M.: Uncertain rule-based fuzzy logic: introduction and new directions. Prentice Hall, Upper Saddle River (2000). McCluskey, E.J.: Minimization of boolean function. Bell Syst. Tech. J. **35**(5), 1417–1444 (1956)
5. Poletajew, B., Slowik, A.: An application of fuzzy logic to traffic lights control and simulation in real time. In: Rutkowski, L., Korytkowski, M., Scherer, R., Tadeusiewicz, R., Zadeh, L.A., Zurada, J.M. (eds.) ICAISC 2016. LNCS (LNAI), vol. 9692, pp. 266–275. Springer, Cham (2016). doi:10.1007/978-3-319-39378-0_23

6. Aksac, A., Uzun, E., Ozyer, T.: A real time traffic simulator utilizing an adaptive fuzzy inference mechanism by tunning fuzzy parameters. Springer Science+ Business Media, pp. 698–720, LLC (2011)
7. Lee, M.-L., Chung, H.-Y., Yu, F.-M.: Modeling of hierarchical fuzzy systems. Department of Electrical Engineering, National Central University, Chung-Li 32054, Taiwan, ROC
8. Renkas, K., Niewiadomski, A.: Hierarchical fuzzy logic systems and controlling vehicles in computer games. J. Appl. Comput. Sci. 22(1), 201–212 (2014)
9. Renkas, K., Niewiadomski, A.: Learning rules for hierarchical fuzzy logic systems using wu & Mendel IF-THEN rules quality measures. In: Rutkowski, L., Korytkowski, M., Scherer, R., Tadeusiewicz, R., Zadeh, L., Zurada, J. (eds.) ICAISC 2016. LNCS, vol. 9692. Springer, Cham (2016)
10. Torra, V.: A review of the construction of hierarchical fuzzy systems. Int. J. Intell. Syst. 17(5), 531–543 (2002)

Hierarchical Fuzzy Logic Systems in Classification: An Application Example

Krzysztof Renkas[(⊠)] and Adam Niewiadomski

Institute of Information Technology, Lodz University of Technology,
ul. Wólczańska 215, 90-924 Łódź, Poland
800561@edu.p.lodz.pl, adam.niewiadomski@p.lodz.pl

Abstract. This paper focuses on problems related to learning rules using numerical data for the *Hierarchical Fuzzy Logic Systems (HFLS)* described in [8]. Using hierarchical structure of *Fuzzy Logic Systems (FLS)* some complex problems could be divided into subproblems with smaller dimensions. "Hierarchical" means that fuzzy sets produced as output of one of fuzzy logic systems are then processed as input of another as the sets of auxiliary variables. The main scope of this paper is to use HFLS in classification problems for different datasets from the UCI Machine Learning Repository (The UC Irvine Machine Learning Repository shared by Center for Machine Learning and Intelligent Systems (University of California, Irvine) available at https://archive.ics.uci.edu/ml/index.html). The proposal presented in this paper operates on a type-1 HFLS, built with the fuzzy logic systems (in the sense of Mamdani). Iris, Abalone, Wine, Wine Quality Red and White datasets were used. Obtained results are described and compared to other classification systems.

Keywords: Hierarchical Fuzzy Logic Systems · Learning fuzzy rules · Rules quality measures · IF-THEN rules · Classification · Fuzzy classification system

1 Introduction

In particular, we are interested in computational intelligence methods based on FLS that make it possible to solve different **complex** problems. By "complex" we mean problems whose rulebases are really large (e.g. 1500 rules with more than two antecedents) and by using HFLS we significantly reduce the number of rules which makes it easier for an expert to express the knowledge processed by systems. Systems based on fuzzy logic [19], which makes decisions based on knowledge containing the rules like *IF ... THEN ...*; with unspecified predicates [17] could be used for many purposes such as controlling, decision making or classification. Our previous work focused on application of our HFLS and learning methods in decision making process for vehicles in computer games based on Tank 1990.

To prove the versatility of our solutions and learning method correctness in this paper we would like to apply these solutions in fuzzy classification problem.

© Springer International Publishing AG 2017
L. Rutkowski et al. (Eds.): ICAISC 2017, Part I, LNAI 10245, pp. 302–314, 2017.
DOI: 10.1007/978-3-319-59063-9_27

The new application proposed here is to learn FLS and HFLS rulebase using numerical data and test their classification accuracy with other attributes in minds such as knowledge base 'size'. Different HFLS structures, *Linguistic Variables (LV)* definitions, Quality Measures (QM) and Learning Ratios (LR) will be used during tests.

The rest of the paper is organized as follows: Subsect. 1.1 treats about our motivation to develop HFLS and HFLS learning methods; Next Subsect. 1.2 describes used datasets; Some literature references with different classification systems including fuzzy ones' description and classification accuracy results' summary for datasets used in this paper are presented in Sect. 2, some literature references with description of our former works are also included. Section 3 contains a brief description our new learning algorithm for HFLS with application of *Wu & Mendel (Wu&M)* IF-THEN rules quality measures. In Sect. 4, tests and the results are described. The last Sect. 5 contains conclusions and some future directions of the research.

1.1 Motivation for Developing Hierarchical FLSs

Below we can see an example of one fuzzy rule that belongs to the FLS rulebase for Abalone dataset:

```
RULE 0: IF gender IS infant AND length IS small AND diameter
IS small AND height IS small AND whole weight IS small AND
shucked weight IS small AND viscera weight IS small AND shell
weight IS small THEN rings IS c;
```

Abalone dataset operates on eight input parameters and one output parameter. The output parameter has 28 possible values. Describing all of input parameters using LV with only three labels (possible values for input parameter), during learning process we have to build about $1.8 \cdot 10^5$ rules using eight antecedents and one consequent. The number of fuzzy rules in final knowledge base could be reduced to more than 2140 rules removing conflicting and zero power rules.

HFLS contains many unit FLSs and outputs of one FLS are then considered as input of another one. The hierarchical structure looks more complicated, but in reality it allows us to simplify a complex problem with at least four input parameters into several subproblems with smaller dimensions. This approach gives us **high level** or **human consistent** fuzzy rules and further knowledge base!

Using HFLS build up with seven unit FLSs (with two input parameters each) for Abalone datasets example final knowledge base could contain only 113 fuzzy rules with maximum two antecedents and one consequent (only 2273 generated rules during the learning process including conflicting and zero power rules). Examples of four rules taken from different unit FLSs rulebases are listed below; example structure of our HFLS for Abalone dataset classification is presented on Fig. 1.

```
FLS1 RULE 3: IF gender IS infant AND length IS small THEN aa IS p3;
FLS2 RULE 1: IF diameter IS small AND height IS small
```

```
THEN ab IS p1;
FLS5 RULE 1: IF aa IS p3 AND ab IS p1 THEN ba IS p3;
FLS7 RULE 22: IF ba IS p3 AND bb IS p3 THEN rings IS i;
```

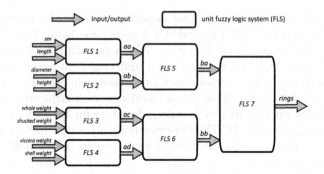

Fig. 1. Example HFLS structure for Abalone dataset classification problem build up with eight input parameters and one output.

Thus HFLS make it possible to divide complex problems into several sub-problems with smaller rulebases and simpler rules, which has a positive impact on performance. This is because of a huge reduction of the number of fired rules and performed mathematical operations during the learning process iteration or each inference. **So it is worth developing HFLS**. In this paper we would like to bring those benefits to play with a classification system based on HFLS using our learning method based on the *Wang & Mendel (W&M)* method adopted for the HFLS dealing with learning with numerical data, which does not contain any data for auxiliary variables. This method was introduced in [10] and improved in [9] applying quality measures of IF-THEN rules in the sense of Wu&M.

1.2 Description of Datasets

In this paper four different datasets are taken into account during our research focused on hierarchical fuzzy classification systems. They are briefly described below (for summary see Table 1.)

Table 1. Summary of chosen datasets classified by our HFLS and other authors' methods described in Subsect. 2.1 respectively.

	Iris	Abalone	Wine	Wine Quality	
Number of instances	150	4 177	178	1 599 (Red)	4 898 (White)
Number of attributes	4	8	13	12	
Number of classes	3	28	3	11	

Iris Plants from Fisher, 1936. Iris is very famous dataset with four numeric, predictive attributes and the class of Iris Plant (Iris Setosa, Iris Versicolour and Iris Virginica). The dataset contains only 150 instances (50 in each of three classes).

Abalone. This dataset contains physical measurements for Abalone age predicting. There are eight attributes plus the output one. The dataset contains 4177 instances.

Wine recognition. This data includes the results of a chemical analysis of wines grown in the same region in Italy but derived from three different cultivars (output parameter). The analysis determined the quantities of 13 constituents (input attributes) found in each of the three types of wines. The dataset contains 178 instances.

Wine Quality (WQ). Two datasets are included, related to the red and white vinho verde wine samples, from the north of Portugal. The output parameter as wine quality depends on 11 physicochemical attributes. The dataset contains 1599 instances in the red dataset and 4898 in the white one.

2   Literature References and Former Works

At the start to better understand the HFLS we could recall to [18]. In [1,3,11,15] authors introduce HFLS for many different problems, i.e. controlling agricultural robots in a natural environment, truck backer-upper system or grouping cars into platoons and controlling the velocity and the gap between cars in single lane platoons.

Authors in [13] say that using fuzzy models all parameters should be learnt in appropriate way by using experimental data and respective learning algorithms. The Wang & Mendel method which is designed as learning algorithm on numerical data for traditional FLS is described in [14]. Authors of [16] use linguistic summaries to generate fuzzy rules.

Our HFLS is described in [8] with comparison to traditional FLS. [10] introduces our new learning method based on the W&M method adopted for HFLS. Application of IF-THEN rules quality measures in the sense of Wu&M as method improvement is described in [9].

2.1   Classification References

Authors in [5] describe fuzzy classification system design using *Particle Swarm Optimization (PSO)* with dynamic parameter adaption through fuzzy logic. Learning ratio is set up to 70% of all instances (30% for testing). The obtained results are shown in Table 2.

Paper [2] describes quality assesment of red wine using logistic classification after clustering the experimental data with the *K*-means and *Expectation Maximization (EM)* algorithms. Additionally, logistic WEKA (Waikato Environment

Table 2. Percentage accuracy obtained by [5] authors during classification experiments using PSO methods and Iris, Abalone, Wine and Wine Quality datasets.

Dataset	PSO simple	FPSO1	FPSO2	FPSO3
Iris [%]	7.00	11.00	11.00	13.00
Abalone [%]	64.00	87.00	73.00	78.00
Wine [%]	52.00	57.00	63.00	57.00
Wine Quality Red [%]	48.00	53.00	51.00	56.00
Wine Quality White [%]	43.00	42.00	45.00	46.00

for Knowledge Analysis) were tested for comparison. Basic logistics classification by WEKA gets 59.78% accuracy. Using K-means clustering classification accuracy of the data was 94.75%, which was 7.32% better than that obtained by fitted logistic classification using EM. All of the obtained results are in Table 3.

Table 3. Percentage accuracy obtained by [2] authors during classification experiments using logistic classification, fitted logistic classification and K-means clustering for WQ Red dataset.

Dataset	EM	Fitted EM	K-means
Wine Quality White [%]	59.79	87.43	94.75

Fuzzy Cooperative Coevolution (Fuzzy CoCo) is presented in [6]. Fuzzy CoCo was applied to flower classification problem (using Iris dataset) obtaining very good classification results coupled with high human interpretability. In this system fuzzy modelling problem is solved by two coevolving cooperative species: encoding values which define completely all the membership functions for all the variables of the system and define a set of fuzzy rules. Fuzzy CoCo classification has high accuracy (about 99%) generating very small rulebase listed below.

```
Rule 1: IF petal lengh IS high THEN output is virginica;
Rule 2: IF sepal width IS low AND petal width IS low
THEN output IS virginica;
Rule 3: IF sepal lengh IS medium AND petal width IS medium
THEN output IS setosa;
Default: ELSE output IS setosa;
```

Authors of [7] introduce fuzzy pattern tree induction as machine learning method for classification. Pattern tree is a tree-like structure, whose inner nodes are marked with fuzzy logical operators or arithmetical operators and whose leaf nodes are associated with fuzzy terms on input attributes. There are two main types of fuzzy pattern tree: bottom-up approach and top-down approach. Classification accuracy results for Iris dataset using different types of fuzzy pattern trees are described in Table 4.

Table 4. Percentage accuracy obtained by [7] authors during classification experiments using fuzzy Pattern Trees (PT) in two different approaches: bottom-up and top-down (TD) for Iris datasets.

Training ratio	PTTD.1 [%]	PTTD.5 [%]	PTTD.25 [%]	Bottom up [%]
0.3	87.00	87.00	87.00	81.00
0.8	87.00	87.00	87.00	87.00

Another approach that could be used in classification is proposed in [4]. Authors of this paper describe Discernibility-Matrix Method Based on the Hybrid of Equivalence and Dominance Relations. The obtained results for classification purposes for different datasets are listed in Table 5.

Table 5. Percentage accuracy obtained by [4] authors during their experiments using Discernibility-Matrix and Attribute Significance methods and Iris, Wine and Wine Quality datasets.

Method	Iris [%]	Wine [%]	Wine Quality [%]
Discernibility-Matrix Method	90.00	47.22	47.63
Attribute Significance Method	89.20	45.00	48.50

In [12] authors describe application of Neuro-Fuzzy (NF) System for classification purposes. Wine Quality Red and White datasets are taken into account with different rule methods. The results are shown in Table 6. Additional classification results are delivered for C4.5 and Multilayer Perceptron (100 epoch) implemented in Weka for comparison.

Table 6. Percentage accuracy obtained by [12] authors during classification experiments using NF system and Wine Quality Red and White datasets. Three variants of system were tested using different rules method: Rectangular, Kosko and Shalaginov. Additional classification experiments classification results using C4.5 and Multilayer Perceptron (100 epochs) implemented in Weka for Wine Quality Red and White datasets are delivered.

Method	Wine Quality Red	Wine Quality White
NF Rectangular [%]	41.34	37.26
NF Kosko [%]	43.09	27.62
NF Shalaginov [%]	68.67	59.06
MLP 1-layer [%]	60.54	52.02
MLP 5-layers [%]	58.54	53.35
MLP 10-layers [%]	58.79	54.53
C4.5 [%]	73.93	58.68

3   Learning Rules Algorithm for Hierarchical Fuzzy Logic Systems

Brief description of our fuzzy rules learning algorithm for HFLS is given below; for detailed description see [9].

STEP 1: for each variable divide input and output domains into fuzzy regions
STEP 2: define groups of variables (including inputs and outputs for each unit FLS).
STEP 3: generate a separate rulebase for each group of variables using all available combinations or choosen strategy when the output variable is an auxiliary
STEP 4: compute the rule's degree or partial Wu&M fuzzy rules quality measure value iterating through learning data for each rule for each of unit rulebases with conflicting rules
STEP 5: compute the final values of Wu&M rules quality measures
STEP 6: remove conflicting rules, leaving rules with the highest value of rule's degree or chosen Wu&M IF-THEN rules quality measures removing also rules with zero quality measure value (zero power fuzzy rules). Additionally we could remove rules with the degree less than some α value, where $0 \leq \alpha \leq 1$.

The mentioned rule's degree is 'simple' and based on membership value - it is some kind of FR quality measure introduced by Wang & Mendel in [14]. Wu & M IF-THEN rules quality measures were used for learning purposes in [16]; in this paper we refer only to three of them: degree of truth, degree of sufficient coverage and degree of usefulness.

4   Tests and Results

4.1   Tests

Both FLS and HFLS were tested for classification purposes taking into account different datasets described in 1.2 and compared to other known methods such as Neuro-Fuzzy System, Fuzzy CoCo System or MLP (see Subsect. 2.1). Moreover, for FLS and HFLS different test cases were prepared taking into account:

- different possible values for LV, e.g. two (yes, no) or three (small, medium, high) labels connected with fuzzy sets
- different HFLS structures, e.g. different number of unit FLSs with different relations between them
- different learning ratio, e.g. 1.0 means that all of input instances would be taken into account during learning process or 0.2 reduces this number to 20%
- different fuzzy rules quality measures, e.g. degree of truth by W&M or Wu&M fuzzy rules quality measures such as degree of sufficient coverage

All of learned FLSs and HFLSs were tested for all input dataset instances collecting different attributes such as number of unit FLSs, number of fuzzy rules (all of generated, conflicting or included in final rulebase), rulebase power or percentage accuracy (percentage number of correct classifications).

Description of some shortcuts used in results description:

- FLS/HFLS - Fuzzy Logic System or Hierarchical FLS
- 3smh - three used values for linguistic variables with labels: small, medium and high
- WM - W&M degree of truth
- LsT, LsC, LsU - rules quality measures in the sense of Wu&M such as degree of truth, degree of sufficient coverage and degree of usefulness
- sp - first layer for HFLS were divided by *sepal* and *petal* into two unit FLSs for Iris dataset.

4.2 Results

Table 7 contains a summary of FLS and HFLS systems learned for Iris dataset with learning ratio 1.0. Different degrees and quality measures were taken into account. All of input parameters were described using only three possible values and the input parameters were dividen in the first HFLS layer by *sepal* and *petal* into two unit FLSs.

Table 7. Summary of FLS and HFLS systems learned for Iris dataset with learning ratio 1.0 for different rule degrees and quality measures.

testCaseName	FLS-3smh				HFLS-sp-3smh			
qualityMeasure	WM	LsT	LsC	LsU	**WM**	**LsT**	LsC	LsU
Number of auxiliary parameters	0	0	0	0	2	2	2	2
Number of unit FLSs	1	1	1	1	3	3	3	3
Number of all generated rules	243	243	243	243	93	93	93	93
Number of final rules	62	62	48	48	39	39	29	29
Number of antecedents	4	4	4	4	2	2	2	2
Rulebase power	0.17	0.82	0.65	0.55	0.47	0.82	0.83	0.71
Percentage accuracy [%]	94.66	94.00	81.33	94.00	**98.00**	**98.00**	93.33	93.33
Mean error	0.20	0.18	0.23	0.19	0.18	0.14	0.16	0.16

In comparison to Table 7 we could see Table 8 which contains summary of FLS and HFLS systems learned for Iris dataset with different learning ratios. Only the degree of truth in the sense of Wu&M were taken into account. Figure 2 shows percentage accuracy comparison for different quality measures and learning ratios.

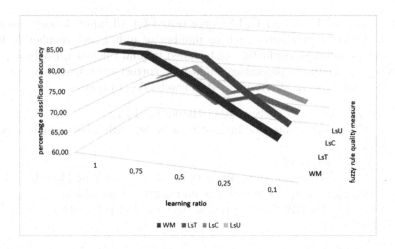

Fig. 2. Percentage classification accuracy for Wine dataset using HFLS with different fuzzy rules quality measures and different learning ratios.

Table 8. Summary of FLS and HFLS systems learned for Iris dataset with different learning ratios for the degree of truth in the sense of Wu&M fuzzy rules quality measure.

testCaseName	FLS-3smh			HFLS-sp-3smh		
learningRatio	0.75	0.25	**0.1**	0.75	0.25	**0.1**
Number of lines	150	150	150	150	150	150
Number of learning lines	112	37	**14**	112	37	**14**
Rulebase power	0.82	0.86	0.90	0.83	0.86	0.84
Percentage accuracy [%]	94.00	93.33	**93.33**	98.00	96.00	**94.00**
Mean error	0.18	0.19	0.19	0.14	0.16	0.17

Table 9 contains the best classification accuracy values obtained during our research (for FLS and HFLS with our learning fuzzy rules method) and other authors' research (mentioned in Subsect. 2.1 including Tables 2, 3, 4, 5 and 6). Of course the collected data is incomplete because not all of datasets used in this paper were used during other, cited authors' research. Also we have not got classification accuracy results for FLS and two datasets (Wine and Wine Quality) because of the number of input parameters for those datasets and very huge number of possible rules' combinations; we are unable to perform learning process. The best values are in bold and the worst are underlined.

Table 9. Summary of the best classification accuracy results [%] obtained during our research and other, cited authors' research (mentioned in Subsect. 2.1. The best accuracy values are bolded and the worst values are underlined (for each dataset). An asterix ('*') means that we were unable to finish learning process due to high complexity of FLS systems.

	Iris	Abalone	Wine	Wine Quality White	Wine Quality Red
our FLS	94.66	22.74	*	*	*
our HFLS	98.00	26.41	**84.27**	48.08	42.28
PSO simple	7.00	64.00	52.00	48.00	43.00
FPSO1	11.00	**87.00**	57.00	53.00	42.00
FPSO2	11.00	73.00	63.00	51.00	45.00
FPSO3	13.00	78.00	57.00	56.00	46.00
Discernibility matrix	90.00		47.22	47.63	
Attribute significance	89.20		45.00	48.50	
EM				59.79	
Fitted EM				87.43	
K-means				**94.75**	
Neuro-Fuzzy Rectangular				37.26	41.34
Neuro-Fuzzy Kosko				27.62	43.09
Neuro-Fuzzy Shalaginov				59.06	68.67
MLP 1-layer				52.02	60.54
MLP 5-layers				53.35	58.54
MLP 10-layers				54.53	58.79
C4.5				58.68	**73.93**
Fuzzy CoCo	**99.12**				
PTTD.1	87.00				
PTTD.5	87.00				
PTTD.25	87.00				
Bottom Up	87.00				

5   Conclusions and Future Work

The tests run prove versatility of the proposed solutions and the correctness of our learning method. Our previous research was focused only on one example problem based on decision making process during tank vehicles controlling process in the game Tank 1990. Application of our solutions for classification purposes gives us proof that the proposed solutions could be successfully used in other complex problems.

We recommend HFLS especially for **complex problems (e.g. at least with four or more parameters)** because we could solve this kind of problems in the easiest way with a highly reduced rulebase which is **human consistent** in comparison to traditional FLS (for example more than 2140 rules with eight antecedents could be reduced to 113 rules with only two antecedents using HFLS) **without any system quality loss registered in our tests**. Moreover, FLS are even unable to handle big dimensional datasets. For more see Subsect. 1.1 where is described our motivation to use HFLS system instead of FLS for Abalone classification problem. Further, taking another dataset as Iris which is not as complex as Abalone but in this case we could also gain more than 60% savings concerning the number of generated rules during learning process; additionally better results in classification accuracy were registered in both cases (see Tables 7, 8 and 9).

Results in Table 7 shows that the appliaction of fuzzy rules quality measures in comparison to the basic rules degree introduced in original Wang & Mendel method do not have any possitive impact on Iris classification accuracy, but it really does for other datasets used during our research (Abalone, Wine and Wine Quality). We dare to say that application of fuzzy rules quality measures in the sense of Wu&M is much more effective in more complex problems or during learning process using much more input instances.

Of course our HFLS and learning method is not a perfect solution for classification purposes because there are more specialized methods that get better classification accuracy, but our systems accuracy is not bad - moreover we coud say that is even good.

Results presented in Table 9 show that our HFLS in comparison to other results presented by authors cited in Subsect. 2.1 gets:

- the best classification accuracy for the Wine dataset
- very high classification accuracy for the Iris dataset with the 98% result (the best one is 99.12% accuracy obtained by Fuzzy CoCo system (see [6]))
- quite good results for the Wine Quality dataset with about 48% classification accuracy for White Wine and 42% for the Red one.
- very poor results for Abalone datasets having only 26.4% accuracy when other methods accuracy is at the level of $64\% - 87\%$

Moreover, we would like to emphasize that our HFLS system obtains quite the same good classification accuracy highly reducing the learning ratio to very small possitive value. Results in Table 8 show that the accuracy is at a high level even for 0.1 learning ratio which means that for Iris dataset just 14 learning instances are enough to provide 94% classification accuracy. Tests on other datasets used in this paper confirm that reducing the learning ratio does not have a negative impact on classification accuracy. For dataset where the number of instances is bigger the safe value of learning ratio is 0.01 for Abalone or White Wine Quality.

During future research we could modify the learning method by using some clustering methods during generating fuzzy sets for linguistic variables using learning data. Well fitted intervals of input domains should have a possitive impact on the whole process of learning and improve obtained results during tests. Also tuning fuzzy sets function using linguistic hedges during learning process could be taken into account during future research.

References

1. Hagras, H., Callaghan, V., Colley, M., Carr-West, M.: A behaviour based hierarchical fuzzy control architecture for agricultural autonomous mobile robots. J. Auton. Robots **13**, 37–52 (2002)
2. Jung, Y.G., Kang, M.S., Heo, J.: Clustering performance comparison using k-means and expectation maximization algorithms. Biotechnol. Biotechnol. Equip. **28**(suppl. 1), S44–S48, pMID: 26019610 (2014)
3. Kim, H.M., Dickerson, J., Kosko, B.: Fuzzy throttle and brake control for platoons of smart cars. Fuzzy Sets Syst. **84**, 209–234 (1996)
4. Li, Y., Zhao, J., Sun, N.-X., Wang, X.-Z., Zhai, J.-H.: Discernibility-matrix method based on the hybrid of equivalence and dominance relations. In: Kuznetsov, S.O., Ślęzak, D., Hepting, D.H., Mirkin, B.G. (eds.) RSFDGrC 2011. LNCS, vol. 6743, pp. 231–239. Springer, Heidelberg (2011). doi:10.1007/978-3-642-21881-1_37
5. Olivas, F., Valdez, F., Castillo, O.: Fuzzy classification system design using PSO with dynamic parameter adaptation through fuzzy logic. In: Castillo, O., Melin, P. (eds.) Fuzzy Logic Augmentation of Nature-Inspired Optimization Metaheuristics. SCI, vol. 574, pp. 29–47. Springer, Cham (2015). doi:10.1007/978-3-319-10960-2_2
6. Peña-Reyes, C.A., Sipper, M.: The Flowering of Fuzzy CoCo: Evolving Fuzzy Iris Classifiers, pp. 304–307. Springer, Vienna (2001)
7. Pophale, S.S., Kinariwala, S.A.: Induction of fuzzy pattern system. Int. J. Innov. Eng. Technol. (IJIET), 198–203
8. Renkas, K., Niewiadomski, A.: Hierarchical fuzzy logic systems and controlling vehicles in computer games. J. Appl. Comput. Sci. (JACS) **22**(1), 201–212 (2014)
9. Renkas K., Niewiadomski A.: Learning rules for hierarchical fuzzy logic systems using Wu & Mendel IF-THEN rules quality measures. In: Rutkowski, L., Korytkowski, M., Scherer, R., Tadeusiewicz, R., Zadeh, L., Zurada, J. (eds.) ICAISC 2016. LNCS, vol. 9692, pp. 299–310. Springer, Cham (2016). doi:10.1007/978-3-319-39378-0_26
10. Renkas, K., Niewiadomski, A., Kacprowicz, M.: Learning rules for hierarchical fuzzy logic systems with selective fuzzy controller activation. In: Rutkowski, L., Korytkowski, M., Scherer, R., Tadeusiewicz, R., Zadeh, L.A., Zurada, J.M. (eds.) ICAISC 2015. LNCS, vol. 9119, pp. 260–270. Springer, Cham (2015). doi:10.1007/978-3-319-19324-3_24
11. Riid, A., Rstern, E.: Fuzzy hierarchical control of truck and trailer. In: BEC 2002: Proceedings of the 8th Biennial Baltic Electronics Conference, pp. 141–144, 6–9 October (2002)
12. Shalaginov, A., Franke, K.: Towards improvement of multinomial classification accuracy of neuro-fuzzy for digital forensics applications. In: Abraham, A., Han, S.Y., Al-Sharhan, S.A., Liu, H. (eds.) Hybrid Intelligent Systems. AISC, vol. 420, pp. 199–210. Springer, Cham (2016). doi:10.1007/978-3-319-27221-4_17
13. Vachkov, G., Fukuda, T.: Structured learning of fuzzy models for reduction of information dimensionality. In: 1999 Proceedings of IEEE International Conference on Fuzzy Systems, FUZZ-IEEE 1999, vol. 2, pp. 963–968, August 1999
14. Wang, L.X., Mendel, J.: Generating fuzzy rules by learning from examples. IEEE Trans. Syst. Man Cybern. **22**(6), 1414–1427 (1992)
15. Wang, L.X.: Modeling and control of hierarchical systems with fuzzy systems. Automatica **33**(6), 1041–1053 (1997)
16. Wu, D., Mendel, J., Joo, J.: Linguistic summarization using if-then rules. In: 2010 IEEE International Conference on Fuzzy Systems (FUZZ), pp. 1–8, July 2010

17. Yager, R.R., Filev, D.P.: Essentials of Fuzzy Modeling and Control. A Wiley-Interscience Publication. John Wiley & Sons (1994)
18. Yager, R.: On a hierarchical structure for fuzzy modeling and control. IEEE Trans. Syst. Man Cybern. **23**(4), 1189–1197 (1993)
19. Zadeh, L.A.: The concept of a linguistic variable and its applications to approximate reasoning (i). Inf. Sci. **8**, 199–249 (1975)

A Bullying-Severity Identifier Framework Based on Machine Learning and Fuzzy Logic

Carmen R. Sedano$^{(\boxtimes)}$, Edson L. Ursini$^{(\boxtimes)}$, and Paulo S. Martins$^{(\boxtimes)}$

University of Campinas, School of Technology, Limeira, Brazil
c154133@g.unicamp.br, {ursini,paulo}@ft.unicamp.br

Abstract. Bullying at schools is a serious social phenomenon around the world that negatively affects the development of children. However, anti-bullying programs should not focus on labeling children as either bullies or victims since they could produce opposite effects. Thus, an approach to deal with bullying episodes, without labeling children, is to determine their severity, so that school staff may respond to them appropriately. Related work about computational techniques to fight against bullying showed promising results but they offer categorical information as a set of labels. This work proposes a framework to determine bullying severity in texts, composed by two parts: (1) evaluation of texts using Support Vector Machine (SVM) classifiers found in the literature, and (2) development of a Fuzzy Logic System that uses the outputs of SVM classifiers as its inputs to identify the bullying severity. Results show that it is necessary to improve the accuracy of SVM classifiers to determine the bullying severity through Fuzzy Logic.

Keywords: Machine learning · Fuzzy logic · Text mining · Bullying

1 Introduction

Bullying is a serious social problem among school children with a high peak during early adolescence (11–14 years old) [1,2]. This problem concerns students, school staff and parents [3] because of its negative consequences for the mental, social and emotional student's health. The entire school environment becomes dangerous when there are not effective interventions against bullying situations and it involves all children, without exception [4]. For example, bullies are likely to suffer depression, to engage in fighting behavior as well as criminal and academic misconduct, to be physically and socially aggressive, to have low empathy and to have an exaggerated air of self-confidence [5]. On the other hand, their victims may suffer from depression, loneliness, anxiety, fear, sadness and low self-esteem [4]. Indeed, students who were either bullies or victims might have health problems and emotional adjustment in adulthood [6].

Therefore, it is important for schools to design and implement prevention and intervention programs to reduce bullying and victimization and the effectiveness of these programs promote positive youth [6]. Additionally, bullying

© Springer International Publishing AG 2017
L. Rutkowski et al. (Eds.): ICAISC 2017, Part I, LNAI 10245, pp. 315–324, 2017.
DOI: 10.1007/978-3-319-59063-9_28

should be viewed along a continuum rather than labeling children as bullies, victims or uninvolved, because labeling overemphasizes the role of individual children without accentuating their capacities and thus it contributes to a negative climate [7]. Therefore, instead of labeling children, an effective anti-bullying program has to be able to respond to signs of any of the categories of involvement [8]: bullying (helping bullies because their behavior is just a response to several causes such as familiar problems [1]) and victimization (helping victims, even by-standers, because typically they do not seek help or prefer to suffer in silence [7]). In [9], the authors propose a Bullying Assessment Matrix which allows school staff to determine the severity of a bullying episode. Indeed, that matrix would avoid labeling students and would help to measure the episode as a whole.

Typically, social scientists of bullying use surveys and questionnaires that may yield limited information due to their cost and potential participant fatigue [10]. In contrast, the computer science study of bullying is emerging with promising results [11]. Nowadays, digital technology forms part of the life of young people and the most prevalent online activity that they are engaged in is the use of social media such as Facebook, Twitter and YouTube. Digital technology may be exploited to determine bullying behavior [9] since the participants of a bullying episode (physical or cyber) often post social media text about their experience [11,12]. The work of Xu [12] shows that social media, with appropriate Machine Learning (ML) and Natural Language Processing (NLP) techniques, can be a valuable tool for the study of bullying.

Following that premise and the proposal of the Bullying Assessment Matrix of [9], this work proposes a bullying-severity identification framework composed by SVM classifiers developed in [12,13] and a Fuzzy Logic System (FLS) in order to determine the severity of bullying episodes occurring in texts.

The remainder of this paper is organized as follows: Sect. 2 presents literature review about bullying and current approaches to reduce bullying episodes at school. It also addresses work on computational techniques to fight against bullying. Section 3 describes our proposed approach used to develop the framework. Section 4 shows the experimental results and related discussion. Finally, Sect. 5 discusses the conclusions and future work.

2 Background and Literature Review

2.1 Definition of Bullying

The most widely accepted definition of bullying was given by the psychologist Dan Olweus [2,14]. He defined that a student is being bullied or victimized when he or she is repeatedly exposed to negative actions on the part of one or more students over time. In fact, there are three key concepts [15] that differentiate bullying from other forms of school violence and conflict: (1) an intent to harm or upset another student, (2) the harmful behavior occurs repeatedly, and (3) the relationship between the bully (or bullies) and the victim(s) is characterized by an imbalance in power. In addition, each bullying episode has a severity level and an impact on the victim [11].

2.2   Types of Bullying

The forms of bullying behaviors are physical aggression (hitting, kicking, punching), verbal (name-calling, threatening), property damage (stealing or damaging the possessions of victims) [6], sexual (inappropriate touching, relational (spreading rumors or exclusionary behavior) and cyberbullying (bullying through electronic devices) [3,15,16]. Regarding the perception of bullying severity, physical bullying had higher values of severity than verbal and relational. However, verbal and relational bullying can be just as harmful as physical bullying [17].

2.3   Participants in a Bullying Episode

Typically, bullies and victims are the key participants in a bullying episode at school [14]. However, bullying is not an isolated event between two individuals [3]. In [3], we found the following roles: the bully, the victim and three types of witnesses such as uninvolved students, bully supporters (students that incite the bully without personally taking action against the victim) and bully intervener (students that defend or console the victim). Additionally, the work of Xu *et al.* [13] augmented two new roles for social media: the reporter, who is someone that may not be present during the episode and the accuser, who accuses someone as being the bully.

2.4   Bullying Assessment Matrix

In [9], the authors consider that all bullying episodes have to be responded appropriately with policies and processes according to their nature. Thus, they propose a Bullying Assessment Matrix to assess severity, impact and frequency of a bullying incident that supports schools in their decisions about how to treat the incident. The bullying incident is considered as moderate, major or severe if the total score varies between 3–5, 6–7 or 8–9 respectively.

2.5   Prior Work in Computer Science

Computational social science is an emerging field has the capacity to collect and analyze large amounts of data, e.g. originating from the Internet [18] and especially from social media networks. Within this context, the computational study of bullying presented in [11–13] shows that social media can enrich it due to their nature (large-scale, near real-time and dynamic data) and their popularity among young people.

Indeed, the social network Twitter, that allows people to post 140-character messages called tweet, produces about 400 million tweets per day and has been used as a data source to answer social science questions [19,20]. Despite the fact that texts posted on Twitter may be unstructured and informal texts with noise as non-standard abbreviations, typographical errors, use of emoticons, irony and sarcasms [20,21], they offer the possibility to identify online (cyberbullying) and offline bullying trends (bullying episodes in the real word) [14].

In [22–25], we found Data Mining and ML techniques to detect cyberbullying using social networks as data sources, including Twitter. In contrast, the work of Mancilla-caceres *et al.* [10] propose the identification of participants of school bullying situations through their social interactions in a computer game instead of using real and social-interaction data from social networks - due to its evaluation complexity and its availability (e.g. Facebook). The thesis of [26] made a comparison between text classification methods and classification using sentiment analysis (emotional vectors) to determine bullying in Twitter conversations. The results showed that using the emotional vectors did not improve the accuracy of the classification (78% versus 75%).

2.6 Support Vector Machine in Text Classification

In [12], there are five SVM classifiers publicly available, but we chose only four of them due to their importance according to the literature presented in Sect. 2. The work of [12] collected 32,477,558 tweets and each tweet was represented by the combination of both unigrams and bigrams (1g2g) which were part of a vocabulary file (4,524 entries). Each SVM classifier handles its own model file where we found that a weight was assigned to each feature of the vocabulary mentioned above. The description and accuracy of the chosen SVM classifiers are shown in Table 1. SVM classifiers in [12] return a real number. For SVM binary classifiers, results less than zero are labeled as No and greater than zero are Yes. For SVM multi-class classifiers, a real number is calculated for each possible class for a tweet, and the final result is the largest value among all of them.

Table 1. SVM classifiers developed by [12]

SVM classif.	Description	Classes	Accuracy
Bullying trace	It identifies if text belongs to a bullying context	Yes No	86%
Teasing trace	It identifies if exists lack of severity of a bullying episode	Yes No	89%
Author role	It identifies participants of a bullying episode	Victim, Defender, Reporter bully, Accuser, Other	61%
Bullying form	It identifies bullying forms of a bullying episode	General, Cyberbullying physical, Verbal	70%

2.7 Fuzzy Logic System

A limitation that led to the creation of fuzzy logic was the poor precision offered by natural languages used to describe or share knowledge that is inherently vague for some contexts [27]. For example, concepts such as cheap, expensive, big and old do not have well-defined boundaries, and thus they may be called fuzzy concepts. The Fuzzy Logic (FL) approach allows us to better deal with the uncertainty present in problems with high complexity, where traditional system modeling techniques do not offer enough precision [27]. In Sect. 1, we have presented the complexity of the bullying problem and we have mentioned that labeling children would not be the correct path to solve the problem. Furthermore, Barton [3] shows that bullies and victims should not be completely (or precisely) regarded as such since it depends on the level of conflict in which they are involved in. Therefore, we propose to include and integrate the capabilities of Fuzzy Logic to determine the severity of bullying episodes according to a number of factors that would be inputs to our FL System.

3   Proposed Approach

Our proposal consists in developing a framework that allows school staff to determine the severity level of a bullying episode in texts, in English written by students aged between 11 and 14 years old. The texts are collected through a social network developed for this project. They are evaluated using a Java Swing application that executes the SVM classifiers and the Fuzzy Logic System in order to obtain the severity level of the text. Figure 1 shows the functionality of the framework.

We divided the development of the framework in four main tasks: (1) Text pre-processing, (2) Text Classification using SVM classifiers of Xu [12], (3) Development of the Fuzzy Logic System, and (4) Development of a Social Network and Java Swing Application.

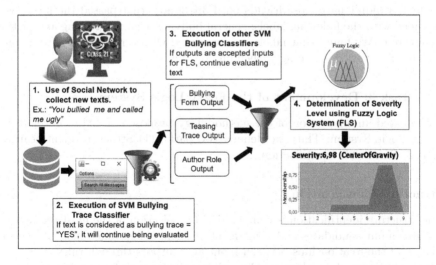

Fig. 1. Functionality of the Bullying-Severity Identifier Framework

3.1   Task 1: Text Pre-processing

We collected 18,504 tweets using *Twitter Streaming API* and *Tweepy* from June to December 2016, and we also added 4,783 of the tweets used by Xu [12] in the training phase of the SVM classifier. These tweets included any of the keywords that belong to a bullying context according to the literature. Additionally, we collected texts from *No-Place4hate.org* and *StampOutBullying.co.uk*, which are stories written by children that were involved in bullying episodes. We analyzed tweets and stories to find high-frequency words, unigrams and bigrams, so that they could assist us in improving the vocabulary used by the SVM classifiers of [12]. We used only tweets (18,504 in total) to test the SVM classifiers in the next task. Since, tweets have an unstructured format, they passed through the Enrichment and Tokenization process defined by Xu [12]. We made some improvements in the Tokenization process such as replacing URLs that started with https:// by HTTPLINK and correctly removing English contractions of words (e.g. he'll → he will).

3.2   Task 2: Text Classification Using SVM Classifiers

To the SVM classifiers of [12], which were developed in Java, we added the following modifications: (1) created a method to read tweets collected in Task 1, (2) created an order of execution of SVM classifiers: It consisted on executing first the SVM Bullying Trace. Only tweets classified as YES can continue being classified by the other SVM classifiers. This scheme allows us to filter out those tweets that are not considered as bullying trace. For example, the tweet I was bullied in elementary school because of my height and then teased for crying about it I didn't have close friends until 3rd grade, that passed the filter, was classified with the following labels: general bullying form, No teasing and victim author role. We found that labels are associated with a real numeric value and that number varied in a range depending on the SVM classifier.

3.3   Task 3: Development of the Fuzzy Logic System

We followed an adaptation of the methodology proposed in [28] to develop the Fuzzy Logic System. Thus, we defined two parts: (1) Structure Identification, and (2) Parameter Identification.

Structure Identification

The structure identification is composed of (1) selection of the most relevant or possible input candidates and assignment of membership functions, (2) Specifications of the relationships between input and outputs through rules.
Input selection and membership: Once all tweets had their labels assigned due to the execution of the SVM classifiers, we analyzed them. Finally, we selected eight labels as input variables and we grouped them by three categories: author role, bullying form, and teasing. Only the Bully and Victim labels were considered out

of the six categories defined in the literature: roles such as Reporter and Assistant were discarded due to their ambiguity; and Accuser and Defender labels were discarded due the complexity involved in inferring their impact on a bullying episode. All of the labels of the *bullying* form (i.e. general, cyberbullying, verbal and physical) where considered due to their relevance according to the literature. The variable teasing was considered important because it allows determination of whether or not the text is a joke - and if so, the severity must decrease.

Next, we defined the Fuzzification process, i.e. for each input variable, we defined its membership functions. Each input variable related to a bullying form (general, physical, cyberbullying, verbal) and author role (bully, victim) had three membership functions: Low, Medium and High. Each input variable related to teasing had two membership functions: Yes and No. The intervals of each membership functions were determined based on the analysis of a set of 18,504 tweets labeled in the previous task. The output variable was called severity and had three membership functions: moderate, major and severe. We used the same names of the Bullying Matrix Assessment [9].

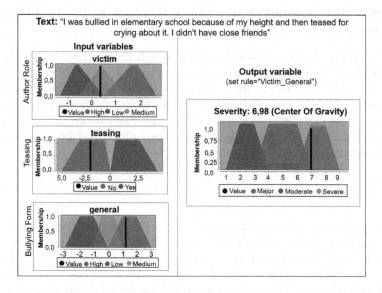

Fig. 2. Determination of the severity of bullying episodes (in text) using Fuzzy Logic

Rule generation: Based on the literature review, we established some criteria for the creation of rules - for example - the importance of each input variable in determining the output variable severity. Thus, a higher importance was assigned to the input variable physical in comparison to the verbal, general and cyber-bullying variables. Regarding roles, victim and bully were considered the most important ones to define the bullying severity. Finally, the "teasing" variable was considered most important when it belongs to the no category in its membership function. We generated eight sets of rules that were modeled in MATLAB software using the Mamdani fuzzy inference system: VictimGeneral, VictimPhysical,

VictimVerbal, VictimCyberbullying, BullyGeneral, BullyPhysical, BullyVerbal, BullyCyberbullying. Each set has 18 rules, the set that will be used must be related to the labels of the tweet. For example, in Fig. 2, the tweet was evaluated with the set of rules VictimGeneral because of its labels (author = victim, bullying form = general and teasing=no) and the output was of severity *severe*.

Parameter Identification

Membership function tuning: The analysis of labeled tweets allowed us to choose the shape of the membership function for each input and output variable. The range of each input variable was determined based on maximum and minimum values extracted from a set of 18,504 tweets.

Parameter adjustment: In [27], we found seven methods used for defuzzification: among them, the Centroid (or Center of Gravity) method is the most widely used. We have also decided to use it because of its accuracy.

3.4   Task 4: Development of a Social Network and Java Swing Application

We developed a social network, in order to collect new texts, with the 140-character limitation - since there is no guarantee that the application of SVM classifiers works with larger texts [12]. This social network was developed in Wordpress (CMS), written in PHP and it uses MySql as the database. To integrate the SVM classifiers with the Fuzzy Logic System, we used the jFuzzyLogic.jar library in our Java Swing Application, which may be used by school staff to manage texts written by their students.

4   Results and Discussion

A set of 18,504 tweets were collected in order to validate the modeled fuzzy logic system, and 11,004 were classified as bullying traces. Once the model processed the tweets, 10,717 were classified as general bullying, 28 as physical bullying, 86 as verbal bullying and 173 as cyberbullying. Finally, 5,949 tweets were classified with the variables of interest, i.e. the variables selected as inputs to the fuzzy logic system.

The resulting tweets were analyzed and the severity of the bullying episodes was determined by agreement of the authors, allowing the comparison to the output of the fuzzy system. The process revealed the difficulty in dealing with the identification of the severity of bullying episodes, even by common sense. The task of determining the severity on a bullying episode turned difficult as each author had a different perception of the same case, even when the same criteria used to create the fuzzy rules were applied.

5   Conclusion

The accuracy of the SVM classifiers used in this work compromised the performance of the fuzzy logic system. However, it is important to emphasize the high complexity involved in addressing social issues from the field of computer science. The analysis performed in this work indicated that the accuracy of the outputs of the SVM classifiers directly affected the fuzzy outputs, meaning that, in order to obtain better results, the SVM classifiers need to be improved.

Currently, we continue working on improving the text classification in order to generate new models used by SVM classifiers. Other factors such as sex, race, social and economic status of the participants of bullying episodes should be considered as well as part of the evaluation of bullying episodes. In future work, other NLP techniques may be used in the preprocessing of texts. Furthermore, improvements in the vocabulary used by the classifiers are also recommended, for example, by assigning proper weights to the most relevant words pertaining to a given bullying context.

References

1. Tusinski, K.: The Causes and Consequences of Bullying (2008)
2. Bauer, N.S., Lozano, P., Rivara, F.P.: The effectiveness of the olweus bullying prevention program in public middle schools: a controlled trial. J. Adolesc. Health **40**, 266–274 (2007)
3. Barton, E.A.: The bully, victim, and witness relationship defined. In: Bully Prevention: Tips and Strategies for School Leaders and Classroom Teachers, pp. 1–18 (2006)
4. Lopes Neto, A.A.: Bullying: comportamento agressivo entre estudantes. J. Pediatr. (Rio. J.) **81**, 164–172 (2005)
5. Dake, J.A., Price, J.H., Telljohann, S.K.: The nature and extent of bullying at school. J. Sch. Health **73**, 173–180 (2003)
6. Evans, C.B.R., Fraser, M.W., Cotter, K.L.: The effectiveness of school-based bullying prevention programs: a systematic review. Aggress. Violent Behav. **19**, 532–544 (2014)
7. Cantillon, P., Hutchinson, L., Wood, D.: Social and Emotional Learning and Bullying Prevention (2009). Accessed. http://casel.org/wp-content/uploads/2016/01/3_SEL_and_Bullying_Prevention_2009.pdf
8. Rettew, D.C., Pawlowski, S.: Bullying. Child Adolesc. Psychiatr. Clin. N. Am. **25**, 235–242 (2016)
9. Bullying Prevention Advisory Group: Bullying prevention and response: A guide for Schools (2015). Accessed. (http://www.education.govt.nz/assets/Documents/School/Bullying-prevention/MOEBullyingGuide2015Web.pdf)
10. Mancilla-Caceres, J.F., Pu, W., Amir, E., Espelage, D.: A computer-in-the-loop approach for detecting bullies in the classroom. In: Proceedings of the 5th International Conference on Social Computing, Behavioral-Cultural Modeling and Prediction, pp. 139–146 (2012)
11. Xu, J.-M., Zhu, X., Bellmore, A.: Fast learning for sentiment analysis on bullying. In: Proceedings of the First International Workshop Issues Sentim. Discov. Opin. Min., WISDOM 2012, pp. 1–6 (2012)

12. Xu, J.: Understanding and Fighting Bullying with Machine Learning (2015)
13. Xu, J., Jun, K., Zhu, X., Bellmore, A.: Learning from bullying traces in social media. In: Proceedings of the 2012 Conference of the North American Chapter of the Association for Computational Linguistics: Human Language Technologies, pp. 656–666 (2012)
14. Olweus, D.: A profile of bullying at school. Educ. Leadersh. **60**, 12–17 (2003)
15. Dulmus, C.N., Sowers, K.M., Theriot, M.T.: Prevalence and bullying experiences of victims and victims who become bullies (bully-victims) at rural schools. Vict. Offender. **1**, 15–31 (2006)
16. California Department of Education: Bullying at School (2003)
17. Perez, V.: Perception of severity, empathy, and disposition to intervene in physical, verbal, and relational bullying among teachers from 5 to 8 grade. In: Psykhe (Santiago), pp. 25–37 (2011) (in Spanish)
18. Lazer, D., Pentland, A., Adamic, L., Aral, S., Barabasi, A.L., Brewer, D., Christakis, N., Contractor, N., Fowler, J., Gutmann, M., Jebara, T., King, G., Macy, M., Roy, D., Al, Vanstyne, M.: Life in the network: the coming age of computational social science. Science **323**(80), 721–723 (2009)
19. Eisenstein, J., O'Connor, B., Smith, N.A., Xing, E.P.: A latent variable model for geographic lexical variation. In: Proceedings of the 2010 Conference Empirical Methods for Natural Language Processing, pp. 1277–1287 (2010)
20. Ajao, O., Hong, J., Liu, W.: A survey of location inference techniques on Twitter. J. Inf. Sci. **41**, 855–864 (2015)
21. Ritter, A., Clark, S., Etzioni, O.: Named entity recognition in tweets: an experimental study. In: Conference Empirical Methods for Natural Language Processing, pp. 1524–1534 (2011)
22. Nalini, K., Sheela, L.J.: Classification of tweets using text classifier to detect cyber bullying. In: Emerging ICT for Bridging the Future - Proceedings of the 49th Annual Convention of the Computer Society of India CSI, vol. 2, pp. 637–645 (2015)
23. Kontostathis, A., Reynolds, K., Garron, A., Edwards, L.: Detecting cyberbullying: query terms and techniques. In: Proceedings of the 5th Annual ACM Web Science Conference on WebSci 2013, pp. 195–204. ACM, New York (2013)
24. Dinakar, K., Reichart, R., Lieberman, H.: Modeling the detection of textual cyberbullying. Assoc. Adv. Artif. Intell., 11–17 (2011)
25. Parime, S., Suri, V.: Cyberbullying detection and prevention: data mining and psychological perspective. In: 2014 International Conference on Circuit, Power and Computing Technologies, ICCPCT 2014, pp. 1541–1547 (2014)
26. Arthur Patch, J.: Detecting Bullying on Twitter using Emotion Lexicons (2015)
27. Sivanandam, S.N., Sumathi, S., Deepa, S.N.: Introduction to Fuzzy Logic Using MATLAB. Springer, Heidelberg (2007)
28. Emami, M.R., Trksen, I.B., Goldenberg, A.A.: Development of a systematic methodology of fuzzy logic modeling. IEEE Trans. Fuzzy Syst. **6**, 346–361 (1998)

Evolutionary Algorithms and Their Applications

On the Efficiency of Successful-Parent Selection Framework in the State-of-the-art Differential Evolution Variants

Petr Bujok[(✉)]

Department of Computer Science, University of Ostrava,
30. dubna 22, 70103 Ostrava, Czech Republic
petr.bujok@osu.cz
http://prf.osu.eu/kip/

Abstract. Successful-parent selection (SPS) framework in differential evolution (DE) is studied. Two SPS versions (SPS1 proposed recently in literature and SPS2 newly proposed in this paper) are applied to seven state-of-the-art DE variants. The algorithms are compared experimentally on CEC 2014 test suite used as a benchmark. The application of SPS1 increases the efficiency of two DE algorithms in over 50% of test problems. An overall comparison shows that the newly proposed SPS2 performs well only in two cases whereas SPS1 outperforms six out of 7 original algorithms.

Keywords: Global optimization · Differential evolution · Successful-parent selection · Experimental comparison · CEC 2014 test suite

1 Introduction

The search of the global optimum of an objective function arises frequently in many fields of human activities. Without loss of generality, this problem can be simply formed as a minimization problem in a very clear form. This single-objective continuous global optimization problem is formed as follows.

The objective function is denoted $f(\boldsymbol{x}), \boldsymbol{x} = (x_1, x_2, \ldots, x_D) \in \mathbb{R}^D$ and the search domain Ω is limited by boundaries given by lower (a_j) and upper (b_j) border, $\Omega = \prod_{j=1}^{D}[a_j, \ b_j]$, $a_j < b_j$, $j = 1, 2, \ldots, D$. Then the global minimum point $\boldsymbol{x}^*$, which satisfies condition $f(\boldsymbol{x}^*) \le f(\boldsymbol{x}), \forall \boldsymbol{x} \in \Omega$ is the solution of the problem.

The successful-parent selection (SPS) mechanism [4] was constructed in order to prevent stagnation of the population. In this paper, another variant of SPS mechanism is also proposed. Both SPS variants are applied to seven state-of-the-art adaptive variants of differential evolution algorithm and experimentally compared on CEC 2014 test suite [6].

The remaining part of the paper is organized as follows. Basics of differential evolution algorithm is explained in Sect. 2. The used state-of-the-art variants

© Springer International Publishing AG 2017
L. Rutkowski et al. (Eds.): ICAISC 2017, Part I, LNAI 10245, pp. 327–336, 2017.
DOI: 10.1007/978-3-319-59063-9_29

of differential evolution and details of the applied mechanism are described in Sects. 3 and 4. CEC 2014 test suite with the experiment settings are introduced in Sect. 5. Results of the experimental comparison of the algorithms are presented in Sect. 6 and some remarks are provided in Sect. 7.

2 Differential Evolution Algorithm

Differential evolution (DE) is a stochastic population based evolutionary technique which searches for the global optimum of the objective function heuristically. It was introduced by Storn and Price approximately twenty years ago [11] in order to solve real-valued problems. Despite its simplicity, DE is a powerful global optimizer that frequently outperforms other well-known optimization algorithms [14,15] and therefore DE is very often applied to many real problems [9]. On the other hand, settings of DE control parameters affect the efficiency of the search substantially. This matter was very intensively studied including the way of their adaptive or self-adaptive setting, for a comprehensive summary of recent results see [2,3,8]. DE algorithm uses a population of

---

Algorithm 1. Differential evolution algorithm

---

initialize population $P = \{x_1, x_2, \ldots, x_N\}$
evaluate $f(x_i)$, $\quad i = 1, 2, \ldots, N$
while stopping condition not reached **do**
   for $i = 1, 2, \ldots, N$ **do**
      create a new trial vector y (mutation and crossover)
      evaluate $f(y)$
      if $f(y) \leq f(x_i)$ **then**
         insert y into Q
      else
         insert x_i into Q
      end if
   end for
   $P \leftarrow Q$
end while

N points - candidates of the solutions in the search domain. The population is developed during the whole search process using evolutionary operators, i.e. *selection, mutation* and *crossover*. Mutation is controlled by F parameter, $F > 0$, and crossover is controlled by parameter CR, $0 \leq CR \leq 1$. The DE algorithm is shown in pseudo-code in Algorithm 1.

The combination of mutation and crossover is called DE *strategy* and it is denoted by the abbreviation of DE/m/n/c. Symbol m specifies the kind of mutation, n is used for the number of differences in mutation, and c indicates the type of crossover. Various values of F and CR can be used in each strategy. DE has a few control parameters only. Besides setting the population size N and defining the stopping condition necessary for all evolutionary algorithms, the selection of DE strategy and setting the values of F and CR is all that must be done.

3   State-of-the-art De Variants

The self-adaptive DE variants (jDE [1], JADE [17], SaDE [10], EPSDE [7] and SHADE [12]) are currently considered as the state-of-the-art DE variants and the performance of novel DE variants is compared with these state-of-the-art DE variants in currently appearing studies. These variants are also included in this experiment along with a variant of composite trial vector generation strategies and control parameters published recently (CoDE) [16] and recently proposed DE with an individual-dependent mechanism (IDE) [13].

A simple and efficient adaptive DE variant (mostly called jDE in literature) was proposed by Brest et al. [1]. It uses the DE/rand/1/bin with an evolutionary self-adaptation of F and CR. The pair of these control parameters is encoded with each individual of the population and survives if an individual is successful, i.e. if it generates such a trial vector which is inserted into next generation. The values of F and CR are initialized randomly for each point x_i in population P and survive with the individuals in the population, but they can be randomly mutated in each generation with given probabilities τ_1 and τ_2.

The differential evolution algorithm with strategy adaptation (SaDE) was introduced by Qin and Suganthan. An enhanced variant of SaDE was proposed later in [10] and it is used in our experimental comparison. Four mutation strategies (rand/1/bin, rand/2/bin, rand-to-best/2/bin, and current-to-rand/1) for creating new trial vectors are stored in a strategy pool. Each strategy has a probability to be selected. These probabilities are updated after each LP generations.

JADE is an algorithm of adaptive differential evolution introduced by Zhang and Sanderson in [17]. The original DE concept is extended with three different improvements - current-to-$p$best mutation strategy, adaptive control of parameters F and CR, and archive A. Archive A is initialized as an empty set. In every generation, parent individuals replaced by a better offspring, $(f(y) \leq f(x_i))$, individuals are put into the archive. After every generation, the archive size is reduced to N individuals by random dropping surplus individuals.

In EPSDE adaptive variant [7], an ensemble of mutation strategies and parameter values is applied. The mutation strategies and the values of control parameters are chosen from pools. The combination of the strategies and the parameters in the pools should have diverse characteristics, so that they can exhibit distinct performance during different stages of evolution when dealing with a particular problem. The triplet of (strategy, F, CR) is encoded along with each individual of population. If the parent vector produces a successful offspring vector, this triplet survives with the trial vector for next generation. Otherwise, the triplet is randomly reinitialized.

The adaptive DE variant with composite trial vector generation strategies and control parameters, CoDE, was presented by Wang et al. [16]. The results showed that CoDE is at least competitive with the algorithms in the comparison. The CoDE combines three well-studied trial vector strategies with three control parameter settings in a random way to generate trial vectors. The strategies are rand/1/bin, rand/2/bin, and current-to-rand/1 and all the three strategies are applied when generating a new vector (select the offspring vector with the least function from the triplet).

DE variant called Success-History Based Parameter Adaptation for Differential Evolution (SHADE) was proposed by Tanabe and Fukunaga in 2013 [12]. It was derived from the original JADE algorithm proposed by Zhang and Sanderson in 2009 [17]. The main extension of SHADE compared to original JADE is in a history-based adaptation of the control parameters F and CR. Both JADE and SHADE variants use an efficient greedy *current-to-pbest* mutation strategy, in the case of SHADE, the parameter of p_i for selecting the best point is generated randomly for each point of the population according to $p_i = rand[2/N, \ 0.2]$.

The last algorithm in the comparison is Differential Evolution With an Individual-Dependent Mechanism (abbreviated IDE) introduced in 2015 by Tang et al. [13]. The algorithm uses an individual dependent mechanism of F and CR control parameters and the selection of the DE mutation strategy with respect to the ranking the individual of the population sorted according to the function values. The selection of the mutation also depends on the stage of the search. The search process of IDE is divided to two stages based on the input parameter g_t – exploration and exploitation. A detailed description of IDE algorithm is provided in [13].

4 DE with Successful-Parent Selection Framework

Although DE belongs to very efficient global optimization methods, even the DE search process stopped by a stagnation in many cases of the problems. The stagnation is situation when individuals of the population are not able to generate better solutions [5]. If the stagnation of the population is detected, there are several ways to help generate better offspring individuals.

In 2015 Guo et al. came up with new approach of smart selection of the individuals for mutation in DE called Successful-Parent Selection (SPS) framework [4]. This framework purports to avoid the stagnation of the population. A very simple idea of SPS is based on two steps – *detecting the stagnation* and *use good archived individual(-s)* to generate better offspring individuals.

Stagnation of population is detected for each individual $\boldsymbol{x}_i$ independently, based on the counts of ongoing failures in previous generations st_i. When an individual is not able to generate a better new trial point $\boldsymbol{y}$ in several consecutive generations (ST), we suppose that it is stagnated. In this case, individuals from the archive of successful parents (SPA) are used in order to generate a new trial point. A pseudo-code of standard DE algorithm extended with SPS framework is pictured in Algorithm 2.

At the beginning, the limit for consecutive failures called stagnation tolerance (ST) is initialized, the counts of individual's failures (st_i) are set to zero and the initialized population is copied to archive of successful parents (SPA). Then from generation to generation, the counter of ith individual failures, st_i, is compared with the stagnation tolerance ST in order to detect individual stagnation. The counter st_i is increased when the ith individual generates an unsuccessful trial point y, $f(y) > f(x_i)$. In case stagnation of x_i is detected, $st_i > ST$, new trial point for the ith individual is created by parent individuals selected from SPA. An archive of successful parents is updated in order to hold 'fresh' old successful

---

Algorithm 2. Successful-Parent Selection Framework

---

initialize stagnation tolerance ST
initialize counters of failures $st_1 = st_2 = \ldots = st_N = 0$
initialize population $P = \{x_1, x_2, \ldots, x_N\}$
evaluate $f(x_i)$, $i = 1, 2, \ldots, N$
$SPA = P$
while stopping condition not reached **do**
  for $i = 1, 2, \ldots, N$ **do**
    if $st_i \leq ST$ **then**
      create a new trial vector y using P
    else
      create a new trial vector y using SPA
    end if
    evaluate $f(y)$
    if $f(y) \leq f(x_i)$ **then**
      insert y into Q
      insert x_i into SPA
      $st_i = 0$
    else
      insert x_i into Q
      $st_i = st_i + 1$
    end if
  end for
  $P \leftarrow Q$
end while

---

individuals. An individual x_i is inserted to SPA when it creates better trial point y, $f(y) \leq f(x_i)$. In such a case, the counter of the ith individual failures is reset to zero, $st_i = 0$.

There are two logical ways how to refresh the individuals in SPA. In the first case, the oldest individual is replaced by newcomer x_i. Thus, kept SPA individuals enable to avoid the stagnation using very recent old good parent solutions. This approach was proposed in paper [4] and applied to state-of-the-art adaptive JADE, SaDE and SHADE variants solving the problems of CEC 2014 test suite with lower levels of dimension. Here we extended the experimental comparison to some other algorithms and the comparison of two SPS methods.

The second possible considered refreshing of SPA is realized as follows. When a new trial point y outperforms parent solution, $f(y) \leq f(x_i)$, this parent individual is placed into SPA in the ith position. Thus, only the ith parent individual in SPA is refreshed. It is clear that some individuals in such an updated SPA could be stored for a long time but older successful parent solutions could generate new individuals which could avoid the stagnation of P. It is clear that a fundamental role in the SPS approach is played by the value of the stagnation limit, ST. Small values bring that new trial points y are created using parents from SPA predominantly and vice versa.

Both of the aforementioned SPS approaches are applied to seven proposed state-of-the-art DE variants from Sect. 3 and experimentally compared. For better reading of the results, the original SPS approach is abbreviated SPS1 and the newly proposed one is labeled by SPS2.

5   Experimental Setting

A test suite of 30 functions was proposed for the special session on Real-Parameter Numerical Optimization, a part of Congress on Evolutionary Computation (CEC) 2014. This session is intended as a competition of optimization algorithms. The functions are described in report [6], including the experimental settings required for the competition. We can expect that this test suite will become one of the most relevant benchmarks required for publishing new single-objective optimization algorithms.

The search range (domain) for all the test functions is $[-100, 100]^D$. The true function value at the global minimum point of each test problem $f(x^*)$ is known. Thus, the function error can be computed as $f(x_{\min}) - f(x^*)$, where $f(x_{\min})$ is the function value of the solution by the algorithm.

The tests were carried out at four levels of dimension, $D = 10, 30, 50, 100$, with 51 times repeated runs per each test function. The run stops if the prescribed value of MaxFES $= D \times 10^4$ is reached or if the minimum function error in the population is less than 1×10^{-8} because such a value of the error is considered sufficient for an acceptable approximation of the correct solution.

The population size was set up to $N = 100$ for all the problems and the levels of dimension. The authors of the SPS framework studied the values of the stagnation limit from $ST = 1$ to $ST = 512$, good results was reached for $ST = 16, 32, 64$. We select a bigger value for our experiments, i.e. $ST = 64$. The remaining control parameters of the algorithms were set up to the recommended values described in Sect. 3. All the algorithms are implemented in Matlab 2010a and all computations were carried out on a standard PC with Windows 7, Intel(R) Core(TM)i7-4790 CPU 3.6 GHz 3.6 GHz, 16 GB RAM.

6   Results

A statistical comparison of SPS1 and SPS2 frameworks efficiency is assessed by Wilcoxon two-sample test at significance level 0.05. The numbers of the test functions where SPS1 or SPS2 framework significantly outperforms the original adaptive DE algorithm are in Tables 1 and 3 for each dimension level. Conversely, the numbers of the test problems in which the original adaptive DE variant performs significantly better than SPS1 or SPS2 are in Tables 2 and 4. In the remaining problems, there is no significant difference between the original DE algorithms and algorithms with proposed frameworks.

Table 1. Counts of the significant wins of the SPS1 framework.

Algorithm	D10	D30	D50	D100	$\sum$	%
CoDE	14	20	18	15	67	56
EPSDE	22	17	17	15	71	59
IDE	5	6	9	11	31	26
JADE	15	9	10	7	41	34
jDE	18	11	6	5	40	33
SaDE	3	0	0	0	3	3
SHADE	19	11	10	8	48	40

Table 2. Counts of the significant losses of the SPS1 framework.

Algorithm	D10	D30	D50	D100	$\sum$	%
CoDE	0	0	1	0	1	1
EPSDE	0	0	1	0	1	1
IDE	8	9	9	6	32	27
JADE	0	2	1	3	6	5
jDE	0	1	2	6	9	8
SaDE	1	0	3	0	4	3
SHADE	0	1	1	4	6	5

Table 3. Counts of the significant wins of the SPS2 framework.

Algorithm	D10	D30	D50	D100	$\sum$	%
CoDE	1	3	1	1	6	5
EPSDE	1	0	2	1	4	3
IDE	1	0	1	1	3	3
JADE	0	1	1	1	3	3
jDE	1	0	0	2	3	3
SaDE	1	2	1	0	4	3
SHADE	1	3	0	0	4	3

In Tables 1, 2, 3 and 4 of results, the total count of the significant wins or losses out of 120 test problems is computed for all dimensions, denoted by symbol '$\sum$'. In the last column, the percentage of overall significant wins or losses for all algorithms is computed. Based on Table 1 of significant wins of SPS1 framework, we can see that the biggest increase of the performance was reached for EPSDE and CODE variants (over 50%). In the case of SaDE variant, SPS1 benefited only in less than 3% of test problems. Analysing the number of the significant losses of SPS1, we can see that the least decrease of efficiency is

Table 4. Counts of the significant losses of the SPS2 framework.

Algorithm	D10	D30	D50	D100	$\sum$	%
CoDE	1	0	0	1	2	2
EPSDE	1	0	2	0	3	3
IDE	16	23	22	22	83	69
JADE	5	1	1	0	7	6
jDE	5	10	15	16	46	38
SaDE	2	1	1	1	5	4
SHADE	2	0	2	1	5	4

Table 5. Mean ranks from Friedman-rank test results for all the algorithms in comparison.

Algorithm	$D = 10$	$D = 30$	$D = 50$	$D = 100$	Mean
EPSDESPS1	8.5	**5.4**	**6.2**	8	7
JADESPS1	**5.3**	7.8	**7.7**	**7.7**	7.1
IDE	**6.4**	**7.2**	9.2	9.2	8
IDESPS1	8.4	7.7	8.1	8.5	8.2
JADESPS2	7.8	9.6	9.7	**7.7**	8.7
JADE	7.9	9.9	9	8.2	8.7
jDESPS1	6.6	9.1	9.6	10.3	8.9
CoDESPS1	15.7	9.4	8	9.8	10.7
IDESPS2	9.5	10.3	11.9	11.3	10.8
jDE	11.7	11.3	10.6	10.7	11.1
SHADESPS1	7.3	12	12.3	13	11.2
SaDESPS1	11.7	12.4	12.4	10.4	11.7
SaDE	12.1	12.7	12.8	10.9	12.1
SaDESPS2	12.1	12.5	12.5	11.7	12.2
jDESPS2	13.8	12.6	12	12.8	12.8
SHADE	10.9	13.7	13.4	13.3	12.8
SHADESPS2	10.5	13.2	14.1	14.1	13
EPSDE	15.4	13	12.3	12.7	13.4
EPSDESPS2	14.9	13	12.6	12.9	13.4
CoDESPS2	17.2	14.1	13.1	14	14.6
CoDE	17.2	14.2	13.6	13.9	14.7

reached for the CODE and EPSDE variants (less than 1% of the test problems). The worst results are obtained for IDE, i.e. performance decreases in 27% of the test problems. Regarding the dimension level, it is obvious that efficiency of SPS1 more frequently decreases with increased D (Tables 1 and 2). On the other hand, using SPS2 framework increases the performance of adaptive DE variants. The number of wins of SPS2 compared with the original DE variants is only about in 3% of the test problems (Table 3). Conversely, SPS2 mostly decreases the performance for IDE and jDE variants, i.e. influences worse results in 69% and 38% of the test problems.

The overall performance of all 21 algorithms was compared using Friedman test for medians of function-error values. The null hypothesis on the equal performance of the algorithms was rejected, the achieved p value for rejection was $p < 5 \times 10^7$. Mean ranks of the algorithms are presented in Table 5. Note that the algorithm winning uniquely in all the problems would have the mean rank 1 and another algorithm being a unique looser in all the problems would have the mean rank 21. In the last column of Table 5, the average mean rank is computed for all dimensions. The table is sorted based on an average mean rank in an ascending way, i.e. algorithms which perform overall better are in higher rows and vice versa. For the best performing algorithm in each dimension, the mean rank is printed bold and underlined, the mean rank for the second is printed bold and the third best variant is underlined. It is obvious that the least average mean rank is for the EPSDE variant with SPS1 framework and the second position is for JADE variant with SPS1 framework. Only in the case of IDE variant, even SPS1 and SPS2 framework is not able to increase the algorithm performance. For four DE variants in comparison, SPS1 framework is able to increase the overall performance and for CODE and JADE even SPS2 increases their performance, based on the average mean rank (last column in Table 5).

7 Conclusion

The experimental comparison has shown that SPS could significantly increase the performance of the state-of-the-art DE variants. A better performance is reached especially with the original SPS framework [4], when the oldest individuals in SPA are updated. The newly proposed SPS framework was successful only in a minor part of the test problems. A big benefit of SPS was observed especially for EPSDE and CODE variants in more than 50% of the test problems. These results are better compared with the results of DE algorithms in the original paper [4]. There is no benefit of SPS applied in SaDE variant. The results of Friedman test show that in two DE out of 7 both SPS1 and SPS2 approaches perform better and in four DE the SPS1 framework performs better than the original algorithms.

References

1. Brest, J., Greiner, S., Bošković, B., Mernik, M., Žumer, V.: Self-adapting control parameters in differential evolution: a comparative study on numerical benchmark problems. IEEE Trans. Evol. Comput. **10**, 646–657 (2006)
2. Das, S., Mullick, S., Suganthan, P.: Recent advances in differential evolution-an updated survey. Swarm Evol. Comput. **27**, 1–30 (2016)

3. Das, S., Suganthan, P.N.: Differential evolution: a survey of the state-of-the-art. IEEE Trans. Evol. Comput. **15**, 27–54 (2011)
4. Guo, S.M., Yang, C.C., Hsu, P.H., Tsai, J.S.H.: Improving differential evolution with successful-parent-selecting framework. IEEE Trans. Evol. Comput. **19**(5), 717–730 (2015)
5. Lampinen, J., Zelinka, I.: On stagnation of the differential evolution algorithm. In: Matousek, R. (ed.) 6th International Conference on Soft Computing Mendel 2000, pp. 76–83 (2000)
6. Liang, J.J., Qu, B.Y., Suganthan, P.N.: Problem definitions and evaluation criteria for the CEC 2014 special session and competition on single objective real-parameter numerical optimization. Tech. rep., Nanyang Technological University, Singapore (2013). http://www.ntu.edu.sg/home/epnsugan/
7. Mallipeddi, R., Suganthan, P.N., Pan, Q.K., Tasgetiren, M.F.: Differential evolution algorithm with ensemble of parameters and mutation strategies. Appl. Soft Comput. **11**, 1679–1696 (2011)
8. Neri, F., Tirronen, V.: Recent advances in differential evolution: a survey and experimental analysis. Artif. Intell. Rev. **33**, 61–106 (2010)
9. Price, K.V., Storn, R., Lampinen, J.: Differential Evolution: A Practical Approach to Global Optimization. Springer, Heidelberg (2005)
10. Qin, A.K., Huang, V.L., Suganthan, P.N.: Differential evolution algorithm with strategy adaptation for global numerical optimization. IEEE Trans. Evol. Comput. **13**(2), 398–417 (2009)
11. Storn, R., Price, K.V.: Differential evolution - a simple and efficient heuristic for global optimization over continuous spaces. J. Global Optim. **11**, 341–359 (1997)
12. Tanabe, R., Fukunaga, A.: Success-history based parameter adaptation for differential evolution. IEEE Congr. Evol. Comput. (CEC) 2013, pp. 71–78, June 2013
13. Tang, L., Dong, Y., Liu, J.: Differential evolution with an individual-dependent mechanism. IEEE Trans. Evol. Comput. **19**(4), 560–574 (2015)
14. Tušar, T., Filipič, B.: Differential evolution versus genetic algorithms in multiobjective optimization. In: Obayashi, S., Deb, K., Poloni, C., Hiroyasu, T., Murata, T. (eds.) EMO 2007. LNCS, vol. 4403, pp. 257–271. Springer, Heidelberg (2007). doi:10.1007/978-3-540-70928-2_22
15. Vesterstrom, J., Thomsen, R.: A comparative study of differential evolution, particle swarm optimization, and evolutionary algorithms on numerical benchmark problems. In: Congress on Evolutionary Computation, CEC 2004, vol. 2, pp. 1980–1987, June 2004
16. Wang, Y., Cai, Z., Zhang, Q.: Differential evolution with composite trial vector generation strategies and control parameters. IEEE Trans. Evol. Comput. **15**, 55–66 (2011)
17. Zhang, J., Sanderson, A.C.: JADE: adaptive differential evolution with optional external archive. IEEE Trans. Evol. Comput. **13**, 945–958 (2009)

State Flipping Based Hyper-Heuristic for Hybridization of Nature Inspired Algorithms

Robertas Damaševičius[1] and Marcin Woźniak[2(✉)]

[1] Department of Software Engineering, Kaunas University of Technology,
Studentu 50, Kaunas, Lithuania
robertas.damasevicius@ktu.lt
[2] Institute of Mathematics, Silesian University of Technology,
Kaszubska 23, 44-100 Gliwice, Poland
marcin.wozniak@polsl.pl

Abstract. The paper presents a novel hyper-heuristic strategy for hybridization of nature inspired algorithms. The strategy is based on switching the state of agents using a logistic probability function, which depends upon the fitness rank of an agent. A case study using two nature inspired algorithms (Artificial Bee Colony (ABC) and Krill Herding (KH)) and eight optimization problems (Ackley Function, Bukin Function N.6, Griewank Function, Holder Table Function, Levy Function, Schaffer Function N.2, Schwefel Function, Shubert Function) is presented. The results show a superiority of the proposed hyper-heuristic (mean end-rank for hybrid algorithm is 1.435 vs. 2.157 for KH and 2.408 for ABC).

Keywords: Nature inspired algorithms · Metaheuristic · Optimization

1 Introduction

Nature inspired algorithms (NIA) are based on the collective behavior of living algorithms such as swarms of insects or flocks of birds. More generally, NIA describe the behavior of communicating agents forming a multi-agent system. While each individual agent has limited processing power, the system (swarm, network) as a whole can perform complex tasks such as resource search. Many modern meta-heuristic algorithms inspired by natural or social phenomena have been proposed such as Artificial Bee Colony (ABC) [1], Firefly Algorithm (FA) [2], Grey Wolf Optimizer (GWO) [3], Krill Herd (KH) [4], and Particle Swarm Optimization (PSO) [5]. Such algorithms are widely applied in different domains, such as scientific computing [6], transportation engineering [7], engineering optimization [4], internet server allocation [8], image processing [9,10], image recognition [11], bioinformatics [13], intelligent systems [14], control systems [15], voice recognition [12,16], and transparent optical networks [17].

Comparison of nature-inspired algorithms for solving different optimization problems has been motivated by the No Free Lunch theorems [18], which prove

© Springer International Publishing AG 2017
L. Rutkowski et al. (Eds.): ICAISC 2017, Part I, LNAI 10245, pp. 337–346, 2017.
DOI: 10.1007/978-3-319-59063-9_30

that for any search or optimization algorithm, any elevated performance over one class of problems is exactly paid for in performance over another class. For example, Çivicioglu and Besdok [19] compare the Cuckoo-search, PSO, Differential Evolution and ABC algorithms. Alqattan and Abdullah [13] compare ABC and PSO. Researchers also proposed their hybrid algorithms by combining two or more metaheurestic algorithms such as MultiStart Hyper-heuristic algorithm (MSH-QAP) [20], which uses of Simulated Annealing (SA), Robust Tabu Search (RTS), Ant Colony Optimization (FAnt), and Breakout Local Search (BLS).

Hybridization is an evolutionary meta-heuristic approach [21]. Zhao et al. [22] introduced the combined algorithm of dynamic multi-swarm particle swarm optimizer (DMS-PSO) and Harmony Search (HS). Yang and Deb [23] proposed a two-stage hybrid search method, Eagle Strategy, which iteratively combined random search using Lèvy walk with the firefly algorithm. A simple way of hybridization is to form a new algorithm by combining other two algorithms. A more general idea is to automatically devise new algorithms by combining the strength and compensating for the weaknesses of heuristic algorithms. The process is called hyper-heuristic, a method that seeks to automate the process of selecting, combining, or adapting several simpler heuristics (or components of such heuristics) to efficiently solve computational search problems [24].

Hyper-heuristic works at a higher level when compared with the typical application of meta-heuristics to optimisation problems i.e. a hyper-heuristic is a (meta-)heuristic which operates on lower level (meta-)heuristics [25] aiming to choose intelligently the right heuristic or algorithm for a given problem. Hyper-heuristics often involves the use of a 'choice function', which trades off the exploitation and exploration activities in choosing the next heuristic to use [26]. The choice function can be random [27], lower-level heuristic performance based [25], evolutionary [28], or greedy [27]. Greedy choice function rewards the low level heuristics with currently best performance. For example, the choice function can be defined as a 3rd-order tensor of the trail of a hyper-heuristic mixing heuristics [29]. Factorization of such a tensor reveals the latent relationships between the low level heuristics and the hyper-heuristic itself.

Related approaches to hyper-heuristics are evolving evolutionary algorithms (EEA) [30], which let an EA discover the rules and knowledge, so that it can find the best EA to optimise the solutions of a problem; Self-modification Cartesian Genetic Programming (SMCGP) [31], which encodes the chosen low-level heuristic, alongside self-modifying operators in a graph structure; Auto-constructive Evolution [32] as reproductive mechanisms, which are evolved and then used to derive problem solutions; and exponentially increasing hyper-heuristic (EIHH) [33], which uses a meta-hyper-heuristic algorithm to search heuristic space for greater performance benefits. Combination of different heuristics also can be defined as the leader-following consensus problem, where leader is a heuristic with best fitness, which is followed by other heuristics. The leader-following configuration can be considered as an energy saving mechanism found in many biological systems, which can enhance the communication and orientation of the flock [34].

We present a hyper-heuristic to combine the behaviour of agents described by two different heuristic algorithms. A case study in combining the ABC [1] and KH [4] is presented. The performance of hyper-heuristic is evaluated with respect to stand-alone ABC and KH using 8 benchmark optimization problems.

2 Hybridization Method

2.1 Description and Pseudocode

The proposed hybridization method assigns a state to each swarming agent and defines an algorithm how this state changes during the execution of the algorithm. We assume that the state of an agent defines its behaviour. If a population of agents is under-performing, their states can be flipped hoping that the change in behaviour would result in higher performance. Thus, the state of agents is not constant and can vary from iteration to iteration. The pseudo-code of the hybridization hyper-heuristic is presented below.

```
(1) BEGIN
(2)   Initialize positions of all agents
(3)   WHILE iterations remain DO
(4)    IF first Iteration THEN
(5)     Initialize the state of agents to one of (Agent1 or Agent2) randomly
(6)    ELSE
(7)     Flip the state of agents using logistic probability
(8)    END IF
(9)    FOREACH agent IN state Agent1
(10)    Calculate new position using one iteration of Algorithm1
(11)   END FOREACH
(12)   FOREACH agent IN state Agent2
(13)    Calculate new position using one iteration of Algorithm2
(14)   END FOREACH
(15)    Merge populations
(16)   END WHILE
(17)   Return the fitness of best performing agent
(18) END
```

Here we assume that there are only two possible states ($Agent_1$ or $Agent_2$), while their corresponding behaviour is defined by $Algorithm_1$ and $Algorithm_2$.

The probability of flipping the state in Line 7 of pseudo-code is calculated using a logistic function with equation:

$$P(r) = \frac{1}{1 + e^{-(r-n/2)}}. \tag{1}$$

here r is the fitness rank of an agent (agents are sorted from best performers to worst performers by the fitness value), and n is the number of agents.

Hereinafter for our further experiments, $Algorithm_1$ is Krill Herding (KH) and $Algorithm_2$ is Artificial Bee Colony (ABC), which are described below.

2.2   Krill Herding (KH)

Krill Herd (KH) [4] algorithm simulates the behavior of Antarctic krill (*Euphausia superba*) individuals found in the Southern Ocean. Krills are able to form a large herd that can reach hundreds of meters in length. When predators (penguins or sea birds) attack krill herds, they take individual krill which leads to reduced krill density. After the attack by predators, formation of krill is a multi-objective process aimed at increasing krill density and reaching food.

The position of krills is defined by movement induced by the presence of other krills, foraging activity, and random diffusion. All individual krill move towards the best possible solution while searching for highest density and food. The smallest distances of each krill from food and from highest density of the herd are considered as the objective function for the krill movement, which finally lead the krills to herd around global minima.

The KH algorithm has been shown of being capable of to solve efficiently a wide range of numerical optimization problems [4].

2.3   Artificial Bee Colony (ABC)

The Artificial Bee Colony (ABC) algorithm [1] is based on a model of foraging behaviour of a honeybee colony described analytically by reactiondiffusion equations developed by Tereshko [35]. This model consists of three essential components: food sources, employed foragers, and unemployed foragers, and defines two leading modes of the honeybee colony behaviour: recruitment to a food source and abandonment of a source. In order to select a food source, a forager bee evaluates several properties related with the food source such as its closeness to the hive, richness of the energy, taste of its nectar, and the ease or difficulty of extracting this energy. An employed forager is employed at a specific food source which she is currently exploiting. She carries information about this specific source and shares it with other bees waiting in the hive. The information includes the distance, the direction and the profitability of the food source. Unemployed foragers search the environment randomly or try to find a food source by means of the information given by the employed bee.

The ABC algorithm implements a population-based search in which artificial bees change their positions in time aiming to discover the places of food sources with high nectar amount and finally the one with the highest nectar. Two types of artificial bees (employed and onlooker bees) choose food sources depending on the experience of themselves and their nest mates, and adjust their positions. Scout bees fly and choose the food sources randomly without using experience. If the nectar amount of a new source is higher than that of the previous one in their memory, they memorize the new position and forget the previous one.

Thus, ABC combines local search, which is carried out by employed bees and on-looker bees, with global search, managed by onlookers and scout bees, therefore attempting to balance the exploration and exploitation policies.

2.4   Summary

The proposed hyper-heuristic hybridization method adapted to hybridization of ABC and KH algorithms is summarized graphically in Fig. 1. The simple mechanism of probabilistic state flipping leads to an emergence of interesting collective behaviour, when under-performing agents try to imitate the behaviour of well-performing agents by switching to their type of behaviour. On the other hand, if a single state attracts a majority of agents, less-performing agents start switching to an alternative state. Such an opportunistic behaviour provides a balance between exploration and exploitation. Exploration helps to come close to the global minimum or good local minimum, while exploitation helps to locate the global minimum more accurately. Thus the hyper-heuristic provides a balancing mechanism between global and local search properties (underlying algorithm can be biased towards local or global search) allowing to diversifying the search in order to avoid getting trapped in a local optimum.

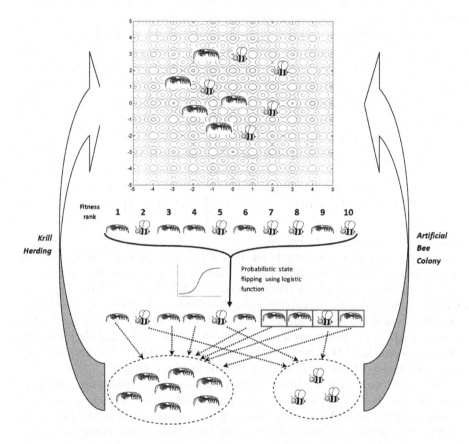

Fig. 1. Conceptual diagram of the proposed hybridization method

3 Benchmarks

To validate the proposed method, we have selected 8 benchmark functions: Ackley Function (Eq. 1), Bukin Function N.6 (Eq. 2), Griewank Function (Eq. 3), Holder Table Function (Eq. 4), Levy Function (Eq. 5), Schaffer Function N.2 (Eq. 6), Schwefel Function (Eq. 7), Shubert Function (Eq. 8):

$$f_1(x) = -20e^{-0.2\sqrt{\frac{1}{d}\sum_{i=1}^{d} x_i^2}} - e^{\frac{1}{d}\sum_{i=1}^{d} cos(2\pi x_i)} + 20 + e \tag{2}$$

$$f_2(x) = 100\sqrt{|x_2 - 0.01x_1^2|} + 0.01|x_1 + 10| \tag{3}$$

$$f_3(x) = \frac{1}{4000}\sum_{i=1}^{n} x_i^2 - \prod_{i=1}^{n} cos(\frac{x_i}{\sqrt{i}}) + 1, \tag{4}$$

$$f_4(x) = -|sin(x_1)cos(x_2)e^{|1 - \frac{\sqrt{x_1^2 + x_2^2}}{\pi}|}|, \tag{5}$$

$$f_5(x) = sin^2(\pi\omega_1) + \sum_{i=1}^{d-1}(\omega_i - 1)^2[1 + 10sin^2(\pi\omega_i + 1)]$$
$$+ (\omega_i - 1)^2[1 + 10sin^2(2\pi\omega_d)], \omega_i = 1 + \frac{x_i - 1}{4} \tag{6}$$

$$f_6(x) = 0.5 + \frac{sin^2(x_1^2 + x_2^2) - 0.5}{[1 + 0.001(x_1^2 + x_2^2)]^2}, \tag{7}$$

$$f_7(x) = 418.9829d - \sum_{i=1}^{d} x_i sin(\sqrt{|x_i|}), \tag{8}$$

$$f_8(x) = (\sum_{j=1}^{5} jcos((j+1)x_1 + j))(\sum_{j=1}^{5} jcos((j+1)x_2 + j)), \tag{9}$$

here d is the number of dimensions.

These functions are commonly used for benchmarking new evolutionary and nature inspired algorithms [36]. All these functions have many local minima.

The algorithms have been implemented in MATLAB in a 2.5GHz Dual Core CPU with 4GB RAM computer, running Windows 8 OS. We have implemented both the hybrid algorithm, which combines KH and ABC as well as stand-alone KH and ABC, for comparison. We used 50 agents, and each algorithm was allowed to run for 200 iterations. We have repeated the process for 100 times each and have performed the statistical analysis of the results.

We have compared the mean ranks of analysed algorithms in each iteration (see Fig. 2). We can see that initially, the ABC is the better method until approximately the 50th iteration. Then the proposed hybrid method overtakes as the best method by rank. On the over hand, KH never scores as the best method, however, it manages to overtake ABC when algorithms are allowed to run for a larger number of iterations. We have performed the analysis of the algorithm performance using the Friedman rank test. The hypothesis that there is no difference between the ranks of the ABC, KH and Hybrid-ABC-KH has

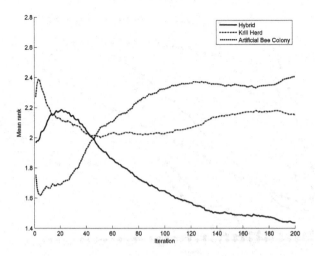

Fig. 2. Ranks vs. Iterations for 8 optimization problems

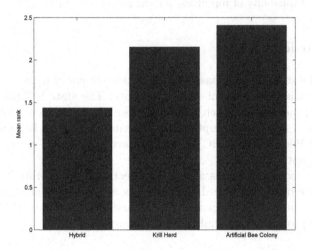

Fig. 3. Mean ranks for 8 optimization problems at final iteration

been rejected. In Fig. 3, we can see the mean ranks of all compared methods at
the final (200th) iteration with statistical confidence limits. We can see that the
proposed method is the best (mean rank 1.435 ± 0.018), KH achieved the second
place (mean rank 2.157 ± 0.018), and ABC is the third (mean rank 2.408 ± 0.022).
Finally, we have calculated probability that the proposed method is the best (i.e.,
achieves top fitness) for each of the analysed optimization problems (Fig. 4). We
can see that while the proposed method does not allow to achieve top fitness in
the beginning, but after approximately 20–30 iteration its performance starts to
increase. In the end, the proposed method achieves top fitness for five benchmark
problems (Bukin, Schwefel, Shaffer No.2, Holder Table, Griewank), but fails to
achieve top fitness for another three problems (Ackley, Levy, Shubert).

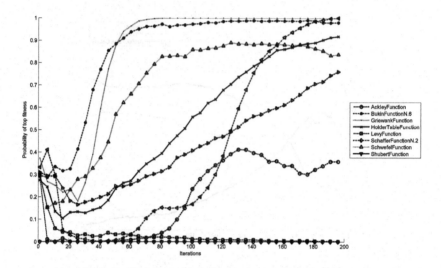

Fig. 4. Probability of top fitness for the proposed hybridization method

4   Conclusions

We proposed a state-flipping based hybridization scheme of nature-inspired algorithms that assigns each agent a variable state. The state is flipped randomly based on the fitness rank of each agent after its current iteration. The application of the proposed method attempts to take merits of the Krill Herding (KH) and the Artificial Bee Colony (ABC) in order to avoid all agents getting trapped in inferior local optimal regions.

The proposed algorithm is able more effectively to generate better quality solutions more often (achieves higher rank as well as has higher probability of achieving top rank when compared to the stand-alone KH and ABC) on several optimization problems with multiple local minima.

References

1. Karaboga, D., Basturk, B.: A powerful and efficient algorithm for numerical function optimization: artificial bee colony (ABC) algorithm. J. Global Optim. **2007**, 459–471 (2007)
2. Yang, X.S., He, X.: Firefly algorithm: recent advances and applications. Int. J. Swarm Intell. **1**(1), 36–50 (2013)
3. Mirjalili, S., Mirjalili, S.M., Lewis, A.: Grey wolf optimizer. Adv. Eng. Softw. **69**, 46–61 (2014)
4. Gandomi, A.H., Alavi, A.H.: Krill herd: a new bio-inspired optimization algorithm. Commun. Nonlinear Sci. Numer. Simul. **17**(12), 4831–4845 (2012)
5. Kennedy, J., Eberhart, R.: Particle swarm optimization. In: Proceedings of the IEEE International Conference on Neural Networks, pp. 1942–1948 (1995)

6. Parpinelli, R.S., Lopes, H.S.: An eco-inspired evolutionary algorithm applied to numerical optimization. In: 3rd World Congress on Nature and Biologically Inspired Computing (NaBIC 2011), pp. 466–471 (2011)

7. Lucic, P., Teodorovic, D.: Bee system: modeling combinatorial optimization transportation engineering problems by swarm intelligence. In: Preprints of the TRISTAN IV Triennial Symposium on Transportation Analysis, pp. 441–445 (2001)

8. Nakrani, S., Tovey, C.: On honey bees and dynamic server allocation in Internet hosting centers. Adap. Behav. 12(2), 223–240 (2004)

9. Grycuk, R., Gabryel, M., Nowicki, R., Scherer, R.: Content-based image retrieval optimization by differential evolution. Proceedings of IEEE Congress on Evolutionary Computation, pp. 86–93 (2016)

10. Korytkowski, M., Rutkowski, L., Scherer, R.: Fast image classification by boosting fuzzy classifiers. Inf. Sci. 327, 175–182 (2016)

11. Polap, D., Wozniak, M., Napoli, C., Tramontana, E., Damasevicius, R.: Is the colony of ants able to recognize graphic objects? In: Proceedings of the 21st International Conference on Information and Software Technologies, ICIST 2015, pp. 376–387 (2015)

12. Polap, D.: Neuro-heuristic voice recognition. In: Proceedings of the 2016 Federated Conference on Computer Science and Information Systems, FedCSIS 2016, Gdańsk, Poland, 11–14 September, 2016, pp. 487–490 (2016)

13. Alqattan, Z.N.M., Abdullah, R.: A comparison between artificial bee colony and particle swarm optimization algorithms for protein structure prediction problem. In: Lee, M., Hirose, A., Hou, Z.-G., Kil, R.M. (eds.) ICONIP 2013. LNCS, vol. 8227, pp. 331–340. Springer, Heidelberg (2013). doi:10.1007/978-3-642-42042-9_42

14. Kapuściński, T., Nowicki, R.K., Napoli, C.: Application of genetic algorithms in the construction of invertible substitution boxes. In: Rutkowski, L., Korytkowski, M., Scherer, R., Tadeusiewicz, R., Zadeh, L.A., Zurada, J.M. (eds.) ICAISC 2016. LNCS, vol. 9692, pp. 380–391. Springer, Cham (2016). doi:10.1007/978-3-319-39378-0_33

15. Cpalka, K., Lapa, K., Przybyl, A.: A new approach to design of control systems using genetic programming. Inf. Technol. Control 44(4), 433–442 (2015)

16. Brester, C., Semenkin, E., Sidorov, M.: Multi-objective heuristic feature selection for speech-based multilingual emotion recognition. J. Artif. Intell. Soft Comput. Res. 6(4), 243–253 (2016)

17. Brasileiro, I., Santos, I., Soares, A., Rabelo, R., Mazullo, F.: Ant colony optimization applied to the problem of choosing the best combination among m combinations of shortest paths in transparent optical networks. J. Artif. Intell. Soft Comput. Res. 6(4), 231–242 (2016)

18. Wolpert, D.H., Macready, W.G.: No free lunch theorems for optimization. IEEE Trans. Evol. Comput. 1(1), 67–82 (1997)

19. Çivicioglu, P., Besdok, E.: A conceptual comparison of the Cuckoo-search, particle swarm optimization, differential evolution and artificial bee colony algorithms. Artif. Intell. Rev. 39(4), 315–346 (2013)

20. Dokeroglu, T., Cosar, A.: A novel multistart hyper-heuristic algorithm on the grid for the quadratic assignment problem. Eng. Appl. Artif. Intell. 52(C), 10–25 (2016)

21. Yang, X.-S.: Nature-Inspired Metaheuristic Algorithms. Luniver Press (2008)

22. Zhao, S.-Z., Suganthan, P.N., Pan, Q.-K., Tasgetiren, M.F.: Dynamic multi-swarm particle swarm optimizer with harmony search. Exp. Syst. Appl. Int. J. 38(4), 3735–3742 (2011)

23. Yang, X.S., Deb, S.: Eagle strategy using Lévy walk and firefly algorithms for stochastic optimization. In: González, J.R., Pelta, D.A., Cruz, C., Terrazas, G., Krasnogor, N. (eds.) Nature Inspired Cooperative Strategies for Optimization (NICSO 2010).SCI, vol. 284, pp. 101–111. Springer, Heidelberg (2010). doi:10.1007/978-3-642-12538-6_9

24. Özcan, E., Bilgin, B., Korkmaz, E.E.: A comprehensive analysis of hyper-heuristics. Intell. Data Anal. **12**(1), 3–23 (2008)

25. Burke, E.K., Hart, E., Kendall, G., Newall, J., Ross, P., Schulenburg, S.: Hyper-heuristics: an emerging direction in modern search technology. In: Glover, F., Kochenberger, G. (eds.) Handbook of Metaheuristics, pp. 457–474. Springer, US (2003)

26. Chakhlevitch, K., Cowling, P.: Hyperheuristics: recent developments. In: Cotta, C., Sevaux, M., Sörensen, K. (eds.) Adaptive and Multilevel Metaheuristics. SCI, vol. 136, pp. 3–29. Springer, Heidelberg (2008). doi:10.1007/978-3-540-79438-7_1

27. Özcan, E., Kheiri, A.: A hyper-heuristic based on random gradient, greedy and dominance. In: Gelenbe, E., Lent, R., Sakellari, G. (eds.) Computer and Information Sciences II, pp. 557–563. Springer, London (2011). doi:10.1007/978-1-4471-2155-8_71

28. Cowling, P., Kendal, G., Han, L.: An investigation of a hyperheuristic genetic algorithm applied to a trainer scheduling problem. In: Proceedings of Congress on Evolutionary Computation (CEC 2002), pp. 1185–1190 (2002)

29. Asta, S., Özcan, E.: A tensor-based selection hyper-heuristic for cross-domain heuristic search. Inf. Sci. Int. J. **299**, 412–432 (2015)

30. Diosan, L., Oltean, M.: Evolutionary design of evolutionary algorithms. Genet. Program Evolvable Mach. **10**(3), 263–306 (2009)

31. Harding, S.L., Miller, J.F., Banzhaf, W.: Developments in cartesian genetic programming: self-modifying CGP. In: Cartesian Genetic Programming, 101–124 (2011)

32. Harrington, K.I., Spector, L., Pollack, J.B., O'Reilly, U.M.: Autoconstructive evolution for structural problems. In: Proceedings of 14th International Conference on Genetic and Evolutionary Computation Conference Companion, pp. 75–82 (2012)

33. Grobler, J., Engelbrecht, A.P., Kendall, G., Yadavalli, V.S.: Heuristic space diversity control for improved meta-hyper-heuristic performance. Inf. Sci. **300**(C), 49–62 (2015)

34. Hammel, D.: Formation flight as an energy saving mechanism. Israel J. Zool. **41**, 261–278 (1995)

35. Tereshko, V.: Reaction-diffusion model of a honeybee colony's foraging behaviour. In: Schoenauer, M., Deb, K., Rudolph, G., Yao, X., Lutton, E., Merelo, J.J., Schwefel, H.-P. (eds.) PPSN 2000. LNCS, vol. 1917, pp. 807–816. Springer, Heidelberg (2000). doi:10.1007/3-540-45356-3_79

36. Li, X., Tang, K., Omidvar, M.N., Yang, Z., Qin, K., China, H.: Benchmark functions for the CEC 2013 special session and competition on large-scale global optimization. Gene **7**, 8 (2013)

Improved CUDA PSO Based on Global Topology

Joanna Kołodziejczyk[1], Dariusz Sychel[2], and Aneta Bera[2(✉)]

[1] Department of Technology, The Jacob of Paradies University,
F. Chopina 52, 66-400 Gorzów Wielkopolski, Poland
jkolodziejczyk@ajp.edu.pl
[2] Faculty of Computer Science and Information Technology,
West Pomeranian University of Technology, Żołnierska 49, 71-210 Szczecin, Poland
{dsychel,abera}@wi.zut.edu.pl

Abstract. We introduce a well-optimized implementation of PSO algorithm based on, Compute Unified Device Architecture (CUDA), using global neighborhood topology with extremely large swarms (greater than 1000 particles). The algorithm optimization is based on effective data organization in GPU memory such as transfer and thread optimization, pinned memory and the zero-copy mechanism usage. Experimental results show that the implementation on GPU is significantly faster than implementation on CPU.

Keywords: Particle Swarm Optimization (PSO) · Global topology · Graphics processing unit (GPU) · CUDA architecture

1 Introduction

The Particle Swarm Optimization algorithm (PSO) can be recommended as an optimizer for complex and large-scale problems because it finds a satisfactory solution in most tests, it is easy to implement, and the number of adjustable parameters is relatively small. Fast convergence and reliability are PSO attributes. However, the PSO often requires a long runtime that prevents from real-time application in dynamically changing environments.

Algorithms can be speedup using parallelization. PSO is intrinsically parallel because each particle in a population moves according to the same rule. Our main idea was to use the graphics processing unit (GPU) which is ideally suited for computations performed across thousands of elements at the same time (in parallel). Software must use a large number of concurrent threads to make a good performance on GPU. To fulfill this requirements we decided to use large number of particles moving over the search space. Among many different PSO variants we choose the one with global neighborhood topology because it is universal. Additionally, we propose an improved data structure that allows for effective CUDA implementation.

© Springer International Publishing AG 2017
L. Rutkowski et al. (Eds.): ICAISC 2017, Part I, LNAI 10245, pp. 347–358, 2017.
DOI: 10.1007/978-3-319-59063-9_31

The performance of the proposed CUDA *Gbest* PSO algorithm was evaluated on standard optimization benchmarks with two criteria: success rate and speedup.

In the paper, we present literature reviews and a discussion on different neighborhood topologies. We also present a detailed description of CUDA PSO improvement (optimization) process. Finally, we discuss experimental results.

2   Previous Reports on GPU PSO

In 2007 Li et al. proposed to implement PSO algorithm on GPU for the first time [11]. Particles were mapped into textures on a graphics card and calculated in parallel. The algorithm was implemented without a CUDA support. In 2009 several authors published results on PSO parallelization based on the CUDA platform [10,15,18,19]. The general conclusion was that GPU accelerated the PSO algorithm when optimizing test functions.

The results from published GPU PSO implementations are hard to compare. Reported speedups and runtimes depend on hardware and software. Authors tested different parallelization models, PSO variants, data structures, and graphics cards. Experimental environments vary in dimension, population, and maximum step sizes, PSO coefficients, initial values, etc. We compare our results to others only if environments are similar.

The GPU PSO presented in the literature can be divided into two categories: with local and global neighborhood topologies. Many implementations used the ring local topology, including: [2,3,12,13,19,20]. The authors of [10], probably applied wheel topology which we guessed from the information about 20 neighbors used.

The global topology PSO on GPU was not very popular in studies. Two papers [4,16] presented the success rate of multi-swarm global topology PSO. The first optimized the entire search space with many parallel swarms and in the second each swarm optimized part of the decomposed search space. Two other papers described a one population PSO. In [15] applied the canonical PSO with a constriction factor. The authors of [17] analyzed the basic PSO with an adaptive inertial factor but did not describe data structures.

Two papers, [5,18], do not reveal the neighborhood topology and other important details about PSO and its CUDA implementation.

Our research fills the gap and tests the basic (non-canonical), one population PSO with the global topology on GPU using the CUDA platform. In the next section, we discuss pros and cons of different neighborhood topologies to motivate our choice.

3   Neighborhood Topologies and Population Size in PSO

The mechanism of particles' cooperation is the key PSO feature resulting in effectiveness of a search space exploration [6,9]. If a particle discovers a local/global optimum it becomes the *best* in its neighborhood. The *Best* particle attracts others to this region. Each particle i in step t explores the search space moving from its old position $\mathbf{x}_i^t$ into a new one:

$$\mathbf{x}_i^{t+1} = \mathbf{x}_i^t + \mathbf{v}_i^{t+1}, \tag{1}$$

where $\mathbf{v}_i^{t+1}$ is the particle velocity and is calculated as follows:

$$\mathbf{v}_i^{t+1} = \omega \mathbf{v}_i^t + \varphi_1 \mathbf{U}_1^t(\mathbf{pb}_i^t - \mathbf{x}_i^t) + \varphi_2 \mathbf{U}_2^t(\mathbf{b}^t - \mathbf{x}_i^t), \tag{2}$$

where ω is the weight of inertia, φ_1 and φ_2 are positive numbers called personal and global best influence respectively, $\mathbf{pb}_i^t$ is the best location found so far of particle i and $\mathbf{b}^t$ is the best position in the neighborhood. $\mathbf{U}_1$ and $\mathbf{U}_2$ are uniformly distributed random numbers.

The neighborhood is the scope of particles influence. The PSO neighborhood is either *Gbest* or *Lbest*.

Definition 1. *Gbest is a neighborhood topology composed of the entire population.*

Definition 2. *Lbest is a neighborhood topology comprising some number of adjacent neighbors in the population.*

The neighborhood topology influences the algorithm performance. Kennedy and Eberhart [9] analyzed the PSO sociometry on standard functions in 30-dimensional spaces comparing *Gbest* and two *Lbest* (circle and wheel) neighborhood topologies. They concluded that particle cooperation was problem dependent i.e. the optimized function influenced the neighborhood. There were no clear rules describing the relationship. Usually, *Gbest* was either best or nearly so in all tests. The authors stressed that *"the main disadvantage of the Gbest PSO is that it is unable to explore multiple optimal regions simultaneously"*.

Kennedy and Eberhart continued their research on the neighborhood association with the optimization problem [7]. They revealed great variation of optimization abilities within topologies. The "von Neumann" *Lbest* neighborhood performs better than *Gbest* or ring *Lbest* topology on some test functions. Yet, no rule for assigning the neighborhood structure to the goal function was discovered. Kennedy and Mendes tested canonical and fully informed PSO with different neighborhood sizes and concluded that *"it will be necessary to find the topologies that are very best suited to each algorithm"* [8].

Bratton and Kennedy defined good practices for PSO application and indicated that the *Gbest* topology performs better in most multidimensional problems, converging faster to the optimum [1]. The convergence rate is a benefit only when the swarm's optimum is global. When particles are attracted to a local optimum the whole population can fix there. Definitely, *Gbest* is the best choice for unimodal functions and a sufficiently good selection for multimodal functions.

Because *Gbest* performs the best or close to the best in solving test functions, we propose to use it. The extremely large population is our strategy to overcome the *Gbest* problem of not exploring many search space areas simultaneously. In sequential PSOs with large populations, the runtime increases and prevents from real-time applications. Parallelism in PSO allows increasing the population size with a slight runtime increase [18]. Tests performed on GPU PSO that used populations up to 2800 particles showed that large populations are necessary for some functions to find the optimum [19].

4   Optimized CUDA *Gbest* PSO

This section presents CUDA *Gbest* PSO implementation details such as data organization and mechanisms used to optimize the algorithm performance. First, we detected time-consuming operations and then proposed data structure reorganization and transfer as well as threads optimizations. At the end of the section, we present an experiment showing a runtime reduction.

4.1   Basic CUDA *Gbest* PSO

The sequential and parallel code identification is the first step in GPU implementations. In the proposed CUDA *Gbest* PSO the following operations run sequentially on the CPU: memory allocation, pre-generated random numbers, the algorithm control, and the best global position update run on the CPU, see Algorithm 1. The swarm initialization and particles update is executed in parallel on the GPU (Algorithm 1).

Algorithm 1. Basic CUDA *Gbest* PSO implementation — CPU/GPU codes separation

---

1: (GPU) Generate vector of random numbers
2: (GPU) Swarm initialization (the entire population in parallel)
3: **for** (CPU) each step t **do**
4:     (GPU) update particles (the entire population in parallel)
5:     **for** (CPU) each particle i **do**
6:         **if** (CPU) $\mathbf{pb}_i$ is better than $\mathbf{b}$ **then**
7:             (CPU) $\mathbf{b} = \mathbf{pb}_i$
8:             (CPU) $\mathbf{b}$ index is i
9:     **if** (CPU) $\mathbf{b}$ index is $\neq -1$ **then**
10:        (CPU) send $\mathbf{b}$ index to GPU
11:        (CPU) load $\mathbf{b}$ from GPU to CPU

The CUDA platform allows writing device (GPU) code in C functions called kernels. Each kernel is executed by many GPU threads in a Single-Instruction-Multiple-Thread (SIMT) fashion. Each thread executes the entire kernel once. We programmed two kernels: one to initialize the population and the second to update particles positions including $f(\mathbf{x})$ fitness function and best particle position $\mathbf{pb}$ evaluation (Algorithm 2). New particle positions are calculated according to Eqs. (1) and (2).

Algorithm 2. The particles update — The GPU kernel

1: update each coordinate of vector $\mathbf{x}_i$
2: calculate fitness function $f(\mathbf{x})$
3: update best particle position **pb**

4.2 Transfer Optimization

A bottleneck in GPU programming is data transfer and communication between the CPU and the GPU. The kernel call procedure runs in the three following steps:

1. transfer input data from host memory to device memories (global, constant and texture memory);
2. run kernel on data;
3. transfer results from device memories to the host memory.

Minimizing data transfer between host and device memories speeds up the algorithm [14].

In the PSO algorithm, the best particle $\mathbf{b}^t$ position is determined by comparing values in a sequence. The process is not suitable for parallelization, and we run it on the CPU resulting in data exchange between device and host memories. We propose a data structure to optimize the number of transferred variables.

In the proposed CUDA *Gbest* PSO data are stored both in GPU (device) and CPU (host) memories (Table 1). Velocities $\mathbf{V}$, positions $\mathbf{X}$, local best positions **Pb** of n particles with m independent variables, fitness values $f(\mathbf{X})$, local best fitness $f(\mathbf{Pb})$ of n particles and the index of the best particle in the entire swarm are stored in separate variables in the device memory. The CPU memory stores: the best particle position **b**, the best particle fitness evaluation $f(\mathbf{b})$ and the local best fitness $f(\mathbf{Pb})$ of n particles.

The data structure affects memory transfers in each step t of the Algorithm 1: n local best fitness values $(f(\mathbf{Pb}))$ are transferred from GPU to CPU and only one value of the global best particle index (**b**) is sent back.

Table 1. CUDA *Gbest* PSO data structure

Memory	GPU (device)	$\mathbf{V}$, $\mathbf{X}$, $\mathbf{Pb}$, $f(\mathbf{X})$, $f(\mathbf{Pb})$
	CPU (host)	$\mathbf{b}$, $f(\mathbf{b})$, $f(\mathbf{Pb})$

4.3 Thread Optimization

To keep the GPU effectively busy, nVidia corporation supplies developers with software called *CUDA Occupancy Calculator* that helps determine device occupancy. The occupancy can be viewed as a percentage of the hardware's ability to process active warps [14]. The *CUDA Occupancy Calculator* is useful for choosing the block size based on shared memory and register requirements to maximize occupancy.

In all experiments presented in the paper we use NVIDIA GeForce GT 540M (ver. CC 2.1) based on Fermi architecture and NVIDIA GeForce 940M (ver. CC 5.0) based on Maxwell architecture. For this graphic board the optimal block size for each kernel was calculated using *CUDA Occupancy Calculator*. The kernel initializing the population was most effectively executed with 608 block size and the kernel updating particles positions with 224 block size for 540M. For 940M both kernel was most effectively executed with 512 block size.

4.4 Pinned Memory and *Zero-copy* mechanism

We use pinned memory and *Zero-copy* features available in CUDA to maximize memory bandwidth.

Pinned memory is a page-locked memory that is never swapped out by the OS. It provides fast access but can reduce system performance because it is a scarce resource.

The extension to pinned memory is the *Zero-copy* mechanism that allows GPU to avoid allocating and copying data between device and host memories. Kernels access the host memory directly. The mechanism allows to improve the data structure (Table 2). Both, sequential and parallel codes access the fitness personal best values $f(\mathbf{Pb})$. We located the vector in the directly accessed memory to speed up the algorithm.

Table 2. CUDA *Gbest* PSO data structure using *Zero-copy* mechanism

Memory	
GPU (device)	$\mathbf{V}$, $\mathbf{X}$, $\mathbf{Pb}$, $f(\mathbf{X})$,
CPU (host)	$\mathbf{b}$, $f(\mathbf{b})$
Shared (host)	$f(\mathbf{Pb})$

The thread and memory optimization resulted in developing extra CUDA *Gbest* PSO code (Algorithm 3) that precedes the Algorithm 1.

Algorithm 3. CUDA *Gbest* PSO code preceding the Algorithm 1

1: (CPU) Get GPU properties
2: (CPU) Set block size
3: **if** (CPU) GPU can map host memory **then**
4:     (CPU) Allocate fitness evaluation of particles best value vector $f(\mathbf{Pb})$ on host using Zero-Copy mechanism

---

4.5 Optimized CUDA *Gbest* PSO experiment

We modified CUDA *Gbest* PSO progressively to observe the influence on runtime. The main goal was to obtain a fast and effective CUDA *Gbest* PSO algorithm with a large number of particles. The sequence of changes was as follows: a) transfer optimization b) threads optimization c) pinned-memory usage d)zero-copy mechanism. Each step was compared to the basic CUDA *Gbest* PSO (Algorithm 1). Figure 1 shows runtime on two-variable Beale's function on different architectures. The PSO was implemented, with $\omega = 0.8$ $\varphi_1 = 2.0$, $\varphi_2 = 0.8$, initial velocities were random numbers in the range $[-4.5, 4.5]$.

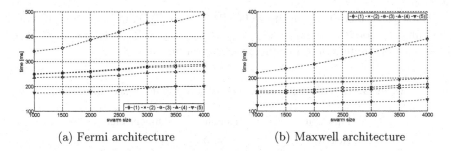

(a) Fermi architecture   (b) Maxwell architecture

Fig. 1. Runtime along increasing swarm size tested on (1) basic CUDA *Gbest* PSO, (2) CUDA *Gbest* PSO with transfer optimization, (3) CUDA *Gbest* PSO with threads optimization, (4) pinned-memory usage (5) zero-copy mechanism

The less time consuming was CUDA *Gbest* PSO implementation with optimized transfer, the optimal block size for each kernel and the zero-copy mechanism. We can reduce the runtime twice on average when compared to the basic CUDA *Gbest* PSO and use the optimized algorithm in further experiments.

5 Experiments, Results, and Discussion

The goal of our experiments was to compare the computational efficiency of optimized CUDA *Gbest* PSO and sequential PSO evaluated by runtime, speedup ratio and success rate. We conducted two experiments: one on three-dimensional benchmarks, and the second on multi-dimensional benchmarks for different swarm sizes. Tests were performed repeatedly because of the stochastic algorithm nature. The results of the tests were combined and then averaged.

For comparison purposes let us introduce the following two definitions.

Definition 3. *The* **speedup** *is the ratio of mean sequential PSO runtime to mean optimized CUDA* Gbest *PSO runtime.*

Algorithms were stopped always and only after 1000 iterations. No other stop condition was used.

Definition 4. *The* **success rate** *is defined to be the ratio of the number of times the algorithm found the global optimum with the 1e-06 tolerance to the number of trials (20 in each experiment).*

Eight standard benchmarks were tested (Table 3): Ackley, Beale, Bohachevsky, Easom, Rastrigin, Schwefel and Three-Hump Camel functions. All were run in 3 dimensions. The Rastrigin and Schwefel both multimodal functions were also run in 6 and 11 dimensions, as to test the algorithm in a more difficult environment. We omit detailed description because all used functions are standard.

Both tested algorithms optimized CUDA *Gbest* PSO and sequential PSO use the following coefficients: $\omega = 0.8$, $\varphi_1 = 2.0$, $\varphi_2 = 0.8$. Particles were initialized in a range of the function domain (ranges given in Table 3).

Table 3. Tested benchmarks, their domains and studied dimensions

Function	Domain	Dimensions
Ackley	$[-30, 30]$	3
Beale	$[-4.5, 4.5]$	3
Bohachevsky	$[-100, 100]$	3
Easom	$[-100, 100]$	3
Michalewicz	$[0, \pi]$	3
Rastrigin	$[-5.12, 5.12]$	3, 6, and 11
Schwefel	$[-500, 500]$	3, 6, and 11
Three-Hump Camel	$[-5, 5]$	3

Numerical experiments on sequential PSO and optimized CUDA *Gbest* PSO were conducted on an Acer Aspire 5742G equipped with an Intel Core i5-480M at 2.667 GHz and 4GB main memory. The operating system was Microsoft Windows 7 (Professional) 64-bit server version. CUDA *Gbest* PSO used GPU from the NVIDA GeForce GT 540M and NVIDIA GeForce 940M card.

5.1 Three-Dimensional Tests

We compared sequential PSO and optimized CUDA *Gbest* PSO time performances. Results (average from 50 runs) presented in Table 4 lead to a natural conclusion that runtime increases with population (swarm) size. The sequential PSO runtime increases more rapidly (sometimes even up to 2 s) than optimized CUDA *Gbest* PSO implementation (at most 0.2 s). Our experiment shows that the large number of particles in CUDA *Gbest* PSO does not significantly influence the runtime. The highest runtime increase was from 184 ms (1000 particles) to 205 ms (4000 particles) using the Michalewicz function for Fermi architecture and from 116 ms (1000 particles) to 131 ms (4000 particles) using Beale function for Maxwell architecture.

The maximum runtime for the optimized CUDA *Gbest* PSO was 214ms (4000 particles). This time is similar to runtimes (few hundreds milliseconds) reported in [2, 4, 12], and is significantly shorter than few seconds reported in [5, 10, 15, 19].

We tested the success rate of sequential PSO and optimized CUDA *Gbest* PSO. It was always 1 for both algorithms meaning that the optimum was found every time with the given precision.

Table 4. Runtime [ms] in three-dimensional domain, where **CPU** — sequential PSO, **GPU 1 (540M), GPU 2 (940M)** — optimized CUDA *Gbest* PSO

	Swarm size					Swarm size			
Test	1000	2000	3000	4000	Test	1000	2000	3000	4000
Ackley CPU	176	350	525	701	Beale CPU	462	923	1377	1844
Ackley GPU 1	175	192	194	199	Beale GPU 1	178	183	189	196
Ackley GPU 2	114	115	125	129	Beale GPU 2	116	117	126	131
Bohachevsky CPU	240	487	723	965	Easom CPU	383	763	1142	1526
Bohachevsky GPU 1	188	192	203	205	Easom GPU 1	191	195	207	208
Bohachevsky GPU 2	113	116	122	127	Easom GPU 2	113	117	118	126
Michalewicz CPU	419	834	1253	1678	Rastrigin CPU	155	308	463	616
Michalewicz GPU 1	184	186	199	205	Rastrigin GPU 1	169	170	177	182
Michalewicz GPU 2	115	117	124	129	Rastrigin GPU 2	110	115	119	122
Schwefel CPU	186	360	540	679	Three-Hump Camel CPU	478	957	1432	1910
Schwefel GPU 1	172	172	184	185	Three-Hump Camel GPU 1	186	190	203	214
Schwefel GPU 2	110	113	121	124	Three-Hump Camel GPU 2	120	117	124	128
Mean CPU	312	623	932	1240	Mean GPU 1	180	186	195	199
					Mean GPU 2	114	117	122	127

We calculated speedup ratios for every benchmark and presented them in Fig. 2. The experiment supports the previously reported conclusion that GPU parallelization of PSO reduces the runtime. In the test with 1000 particles speedup was small (between 1 and 4.1 in 1000 particles) but acceleration increased with population size (between 3.4 and 14.8 in 4000 particles).

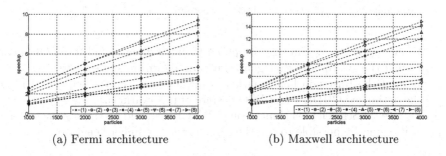

(a) Fermi architecture   (b) Maxwell architecture

Fig. 2. Speedup calculated from runtimes (Table 4) (1) Ackley, (2) Beale, (3) Bohachevsky, (4) Easom, (5) Michalewicz, (6) Rastrigin, (7) Schwefel, (8) Three-Hump Camel

356 J. Kołodziejczyk et al.

5.2 Multi-dimensional Tests

We checked algorithms performances in more complex environments i.e. 6- and 11-dimensional spaces on multimodal functions. The sequential PSO and optimized CUDA *Gbest* PSO runtimes, and speedups averaged from 50 runs are presented in Table 5.

In the case of the Rastrigin function, the maximum speedup was 5.4. This result was similar to those presented in [5,10,15,19,20]. In the Schwefel 11D (500 000 particles) test a speedup > 60 was gained. The highest speedup reported in the literature was 30 (standard benchmarks optimization using GPU PSO implementation). Our result is a new achievement. All results considered, we observed that the speedup increases with population size and function complexity.

The Schwefel test showed that sequential PSO was unable to find a solution in reasonable time, runtime was impractical. In the same test optimized CUDA *Gbest* PSO detected the optimum in only few seconds.

Table 5. Runtimes [ms] and speedup in 6- and 11-dimensional domains, where **CPU** — sequential PSO, **GPU 1**, **GPU 2** — optimized CUDA *Gbest* PSO

	Rastrigin	Swarm size				Schwefel	Swarm size			
Dim	Type	1 000	4 000	50 000	100 000	Type	1 000	50 000	100 000	500 000
6	CPU	455	1 892	14 033	27 527	CPU	415	18 771	37 725	191 265
6	GPU 1	290	444	2 735	5 240	GPU 1	244	2 735	5 032	25 004
6	GPU 2	129	151	1 385	2 240	GPU 2	128	1 335	2 342	9 325
	Speedup 1	1.6	4.3	5.1	5.3	Speedup 1	1.7	6.9	7.45	7.6
	Speedup 2	3.5	12.5	10.1	12.3	Speedup 2	3.2	14.1	16.1	20.5
11	CPU	1 220	4 270	50 587	101 260	CPU	7 398	347 317	657 919	> 1h
11	GPU 1	386	1 013	9 893	19 443	GPU 1	400	10 598	20 128	99 445
11	GPU 2	171	245	4 499	8 989	GPU 2	192	4 250	8 287	33 390
	Speedup 1	3.2	4.2	5.1	5.2	Speedup 1	18.5	32.8	32.7	> 60
	Speedup 2	7.1	17.4	11.2	11.3	Speedup 2	38.5	81.7	79.4	> 60

We increased the swarm size compared to the 3-dimensional tests. Our goal was to find the swarm size that guarantees an algorithm success rate of 1. The optimum of the Rastrigin benchmark was always found by the sequential PSO and the optimized CUDA *Gbest* PSO despite of the swarm size. Table 6 presents success rate for Schwefel benchmark on GPU. CPU should have the same success rate if it would be possible to use the same random number generator.

Table 6. Effectiveness in 6- and 11-dimensional domain Schwefel benchmark

Dimension	1 000	50 000	100 000	500 000
6	1	1	1	1
11	0	0.18	0.72	1

In the optimized CUDA *Gbest* PSO we were always able (based on tested benchmarks) to estimate a swarm size at which the algorithm's success rate equaled 1. Based on this feature and obtained runtimes we claim that the proposed algorithm might be appropriate for a practical real-time implementation.

6 Conclusions

We proposed the optimized CUDA *Gbest* PSO algorithm based on *Gbest* topology utilizing a large number of particles. We applied thread and transfer optimization and the zero-copy mechanism and by that we obtained an algorithm that performs exceptionally good comparing to sequential PSO and other GPU PSO solutions presented in the literature. We evaluated the algorithm using two criteria: success rate and speedup (Sects. 5.1 and 5.2).

Based on experimental results we concluded that the main advantages of optimized CUDA *Gbest* PSO are:

- Short runtime: even complex i.e. multimodal functions can be optimized fast. The algorithm runs in 10 s with 500 000 particles. Experiments show that in 3-dimensional benchmarks, the runtime in a large swarm (4000 particles) is less than 0.22 s for Fermi architecture and less than 0.14 s for Maxwell architecture.
- Speedup in most tests was below 10 for Fermi architecture and below 16 for Maxwell architecture. It can be even more than 60 when the optimization task is complex (Schwefel 11D, 500 000 particles).
- Extremely large populations are recommended. They just make good use of hardware capabilities and guarantee an success rate of 1 (confirmed in all tests).

The optimized CUDA *Gbest* PSO should be tested in higher dimensionality, more complex test functions and on several GPUs.

References

1. Bratton, D., Kennedy, J.: Defining a standard for particle swarm optimization. In: Swarm Intelligence Symposium, SIS 2007, pp. 120–127. IEEE, April 2007
2. Cagnoni, S., Bacchini, A., Mussi, L.: OpenCL implementation of particle swarm optimization: a comparison between multi-core CPU and GPU performances. In: Chio, C., et al. (eds.) EvoApplications 2012. LNCS, vol. 7248, pp. 406–415. Springer, Heidelberg (2012). doi:10.1007/978-3-642-29178-4_41
3. Calazan, R., Nedjah, N., de Macedo Mourelle, L.: Parallel gpu-based implementation of high dimension particle swarm optimizations. In: 2013 IEEE Fourth Latin American Symposium on Circuits and Systems (LASCAS), pp. 1–4, February 2013
4. Calazan, R.M., Nedjah, N., Macedo Mourelle, L.: Swarm grid: a proposal for high performance of parallel particle swarm optimization using GPGPU. In: Murgante, B., Gervasi, O., Misra, S., Nedjah, N., Rocha, A.M.A.C., Taniar, D., Apduhan, B.O. (eds.) ICCSA 2012. LNCS, vol. 7333, pp. 148–160. Springer, Heidelberg (2012). doi:10.1007/978-3-642-31125-3_12

5. Cardenas-Montes, M., Vega-Rodriguez, M.A., Rodriguez-Vazquez, J.J., Gomez-Iglesias, A.: Accelerating particle swarm algorithm with gpgpu. In: 2011 19th International Euromicro Conference on Parallel, Distributed and Network-Based Processing, pp. 560–564, February 2011
6. Kennedy, J., Eberhart, R.: Particle swarm optimization. In: Proceedings of the IEEE International Conference on Neural Networks, vol. 4, pp. 1942–1948, November 1995
7. Kennedy, J., Mendes, R.: Population structure and particle swarm performance. In: Proceedings of the 2002 Congress on Evolutionary Computation, CEC 2002, vol. 2, pp. 1671–1676 (2002)
8. Kennedy, J., Mendes, R.: Neighborhood topologies in fully informed and best-of-neighborhood particle swarms. IEEE Trans. Syst. Man Cybern. Part C Appl. Rev. 36(4), 515–519 (2006)
9. Kennedy, J., Eberhart, R.C.: Swarm Intelligence. Morgan Kaufmann Publishers Inc., San Francisco (2001)
10. Laguna-Sánchez, G.A., Olguín-Carbajal, M., Cruz-Cortés, N., Barrón-Fernández, R., Álvarez-Cedillo, J.A.: Comparative study of parallel variants for a particle swarm optimization algorithm implemented on a multithreading gpu. J. Appl. Res. Technol. 7(3), 292–307 (2009)
11. Li, J., Wan, D., Chi, Z., Hu, X.: An efficient fine-grained parallel particle swarm optimization method based on gpu-acceleration. Int. J. Innov. Comput. Inf. Control 3(6(B)), 1707–1714 (2007)
12. Mussi, L., Daolio, F., Cagnoni, S.: Evaluation of parallel particle swarm optimization algorithms within the cuda architecture. Inf. Sci. 181(20), 4642–4657 (2011). specialIssueonInterpretableFuzzySystems, http://www.sciencedirect.com/science/article/pii/S0020025510004263
13. Mussi, L., Nashed, Y.S., Cagnoni, S.: GPU-based asynchronous particle swarm optimization. In: Proceedings of the 13th Annual Conference on Genetic and Evolutionary Computation, GECCO 2011, NY, USA, pp. 1555–1562 (2011). http://doi.acm.org/10.1145/2001576.2001786
14. nVidia.com: CUDA C Best Practices Guide, DG-05603-001 v6.0 edn. (February 2014)
15. de P. Veronese, L., Krohling, R.: Swarm's flight: accelerating the particles using c-cuda. In: IEEE Congress on Evolutionary Computation, CEC 2009, pp. 3264–3270 (May 2009)
16. Solomon, S., Thulasiraman, P., Thulasiram, R.: Collaborative multi-swarm pso for task matching using graphics processing units. In: Proceedings of the 13th Annual Conference on Genetic and Evolutionary Computation, GECCO 2011, NY, USA, pp. 1563–1570 (2011). http://doi.acm.org/10.1145/2001576.2001787
17. Wachowiak, M.P., Foster, A.E.L.: GPU-based asynchronous global optimization with particle swarm. In: Journal of Physics Conference HPCS 2012, vol. 385 (2012)
18. Wang, W.: Particle swarm optimization on GPU. In: Workshop on GPU Supercomputing. Center for Quantum Science and Engineering National Taiwan University (2009)
19. Zhou, Y., Tan, Y.: GPU-based parallel particle swarm optimization. In: IEEE Congress on Evolutionary Computation, CEC 2009, pp. 1493–1500, May 2009
20. Zhou, Y., Tan, Y.: Particle swarm optimization with triggered mutation and its implementation based on gpu. In: Proceedings of the 12th Annual Conference on Genetic and Evolutionary Computation, GECCO 2010, NY, USA, pp. 1–8 (2010). http://doi.acm.org/10.1145/1830483.1830485

Optimization of Evolutionary Instance Selection

Mirosław Kordos[✉]

Department of Computer Science and Automatics,
University of Bielsko-Biala, Willowa 2, Bielsko-Biała, Poland
mkordos@ath.bielsko.pl

Abstract. Evolutionary instance selection is the most accurate process comparing to other methods based on distance, such as the instance selection methods based on k-NN. However, the drawback of evolutionary methods is their very high computational cost. We compare the performance of evolutionary and classical methods and discuss how to minimize the computational cost using optimization of genetic algorithm parameters, joining them with the classical instance selection methods and caching the information used by k-NN.

1 Introduction

Data preprocessing is frequently the step in machine learning, which in the highest degree determines the success of the entire process. That is because the quality of the possible results is limited by the data quality itself and if the data quality is poor, even the best classification algorithm cannot demonstrate its power. Poor quality data means data with a lot of measurement errors, irrelevant information and other artifacts. The irrelevant information additionally makes the training time long and the model interpretation harder. An important step in data preprocessing is feature and instance selection or in a broader context feature and instance weighting and generation. There are three main purposes of data selection:

- to decrease the data size,
- to reduce the noise in data,
- to select representative data points (prototype) in order to enable us better understand the process.

The first and second objective can be implemented by feature selection and all three by instance selection. This paper is focused on instance selection, although it is likely that some conclusions can be extended onto feature selection and joined instance and feature selection.

The classical instance selection methods select the instances before the classifier training, frequently basing on local predictions made by k-NN (Sect. 2). However, in some cases the instance selection can be also incorporated into the classifier, for instance under some conditions the neural network can reject the instances, which have very high error value as the network response, as they are supposed to be outliers.

© Springer International Publishing AG 2017
L. Rutkowski et al. (Eds.): ICAISC 2017, Part I, LNAI 10245, pp. 359–369, 2017.
DOI: 10.1007/978-3-319-59063-9_32

There are different definitions of what are genetic algorithms and what are evolutionary algorithms. Some of the definitions say that genetic algorithms use binary values and evolutionary algorithms use real values. The values that we use are discretized real numbers, sometimes only with two values (which corresponds to binary) and sometimes with more values. We will use both of the terms while referring to our methods. First we discuss the genetic algorithms with two binary values and how their implementation and parameters influence their performance. We consider generational and steady state genetic algorithms and mixtures of both. We evaluate different crossover schemes, population size and fitness functions. Then we extend that analysis onto multi-valued coding. We discuss how the computational cost can be reduced especially in the case of applying evolutionary algorithms to instance selection (Sect. 3).

There have been some propositions in the literature to use genetic or evolutionary algorithms for instance selections [1–3]. Evolutionary optimization is usually able to find the subset of instances, which is Pareto-optimal in the terms of compression and accuracy. Thus, we cannot improve on that. Instead, we propose two other improvements: reduction of the optimization computational complexity and improvement by including instance weighting. When genetic or evolutionary algorithms are applied to instance selection, their convergence time can be additionally decreased using the results of classical instance selection algorithms while generating the initial population and caching the information from previous model learning (Sect. 4).

Finally we experimentally compare the results of some of the best classical methods and the evolutionary methods on several datasets (Sect. 5) and conclude with our findings regarding the accuracy and computational complexity of those methods (Sect. 6).

2   Classical Instance Selection Methods

The classical instance selection methods usually are based on some local properties of the dataset, frequently the nearest neighbors or Voronoi cells, to assess which instances can be removed as noisy or redundant [4,5]. Their advantage is speed and their disadvantage is the accuracy compared to evolutionary methods.

In the literature frequently DROP-3 is considered as the most effective from the well known instance selection algorithms. However, our test on 10 classification datasets with 200 to 20.000 instances and the tests of Grochowski and Jankowski performed on smaller datasets [7] showed that IB3 performs not much worse. When we proceeded IB3 with ENN, the results were equally good as those of DROP-3 (in fact ENN is also used at the fist stage of DROP-3). Both DROP-3 and ENN with IB3 were able to reduce the number of instances on the datasets used in our experiments on average to 4% with the average accuracy drop from 92% to 87%. However, in our implementation in RapidMiner [6] the selection time of ENN+IB3 was much shorter than that of DROP-3 so for that reason in the final experiments we decided to use ENN followed by IB3.

The purpose of running ENN (Edited Nearest Neighbor) [8] is to increase classification accuracy by noise reduction in the training set. ENN uses k-NN to predict the class of each instance and marks the instances for which the predicted class is different than the real class. In the next step the marked instances are removed from the training set, as they are considered noise. The data size reduction obtained by ENN is usually very weak (at the level of 10% in our experiments) and higher compression indicates that the data quality is poor.

DROP-3 (Decremental Reduction Optimization Procedure v.3) [8] first runs ENN and then the condensation part removes redundant instances. The instances are first sorted by their distance from the nearest enemy (nearest another class instance) and then particular instances get removed if the classification of their neighbors by k-NN does not worsen without them.

IB3 (Instance Based Learning v.3) [9] first selects the instances misclassified by k-NN (as the correctly classified instances are believed not to provide any additional information) and then it further removes from the selected set the instances, which can be removed without the loss of classification accuracy.

The classical instance selection methods were originally designed for classification tasks. However, many of the algorithms can be adapted to regression tasks [10]. There are two ways to accomplish this. The first approach uses output discretization and converting the regression problem to a multi-class classification [11]. The second one replaces the concept of "the same class" by some arbitrary distance threshold in the output space. If a difference between a given instance real value and the value predicted by k-NN is lower than the threshold, then the instance is dealt with in the same way as its class would by correctly predicted in the classification task [12]. Currently our tests with evolutionary instance selection for regression tasks are in progress and they will be probably presented in the next paper.

Frequently a further improvement can be obtained using ensembles of instance selection algorithms [11,13]. However, the results will not be so good as they can be obtained with evolutionary instance selection.

3   Optimization of Genetic Algorithms Parameters

The idea and variants of genetic algorithms are well described in the literature. In [14,15] the reader can find the information. A special feature of genetic algorithms is that they are frequently able to find the optimal solution even without optimal parameters. However, adjusting the parameters allows for significant reduction of computational cost. On the other hand the optimal parameters depend on a given task [16]. Our purpose was to examine how convergence time of genetic algorithms applied to instance selection depends on several parameters. Using only a single processor core the convergence time would be proportional to the number of fitness function evaluations.

We used for the calculations a computer with two Xeon processors, each with 12 physical cores with hyperthreading turned off, thus we had 24 physical cores. So in this case it was optimal to evaluate 24 different solutions simultaneously. For that reason the size of any batch of calculation that we considered was 24 and multiplies of 24.

In the experiments we used chromosomes of different length from 200 to 20.000 (what corresponded to the number of instances in our datasets). The target value (the solution to be found) was obtained by randomly generating zeros and ones at each position or five different values (0, 0.25, 0.5, 0.75, 1), which corresponded to instance weighting. We provide detailed information for optimization of the binary case.

First we evaluated the influence of population size on the required number of fitness function evaluations. The short answer is: optimal population size P is about 100 in general and $P = 96$ for a 24-core system and it only very slightly increases with the chromosome length in the examined range from 200 to 20.000. For a bit smaller populations the average convergence time can be slightly shorter, but the process is less stable: the standard deviation is higher and occasionally the algorithms did not converge at all. The results are shown in Fig. 1.

In the next series of experiments we determined the optimal number of parents of one descendants. It was about 40 for the shortest chromosomes up to 400 for longest ones. An approximate rule is: $3\sqrt{NumInstances}$. The convergence of generational genetic algorithms with that many parents was over 3 times faster than with two parents only. Moreover, it provided a higher diversity in the recombination process, so smaller population was needed. The results are shown in Fig. 2.

It is known from the literature that the steady state genetic algorithms tend to converge faster that generational ones [17,18]. In steady state algorithms, when an offspring is created it immediately replaces the worst individual in the population (or the parent, or the most similar one) and the population immediately gets better as a whole, without the need to wait for the entire next generation. However, in our case the fastest solution was to create and simultaneously evaluate $N = 24$ offspring, because we had 24 cores in our computer. Even if that requires a little bit more calculations, the total time is definitely shorter, because the calculations are performed in parallel. This solution allows to decrease the number of fitness function evaluation by several percent comparing with the generational genetic algorithms with elitism.

The next important point is the fitness function, which can be written in the following form:

$$fitness = \left(\alpha * \frac{accuracy}{avgAccuracy} + (1 - \alpha) * \frac{avgNumVectors}{numVectors} \right)^p \qquad (1)$$

and the exponent p can be gradually increased as the optimization progresses. First, we start from a low p, as $p = 2$ in order not to limit the population diversity and then as the optimization progresses and the individuals tend to be more similar to each other, we gradually increase it to $p = 6$.

Algorithm 1. The genetic algorithm process

---
generate initial currentPopulation of P individuals
calculate fitness for currentPopulation individuals
for n=0 ... numIterations **do**
    apply the crossover operation to generate the newPopulation of N individuals
    ($N <= P$)
    calculate fitness for newPopulation individuals
    if optimal solution found or no further progress **then**
        end process
    end if
    sort together currentPopulation and newPopulation individuals by fitness
    select the best P individuals into currentPopulation
    apply the mutation operation
end for

Algorithm 2. The crossover operation

---
for i=0 ... populationSize **do**
  if RandomDouble(0,1) < crossoverProbability **then**
    for c = 0 ... numCrossoverPoints + 1 **do**
      individual[i][c] = RouletteWheelSelection(RandomDouble(0,sumFitness));
      crossoverPoint[i][c] = RandomInteger(0,numPositions);
    end for
    sort crossoverPoint[i];
    for c = 1 ... numCrossoverPoints + 2 **do**
      for d = crossoverPoint[i][c - 1] ... crossoverPoint[i][c] **do**
        newPopulation[i][d] = currentPopulation[individual[i][c - 1]][d];
      end for
    end for
  end if
end for

---

4 Evolutionary Algorithms for Instance Selection and Instance Weighting

We used two ways to accelerate the process of evolutionary instance selection: starting from a population based on ENN with IB3 selection and cashing the information required for k-NN. The second way can be easily done if the prediction model is k-NN, but the idea can also be partially extended to some other classifiers.

A better way than randomly generate individuals in the initial population is to run the ENN with IB3 instance selection (or another classical algorithm) first and then to use the generated set of prototypes to generate the initial population (Algorithm 3). That allowed us to decrease the calculation time approximately by 30%.

We implemented caching of the information required for k-NN in the following way: first we calculate the distances between each point in the test set and each point in the training set. Then we maintain two arrays for each point in the test set. One array contains the distances to particular points in the training

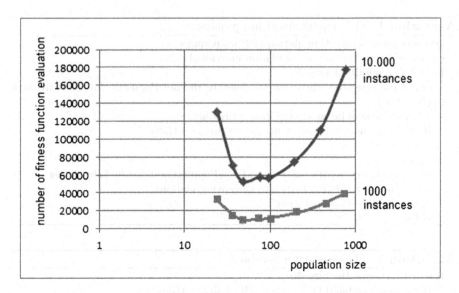

Fig. 1. How the number of fitness function evaluations depends on the population size (number of individuals), with 100 crossover points (101 parents).

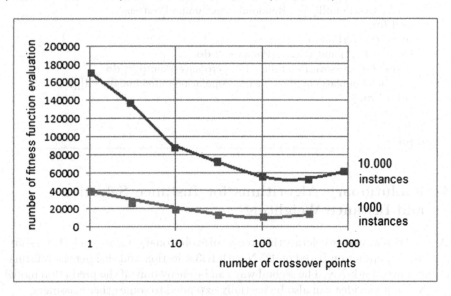

Fig. 2. How the number of fitness function evaluations depends on the number of crossover points (number of parents) with the population size 96 individuals.

set and the other array numbers of that points. Then we sort the first array in the increasing order and modify the second array accordingly to keep track of the point numbers. In the optimization phase it is enough to read from the second array the first k points, which are selected in a given individual and use their classes to predict the class of the point in the test set. In the case of distance-weighted k-NN we additionally read their distances from the first array. As the optimization is converging after some number of iterations it happens that some training set points are included very frequently in the individuals, while others very seldom. At that point we create another set of arrays, which contain only the distances and numbers of the frequent points. The arrays are much smaller than the original arrays (especially for big datasets) and thus searching for the numbers and distances in them is much faster. When a point exists in an individual and does not exist in the small array then it is searched for in the original array. This of course does not decrease the number of fitness function evaluations but frequently more than two orders of magnitude decreases the cost of fitness function evaluation comparing to running the full k-NN each time.

Although it is not so straightforward to extend this idea onto other classifiers, we must remember that k-NN is only the classifier used for instance selection, and the final classifier used for classification can still be a different model. A question may arise if the subset of instances that is optimal for k-NN is also optimal for other classifiers. To answer the questions we performed experiments, when the classifier used to evaluate the fitness function was an MLP network with a hidden single layer trained with the Rprop algorithm. The results showed that the subsets selected with k-NN and MLP contained mostly the same points, however some differences can be sometimes observed. We assume that even if the subset selected by k-NN is not optimal for MLP classifiers, it is enough close to the optimal one and in the cached version of k-NN can be used as much faster fitness function evaluator.

In the case of instance selection, we do not have to find the absolutely best subset, but an enough good suboptimal solution is acceptable. At the end of genetic optimization the improvement progresses very slowly, so by accepting a good suboptimal solution we can save o lot of time.

Algorithm 3. Generation of initial population

---

run the ENN followed by IB3 instance selection to get the set S of selected instances
for i=0 ... populationSize **do**
    for v=0 ... originalNumVectors **do**
        if Instance Selection **then**
            initialPopulation[i][v] = 1 with probability $p1 = 0.1$ if the instance v is in not S
            initialPopulation[i][v] = 0 with probability $p2 = 1 - p1$ if the instance v is in S
        end if
        if Instance Weighting **then**
            initialPopulation[i][v] = Random(0,0.5) if the instance v is in not S
            initialPopulation[i][v] = Random(0.5,1) if the instance v is in S
            initialPopulation[i][v] = Random(0.5,1) with probability $p3 = 0.1$ if v is in not S
            initialPopulation[i][v] = Random(0,0.5) with probability $p3 = 0.1$ if v is in S
        end if
    end for
end for

5   Experiments and Results

To perform the experiments we created a program, which can be downloaded from www.kordos.com/icaisc2017. In k-NN we used $k = 3$ or $k = 1$ if there were too few instances to use $k = 3$. We trained the neural networks for 30 epochs with the R-prop algorithm. We used networks with one hidden layer. The numbers of neurons in the hidden layer was equal to the geometric mean of number of inputs and number of classes. We performed the experiments on 10 classification datasets from the Keel Repository [19]: Ionosphere (351,33,2), Image Segmentation (210, 18, 7), Magic (19020, 20, 2), Thyroid (7200, 21, 3), Page-blocks (5472, 10, 5), WDBC (569, 30, 2), Sonar (208, 60, 2), Satellite Image (6435, 36, 6), Penbased (10992, 16, 10), Pima(768, 8, 2). The numbers in the brackets are (number of instances, number of features, number of classes).

Table 1. Experimental results. EV = evolutionary instance selection, i = percentage of selected instances, a-classification accuracy, KNN, MLP - algorithms used in the fitness function evaluation

Dataset	Alpha	IB3-i	IB3-aKNN	IB3-aMLP	EV-iKNN	EV-aKNN	EV-iMLP	EV-aMLP
Ionosphere 91.15	0.99	15.22	86.87	84.34	1.34	92.05	1.34	92.45
	0.96				1.34	92.25	1.34	92.40
	0.90				1.34	92.15	1.34	92.38
ImgSegm 94.24	0.99	13.88	90.04	91.00	13.71	91.38	13.71	92.49
	0.96				13.71	91.38	13.71	91.55
	0.90				13.71	91.38	13.71	91.49
Magic 85.29	0.99	3.91	82.88	82.51	2.85	84.87	2.68	85.29
	0.96				1.98	84.04	1.96	84.47
	0.90				1.15	82.66	1.05	83.14
Thyroid 98.87	0.99	2.87	88.81	92.97	2.02	96.03	1.81	98.87
	0.96				1.45	95.15	1.32	98.47
	0.90				1.12	92.60	1.02	98.05
PageBlk 96.75	0.99	2.29	93.82	93.62	1.86	97.01	1.80	97.14
	0.96				1.09	96.67	0.98	96.75
	0.90				0.82	95.12	0.83	96.50
WDBC 98.89	0.99	-	-	-	2.85	97.99	3.44	98.66
	0.96				1.20	97.12	1.25	97.89
	0.90				1.15	96.83	1.15	97.15
Sonar 83.19	0.99	27.65	70.61	70.20	3.66	86.14	3.37	85.90
	0.96				2.36	85.01	2.38	85.24
	0.90				2.05	83.85	1.87	83.19
SatImg	0.99	7.47	87.57	84.61	3.13	87.85	4.07	85.01
	0.96				1.52	86.12	1.54	82.52
	0.90				1.15	72.20	0.78	67.78
Penbased	0.99	3.60	97.36	90.35	4.02	97.55	3.20	91.35
	0.98				3.20	97.30	3.25	91.30
	0.96				2.54	91.55	2.20	88.01
Pima 97.15	0.99	-	-	-	1.24	97.35	1.24	98.27
	0.96				1.24	97.27	1.24	98.27
	0.90				1.24	97.29	1.24	98.27

All the experiments were performed in crossvalidation. In case of evolutionary algorithms we used 80% of the training subset for training and 20% for evaluation to select the best individual to be tested. We performed a two-objective optimization, where one objective was to maximize classification accuracy on a test set and the other one was to minimize the number of instances in the training set (Eq. 1). The α parameter was used to assign weights to the two objectives. The power exponent p was gradually increased during the training. We started from $p = 2$ and finish at $p = 6$, as the differences between particular individuals tend to get much smaller as the genetic optimization progresses. If an individual contained fewer than four instances, we always added the missing instances from the most frequent instances.

Many of the evolutionary approaches are able to find the best solution, so there was no point to compare our methods to other evolutionary instance selection methods described in the literature in terms of accuracy. Moreover, the authors usually provided only the results obtained on very small datasets and did not provide enough information to perform such a comparison. However, the advantage of our approach is in the computational time decrease. The whole selection process took from single seconds to minutes depending on the dataset size.

The instance weighting is used in the following way: in k-NN each instance was multiplied by its weight and the weighted number of instances in each class was considered. In the MLP network the error the network makes during learning on each instance was multiplied by the instance weight. We did not allow for subsets with fewer than 4 selected instances. Instance weighting only seldom improved the results for classification, thus we decided not to include this in Table 1. However, based on our previous experience with instance weighting we expect the weighting to work well in regression problems, especially with noisy data [10,12,20].

6   Conclusions

The purpose of the optimization of the process and its parameters was mostly to decrease the computational complexity, while also taking into account the stability (low variance) of the process. We optimized the number of crossover points, the size of the population and the fitness function. We added the caching of the information required by k-NN. We also compared the obtained methods with the results obtained with classical instance selection methods. The results obtained with evolutionary methods are most accurate, but on the other hand most time consuming. The optimization allowed us to shorten the process time by finding the optimal parameters for our tasks, that is: population size: 96 individuals, multi-parent recombinations, with the number of parent 3 times the square root of the number of instances, the fitness function exponent gradually increasing from 2.0 to 6.0. The experiments were first conducted in detail in the "emulation mode" using as the target a randomly generated individual (what was much fasted without the need to use k-NN or to train the neural network) and then confirmed using k-NN and neural networks prediction and number of selected instances as the two objectives of the fitness function. In addition the initial population was generated not totally randomly but was centered about the solution found with the classical methods and randomly perturbated. The total number of iterations that we were able to achieve using 96 individuals and starting from ENN+IB3 based randomized values was approximately $0.5 \cdot numInstances^{0.7}$. Instance weighting did not cause significant improvement in classification tasks. Also setting the balance α between accuracy and number of instances in some cases did not have any influence on the results, as it was enough to have as few instances as four to obtain the best classification accuracy.

The accuracy of the evolutionary methods is their strong point, while the computational complexity, although we were able to significantly lower it is still several hundred times higher than that of classical methods. But it may be acceptable, as in the experiments the whole selection process took from single seconds to minutes depending on the data size.

The direction of our future work is to investigate feature selection and weighting together with instance selection as well for classification as for regression problems.

References

1. Antonelli, M., Ducange, P., Marcelloni, F.: Genetic training instance selection in multiobjective evolutionary fuzzy systems: a coevolutionary approach. IEEE Trans. Fuzzy Syst. **20**(2), 276–290 (2012)
2. Derrac, J., Cornelis, C., Garcia, S., Herrera, F.: Enhancing evolutionary instance selection algorithms by means of fuzzy rough set based feature selection. Inf. Sci. **186**, 73–92 (2012)
3. Tsaia, C.-F., Eberleb, W., Chu, C.-Y.: Genetic algorithms in feature and instance selection. Knowl. Based Syst. **39**, 240–247 (2013)
4. Garcia, S., Derrac, J., Cano, J.R., Herrera, F.: Prototype selection for nearest neighbor classification: taxonomy and empirical study. IEEE Trans. Pattern Anal. Mach. Intell. **34**(3), 417–435 (2012)
5. Olvera-López, J.A., Carrasco-Ochoa, J.A., Martínez-Trinidad, J.F., Kittler, J.: A review of instance selection methods. Artif. Intell. Rev. **34**(2), 133–143 (2010)
6. Hofmann, M., Klinkenberg, R.: RapidMiner: Data Mining Use Cases and Business Analytics Applications. CRC Press, Boca Raton (2013)
7. Jankowski, N., Grochowski, M.: Comparison of instances seletion algorithms I. Algorithms survey. In: Rutkowski, L., Siekmann, J.H., Tadeusiewicz, R., Zadeh, L.A. (eds.) ICAISC 2004. LNCS, vol. 3070, pp. 598–603. Springer, Heidelberg (2004). doi:10.1007/978-3-540-24844-6_90
8. Wilson, D.R., Martinez, T.R.: Reduction techniques for instance-based learning algorithms. Mach. Learn. **38**, 257–286 (2000)
9. Aha, D.W., Kibler, D., Albert, M.K.: Instance-based learning algorithms. Mach. Learn. **6**, 37–66 (1991)
10. Kordos, M., Blachnik, M.: Instance selection with neural networks for regression problems. In: Villa, A.E.P., Duch, W., Érdi, P., Masulli, F., Palm, G. (eds.) ICANN 2012. LNCS, vol. 7553, pp. 263–270. Springer, Heidelberg (2012). doi:10.1007/978-3-642-33266-1_33

11. Arnaiz-González, A., Blachnik, M., Kordos, M., García-Osorio, C.: Fusion of instance selection methods in regression tasks. Inf. Fusion **30**, 69–79 (2016)
12. Kordos, M., Białka, S., Blachnik, M.: Instance selection in logical rule extraction for regression problems. In: Rutkowski, L., Korytkowski, M., Scherer, R., Tadeusiewicz, R., Zadeh, L.A., Zurada, J.M. (eds.) ICAISC 2013. LNCS, vol. 7895, pp. 167–175. Springer, Heidelberg (2013). doi:10.1007/978-3-642-38610-7_16
13. Blachnik, M., Kordos, M.: Bagging of instance selection algorithms. In: Rutkowski, L., Korytkowski, M., Scherer, R., Tadeusiewicz, R., Zadeh, L.A., Zurada, J.M. (eds.) ICAISC 2014. LNCS, vol. 8468, pp. 40–51. Springer, Cham (2014). doi:10.1007/978-3-319-07176-3_4
14. Goldberg, D.D.: Genetic Algorithms in Search, Optimization and Machine Learning. Addison Wesley, Boston (1989)
15. Michalewicz, Z.: Genetic Algorithms + Data Structures = Evolution Programs. Springer, Heidelberg (1992)
16. Lobo, F.G., Lima, C.F., Michalewicz, Z.: Parameter Setting in Evolutionary Algorithms. Studies in Computational Intelligence, vol. 54. Springer, Heidelberg (2007)
17. Zavoianu, Z.C., et al.: Performance comparison of generational and steady-state asynchronous multi-objective evolutionary algorithms for computationally-intensive problems. Knowl. Based Syst. **87**, 47–60 (2015)
18. Cano, J.R., Herrera, F., Lozano, M.: Instance selection using evolutionary algorithms: an experimental study. In: Pal, N.R., Jain, L. (eds.) Advanced Techniques in Knowledge Discovery and Data Mining. Advanced Information and Knowledge Processing, pp. 127–152. Springer, London (2004). doi:10.1007/1-84628-183-0_5
19. Alcala-Fdez, J., et al.: KEEL data-mining software tool: data set repository, integration of algorithms and experimental analysis framework. J. Multiple-Valued Logic Soft Comput. **17**, 255–287 (2011). http://sci2s.ugr.es/keel/datasets.php
20. Rusiecki, A., Kordos, M., Kamiński, T., Greń, K.: Training neural networks on noisy data. In: Rutkowski, L., Korytkowski, M., Scherer, R., Tadeusiewicz, R., Zadeh, L.A., Zurada, J.M. (eds.) ICAISC 2014. LNCS, vol. 8467, pp. 131–142. Springer, Cham (2014). doi:10.1007/978-3-319-07173-2_13

Dynamic Difficulty Adjustment for Serious Game Using Modified Evolutionary Algorithm

Ewa Lach[✉]

Institute of Informatics, The Silesian University of Technology, Gliwice, Poland
Ewa.Lach@polsl.pl

Abstract. Dynamic Difficulty Adjustment (DDA) seeks to adapt the challenge a game poses to a human player. When the game is too easy the player can become bored, when it is too hard - frustrated. In the case of a serious game (educational game), additionally, without a balance between the player competence and the game challenge the game could repeatedly exploit the developed skills, or fail to achieve the pedagogical goals. In this paper evolutionary algorithm (EA) is used to find game settings suitable for the player of a serious math game. To reduce the number of training data and accelerate the search for the 'right' game difficulty level EA modifications are introduced. Various experiments are performed. The obtained results show that proposed methods can substantially decrease the time a human player has to wait for a suitable game level.

Keywords: Dynamic Difficulty Adjustment (DDA) · Game AI · Serious game · Evolutionary algorithm

1 Introduction

Computer games have become an integral part of our social and cultural environment over the past decades. They are now considered a routine part of children's and adolescents' lives [1]. People, especially children, can spend hours playing games without realizing the passage of time. Such a high level of engagement is rare in typical learning experiences. Scholarly interest in educational games also called serious games has grown rapidly in recent years [7] with the realization of the potential of capitalizing on game's entertainment value by offering an instructional content during the gameplay. Various research indicates that serious games render learning easier, more enjoyable, more interesting, and, thus, more effective [6,8–10]. The serious games industry is gaining momentum as games and simulations designed for education and other training purposes continue to gain growing acceptance worldwide. For instance, two major educational technology research journals, Computers & Education and British Journal of Educational Technology contain hundreds of articles on games.

An entertainment has an important role to play in serious games, contributing to their motivational and engaging qualities leading to players voluntarily playing

© Springer International Publishing AG 2017
L. Rutkowski et al. (Eds.): ICAISC 2017, Part I, LNAI 10245, pp. 370–379, 2017.
DOI: 10.1007/978-3-319-59063-9_33

serious games for extended periods of time. Generally, a game in which the challenge level matches the skill of the human player has a greater entertainment value than a game that is either too easy (boring) or too hard (frustrating). The balance between the player competence and the challenge presented by a task is necessary to achieve by the player a state of "flow" [2], the state of mind characterized by focused concentration and elevated enjoyment during performed activities. Therefore the game's tasks should be designed so the player has a reasonable chance of success with intense effort. This is especially important when designing a serious game, otherwise, the game could repeatedly exploit the developed skills, or fail to achieve the pedagogical goals.

To be effective, a cognitive training needs to be adapted to the users' abilities. Unfortunately, fixing a few predefined and static difficulty levels (e.g. easy, medium, hard) is not sufficient. There is a great diversity among players in terms of skills and/or domain knowledge. Even players with a similar level of game playing ability can find different aspects of a game more difficult. Moreover, users may change their performance over time.

Dynamic Difficulty Adjustment (DDA), that seeks to adapt the challenge a game poses to a human player is an emerging and challenging research area of artificial intelligence (AI) in digital games. AI techniques like evolutionary algorithms and artificial neural networks have enormous potential to learn the levels of player's skills for various games. Unfortunately, these methods need lots of training data and a great amount of learning time.

The paper introduces methods reducing the number of training game levels (especially the frustrating ones) for DDA system built with an evolutionary algorithm. A serious game is developed as the test bed to evaluate the efficiency of the proposed methods.

The paper continues by presenting related work in Sect. 2. Section 3 defines the problem. Section 4 introduces the proposed methods. Section 5 presents the serious game, conducted experiments and simulation results. Section 6 concludes the paper.

2 Related Work

The approaches, proposed in work on DDA differ in terms of adjustment methods, as well as adjusted game content. The game difficulty has been set also in various ways: as subjective feedbacks like self-reported fun level [11] or as, more popular, objectively measured performance of a player [12,13,17–19]. In [18] authors showed that objectively measured performance of a player is inversely correlated with the subjectively assessed difficulty.

One approach is to modify the behavior of characters controlled by the computer, the Non-Player Characters (NPCs).

In work [12] NPC is defined with behavior rules, that are adjusted through dynamic scripting. In [14] author focuses on improving previous DDA system with three modifications: dynamic weight clipping, differential learning, and adrenalin rush.

In [20] NPC is described with a set of quantitative attributes, that are adjusted with the use of general quantitative measure resulting from an outcome of player's and opponent's actions.

In [19] good quality, initial NPC strategies are built offline using Reinforcement Learning (RL). Action selection mechanism adjusts the behavior of NPC depending on the current user's skills. During the game, the computer opponent increases or decreases the possibility of 'mistakes', while still learning with RL.

Second approach concerns the personalized procedural content generation (PCG).

In the paper [11] the DDA system automatically generates personalized levels for platform games. The DDA performs an exhaustive search in the space of controllable levels features to find the combination of the features that, taken together with observed gameplay features, maximize the multi-layer perceptron (MLP), which models player experience preferences.

In the paper [16], the racing tracks in a car racing game are built. A multiobjective evolutionary algorithm maximizes the entertainment value of the track relative to a particular player.

In another approach, one can control the game settings in order to make challenges easier or harder.

One of the common directions in commercial and serious games is to rely on the knowledge and experience of human experts to derive the adaptation model tightly dependent on the game. The weights or contributions of individual game components to the game difficulty are manually determined, as well as actions taken to adjust the game [5,15].

In [13] probabilistic technique that dynamically evaluates the difficulty of given obstacles based on user performance is proposed. As the player moves throughout the game world, DDA system, Hamlet, uses statistical metrics to monitor incoming game data. Over time, Hamlet estimates the player's future state from this data. When an undesirable but avoidable state is predicted, the system intervenes and adjusts game settings, such as changing the number of enemies that appear in a certain location or providing a nearby health pack for the player to pick up.

In [17] an artificial neural network (ANN) model describing the relationship between the dynamic game state, player performance, the adaptation decision to be taken and the resultant game difficulty determines both the direction and magnitude of game adaptation.

In [18] player's performance data is used to construct a tensor, which is decomposed to predict the player's performance over time. To receive the desired player performance, the DDA adapts the game contents based on the predicted performance.

All presented approaches have their advantages and limitations. For example, adjusting techniques for NPCs limit the design of computer players. Limitation of using such methods as evolutionary algorithms or artificial neural network lies, on the other hand, in the great amount of training data required and the execution time when the decision space for adaptation becomes large.

3   Problem Definition

The aim of this paper is to present DDA system that can be used to adjust serious game with its educational content to player skills. Especially, we are interested in the first stage of DDA system, in which it finds the difficulty level of the game, which corresponds to the player skills.

Games, on which we concentrate, are made up of levels. Each level consists of the game elements (GE) that can be adjusted to the skills of the player. Some of the game elements are educational game elements (EGE) that player has to solve. The levels could be divided (but this is not required) into smaller parts, Challenge Stages, for which player performance can be calculated. Challenge stage doesn't have to use all GEs of the game level. For each GE, we can identify a set of difficulty points (DP). Each subsequent difficulty point refers to the next difficulty level of the GE.

The DDA system is ignorant of the specific manner in which the game elements at specific difficulty points are designed, allowing for much more system generality. Game Levels are evaluated based on the player performance. The searched situation by DDA is that the player is close to losing, but still wins. So the worse the results the better, with a high penalty for failure. The first goal of the DDA system is to find, as quickly as possible, a vector of game elements difficulty points (VGEDP) for which game challenge is set on the level of the human player abilities. This goal can be described as an optimization problem (where player performance is minimized), which may be solved by evolutionary optimization (e.g. genetic algorithm (GA), that can produce high quality solutions.) With many simultaneously adjusted game elements, the problem, however, is the great amount of training data (VGEDPs), required by the algorithm to find the solution. Because the training is performed by a human player we must find a way to reduce the number of training examples and accelerate the process. This paper focuses on improving the evolutionary algorithm for DDA system by introducing methods to lower the number of training game levels, especially the frustrating ones (too hard for the player).

4   Modified Evolutionary Algorithm

An evolutionary algorithm (EA) [3] begins its search with a population of solutions usually created at random. Next, the procedure enters into an iterative operation of updating the current population to create a new population by the use of three main operators: selection, crossover, and mutation. The operation stops when one or more termination criteria are met.

The EA solutions for our game are represented by bounded integer vectors derived directly from the game (VGEDPs). Each GE can have DP varying from 0 to 10. The DP of 10 means the most difficult. 0 and 1 mean the easiest for no-educational GE and EGE, respectively. If EGE has DP set to 0 then this element is not used in game level as too difficult for the user. To make sure that the EGEs are not specifically removed to improve the player results for each

EGE sets to 0 user gets a penalty. However, this penalty is lower than the one caused by failed EGE. As a result, we have the highest fitness value for unfulfilled EGE, a little lower for unused EGE and still lower for successful EGE.

Proposed in this paper modifications do not change the main algorithm, but work with it to produce results faster and with less data presented to the player.

4.1 Saved Solutions Table [M1]

First, we propose to store all tested solutions with obtained fitness values (player's performance) in the Saved Solutions Table (SST) so they aren't presented to the player again. It is not unheard of that EA proposes the same solutions again and again, especially when population converges to some value. For this solution, an application needs more space but from the point of view of a player, it is a profitable exchange. When the new game level must be evaluated, first it is looked for in the SST, then it is evaluated or the saved fitness is returned.

Next idea uses a domain knowledge: information on parameters' difficulty points for the failed games. If the player has lost for the specified game settings (VGEDP) it is to be expected that he will lose again if all the settings are the same or more difficult. For example, if a player is not able to defeat three opponents, ha can not beat four, when other GE are the same (or more difficult). In SST, in addition to VGEDP and its fitness value, we store an information if the player won or lost. When the new solution must be evaluated, first we check SST if there is a failed solution with all DPs easiest or the same as in the verified vector. If we find such a vector fitness value of found solution is returned.

If the game allows the use of the challenge stages, then we can store the result of the challenge stage in SST and use it the same way as described for the game level. The difference is that same GEs for a challenge stage are set as inactive and we can compare only solutions with the same inactive elements. If the challenge stage is found in SST than the game level will be evaluated only partly - the remaining values will be taken from SST.

4.2 Failed Difficulty Point for the Game Element [M2]

The second method also exchanges less training data for more memory space. This method also uses information about subsequent levels of difficulty for game elements.

We count and store the number of occurrences for each GE with the particular DP and the number of lost game levels when this DP was used. When the new solution must be evaluated we check each GE for using failed DP. DP is believed to be failed if the ratio of losses to the number of occurrences, failure ratio, exceeds a certain threshold lim_F. A failure ratio is checked recursively in the direction of easiest DP starting from DP for checked solution. The occurrences are counted until the DP with a failure ratio below the lim_F is encountered. If the number of counted occurrences is greater than a predetermined threshold lim_{FC} then the value of DP in evaluated vector is reduced.

With this change, impossible solutions should be quickly eliminated.

4.3   Local Optimization Algorithms

There are two phases in the search of an EA. First, the EA exhibits a more global search by maintaining a diverse population. Second, a more local search takes place by bringing the population members closer together. Although the EA degenerates to both these search phases automatically it is believed that more efficient search can be achieved if the local search phase can be executed with a more specialized local search algorithm. Choosing such local search algorithm one must remember that our purpose is to reduce training solutions and that it is not easy as the scale of an optimization problem is set by the dimensionality of the problem, i.e. the number of variables on which the search is performed. In this paper, we choose a simple hill-climbing algorithm where neighbors differ from each other with single DP increased or decreased by one.

Optimization with Threshold [M3]. In the first method, the initial solution for local optimization is chosen when its fitness value is less than certain threshold lim_O. Then, up to $2N$ neighboring solutions are generated and evaluated, where N is the size of VGEDP. The best neighbor is picked and the algorithm continues until the lim_{Oi} iteration is reached or there are no better neighboring solutions. The number of optimization for single generation may not exceed the threshold lim_{Oc}.

The problem with this method is that the EA may not find, fast enough, the solution with fitness value below lim_O. As a result, the second method was proposed.

Optimization with History [M4]. In this method, the fitness value of the optimized solution must be less than the fitness value of the best solution from the previous generation f_{MIN-1}. After that, the algorithm behaves the same as in the previous method. The number of optimization for single generation may not exceed the threshold lim_{Ogc}.

Optimization with Probability [M5]. The last optimization method also uses the minimal fitness value f_{MIN-1}, but with the weight w_O. With the probability p_O a solution is selected from the population for optimization. If its fitness value does not exceed the $w_O * f_MIN - 1$ the solution vector remains unchanged. To reduce the number of evaluated solutions, we check only if a change of 1 (up or down) of one randomly selected DP improves the solution. If the answer is yes the solution is adjusted.

5   Experiments

To evaluate the effect of the proposed methods serious math game for improving the arithmetic skills was employed. The player aim, in this game, is to solve

arithmetic operations, within a limited time with the limited resources, by selecting the results from the available values and shooting them with a cannon. In
the same time, the player must fight off opponents. To win the game the player
must use various skills: knowledge of mathematics, spatial imagination as well as
perception and reflexes. These skills may be at various levels. For example, the
first grade may have little knowledge of mathematics and very quick reflexes.
In turn, a professor of mathematics may have a very long reaction time and
the vast mathematical knowledge. For every mathematical challenge, the player
starts with maximum power and time. The power can be used to kill enemies
and shoot at arithmetic results. This power is also reduced as a result of the
opponents' attacks. A player loses when he loses all the power or runs out of
time. He can also give up.

In this game we can identify eleven game elements: four EGEs corresponding
to the available mathematical operations (addition, subtraction, multiplication,
and division), the number and the size of generated responses, the length of the
guiding beam for the cannon, the number and the movement speed of opponents,
also their strength and the frequency of appearance. Their DPs represent a
solution in AE. The training game level can be divided into four challenge stages
corresponding to 4 mathematical tasks. During a regular game, an arithmetic
operator is selected at random.

To test DDA system we need a player. For this purpose, a computer player
was designed. His behavior is controlled by the variables describing his style,
abilities, and knowledge. We register a time reactions of a random player to
different game events and use their average values in a computer player (e.g.
time reaction to a new opponent). We register also other behaviors which may
be described in a quantitative way (e.g. how many shots he performs over those
necessary to kill the opponent). At this stage, a computer player behavior is
designed for this game only and can not be generalized.

Seven tests were defined, with a different combination of EA modifications
(M1 to M5 from Chap. 4):

T0 - without any modification,
T1 - M1 and M2 ($lim_F = 0.8$, $lim_{FC} = 40$),
T2 - M1, M2 (settings from T1), M3 ($lim_{Oi} = 10$, $lim_{Oc} = 1$, $lim_O = 84$),
T3 - M1, M2 (settings from T1), M4 ($lim_{Ogc} = 1$),
T4 - M1, M2 (settings from T1), M5 ($w_O = 1.1$, $p_O = 0.6$),
T5 - M1, M3, M4, M5 (settings as in tests T2–T4),
T6 - M1, M2, M3, M4, M5 (settings as in tests T1–T4).

First we use classic GA with tournament selection with 2 participants, mutation (mutation operator probability $m_p = 0.3$ and probability for each integer to
be modified by mutation $m_{pp} = 0.3$) and crossover (operator probability $c_p = 0.5$
and equal distribution of individuals elements). The population size is 100. For
test identification we use prefix CGA.

Because we have seen in the obtained results the loss of good solutions in
successive generations we added tests that use EA with an elite-preservation

operator that keep 1 best solution in population. We tray generational (GGA) and steady state (SGA) genetic algorithms [4] with the same setting as for CGA.

Because of premature convergence of SGA to unsatisfactory solutions for a population of 100 individuals we use SGA with a population of 200 individuals.

Each test was repeated from 60 to 100 times and averaged. In the description of the results presented in Sect. 5.1 we refer to average values.

5.1   Results

Figures in Fig. 1a–c presents the best fitness obtained in each generation by different tests. Figure 1d presents the best fitness obtained for different numbers of evaluated solutions.

The results show that all modifications accelerate EA, regardless of its type. Tests T0 are always behind the rest. Tests T6 that uses all modifications simultaneously are always the best. We suspect that this is due to the complementary behavior of the modifications.

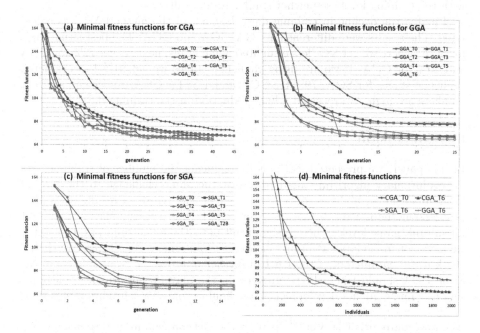

Fig. 1. Experiments results for CGA, GGA and SGA for tests T0–T6

The use of the modification M2 results in a rapid improvement in the early stages of the evolution. The consequence of this is a faster transition from a global to local optimization of EA and getting stuck in a local minimum (tests T1). On the other hand, without M2 other methods starts local optimization from worse evolution points (tests T5). For this reason, each optimization technique is used with M2 (tests T2–T4).

From all minimization techniques M4 (tests T3) get results the closest to the best results (tests T6). We can speculate that this type of modification has the strongest influence on results of tests T6. Optimization M5 is slower but finds better solutions in the end. The method M3 has the worst results. The problem here arises from the use of a threshold, which is often not achieved or is achieved too late. However, when this threshold is reached, we often observed a significant improvement. The results from SGA with M3 must be divided into two parts, due to the clear differences between results of evolution, which reached the required threshold (SG_T2B), and which have not.

At the end, we compare the results of different types of GA for an authentic number of evaluated game levels (Fig. 1d). We can again see that our modifications work. We can also see that the results for SGA and GGA are better than CGA. We can not, however definitely determine which method is better: SGA or GGA. The SGA algorithm quickly generates a better solution. But the 200 individuals in SGA population results in the fact that the GGA finds individuals with tolerable fitness value faster. The player may be prepared to train longer with the training levels somewhat similar to his skills.

Finally, the decline in training data, because of the method M1 was calculated. The results were different for CGA, SGA and GGA. The best situation is for SGA, where we get 88% less evaluated game levels, next we get 72% for GGA and 20% for CGA. If we are using challenge stages we get 91% less evaluated challenge stages for SGA, 79% for GGA and 47% for CGA. As we can see that numbers are high. We received a huge reduction in the training data.

6   Conclusion

In this paper, we presented methods for reducing the number of training data for evolutionary algorithms used to adjust the game challenge to the level of the human player abilities for a serious game. The obtained results show that proposed methods can substantially decrease the time a human player has to wait for suitable game levels. There is significant potential for future work. Currently, we have not made large assumptions about evolutionary algorithms. We should check if the parameters of the EA can further improve the results. Second, the game DDA should track the progression in the player's performance and for a serious game guide its development. We should propose a solution for this task.

Acknowledgment. This work is part of the General Statutory Research Project 02/020/BK_17/0105 conducted at the Institute of Informatics, the Silesian University of Technology.

References

1. Olson, C.K.: Children's motivations for video game play in the context of normal development. Rev. Gen. Psychol. **4**(2), 180–187 (2010)
2. Csikszentmihalyi, M.: Flow: The Psychology of Optimal Experience. Harper & Row, New York (1990)

3. Goldberg, D.E.: Genetic Algorithms for Search, Optimization, and Machine Learning. Addison-Wesley, Reading (1989)
4. Jones, J., Soule, T.: Comparing genetic robustness in generational vs. steady state evolutionary algorithms. In: Proceedings of the 8th Annual Conference on Genetic and Evolutionary Computation GECCO 2006, pp. 143–150 (2006)
5. Osman, Z., Dupire, J., Mader, S., Cubaud, P., Natkin, S.: Monitoring player attention: a non-invasive measurement method applied to serious games. Entertainment Comput. **14**, 33–43 (2016)
6. Kebritchi, M., Hirumi, A., Bai, H.: The effects of modern mathematics computer games on mathematics achievement and class motivation. Comput. Educ. **55**, 427–443 (2010)
7. Hainey, T., Connolly, T., Boyle, E., Wilson, A., Razak, A.: A systematic literature review of games-based learning empirical evidence in primary education. Comput. Educ. **102**, 202–223 (2016)
8. Castellar, E., Looy, J., Szmalec, A., Marez, L.: Improving arithmetic skills through gameplay: assessment of the effectiveness of an educational game in terms of cognitive and affective learning outcomes. Inf. Sci. **264**, 19–31 (2014)
9. Castellar, E., All, A., de Marez, L., Looy, J.: Cognitive abilities, digital games and arithmetic performance enhancement: a study comparing the effects of a math game and paper exercises. Comput. Educ. **85**, 123–133 (2015)
10. Clark, D., Tanner-Smith, E., Killingsworth, S.: Digital games, design, and learning: a systematic review and meta-analysis. Rev. Educ. Res. **86**(1), 79–122 (2016)
11. Shaker, N., Yannakakis, G., Togelius., J.: Towards automatic personalized content generation for platform games. In: Proceedings of Artificial Intelligence and Interactive Digital Entertainment (AIIDE 2010). AAAI Press (2010)
12. Spronck, P., Sprinkhuizen-Kuyper, I., Postma, E.: Difficulty scaling of game AI. In: Proceedings of the 5th International Conference on Intelligent Games and Simulation (GAME-ON 2004), pp. 33–37 (2004)
13. Hunicke, R., Chapman, V.: AI for dynamic difficulty adjustment in games. In: Challenges in Game Artificial Intelligence AAAI Workshop, pp. 91–96 (2004)
14. Joy James Prabhu, A.: Improving dynamic difficulty adjustment to enhance player experience in games. In: Das, V.V., Vijaykumar, R. (eds.) ICT 2010. CCIS, vol. 101, pp. 303–306. Springer, Heidelberg (2010). doi:10.1007/978-3-642-15766-0_44
15. Magerko, B., Stensrud, B., Holt, L.: Bringing the schoolhouse inside the box - a tool for engaging, individualized training. In: Proceedings of the 25th Army Science Conference (2006)
16. Togelius, J., De Nardi, R., Lucas, S.M.: Towards automatic personalised content creation in racing games. In: Proceedings of the IEEE Symposium on Computational Intelligence and Games (2007)
17. Yin, H., Luo, L., Cail, W., Ong, Y., Zhong, J.: A Data-driven approach for online adaptation of game difficulty (2015)
18. Zook, A., Riedl, M.O.: A temporal data-driven player model for dynamic difficulty adjustment. In: Proceedings of the AAAI Conference on Artificial Intelligence and Interactive Digital Entertainment (2012)
19. Andrade, G., Ramalho, G., Corruble, V.: Challenge-sensitive action selection: an application to game balancing. In: IEEE/WIC/ACM International Conference on Intelligent Agent Technology, pp. 194–200 (2005)
20. Lach, E.: Evaluation of automatic calibration method for motion tracking using magnetic and inertial sensors. In: Piętka, E., Badura, P., Kawa, J., Wieclawek, W. (eds.) Information Technologies in Medicine. AISC, vol. 472, pp. 337–348. Springer, Cham (2016). doi:10.1007/978-3-319-39904-1_30

Hybrid Initialization in the Process of Evolutionary Learning

Krystian Łapa[1(✉)], Krzysztof Cpałka[1], and Yoichi Hayashi[2]

[1] Institute of Computational Intelligence,
Częstochowa University of Technology, Częstochowa, Poland
{krystian.lapa,krzysztof.cpalka}@iisi.pcz.pl
[2] Department of Computer Science, Meiji University, Tokyo, Japan
hayashiy@cs.meiji.ac.jp

Abstract. Population-based algorithms are an interesting tool for solving optimization problems. Their performance depends not only on their specification but also on methods used for initialization of initial population. In this paper a new hybridization approach of initialization methods is proposed. It is based on classification of initialization methods that allow various combination of the methods from each category. To test the proposed approach typical problems related to population-based algorithms were used.

Keywords: Population-based algorithms · Population initialization · Initialization methods · Hybrid initialization

1 Introduction

The proper initialization of population in evolutionary algorithms helps to find better solutions to problems and decrease the computation time needed for this process [5]. The initialization process becomes even more important for high dimensional optimization problems (where big set of proper values should be found) [27,43]. Moreover, depending on simulation problem under consideration different initialization methods might result in obtaining better solutions [32].

The evolutionary algorithms are part of computational intelligence methods (see e.g. [8,11,18–21,35,38,50–53,59–62,69–75]) in which specified systems (e.g. fuzzy systems [22,31], neural networks [13,64,65], decision trees [16,54,56]) can learn specified task from data (see e.g. [14,23,34,36,49]) or from experimental observation (see e.g. [24,25]). Computational intelligence methods are used, among others, for: data mining (see e.g. [45,55,57]), modelling (see e.g. [3,4,10,15,17]), classification (see e.g. [28,44,66]) or control (see e.g. [9,30,37,46]).

The initialization methods existing in the literature are focused mostly on uniform distribution of generated numbers in search spaces boundaries (ranges of parameters) under consideration and techniques focusing on searching promising areas of search space (areas that are suspected to give promising results).

© Springer International Publishing AG 2017
L. Rutkowski et al. (Eds.): ICAISC 2017, Part I, LNAI 10245, pp. 380–393, 2017.
DOI: 10.1007/978-3-319-59063-9_34

The other types of techniques are related to the population initialization (e.g. Opposite-Based Learning [47] or Adaptive Randomness [40]). These techniques use already evaluated solutions in assistance to create new solutions and they belong to the computational intelligence methods.

In the literature many attempts for classification of the initialization methods can be found. In [26] the following categories were proposed:

- Randomness. This category contains stochastic and deterministic techniques. In the stochastic techniques initialization depends on stochastic initializers (source of randomness). In this subcategory pseudo-random generators and chaos-based generators can be highlighted. In the deterministic techniques initialization depends on deterministic initializers (source of equal coverage). This subcategory contains mostly sequential generators, the purpose of which is to equally distribute numbers in n-dimensional search space.
- Compositionality. This category consists of the steps needed for the initialization. In a single step subcategory initialization process cannot be divided into separate steps (initialization takes place as a single process). The multi-step subcategory, on the other hand, contains methods with multiple initialization steps (hybrid and multi-step techniques). A good example of methods from this subcategory is an Opposite-Based Learning [47], where first half of the population is initialized randomly and the other half is initialized as opposite (in a terms of numbers) to the first half.
- Generality. This category contains generic and application specified techniques. The generic techniques can be applied for any problems without using knowledge about problem under consideration (the simulation problem is treat as the black box). Due to that it is usually not possible to narrow the search space by omission of areas in which no global optimum can be found. The application specified techniques are opposite techniques which use knowledge about simulation problem. Due to that the initial population can be spread in promising areas of the search space.

The initialization techniques related to the population might also be linked with: (a) automatic selection of initial population size and fitting population size during evolution process, (b) automatic reinitialization of a part of the population in the evolution process, (c) methods using multiple initialization methods in a single population. In this paper a new approach to hybridization of initialization methods is proposed. It is based on classification of initialization methods that allow various combination of the methods from each category. The proposed approach has not been discussed in the literature.

This paper structure is as follows: in the Sect. 2 description of the proposed approach can be found, in the Sect. 3 the obtained results are presented and in the Sect. 4 the conclusions are drawn.

2   Description of Proposed Hybridization Approach

In the proposed approach initialization methods (called later standard initialization methods) were divided in a way to make any hybridization between each possible category. Moreover, the proposed approach is based on test in the standard initialization methods and selection of the fittest in order to create hybrid methods. The proposed classification of the initialization methods contains the following categories (see Fig. 1):

– Number generating methods. This category contains methods focused on generating numbers (it is equal to randomness category proposed by [26]). Those methods can be divided into pseudo-random generators, chaos-based generators and sequential-random generators. An example group of pseudo-random generators are Lagged Fibonacci Generators (LFG) [39]. These methods use numeric values from previous steps to generate pseudo-random numbers according to the following equation:

$$x_i = x_{i-p_1} \Diamond x_{i-p_2} \Diamond x_{i-p_3} \Diamond \ldots \Diamond x_{i-p_k} \bmod m, \tag{1}$$

where $\Diamond$ stands for any binary operator (e.g. addition, multiplication, xor), p_i $(i = 1, ..., k)$ stands for generator delays, k stands for number of generator elements, mod stands for modulo. An example of sequential method is Halton Sequence [7], which works according to the following steps: (a) Divide search space into p (any prime number) equal partitions. (b) Select the partitions separating values as numbers for sequences. (c) Divide again each obtained partitions into p equal partitions and select alternately separating values from each partition. (d) In case of reaching last available value from step (c) repeat step (c). The final sequence of numbers can be made by grouping numbers from several generated sequences obtained from steps (a)-(d). Number generating methods considered in the simulations are presented in Table 1 (methods A01-A15).
– Number transformation methods (see e.g. [42]). The idea of this category of methods relays on transformation of generated numbers in a way to cover search space in a specified way. In [42] authors proved that the transformation of numbers allows us to obtain better results. An example of such method is e.g. Center-Based Initialization [48]. The authors calculated that the average distance between the solutions and the global optimum is smaller if generated numbers are closer to the center of the search space. Due to that only C% of the search space is used by the transformation executed as follows:

$$x_i = \frac{1-C}{2} + C \cdot U(0,1) = U\left(\frac{1-C}{2}, \frac{1+C}{2}\right), \tag{2}$$

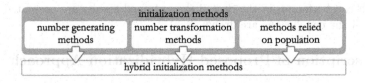

Fig. 1. Proposed initialization methods classification and idea of the hybridization.

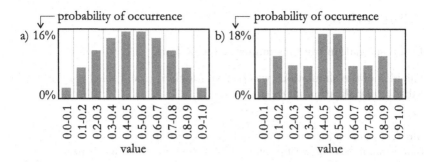

Fig. 2. The distribution of values after using: (a) function (3), (b) function (4).

where $U(a, b)$ stands for any number generator (that returns numbers $\in [a, b]$). Methods from this category might be based on function transformation-for example Hyperbolic Transformation (see Fig. 2.a):

$$x_i = \frac{1}{\pi} \cdot \arccos\left(U\left(-1, 1\right)\right).$$ (3)

Another transformation example is Exponential Transformation (see Fig. 2.b):

$$x_i = \begin{cases} \frac{1}{2} \cdot \exp\left(-\frac{1}{2\sigma^2} \cdot \left(\frac{1}{2} - \alpha\right)^2\right) & \text{for } \alpha < \frac{1}{2} \\ 1 - \frac{1}{2} \cdot \exp\left(-\frac{1}{2\sigma^2} \cdot \left(\alpha - \frac{1}{2}\right)^2\right) & \text{for } \alpha \geq \frac{1}{2}, \end{cases}$$ (4)

where $\alpha = U(0, 1)$. Exponential and Hyperbolic Transformation methods have not been discussed in the literature. Number transformation methods considered in the simulations are presented in Table 1 (methods B01-B10).

- Methods relied on population. This group contains methods that manage initialization of the population. An example method is Simple Sequential Inhibition (SSI, [12]), where newly generated solutions are added into the population only if the smallest distance between rest of solutions is higher than specified threshold value (Δ). In this paper additional modification is proposed (Dynamic SSI), where the parameter Δ is modified by multiplying it by parameter α each time when a solution is rejected. This category of methods also contains Centroidal Voronoi Tessellation methods (CVT, [58]) in which a higher number of solutions are generated (but not evaluated), grouped by any clustering algorithm and centroids of those groups are treat as initial population (see e.g. [58]). Methods relied on population considered in the simulations are presented in Table 1 (methods C01-C10).

Notes on how to combine the initialization methods belonging to the categories listed in Table 1 can be summarized as follows:

- From each category (see Fig. 1) two standard initialization methods will be selected on the basis of the averaged results presented in Table 3.
- Selected methods will be combined (one method from each category) into 8 hybrid methods. Those methods are presented in Table 4.

3   Simulations

Notes on simulation can be summarized as follows:

- The idea of the simulations was to test standard (see Table 1) and hybrid (see Table 4) initialization methods.

Table 1. Selected initialization methods (considered in the simulations). Legend: "#"-method symbol, "src"-literature source, "*"-method proposed in this paper.

#	Method name	Method parameters	Src
A01	Halton sequence	p=(2, 3)	[7]
A02	Halton sequence	p=(2, 3, 5, 7, 11, 13, 17, 19, 23, 29)	[7]
A03	LFG	$\Diamond = +$, $m = 2^{30}$, p= 1 3	[39]
A04	LFG-Knuth-TAOCP	$\Diamond = +$, $m = 2^{30}$, p= 37 100	[39]
A05	LFG-Marsa-LFIB4	$\Diamond = +$, $m = 2^{32}$, p= 55 119 179 256	[39]
A06	LFG-Ziff98	$\Diamond = XOR$, $m = 2^{32}$, p= 471 1586 6988 9689	[39]
A07	Mersenne twister 19937	Default	[6]
A08	Mrg32k3a	Default	[6]
A09	XOR-shift	Default	[41]
A10	PALF	Default	[2]
A11	Sobol sequence	Direction vector: new-joe-kuo-6.21201	[7]
A12	SRNG	Default	[29]
A13	Tent map	$\mu = 1.95$	[27]
A14	Wichmann-Hill's 1982 CMC	Default	[67]
A15	Wichmann-Hill's 2006 CMC	Default	[68]
B01	Box-Muller	Default	[63]
B02	Center-Based	C=70%	[48]
B03	Center-Based	C=80%	[48]
B04	Center-Based	C=90%	[48]
B05	Exponential transformation	default	*
B06	Hyperbolic transformation	default	*
B07	Latin hypercube sampling	$N = 2$, $M = 5$	[33]
B08	Latin hypercube sampling	$N = 3$, $M = 10$	[33]
B09	Latin hypercube sampling	$N = 5$, $M = 12$	[33]
B10	Normal sampling	$\mu = 0, \sigma = 0.2$	[48]
C01	Adaptive randomness	$k = 3$	[40]
C02	Adaptive randomness	$k = 10$	[40]
C03	CVT+k-means	$N = 1000$, $iterations = 50$	[58]
C04	CVT+k-means	$N = 5000$, $iterations = 50$	[58]
C05	Dynamic SSI	$\Delta = 0.32$, $\alpha = 0.999$, distance=Manhattan	*
C06	Dynamic SSI	$\Delta = 0.40$, $\alpha = 0.999$, distance=Euclidean	*
C07	Opposite-based learning	Default	[47]
C08	Quadratic interpolation	Default	[1]
C09	SSI	$\Delta = 0.27$, distance=Manhattan	[12]
C10	SSI	$\Delta = 0.35$, distance=Euclidean	[12]

Table 2. Simulation problems (n stands for dimension). Legend: "#"-method symbol.

#	Problem name	Problem equation	Parameters
F01	Ackley's problem	$-20\exp\left(-\frac{1}{5}\sqrt{\sum\limits_{i=1}^{n} x_i^2/n}\right)+$ $-\exp\left(\sum\limits_{i=1}^{n}\cos\left(2\pi x_i\right)/n\right)+20+e$	$n=30$ $x_i \in [-15,32]$ $\min = \{0,0,...,0\}$
F02	4th De Jong	$\sum\limits_{i=1}^{n} ix_i^4$	$n=30$ $x_i \in [-0.6,1.3]$ $\min = \{0,0,...,0\}$
F03	Levy's function	$\sin^2\left(3\pi x_1\right)+$ $+\sum\limits_{i=1}^{n-1}\left(x_i-1\right)^2\left(1+\sin^2\left(3\pi x_{i+1}\right)\right)+$ $+(x_n-1)^2\left(1+\sin^2\left(2\pi x_n\right)\right)$	$n=30$ $x_i \in [-10,10]$ $\min = \{1,1,...,1\}$
F04	Rastrigin's function	$10n+\sum\limits_{i=1}^{n}\left(x_i^2-10\cos\left(2\pi x_i\right)\right)$	$n=30$ $x_i \in [-2.4,5]$ $\min = \{0,0,...,0\}$
F05	Rosenbrock's valley	$\sum\limits_{i=1}^{n-1}\left[100\left(x_{i+1}-x_i^2\right)^2+\left(1-x_i\right)^2\right]$	$n=30$ $x_i \in [-2,2]$ $\min = \{1,1,...,1\}$
F06	Sphere model	$\sum\limits_{i=1}^{n} x_i^2$	$n=30$ $x_i \in [-2.5,5.1]$ $\min = \{1,1,...,1\}$
F07	Alpine function	$\sum\limits_{i=1}^{n}\left\lvert x_i\sin\left(x_i\right)+0.1x_i\right\rvert$	$n=60$ $x_i \in [-5.5,10]$ $\min = \{0,0,...,0\}$
F08	Inverted cosine wave function	$-\sum\limits_{i=1}^{n-1}\left(\begin{array}{c}\exp\left(\frac{-\left(x_i^2+x_{i+1}^2+0.5x_ix_{i+1}\right)}{8}\right)\cdot\\ \cdot\cos\left(4\sqrt{x_i^2+x_{i+1}^2+0.5x_ix_{i+1}}\right)\end{array}\right)+$ $+n-1$	$n=60$ $x_i \in [-1.2,2.5]$ $\min = \{0,0,...,0\}$
F09	Pathological function	$\sum\limits_{i=1}^{n-1}\left(0.5+\frac{\sin^2\sqrt{100x_i^2+x_{i+1}^2}-0.5}{1+0.001\left(x_i^2-2x_ix_{i+1}+x_{i+1}^2\right)^2}\right)$	$n=60$ $x_i \in [-24,50]$ $\min = \{0,0,...,0\}$
F10	Schwefel's problem 2.21	$\max\left\{\lvert x_i\rvert, 1\le i\le n\right\}$	$n=60$ $x_i \in [-50,99]$ $\min = \{0,0,...,0\}$
F11	Schwefel's problem 2.22	$\sum\limits_{i=1}^{n}\lvert x_i\rvert+\prod\limits_{i=1}^{n}\lvert x_i\rvert$	$n=60$ $x_i \in [-4,10]$ $\min = \{0,0,...,0\}$
F12	Zakharov's function	$\sum\limits_{i=1}^{n} x_i^2+\left(\sum\limits_{i=1}^{n} 0.5ix_i\right)^2+\left(\sum\limits_{i=1}^{n} 0.5ix_i\right)^4$	$n=60$ $x_i \in [-5.5,10]$ $\min = \{0,0,...,0\}$

Table 3. Averaged normalized results obtained for standard initialization methods. Legend: "dev"-standard deviation, "occ"-amount of occurrence of the method in top 3 methods in own group. Top 3 results for each group were marked bold.

#	F01	F02	F03	F04	F05	F06	F07	F08	F09	F10	F11	F12	avg	dev	occ
A01	1.00	**0.27**	0.45	0.46	**0.08**	0.50	1.00	1.00	0.66	1.00	0.57	0.78	0.65	70%	2
A02	0.25	0.29	0.37	0.28	0.24	0.83	0.38	0.73	**0.08**	0.48	0.48	0.58	0.41	51%	1
A03	0.28	0.49	0.40	0.20	**0.10**	0.55	0.29	**0.32**	0.27	0.24	0.08	0.49	0.31	52%	2
A04	0.38	**0.07**	0.65	0.16	0.15	0.46	**0.24**	0.39	0.24	0.14	0.04	**0.23**	**0.26**	48%	3
A05	0.33	0.41	0.49	**0.00**	0.57	0.50	**0.07**	0.38	**0.14**	**0.00**	**0.00**	**0.00**	0.24	36%	6
A06	0.31	0.82	0.18	**0.18**	0.47	**0.33**	0.29	0.37	0.30	**0.13**	0.06	0.41	0.32	53%	3
A07	**0.22**	0.39	**0.32**	0.18	0.23	0.78	0.29	0.47	0.27	0.14	**0.03**	0.39	0.31	49%	3
A08	**0.16**	0.29	0.43	0.16	**0.12**	**0.29**	0.25	0.39	0.21	0.17	0.11	**0.26**	0.24	52%	5
A09	0.93	0.54	0.59	0.17	0.34	**0.15**	0.28	0.41	**0.16**	0.27	0.03	0.53	0.37	48%	2
A10	0.90	0.39	0.60	0.18	0.23	0.81	**0.24**	**0.34**	0.26	0.24	0.04	0.33	0.38	52%	2
A11	0.45	0.58	0.54	**0.16**	0.24	0.56	0.31	0.42	0.27	0.44	0.26	0.92	0.43	62%	1
A12	0.72	0.32	0.42	**0.16**	0.37	0.76	0.28	0.50	0.26	0.18	**0.03**	0.34	0.36	49%	2
A13	0.37	0.44	**0.30**	1.00	0.32	0.50	0.93	**0.19**	0.20	0.84	1.00	1.00	0.59	45%	2
A14	0.55	0.54	0.32	0.20	0.23	0.90	0.27	0.42	0.26	0.15	0.03	0.26	0.34	50%	0
A15	**0.03**	**0.18**	0.53	0.18	0.23	1.00	0.28	0.38	0.26	**0.07**	0.10	0.38	0.30	50%	3
B01	0.28	0.50	0.59	0.31	**0.02**	0.35	0.31	**0.09**	1.00	0.97	0.05	0.17	0.39	49%	2
B02	**0.03**	0.36	0.45	0.26	0.21	0.43	0.25	**0.00**	0.87	0.28	0.08	**0.12**	0.28	52%	3
B03	0.45	**0.22**	**0.32**	0.29	0.34	0.54	**0.19**	0.10	0.46	0.21	0.11	0.22	0.29	52%	3
B04	0.10	0.41	**0.00**	**0.18**	0.39	**0.26**	0.23	0.16	0.33	0.20	0.08	0.29	**0.22**	52%	3
B05	0.18	0.42	**0.30**	**0.21**	0.15	0.44	0.22	0.12	0.37	0.20	0.05	**0.14**	**0.23**	48%	3
B06	0.41	**0.00**	0.43	0.22	**0.06**	**0.14**	0.28	**0.08**	0.51	0.28	0.03	**0.16**	**0.22**	48%	5
B07	**0.06**	0.34	0.39	0.30	1.00	**0.00**	**0.00**	0.52	**0.00**	**0.12**	**0.01**	0.57	0.28	59%	6
B08	**0.00**	**0.34**	0.83	0.34	**0.00**	0.49	0.42	0.53	**0.22**	**0.12**	**0.00**	0.37	0.31	62%	6
B09	0.80	0.35	0.70	0.26	0.43	0.68	**0.13**	0.11	**0.07**	**0.12**	**0.00**	0.18	0.32	57%	4
B10	0.06	0.41	0.45	**0.13**	0.30	0.35	0.25	0.41	0.23	0.16	0.15	0.39	0.27	52%	1
C01	0.44	0.27	1.00	0.21	0.35	**0.17**	0.29	0.54	**0.21**	0.19	0.12	0.32	0.34	54%	2
C02	0.42	0.27	0.39	**0.10**	0.51	**0.30**	0.27	0.54	**0.19**	**0.10**	**0.07**	0.26	**0.29**	50%	5
C03	**0.32**	0.46	0.71	0.29	**0.03**	**0.29**	0.45	**0.10**	0.37	0.56	0.19	**0.08**	0.32	46%	5
C04	0.48	**0.18**	0.48	0.37	**0.10**	0.47	0.52	**0.12**	0.47	0.41	**0.07**	**0.10**	0.32	48%	5
C05	0.61	**0.21**	0.45	0.19	0.33	0.47	0.27	**0.28**	0.23	0.21	0.17	**0.24**	**0.30**	53%	3
C06	**0.33**	0.36	**0.28**	**0.11**	0.27	0.46	**0.22**	0.47	**0.20**	**0.07**	0.12	0.34	**0.27**	50%	6
C07	0.40	**0.05**	0.60	0.24	0.32	0.67	**0.25**	0.38	0.24	0.23	0.09	0.26	0.31	54%	2
C08	0.72	1.00	**0.12**	0.26	**0.16**	0.64	0.43	0.63	0.27	0.52	0.12	0.55	0.45	60%	2
C09	0.76	0.44	0.33	0.19	0.17	0.43	0.27	0.35	0.28	**0.15**	**0.05**	0.33	0.31	49%	2
C10	**0.25**	0.70	**0.26**	**0.08**	0.24	0.82	**0.20**	0.39	0.25	0.17	0.09	0.40	0.32	52%	4

- To test typical problems related to population-based algorithms were used (see Table 2). However, the search space ranges was modified to avoid cases in which a global optimum is placed in the center of the search space (some of the methods are focused on searching center of the search space).
- The genetic algorithm was used as the evolutionary algorithm. It had the following parameters: population size: 100, crossover probability: 0.9, mutation probability: 0.2, mutation range: 0.2, number of iterations (steps): 1000. Simulations for each problem and each function were repeated 200 times and results were averaged.

– The simulation results for standard initialization methods are presented in Table 3. The two top methods from each category were selected as the base for hybrid initialization methods (on the basis of averaged results for all simulation problems F01-F12).
– The simulation results for hybrid initialization methods are shown in Table 5.
– The simulation results from Tables 3 and 5 were normalized in a purpose of clear presentation and proper aggregation of averaged results for all simulation problems. The denormalization values are presented in Table 6.

The conclusions of the simulations:

– It is not ambiguous to choose the best initialization method. This confirms the conclusions of the work [32].
– The best standard initialization methods are: A05, A08, A04, B04, B06, B05, C06, C02, C05 (see Table 3).
– Methods B05, B06, C05 and C06 proposed in the paper cope well in their respective categories.
– The proposed hybrid initialization methods H01, H02, H07 allowed us to obtain better results than standard initialization methods (see Table 5).

Table 4. Hybrid initialization methods.

#	Standard methods used	#	Standard methods used
H01	A05+B04+C02	H05	A08+B04+C02
H02	A05+B04+C06	H06	A08+B04+C06
H03	A05+B06+C02	H07	A08+B06+C02
H04	A05+B06+C06	H08	A08+B06+C06

Table 5. Averaged normalized results obtained for hybrid initialization methods. Legend: "dev"-average standard deviation, "occ"-amount of occurrence of the method in top 3 methods in own group. Top 3 results for each group were marked bold.

#	F01	F02	F03	F04	F05	F06	F07	F08	F09	F10	F11	F12	avg	dev	occ
A05	0.65	0.45	1.00	**0.00**	1.00	0.61	**0.02**	0.73	**0.00**	**0.00**	**0.00**	**0.00**	0.37	36%	6
A08	0.44	0.32	0.88	0.38	0.18	0.40	0.31	0.75	0.15	0.62	0.24	0.44	0.42	52%	0
B04	**0.37**	0.45	**0.00**	0.42	0.69	0.37	0.27	0.35	0.42	0.70	0.18	0.50	0.39	52%	2
B06	0.75	**0.01**	0.89	0.52	**0.08**	**0.25**	0.36	**0.23**	0.85	1.00	0.06	0.27	0.44	48%	4
C02	0.77	0.29	0.80	0.25	0.90	0.41	0.35	1.00	**0.10**	0.35	0.15	0.45	0.48	50%	1
C06	1.00	0.23	0.92	0.44	0.58	0.58	0.33	0.56	0.20	0.73	0.37	0.42	0.53	53%	0
H01	**0.00**	0.55	**0.30**	**0.18**	0.80	0.34	**0.01**	0.44	0.13	**0.01**	0.08	0.34	**0.26**	39%	5
H02	0.96	**0.00**	**0.23**	**0.08**	0.29	0.83	**0.00**	**0.00**	0.27	**0.00**	**0.00**	0.12	**0.23**	35%	8
H03	0.80	0.39	0.56	0.77	0.41	1.00	0.76	0.40	0.55	0.89	1.00	1.00	0.71	99%	0
H04	0.88	0.30	0.77	1.00	**0.09**	0.84	1.00	0.33	0.72	0.69	0.40	0.61	0.64	118%	5
H05	0.69	0.52	0.88	0.39	0.67	0.42	0.23	0.38	0.24	0.56	0.15	0.68	0.48	52%	0
H06	0.73	1.00	0.72	0.49	0.51	**0.13**	0.25	0.30	0.51	0.48	**0.01**	0.52	0.47	47%	2
H07	**0.39**	**0.22**	0.68	0.49	**0.00**	**0.00**	0.28	**0.20**	**0.01**	0.39	0.07	**0.14**	**0.24**	64%	7
H08	0.90	0.63	0.54	0.55	0.22	0.48	0.32	0.24	1.00	0.66	0.19	0.23	0.50	51%	0

Table 6. Denormalization parameters for the results presented in Table 3 and 5.

	Values for Table 3		Values for Table 5	
#	Min	Max	Min	Max
F01	8.4352E-04	9.3715E-04	8.2457E-04	9.0076E-04
F02	2.4189E-17	5.5553E-17	2.3936E-17	5.3209E-17
F03	5.1266E-05	8.8507E-05	5.1266E-05	6.9396E-05
F04	3.0068E+01	6.6583E+01	3.0068E+01	4.5633E+01
F05	2.3167E+01	2.4853E+01	2.3199E+01	2.4123E+01
F06	3.8874E-08	4.6460E-08	3.8013E-08	4.5646E-08
F07	6.8472E-03	1.9001E-02	7.5259E-03	1.5152E-02
F08	1.4555E+01	1.9366E+01	1.4306E+01	1.7161E+01
F09	1.7266E-03	8.9101E-03	2.7394E-03	5.8696E-03
F10	4.8027E-02	6.5068E-01	4.8027E-02	2.1697E-01
F11	3.6007E-02	3.9404E-01	3.6974E-02	1.9817E-01
F12	1.8789E+00	4.9269E+00	1.8789E+00	3.6576E+00

– Some of the proposed hybrid methods yielded worse results than standard initialization methods, therefore, the process of combining methods requires experimental approach (see Table 5).

4 Conclusions

In this paper several hybrid methods for initialization were proposed. Their characteristic is that they combine different (standard) approaches in the field of initialization. They were previously considered generally independently. The effectiveness of combining techniques of initialization has been confirmed in performed simulations. For considered simulation problems the proposed method yielded interesting results.

In further studies in the field of initialization the following actions are taken under consideration: (a) development of methods for automatic configuration of initialization methods (and their parameters), (b) development of solutions for initialization of the population for the evolutionary algorithms using multiple populations.

Acknowledgments. The project was financed by the National Science Centre (Poland) on the basis of the decision number DEC-2012/05/B/ST7/02138.

References

1. Ali, M., Pant, M., Abraham, A.: Unconventional initialization methods for differential evolution. Appl. Math. Comput. **219**(9), 4474–4494 (2013)
2. Aluru, S., Prabhu, G.M., Gustafson, J.: A random number generator for parallel computers. Parallel Comput. **18**, 839–847 (1992)
3. Bartczuk, Ł., Galushkin, A.I.: A new method for generating nonlinear correction models of dynamic objects based on semantic genetic programming. In: Rutkowski, L., Korytkowski, M., Scherer, R., Tadeusiewicz, R., Zadeh, L.A., Zurada, J.M. (eds.) ICAISC 2016. LNCS, vol. 9693, pp. 249–261. Springer, Cham (2016). doi:10.1007/978-3-319-39384-1_22
4. Bartczuk, Ł., Łapa, K., Koprinkova-Hristova, P.: A new method for generating of fuzzy rules for the nonlinear modelling based on semantic genetic programming. In: Rutkowski, L., Korytkowski, M., Scherer, R., Tadeusiewicz, R., Zadeh, L.A., Zurada, J.M. (eds.) ICAISC 2016. LNCS, vol. 9693, pp. 262–278. Springer, Cham (2016). doi:10.1007/978-3-319-39384-1_23
5. Basu, M.: Quasi-oppositional differential evolution for optimal reactive power dispatch. Electr. Power Energy Syst. **78**, 29–40 (2016)
6. Bradley, T., Toit, J.d., Tong, R., Giles, M., Woodhams, P.: Parallelization techniques for random numbers generators. In: GPU Computing Gems Emerald Edition, pp. 231–246 (2011)
7. Cheng, J., Ruzdzel, M.J.: Computational investigation of low-discrepancy sequences in simulation algorithms for Bayesian networks. In: Proceedings of the 16th Annual Conference on Uncertainty in Artificial Intelligence, pp. 72–81 (2000)
8. Cpałka, K.: Design of Interpretable Fuzzy Systems. Springer, Heidelberg (2017)
9. Cpałka, K., Łapa, K., Przybył, A.: A new approach to design of control systems using genetic programming. Inf. Technol. Control **44**(4), 433–442 (2015)
10. Cpałka, K., Rebrova, O., Nowicki, R., Rutkowski, L.: On design of flexible neuro-fuzzy systems for nonlinear modelling. Int. J. Gen. Syst. **42**(6), 706–720 (2013)
11. Cpałka K., Rutkowski, L.: Flexible takagi-Sugeno. Fuzzy systems, neural networks. In: Proceedings of the 2005 IEEE International Joint Conference on IJCNN 2005, vol. 3, pp. 1764–1769 (2005)
12. Diggle, P.J.: Statistical Analysis of Spatial Point Patterns, Mathematics in Biology, 2nd edn. Academic Press, Cambridge (1983)
13. Duda, P., Hayashi, Y., Jaworski, M.: On the strong convergence of the orthogonal series-type kernel regression neural networks in a non-stationary environment. In: Rutkowski, L., Korytkowski, M., Scherer, R., Tadeusiewicz, R., Zadeh, L.A., Zurada, J.M. (eds.) ICAISC 2012. LNCS, vol. 7267, pp. 47–54. Springer, Heidelberg (2012). doi:10.1007/978-3-642-29347-4_6
14. Duda, P., Jaworski, M., Pietruczuk, L.: On pre-processing algorithms for data stream. In: Rutkowski, L., Korytkowski, M., Scherer, R., Tadeusiewicz, R., Zadeh, L.A., Zurada, J.M. (eds.) ICAISC 2012. LNCS, vol. 7268, pp. 56–63. Springer, Heidelberg (2012). doi:10.1007/978-3-642-29350-4_7
15. Dziwiński, P., Avedyan, E.D.: A new approach to nonlinear modeling based on significant operating points detection. In: Rutkowski, L., Korytkowski, M., Scherer, R., Tadeusiewicz, R., Zadeh, L.A., Zurada, J.M. (eds.) ICAISC 2015. LNCS (LNAI), vol. 9120, pp. 364–378. Springer, Cham (2015). doi:10.1007/978-3-319-19369-4_33

16. Dziwiński, P., Avedyan, E.D.: A new approach for using the fuzzy decision trees for the detection of the significant operating points in the nonlinear modeling. In: Rutkowski, L., Korytkowski, M., Scherer, R., Tadeusiewicz, R., Zadeh, L.A., Zurada, J.M. (eds.) ICAISC 2016. LNCS (LNAI), vol. 9693, pp. 279–292. Springer, Cham (2016). doi:10.1007/978-3-319-39384-1_24

17. Dziwiński, P., Avedyan, E.D.: A new method of the intelligent modeling of the nonlinear dynamic objects with fuzzy detection of the operating points. In: Rutkowski, L., Korytkowski, M., Scherer, R., Tadeusiewicz, R., Zadeh, L.A., Zurada, J.M. (eds.) ICAISC 2016. LNCS (LNAI), vol. 9693, pp. 293–305. Springer, Cham (2016). doi:10.1007/978-3-319-39384-1_25

18. Gabryel, M., Cpałka, K., Rutkowski, L.: Evolutionary strategies for learning of neuro-fuzzy systems. In: Proceedings of the I Workshop on Genetic Fuzzy Systems, pp. 119–123. Granada (2005)

19. Gałkowski, T., Rutkowski, L.: Nonparametric fitting of multivariate functions. IEEE Trans. Autom. Control 31(8), 785–787 (1986)

20. Galkowski, T., Starczewski, A., Fu, X.: Improvement of the multiple-view learning based on the self-organizing maps. In: Rutkowski, L., Korytkowski, M., Scherer, R., Tadeusiewicz, R., Zadeh, L.A., Zurada, J.M. (eds.) ICAISC 2015. LNCS (LNAI), vol. 9120, pp. 3–12. Springer, Cham (2015). doi:10.1007/978-3-319-19369-4_1

21. Grycuk, R., Gabryel, M., Korytkowski, M., Scherer, R.: Content-based image indexing by data clustering and inverse document frequency. In: Kozielski, S., Mrozek, D., Kasprowski, P., Małysiak-Mrozek, B., Kostrzewa, D. (eds.) BDAS 2014. CCIS, vol. 424, pp. 374–383. Springer, Cham (2014). doi:10.1007/978-3-319-06932-6_36

22. Hayashi, Y., Tanaka, Y., Takagi, T., Saito, T., Iiduka, H., Kikuchi, H., Bologna, G.: Recursive-rule extraction algorithm with J48graft and applications to generating credit scores. J. Artif. Intell. Soft Comput. Res. 6(1), 35–44 (2016)

23. Jaworski, M., Pietruczuk, L., Duda, P.: On resources optimization in fuzzy clustering of data streams. In: Rutkowski, L., Korytkowski, M., Scherer, R., Tadeusiewicz, R., Zadeh, L.A., Zurada, J.M. (eds.) ICAISC 2012. LNCS, vol. 7268, pp. 92–99. Springer, Heidelberg (2012). doi:10.1007/978-3-642-29350-4_11

24. Jimenez, F., Yoshikawa, T., Furuhashi, T., Kanoh, M.: An emotional expression model for educational-support robots. J. Artif. Intell. Soft Comput. Res. 5(1), 51–57 (2015)

25. Kasthurirathna, D., Piraveenan, M., Uddin, S.: Evolutionary stable strategies in networked games: the influence of topology. J. Artif. Intell. Soft Comput. Res. 5(2), 83–95 (2015)

26. Kazimipour, B., Li, X., Qi, A.K.: A review of population initialization techniques for evolutionary algorithms. In: 2014 IEEE Congress on Evolutionary Computation (CEC), 6–11 July, pp. 2585–2592 (2014)

27. Kazimipour, B., Li, X., Qin, A.K.: Effects of population initialization on differential evolution for large scale optimization. In: 2014 IEEE Congress on Evolutionary Computation (CEC), July 6–11, pp. 2404–2411 (2014)

28. Korytkowski, M., Rutkowski, L., Scherer, R.: Fast image classification, by boosting fuzzy classifiers. Inf. Sci. 327, 175–182 (2016)

29. Knuth, D.E.: The Art of Computer Programming, Volume 2: Seminumerical Algorithms (1997)

30. Lin, C., Dong, F., Hirota, K.: Common driving notification protocol based on classified driving behavior for cooperation intelligent autonomous vehicle using vehicular ad-hoc network technology. J. Artif. Intell. Soft Comput. Res. 5(1), 5–21 (2015)

31. Łapa, K., Przybył, A., Cpałka, K.: A new approach to designing interpretable models of dynamic systems. In: Rutkowski, L., Korytkowski, M., Scherer, R., Tadeusiewicz, R., Zadeh, L.A., Zurada, J.M. (eds.) ICAISC 2013. LNCS, vol. 7895, pp. 523–534. Springer, Heidelberg (2013). doi:10.1007/978-3-642-38610-7_48

32. Maaranen, H., Miettinen, K., Penttinen, A.: On initial populations of a genetic algorithm for continuous optimization problems. J. Global Optim. **37**(3), 405–436 (2007)

33. McKay, M.D., Beckman, R.J., Conover, W.J.: A comparison of three methods for selecting values of input variables in the analysis of output from a computer code. Technometrics **21**(2), 239–245 (1979)

34. Murata, M., Ito, S., Tokuhisa, M., Ma, Q.: Order estimation of japanese paragraphs by supervised machine learning and various textual features. J. Artif. Intell. Soft Comput. Res. **5**(4), 247–255 (2015)

35. Najgebauer, P., Korytkowski, M., Barranco, C.D., Scherer, R.: Novel image descriptor based on color spatial distribution. In: Rutkowski, L., Korytkowski, M., Scherer, R., Tadeusiewicz, R., Zadeh, L.A., Zurada, J.M. (eds.) ICAISC 2016. LNCS (LNAI), vol. 9693, pp. 712–722. Springer, Cham (2016). doi:10.1007/978-3-319-39384-1_63

36. Nikulin, V.: Prediction of the shoppers loyalty with aggregated data streams. J. Artif. Intell. Soft Comput. Res. **6**(2), 69–79 (2016)

37. Nonaka, S., Tsujimura, T., Izumi, K.: Gain design of quasi-continuous exponential stabilizing controller for a nonholonomic mobile robot. J. Artif. Intell. Soft Comput. Res. **6**(3), 189–201 (2016)

38. Nowicki, R., Scherer, R., Rutkowski, L.: A method for learning of hierarchical fuzzy systems. In: Intelligent Technologies-Theory and Applications, pp. 124–129. IOS Press, Amsterdam (2002)

39. Orue, A.B., Montoya, F., Encinas, L.H.: Trifork, a new pseudorandom number generator based on lagged fibonacci maps. J. Comput. Sci. Eng. **1**(10), 46–51 (2010)

40. Pan, W., Li, K., Wang, M., Wang, J., Jiang, B.: Adaptive randomness: a new population initialization method. Math. Probl. Eng. **2014**, 1–14 (2014)

41. Panneton, F., L'Ecuyer, P.: On the xorshift random number generators. ACM Trans. Model. Compu. Simul. **15**(4), 346–361 (2005)

42. Pant, M., Thangaraj, T., Abraham, A.: Particle swarm optimization: performance tuning and empirical analysis. In: Abraham, A., Hassanien, A.-E., Siarry, P., Engelbrecht, A. (eds.) Foundations of Computational Intelligence Volume 3. SCI, vol. 203, pp. 101–128. Springer, Heidelberg (2009)

43. Pant, M., Ali, M., Singh, V.: Differential evolution using quadratic interpolation for initializing the population. In: Advance Computing Conference, pp. 375–380 (2009)

44. Patgiri, C., Sarma, M., Sarma, K.K.: A class of neuro-computational methods for assamese fricative classification. J. Artif. Intell. Soft Comput. Res. **5**(1), 59–70 (2015)

45. Pietruczuk, L., Rutkowski, L., Jaworski, M., Duda, P.: How to adjust an ensemble size in stream data mining? Inf. Sci. **381**, 46–54 (2017)

46. Przybył, A., Jelonkiewicz, J.: Genetic algorithm for observer parameters tuning in sensorless induction motor drive. In: Neural Networks and Soft Computing (Proceedings of the 6th International Conference on Neural Networks and Soft Computing, 2002), pp. 376–381 (2003)

47. Rahnamayan, S., Tizhoosh, H.R., Salama, M.M.A.: A novel population initialization method for accelerating evolutionary algorithms. Comput. Math. Applicat. **53**(10), 1605–1614 (2007)

48. Rahnamayan, S., Wang, G.G.: Toward effective initialization for large-scale search spaces. Wseas Trans. Syst. **8**(3), 355–367 (2009)
49. Rivero, C.R., Pucheta, J., Laboret, S., Sauchelli, V., Patio, D.: Energy associated tuning method for short-term series forecasting by complete and incomplete datasets. J. Artif. Intell. Soft Comput. Res. **7**(1), 5–16 (2017)
50. Rutkowski, L.: Sequential pattern-recognition procedures derived from multiple Fourier-series. Pattern Recogn. Lett. **8**(4), 213–216 (1988)
51. Rutkowski, L., Cpałka, K.: Compromise approach to neuro-fuzzy systems. In: Proceedings of the 2nd Euro-International Symposium on Computation Intelligence. Frontiers in Artificial Intelligence and Applications, vol. 76, pp. 85–90 (2002)
52. Rutkowski, L., Cpałka, K.: Flexible weighted neuro-fuzzy systems. In: Proceedings of the 9th International Conference on Neural Information Processing (ICONIP 2002). Orchid Country Club, Singapore, 18–22 November 2002
53. Rutkowski, L., Cpałka, K.: Neuro-fuzzy systems derived from quasi-triangular norms. In: Proceedings of the IEEE International Conference on Fuzzy Systems, Budapest, 26–29 July, vol. 2, pp. 1031–1036 (2004)
54. Rutkowski, L., Jaworski, M., Pietruczuk, L., Duda, P.: Decision trees for mining data streams based on the gaussian approximation. IEEE Trans. Knowl. Data Eng. **26**(1), 108–119 (2014)
55. Rutkowski, L., Jaworski, M., Pietruczuk, L., Duda, P.: A new method for data stream mining based on the misclassification error. IEEE Trans. Neural Netw. Learn. Syst. **26**, 1048–1059 (2015)
56. Rutkowski, L., Jaworski, M., Pietruczuk, L., Duda, P.: The CART decision tree for mining data streams. Inf. Sci. **266**, 1–15 (2014)
57. Rutkowski, L., Pietruczuk, L., Duda, P., Jaworski, M.: Decision trees for mining data streams based on the mcdiarmid's bound. IEEE Trans. Knowl. Data Eng. **25**(6), 1272–1279 (2013)
58. Saka, Y., Gunzburger, M., Burkardt, J.: Latinized, improved LHS, and CVT point sets in hypercubes. Int. J. Numer. Anal. Model. **4**(3–4), 729–743 (2007)
59. Scherer, R.: Designing boosting ensemble of relational fuzzy systems. Inte. J. Neural Syst. **20**, 381–388 (2010)
60. Scherer, R.: Multiple Fuzzy Classification Systems. Springer, Heidelberg (2012)
61. Scherer, R., Rutkowski, L.: Neuro-fuzzy relational systems. In: 2002 International Conference on Fuzzy Systems and Knowledge Discovery, 18–22 November, Singapore, pp. 44–48 (2002)
62. Scherer, R., Rutkowski, L.: Connectionist fuzzy relational systems. In: Halgamuge, S.K., Wang, L. (eds.) Computational Intelligence for Modelling and Control, Studies in Computational Intelligence, pp. 35–47. Springer, Heidelberg (2005)
63. Shinzato, T.: Box Muller Method (2007)
64. Smoląg, J., Rutkowski, L., Bilski, J.: Systolic array for neural networks. IV KSNiIZ, Zakopane, pp. 487–497 (1999)
65. Szarek, A., Korytkowski, M., Rutkowski, L., Scherer, R., Szyprowski, J.: Application of neural networks in assessing changes around implant after total hip arthroplasty. In: Rutkowski, L., Korytkowski, M., Scherer, R., Tadeusiewicz, R., Zadeh, L.A., Zurada, J.M. (eds.) ICAISC 2012. LNCS, vol. 7268, pp. 335–340. Springer, Heidelberg (2012). doi:10.1007/978-3-642-29350-4_40
66. Villmann, T., Bohnsack, A., Kaden, M.: Can learning vector quantization be an alternative to SVM and deep learning? Recent trends and advanced variants of learning vector quantization for classification learning. J. Artif. Intell. Soft Comput. Res. **7**(1), 65–81 (2017)

67. Wichmann, B.A., Hill, I.D.: Algorithm AS 183: an efficient and portable pseudo-random number generator. Appl. Stat. **31**, 188–190 (1982)
68. Wichmann, B.A., Hill, I.D.: Generating good pseudo-random numbers. Comput. Stat. Data Anal. **51**, 1614–1622 (2006)
69. Zalasiński, M.: New algorithm for on-line signature verification using characteristic global features. Adv. Intell. Syst. Comput. **432**, 137–146 (2016)
70. Zalasiński M., Cpałka, K.: A new method of on-line signature verification using a flexible fuzzy one-class classifier, pp. 38–53. Academic Publishing House EXIT (2011)
71. Zalasiński, M., Cpałka, K.: Novel algorithm for the on-line signature verification using selected discretization points groups. In: Rutkowski, L., Korytkowski, M., Scherer, R., Tadeusiewicz, R., Zadeh, L.A., Zurada, J.M. (eds.) ICAISC 2013. LNCS, vol. 7894, pp. 493–502. Springer, Heidelberg (2013). doi:10.1007/978-3-642-38658-9_44
72. Zalasiński, M., Cpałka, K.: New algorithm for on-line signature verification using characteristic hybrid partitions. In: Wilimowska, Z., Borzemski, L., Grzech, A., Świątek, J. (eds.) Information Systems Architecture and Technology: Proceedings of 36th International Conference on Information Systems Architecture and Technology – ISAT 2015 – Part IV. AISC, vol. 432, pp. 147–157. Springer, Cham (2016). doi:10.1007/978-3-319-28567-2_13
73. Zalasiński, M., Cpałka, K., Hayashi, Y.: A new approach to the dynamic signature verification aimed at minimizing the number of global features. In: Rutkowski, L., Korytkowski, M., Scherer, R., Tadeusiewicz, R., Zadeh, L.A., Zurada, J.M. (eds.) ICAISC 2016. LNCS, vol. 9693, pp. 218–231. Springer, Cham (2016). doi:10.1007/978-3-319-39384-1_20
74. Zalasiński, M., Cpałka, K., Rakus-Andersson, E.: An idea of the dynamic signature verification based on a hybrid approach. In: Rutkowski, L., Korytkowski, M., Scherer, R., Tadeusiewicz, R., Zadeh, L.A., Zurada, J.M. (eds.) ICAISC 2016. LNCS, vol. 9693, pp. 232–246. Springer, Cham (2016). doi:10.1007/978-3-319-39384-1_21
75. Zalasiński, M., Cpałka, K., Rutkowski, L.: A new algorithm for identity verification based on the analysis of a handwritten dynamic signature. Appl. Soft Comput. **43**, 47–56 (2016)

A Tuning of a Fractional Order PID Controller with the Use of Particle Swarm Optimization Method

Krzysztof Oprzędkiewicz[✉] and Klaudia Dziedzic

Department of Automatics and Biomedical Engineering, Faculty of Electrotechnics Automatics,
Informatics and Biomedical Engineering,
AGH University of Science and Technology, Kraków, Poland
kop@agh.edu.pl, klaudia.dziedzic04@gmail.com

Abstract. The paper is devoted to present a new tuning method for Fractional Order PID controller dedicated to temperature control. The proposed method uses Particle Swarm Optimization algorithm. The control plant is described by transfer function with delay. Results of experiments show that the proposed approach assures the good control performance in the sense of known integral cost functions.

Keywords: Digital fractional order PID controller · PSO method · CFE approximation · ORA approximation · Temperature control

1 An Introduction

One of the main areas of application fractional order calculus in automation is a fractional order PID control (FO PID control). Results presented by many Authors [2, 3, 9, 13, 15, 19] show, that FO PID controller is able to assure better control performance than classic integer order PID controller.

An implementation of FO PID controller at each digital platform (PLC, microcontroller) requires us to apply integer order, finite dimensional, discrete approximant. The most known are PSE (Power Series Expansion) and CFE (Continuous Fraction Expansion) (see for example [2, 4, 9, 10, 12, 14]). They allow us to estimate a non integer order element with the use of digital filter. The detailed comparison of both methods was done for example in [4]. In this paper it was marked that the CFE is a more effective method according to the PSE method, but there are some restrictions for the correct choice of the sampling period.

A crucial problem during use a FO PID controller is its correct tune to certain control plant. To do it different methods can be employed (see for example [2, 15]). In the last few years, computational algorithm techniques are being more popularized in tuning controllers due to its combination of low complexity, simplicity and efficiency. Particle Swarm Optimization is one of them, which is based on swarm intelligence. In a PSO system, single solution, called particles, fly in multi-dimensional space adjusting its position according to its own and neighborhoods experience. The performance of each particle is measured by value of fitness function, which is related to the problem. This

© Springer International Publishing AG 2017
L. Rutkowski et al. (Eds.): ICAISC 2017, Part I, LNAI 10245, pp. 394–407, 2017.
DOI: 10.1007/978-3-319-59063-9_35

algorithm is supposed to achieve proper settings of the five parameters k_P, k_I, k_D, α, β to gain optimal control performance for the given plant.

The aim of this paper is to present a method of tuning FO PID controller using Particle Swarm Optimization Algorithm with assumption that tuned FO PID is required to control high order plant, described by transfer function with delay. The presented approach allows us to obtain controller optimal in the sense of known integral cost functions (IAE, ITAE and ISE). The similar approach dedicated to servo drive was presented in [2].

The paper is organized as follows: at the beginning preliminaries are recalled: elementary ideas from Fractional Order Calculus, ORA and CFE approximations and PSO algorithm. Next the considered, closed loop control system with high order plant is remembered. Furthermore the proposed algorithm is presented. Results are illustrated by simulations done with the use of Matlab and associated to real experimental control system.

2 Preliminaries

2.1 Elementary Ideas from Fractional Order Calculus

A non integer order operator is generally described as follows (see for example [6]):

$$
{}_aD_t^\alpha f(t) = \begin{cases} \dfrac{d^\alpha f(t)}{dt^\alpha} & \alpha > 0 \\ 1 & \alpha = 0 \\ \int_a^t f(\tau)(d\tau)^{-\alpha} & \alpha < 0 \end{cases} \tag{1}
$$

In (1) α denotes a non integer order of operation, a, t denote a time interval to calculate operator.

The operator (1) can be described by different definitions given by Grünvald and Letnikov (GL definition), Rieman and Liouville (RL definition) and Caputo (see for example: [6, 18]), but a discretization of the operator (1) can be done the most naturally and easily by the known definition, given by Grünvald and Letnikov:

$$
{}_aD_t^\alpha f(t) = \lim_{h \to 0} h^{-\alpha} \sum_{j=0}^{\left[\frac{t-a}{h}\right]} (-1)^j \binom{\alpha}{j} f(t - jh) \tag{2}
$$

In (2) h is a step of discretization, [...] denotes an integer part, $\binom{\alpha}{j}$ is a generalization of Newton symbol into real numbers:

$$\begin{pmatrix} \propto \\ j \end{pmatrix} = \begin{cases} 1 & for\, j = 0 \\ \dfrac{\propto (\propto -1) \ldots (\propto -j + 1)}{j!} & for\, j > 0 \end{cases} \tag{3}$$

For fractional order systems an idea of transfer function can also be given and its form is analogical as for integer order systems.

Let us consider an elementary non-integer order PID controller described by the transfer function (4):

$$G_c(s) = k_p + \frac{k_I}{s^\alpha} + k_D s^\beta \tag{4}$$

In (4) $\alpha, \beta \in R$ are fractional-orders of integral and derivative actions, k_P, k_I, k_D are coefficients or proportional, integral and derivative actions of a controller.

The analytical form of the step response $y_a(t)$ for the above controller can be calculated with the use of [3, p. 5] and it has the following form:

$$y_a(t) = L^{-1} \left\{ \frac{1}{s} \left(k_p + \frac{k_I}{s^\alpha} + k_D s^\beta \right) \right\} = k_P + k_I \cdot \frac{t^\alpha}{\Gamma(\alpha + 1)} + k_D \frac{t^{-\beta}}{\Gamma(1 - \beta)} \tag{5}$$

where $\Gamma(..)$ denotes complete Gamma function:

$$\Gamma(\alpha) = \int_0^\infty e^{-x} x^{\alpha-1} dx \tag{6}$$

2.2 The Ostaloup Recursive Approximation (ORA)

The method proposed by Oustaloup (see for example [9, 10, 14]) allows us to approximate an elementary non-integer order transfer function s^α in the shape of finite dimensional, integer-order approximation expressed as follows:

$$s^\alpha \cong k_f \prod_{n=1}^{N} \frac{1 + \dfrac{s}{\mu_n}}{1 + \dfrac{s}{\nu_n}} = G_{ORA}(s) \tag{7}$$

In (4) N denotes the order of approximation, μ_n and ν_n are the following coefficients:

$$\begin{aligned} \mu_1 &= \omega_l \sqrt{\eta} \\ \nu_n &= \mu_n \gamma, & n = 1, \ldots, N \\ \mu_{n+1} &= \nu_n \eta, & n = 1, \ldots, N - 1 \end{aligned} \tag{8}$$

where:

$$\gamma = \left(\frac{\omega_h}{\omega_l}\right)^{\frac{\alpha}{N}}$$

$$\eta = \left(\frac{\omega_h}{\omega_l}\right)^{\frac{1-\alpha}{N}} \tag{9}$$

In (9) ω_l and ω_h describe the range of angular frequency, for which parameters are calculated.

A steady-state gain k_f is calculated to assure the convergence the step response of approximation to step response of the real plant in a steady state. Results presented in [12] show, that the order of approximation N assuring its good performance is not too high (really equal 8).

2.3 The CFE Approximation

An alternative form of integer order, finite dimensional approximant for non integer order operator can be obtained with the use of Continuous Fraction Expansion (CFE) approximation (see for example [1, 4, 19]):

$$_0D^\alpha_{kh}f(t) \cong \left(\frac{1+a}{h}\right)^{\pm\alpha} \left(\frac{1-z^{-1}}{1+az^{-1}}\right)^{\pm\alpha} = \left(\frac{1+a}{h}\right)^{\pm\alpha} CFE\{\ldots\} \tag{10}$$

Where:

$$CFE\{\ldots\} = \left(\frac{1-z^{-1}}{1+az^{-1}}\right)^{\pm\alpha} = \frac{v_{\gamma0} + z^{-1}v_{\gamma1}}{w_{\gamma0} + z^{-1}w_{\gamma1}} \tag{11}$$

Coefficients v and w in (11) are equal:

$$v_{\gamma0} = w_{\gamma0} = \frac{2}{a + \gamma + \gamma a - 1}$$

$$v_{\gamma1} = \frac{a - \gamma - \gamma a - 1}{a + \gamma + \gamma a - 1} \tag{12}$$

$$w_{\gamma1} = 1$$

If the Euler approximation is applied, then $a = 0$ and (12) reduces to:

$$v_{\gamma0} = w_{\gamma0} = \frac{2}{\gamma - 1}$$

$$v_{\gamma1} = \frac{-\gamma - 1}{\gamma - 1} \tag{13}$$

$$w_{\gamma1} = 1$$

To calculate the transfer function (10)–(13) the MATLAB function *dfod1* written by I. Petras can be employed (see [16]). This function will also be applied during numerical optimization of FO PID we deal with.

2.4   Particle Swarm Optimization Algorithm

PSO algorithm is a global, stochastic optimization method, which is described in [7, 8]. It is developed from swarm intelligence and it is inspired by behavior of birds flocking and fish schooling. Algorithm relies on the exchange of information between birds in a population. PSO is initialized with random solutions. Each solution regulates its trajectory towards its best result – *pBest* and the best position reached by members of its neighborhood – *gBest*. The fitness function evaluated the performance of particle to decide whether the best fitting solution is reached. In each iteration all solutions update their values by fitness function and according to the following recursive equations (see [7]):

$$p = p + v \tag{14}$$

$$v = c_1 r_1 v + c_2 r_2 (pBest - p) + c_3 r_3 (gBest - p) \tag{15}$$

where the algorithmic parameters are defined as:

- p - position
- v - velocity
- c_1, c_2, c_3 – learning rate
- r_1, r_2, r_3 – a random number from normal distribution $U(0,1)$

In general, the PSO algorithm is depicted by the flowchart shown in Fig. 1.

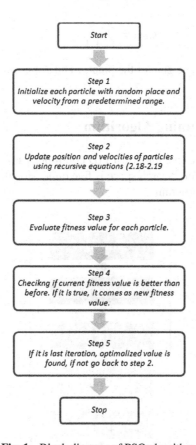

Fig. 1. Block diagram of PSO algorithm.

3   The Closed Loop Control System with High Order Plant and FO PID Controller

The general form of closed loop control system is recalled in Fig. 2. The $G_c(s)$ denotes a transfer function of controller, described by (4). *E, R, U* and *Y* denote Laplace transforms of reference value, error, control signal and process value respectively.

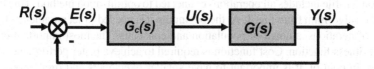

Fig. 2. The closed – loop control system with high order plant.

The considered control plant is described by the following transfer function with delay, describing a huge class of real control plants (for example heat plants):

$$G(s) = \frac{ke^{-\tau s}}{Ts + 1} \qquad (16)$$

Where k is a steady-state gain of the plant, τ and T are dead time and time constant of the plant respectively.

4   The Proposed Tuning Algorithm

The optimal values of the five unknown PID controller parameters k_P, k_I, k_D, α, β are found using PSO algorithm. Search space is five dimensional. Block diagram given in Fig. 3 describes flow of program.

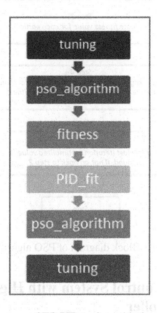

Fig. 3.   The block diagram of FO PID tuning with the use of PSO algorithm.

First tuning m-file is responsible for input parameters. It sends all details about fitness functions, plant parameters and method of approximation to other parts of program. At the end, it creates graphs of results.

Second m-file includes all operations connected to optimization method. This function is based on algorithm from Fig. 1. At the beginning it is needed to set parameters for PSO: number of iterations, amount of population and constants. The most important choice is to select fitness function. Cost function is required to achieve better performance of optimization. In control, it is important to have smaller overshoots and oscillations. In this paper, there is a possibility to choose one from three implemented cost functions:

- Integral of Absolute Error (IAE)

$$I_{IAE} = \int |e(t)| dt \qquad (17)$$

- Integral of Time Absolute Error (ITAE)

$$I_{ITAE} = \int t|e(t)|dt \tag{18}$$

- Integral Square Error (ISE)

$$I_{ISE} = \int e^2(t)dt \tag{19}$$

All particles have their own attributes: position, velocity, *pBest* and *gBest*. These parameters are initialized at the beginning by random numbers in range of their possible values. In each iteration, position and velocity is evaluated by Eqs. (13) and (14). Not only is fitness function changed, but also after comparison with previously best value is found and remembered. Fitness function is calculated in every iteration with the use of Simulink model of closed-loop system shown in Fig. 4. As the output of m-file, we obtain five optimized parameters for PID system.

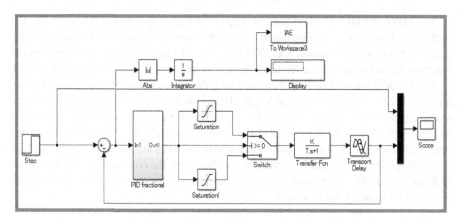

Fig. 4. The Simulink model **PID_fit** calculating cost functions.

Fitness m-file, includes methods of approximation, uses ORA and CFE method. CFE method is based on Ivo Petras files [16, 17]. There are calculated data from actual iteration and later transfer function, which are later send to Simulink model (Fig. 4).

PID_FIT is Simulink model of closed-loop system, containing plant, controller and block calculating a selected cost function.

5   Results of Experiments

Tests were done using approach described in the previous section and Matlab/Simulink model shown in Fig. 4. The model 4 is expected to simulate the real experimental system shown in Fig. 5. It contains the following main parts: experimental heat plant, PLC and PC computers with SCADA and PLC configuration software.

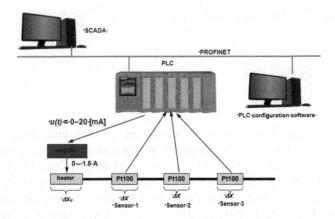

Fig. 5. The experimental control system.

The experimental heat plant has the shape of a thin copper rod 260 [mm] long. For further considerations it will be assumed, that its length is equal 1.0. This implies that localization and length of heater and RTD sensors can be be expressed with respect to 1.0. The rod is heated by an electric heater of the length $\Delta x_0 = 0.14$ localized at one end of rod.

The temperature of the rod is measured with the use of typical Pt100 sensors located in points: 0.29, 0.50 and 0.73 of rod length. The input signal of the system is the standard current signal from range 0–20 [mA]. It is amplified to the range 0–1.5 [A] and next it is the input signal for the heater. Signals from the Pt100 sensors are read directly by analog input module in the PLC. Data from PLC are read by SCADA and stored to further analysis.

For purpose of considerations presented in this paper the control plant was described by transfer function with delay (16) with the parameters (sensor 1 was considered) expressed by transfer function (20):

$$G(s) = \frac{1,4359e^{-3.597s}}{1 + 34,21s} \tag{20}$$

Parameters of the transfer function (20) were identified with the use of Least Square Method by fitting the step response of model (20) to step response of the real plant at the same time grid. From numerical point of view this job consists in minimization of the MSE cost function as a function of model parameters:

$$MSE = \frac{1}{K} \sum_{k=1}^{K} \left(y_e(k) - y_m(k) \right)^2 \tag{21}$$

In (21) $y_e(k)$ and $y_m(k)$ denote the experimental and model step responses respectively, k denotes k-th time moment, equal: kh, where $h = 1$[s] was a sample time during experiments, K is a number of all collected samples, equal 300.

Results of experiments are presented in Figs. 6, 7, 8 and 9. The Fig. 10 presents the step response of the control system with the classic integer order PID controller, tuned by autotune algorithm available at MATLAB.

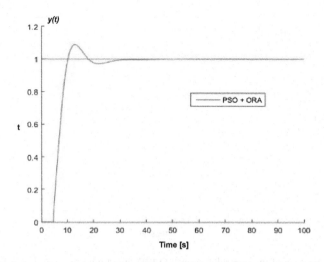

Fig. 6. Step response of the control system. FO PID tuned with the use of cost function IAE. Mp = 8.8731%, IAE = 6.6658, ISE = 5.0194, ITAE = 42.6515, rise time = 4,4968[s], settling time = 25,6976 [s]. Controller parameters: $\alpha = -0.418$, $\beta = 0.357$, $k_p = 3.7050$, $k_I = 0.0538$, $k_D = 0.2593$.

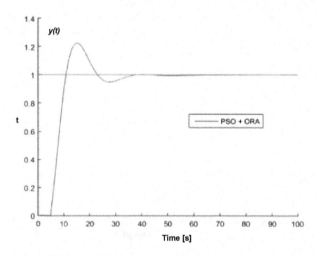

Fig. 7. Step response of the control system. FO PID tuned with the use of cost function ITAE. Mp = 22.4582%, IAE = 8.8608, ISE = 5.9727, ITAE = 87.4357, settling time = 33.5506 [s], rise time = 4.8181[s]. Controller parameters: $\alpha = -0.3418$, $\beta = 0.257$, $k_p = 5.0050$, $k_I = 0.0238$, $k_D = 0.593$.

404   K. Oprzędkiewicz and K. Dziedzic

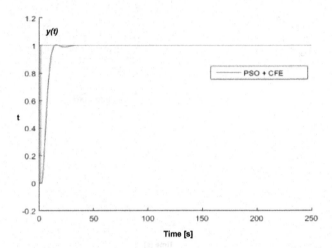

Fig. 8. Step response of the control system. FO PID tuned with the use of cost function IAE. Mp = 3.2651%, IAE = 6.6452, ISE = 7.1111, ITAE = 40.5128, settling time = 20.4345[s], rise time = 7.2748 [s]. Controller parameters: $\alpha = -0.0982$, $\beta = 0.3757$, $k_p = 0.4562$, $k_I = 0.1137$, $k_D = 0.8932$.

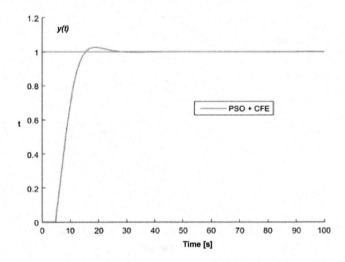

Fig. 9. Step response of the control system. FO PID tuned with the use of cost function ITAE. Mp = 2.7711%, IAE = 7.9193, ISE = 6.1625, ITAE = 47.4985 settling time = 21.7375 [s], rise time = 8.1618 [s]. Controller parameters: $\alpha = -0.1282$, $\beta = 0.057$, $k_p = 0.5350$, $k_I = 0.1838$, $k_D = 0.1293$.

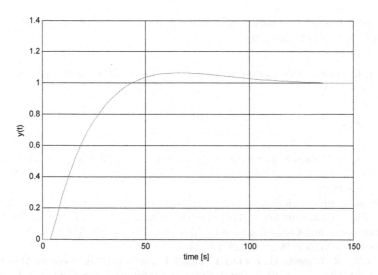

Fig. 10. Step response of the control system with the "classic" Integer Order PID controller tuned with the use of autotune function available at MATLAB. Mp = 6.6259%, IAE = 52.1265, ISE = 39.4198, ITAE = 100.925, settling time = 109.0420 [s], rise time = 27.2994 [s]. Controller parameters: $k_p = 0.6448$, $k_I = 0.3071$, $k_D = 0.6567$.

From Figs. 6, 7, 8, 9 and 10 it can be concluded at once that the proposed, Fractional Order PID controller tuned by PSO method assures a better control performance in the sense all proposed cost functions that classic integer order PID controller.

Next, the use of the discrete-time CFE approximation to construct the FO PID controller gives the better control performance than applying the continuous –time ORA approximation.

It should be also mentioned that duration of calculations necessary to correct tuning the considered FO PID controller was significantly longer that the tuning done with the use of known methods. This was observed during experiments: circa 50 iterations were evaluated about 90[s].

6   Final Conclusions

The final conclusions from the paper can be formulated as follows:

- The proposed PSO based method allows us to effectively tune the considered FO PID controller dedicated to control of high order industrial plant, described by transfer function with delay, independently on approximation method used to construct the FO controller.
- A main disadvantage of the proposed method is associated to the fact that results are hard to repeat or verify, because they are determined by random starting points.
- Further investigations of the presented problems will cover tests done with the use of a real, experimental, PLC control system presented in the previous section.

- An another interesting idea is to implement the proposed tuning algorithm directly at PLC or SCADA platform.

Acknowledgements. This paper was supported by the AGH (Poland) – project no 11.11. 120.815.

References

1. Al Aloui, M.A.: Dicretization methods of fractional parallel PID controllers. In: Proceedings of 6th IEEE International Conference on Electronics, Circuits and Systems, ICECS 2009, pp. 327–330 (2009)
2. Arun, M.K., Biju, U., Rajagopal, N.N., Bagyaveereswaran, V.: Optimal tuning of fractional-order PID controller. In: Suresh, L.P., Panigrahi, B.K. (eds.) Proceedings of the International Conference on Soft Computing Systems. AISC, vol. 397, pp. 401–408. Springer, New Delhi (2016). doi:10.1007/978-81-322-2671-0_38
3. Caponetto, R., Dongola, G., Fortuna, L., Petras, I.: Fractional Order Systems: Modeling and Control Applications. World Scientific Series on Nonlinear Science, Series A, vol. 72. World Scientific Publishing, Singapore (2010)
4. Dorcak, L., Petras, I., Terpak, J., Zbrojovan, M.: Comparison of the methods for discrete approximation of the fractional-order operator. In: Proceedings of the ICCC 2003 Conference, High Tatras, Slovak Republic, pp. 851–856, 26–29 May 2003. arXiv:math/0306017v1
5. Dziedzic, K.: The tuning of fractional-order PID controller to high-order, uncertain parameter plant Master thesis prepared at AGH University in 2016 under supervision of K. Oprzędkiewicz (2016)
6. Kaczorek, T.: Selected Problems in Fractional Systems Theory. Springer, Heidelberg (2011)
7. Kennedy, J., Eberhardt, R.: Particle Swarm Optimization, pp. 1942–1948. IEEE (1995). 0-7803-2768-3/95/$4.00 0
8. Kennedy, J.: Particle swarm optimization. In: Sammut, C., Webb, G.I. (eds.) Encyclopedia of Machine Learning, pp. 760–766. Springer, Heidelberg (2010)
9. Merikh-Bayat, F.: Rules for selecting the parameters of oustaloup recursive approximation for the simulation of linear feedback systems containing $PI^\lambda D^\mu$ controller. Commun. Nonlinear Sci. Numer. Simulat. **17**, 1852–1861 (2012)
10. Merikh-Bayat, F., Mirebahrimi, M., Khalili, M.R.: Discrete-time fractional-order PID controller: Definition, tuning, digital realization and some applications. Int. J. Control Autom. Syst. **13**(1), 81–90 (2015)
11. Oprzędkiewicz, K.: A Strejc model-based, semi- fractional (SSF) transfer function model. Automatyka/Automatics **16**(2), 145–154 (2012). AGH UST 2012, http://journals.bg.agh.edu.pl/AUTOMAT/2012.16.2/automat.2012.16.2.145.pdf
12. Oprzędkiewicz, K., Mitkowski, W., Gawin, E.: An estimation of accuracy of oustaloup approximation. In: Szewczyk, R., Zieliński, C., Kaliczyńska, M. (eds.) Challenges in Automation, Robotics and Measurement Techniques. AISC, vol. 440, pp. 299–307. Springer, Cham (2016). doi:10.1007/978-3-319-29357-8_27
13. Ostalczyk, P.: Discrete Fractional Calculus: Applications in Control and Image Processing. Series in Computer Vision, vol. 4. World Scientific Publishing, Singapore (2016)
14. Oustaloup, A., Levron, F., Mathieu, B., Nanot, F.M.: Frequency-band complex noninteger differentiator: Characterization and synthesis. IEEE Trans. Circuits Syst. I: Fundam. Theory Appl. I **47**(1), 25–39 (2000)

15. Padula, F., Visioli, A.: Tuning rules for optimal PID and fractional-order PID controllers. J. Process Control **21**(1), 69–81 (2011)
16. Petras, I: http://people.tuke.sk/igor.podlubny/USU/matlab/petras/dfod1.m
17. Petras, I: http://people.tuke.sk/igor.podlubny/USU/matlab/petras/dfod2.m
18. Podlubny, I.: Fractional Differential Equations. Academic Press, San Diego (1999)
19. Vinagre, B.M., Podlubny, I., Hernandez, A., Feliu, V.: Some approximations of fractional order operators used in control theory and applications. In: Fractional Calculus and Applied Analysis, vol. 3(3), pp. 231–248 (2000)

Controlling Population Size in Differential Evolution by Diversity Mechanism

Radka Poláková[(✉)]

Centre of Excellence IT4 Innovations, Institute for Research
and Applications of Fuzzy Modeling, Division of University of Ostrava,
30. dubna Street 22, 701 03 Ostrava, Czech Republic
radka.polakova@osu.cz

Abstract. A new mechanism for resizing population in differential evolution algorithm based on diversity of population has been proposed and compared with linear reduction of population size published in 2014. Seven modifications of differential evolution algorithm were chosen for this comparison. Experiments are done on CEC2014 benchmark set. The new diversity-based resizing mechanism frequently improves results of tested variants of differential evolution algorithm more than linear reduction of population size, especially in larger dimensions.

Keywords: Global optimization · Differential evolution · Population size · Population diversity · Experimental comparison

1 Introduction

The differential evolution algorithm is one of the most frequently used evolutionary algorithms for global optimization. It is studied and improved by many groups of researchers. Effectiveness of the algorithm depends on the setting of its control parameters. Many adaptive versions of the algorithm have been proposed because of the fact. An adaptive version called Success-history based differential evolution (SHADE) [14] proposed by Tanabe and Fukunaga is one of them. The authors improved the algorithm by linear reduction of population size. Resulting algorithm called LSHADE [15] was the best DE-version in optimization competition on CEC2014 [6,7]. The algorithm is more efficient in comparison with other DE-versions also because it saves function evaluations.

The idea of saving the evaluations is useful but sometimes adding a point into population may improve the solution, e.g. in case of the premature convergency. If the population diversity (dispersion of population points in search space) is large enough, the algorithm can create many different points as a trial point and it has bigger chance to find the area of global optimum. Thus, if the diversity is large we can save the function evaluations by excluding a point from population. On the other hand, if the diversity is too small (the whole population is located in area of a local minimum), we can try to improve the search by adding a point. The added point should increase the population diversity. These ideas are basic

© Springer International Publishing AG 2017
L. Rutkowski et al. (Eds.): ICAISC 2017, Part I, LNAI 10245, pp. 408–417, 2017.
DOI: 10.1007/978-3-319-59063-9_36

concepts for our mechanism which was introduced in [12]. In this paper, we compare this diversity-based adaptive mechanism with linear reduction of population size proposed in [15]. It is accomplished by implementing both mechanisms in seven effective variants of differential evolution. The comparison is carried out on benchmark set developed for CEC2014 competition [6].

The rest of this paper is organized as follows. In the following section, the differential evolution algorithm is briefly described. In Sects. 3 and 4, linear reduction of population size and our mechanism of population resizing based on population diversity are explained. In Sect. 5, principal ideas of six adaptive versions used in experimental comparison are described. In Sect. 6, experimental settings and summary of the comparison are presented. The last section comes up with conclusions.

2   Differential Evolution

The differential evolution algorithm (DE) is powerful and easy-implemented population based algorithm for global optimization [13]. DE works with population P of NP points. NP is the size of population. The population evolves during the search and the points of population are considered as candidates of solution.

The population is initialized randomly in the search space S, $S = \prod_{j=1}^{D}[a_j, b_j]$, $a_j < b_j$, $j = 1, 2, \ldots, D$. D is dimension of problem. Then a loop repeats until the stopping condition is satisfied. In a body of the cycle, a new generation Q of population P is created by the following way. A new trial point $\boldsymbol{y}_i$ is produced by mutation and crossover operations for each point $\boldsymbol{x}_i \in P$, $i \in \{1, 2, \ldots, NP\}$. If $f(\boldsymbol{y}_i) \leq f(\boldsymbol{x}_i)$, where f is objective function of optimized problem, the trial point $\boldsymbol{y}_i$ is inserted into the new generation Q, otherwise the point $\boldsymbol{x}_i$ enters into Q. After completing the process, Q becomes the current generation of population P. Each trial vector $\boldsymbol{y}_i$ is generated by mutation and crossover operations. As a mutant vector $\boldsymbol{v}_i$ is created by a mutation, then the trial point is computed based on two points, $\boldsymbol{x}_i$ and the mutant $\boldsymbol{v}_i$, by crossover. There are many different versions of mutation operation, e.g. *best/1, randrl/1* [4], *best/2, rand/2,* and *rand/1* [13]. Binomial or exponential crossover can be used. F is parameter of mutation and it affects how far from so called based point of mutation operation the mutant is located. CR is parameter of crossover and it biases the rate of coordinates, which are taken into trial point $\boldsymbol{y}_i$ from the mutant $\boldsymbol{v}_i$.

A combination of a mutation and a crossover is called DE-strategy. It is often denoted by abbreviation $DE/m/n/c$, where m stands for a kind of mutation, n denotes number of differences used in the mutation, and c is employed type of crossover. DE-strategy together with values of F and CR is sometimes called DE-setting.

3   Linear Reduction of Population Size

Linear reduction of population size was proposed in [15]. With this mechanism, DE algorithm starts with a large number of points inside the population in order to emphasize the exploration, $NP = NP_{init}$. It stops with $NP = NP_{min}$ equal to relatively small number. The value of NP is changed linearly according to (1) during the search. The new size of population is computed at the end of each generation G. If the NP_{G+1} is less than current NP, only NP_{G+1} better points from current population form new generation of population. Recommended values are $NP_{init} = 18 * D$ and $NP_{min} = 4$ [15].

$$NP_{G+1} = round\left[NP_{init} - \frac{FES}{MaxFES}\left(NP_{init} - NP_{min}\right)\right]. \qquad (1)$$

The mechanism inspired us to propose our new resizing mechanism.

4 Diversity-Based Resizing Mechanism

There are many adaptive versions of DE algorithm proposed since the algorithm has been introduced [13]. However, there are still optimization problems where even the most effective DE-versions fail in their successful solving due to premature convergency or stagnation [5]. In the case of premature convergency, all points of population are situated inside an area of a local minimum. The stagnation means that no new and better point can be found in the search process. To solve these issues we proposed a new mechanism of resizing the population [12]. Basic ideas are to save the function evaluations when the diversity of population is sufficient and to prevent premature convergency by adding new points into population. The diversity of population is measured by DI defined in (2), where $\bar{x}_j$ is the mean of jth coordinate of the points in the current generation of population (3). DI is square root of average square of distance between a population point and centroid $\bar{x} = (\bar{x}_1, \bar{x}_2, \ldots, \bar{x}_D)$. The diversity in initial population is labelled as DI_{init} and is used as reference value in definition of relative measure RD of the diversity in the current generation of population (4),

$$DI = \sqrt{\frac{1}{NP}\sum_{i=1}^{NP}\sum_{j=1}^{D}(x_{ij} - \bar{x}_j)^2}, \qquad (2)$$

$$\bar{x}_j = \frac{1}{NP}\sum_{i=1}^{NP}x_{ij}, \qquad (3)$$

$$RD = \frac{DI}{DI_{init}}. \qquad (4)$$

Relative number of actually depleted function evaluations is defined by (5).

$$RFES = \frac{FES}{MaxFES}, \qquad (5)$$

where FES is current count of function evaluations and $MaxFES$ is count of function evaluations allowed in the search. The size of population is changed depending on actual relative diversity. We suggested to keep the relative diversity near its required value rRD linearly decreasing from the value 1 at the beginning of the search to value near 0 at the end of the search. We proposed to keep the relative diversity RD somewhere around the required relative diversity rRD as you can see in Fig. 1. It means to change the population size only when the actual relative diversity is less than $0.9 \times rRD$ or larger than $1.1 \times rRD$. It is suggested for first nine tenths of search. In the last tenth, the zero value is strictly required.

RD is computed after each generation. The size of population is increased by 1 and a random point from search space is added when the RD is less than $0.9 \times rRD$. The population size is decreased by 1 and the worst point is excluded when the RD is larger than $1.1 \times rRD$. There are minimal and maximal values of population size used in the approach, $NP_{min} = 8$ and $NP_{max} = 200$. The search process starts with $NP_{init} = 50$.

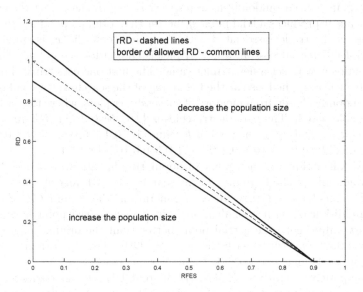

Fig. 1. Required relative diversity rRD and the area of allowed relative diversity

5   Algorithms in Experiments

Seven DE algorithms were chosen for experiments in this paper, original version of differential evolution algorithm, three adaptive versions which are accord- ing to [3] included to the state-of-the-art algorithms, namely CoDE, EPSDE, and jDE. Fifth algorithm placed into our experiments is *b6e6rl* algorithm. It is an effective version of competitive adaptive differential evolution algorithm proposed by Tvrdík [17]. The algorithm and its versions participated in compe- titions organized on congresses on evolutionary computation (CEC) [2,11,18]. Next algorithm employed in the experiment of this paper is SHADE [14], the

412   R. Poláková

first ranked DE-version on CEC2013 [8,9]. Its modification with linear reduction of population size was successful. See CEC2014 [6,7]. Another algorithm included in our tests is recently proposed IDE algorithm [16]. Original version of DE algorithm is used with the most frequent $DE/rand/1/bin$ strategy, input parameters are set as $F = 0.8$ and $CR = 0.5$.

CoDE [19] creates three different trial points by employing three different DE-strategies and the best one is then final trial point. The authors picked out three DE-strategies, namely $DE/rand/1/bin$, $DE/rand/2/bin$, and $DE/current-to-rand/1/-$. A couple of (F, CR) parameters is randomly chosen from three possible pairs $(1, 0.1)$, $(1, 0.9)$, and $(0.8, 0.2)$ for each of the three DE-strategies separately.

EPSDE [10] algorithm uses a pool of DE-strategies, a pool of F-values, and a pool of CR-values. Each point of the population has an associated triplet of parameters initialized randomly from respective pool. If the triplet is successful, which means it produces a trial point y_i better than original point x_i, the triplet continues as the triplet associated with the point (now the x_i is equal to y_i). If the triplet is not successful, a new triplet is chosen for the old x_i. There are two scenarios how to set a new triplet value. The first one uses triplet initialized randomly as it is applied also at the beginning of the search, the second one uses triplet randomly chosen from memory of successful triplets, where they are stored during whole search. The pool of strategies is $\{DE/best/2/bin,\ DE/rand/1/bin,\ DE/current-to-rand/1/-\}$, the pool of F values is $\{0.4, 0.5, 0.6, 0.7, 0.8, 0.9\}$, and the pool of CR values is $\{0.1, 0.2, 0.3, 0.4, 0.5, 0.6, 0.7, 0.8, 0.9\}$.

In jDE algorithm [1], each point of population has its own values of F and CR set randomly at the beginning of the search. The jDE algorithm works with only $DE/rand/1/bin$ strategy. In each generation, the F and CR of each x_i can be re-initialized with a small probability before a trial point is computed. If the new values generate a trial point better than the original one, the new pair of parameters survives with the point x_i. Otherwise, the original couple of parameters is the one associated with the current point.

In competitive adaptive DE [17], K different DE-settings compete. Twelve DE-settings are employed in $b6e6rl$ version [18] used in this experimental comparison. Each of them uses $randrl/1$ mutation [4]. In six of the twelve DE-settings the binomial crossover is used whereas the exponential crossover is applied in the remaining six DE strategies. Two values of F used in this algorithm are 0.5, 0.8. Three different values of CR are employed for each of both crossovers. There are 6 different combinations used with both DE-strategies $(DE/randrl/1/bin,\ DE/randrl/1/exp)$. All twelve DE-settings start with the same probability, $1/K = 1/12$. Then the probabilities are changed according to the success of DE-settings. If a DE-setting has a very small probability, all probabilities are reset to the starting value.

SHADE algorithm [14] is improved version of JADE algorithm [20], with $current-to-pbest/1$ mutation and archive A, into which the old members of population are stored if they are rewritten by their better trial point. Adaptation of F and CR in SHADE is also similar to adaptation of the parameters in JADE.

All successful values of F and also CR are stored during a generation and in following generation, both parameters (F and CR for each x_i) are generated randomly based on respective stored successful values. F is generated from the Cauchy distribution and the first parameter of this distribution is the Lehmer mean of all stored successful values of F, the second distribution parameter is 0.1. CR is generated from the Gauss distribution and the first parameter is arithmetic mean of all stored successful values of CR from last generation. The second parameter of the Gauss distribution is fixed to 0.1. In SHADE algorithm, the adaptation works additionally with two circle memories M_F and M_{CR} of length H, where the first parameters for both distributions computed from successful values of respective parameter in last H generations are stored. When a trial point is needed, algorithm generates index $r \in \{1, \ldots, H\}$ randomly, then F and CR are generated randomly, $F \sim Cauchy(M_{F_r}, 0.1)$, $CR \sim N(M_{CR_r}, 0.1)$. All members of M_F and M_{CR} start with value 0.5.

In IDE algorithm [16], the population is divided into superior and inferior sets (S and I). The proportions of these sets are changing during the search. At the beginning, the size of S is small compared to the size of the I set and it holds almost to the end of the search. Then the superior set grows quickly and at the end of the search, the superior set S is equal to the whole population P. A new variant of mutation operation is used in IDE. The search process is divided into two stages. In the first one, the proposed mutation starts to search mutant from the randomly chosen point (from population), in the second one, it starts to compute the mutant point from the best point of population. Moreover, the proposed mutation works differently for a point from S and for a point from I. Parameters F and CR are set to different values for each point of population dependently to its rank. The smallest values for the best point and the biggest values for the worst point of population. Moreover, there is a point in employed mutation for which each coordinate is reset with small probability to a new value from the range of respective dimension of search space. This prevents the premature convergence of IDE algorithm.

6   Experiments and Results

Seven algorithms mentioned in Sects. 2 and 5 were chosen for the comparison of two approaches to the adaptation of population size described in Sects. 3 and 4. The set of 30 optimization problems with different complexity developed for CEC2014 competition [6] was selected as a benchmark set. Experiments were carried out for all four levels of dimension $D = 10, 30, 50$, and 100 required by [6]. Parameter *MaxFES* for stopping condition was set $MaxFES = D \times 10^4$ for all the test problems. We implemented both investigated mechanisms into all seven algorithms and got fourteen new algorithms. All tested algorithms were implemented in Matlab 2015b and all experiments were carried out in the environment on standard PC with Windows 7. The algorithms, in which the mechanism of resizing the population according to Sect. 4 is implemented, are labelled with suffix "d" ("d" means diversity). The algorithms, in which the mechanism of linear reduction of population size is implemented, are labelled with prefix "L". Thus, we have twenty one algorithms, 7 original algorithms, the ones with implemented resizing population diversity-based mechanism DEd, CoDEd, EPS-DEd, jDEd, *b6e6rld*, SHADEd, and IDEd, and the algorithms LDE, LCoDE, LEPSDE, LjDE, *Lb6e6rl*, LSHADE, and LIDE in which the linear reduction of population size mechanism is implemented. The names of some algorithms are too long for tables, so all SHADE algorithms are labelled as SHA, SHAd, LSHA and all EPSDE algorithms are labelled as EPS, EPSd, and LEPS there. Parameters of all tested algorithms are set according to the recommendations of their authors except for the NP_{min} parameter used in LCoDE, LEPSDE, and LIDE algorithms. NP_{min} is set to 6 for them. For other tested L-versions, NP_{min} is set to value 4. NP_{init} is set to value $18 \times D$ for all L-versions in tests. Original algorithms are run with $NP = 50$. For each algorithm, problem, and dimension, we carried out 25 independent runs. A problem from benchmark set has known solution x^*. The solution of a run is x_{min}. The result of run is the error $f(x_{min}) - f(x^*)$.

We compared results of original version with d-version and also the original version with L-version for each of 30 problems at all levels of dimension for all seven DEs. Then we compared results of L-version and d-version for each of 30 problems in each dimension for all seven algorithms. The differences in efficiency of the algorithms were assessed by Wilcoxon two-sample rank-sum test at 0.05 significance level.

Summarized results of statistical comparison of original DE variants with the corresponding DE variants using a population-size adaptation are shown in the Table 1. The numbers labelled with "$-$" are counts of test problems, where the population-size adaptation version is statistically worse than original algorithm. The numbers labelled with "$+$" are the counts of test problems where the population-size adaptation version is statistically better than original one. In the rows labelled with "$\approx$", you see counts of test problems where both compared algorithms are not significantly different. The results are depicted for each dimension separately and also for all dimensions together. The counts where the population-size adaptive approach wins are underlined. The counts of original algorithm wins are printed in boldface.

We can see that improvement caused by implementation of new resizing mechanism outbalances the impairment for all tested algorithms. The biggest improvement occurred for classical DE, in 90 of 120 cases. The least impact arose for IDE algorithm. The implementation of linear reduction of population size more likely deteriorates the results of algorithms in experiment, with the exception of SHADE and jDE. The diversity-based mechanism improves the efficiency of tested DE versions in more problems than linear reduction of population size. The count of problems with decreased efficiency is less than for linear reduction for all tested DE variants. Each original algorithm has larger number of improvements with new resizing approach than with linear reduction of population size in dimensions $D = 30, 50$, and 100. The number of deteriorations with diversity-based resizing approach is also less than the number of deteriorations with linear reduction of population size in all the dimensions except $D = 10$.

Table 1. Comparison of original algorithms with population-size adaptive versions

	DE vs. DEd					DE vs. LDE					CoDE vs. CoDEd					CoDE vs. LCoDE				
D	10	30	50	100	all	10	30	50	100	all	10	30	50	100	all	10	30	50	100	all
−	7	1	0	0	8	0	**19**	**22**	**15**	**56**	14	8	3	4	29	0	**27**	**28**	**29**	**84**
+	17	23	25	25	90	16	0	1	8	25	11	17	21	22	71	18	0	0	0	18
≈	6	6	5	5	22	14	11	7	7	39	5	5	6	4	20	12	3	2	1	18

	EPS vs. EPSd					EPS vs. LEPS					jDE vs. jDEd					jDE vs. LjDE				
D	10	30	50	100	all	10	30	50	100	all	10	30	50	100	all	10	30	50	100	all
−	**13**	1	2	4	20	2	**17**	**14**	**13**	**46**	**10**	2	5	5	22	1	10	**12**	**19**	42
+	9	18	22	19	68	3	4	12	12	31	8	17	18	17	60	17	10	11	9	47
≈	8	11	6	7	32	25	9	4	5	43	12	11	7	8	38	12	10	7	2	31

	b6e6rl vs. b6e6rld					b6e6rl vs. Lb6e6rl					SHA vs. SHAd					SHA vs. LSHA				
D	10	30	50	100	all	10	30	50	100	all	10	30	50	100	all	10	30	50	100	all
−	**10**	3	4	7	24	0	**12**	**14**	**20**	**46**	8	0	2	2	12	2	4	5	8	19
+	7	16	19	15	57	11	8	10	6	35	9	24	22	18	73	10	19	18	16	63
≈	13	11	7	8	39	19	10	6	4	39	13	6	6	10	35	18	7	7	6	38

	IDE vs. IDEd					IDE vs. LIDE					orig. vs. d-ver.					orig. vs. L-ver.				
D	10	30	50	100	all	10	30	50	100	all	10	30	50	100	all	10	30	50	100	all
−	**14**	5	4	7	30	**16**	**25**	**17**	**23**	**81**	**76**	20	20	27	145	21	**114**	**112**	**127**	**374**
+	4	8	15	12	39	1	1	3	4	9	65	123	142	128	458	76	42	55	55	228
≈	12	17	11	11	51	13	4	10	3	30	69	67	48	53	237	113	54	43	28	238

Table 2. Comparison of L-versions with d-versions

	LDE vs. DEd					LCoDE vs. CoDEd					LEPS vs. EPSd					LjDE vs. jDEd				
D	10	30	50	100	all	10	30	50	100	all	10	30	50	100	all	10	30	50	100	all
−	10	0	0	1	11	**18**	3	0	1	22	**12**	0	2	7	21	**19**	5	3	4	31
+	12	29	28	26	95	11	24	30	29	94	7	23	20	20	70	3	8	16	20	47
≈	8	1	2	3	14	1	3	0	0	4	11	7	8	3	29	8	17	11	6	42

	Lb6e6rl vs. b6e6rld					LSHA vs. SHAd					LIDE vs. IDEd					L-ver. vs. d-ver.				
D	10	30	50	100	all	10	30	50	100	all	10	30	50	100	all	10	30	50	100	all
−	**16**	2	2	3	23	**7**	3	9	10	29	5	0	0	0	5	**87**	13	16	26	142
+	1	16	22	25	64	5	10	12	11	38	9	27	26	28	90	48	137	154	159	498
≈	13	12	6	2	33	18	17	9	9	53	16	3	4	2	25	75	60	40	25	200

The summarized results of L-versions with d-versions comparison are depicted in Table 2. The rows are labelled in the same way as in Table 1. The numbers labelled with "+" are the counts of problems where the d-version of an algorithm is statistically better than its L-version. Counts where the diversity-based approach wins are underlined. Bolded numbers mean that the L-version wins. The d-version has large number of wins for all tested algorithms, when we discuss all dimensions together. The implementation of the new resizing mechanism wins for dimensions 30, 50, and 100 for all algorithms in experiments, nevertheless it does not hold in all cases for $D = 10$.

Success of the population resizing diversity-based mechanism in comparison with linear reduction of population size mechanism is caused by guaranteeing function evaluations saving when the population diversity is large enough and also by increasing of population diversity when it is too small considering the stage of search process.

7   Conclusion

Recently proposed diversity-based approach in differential evolution adapting the size of population is compared with linear reduction of population size in this paper. The resizing mechanism based on population diversity improves efficiency of the tested DE variants more frequently than linear reduction of population size, mainly for higher problem dimensions. When both approaches are compared directly, DE algorithms in which the new diversity-based mechanism is implemented are more effective in almost 60% of test problems whereas algorithms with linear reduction are more effective in less than 17% of test problems. In the future, we would like to modify the approach in order to enhance its efficiency also for the problems of lower dimensions.

Acknowledgments. This work was supported by the project LQ1602 IT4Innovations excellence in science.

References

1. Brest, J., Greiner, S., Bošković, B., Mernik, M., Žumer, V.: Self-adapting control parameters in differential evolution: A comparative study on numerical benchmark problems. IEEE Trans. Evol. Comput. **10**, 646–657 (2006)
2. Bujok, P., Tvrdík, J., Poláková, R.: Differential evolution with rotation-invariant mutation and competing-strategies adaptation. IEEE Congr. Evol. Comput. **2014**, 2253–2258 (2014)
3. Das, S., Suganthan, P.N.: Differential evolution: A survey of the state-of-the-art. IEEE Trans. Evol. Comput. **15**, 27–54 (2011)
4. Kaelo, P., Ali, M.M.: A numerical study of some modified differential evolution algorithms. Eur. J. Oper. Res. **169**, 1176–1184 (2006)
5. Lampinen, J., Zelinka, I.: On stagnation of differential evolution algorithm. In: 6th International Conference on Soft Computing, MENDEL 2000, pp. 76–83 (2000)
6. Liang, J.J., Qu, B., Suganthan, P.N.: Problem definitions and evaluation criteria for the CEC 2014 special session and competition on single objective real-parameter numerical optimization (2013). http://www.ntu.edu.sg/home/epnsugan/
7. Liang, J.J., Qu, B., Suganthan, P.N.: Ranking results of CEC14 special session and competition on real-parameter single objective optimization (2014). http://www3. ntu.edu.sg/home/epnsugan/

8. Liang, J.J., Qu, B., Suganthan, P.N., Hernández-Díaz, A.G.: Problem definitions and evaluation criteria for the CEC 2013 special session on real-parameter optimization (2013). http://www.ntu.edu.sg/home/epnsugan/
9. Loshchilov, I., Stuetzle, T., Liao, T.: Ranking results of CEC 2013 special session and competition on real-parameter single objective optimization (2013). http://www3.ntu.edu.sg/home/epnsugan/
10. Mallipeddi, R., Suganthan, P.N., Pan, Q.K., Tasgetiren, M.F.: Differential evolution algorithm with ensemble of parameters and mutation strategies. Appl. Soft Comput. **11**, 1679–1696 (2011)
11. Poláková, R., Tvrdík, J., Bujok, P.: Controlled restart in differential evolution applied to CEC 2014 benchmark functions. IEEE Congr. Evol. Comput. **2014**, 2230–2236 (2014)
12. Poláková, R., Tvrdík, J., Bujok, P.: Population-size adaptation through diversity-control mechanism for differential evolution. In: 22nd International Conference on Soft Computing MENDEL 2016, Brno, pp. 49–56 (2016)
13. Storn, R., Price, K.: Differential evolution - a simple and efficient heuristic for global optimization over continuous spaces. J. Glob. Optim. **11**, 341–359 (1997)
14. Tanabe, R., Fukunaga, A.: Success-history based parameter adaptation for differential evolution. IEEE Congr. Evol. Comput. **2013**, 71–78 (2013)
15. Tanabe, R., Fukunaga, A.: Improving the search performance of SHADE using linear population size reduction. IEEE Congr. Evol. Comput. **2014**, 1658–1665 (2014)
16. Tang, L., Dong, Y., Liu, J.: Differential evolution with an individual-dependent mechanism. IEEE Trans. Evol. Comput. **19**, 560–574 (2015)
17. Tvrdík, J.: Competitive differential evolution. In: Matoušek, R., Ošmera, P. (eds.) MENDEL 2006, 12th International Conference on Soft Computing, pp. 7–12 (2006)
18. Tvrdík, J., Poláková, R.: Competitive differential evolution applied to CEC 2013 problems. IEEE Congr. Evol. Comput. **2013**, 1651–1657 (2013)
19. Wang, Y., Cai, Z., Zhang, Q.: Differential evolution with composite trial vector generation strategies and control parameters. IEEE Trans. Evol. Comput. **15**, 55–66 (2011)
20. Zhang, J., Sanderson, A.C.: JADE: Adaptive differential evolution with optional external archive. IEEE Trans. Evol. Comput. **13**, 945–958 (2009)

Cosmic Rays Inspired Mutation
in Genetic Algorithms

Wojciech Rafajłowicz[✉]

Department of Computer Engineering, Faculty of Electronics, Wrocław University
of Science and Technology, Wybrzeże Wyspiańskiego 27, 50 370 Wrocław, Poland
wojciech.rafajlowicz@pwr.edu.pl

Abstract. In this paper a new mutation operator is presented. It is
based on simulating cosmic ray impact on living tissue. It was proved that
the proposed mutation method has a compound probability distribution,
which is also derived. Numerical experiments indicate the usefulness of
this concept for problems of moderate sizes.

1 Introduction

Genetic algorithms are a wide group of metaheuristic methods. When electronic
computers became available, one of their early uses were attempts at artificial life
simulation and simulating evolution of living organisms. Later on it was noted
by Rechenberg [15] that evolutionary strategies can be used for optimization of
real value functions. That idea was developed further by Schwefel [17]. One can
trace back the history of genetic and evolutionary algorithms in [5]. All these
nature-inspired methods remain an active field of investigation up to this day.

In particular, genetic algorithms, although criticized, are still developed and
used in many applications [9,14]. The influence of probability distribution of
mutations in evolutionary algorithms has been intensively studied in [1,6,7,10]
and more recently in [11]. The aim of this paper is to provide a preliminary study of
a new probability distribution of mutations for genetic algorithms with the hope to
improve their performance. This distribution is inspired by the cosmic rays influence on living organisms and digital devices. Sources of such radiation are various
to (see [2]) and its effect on electronic devices is widely investigated [8].

We prove that the proposed distribution of mutations is of the compound
type (see [4] page 164 for the definition). Then, we derive its general form and
provide a closed form formula for it in the case when the probability of selecting
genes for mutation can be approximated by the uniform distribution. At the rest
of the paper we provide an excerpt of the simulation results.

In this paper we would investigate a new method of mutation loosely based
on behavior of cosmic rays damaging DNA.

2 Proposed Mutation Model

In a typical mutation each bit of the genotype has an equal probability of being
changed. Their positions are not important. For example the simplest method is

© Springer International Publishing AG 2017
L. Rutkowski et al. (Eds.): ICAISC 2017, Part I, LNAI 10245, pp. 418–426, 2017.
DOI: 10.1007/978-3-319-59063-9_37

as follows: for each bit in the genotype select random number $r \in [0, 1]$. If $r < p$, where p is a probability of a mutation, then change (negate) bit. For probability $p = 0.16$ this method works as shown in Table 1.

Table 1. Example of the performance of the classic mutation algorithm.

Genotype (before)	1	0	0	1	1	0	0	0
r		0.6541	0.8258	0.7749	0.9106	0.1450	0.1233	0.4893	0.4039
Genotype (mutated)	1	0	0	0	0	1	0	0

This method can be considered as a model of information loss in the DNA replication process. We must say that the rate of a biological mutation is much smaller – current estimation for each bit is around $0.5 \cdot 10^{-9}/year$ (see [16]).

In electronics (mostly in aerospace electronics) there is a well known mechanism of space radiation altering bit patterns in memory. Similar, complex mutations were reported in DNA (see [13]) due to alpha particles.

Here we propose to simulate such an impact on a genotype in the following manner. When impact occurs (in the simplest example with some, small probability) then a random point of impact is selected. For each bit in the genotype a probability of change is selected according to a said point distance from the point of impact. The simplest idea is to use a Gaussian curve as this probability. Notice that here the Gaussian curve is normalized so as it has a value at most 1 as its maximum. The idea is shown in Fig. 1.

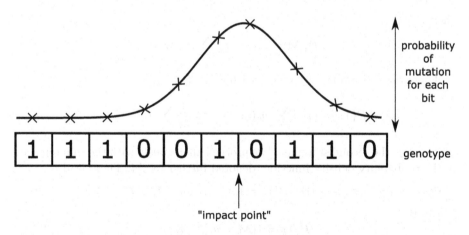

Fig. 1. Simple model of particle hit.

The proposed method of deciding whether k-th bit should be changed or not can be described in terms of a binomial random variable $Y_k \in \{0, 1\}$, but the probability of the success in a given trail depends on another random variable, namely on the uniform distribution of setting the position of maximum of the Gaussian curve.

Let $\mathcal{M}$ be a random variable having this distribution on positive integers up to M (number of bits).

Distributions constructed in this way called in [4] (page 164) compound distributions. It is of interest to derive explicitly an expression for $P\{Y_k = 1\}$.

Theorem 1. *Let for each bit k in a genotype the probability of change is selected according to the Gaussian curve: tu wzr, where m is selected at random at each stage. (a) If m is drawn from positive integers up to $M > 1$, then random variable Y_k has the compound distribution with*

$$P\{Y_k = 1\} = \frac{1}{M} \sum_{j=1}^{M} e^{-\frac{(k-j)^2}{2s^2}}$$

(b) When for large M the probability distribution of random variable $\mathcal{M}$ can be approximated by the uniform distribution on $[0, M]$, then for large M random variable $\mathcal{M}$ can be approximated by a random variable $\tilde{\mathcal{M}}$, having the uniform distribution $[0, M]$ and then

$$P\{Y_k = 1\} = \int_0^M e^{-\frac{(k-j)^2}{2s^2}} dm \frac{1}{M+1} = \frac{\sqrt{\frac{\pi}{2}}s \left(Erf\left(\frac{k}{\sqrt{2}s}\right) - Erf\left(\frac{k-M}{\sqrt{2}s}\right) \right)}{1+M}$$

Proof. To prove the first statement it is expedient to consider the join distribution of Y_k and $\mathcal{M}$ that – by conditioning – can be expressed as follows

$$P\{Y_k - 1, \mathcal{M} = m_j\} = P\{Y_k = 1 | \mathcal{M} = m_j\} \cdot P\{\mathcal{M} = m_j\} = p_k(m_j) \cdot r_j,$$

where

$$P\{\mathcal{M} = m_j\} = r_j \sum_{j=1}^{M} r_j = 1$$

Then, the full probability of $Y_k = 1$ has the form:

$$P\{Y_k = 1\} = \sum_{j=1}^{M} p_k(m_j) \cdot r_j = \frac{1}{M} \sum_{j=1}^{M} e^{-\frac{(k-j)^2}{2s^2}}$$

To prove the second statement, we proceed analogously with the sum replaced by the probability density function, denoted further by $\tilde{\mathcal{M}}$. These lead to

$$\frac{1}{h} P\{Y_k = 1, \tilde{\mathcal{M}} \in (m, m+h)\} = P\{Y_k = 1 | \tilde{\mathcal{M}} \in (m, m+h)\} \cdot P\{\tilde{\mathcal{M}} \in (m, m+h)\} =$$

as $h \to 0$

$$P\{Y_k = 1 | \tilde{\mathcal{M}} = m\} f_{\tilde{\mathcal{M}}}(m)$$

Then, the full probability of $Y_k = 1$ has the form:

$$P\{Y_k = 1\} = \int_0^M p_k(m) dm \frac{1}{M+1}$$

$$= \int_0^M e^{-\frac{(k-j)^2}{2s^2}} dm \frac{1}{M+1} = \frac{\sqrt{\frac{\pi}{2}}s \left(Erf\left(\frac{k}{\sqrt{2}s}\right) - Erf\left(\frac{k-M}{\sqrt{2}s}\right) \right)}{1+M}$$

3 Benchmark Functions

The comparison between different methods is usually done by standard test functions known as benchmarks. Generally these test functions can be found in research papers like [3] or [12]. Functions like presented ones are typical element of many benchmark packages, sometimes in some modified form.

3.1 Simple Unimodal Problem

The simplest possible test function is the distance between zero of the coordinate system and decision variable in n dimensional space.

$$f_1(x) = \sum_{i=1}^{d} x_i^2 \tag{1}$$

As in most cases of genetic optimization the range of possible x must be defined. In this case it is unrestricted by the function so we use $x_i \in (-2, 2)$.

3.2 Ackley Function

An Ackley function is a multimodal problem with clear global minimum and large number of local minimas. The plot in $d = 2$ case is shown in Fig. 2.

$$20 + e - 20e^{-0.2\sqrt{\frac{1}{d}\sum_{i=1}^{d} x_i^2}} - e^{\frac{1}{d}\sum_{i=1}^{n} cos(2\pi x_i)} \tag{2}$$

The box limit for a search was $x \in (-30, 30)$. The minimum if $f(0) = 0$.

3.3 Rosenbrock Function

The Rosenbrock function is a typical example of a difficult unimodal problem due to a minimum in a narrow valley. The function has a minimum in $f(1) = 0$, a hypercube used was $x \in (-2.048, 2048]$. This means that at least 11-bits for each dimension are required to cover this range and be able to represent the minimum point exactly.

$$\sum_{i=1}^{d-1} \left(100 \cdot (x_{i+1} - x_i^2)^2 + (x_i - 1)^2 \right) \tag{3}$$

3.4 Schwefel Function

It is another difficult function with a large number of local minimas.

$$418.9829 \cdot d + \sum_{i=1}^{d} x_i sin\sqrt{|x_i|} \tag{4}$$

The function has a global minimum at $x_i = -420.9687$ with typical range $x_i \in (-512, 512)$ (Fig. 3).

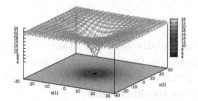

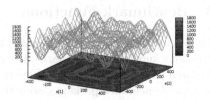

Fig. 2. Ackley function.   **Fig. 3.** Schwefel function.

3.5   Rastrigin Function

It is another example of a multimodal function but its local minimas are regularly distributed. Also the global minimum is at $f(0) = 0$. Typical range of $x \in [-5.12, 5.12]$ (Fig. 4).

$$10 \cdot d + \sum_{i=1}^{d} \left(x_i^2 - 10cos(2\pi x_i) \right) \tag{5}$$

3.6   Solving a Minimization Problem

All the previously described problems require us to find the minimum of the specified function. Additionally most versions of genetic algorithm can work only for $f \geq 0$ and look for the maximum of the function.

A well known solution to this problem is to find $C \geq f(x)\, x \in X$ and find the maximum of

$$g(x) = C - f(x) \tag{6}$$

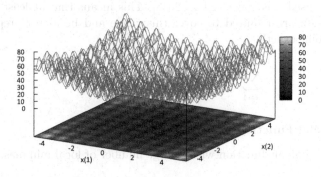

Fig. 4. Rastrigin function

4   Numerical Results

Numerical experiments were carried out in order to compare the proposed mutation method with traditional, simple mutation. As a test functions, benchmarks from the previous section were used. In the first series of simulations the following settings were used: goal functions were **10-dimensional** with 300 bits in the genotype and 150 as a population size.

The results of numerical experiments are shown in Table 2. It can be seen that the simulated cosmic rays impact method shows an improvement in some difficult cases and was not essentially worse in all of them. One can not fail to notice smaller variance of the resulting solution in the proposed method.

Table 2. Numerical experiment results.

	ball		ackley		rosenbrock		schwefel		rastrigin	
	$\bar{f}$	σ	$\bar{f}$	σ	$\bar{f}$	σ	$\bar{f}$	σ	$\bar{f}$	σ
Simple mutation	48.92	0.16	83.47	0.85	9896	1251	8842	90124	9921	162.31
Cosmic ray	48.38	0.15	83.64	1.52	9899	1079	8786	38981	9922	145.3

The largest improvement is achieved with a multimodal Schwefel function (compare Figs. 7 and 6). For the Rosenbrock function behaves similarly, but with smaller improvement (see Fig. 5).

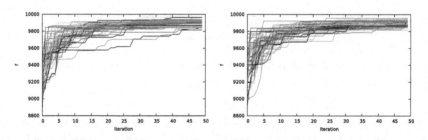

Fig. 5. Behavior of 50 trial runs for the Rosenbrock function using traditional mutation (left) and using simulated impact mutation (right).

It is well known that the problem dimension have drastic impact on a method performance. For this test the Schwefel function was chosen. In this test, due to changing problem size, genotype size n was chosen to be $n = 30 \cdot d$, where d is a problem dimension. In all cases 500 generations were investigated. **The biggest problem size was** $d = 30$. In Fig. 8 vertical axis shows problem size, on the horizontal one a result in both cases is shown. It is clearly visible that simulated cosmic rays impact method results in much smaller rate of decline of the goal function value then traditional one.

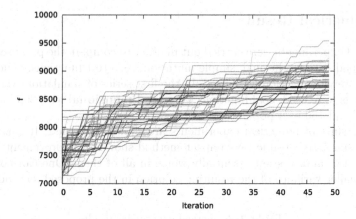

Fig. 6. Behavior of 50 trial runs for the Schwefel function using traditional mutation.

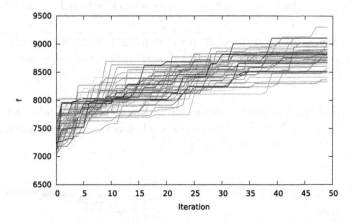

Fig. 7. Behavior of 50 trial runs for the Schwefel function using simulated impact mutation.

One of the most common statistical tests is a test for equality of means. We perform this test using more traditional test statistics values instead more common p-values.

$$H_0: \mu_t = \mu_s,$$
$$H_A: \mu_t < \mu_s,$$

where μ_s is an mean calculated for the space radiation based method and μ_t is a mean in case of the simple mutation. As a significance level, a traditional value $\alpha = 0.05$ was chosen. We must note that by **rejection** of the null hypothesis and subsequent acceptance of the **alternative** one shows the **improvement** in proposed method.

Results for different problem size are shown in Fig. 9. The horizontal axis shows problem size, the vertical one represents test statistics.

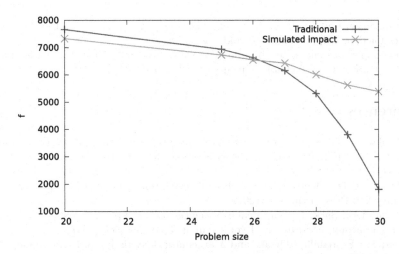

Fig. 8. Results changing for Schwefel function using traditional mutation and using simulated impact mutation with different problem dimension.

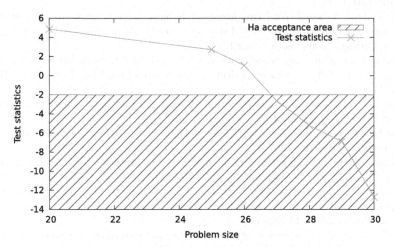

Fig. 9. Test statistics for average difference.

5   Summary

In this paper a new mutation operator was proposed, one based on simulated impact of a space radiation particle and their effect on living tissue.

The proposed mutation distribution is of the compound type and it can be generalized in a number of ways by changing probability distributions that are compounded.

The preliminary computational experiments are successful especially for multimodal function. The proposed method is worth further investigation as a part of a hybrid mutation operator.

Acknowledgements. The author would like to express his thanks to the anonymous referees for their helpful comments and suggestions.

This paper was supported by grant for young research of Faculty of Electronics, Wroclaw University of Technology project number 0402/0174/16

References

1. Chellapilla, K., Fogel, D.B.: Fitness distributions in evolutionary computation: motivation and examples in the continuous domain. BioSystems **54**(1), 15–29 (1999)
2. Dermer, C.D., Menon, G.: High Energy Radiation From Black Holes. Princeton University Press, Princeton (2009)
3. Eiben, A.E., Bäck, T.: Empirical investigation of multi-parent recombination operators in evolution strategies. Evol. Comput. **5**(3), 347–365 (1997)
4. Fisz, M.: Probability Theory and Mathematical Statistics, 3rd edn. Willey, New York (1967)
5. Fogel, D.B.: Evolutionary Computation: The Fossil Record. Wiley-IEEE Press, New York (1998)
6. Fogel, D.B., Ghozeil, A.: Using fitness distributions to design more efficient evolutionary computations. In: 1996 Proceedings of IEEE International Conference on Evolutionary Computation. IEEE (1996)
7. Galar, R.: Evolutionary search with soft selection. Biol. Cybern. **60**, 357–364 (1989)
8. Iniewski, K. (ed.): Radiation Effects is Semiconductors. CRC Press, Boca Raton (2011)
9. Li, M., et al.: Accurate determination of geographical origin of tea based on terahertz spectroscopy. Appl. Sci. **7**(2), 172 (2017)
10. Michalewicz, Z.: Genetic Algorithms + Data Structures = Evolution Programs. Springer, Heidelberg (2013)
11. Obuchowicz, A.: Algorytmy ewolucyjne z mutacj. Informatyka. Akademicka Oficyna Wydaw. EXIT, Warszawa (2013). ISBN: 978-83-7837-020-8
12. Ortiz-Boyer, D., Hervas-Martinez, C., Garcia-Pedrajas, N.: CIXL2: A crossover operator for evolutionary algorithms based on population features. J. Artif. Intell. Res. (JAIR) **25**, 1–48 (2005)
13. Prise, K.M., et al.: A review of studies of ionizing radiation-induced double-strand break clustering. Radiat. Res. **156**(5), 572–576 (2001)
14. Ramadan, B.M.S.M., et al.: Hybridization of genetic algorithm and priority list to solve economic dispatch problems. In: 2016 IEEE Region 10 Conference (TENCON). IEEE (2016)
15. Rechenberg, I.: Evolution Strategy: Optimization of Technical systems by means of biological evolution. Fromman-Holzboog, Stuttgart 104 (1973)
16. Scally, A.: The mutation rate in human evolution and demographic inference. Curr. Opin. Genet. Dev. **41**, 36–43 (2016)
17. Schwefel, H.P.: Evolution strategy and numerical optimization. Technical University of Berlin (1975)

OC1-DE: A Differential Evolution Based Approach for Inducing Oblique Decision Trees

Rafael Rivera-Lopez[1], Juana Canul-Reich[2(✉)], José A. Gámez[3], and José M. Puerta[3]

[1] Instituto Tecnológico de Veracruz, Veracruz, Mexico
`rrivera@itver.edu.mx`
[2] Universidad Juárez Autónoma de Tabasco, Cunduacán, Mexico
`juana.canul@ujat.mx`
[3] Universidad de Castilla-La Mancha, Albacete, Spain
`{jose.gamez,jose.puerta}@uclm.es`

Abstract. This paper describes the application of a Differential Evolution based approach for inducing oblique decision trees in a recursive partitioning strategy. Considering that: (1) the task of finding an optimal hyperplane with real-valued coefficients is a complex optimization problem in a continuous space, and (2) metaheuristics have been successfully applied for solving this type of problems, in this work a differential evolution algorithm is applied with the aim of finding near-optimal hyperplanes that will be used as test conditions of an oblique decision tree. The experimental results show that this approach induces more accurate classifiers than those produced by other proposed induction methods.

Keywords: Machine learning · Classification · Evolutionary algorithms · Recursive partition strategy

1 Introduction

Nowadays, data mining has emerged as a very useful tool for extracting knowledge from any source of information including text files, large data repositories, and data in the cloud. Machine learning provides data mining with useful procedures for constructing models from data. The most representative machine learning approaches are supervised learning, in which a model is learned from labeled data, and unsupervised learning, where a model is obtained from unlabeled data. The main supervised techniques are classification and regression, while clustering is the main unsupervised technique. In particular, decision trees are a classification method that allows to construct simple models with high interpretability level.

Oblique DTs are classifiers that split the data using hyperplanes that are oblique to the axis of the instance space. Oblique DTs are generally more compact and precise than univariate DTs. In this paper, the application of a differential evolution-based approach named OC1-DE for inducing oblique decision trees

© Springer International Publishing AG 2017
L. Rutkowski et al. (Eds.): ICAISC 2017, Part I, LNAI 10245, pp. 427–438, 2017.
DOI: 10.1007/978-3-319-59063-9_38

in a recursive partitioning strategy is described. Experimental results obtained show that OC1-DE induces more accurate classifiers than those produced by other proposed induction methods.

In order to describe the implementation of OC1-DE method, this paper is organized as follows: Sect. 2 describes the characteristics of the decision tree induction process for classifications tasks. The use of metaheuristics for inducing oblique decision trees is described in Sect. 3 and details of differential evolution algorithm are given in Sect. 4. In Sect. 5 the OC1-DE method is outlined and experimental results are presented in Sect. 6. Finally, Sect. 7 holds conclusions and future work of this proposal.

2   Induction of Decision Trees for Classification

A decision tree (DT) is a classification model constructed using a set of training instances in order to define a simple and effective procedure for predicting the class membership of new unclassified instances. A DT is a connected acyclic graph whose nodes contain both test conditions (internal nodes) and class labels (leaf nodes) and whose arcs represent the possible results of each test condition. This graph defines a hierarchical evaluation schema that stands out for its simplicity and its high level of interpretability. These characteristics along with its predictive power allow to place to DT as one of the most widely used classifier. It should be noted that two methods for decision tree induction (CART [4] and C4.5 [36]) have been considered in the top ten most influential data mining algorithms [46].

The great majority of algorithms for inducing decision trees apply a recursive partitioning strategy that implements some splitting criterion in order to separate the training instances. This partitioning procedure is usually complemented with a pruning mechanism in order to avoid overfitting and to improve the performance of the classifier. Furthermore, other strategies such as a global search in the space of DTs have been utilized with the aim of reaching an optimal classifier [19,29,44] although it is known that building optimal DTs is NP-Hard [18].

In accordance to the number of attributes used in test conditions of DT internal nodes, two types of DTs can be induced: univariate and multivariate DTs. ID3 [34], C4.5 [36], J48 [45] and C5.0 are methods for inducing univariate DTs in which a single attribute is evaluated to split the training set. An advantage of univariate DTs is that they are easily interpretable, however when the instances cannot be separated by axis-parallel hyperplanes, induced DTs are more complex as they contain a number of internal nodes [37].

On the other hand, a combination of attributes participates in the test condition at each internal node of a multivariate DT. In case of a linear combination of attributes the induced classifier is known as oblique DT [31] or Perceptron DT [2,29]. Hyperplanes splitting the training instances in this type of DT have an oblique orientation relative to the axes of instance space. Oblique DTs are generally smaller and more accurate than univariate DTs but they are generally more difficult to interpret [5]. In addition, finding the best partition using a single attribute takes low computational effort, however determining the best partition using an oblique hyperplane is a NP-hard problem [16]. CART [4], OC1 [31], LMDT [41] and LTree [10] are methods for inducing oblique DTs.

The training set used for inducing oblique DTs is a group of instances, where each instance can be represented with a real-valued vector $\mathbf{v} = (v_1, v_2, \ldots, v_d, c)^T$ of d attributes and a label c indicating the class membership of the instance. A hyperplane divides the instance space in two halfspaces and it is defined as follows:

$$\sum_{i=1}^{d} x_i v_i + x_{d+1} > 0 \tag{1}$$

where v_i is the value of attribute i, x_i is a real-valued coefficient in the hyperplane and x_{d+1} represents the independent term.

3 Metaheuristics for Inducing Oblique Decision Trees

Metaheuristics are methods that implement both local search procedures and higher level strategies to perform a robust search of a solution space [14]. They have been used for inducing both univariate and multivariate DTs. Metaheuristics can be classified into three categories: single-solution based procedures, evolutionary algorithms and swarm intelligence methods.

Although global search is the most commonly used strategy for inducing DTs in which metaheuristic-based methods are applied, several proposals using metaheuristics through a recursive partitioning strategy have been developed. Single-solution based metaheuristics such as stochastic local search (SLS) [31], simulated annealing (SA) [5,17] and tabu search (TS) [26,32], as well as evolutionary algorithms such as genetic algorithms (GA) [5,6,20,39] and evolutionary strategies (ES) [5,47] have been used for finding good hyperplanes of an oblique DT.

Simulated Annealing of Decision Trees (SADT) method [17] applies SA to find a near-optimal hyperplane for each internal node of an oblique DT. Using a fixed initial hyperplane, SADT iteratively perturbs one coefficient of this hyperplane and each new value is accepted as a coefficient according to the Boltzmann criterion. OC1 method [31] implements a two-step process for searching a better hyperplane: First, it uses a deterministic strategy that adjusts the coefficients of a hyperplane individually, finding a local optimal value for one coefficient at a time. Once this value is reached, OC1 applies SLS in order to jump out of this local optimum value. OC1-AP [30] is a OC1 variant that induces only univariate DTs. OC1 variants such as OC1-SA, OC1-ES and OC1-GA [5] simultaneously modify several coefficients of the best hyperplane produced by OC1-AP, using SA, ES and GA, respectively. Furthermore, TS has been used in two approaches: (1) in conjunction with linear discriminant analysis in the LDTS method [26], and (2) in combination with a discrete support vector machine in the LDSDT$_{TS}$ method [32].

Others have used evolutionary algorithms for inducing oblique DTs: a multi-membered ES is applied for finding near-optimal hyperplanes in the MESODT

method [47] and GA is implemented in two proposals: (1) in the BTGA method
[6] that uses a binary chromosome for representing the hyperplane coefficients,
and (2) in a method based on dipoles[1] [20] that encodes a chromosome by means
of a real-valued vector. An especial GA known as HereBoy algorithm [24] that
evolves a unique chromosome is used in HBDT method [39] for inducing an
oblique DT.

On the other hand, GA [9,15,21,33,43], differential evolution [29] and genetic
programming (GP) [1,3,28] have been used for inducing oblique DTs in a global
search strategy.

4   Differential Evolution Algorithm

Differential evolution (DE) is an efficient evolutionary algorithm designed for
solving optimization problems in continuous spaces [38]. DE encodes candidate
solutions by means of real-valued vectors and it applies a difference vector to
disrupt a population of these solutions. At the begining an initial population of
candidate solutions is generated, then the iterative part of DE follows. At each
iteration of DE a new population is constructed using mutation, crossover and
selection operators. Instead of implementing traditional crossover and mutation
operators, DE applies a linear combination of several candidate solutions selected
randomly to produce a new solution. Finally, when the stop condition is fulfilled
DE return the best candidate solution in the current population. An advantage
of DE is that it utilizes few control parameters: a crossover rate Cr, a mutation
scale factor F, and a population size NP.

DE/A/B/C is the notation commonly adopted for identifying DE variants,
where A represents the selection procedure of candidate solutions that will be
used for constructing a new solution, B is the number of difference vectors used
for the mutation operator, and C identifies the type of crossover operator [7].
The most popular DE variant is the DE/rand/1/bin that consists in combining
three randomly selecting candidate solutions ($\mathbf{x}_a$, $\mathbf{x}_b$ and $\mathbf{x}_c$), using Eq. (2), to
yield a mutated solution $\mathbf{x}_{mut}$.

$$\mathbf{x}_{mut} = \mathbf{x}_a + F\left(\mathbf{x}_b - \mathbf{x}_c\right) \tag{2}$$

Mutated solution is utilized to perturb the values of another candidate solution $\mathbf{x}_{cur}$ using the binomial crossover operator as follows:

$$x_{new,j} = \begin{cases} x_{mut,j} & \text{if } r \leq Cr \vee j = k \\ x_{cur,j} & \text{otherwise} \end{cases} ; j \in \{1, \ldots, d\} \tag{3}$$

where $x_{new,j}$, $x_{mut,j}$ and $x_{cur,j}$ are the values of the j-th position of $\mathbf{x}_{new}$, $\mathbf{x}_{mut}$
and $\mathbf{x}_{cur}$, respectively, and $r \in [0,1)$ and $k \in \{1, \ldots, d\}$ are uniformly distributed
random numbers.

---

[1] A dipole is a pair of instances in training set represented as vectors.

Finally, $\mathbf{x}_{new}$ is selected as a member of the new population if it has a better fitness value than that of $\mathbf{x}_{cur}$.

DE has been used in conjunction with support vector machines [25], neural networks [23], bayesian classifiers [13] and instance based classifiers [11] for implementing classification methods. It has been mainly used for optimizing the parameters of classification methods or for conducting preprocessing tasks in a data mining process.

In the case of DTs, DE is applied to finding the parameter settings of a classification method and for participating in the decision tree induction process. In DEMO algorithm [40] DE finds the J48 parameters in order to yield accurate and small DTs. On the other hand, TreeDE method [42] induces univariate DTs for mapping full trees to vectors and for representing discrete symbols by points in a real-valued vector space. In an improved version of TreeDE method [22] each candidate solution has two types of genes: one express the neighborhood structure between the symbols and the other represents a full tree structure. Finally, in a method for inducing perceptron DTs [29], the coefficients of all possible hyperplanes of an oblique-DT with a fixed size are encoded in a matrix, and then DE evolves the matrix values in order to construct optimized oblique DTs.

5   OC1-DE Method for Inducing Oblique Decision Trees

In this work a method for inducing an oblique DT using DE/rand/1/bin in a recursive partitioning strategy is introduced. This method, named OC1-DE, is similar to OC1 and its variants but it applies DE to finding a near-optimal hyperplane at each internal node of an oblique DT. Since the task of finding a near-optimal hyperplane with real-valued coefficients is an optimization problem in a continuous space, DE operators can be applied without any modification and OC1-DE should induce better oblique DTs.

When a recursive partitioning strategy is implemented for inducing oblique DTs, the classification method starts finding the hyperplane that best splits an instance set into two subsets. This hyperplane will be used as test condition of a new internal node that is added in DT. This procedure is recursively applied using each subset until a leaf node is created due to all instances in the subset have the same class label or a threshold value of unclassified instances is reached. Quality of hyperplane is obtained using a splitting criterion that measures the impurity of a partition or some other discriminant value. Finally, a pruning procedure is applied in order to reduce the overfitting of DT produced and to improve its predictive power.

The DE implementation for finding a near-optimal hyperplane at each internal node of an oblique DT is shown in Algorithm 1. In a similar fashion to OC1-GA method [5] the axis-parallel hyperplane that best splits a set of training instances is obtained (line 2). This hyperplane is copied to 10% of the initial population and remaining hyperplanes are randomly created (line 3). Each random hyperplane is constructed considering that almost two instances with different class label are separated by the hyperplane. This population is evolved through

---

Algorithm 1. Hyperplane selection using the OC1-DE algorithm.

1: $\mathbf{V} \leftarrow$ A set of training instances
2: $\mathbf{x}_{ap} \leftarrow$ Best axis-parallel hyperplane for $\mathbf{V}$
3: $\mathbf{X}_0 \leftarrow$ Initial population of random hyperplanes with 10% of copies of $\mathbf{x}_{ap}$.
4: **for** k in $[1, \ldots, it]$ **do**
5:     $\mathbf{X}_k \leftarrow \emptyset$
6:     **for** cur in $[1, \ldots, NP]$ **do**
7:         $\{\mathbf{x}_a, \mathbf{x}_b, \mathbf{x}_c\} \leftarrow$ Randomly selected candidate solutions of $\mathbf{X}_{k-1}$
8:         $\mathbf{x}_{mut} \leftarrow$ Mutated vector using (2)   ▷ DE mutation operator
9:         $\mathbf{x}_{new} \leftarrow$ Perturbed vector of $\mathbf{x}_{cur}$ using (3)   ▷ DE crossover operator
10:        $\mathbf{x}_{hyp} \leftarrow \begin{cases} \mathbf{x}_{new} & \text{if } fitness(\mathbf{x}_{new}) < fitness(\mathbf{x}_{cur}) \\ \mathbf{x}_{cur} & \text{otherwise} \end{cases}$ ▷ DE selection operator
11:        $\mathbf{X}_k \leftarrow \mathbf{X}_k \cup \{\mathbf{x}_{hyp}\}$
12:     **end for**
13: **end for**
14: $\mathbf{x}_{best} \leftarrow$ Best hyperplane in $\mathbf{X}_k$
15: **return** $\begin{cases} \mathbf{x}_{ap} & \text{if } fitness(\mathbf{x}_{ap}) < fitness(\mathbf{x}_{best}) \\ \mathbf{x}_{best} & \text{otherwise} \end{cases}$

---

several generations using DE operators (lines 4–13) and the best hyperplane in population is selected (line 14). Algorithm returns the hyperplane selected between the best axis-parallel hyperplane and the best oblique hyperplane produced by DE (line 15).

6   Experiments

In order to evaluate the performance of OC1-DE and for comparison with other approaches, two experiments were conducted using several datasets with numerical attributes chosen from UCI repository [27]. Table 1 shows the description of datasets used in experiments. Results were obtained by applying repeated k-fold stratified cross-validation (CV). OC1-DE is compared with OC1 method [31], three OC1 variants proposed in [5] and the HBDT method described in [39].

DE version implemented in Java language is the canonical `DE/rand/1/bin`. The parameters used in the experiments are described in Table 2. The population size is adjusted in accordance with [5] and the number of iterations of DE is small with the aim of reducing the time of experiments due to DE is applied at each internal node of the induced oblique DT.

In the first experiment OC1-DE is compared to OC1 variants applying 5-fold CV. 10 independent runs of each dataset are conducted. For each run, an oblique DT is induced and its test accuracy is calculated. Both average accuracy and average DT size across these 10 runs are obtained. Table 3 shows the experimental results[2]: Columns 2–9 show both average accuracy and average DT size of induced oblique DTs reported by [5]. Results obtained by OC1-DE

---

[2] Highest values for each dataset are in bold.

Table 1. Description of datasets used in the experiments.

Dataset	Inst.	Attr.	Classes	Dataset	Inst.	Attr.	Classes
Breast-w	683	9	2	Australian	690	14	2
Diabetes	768	8	2	Ionosphere	351	34	2
Glass	214	9	7	Wine	178	13	3
Housing	506	12	2	Sonar	208	60	2
Iris	150	4	3	Car	1728	6	4
Vehicle	846	18	4	Cleveland	297	13	5
Vowel	990	10	11	Liver-disorders	345	6	2
Heart-statlog	270	13	2	Page-blocks	5473	10	5

Table 2. Parameters used in experiments with OC1-DE.

Parameter	Value
Mutation scale factor	0.5
Crossover rate	0.9
Size of population	$20\sqrt{d}$
Number of generations	50
Fitness function	Twoing rule [4]
Pruning method	Reduced error pruning [35]

are shown in columns 10–11 of this table. In this table can be observed that accuracies obtained by OC1-DE are better than those reported by OC1 variants, although in general OC1-DE produces oblique-DTs with more leaf nodes than those produced by OC1 variants. Fig. 1(a) shows a comparative plot of the average accuracies obtained.

In the second experiment OC1-DE is compared with HBDT method. 5 independent runs of 10-fold CV are conducted. Results are shown in Table 4: Columns 2–5 show both average accuracy and average DT size of induced oblique DTs reported by [39]. Results obtained by OC1-DE are shown in columns 6–7 of this table. In this table can be observed that both accuracies and DT sizes obtained by OC1-DE are comparable with those reported by HBDT method. OC1 induces oblique DTs with better accuracies for 2 datasets, HDBT constructs DTs with better accuracies for 5 datasets and the accuracies of DTs induced using OC1-DE are better than those reported by [39] for 8 datasets. Fig. 1(b) shows a comparative plot of the average accuracies obtained.

In order to evaluate the performance of OC1-DE a statistical test of the results obtained was realized. Due to the conditions for using parametric test are not guaranteed when analysing results of evolutionary algorithms [12], the Friedman test is applied for detecting the existence of significant differences between the performance of two or more methods and the Nemenyi post-hoc test is utilized for checking these differences. Nemenyi test uses the average ranks

Table 3. Accuracy and size obtained for OC1 variants and OC1-DE.

(1)	(2)	(3)	(4)	(5)	(6)	(7)	(8)	(9)	(10)	(11)
	OC1		OC1-SA		OC1-GA		OC1-ES		OC1-DE	
Dataset	Acc	Size	Acc	Size	Acc	Size	Acc	Size	Acc	Size
Breast-w	96.1	**3.3**	93.5	11.5	94.3	11.1	95.2	5.2	**96.63**	4.92
Diabetes	74.1	**15.3**	73.9	19.0	73.9	20.0	73.7	17.1	**74.33**	40.54
Glass	64.1	**14.5**	65.7	15.2	65.9	19.5	66.4	17.2	**66.78**	16.56
Housing	82.7	**7.3**	82.8	11.6	82.9	13.5	82.8	11.5	**91.50**	10.48
Iris	95.3	3.6	93.8	3.4	96.3	4.6	94.5	3.7	**96.93**	**3.00**
Vehicle	69.4	37.1	70.3	**31.6**	69.6	44.2	69.3	40.0	**70.78**	58.74
Vowel	80.1	64.8	83.1	45.7	82.0	86.2	80.6	82.4	**84.84**	**41.84**

of each classifier and checks for each pair of classifiers whether the difference between their ranks is greater than the critical difference [8] defined as follows:

$$CD = q_\alpha \sqrt{\frac{k(k+1)}{6N}} \qquad (4)$$

where CD is the critical difference, k is the number of methods, N is the number of datasets, and q_α is a critical value associated of the significance level α.

Table 4. Comparison of accuracy and size reported of HBDT and OC1-DE.

(1)	(2)	(3)	(4)	(5)	(6)	(7)
	OC1		HBDT		OC1-DE	
Dataset	Acc	Size	Acc	Size	Acc	Size
Breast-w	95.53	**3.68**	**96.72**	4.04	96.22	5.80
Diabetes	**73.03**	**6.54**	71.51	17.82	71.80	47.36
Glass	62.04	13.12	62.51	**12.74**	**67.66**	18.30
Iris	95.60	3.54	95.60	3.48	**97.20**	**3.00**
Vehicle	68.16	33.53	**74.51**	**16.98**	71.75	65.14
Vowel	74.55	51.68	73.94	**37.84**	**84.58**	44.46
Heart-statlog	76.30	4.70	**79.85**	**4.26**	77.04	15.92
Australian	83.63	**6.10**	82.52	10.26	**84.38**	27.84
Ionosphere	88.26	6.18	88.24	**6.14**	**90.03**	7.82
Wine	90.65	4.58	**94.14**	**3.00**	92.92	5.68
Sonar	79.39	6.76	75.96	**4.26**	**79.56**	9.64
Car	93.42	25.26	95.65	**15.86**	**95.78**	36.80
Cleveland	51.50	**10.28**	**55.50**	10.46	52.39	25.72
Liver-disorders	**67.23**	**5.38**	65.96	12.88	66.03	25.58
Page-blocks	97.05	**23.78**	97.00	26.02	**97.12**	38.12

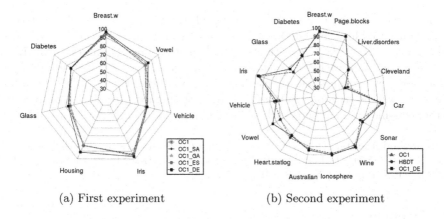

(a) First experiment   (b) Second experiment

Fig. 1. Average accuracies obtained in experiments.

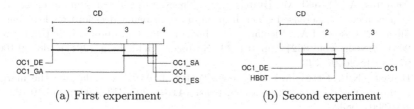

(a) First experiment   (b) Second experiment

Fig. 2. Comparison of classifiers using Nemenyi post-hoc test.

For the first experiment the Friedman statistic for 5 methods using 7 datasets is 15.507 and the p-value obtained is 0.003757 for 4 degrees of freedom (dof) of chi-square distribution. This p-value indicates the existence of statistical differences between these methods and then Nemenyi test post-hoc is executed. Fig. 2(a) shows the critical differences obtained for Nemenyi test and it shown that OC1-DE is the best method compared with OC1 variants. For the second experiment, Friedman statistic for 3 methods using 15 datasets is 6.8136 and the p-value is 0.03315 for 2 dof and it also indicates statistical differences between these methods. Fig. 2(b) shows that OC1-DE is has better performance that OC1 and it have a comparable performance with HDBT.

7   Conclusions

In this paper, OC1-DE method that uses the differential evolution algorithm to find a near-optimal hyperplane with real-valued coefficients at each internal node of an oblique DT is introduced. OC1-DE is a recursively partitioning scheme based on OC1 in which the DE operators can be applied without any modification. OC1-DE was evaluated using a stratified cross-validation procedure with several UCI datasets and statistical tests suggest that OC1-DE achieves a better performance than OC1 variants and HBDT method. In general, OC1-DE

R. Rivera-Lopez et al.

induces more accurate oblique DTs although their sizes are overall larger than those produced for other methods. Based on our results, future work will be oriented to evaluate other DE variants for inducing oblique DTs and to investigate the effect of using several parameter configurations on the OC1-DE performance, also more experiments will be conducted for analyzing the OC1-DE execution time, as well as to compare the OC1-DE performance with those obtained by other classification methods such as random forest and support vector machines.

Acknowledgments. This work has been supported in part by the Mexican Government (CONACyT FOMIX-DICC project No. TAB-2014-C01-245876 and the PROMEP-SEP project No. DSA/103.5/15/6409).

References

1. Agapitos, A., O'Neill, M., Brabazon, A., Theodoridis, T.: Maximum margin decision surfaces for increased generalisation in evolutionary decision tree learning. In: Silva, S., Foster, J.A., Nicolau, M., Machado, P., Giacobini, M. (eds.) EuroGP 2011. LNCS, vol. 6621, pp. 61–72. Springer, Heidelberg (2011). doi:10.1007/978-3-642-20407-4_6
2. Bennett, K.P., Cristianini, N., Shawe-Taylor, J., Wu, D.: Enlarging the margins in perceptron decision trees. Mach. Learn. **41**(3), 295–313 (2000). doi:10.1023/A:1007600130808
3. Bot, M.C.J., Langdon, W.B.: Improving induction of linear classification trees with genetic programming. In: Whitley, L.D., Goldberg, D.E., Cantú-Paz, E., Spector, L., Parmee, I.C., Beyer, H.G. (eds.) GECCO-2000, pp. 403–410. Morgan Kaufmann (2000)
4. Breiman, L., Friedman, J., Olshen, R., Stone, C.: Classification and Regression Trees. Taylor & Francis, Abington (1984)
5. Cantú-Paz, E., Kamath, C.: Inducing oblique decision trees with evolutionary algorithms. IEEE Trans. Evol. Comput. **7**(1), 54–68 (2003). doi:10.1109/TEVC.2002.806857
6. Chai, B.B., Zhuang, X., Zhao, Y., Sklansky, J.: Binary linear decision tree with genetic algorithm. In: ICPR 1996, vol. 4, pp. 530–534 IEEE (1996). doi:10.1109/ICPR.1996.547621
7. Das, S., Suganthan, P.N.: Differential evolution: A survey of the state-of-the-art. IEEE Trans. Evol. Comput. **15**(1), 4–31 (2011). doi:10.1109/TEVC.2010.2059031
8. Demšar, J.: Statistical comparisons of classifiers over multiple data sets. J. Mach. Learn. Res. **7**, 1–30 (2006)
9. Dumitrescu, D., András, J.: Generalized decision trees built with evolutionary techniques. Stud. Inf. Control **14**(1), 15–22 (2005)
10. Gama, J., Brazdil, P.: Linear tree. Intell. Data Anal. **3**(1), 1–22 (1999). doi:10.1016/S1088-467X(99)00002-5
11. García, S., Derrac, J., Triguero, I., Carmona, C.J., Herrera, F.: Evolutionary-based selection of generalized instances for imbalanced classification. Knowl. Based Syst. **25**(1), 3–12 (2012). doi:10.1016/j.knosys.2011.01.012
12. García, S., Fernández, A., Luengo, J., Herrera, F.: A study of statistical techniques and performance measures for genetics-based machine learning: Accuracy and interpretability. Soft Comput. **13**(10), 959 (2008). doi:10.1007/s00500-008-0392-y

13. Geetha, K., Baboo, S.S.: An empirical model for thyroid disease classification using evolutionary multivariate bayseian prediction method. Glob. J. Comput. Sci. Technol. **16**(1), 1–9 (2016)

14. Gendreau, M., Potvin, J.Y.: Handbook of Metaheuristics, vol. 2. Springer, Heidelberg (2010). doi:10.1007/978-1-4419-1665-5

15. Gray, J.B., Fan, G.: Classification tree analysis using TARGET. Comput. Stat. Data Anal. **52**(3), 1362–1372 (2008). doi:10.1016/j.csda.2007.03.014

16. Heath, D.G.: A geometric framework for machine learning. Ph.D. thesis, Johns Hopkins University (1993)

17. Heath, D.G., Kasif, S., Salzberg, S.: Induction of oblique decision trees. In: Bajcsy, R., et al. (ed.) IJCAI 1993, pp. 1002–1007 (1993)

18. Hyafil, L., Rivest, R.L.: Constructing optimal binary decision trees is NP-complete. Inf. Process. Lett. **5**(1), 15–17 (1976). doi:10.1016/0020-0190(76)90095-8

19. Jankowski, D., Jackowski, K.: Evolutionary algorithm for decision tree induction. In: Saeed, K., Snášel, V. (eds.) CISIM 2014. LNCS, vol. 8838, pp. 23–32. Springer, Heidelberg (2014). doi:10.1007/978-3-662-45237-0_4

20. Krętowski, M.: An evolutionary algorithm for oblique decision tree induction. In: Rutkowski, L., Siekmann, J.H., Tadeusiewicz, R., Zadeh, L.A. (eds.) ICAISC 2004. LNCS, vol. 3070, pp. 432–437. Springer, Heidelberg (2004). doi:10.1007/978-3-540-24844-6_63

21. Krętowski, M., Grześ, M.: Evolutionary learning of linear trees with embedded feature selection. In: Rutkowski, L., Tadeusiewicz, R., Zadeh, L.A., Żurada, J.M. (eds.) ICAISC 2006. LNCS, vol. 4029, pp. 400–409. Springer, Heidelberg (2006). doi:10.1007/11785231_43

22. Kushida, J.I., Hara, A., Takahama, T.: A novel tree differential evolution using inter-symbol distance. In: IWCIA 2014, pp. 107–112. IEEE (2014). doi:10.1109/IWCIA.2014.6988087

23. Leema, N., Nehemiah, H.K., Kannan, A.: Neural network classifier optimization using differential evolution with global information and back propagation algorithm for clinical datasets. Appl. Soft Comput. **49**, 834–844 (2016). doi:10.1016/j.asoc.2016.08.001

24. Levi, D.: Hereboy: A fast evolutionary algorithm. In: Lohn, J., et al. (ed.) EH 2000, pp. 17–24. IEEE (2000). doi:10.1109/EH.2000.869338

25. Li, J., Ding, L., Li, B.: Differential evolution-based parameters optimisation and feature selection for support vector machine. Int. J. Comput. Sci. Eng. **13**(4), 355–363 (2016). doi:10.1504/ijcse.2016.080212

26. Li, X.B., Sweigart, J.R., Teng, J.T.C., Donohue, J.M., Thombs, L., Wang, S.M.: Multivariate decision trees using linear discriminants and tabu search. IEEE Trans. Syst. Man Cybern. Part A: Syst. Hum. **33**(2), 194–205 (2003). doi:10.1109/TSMCA.2002.806499

27. Lichman, M.: UCI Machine Learning Repository. University of California, Irvine (2013). http://archive.ics.uci.edu/ml

28. Liu, K.H., Xu, C.G.: A genetic programming-based approach to the classification of multiclass microarray datasets. Bioinformatics **25**(3), 331–337 (2009). doi:10.1093/bioinformatics/btn644

29. Lopes, R.A., Freitas, A.R.R., Silva, R.C.P., Guimarães, F.G.: Differential evolution and perceptron decision trees for classification tasks. In: Yin, H., Costa, J.A.F., Barreto, G. (eds.) IDEAL 2012. LNCS, vol. 7435, pp. 550–557. Springer, Heidelberg (2012). doi:10.1007/978-3-642-32639-4_67

30. Murthy, S.K., Kasif, S., Salzberg, S.: A system for induction of oblique decision trees. J. Artif. Intell. Res. **2**(1), 1–32 (1994). doi:10.1613/jair.63

31. Murthy, S.K., Kasif, S., Salzberg, S., Beigel, R.: OC1: A randomized algorithm for building oblique decision trees. In: Proceedings of AAAI 93, vol. 93, pp. 322–327 (1993)

32. Orsenigo, C., Vercellis, C.: Discrete support vector decision trees via tabu search. Comput. Stat. Data Anal. **47**(2), 311–322 (2004). doi:10.1016/j.csda.2003.11.005

33. Pangilinan, J.M., Janssens, G.K.: Pareto-optimality of oblique decision trees from evolutionary algorithms. J. Glob. Optim. **51**(2), 301–311 (2011). doi:10.1007/s10898-010-9614-9

34. Quinlan, J.R.: Induction of decision trees. Mach. Learn. **1**(1), 81–106 (1986). doi:10.1007/BF00116251

35. Quinlan, J.R.: Simplifying decision trees. Int. J. Hum. Comput. Stud. **27**(3), 221–234 (1987). doi:10.1006/ijhc.1987.0321

36. Quinlan, J.R.: C4.5: Programs for Machine Learning. Morgan Kaufmann, San Francisco (1993)

37. Shali, A., Kangavari, M.R., Bina, B.: Using genetic programming for the induction of oblique decision trees. In: Arif-Wani, M. (ed.) ICMLA 2007, pp. 38–43. IEEE (2007). doi:10.1109/ICMLA.2007.66

38. Storn, R., Price, K.: Differential evolution-a simple and efficient heuristic for global optimization over continuous spaces. J. Glob. Optim. **11**(4), 341–359 (1997). doi:10.1023/A:1008202821328

39. Struharik, R., Vranjkovic, V., Dautovic, S., Novak, L.: Inducing oblique decision trees. In: SISY-2014, pp. 257–262. IEEE (2014). doi:10.1109/SISY.2014.6923596

40. Tušar, T.: Optimizing accuracy and size of decision trees. In: ERK-2007, pp. 81–84 (2007)

41. Utgoff, P.E., Brodley, C.E.: Linear machine decision trees. University of Massachusetts, Amherst, MA, USA, Technical report (1991)

42. Veenhuis, C.B.: Tree based differential evolution. In: Vanneschi, L., Gustafson, S., Moraglio, A., Falco, I., Ebner, M. (eds.) EuroGP 2009. LNCS, vol. 5481, pp. 208–219. Springer, Heidelberg (2009). doi:10.1007/978-3-642-01181-8_18

43. Vukobratović, B., Struharik, R.: Evolving full oblique decision trees. In: CINTI 2015, pp. 95–100. IEEE (2015). doi:10.1109/CINTI.2015.7382901

44. Wang, P., Tang, K., Weise, T., Tsang, E.P.K., Yao, X.: Multiobjective genetic programming for maximizing ROC performance. Neurocomputing **125**, 102–118 (2014). doi:10.1016/j.neucom.2012.06.054

45. Witten, I., Frank, E.: Data Mining: Practical Machine Learning Tools and Techniques. Morgan Kaufmann, San Francisco (2005)

46. Wu, X., Kumar, V., Quinlan, J.R., Ghosh, J., Yang, Q., Motoda, H., McLachlan, G.J., Ng, A., Liu, B., Philip, S.Y.: Top 10 algorithms in data mining. Knowl. Inf. Syst. **14**(1), 1–37 (2008). doi:10.1007/s10115-007-0114-2

47. Zhang, K., Xu, Z., Buckles, B.P.: Oblique decision tree induction using multi-membered evolution strategies. In: Dasarathy, B.V. (ed.) SPIE 2005, vol. 5812, pp. 263–270. SPIE (2005). doi:10.1117/12.596766

An Application of Generalized Strength Pareto Evolutionary Algorithm for Finding a Set of Non-Dominated Solutions with High-Spread and Well-Balanced Distribution in the Logistics Facility Location Problem

Filip Rudziński[✉]

Department of Electrical and Computer Engineering,
Kielce University of Technology, Al. 1000-lecia P.P. 7, 25-314 Kielce, Poland
f.rudzinski@tu.kielce.pl

Abstract. The paper presents an application of generalized Strength Pareto Evolutionary Algorithm (SPEA) in the Logistic Facilities Location (LFL) problem. The task is to optimize a distribution network, i.e. the number of distribution centers and their locations as well as the number of clients served by the particular centers in terms of three following contrary/contradictory criteria: (a) the total maintenance cost of the network, (b) carbon emissions emitted by combustion engines of trucks into the atmosphere (subjects to minimization) and (c) the customer service reliability (subject to maximization). For this purpose, an original multi-objective optimization technique which allow to obtain a set of so-called non-dominated solutions of the considered problem, representing different levels of compromise between the above criteria, is applied. In order to provide a broad, flexible selection of the final solution from the obtained set, the proposed approach aims at finding the set of solutions with high spread and well-balanced distribution in the objective (criteria) space. The functionality of our technique is demonstrated using numerical experiments. Its distinct advantages over alternative approaches are presented in the frame of comparative analysis as well.

Keywords: Multi-objective genetic optimization · Finding set of non-dominated solutions with high spread and well-balanced distribution · Logistic facilities location problem

1 Introduction

One of the currently discussed issues relating to road transport is the growing emissions of carbon dioxide emitted into the atmosphere by combustion engines. The existing models of distribution networks usually do not take into account this aspect, putting the accent only on minimizing the cost of maintenance and maximizing the customer service. The need of a widely understood environmental protection forces the seeking of new models which take into account the

© Springer International Publishing AG 2017
L. Rutkowski et al. (Eds.): ICAISC 2017, Part I, LNAI 10245, pp. 439–450, 2017.
DOI: 10.1007/978-3-319-59063-9_39

level of CO_2 emissions as well. These three criteria of the model evaluation are contrary/contradictory (they cannot be satisfy simultaneously in most cases in practice). Therefore, a multi-objective optimization technique which allow to obtain a set of so-called non-dominated solutions of the problem, representing different levels of compromise between such criteria, should be applied. The effective tools capable to solve this task are multi-objective evolutionary algorithms (MOEA) [1].

However, most popular MOEAs used in the multi-criteria optimization problems do not fulfill a number of requirements raised currently [2]. First, the *accuracy* of all solutions in the set, with respect to the optimal solutions, must be as high as possible. Then, a *spread* of the set (i.e., the distance between the extreme solutions) must be as wide as possible in order to achieve the solutions located on the boundaries of the objective space. Furthermore, the solutions must have a satisfactory *distribution* in the set (i.e., the distances between the nearest neighboring solutions in the objective space must be as similar as possible) in order to provide well-balanced levels of compromise between the values of particular optimization objectives. Finally, the obtained set must contain a balanced number of solutions (neither too many nor too few) in order to give a flexible choice of solutions for a decision maker.

In this paper, we apply a generalization of very known MOEA – the Strength Pareto Evolutionary Algorithm 2 (SPEA2) [3] – referred to as our SPEA3 [4] (see also [5–7] for its recent applications) in the Logistics Facility Location (LFL) problem introduced in [8]. In order to provide a broad flexible selection of the final solution from the obtained set of non-dominated solutions, the proposed approach aims at finding that set with high spread and well-balanced distribution in the objective (criteria) space. The functionality of our technique will be demonstrated using very known two- and three-objective benchmark tests. Then, our SPEA3 will be applied to optimize a distribution network, i.e. the number of distribution centers and their locations as well as the number of clients served by the particular centers in terms of three above-mentioned contrary/contradictory criteria. Distinct advantages our SPEA3 over alternative approaches will be presented in the frame of comparative analysis as well.

2   A Generalization of SPEA2 Approach - an Outline

For clarity of presentation, the SPEA2 approach and its our generalization referred to as SPEA3 are outlined in Table 1. The SPEA2 (and thus also SPEA3) uses an archive (an external set) of final solutions whose size can be regulated depending on the needs and adjusted to the requirements of a decision maker. The proposed generalization of SPEA2 consists in the exchange of its environmental selection procedure - which is responsible for selecting non-dominated solutions from the population to the archive (see Step 3 in Algorithm 1) - for a new original algorithm (Step 3, and Steps 3a-e in Algorithm 2) which aims to determine the final non-dominated solutions with a high spread and well-balanced distribution in the objective space.

Table 1. The SPEA2 algorithm and its generalization referred to as our SPEA3

Algorithm 1 – SPEA2 (Main loop) [3]:		Algorithm 2 – SPEA3 (Main loop):	
Input:	N (*population size*) $\bar{N}$ (*archive size*) T (*maximum number of generations*)	**Input:**	N (*population size*) $\bar{N}$ (*archive size*) T (*maximum number of generations*)
Output:	**A** (*non-dominated set*)	**Output:**	**A** (*non-dominated set*)
Step 1:	*Initialization*: Generate an initial population P_0 and create the empty archive (external set) $\bar{P}_0 = \emptyset$. Set $t = 0$.	**Step 1:** **Step 2:**	} As in Algorithm 1 (SPEA2).
Step 2:	*Fitness assignment*: Calculate fitness values of individuals in P_t and $\bar{P}_t$ (*cf. Section 3.1 in* [3]).	**Step 3:**	*Environmental selection*: If $\bar{P}_t$ is empty then copy three randomly selected individuals from P_t to $\bar{P}_{t+1}$ and go to Step 4, otherwise follow the Steps 3a–e.
Step 3:	*Environmental selection*: Copy all non-dominated individuals in P_t and $\bar{P}_t$ to $\bar{P}_{t+1}$. If size of $\bar{P}_{t+1}$ exceeds N then reduce $\bar{P}_{t+1}$ by means of the truncation operator, otherwise if size of $\bar{P}_{t+1}$ is less than N then fill $\bar{P}_{t+1}$ with dominated individuals in P_t and $\bar{P}_t$ (*cf. Section 3.2 in* [3]).	**Step 3a:** **Step 3b:**	*Make auxiliary archive $\bar{Q}_{t+1}$ by copying all non-dominated individuals in P_t, which are also not weakly dominated by at least one individual in $\bar{P}_t$.* *Copy all individuals in $\bar{P}_t$ to $\bar{P}_{t+1}$ and if size of $\bar{P}_{t+1}$ does not exceed $\bar{N}$ then replenish $\bar{P}_{t+1}$ with auxiliary individuals from $\bar{Q}_{t+1}$ using Sub-algorithm 2a.*
Step 4:	*Termination*: If $t \geq T$ or another stopping criterion is satisfied then set **A** to the set of decision vectors represented by the non-dominated individuals in $\bar{P}_{t+1}$. Stop.	**Step 3c:**	*Replace all individuals in $\bar{P}_{t+1}$ which are dominated by at least one individual in $\bar{Q}_{t+1}$ with their nearest neighbors belonging to $\bar{Q}_{t+1}$. Every time when the nearest neighbor replaces an individual in $\bar{P}_{t+1}$, it must be immediately removed from $\bar{Q}_{t+1}$.*
Step 5:	*Mating selection*: Perform binary tournament selection with replacement on $\bar{P}_{t+1}$ in order to fill the mating pool.	**Step 3d:**	*Minimize the distance differences between individuals in archive $\bar{P}_{t+1}$ using Sub-algorithm 2b.*
Step 6:	*Variation*: Apply recombination and mutation operators to the mating pool and set $\bar{P}_{t+1}$ to the resulting population. Increment generation counter ($t = t + 1$) and go to Step 2.	**Step 3e:**	*Clear all dominated individuals in archive $\bar{P}_{t+1}$ and go to Step 4.*
		Step 4: **Step 5:** **Step 6:**	} As in Algorithm 1 (SPEA2).

Two main activities of Algorithm 2 can be distinguished. The first activity (see Sub-algorithm 2a in Table 2) replenishes the archive by gradually adding auxiliary solutions selected from the population (starting with three randomly selected solutions and ending when the archive contains the desired number of solutions). The second activity (see Sub-algorithm 2b in Table 2) gradually relocates the archived solutions in the objective space in such a way that the distances between them and their nearest neighbors are the greatest possible.

The quality of the obtained set of non-dominated solutions (and, thus, the effectiveness of our approach) can be evaluated using several performance indices (their broad review can be found in [2]) with respect to the Pareto-optimal set or, if it is unknown, with respect to such-called *reference set*, which contains a finite number of Pareto-optimal solutions (or solutions that are as close to them as possible) created in an artificial way. However, in the case of the considered LFL problem (see Sect. 3) both sets are unknown. For this reason, before the target application, the operation of our SPEA3 will be clearly demonstrated using well-known numerical benchmark tests with known Pareto-reference sets (see Table 3

Table 2. The main activities of SPEA3

Sub-algorithm 2a (Replenishment of archive):		Sub-algorithm 2b (Minimization of distance differences between individuals in archive):	
Input:	$\bar{N}$ (*archive size*) $\bar{P}_{t+1}$ (*the main archive*) $\bar{Q}_{t+1}$ (*the auxiliary archive*)	**Input:**	$\bar{P}_{t+1}$ (*the main archive*) $\bar{Q}_{t+1}$ (*the auxiliary archive*)
Output:	$\bar{P}_{t+1}$ (*the main archive*) $\bar{Q}_{t+1}$ (*the auxiliary archive*)	**Output:**	$\bar{P}_{t+1}$ (*the main archive*) $\bar{Q}_{t+1}$ (*the auxiliary archive*)
Step 1:	*Make new archive $\bar{Z}_{t+1}$ as a copy of $\bar{P}_{t+1}$.*	**Step 1:**	*If size of $\bar{P}_{t+1}$ does not exceed 1 or $\bar{Q}_{t+1}$ is empty then stop the algorithm, otherwise set individual counter $i = 1$.*
Step 2:	*If $\bar{Z}_{t+1}$ is empty or size of $\bar{P}_{t+1}$ exceeds $\bar{N}$ then stop the algorithm.*	**Step 2:**	*Select i-th individual $\bar{P}_i$ in archive $\bar{P}_{t+1}$. Make new archive $\bar{Z}_{t+1}$ as a copy of $\bar{Q}_{t+1}$.*
Step 3:	*Determine individual $\bar{Z}_i$ in $\bar{Z}_{t+1}$ with the greatest distance to its nearest neighbor $\bar{Z}_j$ ($i \neq j$; $i,j \in \{1, 2, \dots, |\bar{Z}_{t+1}|\}$.)*	**Step 3:**	*If $\bar{Z}_{t+1}$ is empty then go to Step 8.*
Step 4:	*Select an auxiliary individual $\bar{Q}_{AUX}$ in $\bar{Q}_{t+1}$ for which an absolute value of the difference between distances from $\bar{Q}_{AUX}$ to $\bar{Z}_i$ and to $\bar{Z}_j$ is the smallest.*	**Step 4:**	*Among members of $\bar{Z}_{t+1}$ determine individual $\bar{Z}_{NN}$ which is the nearest neighbor of $\bar{P}_i$.*
Step 5:	*If the distance between $\bar{Z}_i$ and $\bar{Q}_{AUX}$ is greater than the distance between $\bar{Z}_i$ and $\bar{Z}_j$ then remove $\bar{Z}_i$ from $\bar{Z}_{t+1}$ and go to Step 2.*	**Step 5:**	*Among members of $\bar{P}_{t+1}$ determine individual $\bar{P}_j$ which is the nearest neighbor of $\bar{Z}_{NN}$, such that $i \neq j$.*
Step 6:	*If the distance between $\bar{Z}_j$ and $\bar{Q}_{AUX}$ is greater than the distance between $\bar{Z}_i$ and $\bar{Z}_j$ then remove $\bar{Z}_j$ from $\bar{Z}_{t+1}$ and go to Step 2.*	**Step 6:**	*If the distance between $\bar{Z}_{NN}$ and $\bar{P}_j$ is greater than the distance between $\bar{P}_i$ and $\bar{P}_j$ then replace $\bar{P}_i$ with $\bar{Z}_{NN}$ in $\bar{P}_{t+1}$.*
Step 7:	*Move auxiliary individual $\bar{Q}_{AUX}$ from $\bar{Q}_{t+1}$ to $\bar{P}_{t+1}$. Clear $\bar{Z}_{t+1}$ (or alternatively, remove only $\bar{Z}_i$ and $\bar{Z}_j$ from $\bar{Z}_{t+1}$, in order to fill $\bar{P}_{t+1}$ with more auxiliary individuals) and go to Step 2.*	**Step 7:**	*Remove $\bar{Z}_{NN}$ from $\bar{Z}_{t+1}$ and go to Step 3.*
		Step 8:	*Increment individual counter ($i = i + 1$). If i exceeds the size of $\bar{P}_{t+1}$ then stop the algorithm, otherwise go to Step 2.*

for details). The SCH1 test (called also the Shaffer test) and the DTLZ2 test (proposed by Deb, Thiele, Laumanns and Zitzler in [9] will be considered. Their reference sets are available on the jMetal Web Site (http://jmetal.sourceforge. net/problems.html)). The accuracy, spread and distribution of the obtained sets are evaluated with the use of two most often applied indices, i.e. the Generational Distance (GD) [10] (the accuracy measure) and the generalization of Δ metric proposed in [1] (the spread and distribution measure).

Figures 1 and 2 show the exemplary final approximations of Pareto-optimal fronts of the considered problems listed in Table 3 and obtained using our SPEA3 and two most popular MOEAs, i.e. SPEA2 [3] and ε-NSGA-II [11]. The experiments have been performed with the following parameters: the population of 100 individuals, the crossover and mutation probabilities equal to 0.8 and 0.2, respectively, the tournament selection (with the selection pressure equals to 2), and the maximum size of the final solution archive equals to 40 (for SPEA2 and our SPEA3) as well as the grid size equal to 100 (for ε-NSGA-II). The optimization process lasted for 10000 generations. It can clearly be seen that our approach determines the final solutions with the best spread and distribution in comparison to the remaining techniques, in the case of both considered

Table 3. Formulation of the benchmark problems

Test name	Objective functions	Search space	Optimal solutions
SCH1	$f_1(x) = x^2 \quad f_2(x) = (x - 2)^2$	$x \in [-10^3, 10^3]$	$x \in [0, 2]$
DTLZ2	$f_1(\boldsymbol{x}) = (1 + g(\boldsymbol{x})) \cos(x_1\pi/2) \cos(x_2\pi/2)$ $f_2(\boldsymbol{x}) = (1 + g(\boldsymbol{x})) \cos(x_1\pi/2) \sin(x_2\pi/2)$ $f_3(\boldsymbol{x}) = (1 + g(\boldsymbol{x})) \sin(x_1\pi/2),$ where $g(\boldsymbol{x}) = \sum_{i=2}^{12}(x_i - 0.5)^2$	$x_i \in [0, 1]$ $i = 1, 2, \ldots, 12$	$x_1, x_2 \in [0, 1]$ $x_i = 0.5,$ $i = 3, 4, \ldots, 12$

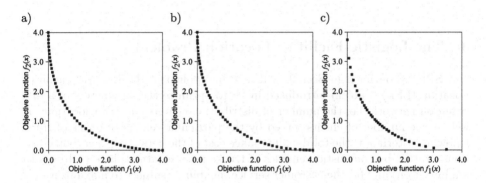

Fig. 1. Final approximations of the Pareto-optimal solution sets for our SPEA3 (a), SPEA2 (b), and ε-NSGA-II (c) algorithms (*SCH1* benchmark test)

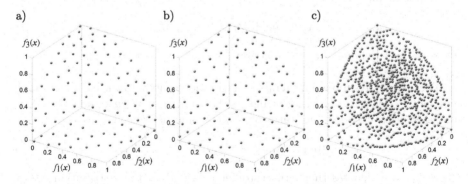

Fig. 2. Final approximations of the Pareto-optimal solution sets for our SPEA3 (a), SPEA2 (b), ε-NSGA-II (c) algorithms (*DTLZ2* benchmark test)

benchmark tests. A comparative analysis of our approach with the alternative techniques confirms the above findings. Table 4 contains the averaged results of 10-fold validations of each of the possible pairs of the considered benchmark tests and algorithms. In terms of determining the solutions with a high spread and well-balanced distribution, our approach is superior for both benchmark tests.

Table 4. A comparative analysis of our approach with alternative techniques

Test name	Algorithm	Generational distance (GD)		Spread and distribution measure (Δ)	
		Average	Std. deviation	Average	Std. deviation
SCH	Our SPEA3	1.3451E-3	6.1730E-5	**3.9995E-2**	**1.2251E-2**
	SPEA2	1.3380E-3	9.6155E-5	1.2570E-1	1.4079E-2
	NSGA-II	**9.5733E-4**	**5.3899E-5**	6.1462E-1	4.9235E-2
	ε-NSGA-II	1.5788E-3	1.2666E-4	5.0620E-1	2.3372E-2
DTLZ2	Our SPEA3	4.679E-3	2.031E-4	**1.117E-1**	**6.527E-2**
	SPEA2	1.521E-2	3.205E-3	1.33E-1	1.281E-2
	NSGA-II	5.144E-3	1.767E-3	5.729E-1	5.746E-2
	ε-NSGA-II	**3.031E-4**	**1.451E-5**	2.48E-1	6.733E-3

3   The Logistic Facilities Location Problem

Our SPEA3, outlined in Sect. 2, will now be applied in the Logistic Facilities Location (LFL) problem introduced in [8] to optimize the parameters of a distribution network, i.e. the number of distribution centers and their locations as well as the number of clients served by the particular centers in terms of three following criteria: (a) the total maintenance cost of the network, (b) carbon emissions emitted by combustion engines of trucks into the atmosphere (subjects to minimization) and (c) the customer service reliability (subject to maximization). In this model, the distribution centers are placed in the locations with numbers $j \in \{1, 2, \ldots, J\}$, where J means the number of locations. The status of j-th location is defined as follows:

$$X_j \in \{0, 1\}, \forall j = 1, 2, \ldots, J, \tag{1}$$

where $X_j = 1$ denotes that the distribution center in j-th location is open and $X_j = 0$ otherwise.

The goods are supplied from open distribution centers to the customers with numbers $i \in \{1, 2, \ldots, I\}$, where I means the number of all customers served by all distribution centers. The status of i-th customer is defined as follows:

$$Y_{ij} \in \{0, 1\}, \forall i = 1, 2, \ldots, I, j = 1, 2, \ldots, J, \tag{2}$$

where $Y_{ij} = 1$ denotes that i-th customer is served by the distribution center opened in j-th location and $Y_{ij} = 0$ otherwise.

Assuming that a single customer can be served only by one distribution center:

$$\sum_{j=1}^{J} Y_{ij} = 1, \forall i = 1, 2, \ldots, I, \tag{3}$$

and i-th customer cannot be served by closed distribution center:

$$Y_{ij} \leq X_j, \forall i = 1, 2, \ldots, I, j = 1, 2, \ldots, J, \tag{4}$$

the task is to find a set of variables $X_j^{(opt)}$ and $Y_{ij}^{(opt)}$, $i = 1, 2, \ldots, I$, $j = 1, 2, \ldots, J$ taking into account limitations (1–4), which represents optimal - from the point of view of the considered quality criteria (see below) - placement of a number of $J^{(opt)} \leq J$ distribution centers in possible locations and assignment each of the centers to certain number $I_j^{(opt)} \leq I$ ($\sum_{j=1}^{J} I_j^{(opt)} = I$) of customers.

In order to solve the above defined task, a multi-objective optimization technique which minimize three objective function:

$$f_1 = Q_{TC}, \; f_2 = Q_{CO_2}, \text{ and } f_3 = 1 - Q_{SR}, \tag{5}$$

must be applied, where Q_{TC}, Q_{CO_2}, and Q_{SR} are the quality criteria defined as follows:

- the total cost of the distribution network Q_{TC} being the sum of fixed costs to maintain the distribution centers Q_L and cost of transport services Q_T:

$$Q_{TC} = Q_L + Q_T = \sum_{j=1}^{J} f_j X_j + \sum_{j=1}^{J} \sum_{i=1}^{I} c_{ij} h_i d_{ij} Y_{ij}, \tag{6}$$

- the total level of carbon dioxide emissions Q_{CO_2} into the atmosphere from trucks:

$$Q_{CO_2} = \sum_{j=1}^{J} \sum_{i=1}^{I} d_{ij} [\epsilon_{vf} - \epsilon_{ve} \frac{h_i}{w} + \epsilon_{ve} \frac{h_i}{w}] Y_{ij}, \tag{7}$$

- indicator of the service reliability $Q_{SR} \in [0, 1]$, which should be understood as a statistical chance of delivery of the goods to the customer in the required time ($Q_{SR} \approx 1$ means high quality of service):

$$Q_{SR} = \min_{\substack{i=1,2,\ldots,I \\ j=1,2,\ldots,J}} \{(1 - F_v \frac{d_{ij}}{T_i}) Y_{ij}; \forall Y_{ij} \neq 0\}, \tag{8}$$

and, moreover, c_{ij} is the unit cost of delivery of the goods from a distribution center in j-th location to i-th customer, d_{ij} is the distance between i-th customer and j-th location, f_j is fixed cost of maintaining the distribution center in j-th location, h_i is demand for the goods of i-th customer, T_i is the required maximum time of delivery of the goods to i-th customer, ϵ_{ve} and ϵ_{vf} are the amount of CO_2 emissions for empty and fully loaded trucks, respectively, and w is the permissible payload of the truck. The speed of truck delivering goods from the distribution center to customers is modeled as a random variable v with a certain probability distribution function F_v [8].

4 Experimental Results

The parameters of the distribution network formulated in Sect. 3 are adjusted as in [8] in order to permit a direct comparison of the obtained results with the results of [8]. The costs of maintaining distribution centers (f) and customers'

Table 5. The costs of maintaining the distribution centers

j	1	2	3	4	5	6	7	8	9	10
f_j [\$]	78500	46900	92200	58700	89800	46300	85400	67100	76000	85300

demands for the goods (h) as well as the distances between customers and locations of distribution centers (d) are presented in Tables 5 and 6, respectively. The remaining parameters of the network are the following:

- the number of possible locations: $J = 10$,
- the number of customers: $I = 25$,
- the unit cost of delivery of the goods from a distribution center in j-th location to i-th customer: $c_{ij} = 1\$ $ ($i = 1, 2, \ldots, I$, $j = 1, 2, \ldots, J$),
- the amount of CO_2 emissions for empty and fully loaded trucks: $\epsilon_{ve} = 0.772\,\text{kg/km}$, $\epsilon_{vf} = 1.096\,\text{kg/km}$,
- the required maximum time of delivery of the goods to i-th customer: $T_i = 2h$ ($i = 1, 2, \ldots, I$),
- the permissible payload of the truck: $w = 25ton$,
- the average speed of the truck: $80\,\text{km/h}$
- the maximal fixed cost of maintaining the distribution center $Q_{TCMAX} = 700000\$$,
- the minimal indicator of the service reliability: $Q_{SRMIN} = 0.9$.

The Gaussian probability distribution function F_v of the random variable v has been assumed:

$$F_v(v) = \begin{cases} exp(-\frac{(v-60)^2}{2\sigma^2}), & \text{for } v < 60, \\ 0, & \text{elsewhere,} \end{cases} \quad \text{where } \sigma = 10. \tag{9}$$

Figure 3 presents final approximations of the Pareto-optimal sets for our SPEA3 (Fig. 3a), SPEA2 (Fig. 3b), and ε-NSGA-II (Fig. 3c) techniques. The particular solutions of obtained distribution networks are labelled by the numbers of distribution centers opened in them. Similarly as in the case of the experiments presented in Sect. 2, this time is also evident that our SPEA3 gives the final set with much better spread and distribution of the solutions than the remaining methods. In the case of our SPEA3, a "chain" of solutions (for $Q_{SR} = 0.95$) with well balanced distribution is clearly visible (see Fig. 3a). The remaining solutions (for $Q_{SR} > 0.95$) form a kind of wall of solutions also with the best spread and distribution. Among the alternative techniques, similar "chain" (but with not so good distribution) is visible only for ϵ-NSGA-II. The accuracies of solutions obtained with the use of all techniques are comparable.

Table 7 contains four exemplary configurations of obtained distribution networks. Three of them have been selected from the approximation of the Pareto-optimal set obtained using our SPEA3 (Fig. 3a). For the comparative purpose, the last one configuration (created using ϵ-constraint method) has been cited from [8]. The distribution center in j-th location is indicated by F_j, whilst C_i

Table 6. The demands for the goods of the customers as well as distances between the customers and the locations of distribution centers

i	h_i [$\times 10^3\ kg$]	j									
		1	2	3	4	5	6	7	8	9	10
		d_{ij} [km]									
1	166	85	47	100	86	166	262	13	272	40	251
2	156	241	276	42	152	122	53	173	182	65	156
3	88	297	147	209	124	11	88	241	104	25	154
4	59	111	222	158	242	246	57	38	247	192	75
5	163	269	155	164	182	229	257	115	26	221	100
6	191	252	112	249	53	39	264	14	207	212	132
7	79	114	294	120	133	98	48	95	269	75	94
8	141	123	213	44	262	139	25	10	226	211	65
9	99	204	168	256	168	271	126	108	147	77	279
10	170	141	77	130	211	121	55	257	176	113	67
11	159	66	157	131	223	60	255	204	41	258	22
12	50	183	163	49	27	232	230	127	18	176	53
13	176	219	161	76	276	228	267	210	56	222	21
14	199	234	151	128	184	257	202	158	90	212	115
15	113	171	267	253	270	282	245	61	101	27	79
16	180	42	128	103	163	278	90	102	258	103	37
17	126	153	257	116	209	189	136	143	285	26	84
18	48	135	177	264	141	132	224	111	259	140	150
19	56	147	69	26	21	267	70	259	214	262	282
20	93	42	119	295	194	269	145	55	187	70	159
21	155	218	183	177	131	74	129	48	183	288	29
22	169	11	266	75	32	245	43	264	29	106	179
23	162	176	201	195	131	42	226	73	196	258	26
24	77	292	10	251	251	15	164	283	97	242	181
25	116	237	240	15	85	197	147	292	225	171	90

denoted the i-th customer. The first distribution network with four distribution centers is characterized by a relatively small total cost of maintenance and high quality of the customer service, but at the expense of high emissions of carbon dioxide. The next solution with five distribution centers has a relatively good level of compromise between all considered evaluation criteria. The last our distribution network with seven distribution centers dominates the alternative solution from [8], i.e. it is better in terms of all evaluation criteria.

a)   b)   c)

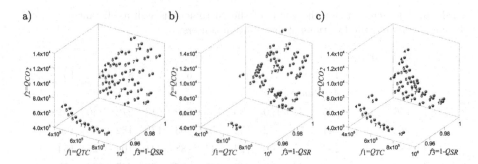

Fig. 3. Final approximations of the Pareto-optimal solution sets for our SPEA3 (a), SPEA2 (b), and ε-NSGA-II (c) algorithms (*LFL* problem)

Table 7. Comparison of the selected solutions with different levels of compromise between the evaluation criteria Q_{TC}, Q_{CO_2}, and Q_{SR}, obtained with the use of our SPEA3 and ε-constraint techniques

	Q_{TC} [$]	Q_{CO_2} [kg]	Q_{SR}	$J^{(OPT)}$	Configuration of the distribution network
our SPEA3	455483	7381.91	0.998	4	F_3 (C_{18}, C_5, C_{24}),
					F_7 (C_4, C_{17}, C_{21}, C_{13}, C_{11}),
					F_8 (C_1, C_{19}, C_8, C_6, C_0, C_{14}, C_{16}, C_2),
					F_9 (C_9, C_{10}, C_7, C_{22}, C_{15}, C_{12}, C_{20}, C_3, C_{23})
	494350	5885.52	0.993	5	F_5 (C_{18}, C_6, C_9, C_1, C_{17}),
					F_6 (C_7, C_0, C_5, C_3, C_{19}),
					F_7 (C_4, C_{21}, C_{13}, C_{23}, C_{11}),
					F_8 (C_8, C_2, C_{16}, C_{14}),
					F_9 (C_{10}, C_{20}, C_{22}, C_{12}, C_{24}, C_{15})
	606488	5304.11	0.998	7	F_0 (C_{21}, C_{19}),
					F_1 (C_{23}, C_{18}, C_{17}),
					F_5 (C_6, C_9, C_1),
					F_6 (C_5, C_3, C_7, C_0),
					F_7 (C_{13}, C_{11}, C_4),
					F_8 (C_8, C_2, C_{16}, C_{14})
					F_9 (C_{10}, C_{20}, C_{22}, C_{12}, C_{24}, C_{15})
ε-constraint [8]	658752	5313.7	0.9929	8	F_1 (C_{20}, C_{22}),
					F_2 (C_{24}),
					F_3 (C_{19}, C_{25}),
					F_4 (C_2, C_7, C_{10}),
					F_5 (C_1, C_4, C_6, C_8, C_{18}),
					F_6 (C_5, C_{12}, C_{14}),
					F_7 (C_3, C_9, C_{15}, C_{17}),
					F_8 (C_{11}, C_{13}, C_{16}, C_{21}, C_{23})

5 Conclusions

The application of generalized Strength Pareto Evolutionary Algorithm - referred to as our SPEA3 - for finding a set of non-dominated solutions with high-spread and well-balanced distribution in the Logistic Facilities Location (LFL) problem have been presented. The main goal of the problem is to optimize a distribution network, i.e. the number of distribution centers and their locations as well as the number of clients served by the particular centers in terms of three following contrary/contradictory criteria: (a) the total maintenance cost of the network, (b) carbon emissions emitted by combustion engines of trucks into the atmosphere (subjects to minimization) and (c) the customer service reliability (subject to maximization). First, very known two- and three-objective benchmark tests have been used to demonstrate the functionality of our SPEA3. Then, this technique has been applied to the target problem. The obtained results and comparative analysis with alternative approaches show the superiority of SPEA3. It generates better - in terms of the spread and distribution - non-dominated solutions than the alternative techniques, whilst it still remains competitive in terms of the accuracy of those solutions. In conclusion, our SPEA3 provides a broad, flexible selection of the final solution from the obtained set.

References

1. Deb, K.: Multi-Objective Optimization using Evolutionary Algorithms. Wiley, New York (2001)
2. Okabe, T., Jin, Y., Sendhoff, B.: A critical survey of performance indices for multi-objective optimisation. In: Proceedings of 2003 Congress on Evolutionary Computation, pp. 878–885. IEEE Press (2003)
3. Zitzler, E., Laumanns, M., Thiele, L.: SPEA2: improving the strength pareto evolutionary algorithm for multiobjective optimization. In: Proceeedings of Evolutionary Methods for Design, Optimisation, and Control, CIMNE, Barcelona, Spain, pp. 95–100 (2002)
4. Rudziński, F.: Finding sets of non-dominated solutions with high spread and well-balanced distribution using generalized strength Pareto evolutionary algorithm. In: Proceedings of the 16th World Congress of the International Fuzzy Systems Association (IFSA) and the 9th Conference of the European Society for Fuzzy Logic and Technology (EUSFLAT), IFSA-EUSFLAT 2015 (part of Advances in Intelligent System Research, vol. 89), pp. 178–185. Atlantis Press (2015)
5. Gorzałczany, M.B., Rudziński, F.: An improved multi-objective evolutionary optimization of data-mining-based fuzzy decision support systems. In: Proceedings of IEEE World Congress on Computational Intelligence, 25–29 July 2016, Vancouver, Canada, pp. 2227–2234 (2016)
6. Gorzałczany, M.B., Rudziński, F.: A multi-objective-genetic-optimization-based data-driven fuzzy classifier for technical applications. In: Proceedings of IEEE International Symposium on Industrial Electronics, June 8–10 2016, Santa Clara CA, USA, pp. 78–83 (2016)

7. Gorzałczany, M.B., Rudziński, F.: Classification of splice-junction DNA sequences using multi-objective genetic-fuzzy optimization techniques. In: Proceedings of the 16th International Conference on Artificial Intelligence and Soft Computing ICAISC 2017, June 11–15, Zakopane, in this volume (2017)

8. Xifeng, T., Zhang, J., Xu, P.: A multi-objective optimization model for sustainable logistics facility location. Transp. Res. (D) **22**, 45–48 (2013)

9. Deb, K., Thiele, L., Laumanns, M., Zitzler, E.: Scalable multi-objective optimization test problems. In: CEC 2002, pp. 825–830. IEEE Press (2002)

10. Van Veldhuizen, D.A.: Multiobjective Evolutionary Algorithms: Classifications, Analyses, and New Innovations. Air Force Institute of Technology, Wright Patterson AFB (1999)

11. Kollat, J.B., Reed, P.M.: Comparing state-of-the-art evolutionary multi-objective algorithms for long-term groundwater monitoring design. Adv. Water Resour. **29**(6), 792–807 (2006)

Efficient Creation of Population of Stable Biquad Sections with Predefined Stability Margin for Evolutionary Digital Filter Design Methods

Adam Slowik[✉]

Department of Electronics and Computer Science,
Koszalin University of Technology, Sniadeckich 2 Street, 75-453 Koszalin, Poland
aslowik@ie.tu.koszalin.pl

Abstract. Evolutionary methods are very often used to digital filters design (especially for digital filters with non-standard amplitude characteristics). In practical digital filter implementation, we can use a cascade of biquad sections. Each biquad section will be stable if all poles of transfer function are located into unitary circle in the z-plane. To generate $k - th$ stable biquad section, we can use an existing equations which are generated for the case when the coefficient $a_{k,0} = 1$. The problem is complicated if we want to have the vary values of all filter coefficients from the continuous range $[-1; 1]$ or from the discrete range $[-1; 1 - 2^{-M}]$ (if filter will be implemented into $Q.M$ fixed-point format). In this paper we have presented an efficient method for generation of stable biquad sections. The proposed method can be used in any evolutionary digital filter design method for increase its efficiency. Using proposed approach we can fast generate the population of stable biquad sections with prescribed stability margin. Due to presented approach, we can also very fast evaluate the stability of the given biquad section (the methods for polynomial roots generation are not needed).

1 Introduction

In digital filters design very often are used biquad sections. If we have to design a digital filter with typical amplitude characteristics we can use Butterworth, Cauer of Chebyshev approximations [1]. But the problem is more complicated if our filter will possess a non-standard amplitude characteristics. In this case to obtain a digital filter with arbitrary amplitude characteristics we can use Yule-Walker [2,3] method or one of evolutionary computation techniques [4,5]. Also, we can use the hybridization of Yule-Walker method with evolutionary method for more efficient design of digital filters with arbitrary amplitude characteristics [6]. This hybridization depends on initialization of the initial population in the evolutionary algorithm by the solution obtained using Yule-Walker method. The problem is more complicated if we want to design a IIR (Infinite Impulse Response) digital filter which can be directly implemented into the hardware without rounding errors. Then the coefficients obtained using Yule-Walker method must be scaled into the range $[-1; 1]$ and quantized in given

© Springer International Publishing AG 2017
L. Rutkowski et al. (Eds.): ICAISC 2017, Part I, LNAI 10245, pp. 451–460, 2017.
DOI: 10.1007/978-3-319-59063-9_40

Q.M fixed-point format. After these operations the IIR digital filter properties will be probably changed, because IIR digital filters are very sensitive on changing of coefficient values. Therefore, the application of other global optimization techniques is strongly recommended in this case. The evolutionary methods are widely used for digital filter design problems. In paper [7] the differential evolution algorithm was proposed for IIR (Infinite Impulse Response) digital filter design. In the article [8] the finite word-length digital filter was designed using an annealing algorithm, in the paper [9] the finite word-length design for IIR digital filters based on the modified least-square criterion in the frequency domain was shown. In the paper [10], the differential evolution algorithm was used for design of IIR digital filters with non-standard amplitude characteristics, in work [11] the continuous ant colony optimization algorithm was used for the same problem. In the paper [12] the application of evolutionary algorithm to design of minimal phase and stable digital filters with non-standard amplitude characteristics and finite bits word length was shown. In the article [6] the hybridization of evolutionary algorithm with Yule Walker method to design minimal phase and stable digital filters with arbitrary amplitude characteristics was presented. In the algorithms described in papers [6–12], the initial population of individuals was created randomly. Therefore, the some of created individuals were non-acceptable solutions (the created digital filters were unstable). In hardware implementation (for example in DSP processors, such as the TMS320C5000 [13,14]) the digital filters are represented by biquad sections [15]. Therefore, in this paper, we show how to efficiently generate a population of stable biquad sections with prescribed stability margin in evolutionary algorithm. The generation of stable biquad sections with prescribed stability margin improve the quality of evolutionary algorithm search process. It is especially important in the algorithms where the re-initialization of population is occurred (as for example micro genetic algorithm [16]). It is significant to point out, that due to our approach, we can also very fast evaluate the stability of the given biquad section (the methods for polynomial roots generation are not needed). The approach presented in this paper is a continuation (and extension) of the approach presented in the paper [18].

2   IIR Digital Filter Biquad Section

In general the IIR (Infinite Impulse Response) digital filter possesses the transfer function as follows:

$$H(z) = \frac{b_0 + b_1 \cdot z^{-1} + b_2 \cdot z^{-2} + ... + b_{n-1} \cdot z^{n-1} + b_n \cdot z^{-n}}{a_0 + a_1 \cdot z^{-1} + a_2 \cdot z^{-2} + ... + a_{n-1} \cdot z^{n-1} + a_n \cdot z^{-n}} \tag{1}$$

where: b_i and a_i (for $i = 0, 1, ..., n$) are the filter coefficients, n is a filter order.

One of the practical realization of the transfer function (1) is a cascade connection of biquad sections. Therefore, the transfer function $H(z)$ can be also represented by the following equation:

$$H(z) = H_1(z) \cdot H_2(z) \cdot H_3(z) \cdot ... \cdot H_{k-1}(z) \cdot H_k(z) \tag{2}$$

where: k is a number of biquad sections, $H_k(z)$ is a transfer function for $k - th$ biquad section.

Each $k - th$ biquad section is represented by the transfer function as follows:

$$H_k(z) = \frac{b_{k,0} + b_{k,1} \cdot z^{-1} + b_{k,2} \cdot z^{-2}}{a_{k,0} + a_{k,1} \cdot z^{-1} + a_{k,2} \cdot z^{-2}} = \frac{b_{k,0} \cdot z^2 + b_{k,1} \cdot z^1 + b_{k,2}}{a_{k,0} \cdot z^2 + a_{k,1} \cdot z^1 + a_{k,2}} \tag{3}$$

The $k - th$ biquad section is stable if and only if all poles of its transfer function $H_k(z)$ are located inside the unitary circle in the z-plane.

3   Q.M Fixed-Point Format

In many DSP (Digital Signal Processing) systems the numbers are represented by $Q.M$ fixed-point format. If we want to implement the designed digital filters into the $Q.M$ DSP system without any additional errors (i.e. filter coefficients rounding error) the value of $b_{k,0}$, $b_{k,1}$, $b_{k,2}$, $a_{k,0}$, $a_{k,1}$, $a_{k,2}$ coefficients must be taken from the predefined set X. The set X of potential values for these coefficients (in $Q.M$ fixed-point format) is defined as follows:

$$X = \left[\frac{(-1) \cdot 2^M}{2^M} ; \frac{2^M - 1}{2^M} \right] \tag{4}$$

In the 2's complement fractional representation, $M + 1$ bits (in fixed-point format $Q.M$) binary word can represent 2^{M+1} equally spaced numbers from $\frac{(-1) \cdot 2^M}{2^M} = -1$ to $\frac{2^M - 1}{2^M} = 1 - 2^{-M}$. The binary word BW which consists of $M + 1$ bits (bw_i):

$$BW = [bw_M, bw_{M-1}, bw_{M-2}, ..., bw_1, bw_0] \tag{5}$$

we interpret as a fractional number x:

$$x = -(bw_M) + \sum_{i=0}^{M-1} \left(2^{i-M} \cdot bw_i \right) \tag{6}$$

If we assure that the value of filter coefficients $b_{k,0}$, $b_{k,1}$, $b_{k,2}$, $a_{k,0}$, $a_{k,1}$, $a_{k,2}$ will be the elements from the set X then the digital filter will be resistive on rounding error after its implementation into $Q.M$ DSP system.

4   Biquad Section Stability and Stability Margin

Generally in the literature [15], we can find that the biquad section $H_k(z)$ with the values of $b_{k,0}$, and $a_{k,0}$ coefficients equal to one, and described as follows:

$$H_k(z) = \frac{1 + b_{k,1} \cdot z^{-1} + b_{k,2} \cdot z^{-2}}{1 + a_{k,1} \cdot z^{-1} + a_{k,2} \cdot z^{-2}} \tag{7}$$

is stable if and only if:

$$|a_{k,2}| < 1 \tag{8}$$

$$a_{k,1} < 1 + a_{k,2} \tag{9}$$

$$a_{k,1} > -1 - a_{k,2} \tag{10}$$

In practical realization of biquad sections, the stability margin has been introduced. Due to stability margin we can increase the certainty that given biquad section will be stable after its hardware implementation. In this case, we design a biquad section in which all poles of its transfer function are located inside the circle having the radius equal to: $1 - w$ (where w is a value of stability margin). Additionally, we have assumed that the value of coefficients $b_{k,0}$ and $a_{k,0}$ can be arbitrary from the range $[-1; 1]$. In the next section, we present a simple mathematical equations for a fast generation of stable biquad sections with prescribed value of stability margin.

5   Proposed Approach

Let's consider the biquad section $H_k(z)$ described by Eq. (3). Assume that the values for all coefficients $a_{k,i}$ and $b_{k,i}$ (for $i=0,1,2$) are from the range $[-1; 1]$. Also, we must remember, that if our biquad section will be implemented into DSP system with given $Q.M$ fixed-point format, then the values for all coefficients $a_{k,i}$ and $b_{k,i}$ (for $i=0,1,2$) must be from the range X (see Eq. 4). In the Fig. 1, we have presented the stability region (white color), non-stability region (black color) and the stability region with prescribed stability margin (gray color) for different values of parameter $a_{k,0}$ and different values of parameter w (stability margin). The horizontal axis represents the values of coefficient $a_{k,2}$, the vertical axis represents the values of coefficient $a_{k,1}$. The left-bottom corner is a point $(a_{k,2} = 1, a_{k,1} = -1)$. The right-upper corner is a point $(a_{k,2} = -1, a_{k,1} = 1)$.

Based on the Fig. 1, we can see that the stability region with prescribed stability margin (gray color) always is limited: in the right side of the axis $a_{k,2}$ by the line $a_{k,2} = (-1) \cdot (w - 1)^2 \cdot a_{k,0}$; in the left side of the axis $a_{k,2}$ by the line $a_{k,2} = (w - 1)^2 \cdot a_{k,0}$; in the upper side of the axis $a_{k,1}$ by the line $a_{k,1} = (-2 \cdot w + 2) \cdot a_{k,0}$; and in the bottom side of the axis $a_{k,1}$ by the line $a_{k,1} = (2 \cdot w - 2) \cdot a_{k,0}$. If we assume that the value of $a_{k,0}$ is positive, then the upper relation between $a_{k,1}$ and $a_{k,2}$ can be represented by the line undergoing by two points: $((-1) \cdot (w - 1)^2 \cdot a_{k,0}, 0)$ and $((w - 1)^2 \cdot a_{k,0}, (-2 \cdot w + 2) \cdot a_{k,0})$. Then the $a_{k,1}$ as a function of $a_{k,2}$ can be described as follows:

$$a_{k,1} = A \cdot a_{k,2} + B \tag{11}$$

where:

$$A = \frac{(-2 \cdot w + 2) \cdot a_{k,0}}{(w - 1)^2 \cdot a_{k,0} + (w - 1)^2 \cdot a_{k,0}} \cdot a_{k,2} \tag{12}$$

$$B = -\frac{((-2 \cdot w + 2) \cdot a_{k,0}) \cdot (-(w - 1)^2 \cdot a_{k,0})}{(w - 1)^2 \cdot a_{k,0} + (w - 1)^2 \cdot a_{k,0}} \tag{13}$$

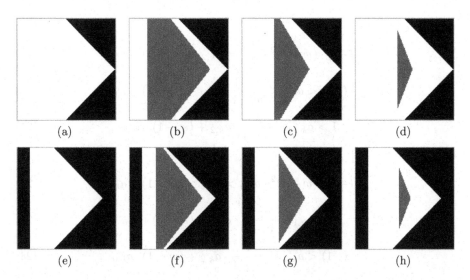

Fig. 1. The stability region (white color), non-stability region (black color), and stability region with prescribed stability margin (gray color) for different values of parameter $a_{k,0}$ and w: $a_{k,0} = 1$ and $w = 0$ (a), $a_{k,0} = 1$ and $w = 0.2$ (b), $a_{k,0} = 1$ and $w = 0.4$ (c), $a_{k,0} = 1$ and $w = 0.6$ (d), $a_{k,0} = 0.75$ and $w = 0$ (e), $a_{k,0} = 0.75$ and $w = 0.2$ (f), $a_{k,0} = 0.75$ and $w = 0.4$ (g), $a_{k,0} = 0.75$ and $w = 0.6$ (h)

after some simple mathematical transformations, we obtain:

$$a_{k,1} = \frac{(-1)}{w-1} \cdot a_{k,2} - (w-1) \cdot a_{k,0} \tag{14}$$

The bottom relation between $a_{k,1}$ and $a_{k,2}$ can be represented by the line undergoing by two points: $((-1) \cdot (w-1)^2 \cdot a_{k,0}, 0)$ and $((w-1)^2 \cdot a_{k,0}, (2 \cdot w - 2) \cdot a_{k,0})$. Then the $a_{k,1}$ as a function of $a_{k,2}$ can be described as follows:

$$a_{k,1} = A \cdot a_{k,2} + B \tag{15}$$

where:

$$A = \frac{(2 \cdot w - 2) \cdot a_{k,0}}{(w-1)^2 \cdot a_{k,0} + (w-1)^2 \cdot a_{k,0}} \cdot a_{k,2} \tag{16}$$

$$B = -\frac{((2 \cdot w - 2) \cdot a_{k,0}) \cdot (-(w-1)^2 \cdot a_{k,0})}{(w-1)^2 \cdot a_{k,0} + (w-1)^2 \cdot a_{k,0}} \tag{17}$$

after some simple mathematical transformations, we obtain:

$$a_{k,1} = \frac{1}{w-1} \cdot a_{k,2} + (w-1) \cdot a_{k,0} \tag{18}$$

If we do similar computation for the negative value of $a_{k,0}$, we can see that the all poles of transfer function (for $k - th$ biquad section) will be located into the stability area with prescribed stability margin if and only if:

- for $a_{k,0} \in [+1; 0)$

$$(-1) \cdot (w - 1)^2 \cdot a_{k,0} < a_{k,2} < (w - 1)^2 \cdot a_{k,0} \tag{19}$$

$$(-1) < a_{k,1} < \frac{(-1)}{w - 1} \cdot a_{k,2} - (w - 1) \cdot a_{k,0} \tag{20}$$

$$1 > a_{k,1} > \frac{1}{w - 1} \cdot a_{k,2} + (w - 1) \cdot a_{k,0} \tag{21}$$

- for $a_{k,0} \in (0; -1]$

$$(-1) \cdot (w - 1)^2 \cdot a_{k,0} > a_{k,2} > (w - 1)^2 \cdot a_{k,0} \tag{22}$$

$$1 > a_{k,1} > \frac{(-1)}{w - 1} \cdot a_{k,2} - (w - 1) \cdot a_{k,0} \tag{23}$$

$$(-1) < a_{k,1} < \frac{1}{w - 1} \cdot a_{k,2} + (w - 1) \cdot a_{k,0} \tag{24}$$

Using the Eqs. (19–24) in evolutionary algorithm, we can very fast generate the population of stable biquad sections with a prescribed stability margin or very fast evaluate the stability of given biquad section (in this case in the Eqs. (19)–(24) we must assume that the value of stability margin is equal to zero ($w = 0$)). Due to Eqs. (19–24) the population of stable biquad sections with prescribed stability margin can be generated very fast in evolutionary algorithm. In the Table 1 (part A), the pseudo-code for simple generation of stable $k - th$ biquad section (in continuous domain of digital filter coefficients) with prescribed stability margin is presented (the symbol r in Table 1 is a random value from the range $(0; 1)$).

Table 1. The pseudo-code for simple generation of stable $k - th$ biquad section coefficients $a_{k,i}$ (for $i = 0, 1, 2$) with prescribed stability margin w in: continuous coefficient domain $[-1; 1]$ (A), discrete coefficient domain $[-1; 1 - 2^{-M}]$ (in given $Q.M$ fixed-point format) (B)

A. In continuous coefficient domain	B. In discrete coefficient domain
if $r < 0.5$ then $a_{k,0} = r$	if $r < 0.5$ then $a_{k,0} = r * (1 - 2^{-M+1}) + 2^{-M}$
else $a_{k,0} = r - 1$; endif	else $a_{k,0} = r * (-2^{-M} + 1) - 1$; endif
$a_{k,2} = (w - 1)^2 * (2 * r * a_{k,0} - a_{k,0})$;	$a_{k,2} = (w - 1)^2 * (2 * r * a_{k,0} - a_{k,0})$;
$t_1 = \frac{1}{w-1} * a_{k,2} + (w - 1) * a_{k,0}$;	$t_1 = \frac{1}{w-1} * a_{k,2} + (w - 1) * a_{k,0}$;
if $(t_1 > 1)$ then $t_1 = 1$; endif	if $(t_1 > (1 - 2^{-M}))$ then $t_1 = 1 - 2^{-M}$; endif
if $(t_1 < -1)$ then $t_1 = -1$; endif	if $(t_1 < -1)$ then $t_1 = -1$; endif
$a_{k,1} = 2 * r * t_1 - t_1$;	$a_{k,1} = 2 * r * t_1 - t_1$;

If we want to generate (in evolutionary algorithm) a population of stable biquad sections with coefficients in given $Q.M$ fixed-point format, then the pseudo-code from Table 1 (part B) must be used and next the obtained values

$a_{k,i}$ and $b_{k,i}$ (for $i = 0, 1, 2$) must be transformed into the discrete values from the range X (see Eq. 4). This transformation for $a_{k,i}$ coefficients (for $i=0,1,2$) can be done using following formula:

$$ind_a = round \left(\frac{1 - 2^{M+1}}{2 - 2^{-M}} \cdot (a_{k,i} + 1) + 2^{M+1} \right) \tag{25}$$

where: $round(.)$ is a function which round the input argument to the nearest integer value, ind_a is a number of element in set X. The value of the element having index ind_a in set X is a new value in $Q.M$ fixed-point format for a given $a_{k,i}$ coefficient.

If we have value of ind_a, we can also very fast compute, the new value of $a_{k,i}$ coefficient in given $Q.M$ fixed-point format $(a_{k,i}^{Q.M})$ using equation:

$$a_{k,i}^{Q.M} = 1 - ind_a \cdot 2^{-M} \tag{26}$$

6   Description of Experiments

In order to test our approach, two procedures of random generation of population consists of $PopSize = 100$ individuals for evolutionary algorithm were created. Each individual consists of K stable biquad section with prescribed stability margin w. In the first procedure (named "Our approach" in Table 2), the coefficient $a_{k,i}$ (for $i=0, 1, 2$) values stored in individuals are generated randomly with the use of equations (19-24). In the second procedure (named "Standard approach" in Table 2), the coefficient $a_{k,i}$ (for $i=0, 1, 2$) values stored in individuals are generated randomly with uniform distribution. The experiments were performed using DELL Latitude E6420 computer and the FreeMat software [17] in version 4.2. The FreeMat is a free environment for rapid engineering and scientific prototyping and data processing [17]. It is similar to commercial system such as MATLAB [17].

In the first experiment, we have assumed that the value of coefficients $a_{k,i}$ (for $i=0, 1, 2$) are from the continuous range $[-1; 1]$. The symbols in Table 2 are as follows: K - the number of biquad sections in each individual in population, w - the value of stability margin, $StableM$ - the average number (with standard deviation) of stable biquad section for which all poles of the transfer function $H_k(z)$ are located in the circle having the radius $1 - w$ in the z-plane, Max - the obtained maximal value of transfer function pole. The average values (with standard deviations) were computed after 10-fold repetition of each procedure. The pseudo-code from Table 1 (part A) has been used as a "Our approach". The results obtained for this experiment has been presented in Table 2 (part A).

In Table 2 (part A), we can see that the biquad sections which were generated using our approach ("Our approach") are in all cases stable. Also in all cases the poles of transfer function are located inside the circle with the radius $1 - w$ in the z-plane. The correctness of generated individuals in evolutionary algorithm population is equal to 100%. In the case of procedure of standard individuals generation ("Standard approach") the correctness is from the range 15% up to 30%.

Table 2. The comparison of randomly generated populations of biquad sections with the values of $a_{k,i}$ coefficients from the continuous domain $[-1;1]$ (A), from the discrete domain $[-1;1-2^{-15}]$ (Q.15 fixed-point format) (B) - after 10-fold repetition of each procedure

| | | A. continuous domain $[-1;1]$ | | | | B. discrete domain $[-1;1-2^{-15}]$ | | | |
| | | Our approach | | Standard approach | | Our approach | | Standard approach | |
K	w	StableM	Max	StableM	Max	StableM	Max	StableM	Max
1	0.0	100 ± 0	0.98	29.90 ± 4.72	2678	100 ± 0	0.99	28.60 ± 4.32	4757
	0.1	100 ± 0	0.89	21.90 ± 3.69	40436	100 ± 0	0.89	18.40 ± 2.83	412
	0.2	100 ± 0	0.79	19.70 ± 4.08	1804	100 ± 0	0.79	16.30 ± 3.71	3524
2	0.0	200 ± 0	0.96	62.30 ± 4.54	932	199.80 ± 0.63	1.00	59.60 ± 6.36	312
	0.1	200 ± 0	0.87	44.20 ± 5.86	3300	200 ± 0	0.89	41.70 ± 5.12	1098
	0.2	200 ± 0	0.78	32.00 ± 5.73	866	199.90 ± 0.31	0.80	31.80 ± 4.44	2465
3	0.0	300 ± 0	0.99	92.10 ± 9.58	7377	299.80 ± 0.42	1.00	87.30 ± 7.95	4305
	0.1	300 ± 0	0.89	68.50 ± 5.50	1124	300 ± 0	0.89	62.30 ± 8.69	403
	0.2	300 ± 0	0.79	50.20 ± 5.67	2477	299.80 ± 0.63	0.80	50.20 ± 5.67	5062
4	0.0	400 ± 0	0.99	111.60 ± 10.16	1881	399.50 ± 0.97	1.00	120.80 ± 7.91	873
	0.1	400 ± 0	0.89	88.30 ± 6.94	8118	400 ± 0	0.89	90.50 ± 7.56	2291
	0.2	400 ± 0	0.79	62.50 ± 4.22	2751	399.89 ± 0.31	0.81	63.00 ± 5.96	1012
5	0.0	500 ± 0	0.99	141.60 ± 7.51	2271	499.60 ± 0.51	1.00	142.10 ± 9.74	10639
	0.1	500 ± 0	0.89	106.00 ± 9.92	1674	499.89 ± 0.31	0.90	115.10 ± 9.40	3792
	0.2	500 ± 0	0.79	86.20 ± 10.23	2914	499.70 ± 0.48	0.83	83.30 ± 9.90	11979
100	0.0	10000 ± 0	0.99	2894.00 ± 44.79	56414	9994.00 ± 2.10	1.00	2915.00 ± 31.65	32357
	0.1	10000 ± 0	0.89	2222.80 ± 51.34	94488	9997.50 ± 1.84	1.00	2200.30 ± 27.25	32076
	0.2	10000 ± 0	0.79	1621.10 ± 26.41	46384	9997.90 ± 1.28	0.86	1608.90 ± 39.37	13719

In the second experiment we want to show the computational time which is required for the generation of single biquad section using procedures: "Our approach" and "Standard approach". We have generate 500 biquad sections in 10-fold repetition. The average time (with standard deviation) for generation of single biquad section was equal: 0.1898 ± 0.0144 [ms] for "Our approach" and 0.1066 ± 0.0175 [ms] for "Standard approach". We can see that the "Our approach" procedure is about 44% more time expensive than the "Standard approach" procedure.

In the third experiment, we have assumed that the biquad section will be implemented into the DSP system with $Q.15$ fixed-point format ($M = 15$). Therefore, the value of coefficients $a_{k,i}$ (for $i=0, 1, 2$) are from the range $[-1;1-2^{-15}]$. The pseudo-code from table 1 (part B) has been applied and additionally, after generation of the values for biquad section coefficients, the generated coefficients were transformed into $Q.15$ fixed-point format using Eqs. (25) and (26). Before this experiments, the vector X (see Eq. 4) consists of $2^{M+1} = 2^{16} = 65536$ elements from the range $[-1;1-2^{-15}]$ was created and stored in the computer memory. The value of the first element from the set X is equal to $1-2^{-15}$ ($X(1) = 1-2^{-15}$), and the value of the last element from the set X is equal to -1 ($X(65536) = -1$). The average values (with standard deviations) were computed after 10-fold repetition of each procedure. The results obtained for this experiment has been presented in Table 2 (part B).

In Table 2 (part B), we can see that the biquad sections which were generated using our approach ("Our approach") are in almost all cases stable with prescribed stability margin. The correctness of generated individuals in evolutionary algorithm population is higher than 99%. In the case of procedure of standard individuals generation ("Standard approach") the correctness is from the range 15% up to about 30%.

In the fourth experiment we want to show the computational time which is required for the generation of single biquad section using procedures: "Our approach" and "Standard approach" (in Q.15 fixed-point format). We have generate 500 biquad sections in 10-fold repetition. The average time (with standard deviation) for generation of single biquad section was equal: 0.2136 ± 0.0264 [ms] for "Our approach" and 0.1280 ± 0.0148 [ms] for "Standard approach". We can see that the "Our approach" procedure is about 60% more time expensive than the "Standard approach" procedure.

In the fifth experiment, we have studied the computational time which is required to decide whether given biquad section is stable with prescribed stability margin. To perform this study, the population consisting of 100 individuals has been generated randomly with uniform distribution in 10-fold repetition. Each individual consists of one biquad section. Each biquad section from the population was validate on stability criteria (with stability margin equal to 0.1). We have used the Eqs. (19)–(24) for "Our approach", and $roots(.)$ and $abs(.)$ functions from FreeMat [17] software (in version 4.2) for "Standard method of validation". The average computational time which was required for stability validation of single individual in population was equal: 9.7700 ± 0.1220 [ms] for "Our approach" and 14.5060 ± 1.1870 [ms] for "Standard method of validation". We can see that the individual validation using "Our approach" procedure is about 40% faster than the individual validation using "Standard method of validation".

7   Conclusions

Due to approach presented in this paper, we can generate the population of stable biquad sections (with prescribed stability margin) for application in evolutionary algorithms (with 100% of correctness for continuous values of biquad section coefficients and in over 99% of correctness for discrete values of biquad section coefficients). The proposed approach can highly increase the efficiency of evolutionary methods for digital filters design problem. Especially, in the case when in the evolutionary algorithm, the block of re-initialization of individuals in population exists. Due to our approach we can efficiently generate the population of stable biquad sections with prescribed stability margin and we can efficiently validate the stability criteria for each individual in population. Also, it is worth to notice that our approach can be used for generation of minimal phase biquad section too.

References

1. Lyons, R.G.: Understanding Digital Signal Processing, 3rd edn. Prentice Hall, Upper Saddle River (2010)
2. Orfanidis, S.J.: Introduction to Signal Processing. Prentice-Hall, Englewood Cliffs (1995)
3. Ding, H., Lu, J., Qiu, X., Xu, B.: Anadaptive speech enhancement method for siren noise cancellation. Appl. Acoust. **65**, 385–399 (2004)
4. Michalewicz, Z.: Genetic Algorithms + Data Structures = Evolution Programs. Springer, Heidelberg (1992)
5. Goldberg, D.: Genetic Algorithms in Search, Optimization, and Machine Learning. Addison-Wesley Publishing Company Inc., Boston (1989)
6. Slowik, A.: Hybridization of evolutionary algorithm with yule walker method to design minimal phase digital filters with arbitrary amplitude characteristics. In: Corchado, E., Kurzyński, M., Woźniak, M. (eds.) HAIS 2011. LNCS (LNAI), vol. 6678, pp. 67–74. Springer, Heidelberg (2011). doi:10.1007/978-3-642-21219-2_10
7. Karaboga, N.: Digital IIR filter design using differential evolution algorithm. EURASIP J. Appl. Signal Process. **2005**(8), 1269–1276 (2005)
8. Benvenuto, N., Marchesi, M., Orlandi, G., Piazza, F., Uncini, A.: Finite wordlength digital filter design using an annealing algorithm. In: International Conference on Acoustics, Speech, and Signal Processing, vol. 2, pp. 861–864 (1989)
9. Nakamoto, M., Yoshiya, T., Hinamoto, T.: Finite word length design for IIR digital filters based on the modified least-square criterion in the frequency domain. In: International Symposium on Intelligent Signal Processing and Communication Systems, ISPACS, pp. 462–465 (2007)
10. Slowik, A., Bialko, M.: Design of IIR digital filters with non-standard characteristics using differential evolution algorithm. Bull. Pol. Acad. Sci. Techn. Sci. **55**(4), 359–363 (2007)
11. Slowik, A., Bialko, M.: Design and optimization of IIR digital filters with non-standard characteristics using continuous ant colony optimization algorithm. In: Darzentas, J., Vouros, G.A., Vosinakis, S., Arnellos, A. (eds.) SETN 2008. LNCS, vol. 5138, pp. 395–400. Springer, Heidelberg (2008). doi:10.1007/978-3-540-87881-0_39
12. Slowik, A.: Application of evolutionary algorithm to design of minimal phase digital filters with non-standard amplitude characteristics and finite bits word length. Bull. Pol. Acad. Sci. Techn. Sci. **59**(2), 125–135 (2011)
13. TMS320C54x DSP Library Programmers Reference, Texas Instruments (2001)
14. TMS320C55x DSP Library Programmers Reference, Texas Instruments (2002)
15. STMicroelectronics, "AN2874 Applications note", February 2009
16. Tiwari, S., Koch, P., Fadel, G., Deb, K.: Amga: an archive-based micro genetic algorithm for multi-objective optimization. In: Proceedings of the 10th Annual Genetic and Evolutionary Computation Conference, USA, pp. 729–736 (2008)
17. http://freemat.sourceforge.net/
18. Slowik, A.: On fast randomly generation of population of minimal phase and stable biquad sections for evolutionary digital filters design methods. In: Nguyen, N.T., Trawiński, B., Fujita, H., Hong, T.-P. (eds.) ACIIDS 2016. LNCS (LNAI), vol. 9621, pp. 511–520. Springer, Heidelberg (2016). doi:10.1007/978-3-662-49381-6_49

Computer Vision, Image and Speech Analysis

Contiguous Line Segments in the Ulam Spiral: Experiments with Larger Numbers

Leszek J. Chmielewski$^{(\boxtimes)}$, Maciej Janowicz, Grzegorz Gawdzik, and Arkadiusz Orłowski

Faculty of Applied Informatics and Mathematics – WZIM, Warsaw University of Life Sciences – SGGW, ul. Nowoursynowska 159, 02-775 Warsaw, Poland
{leszek_chmielewski,arkadiusz_orlowski}@sggw.pl
http://www.wzim.sggw.pl

Abstract. In our previous papers we have investigated the directional structure and the numbers of straight line segments in the Ulam spiral. Our tests were limited to primes up to 25 009 991 due to memory limits. Now we have results for primes up to about 10^9 for the previously used directional resolution, and for the previous maximum number but with greatly increased directional resolution. For the extended resolution, new long segments have been found, among them the first one with 14 points. For larger numbers and the previous resolution, the new segments having up to 13 points were found, but the longest one is still the one with 16 points. It was confirmed that the relation of the number of segments of various lengths to the corresponding number of primes for a given integer, for large numbers, is close to linear in the double logarithmic scale.

Keywords: Ulam spiral · Ulam square · Prime numbers · Line segments · Number of segments · Large numbers

1 Introduction

Studies on the set of prime numbers are important due to many reasons, including for example the design of ciphers. One of the ways through which this set can be investigated is the Ulam spiral [9]. It is a two-dimensional pattern formed by prime numbers distinguished in the set of natural numbers written down in a square grid as a spiral going from the center around to infinity [11]. This makes it possible to investigate some of the properties of this set with the use of the image processing methods. The Ulam spiral will be considered together with the square in which it is embedded, which we shall call the Ulam square. This square will be considered as an image composed of pixels, called also points.

There is a common opinion that the diagonals in the Ulam spiral are important, although the sources which can be found belong to the class of volatile publications [7,8]. This drew our attention to the question on contiguous line segments within the Ulam square.

© Springer International Publishing AG 2017
L. Rutkowski et al. (Eds.): ICAISC 2017, Part I, LNAI 10245, pp. 463–472, 2017.
DOI: 10.1007/978-3-319-59063-9_41

Our previous experience described in [2,3,5,6] suggests that there is a gap in lengths of the set of continuous line segments. Up till now we have investigated the Ulam square of dimension up to 5001×5001 which corresponds to a largest prime equal to 25 009 991. The directions of segments considered were those possible to be expressed as a quotient of integers up to 10. Under these conditions there are numerous segments of lengths from 2 to 10 points, less than ten segments 11 and 12 points long, and single segments 13 and 16 points long. There are neither segments longer than 16 nor segments of lengths 14 and 15 points found.

Now we have extended the search to the Ulam square up to 31623×31623 which corresponds to the largest prime equal to $1\,000\,014\,121 > 10^9$ if the same directions as previously are considered. We will show in this paper that there are still no segments of 14 and 15 points and that the 16 point segment is the longest one. This means that there are longer segments neither on lines inclined by a multiple of $45°$ on those with directions expressed by quotients of small integers. This can have some consequences for the research on prime numbers, although it is not clear yet in exactly which way.

We have also studied the segments with directions expressible by quotients of larger integers, up to 50. This was done for squares up to 5001×5001, as previously studied, due to the technical limitations we have at present. We will show here that under these conditions the 16 point segment is still the longest one, and there are still no segments of length 15. However, a first single segment of length 14 has been found. Also some new segments of other large length have been detected.

The next parts of this paper are organized as follows. In Sect. 2 we shall very briefly mention the method of finding line segments, described elsewhere in more detail. In Sect. 3 we shall present the results obtained for larger numbers, with the set of directions as previously considered. In Sect. 4 the results for an extended set of directions will be shown, but for numbers as those considered in previous papers. The paper will be concluded in Sect. 5.

2   Method in Brief

The method was described in detail in [2] and its main elements were repeated in [3,5,6], so here let us only provide the main functional information.

The central part of the Ulam spiral is shown in Fig. 1a. The origin of the coordinate system Opq is located in the center containing number 1. Let us denote the coordinates of the number in the Ulam square, say number five, as $p(5)$ and $q(5)$. Let us consider three points corresponding to numbers $23, 7, 19$ which form a contiguous segment. Its slope can be described by the differences in coordinates between its ends: $\Delta p = p(19) - p(32) = 2$ by $\Delta q = q(19) - q(32) = 2$. Now let us pay attention to the table shown in Fig. 1b called the *direction table*

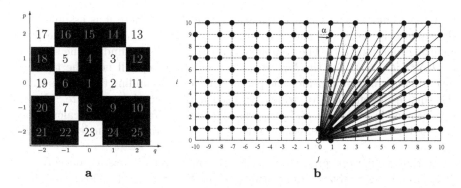

Fig. 1. (a) The central part of the Ulam spiral for dimensions 5 × 5. Coordinates are (p, q). Primes: black on white, other numbers: grey on black. (b) Directional vectors represented by the *direction table* D with elements D_{ij} containing increments $(\Delta p, \Delta q) = (i, j)$. Each vector has the initial point at the empty circle $(0, 0)$ and the terminal point in one of the black circles (i, j), $\neg(i = 0 \wedge j = 0)$. Angle α is the angle between the line segment and the vertical axis. Figures from [2], Copyright 2013 Machine Graphics & Vision. Reprinted with permission.

denoted D, with elements D_{ij}. If we denote $\Delta p = i$ and $\Delta q = j$ we can see that this segment, inclined at the angle $\alpha = -45°$, can be described, after reducing the directional vector from $(-2, 2)$ to $(-1, 1)$, by the point $(i, j) = (-1, 1)$ in the direction table. Its offset can be described as q of its section with axis Oq, that is $q(19) = -2$. The dimensions of table D_{ij} make it possible to represent slopes expressed by pairs $i \in [0 : N]$ and $j \in [-N, N]\backslash\{0\}$. By example, let us consider points $23, 2, 13$. They form a segment inclined by $(i, j) = (2, 1)$, with offset $q(2) = 1$. The points in this segment are the closest possible at this direction, although they are not neighbors in the normal sense. However, they are the closest possible at the direction considered; hence, the segment formed by them will be denoted as *contiguous*. The angle can take as many different values as is the number of black circles in Fig. 1b with thin lines.

The direction table can be used as the accumulator in the Hough transform for straight lines passing exactly through the points in the Ulam spiral. The neighborhood from which a second point is taken for each first point in the two-point Hough transform is related to the dimensions of D_{ij} (with avoiding doubling the pairs). During the accumulation process, in each D_{ij} a one-dimensional data structure is formed. For each vote the line offset and the locations of voting points are stored and the pairs and their primes are counted.

After the accumulation process the accumulator can be analyzed according to the need. Here, as it was in [3], we shall be interested in finding the numbers of contiguous segments of different lengths.

The object of our interest is well defined and is constant in time, so the calculations have to be carried out only once. However, let us remind after [5] that the complexity of the accumulation is $O(PN^2)$, where it is assumed that the square size is $S \times S$ and it contains $P = \pi(S^2)$ primes ($\pi(\cdot)$ is the prime counting function [10]). The complexity of the analysis for finding the segments is $O(N^2S^2log(S))$. For the largest squares analyzed up till now the time was several hours, but its considerable part was devoted to finding the primes within the square dimensions, for which we used a very simple procedure. Memory requirement is $O(N^2S^2)$ and this was the limiting factor in the calculations. We used a 64-bit machine with 64 GB of RAM, programmed single-threaded in GNU C.

In the calculations presented in the previous papers it was assumed $N = 10$ or less, and the analyzed squares were up to $S \times S = 5001 \times 5001$ containing 1 566 540 primes, the largest of which was 25 009 991. In the present paper we shall relax both limits, one at a time. The largest square we shall consider will be 31623×31623 which corresponds to the largest prime equal to 1 000 014 121 > 10^9. This is still very little in relation to the largest primes known at present, with the number of digits exceeding 20 millions [1]. However, a step of several orders of magnitude has been made, and next steps can be made should it be necessary.

3   Larger Numbers, Formerly Used Directions

For the direction table of dimensions $[0, 10] \times [-10, 10]$ as specified before, in [3] the squares of dimensions up to 5001×5001 were analyzed. Now we extend this dimension up to 31623×31623. In the graph in Fig. 2 the distribution of the numbers of segments versus their lengths is shown for the previously and newly investigated sizes of the Ulam square. The numbers of short segments grow together with the size of the square. This could have been expected due to that, informally speaking, in a larger square there are simply more points. However, the numbers of the longest segments increase only by a small difference, or even remain constant, as it is in the case of the globally longest segment found until now. For primes considered now, there are no segments of lengths 14 and 15, and the segment with 16 points remains single. A new 13 point segment was found among the numbers above 6.6×10^8.

Let us look now at the graphs of the numbers of segments versus the value of the largest prime number, and the number of prime numbers, as it was first done in [5], but with new data for larger primes added, in Fig. 3. What is interesting is that the tendency of the graphs to become a straight line for larger arguments starts to be seen also for the longest segments, like for example the 11 points segments, and also the 12 points ones, however less clearly. It is possible that for larger numbers also the 13 points segments could follow this tendency. Little can be said on the 16 points one, which is still single. The intriguing gap in length between the segments with 13 and 16 points stays up to the largest prime slightly over 10^9.

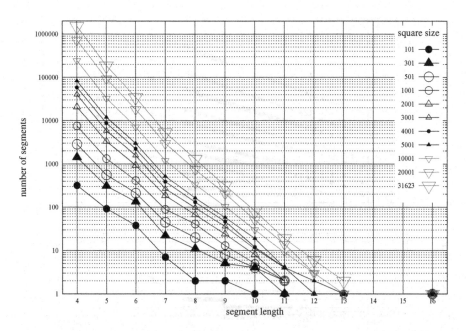

Fig. 2. Number of segments versus length of the segment for various sizes of the Ulam square, for direction array $N = 10$. If the number of lines is zero the data point is absent. The newly investigated square sizes graphed in red (■). (Color figure online)

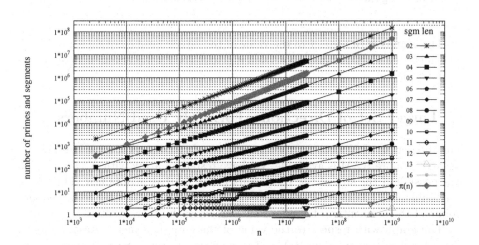

Fig. 3. Number of primes less than or equal to n, that is, the $\pi(n)$ function, shown in red (■)), and the numbers of segments of lengths encountered in the images considered, versus n, for direction array $N = 10$. If the number of segments is zero the data point is absent. Dark green (■) used to enhance discernibility. In each graph, three rightmost points were added w.r.t. [5]. (Color figure online)

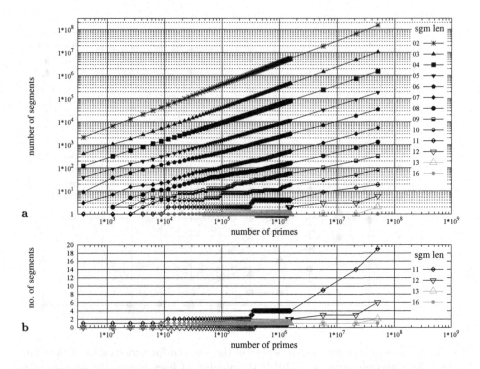

Fig. 4. Number of segments versus number of primes in the square, for direction array $N = 10$. (**a**) Results for all the segments; (**b**) results for the longest segments. In each graph, three rightmost points were added w.r.t. data shown in [5].

The graphs with respect to the value of the largest prime are shown in Fig. 4. The linear tendency is conspicuous and it is followed for larger numbers of primes without change.

4   Larger Set of Directions, Formerly Used Numbers

The restriction of $N = 10$ for the dimensions of the direction table can now be relaxed as more RAM is available. We have tested an array with $N = 50$, which was possible if the size of the Ulam square $S = 5001$ is maintained. This made it possible for the directions to be more numerous, but also for the subsequent points of the segments to be more remote from each other.

The graph with respect to the value of the largest prime (corresponding to Fig. 4 for direction table with $N = 10$) is shown in Fig. 5. Several observations can be made.

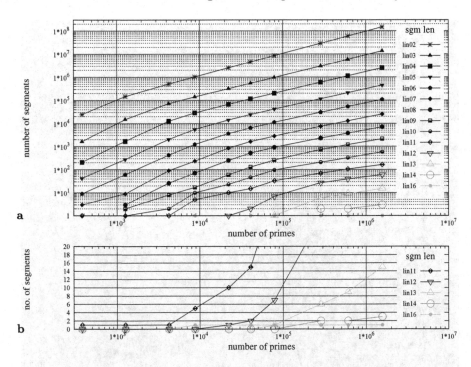

Fig. 5. Number of segments versus number of primes in the square, for direction array $N = 50$. **(a)** Results for all the segments; **(b)** results for the longest segments. In each graph, three rightmost points were added w.r.t. [5], Fig. 3.

The most apparent change is that there are three new long segments of 14 points. Two of them emerged near 3×10^5, and the third one for the primes over 10^6. The first two are shown in Fig. 6a, together with segment 16 for comparison of scale and range. In the new segments the distances between points are very long in comparison to the formerly found ones. Their directions are $(39, -37)$, $(43, 33)$ for the segments visible in Fig. 6 and $(47, -31)$ for the one farther from the spiral center.

In Fig. 7 the segments with 13 points visible in a 2001×2001 square are shown. One of them was found with directions at $N = 10$, the others were found with directions at $N = 50$. As in the case of 14 points segments, in the newly found segments the distances between their points are larger than in those found previously.

Consequently, it should be noted that the gap between the segments with 13 and 16 points has been partly filled.

Fig. 6. The Ulam square of size 2001×2001 for direction array $N = 50$ with segments of length 14 in red (■), direction $(39, -37)$ and magenta (■), direction $(43, 33)$. For comparison of scale and range the segment of length 16 also shown in yellow ( ), direction $(3, 1)$. Grey pixels (■): primes, black pixels: other numbers. Color pixels and center of the square (marked white) replaced with squares 7×7 for better visibility. (Color figure online)

The last observation is that the close-to-linear shape of the graphs became less apparent in the left-hand side and the central part of the graphs. The linearity seems to hold for numbers over 10^5, quite similarly as it was in the case of less directions. Extending the calculations to larger numbers could reveal the validity of the experimental asymptotic tendency in this case.

Fig. 7. The Ulam square of size 2001 × 2001 for direction array $N = 50$ with the new segments of length 13 in blue (■), green (■), red (■) and magenta (■). The segment present already for $N = 10$ shown in yellow ( ). Grey pixels (■): primes, black pixels: other numbers. Color pixels and center of the square (marked white) replaced with squares 7 × 7 for better visibility. (Color figure online)

5 Conclusion

The Ulam spiral containing prime numbers up to slightly above 10^9 was investigated from the point of view of the existence of contiguous straight line segments. This paper was a continuation of the previous works in which the Ulam square of size up to 5001 × 5001 were analyzed, for directions expressible by quotients of integers up to 10. In this paper we have studied the numbers in squares up to 31623 × 31623 which corresponds to the largest prime equal to $1\,000\,014\,121 > 10^9$, which is 40 times more than previously. Alternatively, square

of the previous size was analyzed, but the directions extended to those expressible by quotients of integers up to 50.

For larger numbers it was confirmed that the longest segment has 16 points
and that there are no segments of length 14 and 15. Previously, the segment
with 13 points was single, but now another such segment has been found at the
numbers above 6.6×10^8. For the extended directional resolution, new long segments have been found, among them the first one with 14 points. This partially
fills the previously observed gap between the segments having 13 and 16 points.

For large numbers, it was confirmed that the relation of the number of segments of various lengths to the corresponding number of primes is close to linear
in the double logarithmic scale. This holds for primes over 10^5 and segments of
lengths not larger than 11.

The data obtained in this and our other papers on the analysis of regularities
in the Ulam spiral in a downloadable form can be found in the web page [4].

References

1. Caldwell, C.K. (ed.): The Prime Pages (2016). primes.utm.edu. Accessed 15 Oct
   2016
2. Chmielewski, L.J., Orłowski, A.: Hough transform for lines with slope
   defined by a pair of co-primes. Mach. Graph. Vis. **22**(1/4), 17–25 (2013).
   mgv.wzim.sggw.pl/MGV22.html
3. Chmielewski, L.J., Orłowski, A.: Finding line segments in the Ulam square with the
   Hough transform. In: Chmielewski, L.J., Datta, A., Kozera, R., Wojciechowski, K.
   (eds.) ICCVG 2016. LNCS, vol. 9972, pp. 617–626. Springer, Cham (2016). doi:10.
   1007/978-3-319-46418-3_55
4. Chmielewski, L.J., Orłowski, A.: Prime numbers in the Ulam square (2016).
   www.lchmiel.pl/primes. Accessed 14 Oct 2016
5. Chmielewski, L.J., Orłowski, A., Gawdzik, G.: Segment counting versus prime
   counting in the Ulam square. In: Nguyen, N.T., Tojo, S., Nguyen, L.M., Trawiński,
   B. (eds.) ACIIDS 2017. LNCS, vol. 10192, pp. 227–236. Springer, Cham (2017).
   doi:10.1007/978-3-319-54430-4_22
6. Chmielewski, L.J., Orłowski, A., Janowicz, M.: A study on directionality in the
   Ulam square with the use of the Hough transform. In: Kobayashi, S., Piegat, A.,
   Pejaś, J., El Fray, I., Kacprzyk, J. (eds.) ACS 2016. AISC, vol. 534, pp. 81–90.
   Springer, Cham (2017). doi:10.1007/978-3-319-48429-7_8
7. GeMir: Quadratic polynomials describe the diagonal lines in the Ulam-Spiral.
   Mathematics Stack Exchange, 15 July 2015. math.stackexchange.com/q/1347560.
   Accessed 28 Jan 2017
8. PrimeCrank: Ulam spiral explained (sort of...). Prime Numbers on bark-
   erhugh.blogspot.com, 22 February 2012. barkerhugh.blogspot.com/2012_02_01_
   archive.html. Accessed 28 Jan 2017
9. Stein, M.L., Ulam, S.M., Wells, M.B.: A visual display of some properties of the distribution of primes. Am. Math. Mon. **71**(5), 516–520 (1964). doi:10.2307/2312588
10. Weisstein, E.W.: Prime counting function. From MathWorld-A Wolfram Web
    Resource (2016). mathworld.wolfram.com/PrimeCountingFunction.html. Accessed
    15 Oct 2016
11. Weisstein, E.W.: Prime spiral. From MathWorld-A Wolfram Web Resource (2016).
    mathworld.wolfram.com/PrimeSpiral.html. Accessed 15 Oct 2016

Parallel Realizations of the Iterative Statistical Reconstruction Algorithm for 3D Computed Tomography

Robert Cierniak[1]([⊠]), Jarosław Bilski[1], Jacek Smoląg[1], Piotr Pluta[1], and Nimit Shah[2]

[1] Institute of Computational Intelligence, Czestochowa University of Technology, Armii Krajowej 36, 42-200 Czestochowa, Poland
robert.cierniak@iisi.pcz.pl
[2] Department of Electrical Engineering, Maharaja Sayajirao University of Baroda, Vadodara, India

Abstract. The presented paper describes a parallel realization of an approach to the reconstruction problem for 3D spiral x-ray tomography. The reconstruction problem is formulated taking into consideration the statistical properties of signals obtained by x-ray CT and the analytical methodology of image processing. The concept shown here significantly accelerates calculations performed during iterative reconstruction process in the formulated algorithm. Computer simulations have been performed which prove that the reconstruction algorithm described here, does indeed significantly outperform conventional analytical methods in the quality of the images obtained.

Keywords: Image reconstruction from projections · X-ray computed tomography · Statistical reconstruction algorithm · Parallel computation

1 Introduction

Even though computed tomography was invented many years ago, it continues to be a very attractive field of research. Every new generation of CT devices stimulates the development of reconstruction algorithms adapted for the new design. The use of the algebraic method in the first historical CT apparatus was presumably because there was no alternative at the time. After this "early mistake", the next generation of CT systems used only reconstruction algorithms based on analytical image processing methods. The main reason for this was the huge size of the matrices which appear in the algebraic reconstruction problem and the calculation complexity of the reconstruction method based on this methodology that this caused. The analytical (or transformation) methodology drastically simplifies the number of calculations needed and so is more appealing. It is possible to improve the resistance of tomographic images to the measurement noise which occurs during image reconstruction by using appropriate

L. Rutkowski et al. (Eds.): ICAISC 2017, Part I, LNAI 10245, pp. 473–484, 2017.
DOI: 10.1007/978-3-319-59063-9_42

statistical signal processing. This means that it is possible to decrease the radiation intensity applied, and so decrease the dose absorbed by patients. Recently, some commercial solutions of such systems have been developed, which perform reconstruction processing iteratively to decrease the noise in the images. Their practical usefulness has been confirmed by many papers published in radiological journals. These systems take into consideration the probabilistic conditions present in the measurement systems of CT scanners in order to limit the influence of noise on the images obtained from the measurements. The most interesting approach, called MBIR (Model-Based Iterative Reconstruction), is presented in such papers as [1,2]. The huge number of coefficients in this model means that it is impossible to keep all of them in memory at the same time and the requirement for the simultaneous calculation of all voxels in the range of the reconstructed 3D image make the reconstruction problem extremely complex. Although, there have been attempts to decrease the calculation complexity of this approach, as presented for example in the paper [3], they have, as yet, only met with limited results. Therefore, there is still room for improvement of such systems. It would be much more profitable to construct a statistical reconstruction method which would take into consideration the statistical conditions of the measurement process and the geometry of the projections performed, thereby eliminating most of the disadvantages of the algebraic scheme of signal processing methodology. We could avoid the above mentioned difficulties connected with using an algebraic methodology by using an analytical strategy for the reconstructed image processing. In previous papers, we have shown how to formulate the analytical reconstruction problem consistent with the ML methodology for scanners with parallel geometry [4–6], for fan-beams [8], and finally we have proposed a scheme of reconstruction method for the spiral cone-beam scanner [9]. Our approach has some significant advantages compared with algebraic methodology. Firstly, in our method, we establish certain coefficients, but this is performed in a much easier way than in comparable methods. Secondly, we perform the reconstruction process in only one plane in 2D space, greatly simplifying the problem. In this way, the reconstruction process can be performed for every cross-section image separately. After this, it is possible to reconstruct the whole 3D volume image from the set of previously reconstructed 2D images. And finally, because of the analytical methodology of the reconstruction process, we can perform most of the computationally expensive operations in the frequency domain (2D convolutions). Because it is a very much less computationally demanding approach, by using FFT, we make our reconstruction method independent of the dimensions of the reconstructed image, to an acceptable degree. The main motivation for this paper is to present a feasible, practical solution for 3D spiral tomographic scanners based on the analytical statistical reconstruction approach mentioned above with parallel realization of the most computational demanding elements of the iterative reconstruction procedure.

2   3D Reconstruction Algorithm for the Spiral Cone-Beam Scanner

The 2D analytical approximate reconstruction problem was originally formulated for a parallel scanner geometry [5–8,10,11]. However, the concept can also form the starting point for the design of a 3D reconstruction algorithm for a spiral cone-beam scanner geometry. Because the basic methodology is very strongly associated with the parallel geometry of x-ray beams, we preferred to choose an appropriate signal processing strategy from among reconstruction algorithms which rely on rebinning. This could be, for example, the reconstruction idea involved in the SSR (Single Slice Rebinning) method [12] or the principles behind the ASSR (Advanced Single Slice Rebinning) algorithm [13]. Because of the practical advantages of the ASSR algorithm, we favored this latter approach. The general scheme of the reconstruction procedure we propose is depicted in Fig. 1.

A three-dimensional view of the scanner is given in Fig. 2.

The system consists of an x-ray tube and a rigidly coupled, partial cylindrical screen with a multi-row matrix of detectors. During a scan, this assembly rotates around the z-axis, the principal axis of the system, and at the same time, the patient table moves into the gantry. The moving projection system thus traces a spiral path around the z-axis. The projection function measured at the screen in a cone-beam system can be represented by $g^h\left(\beta, \alpha^h, \dot{z}\right)$, where β is the angle between a particular ray in the beam and the axis of symmetry of the moving 'projection system; α^h is the angle at which the projection is made, i.e. the angle between the axis of symmetry of the rotated projection system and the y-axis; $\dot{z}$ is the z-coordinate relative to the current position of the moving projection system. R is the radius of the circle defining the space in which the scan is carried out and R_f is the radius of the circle described by the focus of the tube. Unlike β, which is an angle, $\dot{z}$ represents a distance on the screen. It is the distance between the point where a ray strikes the screen and the vertical plane of symmetry of the projection system. Assuming that the tube rotating around the test object starts at a projection angle $\alpha^h = 0$, the vertical plane of symmetry of the projection system (and the focus of the tube) moves along the z-axis and its current location along the axis is defined by the relationship:

$$z_0 = \lambda \cdot \frac{\alpha^h}{2\pi}, \tag{1}$$

where: λ is the relative travel of the spiral described by the tube around the test object.

After selecting the angle of rotation α_p^h of the spiral projection system so that the central ray of the beam intersects the z-axis at the midpoint of reconstructed slice, we then have to determine the inclination of the plane of the slice, represented by the symbol v. The image is reconstructed from projections made in this case by only those selected rays in the conical beam, which lie exactly in the plane at three positions, that is for: $\alpha^h = \alpha_p^h$, $\alpha^h = \alpha_p^h + \alpha_*^h$ and $\alpha^h = \alpha_p^h - \alpha_*^h$. Other projections, necessary for the reconstruction, will be obtained using interpolation based on measurements performed at positions other than the desired

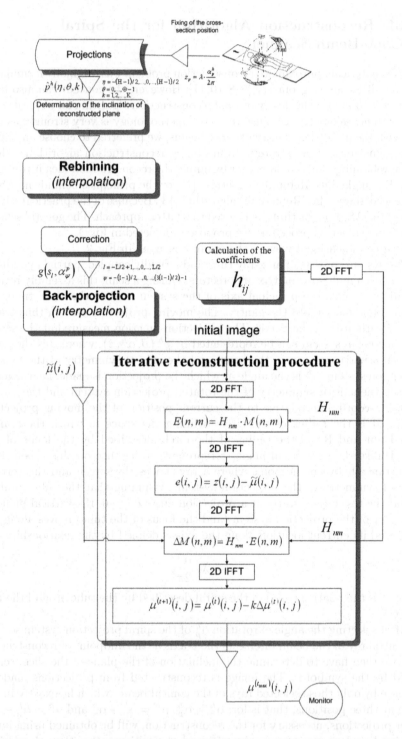

Fig. 1. An image reconstruction algorithm for the cone-beam geometry scanner

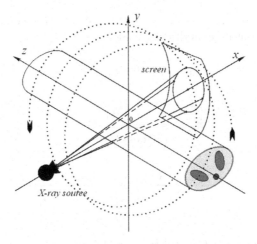

Fig. 2. The projection system of a cone-beam scanner – a three-dimensional perspective view

positions of the tube in relation to the z axis. If the inclination angle meets the following condition:

$$v = \arctan \left(\frac{\lambda \cdot \arccos \left(\frac{1}{2} \left(1 + \cos \left(\frac{\pi}{2} \right) \right) \right)}{2\pi R_f \sin \left(\arccos \left(\frac{1}{2} \left(1 + \cos \left(\frac{\pi}{2} \right) \right) \right) \right)} \right) \tag{2}$$

then errors of interpolation will be minimized.

In the next step of the reconstruction algorithm, we reformulate the 3D reconstruction problem as a 2D problem, taking into consideration a reconstruction procedure for parallel virtual rays lying in the reconstruction plane. These virtual rays, which establish the projections necessary for the reconstruction procedure, are approximated by rays obtained by interpolation of real rays obtained from a physical scan. This operation is called longitudinal approximation. This approximation uses the fact that both the interpolated ray and the desired parallel ray lie in the same plane parallel to the z-axis. In the longitudinal approximation, we firstly specify all the parameters of the parallel projections $g^p \left(s_l, \alpha_\psi^p \right)$ needed for the reconstruction procedure. We assume that the virtual detectors are equidistant on a flat screen and fixed at the points $s_l = l \cdot \Delta_s$, where $l = -L/2 - 0.5, \ldots, L/2 - 0.5$, and L is a even number of detectors. Every virtual parallel projection is performed in an equiangular way at specific angles $\alpha_\psi^p = \psi \cdot \Delta_\psi$, where $\psi = 0, \ldots, \Psi - 1$, and Ψ is the number of projections carried out. Based on the set of values of parameters $\left(s_l, \alpha_\psi^p \right)$, we can establish the projection angles α^h at which the helical projections should be performed, according to the following geometrical relation:

$$\alpha_{l\psi}^h = \alpha_\psi^p - \arcsin \frac{s_l}{R_f} + \alpha_p^h. \tag{3}$$

The following pair of equations allows us to calculate the remaining coordinates of this projection onto a cylindrically shaped screen:

$$\beta_{l\psi} = -\arctan\frac{w}{R_f + R_d} \tag{4}$$

and

$$\dot{z}_{l\psi} = \frac{vR_f}{\sqrt{(R_f + R_d)^2 + w^2}}. \tag{5}$$

where:

$$w = -\frac{R_f + R_d}{R_f} \cdot \frac{s_l}{\cos\left(\alpha^h - \alpha_p^h - \alpha_\psi^p\right)} \tag{6}$$

and

$$v = \frac{R_f + R_d}{R_f}\left(\frac{s_l \cdot \cos\left(\alpha^h - \alpha_p^h\right) \cdot \tan\upsilon}{\cos\left(\alpha^h - \alpha_p^h - \alpha_\psi^p\right)} - \lambda\frac{\alpha^h - \alpha_p^h}{2\pi}\right). \tag{7}$$

It is highly unlikely that any physical ray will be consistent with the line described by the parameters specified by (3), (5) and (6). That is why an additional interpolation operation is necessary to establish a projection value $\dot{g}^h\left(\beta_{l\psi}, \alpha_{l\psi}^h, \dot{z}_{l\psi}\right)$. This can be done using trilinear interpolation based on the eight projections nearest to the desired ray.

It is worth noting that the interpolated ray passes through the various tissues along a path longer than that of the approximated parallel ray. We therefore have to make a correction for this effect. This correction can be performed by multiplying the interpolated projection by the following factor:

$$CORR_1 = \cos\varepsilon \tag{8}$$

$$= \frac{(R_f + R_d)\cos\left(\alpha^h - \alpha_p^h - \alpha_\psi^p\right)\cos\upsilon}{\sqrt{w^2 + v^2 + (R_f + R_d)^2} \cdot \sqrt{\sin^2\alpha_\psi^p + \cos^2\upsilon\sin^2\alpha_\psi^p}}$$

$$+ \frac{w\sin\left(\alpha^h - \alpha_p^h - \alpha_\psi^p\right)\cos\upsilon - v\sin\alpha_\psi^p\sin\upsilon}{\sqrt{w^2 + v^2 + (R_f + R_d)^2} \cdot \sqrt{\sin^2\alpha_\psi^p + \cos^2\upsilon\sin^2\alpha_\psi^p}},$$

where Eqs. 6 and 7 still hold true.

To avoid having to transfer the results of the reconstruction procedure from the local coordinate system of the reconstructed cross-section to the global coordinate system, the projection values should be corrected by a second factor, which can be expressed in the following way:

$$CORR_2 = \frac{\cos\upsilon}{\sqrt{\cos^2\alpha^p + \cos^2\upsilon\sin^2\alpha_\psi^p}}. \tag{9}$$

Taking all the above considerations together, it is possible to write the approximated parallel projection as:

$$g^p(s, \alpha^p) = \dot{g}^h\left(\beta, \alpha^h_{s,\alpha^p}, \dot{z}\right) \cdot CORR_1 \cdot CORR_2, \tag{10}$$

where $\dot{g}^h\left(\beta, \alpha^h_{s,\alpha^p}, \dot{z}\right)$ is an interpolated value of the helical projection for parameters specified by relations (3), (4) and (5).

After the operations described above, we have a set of the parallel projections $g^p\left(s_l, \alpha^p_\psi\right)$ ready to be applied to a reconstruction method designed for this type of scanner geometry, as it has been described in e.g. [11]. This method is formulated in the folowing way

$$\mu_{\min} = \arg\min_\mu \left(\sum_{i=1}^{I} \sum_{j=1}^{J} \left(\sum_{\bar{i}} \sum_{\bar{j}} \mu(\bar{i}, \bar{j}) \cdot h_{\Delta i, \Delta j} - \tilde{\mu}(i,j) \right)^2 \right), \tag{11}$$

where the function $\mu(x, y)$ denotes the unknown image representing a cross-section of an examined body, and the function $\tilde{\mu}(i, j)$ denotes the image obtained after the back-projection operation using samples of the projections $g^p\left(s_l, \alpha^p_\psi\right)$.

We propose the gradient descent method to solve the optimization problem described by the formula (11) because of its simplicity. In this case, the pixels in the reconstructed image will take values according to the following very easy iterative procedure

$$\mu^{*(t+1)}(i, j) = \mu^{*(t)}(i, j) - \sum_{\bar{i}=1}^{I} \sum_{\bar{j}=1}^{J} e^{(t)}(\bar{i}, \bar{j}) h_{\Delta i, \Delta j}, \tag{12}$$

where

$$e^{(t)}(\bar{i}, \bar{j}) = \sum_{i} \sum_{j} \hat{\mu}^{*(t)}(i, j) \cdot h_{\Delta i, \Delta j} - \tilde{\mu}(\bar{i}, \bar{j}). \tag{13}$$

3   Experimental Results

In our experiments, we have adapted the well-known 3D Shepp-Logan mathematical phantom of the head. First, before the virtual parallel projections were obtained, we determined the pitch (at $\lambda = 2$) and we pointed those places on the z axis at which the plane of reconstruction intersects the axis (we chose: $z_p = -22.5$; see Fig. 3a). After this decision, for each position z_p, the angle v was calculated using the formula (2).

During the simulations, for parallel projections, we fixed L = 1024 virtual measurement points (detectors) on the screen. The number of parallel views was chosen as $\Psi = 1610$ per half-rotation and the size of the processed image was fixed at I × I = 1024 × 1024 pixels.

After making these assumptions, it is then possible to conduct the virtual measurements and complete all the required parallel projections which relate to the cone-beam lines, according to formulas (3)–(7).

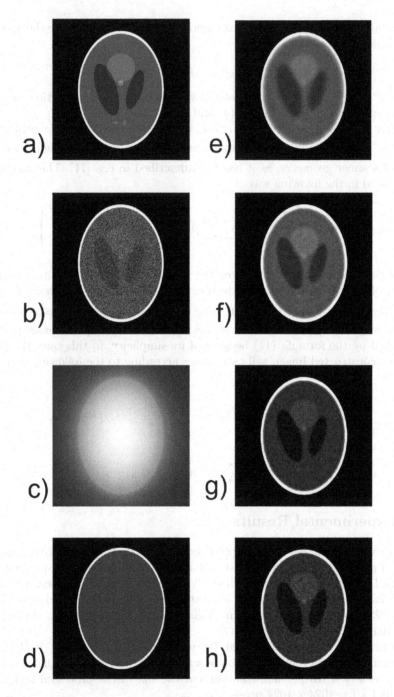

Fig. 3. Views of the images: original image (a); reconstructed image using the standard FBP method with Shepp-Logan kernel (b); image obtained after back-projection opeartion (c); starting image for the iterative reconstruction procedure (d); reconstructed images using the method described in this paper after: $t = 10^3$ (e), $t = 3 \cdot 10^3$ (f), $t = 5 \cdot 10^3$ (g), $t = 3 \cdot 10^3$ (h) iterations.

Then, through longitudinal approximation and suitable corrections, the back-projection operation can be carried out to obtain an image $\tilde{\mu}(i,j)$ (see Fig. 3c) which can be used as a referential image for the optimization procedure to be realized iteratively. There are depicted the reconstructed images after 1000, 3000, 5000 and 10000 iterations in the Figs. 3e–h, respectively.

Coefficients $h_{\Delta i, \Delta j}$ necessary for the optimization procedure can be pre-calculated before the reconstruction process is started, and in our experiments, these coefficients were fixed for the subsequent processing. The image obtained after the back-projection operation was then subjected to a process of reconstruction (optimization) using a gradient descent method, whose procedure is described by relations (12) and (13), wherein convolution operations were performed in the frequency domain. For comparison, a view of the reconstructed image using a traditional FBP algorithm is also presented (see Fig. 3b).

The motivation for this paper was to present a feasible, practical solution for 3D spiral tomographic scanners based on the analytical statistical reconstruction approach mentioned above with parallel realization of the most computational demanding elements of the iterative reconstruction procedure. During our experiments, we have carried out calculations necessary to realize the iterative reconstruction procedure using three approaches to hardware implementation which are described in Table 1.

Table 1. Hardware configurations

Sequential implementation (without optimization)	GPU	Assembler (with parallelization of the calculations)
IvyBridge 3570, RAM 16 GB	Latest 6th Gen. Intel Core i7 processor, Chipset Intel HM17, RAM DDR4-2133, 32 GB, GTX 960 M Engine Specs: 640CUDA Cores 1096 + BoostBase Clock (MHz) GTX 960 M Memory Specs:2500 MHzMemory ClockGDDR5Memory Interface128-bitMemory Interface Width 80	IvyBridge 3570, RAM 16GB

There are presented results obtained using the above mentioned realizations in the Table 2.

Table 2. Comparison of the computation times for the different realizations of the iterative reconstruction procedure.

Nuber of iterations	Sequential realization	GPU	Assembler realization
1000	2820000 ms	22625,2 ms	814000 ms
3000	8460000 ms	67922,38 ms	2442000 ms
5000	14100000 ms	113207 ms	4070000 ms
10000	28200000 ms	226322,1 ms	8140000 ms

4   Conclusion

In this paper, it has been proven that this statistical approach, which was originally formulated for parallel beam geometry, can be adapted for helical scanner geometry. We have presented a fully feasible statistical reconstruction algorithm for helical cone-beam projections. Simulations have been performed, which prove that our reconstruction method is very fast (thanks to the use of FFT algorithms) and gives satisfactory results with suppressed noise, even without the introduction of any additional regularization term. The computational complexity for 2D reconstruction geometries (e.g. parallel rays), is proportional to $I^2 \Psi \times N$ for each iteration of the algebraic reconstruction procedure (if we use the approximate algorithmic solution presented in [3]), but our original analytical approach only needs approximately $4I^2 \log_2 (2I)$ operations. For the 3D reconstruction problem these proportions change to $I^2 \times numbers\_of\_reconstructed\_cross-sections \times \Psi \times N$ while our analytical approach still only requires $4I^2 \log_2 (2I)$ i.e. there is no increase. In comparison to the ASSR algorithm for spiral cone-beam geometry, our reconstruction method (presented in this paper) differs only in the final stage: where FBP (filtration back-projection) is performed in the traditional ASSR approach, and an iterative reconstruction procedure after the back-projection operation in our approach. Simulations have been performed, which prove that our reconstruction method leads to a reconstructed image with suppressed noise even without introducing any additional regularization term. These simulations shown that our reconstruction method is extremely fast, and the whole iterative reconstruction process can be competed within 1 min, what is possible thanks to the use of the GPU. Also soft computing techniques can find their application in this parallelization (see e.g. [14–20]).

References

1. DeMan, B., Basu, S.: Distance-driven projection and backprojection in three dimensions. Phys. Med. Biol. **49**, 2463–2475 (2004)
2. Thibault, J.-B., Sauer, K.D., Bouman, C.A., Hsieh, J.: A three-dimensional statistical approach to improved image quality for multislice helical CT. Med. Phys. **34**(11), 4526–4544 (2007)

3. Zhou, Y., Thibault, J.-B., Bouman, C.A., Sauer, K.D., Hsieh, J.: Fast model-based x-ray CT reconstruction using spatially non-homogeneous ICD optimization. IEEE Trans. Image Process. **20**(1), 161–175 (2011)
4. Cierniak, R.: A novel approach to image reconstruction from discrete projections using hopfield-type neural network. In: Rutkowski, L., Tadeusiewicz, R., Zadeh, L.A., Żurada, J.M. (eds.) ICAISC 2006. LNCS, vol. 4029, pp. 890–898. Springer, Heidelberg (2006). doi:10.1007/11785231_93
5. Cierniak, R.: A new approach to tomographic image reconstruction using a Hopfield-type neural network. Int. J. Artif. Intell. Med. **43**(2), 113–125 (2008)
6. Cierniak, R.: A new approach to image reconstruction from projections problem using a recurrent neural network. Int. J. Appl. Math. Comput. Sci. **183**(2), 147–157 (2008)
7. Cierniak, R.: A novel approach to image reconstruction problem from fan-beam projections using recurrent neural network. In: Rutkowski, L., Tadeusiewicz, R., Zadeh, L.A., Zurada, J.M. (eds.) ICAISC 2008. LNCS, vol. 5097, pp. 752–761. Springer, Heidelberg (2008). doi:10.1007/978-3-540-69731-2_72
8. Cierniak, R.: New neural network algorithm for image reconstruction from fan-beam projections. Neurocomputing **72**, 3238–3244 (2009)
9. Cierniak, R.: A three-dimentional neural network based approach to the image reconstruction from projections problem. In: Rutkowski, L., Scherer, R., Tadeusiewicz, R., Zadeh, L.A., Zurada, J.M. (eds.) ICAISC 2010. LNCS, vol. 6113, pp. 505–514. Springer, Heidelberg (2010). doi:10.1007/978-3-642-13208-7_63
10. Cierniak, R.: Neural network algorithm for image reconstruction using the grid-friendly projections. Australas. Phys. Eng. Sci. Med. **34**, 375–389 (2011)
11. Cierniak, R.: An analytical iterative statistical algorithm for image reconstruction from projections. Appl. Math. Comput. Sci. **24**(1), 7–17 (2014)
12. Bruder, H., Kachelrieß, M., Schaller, S., Stierstorfer, K., Flohr, T.: Single-slice rebinning reconstruction in spiral cone-beam computed tomography. IEEE Trans. Med. Imag. **9**(9), 873–887 (2000)
13. Kachelrieß, M., Knaup, M., Kalender, W.A.: Extended parallel back projection for standard three-dimensional and phase-correlated four-dimensional axial and spiral cone-beam CT with arbitrary pitch, arbitrary cone-angle, and 100% dose usage. Med. Phys. **31**, 1623–1641 (2004)
14. Chu, J.L., Krzyżak, A.: The recognition of partially occluded objects with support vector machines, convolutional neural networks and deep belief networks. J. Artif. Intell. Soft Comput. Res. **4**(1), 5–19 (2014)
15. Bas, E.: The training of multiplicative neuron model artificial neural networks with differential evolution algorithm for forecasting. J. Artif. Intell. Soft Comput. Res. **6**(1), 5–11 (2016)
16. Aghdam, M.H., Heidari, S.: Feature selection using particle swarm optimization in text categorization. J. Artif. Intell. Soft Comput. Res. **5**(4), 231–238 (2015)
17. El-Samak, A.F., Ashour, W.: Optimization of traveling salesman problem using affinity propagation clustering and genetic algorithm. J. Artif. Intell. Soft Comput. Res. **5**(4), 239–245 (2015)
18. Leon, M., Xiong, N.: Adapting differential evolution algorithms for continuous optimization via greedy adjustment of control parameters. J. Artif. Intell. Soft Comput. Res. **6**(2), 103–118 (2016)

19. Miyajima, H., Shigei, N., Miyajima, H.: Performance comparison of hybrid electromagnetism-like mechanism algorithms with descent method. J. Artif. Intell. Soft Comput. Res. 5(4), 271–282 (2015)
20. Rutkowska, A.: Influence of membership function's shape on portfolio optimization results. J. Artif. Intell. Soft Comput. Res. 6(1), 45–54 (2016)

Efficient Real-Time Background Detection Based on the PCA Subspace Decomposition

Bogusław Cyganek[1(✉)] and Michał Woźniak[2]

[1] AGH University of Science and Technology, Al. Mickiewicza 30, 30-059 Kraków, Poland
cyganek@agh.edu.pl
[2] Wroclaw University of Science and Technology, Wybrzeże Wyspiańskiego 27,
50-370 Wrocław, Poland
Michal.Wozniak@pwr.wroc.pl

Abstract. We investigate performance of the classical PCA based background subtraction procedure and compare it with the robust PCA versions which are computationally demanding. We show that the simple PCA based version endowed with the fast eigen-decomposition method allows real-time operation on VGA video streams while offering accuracy comparable with some of the robust versions.

Keywords: Background subtraction · Computer vision · Real-time computations

1 Introduction

Artificial intelligent systems, such as autonomous cars or surveillance systems, frequently rely on processing of streams of video data [5–7, 15, 21, 22, 25, 26]. In this context the background subtraction (BS) belong to the classical building blocks of video analytics methods. Background subtraction, sometimes called background extraction, denotes a method of segmenting a video frame into a part of moving objects and the static background of a scene. There is a significant amount of works on BS methods. A valuable source of information is the website [12]. A comparative study of BS methods is presented by Benezeth *et al.* [2]. In the review paper, Bouwmans discusses the traditional and recent approaches of background modeling for foreground detection [4].

The simplest BS algorithm subtracts two frames, and then does the thresholding on values different from 0. However, this does not work well in practice due to the local variations of pixels and noise [18]. Therefore many other methods were devised, such as the one building a statistical model of the static background with the mixture of Gaussians, as proposed by Stauffer and Grimson [20], the codebook method by Kim *et al.* [13], fuzzy logic based [14, 29], or the eigen-background proposed by Oliver *et al.* [17], to name a few. There are also novel robust versions of the eigen-background algorithm which are of the special interest [27]. The work by Guyon *et al.* presents a systematic evaluation and comparative analysis of the Robust PCA (RPCA) for BS [10]. Similarly, Papusha discusses a fast automatic background extraction via RPCA [19].

© Springer International Publishing AG 2017
L. Rutkowski et al. (Eds.): ICAISC 2017, Part I, LNAI 10245, pp. 485–496, 2017.
DOI: 10.1007/978-3-319-59063-9_43

In this paper we investigate performance of the eigen-background method compared to their novel versions based on RPCA. The latter try to find a low rank subspace of a scene, whereas the moving foreground objects are among the correlated sparse outliers. However, despite better accuracy obtained with the robust versions, their run time performance is rather prohibitive in the case of systems demanding a real-time perform- ance. As written in the paper by Guyon *et al.* regarding performance of the RPCA methods [10]: *"the current implementations are faraway to achieve real-time. Indeed, the computing of the backgrounds take few hours for a training sequence with 200 frames for each algorithm. This time can be reduced by C/Cuda implementation"*. In the similar way, Papusha concludes that *"A criticism of Robust PCA is that it is not real time"* [18]. Therefore, we try to answer the question on performance of the basic PCA based BS method. We show that it can obtain accuracy comparable with some versions of the RPCA while being *real-time,* as shown by experiments.

2   Subspace Based Background Subtraction Method

The main idea of the analyzed method is to build a scene model, as proposed by Olivier *et al.* [17], and then to apply the fast eigen-decomposition based on the power method, as proposed in this paper and discussed in the next section. Finally, the consecutive model gets updated by the weighted covariance matrix update method.

2.1   PCA Background Subtraction

The main concept behind the subspace BS is to describe static regions of the scene with the eigenspace. This way, the scene model, which comprises W frames from the input video stream, is constructed. On the other hand, new frames with content different from the model cannot be perfectly reconstructed by this model. This feature is used to discover different regions, such as moving objects. In other words, having a frame I_k, it is first projected onto the eigenspace. This way, the reconstructed frame I_{k+1} is obtained. The last step is to compute a pixel-by-pixel difference between these two, which is usually done by a constant thresholding of a difference of the pixels. In this section the main steps of this procedure are discussed.

The PCA model is created computing parameters $\mathbf{T}$ and Λ of the following

$$\mathbf{T} \underbrace{E\left((\mathbf{x} - \mathbf{m}_x)(\mathbf{x} - \mathbf{m}_x)^T\right)}_{\Sigma_x} \mathbf{T}^T = \Lambda, \tag{1}$$

where $\mathbf{x}$ stands for a data vector of dimension L, $\mathbf{m}_x$ is an average of all data points, E denotes expected value, and finally $\Sigma_x = \mathbf{\bar{X}}\mathbf{\bar{X}}^T$ is a covariance matrix. In the case of video processing, each frame of resolution $r \times c$ is converted into a vector $\mathbf{x}$ of size $L = rc$. Usually, this is done in a row-scan order. However, for even small resolution of the processed frames, direct computation of (1) can be time and memory limited. This happens because the covariance matrix Σ_x has dimensions $L \times L$. For example, from frames of resolution 160×120, the matrix 19200×19200 is obtained which decomposition surely forbids the real-time performance. However, as shown by Olivier *et al.*, by a simple matrix manipulation a much smaller eigen-decomposition problem needs to be solved [17]. This, as well as application of the fast eigen-decomposition method, constitute the core of our approach to BS problem.

From the W frames $\mathbf{x}_i$ the zero-mean vectors $\bar{\mathbf{x}}_i$ are computed by subtracting $\mathbf{m}_x$. Then, a model is built out of W frames in a form of an $L \times W$ matrix $\bar{\mathbf{X}}$. Now, the key step is to compute a product $\bar{\mathbf{X}}^T\bar{\mathbf{X}}$ of dimensions $W \times W$, which is much smaller than dimensions of the original covariance matrix $\Sigma_x = \bar{\mathbf{X}}\bar{\mathbf{X}}^T$. Thanks to this, $\bar{\mathbf{X}}^T\bar{\mathbf{X}}$ can be efficiently eigen-decomposed, as follows

$$\underbrace{\bar{\mathbf{X}}^T\bar{\mathbf{X}}}_{Q}\ \mathbf{e}_k = \lambda_k\, \mathbf{e}_k, \tag{2}$$

where $\mathbf{e}_k$ are eigenvectors and λ_k eigenvalues of the matrix $\mathbf{Q} = \bar{\mathbf{X}}^T\bar{\mathbf{X}}$, respectively. Thanks to the symmetry of $\bar{\mathbf{X}}^T\bar{\mathbf{X}}$, $\mathbf{e}_k$ can be efficiently computed with the fixed-point method, as will be discussed. However, $\mathbf{e}_k$ are not the final eigenvectors we need. To obtain eigenvectors of Σ_x, let us left multiply both sides of (2) by $\bar{\mathbf{X}}$. This yields

$$\underbrace{\bar{\mathbf{X}}\bar{\mathbf{X}}^T}_{\Sigma_x}\ \underbrace{\left(\bar{\mathbf{X}}\,\mathbf{e}_k\right)}_{\mathbf{p}_k} = \lambda_k\ \underbrace{\left(\bar{\mathbf{X}}\,\mathbf{e}_k\right)}_{\mathbf{p}_k} \tag{3}$$

From (3) it can be seen that the vectors

$$\mathbf{p}_k = \bar{\mathbf{X}}\,\mathbf{e}_k, \tag{4}$$

are eigenvectors of dimensions $L \times 1$ of the covariance matrix Σ_x. Thus, if $\mathbf{e}_k$ are computed from (2), then based on (4) the matrix of eigenvectors of Σ_x can be computed is a straightforward way, as follows

$$\mathbf{P} = \bar{\mathbf{X}}\,\mathbf{E}. \tag{5}$$

Algorithm 1. Subspace background detection method.

Input – Video stream, a number of model frames W, a number of leading eigenvectors k_{max}, threshold τ

Output – A binary foreground image $\mathbf{M}$

1   From W consecutive video frames build a scene model by PCA computing with k_{max} leading eigenvectors;

2   For a frame I_k compute its projection I_{k+1} onto the eigenspace from step 1;

3   Reconstruct the projected image I_{k+1} to I_{k+2};

4   Compute the background map $\mathbf{M}_k$ as follows:

$$\mathbf{M}_k = \begin{cases} 0, & for \ \left| I_{k+1} - I_{k+2} \right| < \tau \\ 1, & otherwise \end{cases} \tag{6}$$

5   Update the covariance matrix of the window W_{t+1} and repeat;

In (5) $\mathbf{E}$ denotes a matrix with columns of $\mathbf{e}_k$. Finally, let us notice that eigenvectors in $\mathbf{P}$ need to be normalized to form an orthonormal basis. This way, the matrix $\mathbf{T}$ in (1) is obtained, as follows

$$\mathbf{T} = \bar{\mathbf{P}}^T, \tag{7}$$

where $\bar{\mathbf{P}}$ is a column-wise normalization of the matrix $\mathbf{P}$. Thanks to the above procedure, which requires eigenvalue decomposition of a small matrix $\bar{\mathbf{X}}^T\bar{\mathbf{X}}$ instead of $\bar{\mathbf{X}}\bar{\mathbf{X}}^T$, a much faster procedure is obtained. Finally, the background subtraction algorithm is summarized in Algorithm 1.

2.2   Eigen-Decomposition Method for Efficient Background Subtraction

A computational framework presented in the previous section has yet another useful feature. Namely, computations of the eigenvectors are done on a symmetric positive definite matrix $\mathbf{Q}$, as shown in (2). For this group of matrices there exist eigen-decomposition methods which are less computationally demanding than the methods used in a general case [8, 9]. One of them, called the fixed-point eigen-decomposition, is based on power iterations [16]. Its operation is outlined in Algorithm 2.

Algorithm 2. Fixed-point eigen-decomposition method.

Input – A symmetric matrix $\mathbf{Q}$, a number of eigenvectors k_{max}, a threshold ρ_{th}

Output – k_{max} leading eigenvectors of $\mathbf{Q}$

1   Init $\mathbf{e}_0^{(0)}$ - first-run randomly, next based on a previous run; $k \leftarrow 0$

2   **for** $k < k_{max}$

3        $i \leftarrow 1$

4   **do**

5            $\mathbf{e}_k^{(i)} \leftarrow \mathbf{Q}\,\mathbf{e}_k^{(i-1)}$

         Orthogonalize $\mathbf{e}_k^{(i)}$ to eigenvectors $\left\{\mathbf{e}_{0 \le j < k}\right\}$ (Gram-Schmidt):

6            $\mathbf{e}_k^{(i)} \leftarrow \mathbf{e}_k^{(i)} - \sum_{j=0}^{k-1}\left(\mathbf{e}_k^{T(i)}\mathbf{e}_j\right)\mathbf{e}_j$

7   Normalize $\mathbf{e}_k^{(i)}$: $\mathbf{e}_k^{(i)} \leftarrow \mathbf{e}_k^{(i)}\big/\left\|\mathbf{e}_k^{(i)}\right\|_2$

8            $\rho = \left|\mathbf{e}_k^{T(i-1)}\mathbf{e}_k^{(i)} - 1\right|$

9            $i \leftarrow i + 1$

10   **while** $\rho > \rho_{th}$

11   **end for**

The above algorithm is iterative, but in practice it converges fast. This feature is obtained due to initialization of the initial eigenvalues with the ones from the previous run for the window W as shown in Algorithm 1.

3   Experimental Results

The methods were implemented in C++ in the Microsoft Visual 2015 IDE. Code was parallelized with the OpenMP library [28]. In the presented system this is used only in the matrix multiplication in (2) and (5), however. The experiments were run on a laptop computer equipped with the Intel® Xeon® E-1545 CPU @2.9 GHz, 64 GB RAM, and OS 64-bit Windows 10.

In the experiments the frequently compared *Wallflower* dataset was used [11, 24]. It consists of a number of color video sequences of small resolution of 160×120. It contains also manually labeled ground-truth frames which are used for quantitative evaluation. In our method color frames are converted to monochrome.

The presented experiments were carried on to answer the following questions:

1. What is accuracy of the subspace based background computations?
2. How the PCA accuracy compares to the robust PCA?
3. Does it allow real-time operations?

To answer these questions, four test videos were used from which 200 frames were used to build the scene model, and then one frame was used to compute its background. The results were always compared with the ground-truth map for that frame, provided in the *Wallflower* dataset.

The background maps were further median filtered to remove sparse points. The quantitative results were measured in terms of the true-positive (TP), true-negative (TN), false-positive (FP), and false-negative (FN) parameters. From these, the so called F value was computed. We used the procedures described in [5]. Finally, the F value obtained for the presented PCA method was checked with variants of the robust PCA. In each case, the worst and the best F value for the robust PCA is presented, based on the publication Guyon *et al.* [10]. The results for the *WaivingTrees*, *Bootstrap*, *Camouflage*, as well as *ForegroundAperture* videos, are shown in Figs. 1, 2, 3 and 4, respectively.

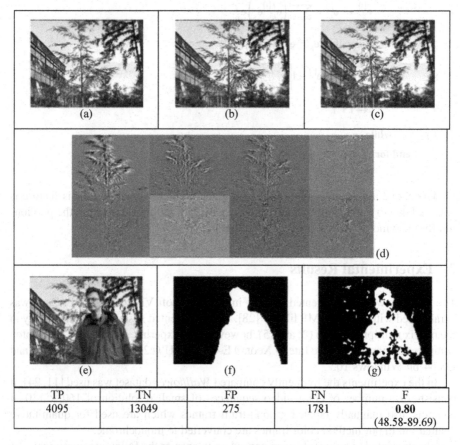

TP	TN	FP	FN	F
4095	13049	275	1781	**0.80** (48.58-89.69)

Fig. 1. Results obtained for the *WaivingTrees* test sequence from the *Wallflower* dataset. Examples from the training chunk of frames (a)–(c). Eigenimages corresponding to the largest eigenvalues (d). Test frame (e), background ground-truth for the test frame (f), obtained background (g). Below there are results TP/TN/FP/FN as well as the F score. In the parenthesis the worst and the best F value for the robust PCA methods (TFOCS, RSL).

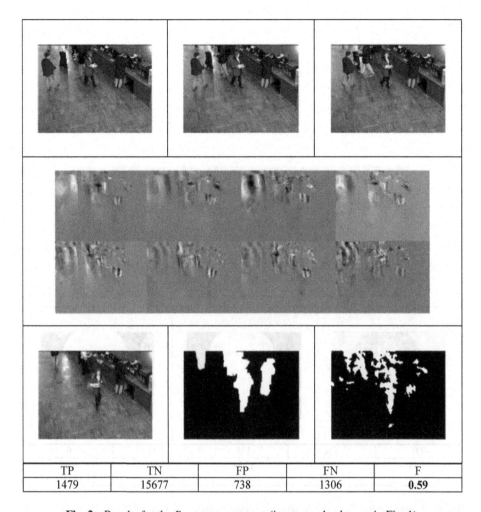

TP	TN	FP	FN	F
1479	15677	738	1306	**0.59**

Fig. 2. Results for the *Bootstrap* sequence (images and values as in Fig. 1).

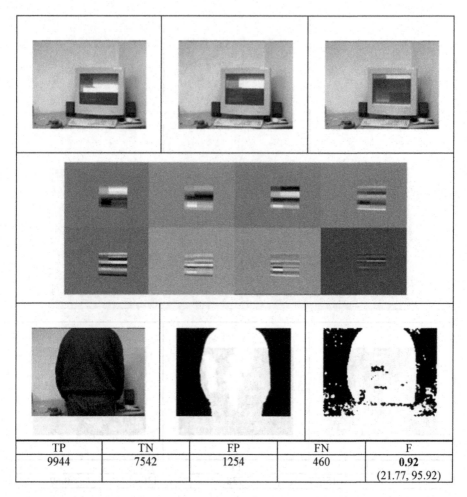

TP	TN	FP	FN	F
9944	7542	1254	460	**0.92** (21.77, 95.92)

Fig. 3. Results for the *Camouflage* sequence (images and values as in Fig. 1). The min. F value for the TFOCS method, the max. F value for the SUB method.

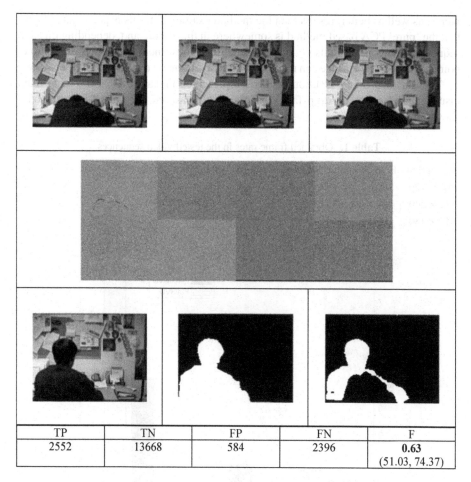

TP	TN	FP	FN	F
2552	13668	584	2396	**0.63** (51.03, 74.37)

Fig. 4. Results for the *ForegroundAperture* sequence (images and values as in Fig. 1). The min. F value for the TFOCS method, the max. F value for the RSL method.

The background map shown in Fig. 1(g) exhibits some irregularities pertinent to the PCA based method. However, in this case the TFOCS (Templates for First-Order Conic Solvers) method shows much worse performance [1]. On the other hand, the RSL (Robust Subspace Learning) method is better in this case [23].

In the case of the Camouflage sequence, the background map – shown in Fig. 3(g) – the obtained F value is higher than majority of the robust methods presented in the paper by Guyon *et al.*; It is only worse than the SUB, which denotes the basic average algorithm [10].

The background map in Fig. 4(g) of the *ForegroundAperture* is due to a part of the foreground object being present in the background model, which is well visible in Fig. 4(a)–(c).

Interestingly enough, in each of the above cases, except the *Boostrap* sequence for which the values were not available, there are versions of the robust PCA which give worse, as well as better, results than the method examined in this paper. This indicates that the 'pure' PCA based method is somewhere in the middle and, depending on a real sequence, it can give results comparable with the robust methods. However, the main point is that its performance can reach real-time operation on the CPU, as will be shown.

The obtained time execution results are presented in Table 1. On the other hand, visualization of the speed-up factor for the two measured decomposition method is shown Fig. 5.

Table 1. Obtained frame rates in the tested video sequences.

Resolution	Frames	Time (fixed-point SVD)	Frame rate	Time (full SVD)	Frame rate
160×120	200	0.197	1015	0.635	314
352×240	200	0.894	223	2.1	95
640×480	200	4.103	48	8.292	24

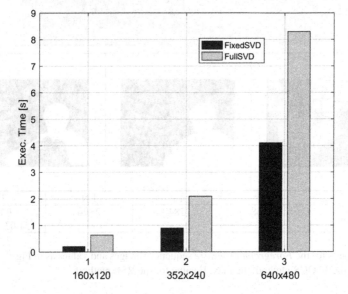

Fig. 5. Timing bars of the two versions of the eigen-decomposition for scene model building in the subspace background computation. Three sets of video used with different resolutions. In all experiments 200 images were employed.

From these it can be observed that in all cases we obtained a real-time processing conditions which we set as 24 frames per second. Moreover, in the case of the fast fixed point eigen-decomposition method, the speed-up factor is up to 2 times.

In many practical cases, in which a background subtraction is a supporting module for object detection etc., a useful strategy is to process low resolution video and interpolate the background map to the requested dimensions. Thanks to this, even on CPU platforms it is possible to employ CPU to other tasks than a sole background computation.

4   Conclusions

In the paper, the PCA based background method is evaluated in the context of robust PCA approaches. We tried to shed more light onto performance of this basic algorithm, especially in the framework of embedded systems and robotics which require real-time performance. We tried to answer a number of questions which can help in choosing right method in real applications. The first parameter, which is background map accuracy, achieves values which are in-between, but sometimes outperform, the RPCA methods. On the other hand, we show that in the C++ implementation on the CPU it is possible to process a VGA video stream in real-time.

However, it should be remembered that in practice, a processed video can contain much different statistical properties than the sequences used in the evaluation. Therefore, a background method, as well as its parameters should be chosen to the particular application. Nevertheless, as we show in this paper, the subspace methods exhibit a number of valuable properties, at the same time being able to achieve real-time performance on CPU platforms.

Acknowledgement.   The authors would like to express their gratitude to prof. Ryszard Tadeusiewicz and prof. Janusz Kacprzyk for their great scientific influence and continuous support.

This work was supported by the Polish National Science Center NCN under the grant no. 2014/15/B/ST6/00609.

This work was also supported by the statutory funds of the Department of Systems and Computer Networks, Faculty of Electronics, Wroclaw University of Science and Technology.

References

1. Becker, S., Candes, E., Grant, M.: TFOCS: flexible first-order methods for rank minimization. In: Low-Rank Matrix Optimization Symposium, SIAM Conference on Optimization (2011)
2. Benezeth, Y., Jodoin, P-M., Emile, B., Laurent, H., Rosenberger, C.: Comparative study of background subtraction algorithms. SPIE J. Electron. Imaging **19**(3), 033003 (2010)
3. Bingham, E., Hyvärinen, A.: A fast fixed-point algorithm for independent component analysis of complex valued signals. Int. J. Neural Syst. **10**(1) (2000). World Scientic Publishing Company
4. Bouwmans, T.: Traditional and recent approaches in background modeling for foreground detection: an overview. Comput. Sci. Rev. **11**, 31–66 (2014)
5. Cyganek, B.: Object Detection and Recognition in Digital Images: Theory and Practice. Wiley, Hoboken (2013)
6. Cyganek, B., Gruszczyński, S.: Hybrid computer vision system for drivers' eye recognition and fatigue monitoring. Neurocomputing **126**, 78–94 (2014)
7. Cyganek, B.: An analysis of the road signs classification based on the higher-order singular value decomposition of the deformable pattern tensors. In: Blanc-Talon, J., Bone, D., Philips, W., Popescu, D., Scheunders, P. (eds.) ACIVS 2010. LNCS, vol. 6475, pp. 191–202. Springer, Heidelberg (2010). doi:10.1007/978-3-642-17691-3_18

8. Demmel, J.W.: Applied Numerical Linear Algebra. Siam (1997)
9. Golub, G.H., van Loan, C.F.: Matrix Computations. Johns Hopkins Studies in the Mathematical Sciences. Johns Hopkins University Press (2013)
10. Guyon, C., Bouwmans, T., Zahzah, E.: Robust Principal Component Analysis for Background Subtraction: Systematic Evaluation and Comparative Analysis. Principal Component Analysis, Edited by Sanguansat, P. InTech (2012)
11. http://research.microsoft.com/en-us/um/people/jckrumm/wallflower/testimages.htm
12. https://sites.google.com/site/backgroundsubtraction/Home
13. Kim, K., Chalidabhongse, T.H., Harwood, D., Davis, L.: Real- time foreground-background segmentation using codebook model. Real-Time Imaging 11, 172–185 (2005)
14. Kim, W., Kim, C.: Background subtraction for dynamic texture scenes using fuzzy color histograms. IEEE Signal Process. Lett. 3(19), 127–130 (2012)
15. Korytkowski, M., Rutkowski, L., Scherer, R.: Fast image classification by boosting fuzzy classifiers. Inf. Sci. 327, 175–182 (2016)
16. Marot, J., Fossati, C., Bourennane, S.: About advances in tensor data denoising methods. EURASIP J. Adv. Sig. Process. 2008, 12 (2008)
17. Oliver, N.M., Rosario, B., Pentland, A.P.: A Bayesian computer vision system for modeling human interactions. IEEE Trans. Pattern Anal. Mach. Intell. 22(8), 831–843 (2000)
18. Piccardi, M.: Background subtraction techniques: a review. In: IEEE International Conference on Systems, Man and Cybernetics, vol. 4, pp. 3099–3104 (2004)
19. Papusha, I.: Fast Automatic Background Extraction Via Robust PCA. Stanford Electrical Engineering Department, Stanford (2011). http://web.stanford.edu/class/ee364b/projects/2011projects/reports/papusha.pdf
20. Stauffer C., Grimson W. E. L.: Adaptive background mixture models for real-time tracking. In: Proceedings of the International Conference on Computer Vision and Pattern Recognition, vol. 2. IEEE, NJ (1999)
21. Tadeusiewicz, R.: Introduction to intelligent systems. In: Wilamowski, B.M., Irvin, J.D. (eds.) The Industrial Electronics Handbook – Intelligent Systems, pp. 1–12. CRC Press, Boca Raton (2011)
22. Tadeusiewicz, R.: Neural networks in mining sciences – general overview and some representative examples. Archives of Mining Sciences (Archiwum Górnictwa), vol. 60, no. 4, pp. 971–984 (2015). ISSN 0860-7001
23. Torre, F.D.L., Black, M.: A framework for robust subspace learning. Int. J. Comput. Vis. 54, 117–142 (2003)
24. Toyama, K., Krumm, J., Brumitt, B., Meyers, B.: Wallflower: principles and practice of background maintenance. In: Seventh International Conference on Computer Vision, Kerkyra, Greece, pp. 255–261, IEEE Computer Society Press, September 1999
25. Woźniak, M.: A hybrid decision tree training method using data streams. Knowl. Inf. Syst. 29(2), 335–347 (2011)
26. Woźniak, M., Grana, M., Corchado, E.: A survey of multiple classifier systems as hybrid systems. Inform. Fusion 16(1), 3–17 (2014)
27. Wright, J., Peng, Y., Ma, Y., Ganesh, A., Rao, S.: Robust principal component analysis: Exact recovery of corrupted low-rank matrices by convex optimization. Neural Information Processing Systems, NIPS (2009)
28. http://www.openmp.org/
29. Zadeh, L.A., Kacprzyk, J.: Fuzzy Logic for the Management of Uncertainty. John Wiley & Sons, Inc., New York (1992)

The Image Classification with Different Types of Image Features

Marcin Gabryel[1]([✉]) and Robertas Damaševičius[2]([✉])

[1] Institute of Computational Intelligence, Częstochowa University of Technology,
Al. Armii Krajowej 36, 42-200 Częstochowa, Poland
marcin.gabryel@iisi.pcz.pl
[2] Software Engineering Department, Kaunas University of Technology,
Studentu 50, Kaunas, Lithuania
robertas.damasevicius@ktu.lt

Abstract. In this paper we present a modified Bag-of-Words algorithm used in image classification. The classic Bag-of-Words algorithm is used in natural language processing. A text (such as a sentence or a document) is represented as a bag of words. In image retrieval or image classification this algorithm also works on one characteristic image feature and most often it is a descriptor defining the surrounding of a keypoint obtained by using e.g. the SURF algorithm. The modification which we have introduced involves using two different types of image features – the descriptor of a keypoint and also the colour histogram, which can be obtained from the surrounding of a keypoint. This additional feature will make it possible to obtain more information as the commonly used SURF algorithm works only on images with greyscale intensity. The experiments which we have conducted show that using this additional image feature significantly improves image classification results by using the BoW algorithm.

1 Introduction

The literature on the subject offers a lot of information on the application of the Bag-of-Words algorithm, which is also known as the Bag-of-Visual-Words or the Bag-of-Features algorithm. The Bag-of-Words algorithm is one of the most popular and widely spread algorithms used for indexation and image retrieval. The Bag-of-Words model, i.e. an algorithm version applied in computer vision, comprises three main steps: (i) feature extraction, (ii) building visual vocabulary and (iii) storing vocabulary in a database. In the first step image characteristic features are obtained, which are most often keypoint descriptors, i.e. colour- or shape-based features or fragments of an image. The next step involves creating a visual dictionary, where specific words are obtained by clustering of features. The k-means algorithm is the most frequently used clustering algorithm. Obtained words are assigned to images or image classes, which results in creating a dictionary. The last step of the algorithm involves storing of the dictionary in a database. Relevant indexation enables users to make a swift search for the image class or similar images to which a given query image belongs.

L. Rutkowski et al. (Eds.): ICAISC 2017, Part I, LNAI 10245, pp. 497–506, 2017.
DOI: 10.1007/978-3-319-59063-9_44

In our method we introduced a number of innovative solutions improving general operational efficiency of the BOW algorithm. The first innovation which we put forward is creating a visual words dictionary by using the clustering algorithm which in itself is responsible for selecting the required number of clusters. This solution results in significant automation of image database creation. Another innovation is presenting an image by using two independent features: descriptors defining the surrounding of keypoints and the histogram defining colours from the surrounding of keypoints. In our algorithm we proposed using the majority voting instead of comparing histograms with information on frequency of particular features present in an image. The majority vote method involves computation of representation of keypoint clusters and colour histogram clusters.

Currently, image recognition and classification present one of the most complex problems, which requires extensive expertise of a variety of fields, where elements of computational intelligence [3,17], fuzzy systems [11,18], neural networks [4], evolutionary algorithms [10,20,22], multi-objective optimization [6], mathematics [21,23] and data mining [19] could be used successfully. This kind of knowledge is used in deep-learning neural networks or, indeed, in the Bag-of-Words algorithms. In the literature on the subject there are a number of papers which present image classification in which the BoW model is used. In the paper [7] the authors present the classical system called bag-of-keypoints, which is presented earlier in this section. In [13] the authors present the image of a scene using a collection of local regions, denoted as codewords obtained by unsupervised learning. In the paper [16] given a set of images containing multiple object categories, the authors seek to discover those categories and their image locations without supervision. In [15] the authors propose a Term-Frequency and Inverse Document Frequency weighting scheme to characterize the importance of visual words in the BoW model. The authors also discuss a method to fuse different Bag-of-Words obtained with different vocabularies. The paper [12] presents a method for recognizing natural scene categories based on approximate global geometric correspondence.

The article is divided into the following parts. In Sect. 2 we can find familiar algorithms such as Speeded Up Robust Features (SURF) and the modified k-means algorithms. Those are also the ones which we use in our method. In the following section there is also a description of our original idea of retrieving images and creating new databases algorithms with the use of the k-means algorithm with different types of image. In the last section we present the results of experiments as well as a summary of our work.

2 Algorithms Used in the Proposed Approach

2.1 SURF

The SURF method (Speeded Up Robust Features) [2] is a fast and robust algorithm for local, similarity invariant representation and comparison of images. It is partly inspired by the SIFT descriptor [14]. Interest points of a given image are defined as salient features from a scale-invariant representation. The SURF algorithm is in itself based on two consecutive steps (feature detection and description). The main element of this algorithm is the structure named integral images, which allows us to significantly reduce the number of operations. Next, SURF uses a blob detector based on the Hessian matrix to find points of interest. The determinant of the Hessian matrix is used as a measure of local change around the point and points are chosen where this determinant is maximal. In order to achieve rotational invariance, the orientation of the point of interest needs to be found. The Haar wavelet responses in both x- and y-directions within a circular neighbourhood around the point of interest are computed. To describe the region around a point, a square region is extracted, centered on the interest point and oriented along the orientation as selected above. The interest region is split into smaller 4x4 square sub-regions and they are described by a 64-number vector. In our approach we additionally generate the RGB colour histogram from the interest region. Sample keypoints together with the obtained histogram are presented in Fig. 1.

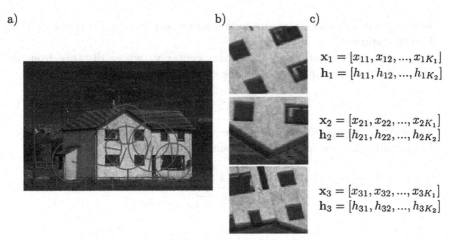

a)   b)   c)

$$\mathbf{x}_1 = [x_{11}, x_{12}, ..., x_{1K_1}]$$
$$\mathbf{h}_1 = [h_{11}, h_{12}, ..., h_{1K_2}]$$

$$\mathbf{x}_2 = [x_{21}, x_{22}, ..., x_{2K_1}]$$
$$\mathbf{h}_2 = [h_{21}, h_{22}, ..., h_{2K_2}]$$

$$\mathbf{x}_3 = [x_{31}, x_{32}, ..., x_{3K_1}]$$
$$\mathbf{h}_3 = [h_{31}, h_{32}, ..., h_{3K_2}]$$

Fig. 1. (a) A sample image for which the SURF algorithm is operated, (b) sample keypoints found together with their local regions, (c) descriptor of keypoints $\mathbf{x}_i$ and colour histogram obtained for a sample keypoint $\mathbf{h}_i$ (Color figure online)

2.2 Modified k-means Algorithm

The k-means algorithm is the most frequently used clustering algorithm used in the BoW. Its only drawback involves having to define the initial number of classes c. In this section we present an automatic selection mechanism of the number of classes during the operation of this algorithm. We have used the growing method used in the Growing Self-Organizing Map (GSOM) algorithm [8]. In that method a cluster is divided when the amount of its data exceeds a certain threshold value Θ. Operation of the said algorithm starts with setting the threshold value Θ and defining two clusters ($c = 2$). In the subsequent steps the algorithm works as a classic k-means with the only difference being that at the end of each iteration the number of points belonging to each cluster τ_j, $j = 1, ..., c$, is checked. If the number τ_j exceeds the threshold already set at Θ, then another cluster $c + 1$ is created. The algorithm is presented below in detail.

Let $\mathbf{X} = \mathbf{x}_1, ..., \mathbf{x}_n$ be a set of points in d-dimensional space, and $\mathbf{V} = \mathbf{v}_1, ..., \mathbf{v}_c$ be cluster centers, where n is the number of samples, $\mathbf{x}_i = [x_{i1}, ..., x_{id}]$, c is the number of clusters, and $\mathbf{v}_j = [v_{j1}, ..., v_{jd}]$.

1. Let the number of clusters $c = 2$. Determine Θ.
2. Randomly select c cluster centers $\mathbf{v}_j$, $j = 1, ..., c$, for example:

$$v_{ji} = rand(\min(x_{ij}), \max(x_{ij})), \tag{1}$$

   where $rand(a, b)$ is a random number generated from the interval $[a; b]$.
3. Calculate the distance d_{ij} between each data point $\mathbf{x}_i$ and cluster centers $\mathbf{v}_j$:

$$d_{ij} = \|\mathbf{x}_i - \mathbf{v}_j\|, \tag{2}$$

   where $\| \cdot \|$ is a distance measure between two vectors (e.g. Euclidian or Manhattan distance).
4. Assign the data point $\mathbf{x}_i$ to the cluster center $\mathbf{v}_s$ whose distance from the cluster center is a minimum of all the cluster centers

$$\mathbf{x}_i \in \mathbf{v}_s \rightarrow d_{is} \le d_{is}, m = 1, ..., c \tag{3}$$

   and increase counter of winnings $\tau_s = \tau_s + 1$.
5. Recalculate the new cluster center using:

$$\mathbf{v}_i = \frac{1}{c_i} \sum_{j=1}^{c_i} \mathbf{x}_j \tag{4}$$

   where c_i represents the number of data points in i-th cluster.
6. If in the center s the number τ_s is greater than the threshold value Θ, create a new cluster, $c := c + 1$ and

$$\mathbf{v}_c = \mathbf{x}_{rand(j)} \tag{5}$$

   where $rand(j)$ generates a random index of point $\mathbf{x}$ belonging to center $\mathbf{v}_s$.
7. Remove clusters for which $\tau_s = 0$. Refresh the number of clusters c.
8. If no data point was reassigned, then stop; otherwise, repeat starting from step 3.

As a result of the algorithm operation we obtain c clusters with the centers in points $\mathbf{v}_j$, $j = 1, ..., c$.

3 Proposed Approach

In this section we present the modified Bag-of-Words algorithm, which uses key-point descriptors and the colour histogram of their region as image characteristic features. Both sets of features are clustered by the k-means algorithm, and next they are stored in the database. Searching for the query image class involves the majority vote method, in which the number of classes to which a particular set of features belongs is counted.

We are considering a set of given images $\mathbf{I}_i$, where $i = 1, ..., I_L$, I_L is the number of all images. Each image $\mathbf{I}_i$ has a class $c(\mathbf{I}_i)$ assigned to it, where $c(\mathbf{I}_i) \in \Omega$, $\Omega = \{\omega_i, ..., \omega_C\}$ is a set of all classes and C is the number of all classes.

The algorithm for building a dictionary presented herein comprises the following steps:

1. Find the key points with the SURF algorithm for all images I_L, $\mathbf{x}_i = [x_{i1}, x_{i2}, ..., x_{iK_1}]$, $i = 1, ..., L$, L – the total number of all characteristic points, $K_1 = 64$ – the dimension of the vector describing a characteristic point.
2. Obtaining the RGB colour histogram for each keypoint region $\mathbf{h}_i = [h_{i1}, h_{i2}, ..., h_{iK_2}]$, $i = 1, ..., L$, K_2 – the dimension of the vector describing the three RGB colour histogram. Each of the colours is defined by 8 elements, thus $K_2 = 24$.
3. Starting operation of the k-means clustering algorithm for key points $\mathbf{x}_i$. We have used the algorithm version presented in Sect. 2.2. As a result we obtain c_1 clusters with the centers in points $\mathbf{v}_j$, $j = 1, ..., c_1$, which are treated as words in the BoF dictionary.
4. Starting operation of the k-means clustering algorithm for histograms $\mathbf{h}_i$. As a result we obtain c_2 clusters with the centers in points $\mathbf{w}_j$, $j = 1, ..., c_2$, which are treated as additional words in the BoF dictionary.
5. For key points the value of the number of classes i of cluster j is calculated and defined as k_{ji}^p. This value is computed by counting the points $\mathbf{x}_n$ which belong to the center j provided that $\mathbf{x}_n \in \mathbf{I}$ and $c(\mathbf{I}) = \omega_i$:

$$k_{ji}^p = \sum_{n=1}^{L} \delta_{nj}^p(i), j = 1, ..., c_1, i = 1, ..., C, \tag{6}$$

where

$$\delta_{nj}^p = \begin{cases} 1 \text{ if } d_{nj} < d_{nm} \text{ for } \mathbf{x}_n \in \mathbf{I} \text{ and } c(\mathbf{I}) = \omega_i, m = 1, ..., c_1, j \neq m \\ 0 \qquad\qquad\qquad\qquad \text{otherwise} \end{cases} \tag{7}$$

The variable $\delta_{nj}^p(i)$ is an indicator if cluster $\mathbf{v}_j$ is the closest vector (a winner) for any sample $\mathbf{x}_n$ from an image $\mathbf{I}$ and $c(\mathbf{I}) = \omega_i$.

6. Similar computations are carried out in the case of coefficients k_{ji}^h for histograms $\mathbf{h}_i$ and centers $\mathbf{w}_j$. This value is computed by counting the points $\mathbf{h}_n$ which belong to the center $\mathbf{v}_j$ provided that $\mathbf{h}_n \in \mathbf{I}$ and $c(I) = \omega_i$:

$$k_{ji}^h = \sum_{n=1}^{L} \delta_{nj}^h(i), j = 1, ..., c_2, i = 1, ..., C, \tag{8}$$

where

$$\delta_{nj}^{h} = \begin{cases} 1 & \text{if } d_{nj} < d_{nm} \text{ for } \mathbf{h}_n \in \mathbf{I} \text{ and } c(\mathbf{I}) = \omega_i, m = 1, ..., c_2, j \neq m \\ 0 & \text{otherwise} \end{cases} \tag{9}$$

The variable $\delta_{nj}^{h}(i)$ is an indicator if cluster $\mathbf{w}_j$ is the closest vector (a winner) for any sample $\mathbf{h}_n$ from an image $\mathbf{I}$ and $c(\mathbf{I}) = \omega_i$.

7. Sorting the results k_{ji}^{p} and k_{ji}^{h} so that

$$k_{js_j^p(i)}^{p} \geq k_{js_j^p(i+1)}^{p} \tag{10}$$

and

$$k_{js_j^h(i)}^{h} \geq k_{js_j^h(i+1)}^{h} \tag{11}$$

where $s_j^p(i)$ and $s_j^h(i)$ are functions returning sorted indexes i depending on the cluster j for keypoint and histogram clusters respectively.

8. Storing the obtained centres $\mathbf{v}_j$, $\mathbf{w}_j$ and coefficients $k_{js_j^p(i)}^{p}$, $k_{js_j^h(i)}^{h}$ in the database.

The classification process, i.e. the testing process whether a given query image $\mathbf{I}_q$ belongs to a particular class is conducted as follows:

1. Using a feature extraction algorithm on query image $\mathbf{I}_q$ in order to obtain values of characteristic features $\mathbf{x}_i^{q}, i = 1, ..., L_q$, L_q — the number of obtained features.
2. Generating RGB colour histograms $\mathbf{h}_i^{q}, i = 1, ..., L_q$ from the keypoints' surrounding obtained in point 1.
3. Defining the value of parameter τ_c, which specifies the value of the number of the best classes included in the majority vote. Condition $1 \leq \tau_c \leq C$ needs to be accounted for.
4. The classification process involves assigning of the obtained descriptors and histograms to their relevant clusters according to the following formula:

$$k_i^{q} = \sum_{n=1}^{L_q} \alpha_n(i) + \sum_{n=1}^{L_q} \beta_n(i), \ i = 1, ..., \tau_c \tag{12}$$

where

$$\alpha_n(i) = \begin{cases} k_{js_j^p(i)}^{p} & \text{if } d_{jn}^{qp} < d_{mn}^{qp}, \ j \neq m \\ 0 & \text{otherwise} \end{cases}, \tag{13}$$

$$\beta_n(i) = \begin{cases} k_{js_j^h(i)}^{h} & \text{if } d_{jn}^{qh} < d_{mn}^{qh}, \ j \neq m \\ 0 & \text{otherwise} \end{cases}, \tag{14}$$

and for keypoints groups:

$$d_{jn}^{qp} = \|\mathbf{v}_i - \mathbf{x}_n^{q}\|, j = 1, ..., c_1, \tag{15}$$

$$d_{mn}^{qp} = \|\mathbf{v}_m - \mathbf{x}_n^{q}\|, j = 1, ..., c_1. \tag{16}$$

It is similar for histogram groups:

$$d_{jn}^{qh} = \|\mathbf{w}_i - \mathbf{h}_n^q\|, j = 1, ..., c_2, \tag{17}$$

$$d_{mn}^{qh} = \|\mathbf{w}_m - \mathbf{h}_n^q\|, j = 1, ..., c_2. \tag{18}$$

5. Assigning to class $c(\mathbf{I}_q)$ is done by the majority vote checking the maximum value k_i^q:

$$c(\mathbf{I}_q) = \operatorname*{argmax}_{i=1,...,c} k_i^q \tag{19}$$

The algorithm's experimental research results are presented in the next section.

4   Experimental Research

In this section we present the research experiments showing the operational efficiency of the discussed Bag-of-Words algorithm in relation to its parameter values. The research was carried out on the commonly used Caltech 101 data set [9], from which six image classes were randomly selected. Each class was divided into two parts at the proportion of 80/20. The first one was used to build the dictionary which was stored in the database (training data). The other part was the testing data, which was used to verify the operational efficiency of the algorithm. The algorithm was implemented in Java with the use of JavaCV library [1]. This library offers functions implemented in OpenCV [5]. The recall value expressed as a percentage was used to calculate the classification efficiency according to the following formula:

$$Recall = \frac{TruePositiveCount}{TruePositiveCount + FalseNegativeCount} \cdot 100\% \tag{20}$$

Table 1 presents the results of the experiments for the learning and testing data with the use of different parameter values. In the first column the changes in the threshold values τ_{max} used in the k-means algorithm are marked. In the following column the features selected for the majority vote are marked. For the experiments marked as "kp" only keypoint descriptors are taken into consideration. Formula 12 was simplified and took the following form (the part responsible for the colour histograms was removed):

$$k_i^q = \sum_{n=1}^{L_q} \alpha_n(i), \ i = 1, ..., \tau_c \tag{21}$$

For the experiments marked as "h" only colour histograms from the keypoint regions were taken into consideration. The formula (12) then presents as follows:

$$k_i^q = \sum_{n=1}^{L_q} \beta_n(i), \ i = 1, ..., \tau_c \tag{22}$$

504   M. Gabryel and R. Damaševičius

Table 1. Operational results of the Bag-of-Words algorithm presented for different parameters. The results show classification efficiency expressed as a percentage.

τ_{max}	features	τ_c 1 train	1 test	2 train	2 test	3 train	3 test	4 train	4 test	5 train	5 test	6 train	6 test
100	kp	78.84	49.65	83.42	51.05	83.60	**55.94**	88.01	53.85	94.53	41.26	**98.94**	41.96
	h	89.42	**72.73**	89.07	69.93	91.36	64.34	95.24	64.34	97.71	65.03	**99.12**	65.03
	b	91.36	**73.43**	93.30	69.23	94.53	67.83	95.94	61.54	97.71	49.65	**99.82**	51.05
50	kp	92.06	55.24	94.00	**59.44**	94.36	56.64	96.65	57.34	99.47	46.15	**100.00**	44.06
	h	91.53	**76.92**	95.24	72.03	96.30	68.53	97.71	67.83	99.47	68.53	**100.00**	68.53
	b	95.59	**76.22**	97.71	76.22	98.06	72.03	98.94	72.73	**100.00**	65.03	100.00	65.03
25	kp	99.12	64.34	99.65	**65.03**	99.82	65.03	**100.00**	61.54	100.00	56.64	100.00	55.24
	h	96.12	**75.52**	98.24	75.52	99.12	74.13	99.47	73.43	**100.00**	73.43	100.00	73.43
	b	98.94	74.83	99.65	**78.32**	100.00	76.92	100.00	76.92	**100.00**	73.43	100.00	72.73
10	kp	**100.00**	**69.23**	100.00	65.03	100.00	66.43	100.00	64.34	100.00	62.24	100.00	62.24
	h	98.24	**79.02**	99.65	77.62	99.82	76.92	**100.00**	76.92	100.00	76.92	100.00	76.92
	b	99.82	79.72	**100.00**	**81.12**	100.00	79.72	100.00	79.02	100.00	79.02	100.00	79.02

For the "b" marking the formula (12) does not change, and then both image features are accounted for.

The following columns present the percentage values of the classification efficiency in relation to the τ_c coefficient for the two parts of the image set, i.e. the training and testing data. The best results obtained for the learning and testing data are shown in bold. The results show it clearly that using colour histograms is more effective than using keypoint descriptors. However, using a combination of these two features provides the best results. Moreover, the τ_{max} coefficient affects the number of clusters obtained in the k-means algorithm. The smaller the value of this coefficient, the better the results obtained.

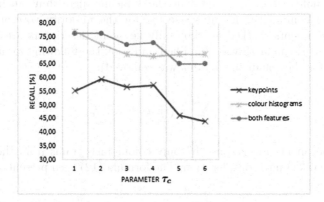

Fig. 2. Diagrams presenting the efficiency classification as percentage for the testing data for $\tau_{max} = 50$ in relation to different values τ_c.

The τ_c parameter affects the efficiency of the algorithm. For the test data it is important that it should have the smallest possible value.

In Fig. 2 the sample results in the form of a diagram are presented. The following diagrams show the percentage value of the efficiency for the testing data for $\tau_{max} = 50$ in relation to different values τ_c.

5   Conclusions

This paper presents several changes in the Bag-of-Words algorithm. We show it is possible to use more than just one image feature. We use the k-means algorithm, which automatically selects the number of clusters treated as elements of the dictionary. Image classification is based on the majority vote method, where the number of classes to which a given image feature is assigned is computed. All these factors have contributed to an improved operational performance of the Bag-of-Words algorithm, which has been confirmed by the experimental research conducted.

References

1. Audet, S.: JavaCV (2017). http://bytedeco.org/. Accessed 1 Feb 2017
2. Bay, H., Tuytelaars, T., Gool, L.: SURF: Speeded Up Robust Features. In: Leonardis, A., Bischof, H., Pinz, A. (eds.) ECCV 2006. LNCS, vol. 3951, pp. 404–417. Springer, Heidelberg (2006). doi:10.1007/11744023_32
3. Bertini Junior, J.R., Nicoletti, M.C.: Enhancing constructive neural network performance using functionally expanded input data. J. Artif. Intell. Soft Computing Res. 6(2), 119–131 (2016)
4. Bilski, J., Wilamowski, B.M.: Parallel learning of feedforward neural networks without error backpropagation. In: Rutkowski, L., Korytkowski, M., Scherer, R., Tadeusiewicz, R., Zadeh, L.A., Zurada, J.M. (eds.) ICAISC 2016. LNCS, vol. 9692, pp. 57–69. Springer, Cham (2016). doi:10.1007/978-3-319-39378-0_6
5. Bradski, G.: The OpenCV Library. Dr. Dobb's Journal of Software Tools (2000)
6. Christina, B., Eugene, S., Maxim, S.: Multi-objective heuristic feature selection for speech-based multilingual emotion recognition. J. Artif. Intell. Soft Comput. Res. 6(4), 243 (2016)
7. Csurka, G., Dance, C.R., Fan, L., Willamowski, J., Bray, C.: Visual categorization with bags of keypoints. In: Workshop on Statistical Learning in Computer Vision, ECCV, pp. 1–22 (2004)
8. Dittenbach, M., Merkl, D., Rauber, A.: The growing hierarchical self-organizing map. In: Proceedings of the IEEE-INNS-ENNS International Joint Conference on Neural Networks, IJCNN 2000, vol. 6, pp. 15–19 (2000)
9. Fei-Fei, L., Fergus, R., Perona, P.: Learning generative visual models from few training examples: an incremental Bayesian approach tested on 101 object categories. In: Conference on Computer Vision and Pattern Recognition Workshop, CVPRW 2004, pp. 178–178 (2004)
10. Kasthurirathna, D., Piraveenan, M., Uddin, S.: Evolutionary stable strategies in networked games: the influence of topology. J. Artif. Intell. Soft Comput. Res. 5(2), 83–95 (2015)

11. Korytkowski, M.: Novel visual information indexing in relational databases. Integr. Comput.-Aid. Eng. **24**(2), 119–128 (2017)
12. Lazebnik, S., Schmid, C., Ponce, J.: Beyond bags of features: spatial pyramid matching for recognizing natural scene categories. In: 2006 IEEE Computer Society Conference on Computer Vision and Pattern Recognition, vol. 2, pp. 2169–2178 (2006)
13. Li, F.F., Perona, P.: A Bayesian hierarchical model for learning natural scene categories. In: Proceedings of the 2005 IEEE Computer Society Conference on Computer Vision and Pattern Recognition (CVPR 2005), vol. 2, pp. 524–531. IEEE Computer Society (2005)
14. Lowe, D.: Distinctive image features from scale-invariant keypoints. Int. J. Comput. Vis. **60**(2), 91–110 (2004)
15. Moulin, C., Barat, C., Ducottet, C.: Fusion of tf.idf weighted bag of visual features for image classification. In: 2010 International Workshop on Content-Based Multimedia Indexing (CBMI), pp. 1–6 (2010)
16. Sivic, J., Russell, B., Efros, A., Zisserman, A., Freeman, W.: Discovering objects and their location in images. In: Tenth IEEE International Conference on Computer Vision, ICCV 2005, vol. 1, pp. 370–377 (2005)
17. Starczewski, A., Krzyżak, A.: A modification of the Silhouette index for the improvement of cluster validity assessment. In: Rutkowski, L., Korytkowski, M., Scherer, R., Tadeusiewicz, R., Zadeh, L.A., Zurada, J.M. (eds.) ICAISC 2016. LNCS, vol. 9693, pp. 114–124. Springer, Cham (2016). doi:10.1007/978-3-319-39384-1_10
18. Starczewski, J.: Advanced Concepts in Fuzzy Logic and Systems with Membership Uncertainty, Studies in Fuzziness and Soft Computing, vol. 284. Springer, Heidelberg (2013)
19. Staszewski, P., Woldan, P., Korytkowski, M., Scherer, R., Wang, L.: Query-by-example image retrieval in microsoft SQL server. In: Rutkowski, L., Korytkowski, M., Scherer, R., Tadeusiewicz, R., Zadeh, L.A., Zurada, J.M. (eds.) ICAISC 2016. LNCS, vol. 9693, pp. 746–754. Springer, Cham (2016). doi:10.1007/978-3-319-39384-1_66
20. Woźniak, M.: Novel image correction method based on swarm intelligence approach. In: Dregvaite, G., Damasevicius, R. (eds.) ICIST 2016. CCIS, vol. 639, pp. 404–413. Springer, Cham (2016). doi:10.1007/978-3-319-46254-7_32
21. Wozniak, M., Polap, D.: On manipulation of initial population search space in heuristic algorithm through the use of parallel processing approach. In: 2016 IEEE Symposium Series on Computational Intelligence, pp. 1–6. IEEE (2016)
22. Wozniak, M., Polap, D., Napoli, C., Tramontana, E.: Graphic object feature extraction system based on cuckoo search algorithm. Expert Syst. Appl. **66**, 20–31 (2016)
23. Zalasiński, M., Cpałka, K.: New algorithm for on-line signature verification using characteristic hybrid partitions. In: Wilimowska, Z., Borzemski, L., Grzech, A., Świątek, J. (eds.) ISAT 2015. AISC, vol. 432, pp. 147–157. Springer, Cham (2016). doi:10.1007/978-3-319-28567-2_13

Local Keypoint-Based Image Detector with Object Detection

Rafał Grycuk[1]($^{\boxtimes}$), Magdalena Scherer[2], and Sviatoslav Voloshynovskiy[3]

[1] Institute of Computational Intelligence, Częstochowa University of Technology,
Al. Armii Krajowej 36, 42-200 Częstochowa, Poland
rafal.grycuk@iisi.pcz.pl
[2] Faculty of Management, Częstochowa University of Technology,
al. Armii Krajowej 19, 42-200 Częstochowa, Poland
[3] Computer Science Department, University of Geneva,
7 Route de Drize, Geneva, Switzerland
http://iisi.pcz.pl
http://sip.unige.ch

Abstract. Accurate and efficient image content description is crucial for image retrieval systems. In the paper we propose a novel method to describe images by a combination of the SURF local keypoint detector and the Canny edge detector. Then, a crawler is used to detect objects. The experiments performed on state-of-the-art image dataset showed that the method generates less data than standalone local keypoint detectors.

Keywords: Content-based image retrieval · Crawler · Edge detection · Bag of words · Image descriptor · Object extraction

1 Introduction

Emergence of content-based image retrieval (CBIR) in the 1990s enabled automatic retrieval of images and allowed to depart from searching collections of images by keywords and meta tags or just by browsing them. Generally, CBIR consists on finding images similar to the query image or images of a certain class [8,22,23,26,29–31,42] or classification [1,20,21,27,41,44,45,47,51,53] of the query image. To analyse images we have to compute visual features and then match them between the query image and the image database. Identifying features and objects in images is still a challenge as the same objects and scenes can be viewed under different imaging conditions. Features can be based on color representation [17,25,40], textures [6,10,19], shape [18,24,50] or edge detectors [54]. Another group of methods are local invariant features [33,35,37,38,46] which detect keypoints and generate descriptors, i.e. SURF [3], SIFT [33] or ORB [43].

In this paper we propose a novel method, which describes images by a combination of the SURF local keypoint detector and the Canny edge detector.

© Springer International Publishing AG 2017
L. Rutkowski et al. (Eds.): ICAISC 2017, Part I, LNAI 10245, pp. 507–517, 2017.
DOI: 10.1007/978-3-319-59063-9_45

The paper is organized as follows. Section 2 describes the SURF algorithm and Sect. 3 the Canny edge detector. The proposed method of image description is presented in Sect. 4 and numerical experiments on the PASCAL Visual Object Classes (VOC) 2012 dataset [9] in Sect. 5.

2 Speeded-Up Robust Features (SURF)

Speeded-Up Robust Features is a method which allows to detect and match local features of an image [4]. SURF is an improved version of SIFT (Scale-invariant feature transform) [33]. It is faster and provides similar functionality. SURF keypoints are composed of two vectors. First one provides the following information: position (x,y), scale (detected scale), response (response of the detected feature, strength), orientation (orientation, measured anti-clockwise from +ve x-axis), laplacian (sign of laplacian). Second one is a descriptor which contains 64 numbers. An important advantage of SURF is that it generates less data then SIFT (SIFT has longer descriptor of 128 length), which speeds-up further processing. The method has also a parallel version [48,49], thus it generates the results much faster. It is widely used in e.g. image description [32], segmentation [11], image recognition, object tracking, image analysis [13,16] content-based image retrieval [12,14]. The algorithm consists of four main stages [3]:

1. Computing Integral Images,
2. Fast-Hessian Detector:
   – The Hessian,
   – Constructing the Scale-Space,
   – Accurate Interest Point Localization,
3. Interest Point Descriptor:
   – Orientation Assignment,
   – Descriptor Components,
4. Generating vectors describing the interest points.

Figure 1 shows an example of the SURF method with two images containing similar objects. The lines between these two images represent the corresponding keypoints found on both images. The rectangle on the observed image, pinpoints the object location.

Fig. 1. The SURF algorithm example with keypoints detection and matching.

3   Canny Edge Detection

The Canny edge detector [5] is one of the most commonly used image processing methods for detecting edges [2,28,52]. The algorithm takes a gray scale image and returns an image with the positions of tracked intensity discontinuities [7,39]. The algorithm consists of four main steps [16,34]:

1. Noise reduction. The image is smoothed by applying an appropriate Gaussian filter.
2. Finding the intensity gradient of the image. During this step the edges should be marked where gradients of the image have large magnitudes.
3. Non-maxima suppression. If the gradient magnitude at a pixel is larger than those at its two neighbours in the gradient direction, mark the pixel as an edge. Otherwise, mark the pixel as the background.
4. Edge tracking by hysteresis. Final edges are determined by suppressing all edges that are not connected to genuine edges.

Fig. 2. The edge linking process. A - input image, B - edge detection, C - edge linking.

The result of the Canny edge detector is determined by two input parameters [16,34]:

- The width of the Gaussian filter used in the first stage directly affects the results of the Canny method,
- The thresholds used during edge tracking by hysteresis. It is difficult to give a generic threshold that works well on all images.

The algorithm basically finds the pixel intensity changes (gradients). The following formulas describe the approximation gradient. Before this step non-important edges need to be removed. This process is performed by the Gaussian function for calculating transformation for each image pixel [34]

$$G(x,y) = \frac{1}{\sqrt{2\pi\sigma^2}} e^{-\frac{x^2+y^2}{2\sigma^2}}, \tag{1}$$

where x is a distance from pixel position in the horizontal axis, y is a distance from pixel position in the vertical axis, σ is a standard deviation of Gaussian distribution. The process of edge detection is performed by Sobel filters [5,34]

$$K_{Gx} = \begin{bmatrix} -1 & 0 & 1 \\ -2 & 0 & 2 \\ -1 & 0 & 1 \end{bmatrix}, \tag{2}$$

$$K_{Gy} = \begin{bmatrix} 1 & 2 & 1 \\ 0 & 0 & 0 \\ -1 & -2 & -1 \end{bmatrix}, \tag{3}$$

In order to determine the edge strengths we need to use the Euclidean (4) or Manhattan (5) distance measures [34]:

$$|G| = \sqrt{G_x^2 + G_y^2}, \tag{4}$$

$$|G| = |G_x^2| + |G_y^2|, \tag{5}$$

where G_x is horizontal direction gradient and G_y is vertical direction gradient. The edge direction (angle) is determined by the following formula [5,34]

$$\Theta = \arctan\left(\frac{G_x}{G_y}\right). \tag{6}$$

4   Proposed Method for Image Description

In this section we describe the novel method for image description, which can be successfully used in content-based image retrieval. The presented algorithm is based on two well-known methods: Canny edge detection (see Sect. 3) and SURF (see Sect. 2). Both methods are widely used in computer vision. The main problem in the image description is object detection. We need to extract objects and perform mathematical transformation to obtain the mathematical description of the image. Many image databases contain images with several objects with non-homogeneous background. The more homogeneous object is, then more accurate results are. Therefore we decided to develop a method which extracts only important features of the object.

Our method takes an image as input and returns two sets of data:

– Extracted objects - objects extracted by the crawler method described in [15],
– Keypoints dictionary - list of key value pairs, where key contains image id, and value which contains list of keypoints. Such a structure provides access to keypoints by the image id.

In the first step we perform the edge detection by the Canny algorithm. Unfortunately, the detected edges are rarely complete and the object contours do not allow to run the crawler (see Fig. 2B). To eliminate this issue we use the edge linking algorithm, which completes the missing edges (see Fig. 2C). Subsequently, we detect the keypoints (features) on the input image. Afterwards, we run the crawler algorithm. Crawler moves between pixels until reaches the start pixel or all pixels are visited. The movements of this method are presented in Fig. 4(b) (for more details see [15]). The next step in our algorithm extracts the obtained objects. Then, for each keypoint we check the position. If it belongs to any of the obtained objects (i.e. it is located in object's ROI, Region of Interest) we add it to the keypoint dictionary. This stage allows to remove the insignificant keypoints, see Fig. 3. The algorithm steps are presented in both forms: pseudo-code (Algorithm 1) and block diagram (Fig. 4(a)).

Fig. 3. The insignificant keypoints removal. The top image contains all detected pixels. In the bottom pixels non important pixels are removed. As can be seen, pixels describe only the object itself and any background noise is removed.

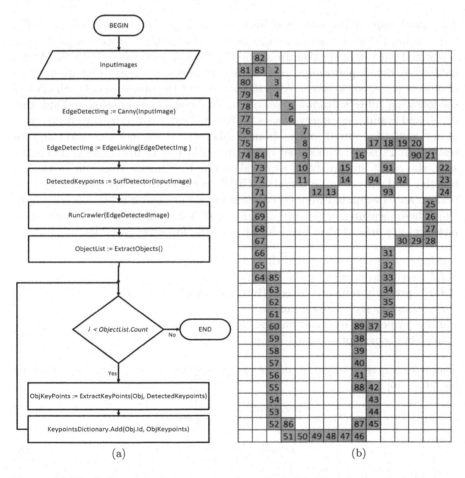

(a) (b)

Fig. 4. (a) Block diagram of the presented method. (b) The crawler method movements.

INPUT: *InputImage*
OUTPUT: *KeypointsDictionary, ObjectsList*
EdgeDetectImg := Canny(InputImage);
EdgeDetectImg := EdgeLinking(EdgeDetectImg);
DetectedKeypoints := SurfDetector(InputImage);
  RunCrawler(EdgeDetectedImage);
ObjectsList := ExtractObjects();
foreach *Obj ∈ ObjectsList* **do**
  | *ObjKeypoints := ExtractKeyPoints(Obj, DetectedKeypoints);*
  | *KeypointsDictionary.Add(Obj.Id, ObjKeypoints);*
end

Algorithm 1. Steps of the proposed algorithm.

5 Experimental Results

The simulation environment was build on our own software, written in .NET, C#. During the experiments we also used the Emgu CV library. The tests were performed with images taken from the PASCAL Visual Object Classes (VOC) dataset [9]. Before performing experiments, we selected images with various types of objects (cars, cows, bikes). Afterwards, we split the image sets of each class in the following subsets: training - set of images for image description and indexing (90%) and evaluation - query images for testing (10%). To verify the effectiveness of our method we used images from a different set for index creation and for testing. In Table 1 we presented the retrieved factors of multi-query. As can be seen, the result are satisfying which allows us to conclude that, our method is effective and proves to be useful in CBIR techniques. For the purposes of the performance evaluation we used two well known measures; *precision* and *recall* [36]. These measures are widely used in CBIR for evaluation. The classification measures are presented in Fig. 5 and it uses the following variables:

- *AI* - appropriate images and they should be returned,
- *RI* - returned images by the system,
- *Rai* - properly returned images (intersection of *AI* and *RI*),
- *Iri* - improperly returned images,
- *anr* - proper not returned images,
- *inr* - improper not returned images.

These measures allows to define *precision* and *recall* by the following formulas [36]

$$precision = \frac{|rai|}{|rai + iri|}, \tag{7}$$

$$recall = \frac{|rai|}{|rai + anr|}. \tag{8}$$

Table 2 shows the visualization of experiment results from a single image query. As can be seen, most images were correctly retrieved (20). Only three of

Table 1. Results of the experiments (MultiQuery). Due to limited space only a small part of the query results are presented.

Image Id	RI	AI	rai	iri	anr	Precision	Recall
29036	35	30	26	9	4	0.74	0.87
29045	29	30	23	6	7	0.79	0.77
29060	24	30	22	2	8	0.92	0.73
29091	31	30	25	6	5	0.81	0.83
354002	30	30	26	4	4	0.87	0.87
car-111	26	30	20	6	10	0.77	0.67
car-115	34	30	26	8	4	0.76	0.87
car-116	26	30	24	2	6	0.92	0.8
car-117	30	30	24	6	6	0.8	0.8
car-119	25	30	20	5	10	0.8	0.67
car-120	22	30	20	2	10	0.91	0.67
car-121	32	30	24	8	6	0.75	0.8
car-130	26	30	20	6	10	0.77	0.67
car-134	27	30	24	3	6	0.89	0.8
car-135	26	30	25	1	5	0.96	0.83
car-139	35	30	26	9	4	0.74	0.87
car-145	27	30	23	4	7	0.85	0.77
car-146	35	30	26	9	4	0.74	0.87
car-147	30	30	24	6	6	0.8	0.8
car-154	33	30	26	7	4	0.79	0.87
car-157	33	30	26	7	4	0.79	0.87
car-165	28	30	22	6	8	0.79	0.73
car-172	23	30	20	3	10	0.87	0.67
car-2	27	30	21	6	9	0.78	0.7
car-40	29	30	20	9	10	0.69	0.67
car-75	27	30	25	2	5	0.93	0.83
car-87	29	30	22	7	8	0.76	0.73
car-95	29	30	22	7	8	0.76	0.73
car-96	26	30	25	1	5	0.96	0.83
car-97	28	30	21	7	9	0.75	0.7
cow-111	32	30	26	6	4	0.81	0.87
cow-150	28	30	22	6	8	0.79	0.73
cow-172	26	30	23	3	7	0.88	0.77
cow-181	23	30	20	3	10	0.87	0.67
cow-191	28	30	25	3	5	0.89	0.83
cow-210	23	30	22	1	8	0.96	0.73
cow-280	27	30	25	2	5	0.93	0.83
cow-292	25	30	22	3	8	0.88	0.73
cow-302	24	30	22	2	8	0.92	0.73

Table 2. Query results. Example images from the experiment. The image with border is the query image.

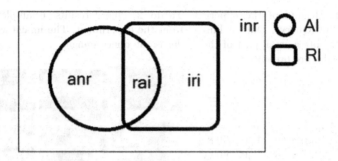

Fig. 5. Performance measures diagram. [14]

them are improperly recognized. The image with a border is the query image. The *precision* value for this experiment equals 0.87 and for *recall* 0.67.

The presented experiments proved that our method is effective and correctly retrieves images.

6   Conclusion

The presented method is a novel approach to content-based image retrieval. Our method involves two well-known algorithms (the Canny edge detector and SURF) and uses the crawler method to extract objects. The effectiveness of our method has been proven by the performed experiments and the method can be used in various content-based image retrieval tasks. The applied SURF method ensure invariance to different geometric modifications (e.g. scale changing and rotation).

References

1. Akusok, A., Miche, Y., Karhunen, J., Bjork, K.M., Nian, R., Lendasse, A.: Arbitrary category classification of websites based on image content. IEEE Comput. Intell. Mag. **10**(2), 30–41 (2015)
2. Bao, P., Zhang, L., Wu, X.: Canny edge detection enhancement by scale multiplication. IEEE Trans. Pattern Anal. Mach. Intell. **27**(9), 1485–1490 (2005)
3. Bay, H., Ess, A., Tuytelaars, T., Van Gool, L.: Speeded-up robust features (SURF). Comput. Vis. Image Underst. **110**(3), 346–359 (2008)
4. Bay, H., Tuytelaars, T., Gool, L.: SURF: Speeded Up Robust Features. In: Leonardis, A., Bischof, H., Pinz, A. (eds.) ECCV 2006. LNCS, vol. 3951, pp. 404–417. Springer, Heidelberg (2006). doi:10.1007/11744023_32
5. Canny, J.: A computational approach to edge detection. IEEE Trans. Pattern Anal. Mach. Intell. **8**(6), 679–698 (1986)
6. Chang, T., Kuo, C.C.: Texture analysis and classification with tree-structured wavelet transform. IEEE Trans. Image Process. **2**(4), 429–441 (1993)
7. Ding, L., Goshtasby, A.: On the canny edge detector. Pattern Recogn. **34**(3), 721–725 (2001)

8. Drozda, P., Sopyła, K., Górecki, P.: Online crowdsource system supporting ground truth datasets creation. In: Rutkowski, L., Korytkowski, M., Scherer, R., Tadeusiewicz, R., Zadeh, L.A., Zurada, J.M. (eds.) ICAISC 2013. LNCS, vol. 7894, pp. 532–539. Springer, Heidelberg (2013). doi:10.1007/978-3-642-38658-9_48

9. Everingham, M., Van Gool, L., Williams, C.K.I., Winn, J., Zisserman, A.: The pascal visual object classes (VOC) challenge. Int. J. Comput. Vis. **88**(2), 303–338 (2010)

10. Francos, J., Meiri, A., Porat, B.: A unified texture model based on a 2-d wold-like decomposition. IEEE Trans. Sig. Process. **41**(8), 2665–2678 (1993)

11. Grycuk, R., Gabryel, M., Korytkowski, M., Romanowski, J., Scherer, R.: Improved digital image segmentation based on stereo vision and mean shift algorithm. In: Wyrzykowski, R., Dongarra, J., Karczewski, K., Waśniewski, J. (eds.) PPAM 2013. LNCS, vol. 8384, pp. 433–443. Springer, Heidelberg (2014). doi:10.1007/978-3-642-55224-3_41

12. Grycuk, R., Gabryel, M., Korytkowski, M., Scherer, R.: Content-based image indexing by data clustering and inverse document frequency. In: Kozielski, S., Mrozek, D., Kasprowski, P., Małysiak-Mrozek, B., Kostrzewa, D. (eds.) BDAS 2014. CCIS, vol. 424, pp. 374–383. Springer, Cham (2014). doi:10.1007/978-3-319-06932-6_36

13. Grycuk, R., Gabryel, M., Korytkowski, M., Scherer, R., Voloshynovskiy, S.: From single image to list of objects based on edge and blob detection. In: Rutkowski, L., Korytkowski, M., Scherer, R., Tadeusiewicz, R., Zadeh, L.A., Zurada, J.M. (eds.) ICAISC 2014. LNCS, vol. 8468, pp. 605–615. Springer, Cham (2014). doi:10.1007/978-3-319-07176-3_53

14. Grycuk, R., Gabryel, M., Scherer, R., Voloshynovskiy, S.: Multi-layer architecture for storing visual data based on WCF and microsoft SQL server database. In: Rutkowski, L., Korytkowski, M., Scherer, R., Tadeusiewicz, R., Zadeh, L.A., Zurada, J.M. (eds.) ICAISC 2015. LNCS, vol. 9119, pp. 715–726. Springer, Cham (2015). doi:10.1007/978-3-319-19324-3_64

15. Grycuk, R., Gabryel, M., Scherer, M., Voloshynovskiy, S.: Image descriptor based on edge detection and Crawler algorithm. In: Rutkowski, L., Korytkowski, M., Scherer, R., Tadeusiewicz, R., Zadeh, L.A., Zurada, J.M. (eds.) ICAISC 2016. LNCS, vol. 9693, pp. 647–659. Springer, Cham (2016). doi:10.1007/978-3-319-39384-1_57

16. Grycuk, R., Scherer, R., Gabryel, M.: New image descriptor from edge detector and blob extractor. J. Appl. Math. Comput. Mech. **14**(4), 31–39 (2015)

17. Huang, J., Kumar, S., Mitra, M., Zhu, W.J., Zabih, R.: Image indexing using color correlograms. In: Proceedings of 1997 IEEE Computer Society Conference on Computer Vision and Pattern Recognition, pp. 762–768, June 1997

18. Jagadish, H.V.: A retrieval technique for similar shapes. SIGMOD Rec. **20**(2), 208–217 (1991)

19. Jain, A.K., Farrokhnia, F.: Unsupervised texture segmentation using Gabor filters. Pattern Recogn. **24**(12), 1167–1186 (1991)

20. Jégou, H., Douze, M., Schmid, C., Pérez, P.: Aggregating local descriptors into a compact image representation. In: 2010 IEEE Conference on Computer Vision and Pattern Recognition (CVPR), pp. 3304–3311. IEEE (2010)

21. Jégou, H., Perronnin, F., Douze, M., Sanchez, J., Perez, P., Schmid, C.: Aggregating local image descriptors into compact codes. IEEE Trans. Pattern Anal. Mach. Intell. **34**(9), 1704–1716 (2012)

22. Kanimozhi, T., Latha, K.: An integrated approach to region based image retrieval using firefly algorithm and support vector machine. Neurocomputing **151**, 1099–1111 (2015). Part 3(0)
23. Karakasis, E., Amanatiadis, A., Gasteratos, A., Chatzichristofis, S.: Image moment invariants as local features for content based image retrieval using the bag-of-visual-words model. Pattern Recogn. Lett. **55**, 22–27 (2015)
24. Kauppinen, H., Seppanen, T., Pietikainen, M.: An experimental comparison of autoregressive and Fourier-based descriptors in 2d shape classification. IEEE Trans. Pattern Anal. Mach. Intell. **17**(2), 201–207 (1995)
25. Kiranyaz, S., Birinci, M., Gabbouj, M.: Perceptual color descriptor based on spatial distribution: a top-down approach. Image Vision Comput. **28**(8), 1309–1326 (2010)
26. Korytkowski, M.: Novel visual information indexing in relational databases. Integr. Comput.-Aid. Eng. **24**(2), 119–128 (2017)
27. Korytkowski, M., Rutkowski, L., Scherer, R.: Fast image classification by boosting fuzzy classifiers. Inf. Sci. **327**, 175–182 (2016)
28. Li, X., Jiang, J., Fan, Q.: An improved real-time hardware architecture for canny edge detection based on FPGA. In: 2012 Third International Conference on Intelligent Control and Information Processing (ICICIP), pp. 445–449. IEEE (2012)
29. Lin, C.H., Chen, H.Y., Wu, Y.S.: Study of image retrieval and classification based on adaptive features using genetic algorithm feature selection. Expert Syst. Appl. **41**(15), 6611–6621 (2014)
30. Liu, G.H., Yang, J.Y.: Content-based image retrieval using color difference histogram. Pattern Recogn. **46**(1), 188–198 (2013)
31. Liu, S., Bai, X.: Discriminative features for image classification and retrieval. Pattern Recogn. Lett. **33**(6), 744–751 (2012)
32. Lowe, D.G.: Object recognition from local scale-invariant features. In: The Proceedings of the Seventh IEEE International Conference on Computer Vision, vol. 2, pp. 1150–1157. IEEE (1999)
33. Lowe, D.G.: Distinctive image features from scale-invariant keypoints. Int. J. Comput. Vis. **60**(2), 91–110 (2004)
34. Luo, Y.M., Duraiswami, R.: Canny edge detection on nvidia cuda. In: IEEE Computer Society Conference on Computer Vision and Pattern Recognition Workshops, CVPRW 2008, pp. 1–8. IEEE (2008)
35. Matas, J., Chum, O., Urban, M., Pajdla, T.: Robust wide-baseline stereo from maximally stable extremal regions. Image Vis. Comput. **22**(10), 761–767 (2004). British Machine Vision Computing 2002
36. Meskaldji, K., Boucherkha, S., Chikhi, S.: Color quantization and its impact on color histogram based image retrieval accuracy. In: First International Conference on Networked Digital Technologies, NDT 2009, pp. 515–517, July 2009
37. Mikolajczyk, K., Schmid, C.: Scale and affine invariant interest point detectors. Int. J. Comput. Vis. **60**(1), 63–86 (2004)
38. Nister, D., Stewenius, H.: Scalable recognition with a vocabulary tree. In: Proceedings of the 2006 IEEE Computer Society Conference on Computer Vision and Pattern Recognition, CVPR 2006, vol. 2, pp. 2161–2168. IEEE Computer Society, Washington, DC (2006)
39. Ogawa, K., Ito, Y., Nakano, K.: Efficient canny edge detection using a gpu. In: 2010 First International Conference on Networking and Computing (ICNC), pp. 279–280. IEEE (2010)
40. Pass, G., Zabih, R.: Histogram refinement for content-based image retrieval. In: Proceedings 3rd IEEE Workshop on Applications of Computer Vision, WACV 1996, pp. 96–102, December 1996

41. Patgiri, C., Sarma, M., Sarma, K.K.: A class of neuro-computational methods for assamese fricative classification. J. Artif. Intell. Soft Comput. Res. 5(1), 59–70 (2015)
42. Rashedi, E., Nezamabadi-pour, H., Saryazdi, S.: A simultaneous feature adaptation and feature selection method for content-based image retrieval systems. Knowl.-Based Syst. 39, 85–94 (2013)
43. Rublee, E., Rabaud, V., Konolige, K., Bradski, G.: Orb: an efficient alternative to sift or surf. In: 2011 IEEE International Conference on Computer Vision (ICCV), pp. 2564–2571, November 2011
44. Scherer, R.: Multiple Fuzzy Classification Systems. Springer, Berlin (2012)
45. Shrivastava, N., Tyagi, V.: Content based image retrieval based on relative locations of multiple regions of interest using selective regions matching. Inf. Sci. 259, 212–224 (2014)
46. Sivic, J., Zisserman, A.: Video google: a text retrieval approach to object matching in videos. In: Proceedings of Ninth IEEE International Conference on Computer Vision, vol. 2, pp. 1470–1477, October 2003
47. Stanovov, V., Semenkin, E., Semenkina, O.: Self-configuring hybrid evolutionary algorithm for fuzzy imbalanced classification with adaptive instance selection. J. Artif. Intell. Soft Comput. Res. 6(3), 173–188 (2016)
48. Šváb, J., Krajník, T., Faigl, J., Přeučil, L.: FPGA based speeded up robust features. In: IEEE International Conference on Technologies for Practical Robot Applications, TPRA 2009, pp. 35–41. IEEE (2009)
49. Terriberry, T.B., French, L.M., Helmsen, J.: GPU accelerating speeded-up robust features. In: Proceedings of 3DPVT, vol. 8, pp. 355–362 (2008)
50. Veltkamp, R.C., Hagedoorn, M.: State of the art in shape matching. In: Lew, M.S. (ed.) Principles of Visual Information Retrieval, pp. 87–119. Springer, London (2001)
51. Villmann, T., Bohnsack, A., Kaden, M.: Can learning vector quantization be an alternative to SVM and deep learning? - recent trends and advanced variants of learning vector quantization for classification learning. J. Artif. Intell. Soft Comput. Res. 7(1), 65–81 (2017)
52. Wang, B., Fan, S.: An improved canny edge detection algorithm. In: 2009 Second International Workshop on Computer Science and Engineering, pp. 497–500. IEEE (2009)
53. Yang, J., Yu, K., Gong, Y., Huang, T.: Linear spatial pyramid matching using sparse coding for image classification. In: IEEE Conference on Computer Vision and Pattern Recognition, CVPR 2009, pp. 1794–1801, June 2009
54. Zitnick, C.L., Dollár, P.: Edge boxes: locating object proposals from edges. In: Fleet, D., Pajdla, T., Schiele, B., Tuytelaars, T. (eds.) ECCV 2014. LNCS, vol. 8693, pp. 391–405. Springer, Cham (2014). doi:10.1007/978-3-319-10602-1_26

Heavy Changes in the Input Flow for Learning Geography of a Robot Environment

Georgii Khachaturov[✉], Josué Figueroa-González,
Silvia B. González-Brambila, and Juan M. Martínez-Hernández

Metropolitan Autonomous University - Azcapotzalco,
San Pablo Av. 180, Reynosa Tamaulipas, 02200 Mexico City, Mexico
{xgeorge,jfgo,sgb}@correo.azc.uam.mx, martinez.juan.hdz@gmail.com

Abstract. A novel approach to generation of geographic knowledge from robot views is presented. It is implemented in a pilot software where a virtual robot operates in a static 2D-environment. The robot sensor scans with rays an angular field of view and produces a 1D view of distances to the closest obstacle. By processing such views, 'heavy changes' are detected to trigger switching local maps in an atlas that represents geography of the robot environment. To detect heavy changes, firstly, each plot is transformed to a string of singular points; then, in time-scale, a pair of such strings is subjected to a treatment based on application of the distance of Levenshtein, which leads to so-called Editorial Prescription (EP); a heavy change is detected if EP shows a considerable distinction between strings. This approach is applied in automatic construction of an atlas for non-Cartesian navigation, while robot explores the scene.

Keywords: Detection of heavy changes · Levenshtein distance · Map switching · Non-Cartesian navigation · Processing range images

1 Introduction

The particular problem treated in this paper is related to a wider research: the one of non-Cartesian robot navigation. Its specific is illustrated below with some views generated by a pilot software that simulates evolution of a virtual robot in an environment, guided by supervisor. The environment is static, bi-dimensional, and populated with obstacles. Figure 1a shows an instance of such environment; it is a supervisor's view of a scene; the robot itself is shown as a dark triangle at the bottom. If the robot would be supplied with a color camera, it could see the same scene as Fig. 1b shows. However, it has a different kind of sensor: the one that scans all directions in a field of view and measures a distance to the first obstacle along each direction; each distance is subjected to a transformation so that the final view obtained by the robot for the same scene is the one shown in Fig. 1c. Note that views like Fig. 1c are called below 'angle-distance plots', but this should not to confuse the readers: they show not crude, but somehow transformed distances.

© Springer International Publishing AG 2017
L. Rutkowski et al. (Eds.): ICAISC 2017, Part I, LNAI 10245, pp. 518–529, 2017.
DOI: 10.1007/978-3-319-59063-9_46

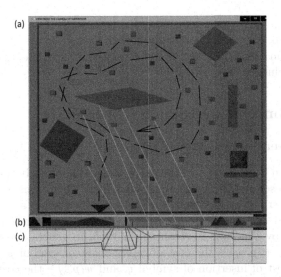

Fig. 1. Three views for the same "robot-in-environment" scene: **(a)** A supervisor's view; the robot position and orientation is shown by a dark triangle at the bottom; the dashed line is commented in Sect. 5. **(b)** A robocentric view by a color camera. **(c)** A robocentric view by an 'angle-distance plot'. The lines between (a) and (b), and between (b) and (c) connect the same objects and the same details shown in the views. (Color figure online)

A single view like the one in Fig. 1a provides the supervisor with full information about geography of the environment. But it is unavailable to robot, so we try to help the robot to reach an equivalent understanding of the geography, but by processing a set of views similar to that in Fig. 1c.

The process of learning geography is based on driving the robot along a continuous trajectory in the scene, combined with interpretation of the input flow of robot views. The process starts from an initial robot position in the scene, like the one of Fig. 1a.

This work presents an approach to a particular issue of processing the input flow: the detection of '*heavy changes*' or, in other words, those changes in the flow, which can be used as milestones for the non-Cartesian navigation. Note that robot should run a permanent process of self-localization for identifying the map in the atlas to which the robot's current state belongs (so called *current map*). While no heavy change occurs, the robot stays in the same current map; otherwise it must switch the current map to other map of the atlas or, if no appropriate map exists, firstly to introduce a new map into the atlas and only then switch the current map to it. That is, the heavy changes are the main interface between the robot sensor system and an inner representation of geography.

Our approach is implemented in the scope of the pilot software mentioned above and its principles are verified by experiments with this software.

The rest of paper is organized as follows: Sect. 2 offers a brief review of previous works and some concepts used in this paper. Section 3 describes a processing scheme for detection of heavy changes. Section 4 develops further-more the draft of Sect. 3, namely it deals with specific issues of the Levenshtein distance. Section 5 describes the pilot software and experiments on it. Section 6 contains a conclusion.

2   Used Concepts and Related Works

2.1   The Levenshtein Distance

The Levenshtein distance [1] is a measure of the difference between two strings and it can be described as the minimum number of operations over single letters needed to convert or change one word into another. Such a step-by-step con-version is called *editorial prescription*. The process for obtaining this measure was developed by Vladimir Levenshtein in 1965. In general case it depends on three functions: $w(p, q)$ – the cost of replacement of symbol p with symbol q, $w(\varepsilon, q)$ – the cost of insertion of symbol q, and $w(p,\varepsilon)$ – the cost of deletion of symbol p. Let string a should be converted to string b, then the following formula (1) determines recursively all elements of a rectangle matrix

$$D_{a,b}(i,j) = \begin{cases} 0 \text{ if } j = i = 0; \text{ else}: \\ D_{a,b}(i-1,0) + w(\varepsilon, a_i) \parallel D_{a,b}(0, j-1) + w(b_j, \varepsilon) \\ \text{for } j = 0 \parallel i = 0; \text{ otherwise}: \\ \min \begin{cases} D_{a,b}(i-1,j) + w(\varepsilon, b_j) \\ D_{a,b}(i, j-1) + w(a_i, \varepsilon) \\ D_{a,b}(i-1,j-1) + w(a_i b_j) * 1_{(a_i \neq b_j)}, \end{cases} \end{cases} \tag{1}$$

where i and $j \geq 0$ run indexes of symbols in respective strings, $1_{(a_i \neq b_j)}$ is so called *indicator function* equal to 0 when $a_i = b_j$ and to 1 otherwise. In particular case, when $w(\varepsilon, q) = 1$, $w(p, \varepsilon) = 1$, $w(p, p) = 0$, and $w(p, q) = 1$ as p $\neq$ q, this matrix defines the Levenshtein distance between a and b. The Levenshtein distance is widely known thanks to the applications used in detecting plagiarism in text or codes [2], and recently in dialect analysis [3]; some uses of this technique related with image processing can be found in [4]. Here it is applied to the strings constructed as a coded form of the angle-distance plot. Such strings are quite short so we do not meet difficulties specific for the detecting plagiarism in large texts.

2.2   Non-Cartesian Navigation

Navigation for humans and animals does not assume pointing the goal as a point in a Cartesian space. The attempts to understand how the human brain tackles the navigation tasks have been undertaken in numerous works, during several decades mainly in psychology. The first mathematically strict model of non-Cartesian navigation was proposed by Khachaturov [5]. Other works on this topic are [6–8].

The model presented in [5] (so called GT-model) is formed by two graphs: G which stands for the robot sensor-motor knowledge, and T represents a geography or administrative system imposed on a corresponding environment. In formal terms, a *GT-model* is defined as pair $\{G,\ T\}$, where the components satisfy the following properties:

- $G = (V_G, E_G)$ is a directed graph with non-negative weights assigned to its edges;
- $T = (V_T,\ E_T)$ is a directed graph with an unique source-node whereas the set of its terminal nodes coincides with V_G.

Interpretation of nodes of V_G and edges of E_G follows. Let S denote the space of all robot states inside an environment. It is assumed that *robot views* are put in one-to-one correspondence with points of S. A set of close in a metric robot views generates a neighborhood of close states in S. Each node of V_G stands for a domain of S and the whole set of nodes V_G represents a sampling of S generated by equivalent in a certain meaning robot views.

An edge $l \in E_G$, where $l = (a,b)$ and $a,\ b \in V_G$, represents a control rule that drives robot to change any view associated with a to a view associated with b. Graph T represents a system of sets composed by nodes of G. This system is determined by the following rule: *Any node α of V_T corresponds to set $V_\alpha \subset V_G$ defined as the set of all terminal nodes of all paths in T that start from α.* Intuitively, α is the name of a geographic object represented by the set V_α – *the domain of α.* An edge of E_T from node β to node α stands for inclusion of respective domains, $V_\beta \supset V_\alpha$, so the first geographic object of an edge is wider than the second.

Since a terminal node of T corresponds to a node of V_G, it represents some close states of a robot inside its environment. Unlike that, for a non-terminal $\alpha \in V_T$, its domain V_α can contain far robot states. Graph T organizes such domains into a system quite similar to a real system of geographical concepts. In particular, the domain of the source-node of V_T is the whole V_G. If there is a path in T from α to β, then the domain of α obviously contains the domain of β. The simplest kind for T is a tree. If T is a tree, it represents so-called *tree decomposition* of G, [9].

In these terms, formalization of a non-Cartesian navigation problem is as follows. Let $\alpha, \beta \in V_T$ and $V_\alpha,\ V_\beta \subset V_G$ be domains of α and β, respectively. The problem is: *Find a path in G with minimal summary weight that begins inside V_α and ends inside V_β.* This problem is called the *extended shortest path problem (espp).* So an *espp* means search of a best route that connects two sets, however not arbitrary sets but only those represented by some nodes of the geographic graph T. It was shown in [5], that under some natural assumptions a navigational problem can be solved by a dynamic programming algorithm with a relatively low computational complexity.

2.3 Geography of the Lowest Level

In spite of the theoretical advantages of the GT-model mentioned in the previous section, no progress in its practical implementation can be found in literature since the publication of [5]. The main obstacle for that consists in the necessity of novel methods for automatic learning the lowest, non-verbalizable level of geography. To clarify this issue, consider any geographic item that has an explicit name like Poznan, Broadway, Asia, etc. It is technically clear how to insert such an item into the geographic graph of GT-model. In contrast to that, each item of the lowest geographic level should be represented by a map that does not have neither a distinctive shape, nor an attributed name.

Moreover, in contrast to the verbalizable levels of the geographic graph of the GT-model, the lowest-level geography depends essentially on the sensor system. Note that the higher levels can be the same, say, for a robot, a blind person, as well as for a human who does not suffer any illness of sight. On the contrary, the lowest level geography depends on available sensors: for a blind person – on the stick which he/she uses for exploration of the environment, and for a normal person – on his/her vision. This is why a special technique for learning geography of the lowest level should be developed and studied, and why the detection of heavy changes is important.

2.4 Related Works

A brief review of some related research lines follows.

Simultaneous Localization and Map Building problem (SLAM) [10–13]: the aim of it this approach is to convert a set of robocentric views into a single map. This 'map' is understood not as in our work, but in a traditional, Cartesian meaning so that a human could read it and use. The SLAM techniques involve statistical methods including extended Kalman filter [14] and Rao–Blackwellized particle filters [15]. They allow feeding the map creation while the robot moves smoothly. The ideas of SLAM seem very fruitful to be combined in future with our approach to cope with the robot dynamics.

Path Planning for Autonomous Vacuum Cleaner Robots: The user of a cleaner robot must be certain that in a certain time with a high probability the robot will clean every corner of the workspace. It can be reached avoiding the requirement that the robot any moment be aware of where it is located. So the efficiency of sweeping workspace, sufficient for practice can be provided without construction of a computational model of the workspace. It is reached by applying some context-specific heuristics and sensors in combination with statistical principles [16].

3   A Draft Scheme for Detection of Heavy Changes

3.1 Sensor System for Learning Geography of the Virtual Robot

The main goal of developing the virtual robot briefly presented in the beginning of paper is to study principles of automatic learning non-Cartesian geography.

Image processing and recognition play an auxiliary role. This is why the sensor system of robot was intentionally designed as simple as possible. In particular, each robot view, like one in Fig. 1c, is just a real function of one variable.

It should be mentioned from the beginning that our processing scheme of the robot views depends essentially on the dimension of the views. A discussion about extension in future of this scheme, which is applied here to the 1D-images, to higher dimensions and other kind of sensors, such as conventional video cameras or ultrasonic sensors, lays beyond the scope of this paper.

While using a virtual reality platform it is easy to generate a complex scene and extract any kind of information related to the scene for different kinds of virtual camera. Our virtual robot is developed by means of OpenGL and the screenshots of Figs. 1 are just different ways to render the same scene.

3.2   Main Idea for Detection of Heavy Changes

Intuitively it is clear that the visual events specifically important for robot navigation are related to the facts of appearance/disappearance of objects or gaps in sight and also to a qualitative change in appearance of an object. So the events of such kinds should be primarily detected and then used for indexing the maps of a geographic database. On the other hand, the value of argument where the image of an obstacle begins or ends in a view strongly corresponds to a discontinuity of the first derivative of the plot; then, a discontinuity of the second derivative has a strong correlation with an angle of the object shape. It can be easily seen by comparison of the corresponding elements of Figs. 1b and c connected by the association lines. This observation suggests us to perform a transformation of each angle-distance plot into *string formed by singular points constructed by discontinuities of the first and the second derivatives* of the plot. The transformation 'plot-to-string' is the first step in detection of heavy changes. Its details are presented in the upper block-diagram of Fig. 2.

In this transformation, the type of any singular point belongs to the following short alphabet: {STEP_UP, STEP_DOWN, ANGLE_TOWARD_ROBOT, ANGLE_OUTWARD_ROBOT} with an obvious intuitive meaning of each option.

A subsequent processing step performs temporal analysis of the flow of such strings. The idea to *use for this step a technique based on the Levenshtein distance* suggests itself: when two strings of the singular points are represented in an alphabet, one can find their editorial prescription. And if, for example, such two strings generated for some close moments completely match each other, it is naturally to claim that no heavy change occurred; otherwise, a further analysis of the vector of editorial prescription should follow to classify possible occurrence of a heavy change. A graphic representation of this step with more details is shown in the lower block-diagram of Fig. 2.

1. Transformation "ADP to String of SPs"

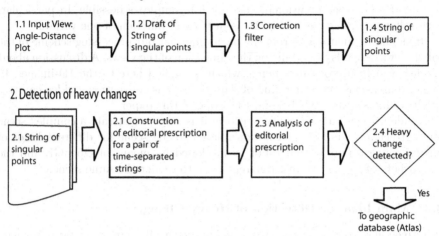

2. Detection of heavy changes

Fig. 2. Block-diagram of the two main steps of detection of heavy changes.

4   Specific Issues of Detection of Heavy Changes

There are two specific issues in the just presented scheme. The former is related to the errors in classifying a singular point, blocks 1.2 and 1.3 of Fig. 2: it turns out to be that such errors cannot be avoided completely. Indeed, if the robot would drive around a pyramid, its shape in the view changes and finally a shape of angle in the angle-distance plot will be transformed to a shape of step-function; consequently the errors are inevitable for some critical region of arguments. Nevertheless, some thresholds of the algorithm of extraction of singular points can be optimized to reduce probability of the errors of this kind. This optimization was provided, [17], and the probability of errors from its initial value 8% for some intuitively chosen thresholds was reduced in result to 0.2–0.3%. Other kind of possible errors is generated by the detection of two or more very close singular points that, in fact, are yielded by a single singular point. The correction filter (Fig. 2, block 1.3) was introduced just to reduce errors of this kind. For example, the filter merges two STEPs of the same kind (see the alphabet in the previous section) if distance between them is very short.

The latter is related to the application of the Levenshtein distance and is a consequence of the fact that as the alphabet for singular points is rather small as well as the strings of singular points are respectively short, which makes the errors in interpretation of editorial prescription (block 2.3 of Fig. 2) rather probable. In the rest of this section, we describe some problem-specific expedients to reduce significantly the probability of the last kind of errors.

4.1 An Enriched Description of Symbols for Computing the Levenshtein Distance

A close look at the process of computation due to formula (1), shows that what-soever specific of a particular problem is hidden in computation of the function $1_{(a_i \neq b_j)}$ and the costs $w(p,q)$, $w(\varepsilon,q)$, and $w(p,\varepsilon)$. As to function $1_{(a_i \neq b_j)}$ if one just compares literal coincidence of the symbols a_i and b_j then no problem-specific is taken in consideration. However, if each symbol is provided with an enriched problem-specific description, it allows us to improve the proper defin-ition of $1_{(a_i \neq b_j)}$. For the problem under consideration, additionally to its type from the above alphabet, each singular point can be accompanied, for instance, with the value of distance. Since the strings to be compared typically correspond to the two robot states, one before and another after application of a robot con-trol, it is technically possible to evaluate whether the distance associated with b_j can match to the distance of a_i. This idea was implemented in our software, which significantly improved the function $1_{(a_i \neq b_j)}$. See the top-right of Fig. 3 for an instance of computation of $1_{(a_i \neq b_j)}$.

One more expedient to reduce errors consists in the introduction of an arti-ficial symbol of the kind LEG between each pair of usual singular points. Its description, in particular, includes the leg-length, that is, the distance between two successive singular points. This allows us to improve furthermore the idea of Sect. 4.1: using lengths of legs, it is analyzed how probable is that under a certain control two legs can match each other.

4.2 Interpretation of an Editorial Prescription

We construct Editorial Prescription (EP) by the well-known Wagner-Fischer algorithm [18]. Finally, an EP is represented as a string in the following alphabet: {MATCH, REPLACE, INSERT, and DELETE }. Evidently, if all symbols of an EP are MATCHes, then no heavy change occurs. If EP contains INSERT or DELETE, it mostly means a heavy change. But some situations regarded in Block 2.4 of Fig. 2 require a problem-specific heuristic. Just two examples of such heuristics for computation of $1_{(a_i \neq b_j)}$ follow:

A limited equivalence between MATCH and REPLACE is allowed. For exam-ple, an STEP_UP is regarded as equivalent to ANGLE_TOWARD_ROBOT for singular points with distances quite close to the far distance-limit of angle-distance plots. This decision naturally follows from the manner of how OpenGL constructs the content of z-buffer [19]: when some real distances are far, they are compressed in z-buffer to very close values; so it is easy to confuse a far STEP with a far ANGLE if they are of appropriate types. Partly, interpretation of an EP depends on a priori knowledge of the applied robot control. For example, if the control command is ROTATE, it is expected that distance to any singu-lar points will stay practically the same after application of the command. A rotation can lead to an appearance or disappearance of a singular point at the periphery of the field of view. If this occurs, we assume that a disappearance of singular point is not a heavy change, but an appearance is a heavy change.

The above heuristics related to the indicator function were combined with a problem-specific choice of the costs in formula (1). It was found experimentally that for our case they should not all be the same. We use the cost of replacement $w(p,q)$ two times bigger than the costs of deletion $w(\varepsilon,q)$ and insertion $w(p,\varepsilon)$, which are set as equal.

5   Experiments

All experiments were accomplished using a pilot software mentioned above. A detailed description of the software lay beyond the scope of this paper. Just a concise list of its functional components follows:

– Interactive creation of obstacles of the robot environment (based on [20]);
– Visualization of the environment and different kinds of robot views (based on [20]);
– Intuitive supervisor graphic interface for controlling robot;
– Training associative memory (Kohonen) to be able to reconstruct a robot control that drives the robot from one view to another;
– Extraction of singular points and transformation of angle-distance plots to a string (see Sect. 3.1 and beginning of Sect. 4);
– Construction of Editorial Prescription (EP) for two strings (see Sects. 2–4);
– Detection of a heavy change by interpretation of a EP (see Sects. 3, 4);
– Support of the self-localization task;
– Manipulations with the atlas of geographic maps;
– Serialization/Deserialization.

Using software comprises three stages: (i) generation of obstacles inside the workspace; (ii) training the associative memory; (iii) learning geography of the workspace. Some details of this process follow.

Supervisor puts on the scene any number of obstacles (pyramids or cubes), controlling their size, shape, position, and orientation. Then he/she drives the robot randomly around the scene; this leads to automatic training of matrices of the Kohonen associative memory (the data generated at this stage are needed for automatic estimation of a robot control command that would drive the robot from one view to some other close view). When the memory is trained, the program automatically initializes the robot position as in Fig. 1a and sets all content of the geographic atlas to a single initial map.

Starting from this point, supervisor generates discrete commands to drive randomly the robot around the scene by means of an intuitive graphic interface. The robot executes the command sequence and automatically learns geography of the workspace: each robot control introduced by supervisor invokes the above scheme of detection of heavy changes; any detected heavy change means that the current map must be changed; the new current map is chosen from the neighbors of the old one or, if it is impossible, a new map is introduced into the atlas and then assigned as the current map.

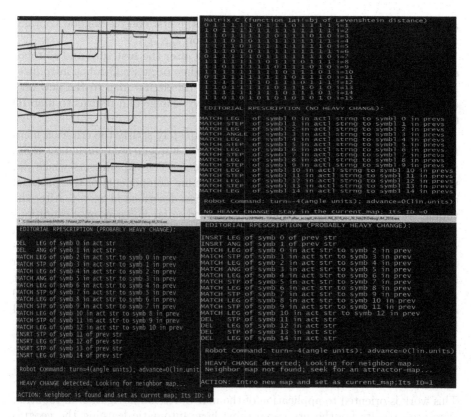

Fig. 3. A short sequence of screen-shots of a learning geography experiment (see the main text).

The correctness of the principles of detection of heavy changes presented in this work should follow from the fact that the robot would demonstrate the ability to learn correctly the geography. The *criterion of correctness* of all taken decisions is as follows: when a robot trajectory returns to a past point with the same orientation, its current map must return to an old map that robot had as the current while visiting this point the last time.

A learning geography experiment is illustrated in Fig. 3. The beginning of the learning process corresponds to the frames of Fig. 1. Then the robot was ordered to perform three turns without any change of position: two successive counter-clockwise turns and a reverse turn with the same angle. The three angle-distance plots show the dynamics of the robot views. Note that each of the three screens shows simultaneously a previous and actual views, respectively, as a stippled and continuous line-strip. Three screenshots of the console window, from the top-right to the bottom-left, correspond to respective robot views and show the response of the learning algorithm; each of them contains an editorial prescription, a result of detection of heavy change, if applicable, and a subsequent action with geographic database. Additionally, the first console-window contains an example of computation of indicator function $1_{(a_i \neq b_j)}$.

It can be seen that no heavy change occurs after the first turn and so robot stays in the same map, ID 0; then a heavy change occurs and robot introduces a new map, ID 1, into the atlas, and changes the actual map to the new one; after the reverse turn, robot again detects heavy change and returns to map ID 0 chosen as a neighbor of map ID 1.

The results of this simplest experiment completely satisfy the criterion mentioned above. A series of such experiments was implemented, including those with a much longer sequence of robot commands. The experiments which did not pass the above criterion were used for tuning components of the package; especially, the tuning is concerned with the issues described in Sect. 4. After the tuning, all succeeding experiments accomplished so far under normal conditions satisfy the criterion. The 'normal conditions' mean that any view contains a sufficient number of singular points and that the possibility of two similar views for distant robot states on a robot path is excluded. The dashed line in Fig. 1a shows a typical example of a long robot trajectory fulfilled in an experiment after the tuning. The whole route up to closing the first cycle corresponds to 60 commands of supervisor. The robot detects correctly the cycles of the path and interacts with the atlas in full correspondence with the expectations.

6   Conclusions

This work is oriented to application of the non-Cartesian navigation in robotics. Presented results for the first time show how automatic learning the geography of a robot environment by means of processing the flow of robot views can be implemented. The presented here approach is based on detection of heavy changes in the input flow and their use for interaction with a geographic database. In its turn, the detection of heavy changes is provided by transformation of each view to a string with subsequent comparison of such strings by a technique based on application of the Levenshtein distance. This scheme is implemented in the scope of a pilot software that simulates actions of a virtual robot in 2D-environment and integrates all components of the approach. The first series of experiments with this software allows us to claim that main principles presented here are correct and lead to automatic learning the geography.

Main research lines for future work are: replacement of the 1D robot sensor to other kinds of sensors; extension of the technique from the 1D to 2D views; extension to a non-static environment; the connection of this approach with the possibility of setting navigation problems in natural terms, etc. As to some challenging goals oriented to practice, we can mention creation of a drone that would be able to navigate automatically in 3D workspace, and connecting our approach with techniques based on GPS.

Acknowledgment. UAM-Azcapotzalco supports this work in the scope of project "Artificial Cognitive Vision".

References

1. Levenshtein, V.I.: Binary codes capable of correcting deletions, insertions, and reversals. Dokl. Akad. Nauk SSSR **163**, 845–848 (1965)
2. Su, Z., Ahn, B.R., Eom, K.Y., Kang, M.K., Kim, J.P., Kim, M.K.: Plagiarism detection using the Levenshtein distance and Smith-Waterman algorithm. In: Innovative Computing Information and Control, ICICIC 2008, pp. 569–569. IEEE Press (2008)
3. Heeringa, W.J.: Measuring dialect pronunciation differences using Levenshtein distance. Doctoral dissertation. University Library Groningen (2004)
4. Schimke, S., Vielhauer, C., Dittmann, J.: Using adapted levenshtein distance for on-line signature authentication. In: Pattern Recognition, 2004 ICPR 2004, pp. 931–934. IEEE Press (2004)
5. Khachaturov, G.: An approach to trip-and route-planning problems. Cybern. Syst. **33**, 43–67 (2002)
6. Gomi, T.: Non-cartesian robotics. Robot. Autonom. Syst. **18**, 169–184 (1996)
7. Gomi, T.: Aspects of non-cartesian robotics. Artif. Life Robot. **1**, 95–103 (1997)
8. Vukobratović, M.: How to control robots interacting with dynamic environment. J. Intell. Rob. Syst. **19**, 119–152 (1997)
9. van Leeuwen, J.: Graph algorithms. In: Handbook of Theoretical Computer Science: Algorithms and Complexity, vol. A, pp. 525–631 (1990)
10. Smith, R., Self, M., Cheeseman, P.: A stochastic map for uncertain spatial relationships. In: Robotics Research: The Fourth International Symposium, pp. 467–474 (1988)
11. Dissanayake, M.G., Newman, P., Clark, S., Durrant-Whyte, H.F., Csorba, M.: A solution to the simultaneous localization and map building (SLAM) problem. IEEE Trans. Robot. Autom. **17**, 229–241 (2001)
12. Davison, A.J., Murray, D.W.: Simultaneous localization and map-building using active vision. IEEE Trans. Pattern Anal. Mach. Intell. **24**, 865–880 (2002)
13. Guivant, J., Nebot, E., Baiker, S.: Autonomous navigation and map building using laser range sensors in outdoor applications. J. Robotic Syst. **17**, 565–583 (2000)
14. Castellanos, J.A., Martinez-Cantin, R., Tardós, J.D., Neira, J.: Robocentric map joining: improving the consistency of EKF-SLAM. Robot. Autonom. Syst. **55**, 21–29 (2007)
15. Grisetti, G., Tipaldi, G.D., Stachniss, C., Burgard, W., Nardi, D.: Fast and accurate SLAM with Rao-Blackwellized particle filters. Robot. Autonom. Syst. **55**, 30–38 (2006, 2007)
16. Hasan, K.M, Reza, K.J., Abdullah-Al-Nahid.: Path planning algorithm development for autonomous vacuum cleaner robots. In: Informatics, Electronics Vision 2014, pp. 1–6. IEEE Press (2014). doi:10.1109/ICIEV.2014.6850799
17. Khachaturov, G., Espinosa de los Monteros, J.A., Figueroa, J.: Applying spatiotemporal analysis to angle-distance views for detection of relevant events. In: Information Technology: Proceedings of Conference on Informatics and Computer Science (CNCIIC-ANIEI 2016), Mexico, pp. 205–214 (2016)
18. Navarro, G.: A guided tour to approximate string matching. ACM Comput. Surv. **33**, 31–88 (2001). doi:10.1145/375360.375365
19. Joy, K.: The depth-buffer visible surface algorithm. Visualization and Graphics Research Group, Department of Computer Science, University of California
20. Martínez-Hernández, J.M.: Development of a virtual robot and its environment based on a non-Cartesian navigation model. Terminal project for graduation in Computer Engineering from UAM-Azcapotzalco (2010) (in Spanish)

Constant-Time Fourier Moments for Face Detection — Can Accuracy of Haar-Like Features Be Beaten?

Przemysław Klęsk$^{(\boxtimes)}$

Faculty of Computer Science and Information Technology, West Pomeranian University of Technology, ul. Żołnierska 49, 71-210 Szczecin, Poland
pklesk@wi.zut.edu.pl

Abstract. We demonstrate a technique allowing for constant-time calculation of low order Fourier moments, applicable in detection tasks. Real and imaginary parts of the moments can be used as features for machine learning and classification of image windows. The technique is based on a *set* of special *integral images*, prepared prior to the scanning procedure. The integral images are constructed as cumulative inner products between the input image and suitable trigonometric terms. Additional time invested in the preparation of such integral images is amortized later at the stage of scanning. Then, the extraction of each moment requires only 21 operations, regardless of the number of pixels in the detection window, and thereby is an $O(1)$ calculation.

As an application example, face detection experiments are carried out with detectors based on Haar-like features serving as opponents to the proposed Fourier-based detectors.

1 Introduction

Constant-time computational complexity is the most attractive complexity for a computer scientist. Unfortunately, favourable opportunities to apply algorithms of that complexity are rare — typically, they pertain to some selected data structures e.g. hash tables, Union-Find[1] [2] and constitute a narrow fragment of a larger software. Often, one deals in fact with a so-called *amortized* constant-time complexity. This means that in the company of essential operations, performed are also some auxiliary operations meant to guarantee the speed for the future.

Not so long ago an algorithmic idea of that class has appeared in the field of computer vision and works remarkably well — namely, the idea of Haar-like features due to Viola and Jones (2001, 2004) [9,10]. Haar-like features are now commonly applied to detect objects (faces, people, vehicles, road signs, etc.) in digital

---

This work was financed by the National Science Centre, Poland. Research project no.: 2016/21/B/ST6/01495.

[1] For strictness: the 'Find' operation in this data structure is of amortized complexity $O(\log^* n)$ — iterated logarithm of n. Wherein $\log_2^* n$ is not greater than 5 for all quantities n observable in the universe; in particular, $\log_2^* 2^{65536} = 5$.

© The Author(s) 2017
L. Rutkowski et al. (Eds.): ICAISC 2017, Part I, LNAI 10245, pp. 530–543, 2017.
DOI: 10.1007/978-3-319-59063-9_47

images [1,8]. One should be aware that the fast performance of Haar-like features is *not* owed to the nature of these features as such; they are simple differential features that can be viewed as rough contours (e.g. difference in average pixel intensity between forehead and eyes regions). Instead, the fast performance is in fact a consequence of a computational trick known as **integral image**. For an image $i(x, y)$ the elementary integral image is: $ii(x, y) = \sum_{1 \leqslant j \leqslant x} \sum_{1 \leqslant k \leqslant y} i(j, k)$. Once such a cumulant is prepared, the sum of intensities over any image window can be calculated in constant time — $O(1)$ — regardless of the number of pixels, using 2 subtractions and 1 addition. This allows for very fast feature extraction.

There exist a few modifications of that idea. For example, a cumulant of squares $ii(x, y) = \sum_{1 \leqslant j \leqslant x} \sum_{1 \leqslant k \leqslant y} i^2(j, k)$ is useful for calculations of variance. In turn, cumulants of so-called vote matrices allow for extraction of HOG[2] features [5,7]. Yet, other propositions of that kind are scarce and, in generality, approaches which would allow for constant-time extraction of more advanced features, exhibiting better approximation properties, are not known.

In this paper we demonstrate that it is possible to prepare a *set* of cumulants of form: $ii(x, y) = \sum_{1 \leqslant j \leqslant x} \sum_{1 \leqslant k \leqslant y} i(j, k) \cdot \cos f(j, k, \cdots)$ and $ii(x, y) = \sum_{1 \leqslant j \leqslant x} \sum_{1 \leqslant k \leqslant y} i(j, k) \cdot \sin f(j, k, \cdots)$, with f being a suitably chosen function, and then to use the cumulants to extract Fourier moments of low orders in constant-time, using 21 operations, regardless of size and position of detection window.

We omit the topic of classifiers cascade in the paper.

2   Haar-Like Features — Short Review

In this section we briefly remind Haar-like features and point out their connection to Haar wavelets.

Recall the mother Haar wavelet $\psi(x)$ defined to yield: 1 for $0 \leqslant x < \frac{1}{2}$, -1 for $\frac{1}{2} \leqslant x < 1$, and 0 otherwise. The descendant wavelets are generated as follows:

$$\psi_{j,k}(x) = \psi(2^{j-1}x - k), \qquad j = 2, 3, \ldots; \quad k = 0, 1, \ldots 2^{j-1} - 1. \qquad (1)$$

Thus, descendants are narrowed and shifted versions of the mother wavelet. For any continuous function f (to be approximated) the orthogonality of wavelets

$$\forall (j, k) \neq (l, m) \quad \langle \psi_{j,k}, \psi_{l,m} \rangle = \int_0^1 \psi_{j,k}(x)\psi_{l,m}(x)\, dx = 0, \qquad (2)$$

allows to write down the following expansion

$$f(x) = c_0 \cdot 1 + \sum_{j=1}^{\infty} \sum_{k=0}^{2^{j-1}-1} c_{j,k}\psi_{j,k}(x), \qquad (3)$$

where the best coefficients can be found through inner products of Haar bases and the target function: $c_{j,k} = 1/\|\psi_{j,k}\|^2 \langle f, \psi_{j,k} \rangle$ and $c_0 = \langle f, 1 \rangle$.

---

[2] Histogram of Oriented Gradients.

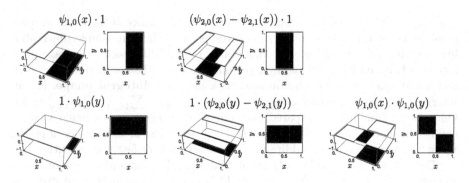

Fig. 1. Five 2D templates for Haar-like features defined in terms of 1D wavelets.

Viola and Jones [9,10] proposed two-dimensional templates resembling Haar wavelets. The templates can be mapped to features by anchoring them within an image window at different positions and scales, and then calculating the difference in average intensity of pixels under white ($+1$) and black (-1) regions. We depict the templates and their connection to wavelets in Fig. 1. The intention of Viola and Jones was to generate a massive multitude of features (e.g. $\sim 10^5$), so that some of them might happen to represent good characteristics of target objects (e.g. for faces: differences between forehead and eyes, nose and cheeks, etc.). Therefore, the way to implement how Haar-like features are actually embedded inside a window (i.e. setting up their positions and scales) is fairly arbitrary. One may allow for overlapping of feature supports and neglect orthogonality. On the other hand, we remark that it is straightforward to define orthogonal two-dimensional wavelets via products $\psi_{j,k;l,m}(x,y) = \psi_{j,k}(x) \cdot \psi_{l,m}(y)$, and to write down a polynomial in wavelets to approximate some fragment of image function $i(x,y)$. Note that in the formula for coefficients, $c_{j,k;l,m} = 1/\|\psi_{j,k;l,m}\|^2 \langle i, \psi_{j,k;l,m}\rangle$, the expression $\langle i, \psi_{j,k;l,m}\rangle$ is then equivalent to taking the white-black difference, as in the definition of Haar-like features, whereas the normalization constant $1/\|\psi_{j,k;l,m}\|^2$ plays the role of averaging (provided that white and black supports are of the same size).

3 Constant-Time Fourier Moments via Integral Images

Consider the following approximation, by a partial Fourier sum, of an image fragment restricted to a rectangle spanning from (x_1, y_1) to (x_2, y_2):

$$i(x,y) \approx \sum_{-n \leq k_x \leq n} \sum_{-n \leq k_y \leq n} c_{x_1,y_1}^{k_x,k_y} e^{2\pi i \left(k_x \frac{x-x_1}{N_x} + k_y \frac{y-y_1}{N_y}\right)}, \qquad \begin{matrix} x_1 \leq x \leq x_2 \\ y_1 \leq y \leq y_2 \end{matrix}, \quad (4)$$

where: n is the harmonic order of approximation (variable-wise), $i = \sqrt{-1}$ is the imaginary unit (please note the calligraphic difference from i denoting the image), the coefficients c are complex numbers, and $N_x = x_2 - x_1 + 1$,

$N_y=y_2-y_1+1$ are rectangle widths in pixels. The superscripts k_x, k_y of c coefficients indicate the particular harmonic indexes. The subscripts represent the boundaries of the rectangle. In the current context the boundaries are fixed, but shall vary later when partitioning of the detection window becomes involved. To avoid confusion, we explain that throughout the paper N_x, k_x and similar subscript notations should *not* be treated as functions of the specific subscript value, but instead as an indication of what coordinate the quantity is associated with.

Due to orthognality of Fourier bases, the optimal complex coefficients from (4) can be derived as

$$c_{\substack{x_1,y_1\\x_2,y_2}}^{k_x,k_y} = \frac{1}{N_x N_y} \sum_{x_1 \leqslant x \leqslant x_2} \sum_{y_1 \leqslant y \leqslant y_2} i(x,y) e^{-2\pi i \left(k_x \frac{x-x_1}{N_x} + k_y \frac{y-y_1}{N_y} \right)}. \tag{5}$$

From now on, we shall refer to the coefficients as Fourier moments, and we intend to use their real and imaginary parts as features for learning and detection.

Let us introduce **two sets of integral images:**

$$\left\{ ii_{\substack{\cos\\N_x,N_y}}^{k_x,k_y} \right\}, \quad \left\{ ii_{\substack{\sin\\N_x,N_y}}^{k_x,k_y} \right\},$$

related to cosine and sine functions, respectively, and constructed as follows:

$$ii_{\substack{\cos\\N_x,N_y}}^{k_x,k_y} (x,y) = \sum_{1 \leqslant j_x \leqslant x} \sum_{1 \leqslant j_y \leqslant y} i(j_x,j_y) \cos\left(-2\pi \left(\frac{k_x j_x}{N_x} + \frac{k_y j_y}{N_y} \right) \right), \tag{6}$$

$$ii_{\substack{\sin\\N_x,N_y}}^{k_x,k_y} (x,y) = \sum_{1 \leqslant j_x \leqslant x} \sum_{1 \leqslant j_y \leqslant y} i(j_x,j_y) \sin\left(-2\pi \left(\frac{k_x j_x}{N_x} + \frac{k_y j_y}{N_y} \right) \right), \tag{7}$$

where indexes (k_x, k_y) iterate over the set:

$$\{(k_x,k_y): \ -n \leqslant k_x \leqslant -1, -n \leqslant k_y \leqslant n\} \cup \{(0,k_y): \ -n \leqslant k_y \leqslant -1\} \cup \{(0,0)\}. \tag{8}$$

We remark that a single integral image of form (6) or (7) can be calculated by induction in linear time with respect to the total number of pixels in the input image (i.e. with one pass).

Let us now define the **growth operator** for any integral image ii taken from either of the sets $\{ii_{\cos}\}, \{ii_{\sin}\}$:

$$\mathop{\Delta}_{\substack{x_1,y_1\\x_2,y_2}} (ii) = ii(x_2,y_2) - ii(x_1-1,y_2) - ii(x_2,y_1-1) + ii(x_1-1,y_1-1). \tag{9}$$

Note that Δ returns a subsum over given cuboid in constant time using 2 subtractions and 1 addition, instead of $\Theta(N_x N_y)$ operations.

The following proposition constitutes the main contribution of the paper.

Proposition 1. *Suppose the two sets of integral images:*

$$\left\{ ii_{\substack{\cos\\N_x,N_y}}^{k_x,k_y} \right\}, \quad \left\{ ii_{\substack{\sin\\N_x,N_y}}^{k_x,k_y} \right\},$$

defined as in (6) and (7), respectively, have been calculated prior to the detection procedure. Then, for any rectangle of widths N_x, N_y in the image, the real and imaginary parts of each of its Fourier moments can be calculated in constant time — $O(1)$ — as follows:

$$Re\left(c_{\substack{x_1,y_1\\x_2,y_2}}^{k_x,k_y}\right) = \frac{1}{N_x N_y}\left(\cos\left(2\pi\left(\frac{k_x x_1}{N_x} + \frac{k_y y_1}{N_y}\right)\right) \underset{\substack{x_1,y_1\\x_2,y_2}}{\triangle}\left(ii_{\cos}^{k_x,k_y}{}_{N_x,N_y}\right)\right.$$

$$\left. - \sin\left(2\pi\left(\frac{k_x x_1}{N_x} + \frac{k_y y_1}{N_y}\right)\right) \underset{\substack{x_1,y_1\\x_2,y_2}}{\triangle}\left(ii_{\sin}^{k_x,k_y}{}_{N_x,N_y}\right)\right), \quad (10)$$

$$Im\left(c_{\substack{x_1,y_1\\x_2,y_2}}^{k_x,k_y}\right) = \frac{1}{N_x N_y}\left(\sin\left(2\pi\left(\frac{k_x x_1}{N_x} + \frac{k_y y_1}{N_y}\right)\right) \underset{\substack{x_1,y_1\\x_2,y_2}}{\triangle}\left(ii_{\cos}^{k_x,k_y}{}_{N_x,N_y}\right)\right.$$

$$\left. + \cos\left(2\pi\left(\frac{k_x x_1}{N_x} + \frac{k_y y_1}{N_y}\right)\right) \underset{\substack{x_1,y_1\\x_2,y_2}}{\triangle}\left(ii_{\sin}^{k_x,k_y}{}_{N_x,N_y}\right)\right). \quad (11)$$

As one can note both parts, real (10) and imaginary (11), require a calculation of two growth operations and two trigonometric functions. It is easy check that this comprises a total of **21 operations**: 8 additions (or subtractions), 8 multiplications, 3 divisions, and 2 trigonometric functions for either of the two formulas. Note that it is sufficient to calculate the argument under trigonometric functions only once. Furthermore, it is worth noting that this argument depends on the offset (x_1, y_1) of the rectangle, but does *not* depend on the rectangle contents — the pixels, thereby making the overall calculation a **constant-time calculation**. The proof of the proposition is a straightforward derivation.

Proof. Rewriting the moments from (5) using Euler's identity leads to:

$$c_{\substack{x_1,y_1\\x_2,y_2}}^{k_x,k_y} = \frac{1}{N_x N_y}\sum_{x_1\leqslant x\leqslant x_2}\sum_{y_1\leqslant y\leqslant y_2} i(x,y)\left(\cos\left(-2\pi\left(k_x\frac{x-x_1}{N_x} + k_y\frac{y-y_1}{N_y}\right)\right)\right.$$

$$\left. + i\sin\left(-2\pi\left(k_x\frac{x-x_1}{N_x} + k_y\frac{y-y_1}{N_y}\right)\right)\right). \quad (12)$$

The argument of the trigonometric functions can be parted into a group of terms independent from the pixel index (x, y) and a group dependent on it as follows:

$$\alpha = 2\pi\left(k_x x_1/N_x + k_y y_1/N_y\right),$$
$$\beta(x,y) = -2\pi\left(k_x x/N_x + k_y y/N_y\right).$$

Now, one can apply in (12) the trigonometric identities for $\cos(\alpha+\beta)$ and $\sin(\alpha+\beta)$. Simultaneously, the $\cos\alpha$ and $\sin\alpha$ terms can be pulled out as factors in front of the summations as they are independent of the pixel index (x, y). Finally, by splitting the expression into real and imaginary parts one obtains:

$$\text{Re}\left(c_{\substack{x_1,y_1 \\ x_2,y_2}}^{k_x,k_y}\right) = \frac{1}{N_x N_y}\left(\cos\alpha \sum_{\substack{x_1\leqslant x\leqslant x_2 \\ y_1\leqslant y\leqslant y_2}} i(x,y)\cos\beta(x,y) - \sin\alpha \sum_{\substack{x_1\leqslant x\leqslant x_2 \\ y_1\leqslant y\leqslant y_2}} i(x,y)\sin\beta(x,y)\right),$$

$$\underbrace{\qquad\qquad\qquad\qquad}_{ii_{\cos}^{\substack{x_1,y_1 \\ x_2,y_2}}\left(\substack{k_x,k_y \\ N_x,N_y}\right)} \qquad\qquad \underbrace{\qquad\qquad\qquad\qquad}_{ii_{\sin}^{\substack{x_1,y_1 \\ x_2,y_2}}\left(\substack{k_x,k_y \\ N_x,N_y}\right)}$$

$$\text{Im}\left(c_{\substack{x_1,y_1 \\ x_2,y_2}}^{k_x,k_y}\right) = \frac{1}{N_x N_y}\left(\sin\alpha \sum_{\substack{x_1\leqslant x\leqslant x_2 \\ y_1\leqslant y\leqslant y_2}} i(x,y)\cos\beta(x,y) + \cos\alpha \sum_{\substack{x_1\leqslant x\leqslant x_2 \\ y_1\leqslant y\leqslant y_2}} i(x,y)\sin\beta(x,y)\right). \quad (13)$$

$$\underbrace{\qquad\qquad\qquad\qquad}_{ii_{\cos}^{\substack{x_1,y_1 \\ x_2,y_2}}\left(\substack{k_x,k_y \\ N_x,N_y}\right)} \qquad\qquad \underbrace{\qquad\qquad\qquad\qquad}_{ii_{\sin}^{\substack{x_1,y_1 \\ x_2,y_2}}\left(\substack{k_x,k_y \\ N_x,N_y}\right)}$$

The underbraces show how the expensive summations over pixels get replaced by cheap (constant-time) growths of integral images, yielding (10), (11). □

The form of indexes set (8) is implied by the known symmetry property i.e. complex conjugacy of opposed Fourier coefficients:

$$\text{Re}\left(c_{\substack{x_1,y_1 \\ x_2,y_2}}^{-k_x,-k_y}\right) = \text{Re}\left(c_{\substack{x_1,y_1 \\ x_2,y_2}}^{k_x,k_y}\right), \qquad \text{Im}\left(c_{\substack{x_1,y_1 \\ x_2,y_2}}^{-k_x,-k_y}\right) = -\text{Im}\left(c_{\substack{x_1,y_1 \\ x_2,y_2}}^{k_x,k_y}\right), \quad (14)$$

and also by the fact that the zeroth order moment is a real number — $\text{Im}(c^{0,0}) = 0$. Hence, it suffices to calculate roughly only a half of all moments. More precisely, the effective number of distinct moments is

$$1/2\left((2n+1)^2 - 1\right) + 1, \quad (15)$$

which yields $2n^2 + 2n + 1$ and corresponds to the size of set (8). In fact, any set of $2n^2 + 2n + 1$ coefficients will do to uniquely reconstruct all coefficients.

As regards the needed number of integral images, it is equal to the double of expression (15) yielding: $(2n+1)^2 + 1$, since required are two kinds of integral images, related to cosine and sine functions, for each (k_x, k_y) pair. Hence, the calculation of all cumulants is potentially expensive. That is why, when using Proposition 1, one should in practice limit himself to low harmonic orders, so that the time invested in the preparation of integral images is reasonably small.

4   Window Paritioning — Piecewise Approximations

Apart from n let us now introduce an additional integer parameter $p > 0$, responsible for the partitioning of detection window and affecting the final number of features. Let the window be partitioned into a regular grid of rectangles: $p \times p$. The moments shall be extracted from each rectangle independently and their concatenation shall form the final vector of features. This approach can be understood as a piecewise Fourier approximation of the window under detection.

Consider a single image pass with a detection window of size $w_x \times w_y$. The partitioning leads to a grid of pieces with widths equal to: $N_x = \lfloor w_x/p \rfloor$,

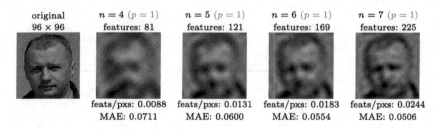

original 96 × 96	n = 4 (p = 1) features: 81	n = 5 (p = 1) features: 121	n = 6 (p = 1) features: 169	n = 7 (p = 1) features: 225
	feats/pxs: 0.0088 MAE: 0.0711	feats/pxs: 0.0131 MAE: 0.0600	feats/pxs: 0.0183 MAE: 0.0554	feats/pxs: 0.0244 MAE: 0.0506

Fig. 2. Reconstructions for successive harmonic orders $n = 4, \ldots, 7$ (fixed $p = 1$).

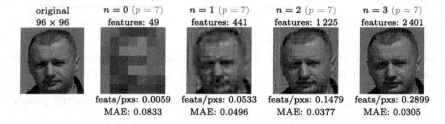

original 96 × 96	n = 0 (p = 7) features: 49	n = 1 (p = 7) features: 441	n = 2 (p = 7) features: 1 225	n = 3 (p = 7) features: 2 401
	feats/pxs: 0.0059 MAE: 0.0833	feats/pxs: 0.0533 MAE: 0.0496	feats/pxs: 0.1479 MAE: 0.0377	feats/pxs: 0.2899 MAE: 0.0305

Fig. 3. Reconstructions for successive harmonic orders $n = 0, \ldots, 3$ (fixed $p = 7$).

$N_y = \lfloor w_y/p \rfloor$. We denote the corresponding division remainders as: $m_x = w_x$ mod p, $m_y = w_y$ mod p. Now, for a window starting at a point (x_1, y_1) and for fixed numbers N_x, N_y we define the collection of features $\left\{ f_{\substack{x_1,y_1 \\ N_x,N_y}} (\cdots) \right\}$ as follows:

$$
f_{\substack{x_1,y_1 \\ N_x,N_y}} (k_x, k_y, p_x, p_y, r) = \begin{cases} \mathrm{Re}\left(c^{k_x,k_y}_{\substack{x_1'+p_x N_x, y_1'+p_y N_y \\ x_1'+(p_x+1)N_x-1, y_1'+(p_y+1)N_y-1}} \right), r = 1; \\ \mathrm{Im}\left(c^{k_x,k_y}_{\substack{x_1'+p_x N_x, y_1'+p_y N_y \\ x_1'+(p_x+1)N_x-1, y_1'+(p_y+1)N_y-1}} \right), r = 0; \end{cases} \tag{16}
$$

where: $(x_1', y_1') = (x_1 + \lfloor m_x/2 \rfloor, y_1 + \lfloor m_y/2 \rfloor)$ represents a shifted starting point taking into account small corrections due to the partitioning remainders[3]; indexes k_x, k_y iterate over the set defined in (8); $0 \le p_x, p_y \le p-1$ represent the index (and hence the offset) of a particular rectangle; and r is a flag switching between real and imaginary parts. Since $\mathrm{Im}(c^{0,0}) = 0$, then $f(0, 0, p_x, p_y, 0)$ is also zero for any p_x, p_y pair, and therefore should not be taken as an actual feature. Finally, the total number of features is: $d(n, p) = (2n + 1)^2 p^2$.

Figures 2 and 3 show example reconstructions of an image from Fourier moments. Reconstructions are carried out according to formula (4) (piecewise reconstructions for $p > 1$). Obviously, image reconstruction as such is not a needed step in a detection procedure. Yet, the quality of reconstructions helps to understand the descriptive capability of the features. Under each reconstruction

---

[3] This operation centers the grid of rectangles within the scanning window.

we report the mean absolute error (MAE) and the ratio of the number of features to the number of pixels (feats/pxs).

5   Face Detection Experiments

Taking advantage of Proposition 1, we have trained four variants of face detectors based on Fourier moments. The variants correspond to different settings of n and p parameters: (1) $n = 2$, $p = 5$ (625 features), (2) $n = 2$, $p = 7$ (1 225 features), (3) $n = 3$, $p = 5$ (1 225 features), (4) $n = 3$, $p = 7$ (2 401 features).

To compare accuracy, we have introduced opponents for our Fourier-based detectors — namely, additional detectors trained on the same learning material but using Haar-like features. Our intention was to impose a similar feature space parameterization in both approaches, but to slightly favour the Haar-like features in terms of their quantity. To achieve this, we were using 5 Haar templates discussed earlier (Fig. 1) and we were anchoring the Haar-like features within the detection window on $p \times p$ grids ($p = 5, 7$) — hence, the grids were of the same sizes as in the case of Fourier moments. Lengths of Haar-like features were scaled independetly along each axis, and the number of scales was controlled by an additional parameter $q > 0$. More precisely, for a window of size $w_x \times w_y$, the lengths of features were changing according to: $w_x \lambda^{s_x}$ and $w_y \lambda^{s_y}$ with $1 \leqslant s_x, s_y \leqslant q$ and the scaling factor chosen to be $\lambda = \sqrt{2}/2$. Hence, the final number of generated Haar-like features became $d_{\mathrm{HF}}(q, p) = 5q^2p^2$. We remind that the corresponding number for Fourier moments is $d_{\mathrm{FM}}(n, p) = (2n + 1)^2 p^2$.

A learning material of moderately large size was used. It contained 7 258 positive examples (windows with faces marked manually from 3 000 images) and 100 000 negative examples (windows sampled randomly from non-face images). Accuracy measures were evaluated on test data consisting of 500 images with 1 000 faces and a total of 70 252 859 windows. In order to produce ROC curves for detectors on the basis of test material, a test set with a limited number of negatives was randomly selected (to fit in RAM memory). We imposed $2 \cdot 10^6$ negative windows in that set, thereby making the precision along the FAR[4] axis at the level of $5 \cdot 10^{-7}$. Details of the experimental setup are listed in Table 1.

We have applied a boosted learning algorithm known as RealBoost+bins, see e.g. [6], with additional weight trimming [3]. In this variant, an ensemble consists of partial (weak) classifiers that are based on selected single features each. Classifiers' responses are real-valued, equal to half the *logit* transform, and a binning mechanism is introduced to store those responses. We have set up 8 bins of equal widths per feature. Finally, $T = 256$ or $T = 512$ rounds of boosting were carried out, yielding ensembles with at most T distinct features selected.

The software has been programmed in C#, with key procedures (e.g. integral images, features extraction) implemented for efficiency in C++ as dll libraries.

We start the review of results by showing in Fig. 4 some example outcomes produced by a Fourier-based detector (variant: $n = 3$, $p = 7$). The left-hand

---

[4] False Alarm Rate.

Table 1. Setup for face detection experiments.

Train data		
Quantity/parameter	Value	Additional information
No. of images with faces	3 000	Photos downloaded from *Google Images* for queries: person, people, group of people, family, children, sportsmen, students, etc.
No. of images without faces	300	As above, queries: view, landscape, street, cars, etc.
No. of positive examples	7 258	Face windows marked manually
No. of negative examples	100 000	Imposed quantity; examples sampled at random positions and scales within images without faces
Train set size	107 258	Positive and negative examples in total
Test data		
No. of images with faces	500	Queries as for train data (other images)
No. of images without faces	300	Queries as for train data (other images)
No. of positive examples	1 000	Face windows marked manually
No. of negative examples	2 000 000	Imposed quantity; examples sampled at random
Test set size	2 001 000	Positive and negative windows in total
Detection procedure (scanning with a sliding window)		
Image height	480	Before detection, images scaled to the height 480, keeping original height: width proportion
No. of detection scales	8	Images scanned with 8 different sizes of window
Window growing coefficient	1.2	Window widths and heights increase by $\approx$20% per scale
Smallest window size	48×48	Faces smaller than $\approx$10% of image height not to be detected
Largest window size	172×172	Faces larger than $\approx$36% of image height not to be detected
Window jumping coefficient	0.05	Window jumps equal to $\approx$5% of its width and height

side images contain all single positive indications. Their counterparts on the right-hand side are postprocessed outcomes, i.e. after grouping of windows clusters has been performed. Figure 5 shows some examples of false alarms, interesting because of their resemblance to faces (we encourage to zoom the document).

ROC curves for all detectors are presented in Fig. 6. To distinguish the curves better, logarithmic scale was imposed on FAR axis. Operational decision

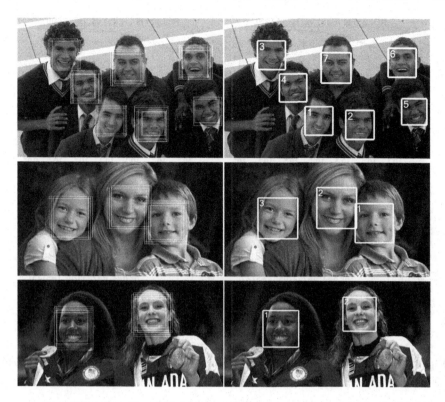

Fig. 4. Examples of single outcomes returned by the Fourier-based detector: before grouping positive windows (left-hand side) and afterwards (right-hand side).

Fig. 5. Examples of some false alarms resembling faces.

thresholds for detectors were taken as averages of threshold values registered for two left-most points on ROCs, with the smallest FAR values ($\approx 5 \cdot 10^{-7}$).

Detailed accuracy results are reported in Table 2. In particular, AUC (area under ROC) measures are stated. In learning tasks with strongly imbalanced classes, it is the left-most part of the ROC curve that is of crucial importance. Therefore, we decided to report normalized AUCs obtained up to several first orders of magnitudes of FAR. More precisely, AUC_α should be understood as $1/\alpha \int_0^\alpha s(f) \, df$ where s and f represent sensitivity and FAR respectively.

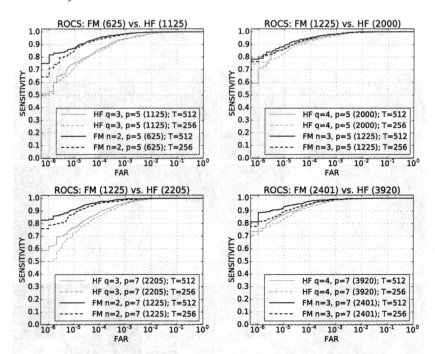

Fig. 6. Comparison of ROC curves for detectors based on Fourier moments (FM, black) and Haar-like features (HF, gray). Logarithmic scale imposed on FAR axis.

Table 2. Accuracy measures of detectors on test data (best variant in gray).

name / description	AUC$_\alpha$			sensiti-vity	FAR per images	FAR per windows	accuracy per windows
	$\alpha=10^{-5}$	$\alpha=10^{-4}$	$\alpha=10^{-3}$				
HF $q=3,p=5$ (1125); $T=512$	0.6761	0.8123	0.9156	0.699	0.098	$6.975\cdot10^{-7}$	0.999995018084708
FM $n=2,p=5$ (625); $T=512$	0.8401	0.9236	0.97144	0.889	0.092	$6.548\cdot10^{-7}$	0.999997765249097
HF $q=4,p=5$ (2000); $T=512$	0.8021	0.9082	0.9624	0.849	0.086	$6.121\cdot10^{-7}$	0.999997238589628
FM $n=3,p=5$ (1225); $T=512$	0.8475	0.9285	0.9703	0.872	0.054	$3.843\cdot10^{-7}$	0.999997793717795
HF $q=3,p=7$ (2205); $T=512$	0.7075	0.8376	0.9320	0.741	0.084	$5.978\cdot10^{-7}$	0.999995715550288
FM $n=2,p=7$ (1225); $T=512$	0.8800	0.9480	0.9826	0.924	0.058	$4.128\cdot10^{-7}$	0.999998505420626
HF $q=4,p=7$ (3920); $T=512$	0.8188	0.9141	0.9729	0.857	0.052	$3.701\cdot10^{-7}$	0.999997594441206
FM $n=3,p=7$ (2401); $T=512$	0.8965	0.9538	0.9845	0.945	0.052	$3.701\cdot10^{-7}$	0.999998847038424

Test results clearly show that detectors based on Fourier moments surpass in accuracy their counterparts based on Haar-like features, even though there were fewer features at disposal at the learning stage. It is an experimental evidence that Fourier moments have better approximation properties for the face detection problem. Naturally, the best (and definitely satisfactory) accuracy was achieved by the variant with the most rich space consisting of 2 401 features ($n = 3$, $p = 7$). Please note that the number $2.4 \cdot 10^3$ is decidedly smaller than the total of $1.8 \cdot 10^5$ features originally used by Viola and Jones in their experiment [9].

Table 3. Time performance for a 480×480 image (parallel computations on: Intel i7 Q 720 4×2-core 1.60 GHz CPU).

Quantity (or operations)	Fourier moments ($T = 512$)	Haar-like features ($T = 512$)
	(456 distinct feats.)	(472 distinct feats.)
No. of analyzed windows	108 873	108 873
No. of prepared integral images	400	1
	(50 images per each of 8 scales)	
Preparation time for integral images	1 015 ms	9 ms
Preparation time per 1 integral image	2.54 ms	9 ms
Total time of detection procedure	2 852 ms	1 032 ms
Time per 1 window	27.20 µs (amortized: 16.87 µs)	9.48 µs
Time per 1 window and 1 feature	57.45 ns (amortized: 37.00 ns)	20.08 ns

Finally, Table 3 reports the time performance we achieved on our machine (Intel i7 Q 720 4 × 2-core 1.60 GHz CPU). Please remember that cascades of classifiers were purposely not involved in the experiment. The observed amortized extraction time for a single Fourier feature was approximately 37 ns — about two times longer than for a Haar-like feature. This is related to the 21 operations we declared in Proposition 1, and is roughly proportional to the number of operations needed for Haar-like features (8 operations for an 'edge' template, 12 for a 'diagonal' template). We remark that by involving a cascade of classifiers and a processor with more cores/threads (to save time for preparation of integral images) the real-time regime could be achieved without difficulty in both approaches (FM and HF).

6 Conclusions

We have proposed a computational technique, based on special integral images, for constant-time extraction of low order Fourier moments. The technique is suitable for detection tasks. Our experiments have shown that fairly small sets of Fourier features — real and imaginary parts of moments — can lead to face detectors superior in accuracy than Haar-based detectors. The proposed approach could be beneficial in other machine learning applications where accuracy is of primary importance (e.g. medical diagnosis, image-based fault dection in production, landmine detection [4,5]), and where one is willing to invest some additional time in the preparation of special integral images in order to improve accuracy.

References

1. Charles, J., et al.: Automatic and efficient human pose estimation for sign language videos. Int. J. Comput. Vis. **110**(1), 70–90 (2014)
2. Cormen, T.H., et al.: Introduction to Algorithms, 3rd edn. MIT Press, Cambridge (2009)
3. Friedman, J., Hastie, T., Tibshirani, R.: Additive logistic regression: a statistical view of boosting. Ann. Stat. **28**(2), 337–407 (2000)
4. Klęsk, P., Kapruziak, M., Olech, B.: Fast Extraction of 3D Fourier Moments via Multiple Integral Images: An Application to Antitank Mine Detection in GPR C-scans. In: Chmielewski, L.J., Datta, A., Kozera, R., Wojciechowski, K. (eds.) ICCVG 2016. LNCS, vol. 9972, pp. 206–220. Springer, Cham (2016). doi:10.1007/978-3-319-46418-3_19
5. Klęsk, P., Godziuk, A., Kapruziak, M., Olech, B.: Fast Analysis of C-scans from Ground Penetrating Radar via 3D Haar-like Features with Application to Landmine Detection. IEEE Trans. Geosci. Remote Sens. **53**(7), 3996–4009 (2015)
6. Rasolzadeh, B., et al.: Response binning: improved weak classifiers for boosting. In: IEEE Intelligent Vehicles Symposium, pp. 344–349 (2006)
7. Said, Y., Atri, M., Tourki, R.: Human detection based on integral Histograms of Oriented Gradients and SVM. In: Communications, Computing and Control Applications (CCCA 2011), pp. 1–5. IEEE (2011)
8. Tresadern, P.A., Ionita, M.C., Cootes, T.F.: Real-Time Facial Feature Tracking on a Mobile Device. Int. J. Comput. Vis. **96**(3), 280–289 (2012)
9. Viola, P., Jones, M.: Rapid Object Detection using a Boosted Cascade of Simple Features. In: Conference on Computer Vision and Pattern Recognition (CVPR 2001), pp. 511–518. IEEE (2001)
10. Viola, P., Jones, M.: Robust Real-time Face Detection. Int. J. Comput. Vis. **57**(2), 137–154 (2004)

Neural Video Compression Based on PVQ Algorithm

Michał Knop[1,2(✉)], Tomasz Kapuściński[1,2], and Rafał Angryk[3]

[1] Institute of Computational Intelligence, Czestochowa University of Technology,
Al. Armii Krajowej 36, 42-200 Czestochowa, Poland
{michal.knop,tomasz.kapuscinski}@iisi.pcz.pl
[2] Institute of Information Technology, Radom Academy of Economics,
Domagalskiego Street 7a, 26-600 Radom, Poland
[3] Department of Computer Science, Georgia State University,
P.O. Box 5060, Atlanta, GA 30302-5060, USA
angryk@cs.gsu.edu
http://www.iisi.pcz.pl
http://wsh.pl/
http://grid.cs.gsu.edu/~rangryk/

Abstract. In this paper we present a video compression algorithm based on predictive vector quantization, which is a combination of vector quantization and differential pulse code modulation. We optimized the algorithm using chroma subsampling which reduces the amount of information that needs to be processed. This allowed us to combine two color channels into one and thereby reduce the number of predictors and codebooks. Furthermore, we introduced inter-frames which only store regions that changed compared to previous frames, further decreasing the size of compressed data.

Keywords: Video compression · Image compression · PVQ

1 Introduction

Video and image data compression has become an increasingly important issue in all areas of computing and communications. Various techniques for encoding of data can be used to eliminate the redundancy of the color information in images and video frames. Most video codecs and algorithms combine spatial compensation of images as well as compensation of movement in time. Currently, there are many compression standards. They can be found in a wide range of applications such as cable and land-based transmission channels, video services over the satellite, video streaming in Internet or local area network and storage formats. The most popular of these algorithms are MPEG, JPEG and H.26x. The MPEG standard describes a family of compression algorithms of audiovisual data, more details can be find in [5]. The well known members of MPEG family are MPEG-1, MPEG-2, and MPEG-4. H.261 is the first of the entire family H.26x of video compression standards. It has been designed for handling video

L. Rutkowski et al. (Eds.): ICAISC 2017, Part I, LNAI 10245, pp. 544–551, 2017.
DOI: 10.1007/978-3-319-59063-9_48

transmission in real time. More information about the family H.26x can be found in [1]. JPEG and JPEG2000 standards are used for image compression with an adjustable compression rate. They are also used for video compression. These methods compress each movie frames individually.

In the proposed approach we used a Predictive Vector Quantization (PVQ) algorithm to compress a sequence of video frames, which combines two methods: Vector Quantization (VQ) and Differential Pulse Code Modulation (DPCM) [2,4]. In order to reduce the amount of information needed to store the video stream we used two types of frames: key frames (intra-frame) and predictive frames (inter-frame). These frames were joined to groups containing one key frame and a series of predictive frames called GOP [14,17].

The rest of the paper is organized as follows. Section 2 describes components of our algorithm. It includes a description of neural image compression and the encoding color information. In Sect. 3 we discuss our approach to neural video compression. Next in Sect. 4 the experimental results are presented. The final section covers conclusions and the plans for future works.

2   Related Works

2.1   Predictive Vector Quantization

Predictive Vector Quantization is a neural algorithm that extends differential pulse code modulation (DPCM) scheme with vector quantization method (VQ) [6,7]. In PVQ neural predictor is responsible for vector quantization and the codebook is fulfilled by the DPCM function. The successive input vectors $V(t)$ are the macroblocks of the same dimensionality obtained from a frame, where t are indices of consecutive macroblocks. The predictor's input is a preceding macroblock, which after processing results in predicted vector $\overline{V}(t)$. The difference between current macroblock and predictor output $E(t) = V(t) - \overline{V}(t)$ is then calculated and used to select the best approximation g_j using neural quantizer from the codebook $G = [g_0, g_1, \ldots, g_J]$. Approximation g_j is then added to the difference which gives reconstructed input vector $\tilde{V}(t) = \overline{V}(t) + g_j$. This vector is later used as the next input to the predictor. The codebook index j is stored in a stream. In order to decompress data, the predictor is again used to calculate predicted vectors while stored codebook indices are used to select their corresponding approximations from codebook that are then combined with predictor outputs to reconstruct the original image.

2.2   Encoding Color Information

Most of modern hardware uses RGB color model to display images and video. Unfortunately, this color model is not an efficient way of storing and processing color information. Common standards for image and video compression such as JPEG and MPEG, and color encoding systems like PAL and NTSC take advantage of the way human vision works. Human eyes are very sensitive to small changes in brightness but they are far less sensitive to changes in chrominance [15]. For this reason we can use YC_bC_r color space. In this color space, Y is the luminance channel which describes brightness and C_b and C_r channels are chrominances which contain the remaining color information. In our method we used a modified conversion to luminance and chrominances which covers full range of values from 0 to 255 like RGB color values do. Conversion from RGB color space to YC_bC_r color space is presented in Eq. 1 while conversion from YC_bC_r space to RGB color space is presented in Eq. 2 [9, 13].

$$\begin{bmatrix} Y \\ C_b \\ C_r \end{bmatrix} = \begin{bmatrix} 0 \\ 128 \\ 128 \end{bmatrix} + \begin{bmatrix} 0.299 & 0.587 & 0.114 \\ -0.169 & -0.331 & 0.500 \\ 0.500 & -0.419 & -0.081 \end{bmatrix} * \begin{bmatrix} R \\ G \\ B \end{bmatrix} \tag{1}$$

$$\begin{bmatrix} R \\ G \\ B \end{bmatrix} = \begin{bmatrix} 1.000 & 0.000 & 1.400 \\ 1.000 & -0.343 & -0.711 \\ 1.000 & 1.765 & 0.000 \end{bmatrix} * \begin{bmatrix} Y \\ C_b - 128 \\ C_r - 128 \end{bmatrix} \tag{2}$$

Since human eye is less sensitive to changes in chrominance, we can reduce an amount of information needed to store C_b and C_r channels without significant loss of quality. This operation is known as chroma subsampling and is used in modern lossy compression algorithms [10]. Commonly used variants, such as 4:2:2 and 4:2:0, are presented in Fig. 1.

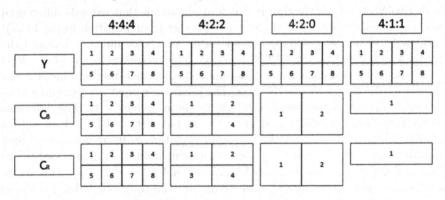

Fig. 1. Chroma subsampling

3   Proposed Method

In this paper, we propose a method of video compression based on the solutions presented in [3, 8, 11–13]. Compared to the previous methods based on the full color chrominance (4:4:4), in our method we use 4:2:0 chrominance in order to reduce the amount of data needed to record frames [15]. In this approach, only the Y channel is stored completely, while the C_b and C_r channels are reduced to half their original resolution. Moreover, C_b and C_r channels are combined into one channel before compression. This approach allowed us to reduce the number

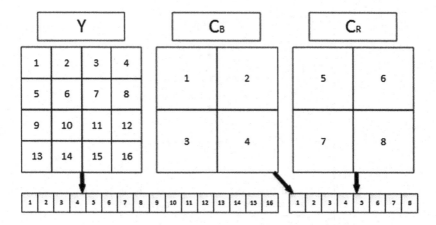

Fig. 2. Scheme encoding YC_bC_r to two channels

of sets of codebooks and predictors to two, one for the luminance channel the other one for the combined chrominance channels (Fig. 2).

Unlike the previous method that uses key frame detection algorithms, this method is based on a fixed size GOP (group of pictures) [14,17]. Each GOP consists of a key frame (intra-frame) and subsequent predictive frames (inter-frame). Each key frame is composed of complete frame data that encode entire frame, and two predictors and codebooks which are used to decode the frame. Predictive frames do not contain predictors and codebooks which are taken from the previously encountered key frame, and instead of encoding the entire frame, they can encode smaller regions within the frame, taking the remaining information from previous frame. This allows us to significantly reduce the size of each predictive frame by storing only changes between subsequent frames (Fig. 3).

4 Experimental Results

For the purpose of our research, we conducted two experiments to test the efficiency of our algorithm. We used sequences of frames from a publicly available video *"Elephants dream"* in resolution 1280×720 [16]. In all our tests we converted the frames into YC_bC_r color space. In 4:4:4 chroma subsampling, frames were encoded using three sets of predictors and codebooks. This corresponds to the method used in our previous works. In 4:2:0 chroma subsampling, C_b and C_r channels were first reduced by averaging 2×2 squares and then combined into one 8-element channel. Due to this modification algorithm only needed one standard set of predictor and codebook for Y channel and one reduced set of predictor and codebook for combined C_bC_r channel. This allowed us to reduce the number of indices stored per macroblock from 3 to 2 and reduced computational complexity of the compression.

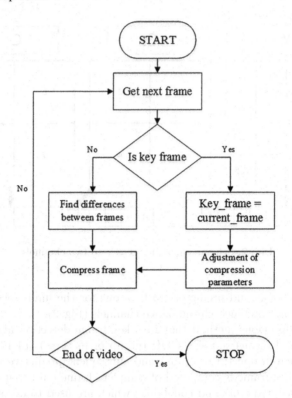

Fig. 3. Video compression algorithm

In the first experiment, we tested the efficiency of our algorithm in two types of chroma subsampling modes, 4:4:4 and 4:2:0. In both cases we used preset macroblock size 4×4. The experiment showed that introduction of chroma subsampling did not significantly degrade the video quality which can be seen in Fig. 4.

We compared frames compressed using 4:4:4 chroma subsampling to frames compressed using 4:2:0 subsampling using Peak Signal-to-Noise Ratio (PSNR). Figure 5 shows that both methods result in very similar PSNR values.

In the second experiment we compared the video quality after compression using predictive frames. Instead of key detection algorithm we used fixed GOP with a length of 30 frames. The experiment showed that there is no significant loss of quality in the example that uses predictive frames instead of encoding full frame what can be seen in Fig. 6. The quality was again compared with the version without predictive frames using PSRN (Fig. 7). Introduction of 4:2:0 subsampling and predictive frames resulted in about 48% reduction of compressed data, which means considerably increased compression ratio.

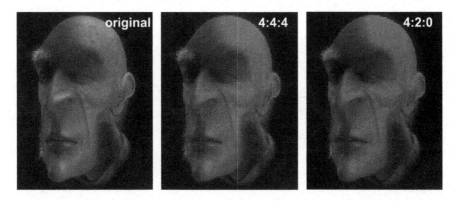

Fig. 4. Differences between images before compression, after compression in 4:4:4 and 4:2:0 subsampling

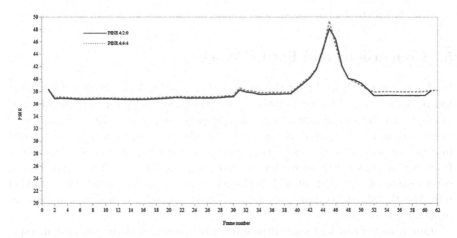

Fig. 5. PSNR of compressed video frames in 4:4:4 and 4:2:0 subsampling

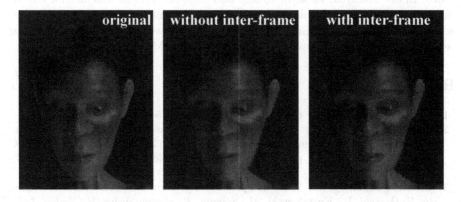

Fig. 6. Differences between images before compression, after compression with and without inter-frame

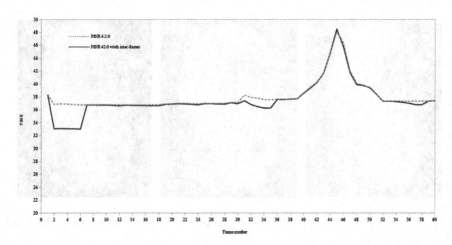

Fig. 7. PSNR of compressed video frames in 4:2:0 and 4:2:0 with inter-frame

5 Conclusions and Future Work

In this paper we showed that the presented method can be successfully used in video compression. The combination of 4:2:0 color subsampling and introduction of predictive frames allowed us to significantly reduce the size of compressed video with respect to previous approach. Thanks to encoding of YC_bC_r channels into two we were able to reduce the number of predictors and codebooks to two from original three. Predictive frames made it possible to eliminate unnecessary compression of regions that didn't change between consecutive frames, further reducing the size of compressed data. Those changes also resulted in reduced processing time.

Our future work will concentrate on improvements of the algorithm, especially optimizing the creation of predictive frames and enhancing the quality of video. In our research, we want to find and compare various schemes for encoding reduced YC_bC_r channels to see which one of theme can improve quality and which one can improve compression ratio. Finally we will try to compare our results with known compression standards, such as MPEG.

References

1. CCITT: Video codec for audio visual services at px 64 kbits/s (1993)
2. Cierniak, R.: An image compression algorithm based on neural networks. In: Rutkowski, L., Siekmann, J.H., Tadeusiewicz, R., Zadeh, L.A. (eds.) ICAISC 2004. LNCS, vol. 3070, pp. 706–711. Springer, Heidelberg (2004). doi:10.1007/ 978-3-540-24844-6_108
3. Cierniak, R., Knop, M.: Video compression algorithm based on neural networks. In: Rutkowski, L., Korytkowski, M., Scherer, R., Tadeusiewicz, R., Zadeh, L.A., Zurada, J.M. (eds.) ICAISC 2013. LNCS, vol. 7894, pp. 524–531. Springer, Heidelberg (2013). doi:10.1007/978-3-642-38658-9_47

4. Cierniak, R., Rutkowski, L.: On image compression by competitive neural networks and optimal linear predictors. Signal Process. Image Commun. **15**(6), 559–565 (2000)
5. Clarke, R.J.: Digital Compression of Still Images and Video. Academic Press Inc., London (1995)
6. Gersho, A., Gray, R.M.: Vector Quantization and Signal Compression. Kluwer Academic Publishers, Norwell (1991)
7. Gray, R.: Vector quantization. IEEE ASSP Magaz. **1**(2), 4–29 (1984)
8. Grycuk, R., Knop, M.: Neural video compression based on SURF scene change detection algorithm. In: Choraś, R.S. (ed.) Image Processing and Communications Challenges 7. AISC, vol. 389, pp. 105–112. Springer, Cham (2016). doi:10.1007/978-3-319-23814-2_13
9. ITU-R BT.709–6: Parameter values for the HDTV standards for production and international programme exchange (2015)
10. Kerr, D.A.: Chrominance subsampling in digital images. The Pumpkin (3), January 2012
11. Knop, M., Cierniak, R., Shah, N.: Video compression algorithm based on neural network structures. In: Rutkowski, L., Korytkowski, M., Scherer, R., Tadeusiewicz, R., Zadeh, L.A., Zurada, J.M. (eds.) ICAISC 2014. LNCS, vol. 8467, pp. 715–724. Springer, Cham (2014). doi:10.1007/978-3-319-07173-2_61
12. Knop, M., Dobosz, P.: Neural Video Compression Algorithm. In: Choraś, R.S. (ed.) ICAISC 2014. AISC, vol. 313, pp. 59–66. Springer, Cham (2015). doi:10.1007/978-3-319-10662-5_8
13. Knop, M., Kapuściński, T., Mleczko, W.K., Angryk, R.: Neural video compression based on RBM scene change detection algorithm. In: Rutkowski, L., Korytkowski, M., Scherer, R., Tadeusiewicz, R., Zadeh, L.A., Zurada, J.M. (eds.) ICAISC 2016. LNCS (LNAI), vol. 9693, pp. 660–669. Springer, Cham (2016). doi:10.1007/978-3-319-39384-1_58
14. Setton, E., Girod, B.: Video streaming with SP and SI frames. In: Proceedings of Visual Communication and Image Processing (2005)
15. Winkler, S., Kunt, M., van den Branden Lambrecht, C.J.: Vision and video: models and applications. In: van den Branden Lambrecht, C.J. (ed.) Vision Models and Applications to Image and Video Processing, pp. 201–229. Springer, Heidelberg (2001)
16. Xiph.org: Video test media. https://media.xiph.org/video/derf/ Accessed 10 Mar 2016
17. Zhang, R., Regunathan, S.L., Rose, K.: Video coding with optimal inter/intra-mode switching for packet loss resilience. IEEE J. Sel. Areas Commun. **18**(6), 966–976 (2000)

Taming the HoG: The Influence of Classifier Choice on Histogram of Oriented Gradients Person Detector Performance

Michał Olejniczak and Marek Kraft[(✉)]

Institute of Control and Information Engineering,
Poznan University of Technology, Piotrowo 3A, 60-965 Poznań, Poland
marek.kraft@put.poznan.pl

Abstract. Histogram of oriented gradients (HoG) is a common choice for hand-crafted feature used in a wide range of machine vision task. It functions as a part of a processing pipeline, in which it's followed by a classifier. The canonical approach proposed by the authors of HoG is the use of a linear support vector machine (SVM). This approach is usually followed by the majority of adopters with good results. However, a range of classifiers have proven to perform better than linear SVM in a variety of applications. In this paper, we investigate the pairing between HoG and a range of classifiers in order to find one with the best performance in terms of accuracy and processing speed for the task of human silhouete detection.

Keywords: Object detection · Machine learning · Histogram of oriented gradients

1 Introduction

Object detection in images and videos has received a lot of attention in the computer vision community in recent years. Since people constitute a very important elements of machine's environment in a vast range of applications, significant effort has been made to develop reliable methods for the detection of human silhouette. One of the most referenced, know and applied techniques to achieve this goal was described in the seminal paper by Dalal and Triggs [5]. The method uses the Histogram of Oriented Gradients (HoG) descriptor to generate a representation that captures the characteristics of the image patch in the detection window in combination with the linear support vector machine (SVM) classifier. The linear SVM+HoG approach became a de facto standard across many computer vision tasks, since its introduction brought forth a significant improvement of performance [6,8,10,12]. But is the linear SVM the only viable choice for classifier to be applied with HoG?

The motivation behind this work was the study performed by Delgado et al. [7]. The authors evaluated 179 classifiers coming from 17 families, using 121 datasets. The random forest is clearly the best all-round family of classifiers

© Springer International Publishing AG 2017
L. Rutkowski et al. (Eds.): ICAISC 2017, Part I, LNAI 10245, pp. 552–560, 2017.
DOI: 10.1007/978-3-319-59063-9_49

achieving 94.1% of the maximum accuracy overcoming 90% in the 84.3% of the datasets (3 out of 5 bests classifiers are RF). Moreover, in this comprehensive evaluation, the kernel versions of support vector machines (SVMs) in general outperform the linear SVM, which is the most common choice for the classifier that is paired with HoG [2]. In this paper, we present the results o the evaluation of the influence of classifier choice on HoG-based object detector performance. Specifically, we compare the approaches based on Linear SVM, kernel SVM, decision trees and random forest.

2   The HoG Region Descriptor

Feature extraction algorithms convert a fixed size patch to a higher level description – a feature vector of a given size. The goal of this process is to create a general, informative and non-redundant characterization. The idea behind the HoG descriptor is, that the appearance of an object can be described using the spatial distributions of edge gradient directions.

The steps for calculating the HOG descriptor for a single image patch are illustrated in Fig. 1.

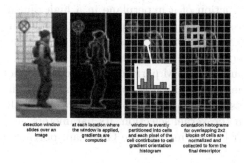

detection window   at each location where   window is evenly   orientation histograms
slides over an   the window is applied,   partitioned into cells   for overlapping 2x2
image   gradients are   and each pixel of the   blocks of cells are
                          computed   cell contributes to cell normalized and
                                                          gradient orientation   collected to form the
                                                          histogram   final descriptor

Fig. 1. Formation of the histogram of oriented gradients descriptor.

The first step in the description process is the edge detection. It is performed by applying the 1-D centered, point discrete derivative masks (see Eq. (1) in both directions by convolution.

$$g_x = [-1, 0, 1] \quad \text{and} \quad g_y = [-1, 0, 1]^T \tag{1}$$

The partial gradients g_x and g_y are then used to compute the gradient magnitude g and gradient direction θ. Please note, that the resulting orientation has a range of 0° to 180° (see Eq. (2)).

$$g = \sqrt{g_x^2 + g_y^2} \quad , \quad \theta = arctan\frac{g_y}{g_x} \tag{2}$$

The image patch is subsequently divided into regular cells. In the original paper, 8 × 8 pixel cells were used for 128 × 64 pixel patches, resulting in a

16×8 cell grid. Each one of the 64 pixels within a cell contributes to a nine bin histogram of gradient orientations. The votes are additionally weighted by their corresponding gradient magnitude and each individual vote is split proportionally between two nearest bins. The normalization is performed on feature histograms grouped in blocks. The blocks have a size of 2×2 cells and a one cell overlap. The described image patch is therefore divided into 7×15 blocks, for a total of 105 blocks. With each block containing the 9-bin histograms of 4 cells, the total number of descriptor elements per block is 36. The complete descriptor is therefore composed of 3780 values.

3   The Tested Classifiers

3.1   Support Vector Machines

Support vector machines are a supervised learning method useful in classification and regression. When presented with a sufficient number of annotated training samples training examples, the SVM training algorithm comes up with a model capable of assigning examples from outside of the training to one of two categories, making it a non-probabilistic binary linear classifier. The learning process results in the construction of a hyperplane in a high dimensional space. Good class separation is assured by the hyper-plane that has the largest distance to the nearest training data points of any class. The larger the margin is, the smaller the generalization error of the classifier.

The basic version of the SVM performs linear classification [3], but an extension called the kernel trick facilitates nonlinear classification by implicitly mapping the SVM inputs into high-dimensional feature spaces [4]. Performance level of SVM depends on the kernel selection, its parameters, and soft margin parameter C. Choices of kernels include the Gaussian, polynomial and radial basis and sigmoid functions.

3.2   Decision Tree Classifier

Decision trees are a non-parametric supervised learning method used in classification and regression. In order to predict a class, a set of simple decision rules is inferred from the data in the process of learning [16]. Main advantages of decision trees are:

- they are simple to understand and interpret,
- can provide meaningful output even when trained with a small set of samples,
- classification using decision trees is fast,
- can effortlessly be combined them with other decision techniques.

Their performance is not on par with the performance of more complex classifiers, but the simplicity and speed may make them a viable candidate for salient image region detection. Therefore, a decision was made to include them in the presented evaluation.

3.3 Random Forest Classifier

The random forest [1] is an ensemble approach that can also be thought of as a form of nearest neighbor predictor. Ensembles are used to improve performance utilizing a divide-and-conquer approach. The fundamental principle behind this is that a cluster of "weak learners" can come together to form a "strong learner" [18]. Each of these classifiers, on his own, is considered a "weak learner", while all of them taken together create a "strong learner". The random forest is composed of a set of decision trees, each of which is such a weak learner. The individual trees are trained using a subset of the complete training set. Each split in each tree is made on a randomly selected subset of features. The final classification is performed based on the votes cast by individual trees. Random decision forests correct for decision trees' habit of overfitting to their training set, achieving significantly better performance. Although composed of a larger number of weak classifiers, random forest classifier is still relatively fast. Moreover, it is capable of dealing with unbalanced and missing data.

4 Evaluation Methodology

The implementation and evaluation of the learning algorithms was performed using the scikit-learn Python library [15]. The data coming from the CVC-03 [13] and CVC-04 [17] virtual pedestrian datasets was used for testing. The datasets were generated using the Half-Life 2 game graphics engine, but their usefulness for training the classifiers that will subsequently be used for real-world data has been confirmed [13,17]. The CVC-03 dataset was used for classifier learning. It consists of 1678 virtual pedestrians (with their corresponding horizontal mirrors) and 2048 pedestrian-free background images to extract negatives for training. The CVC-04 dataset of 1208 virtual pedestrians (with their corresponding horizontal mirrors) and 6828 pedestrian-free background images was used for testing. The size of a single sample is 96 × 48 pixels, so the HoG cells have the size of 6 × 6 pixels (Fig. 2).

Preliminary classifier were trained using two sets – one containing the positive and one containing the negative samples. Each set contains 3356 elements. A single element of each set (sample) is a 96 × 48 pixel window either containing (positive) or not containing a pedestrian (negative). After that we performed exhaustive search on background images to find hard examples. Finding hard examples is essentially hard negative mining process on pedestrian-free images. The false positives found during this step are included in the final training set. It was a two step process, first we used Naive Bayes classifier to pre-decide if there was any sense to use a more advanced classifier (if the probability was below 80%, the sample was omitted). It was used purely to reduce overall hard negative mining time, as the second step was proper detection with preliminary classifier. To ensure that our training vector won't be larger than 6712 samples for the sake of class balance, a subsampling method based on cluster centroids was used. The method undersamples the majority class by replacing a cluster of majority samples by the cluster centroid of the KMeans algorithm. This algorithm keeps

Fig. 2. Sample images from the training set. Top row – pedestrians, middle row – background, negative set, bottom row – background, hard examples.

N majority samples by fitting the KMeans algorithm with N clusters to the majority class and using the coordinates of the N cluster centroids as the new majority samples. The subsampled vector was used to compute the final version of our classifiers.

As there are multiple hyperparameters associated with the evaluated machine learning algorithms, hyperparameter tuning using grid search was performed to ensure the results are close to the theoretical maximum of accuracy. Grid search is an exhaustive search through an earlier created subset of the hyperparameter space of a learning algorithm [9]. A grid search algorithm must be controlled by some performance metric, typically measured by cross-validation on the training set or evaluation on a held-out validation set. In the case of the described experiments, 3-fold cross validation was performed. The parameter values used in grid search are given below. The finally selected values of each parameter (ones associated with the highest classifier accuracy) are underlined.

Linear SVM:

– C values:

    {0.01, 0.1, 1, 10, 100, 1000}

Nonlinear SVM:

– C values:

```
{0.01, 0.1, 1, 10, 100, 1000}
```

- γ values:

```
{0.001, 0.0001}
```

- kernel types:

```
{radial basis function, polynomial, sigmoid}
```

- polynomial degree (for polynomial kernel function only):

```
{2, 3, 4}
```

Decision tree:

- node split selection strategy:

```
{best, best random}
```

- maximum tree depth:

```
{10, 50, 100, no limit}
```

- minimum samples required for split:

```
{2, 3, 6}
```

Random forest:

- maximum depth

```
{10, 50, 100, None}
```

- maximum estimator number:

```
{5, 10, 15, 30, 40, 50, 60}
```

- minimum samples required for split:

```
{2, 3, 6, 10, 12}
```

Random forest class predictions were made with majority voting. Validation was performed on 2416 pedestrian samples and 8533 background samples. Precision, recall and f1-score for both the *pedestrian* and *background* classes were selected as the performance measures.

5 Results and Discussion

Tables 1, 2, 3 and 4 summarize the results for the linear SVM, kernel SVM, decision tree and random forest classifier, respectively. The average values in the bottom row of the tables are weighted to reflect the differences in the sizes of test sets for both classes.

Table 1. Linear SVM.

	Precision	Recall	f1-score	Support
Background	0.974	0.998	0.986	8533
Pedestrian	0.992	0.906	0.947	2416
Avg/total	0.978	0.978	0.977	10949

Table 2. Kernel SVM.

	Precision	Recall	f1-score	Support
Background	0.974	0.998	0.986	8533
Pedestrian	0.991	0.907	0.947	2416
Avg/total	0.978	0.978	0.977	10949

Table 3. Decision tree.

	Precision	Recall	f1-score	Support
Background	0.927	0.861	0.893	8533
Pedestrian	0.608	0.761	0.676	2416
Avg/total	0.857	0.839	0.845	10949

Table 4. Random forest.

	Precision	Recall	f1-score	Support
Background	0.984	0.993	0.988	8533
Pedestrian	0.976	0.941	0.958	2416
Avg/total	0.982	0.982	0.982	10949

In terms of accuracy, there is no gain from using the kernel SVM over the linear SVM. In this particular case, the general claims from [7] do not hold. As stated in [2], the HoG descriptor preserves the second-order interactions between the pixels, so the lack of gain from using a higher order transformation of the input features seems to be the logical consequence of the fact. Moreover, the use of kernel SVM makes the training significantly longer, and requires the tuning of multiple hyperparameters.

The decision trees, being a rather simplistic approach, have achieved significantly worse accuracy scores. This was an expected outcome. However, relatively high precision and recall scores of a single decision tree for the *pedestrian*, paired with its ease of use and high computational speed makes it a viable candidate for a preliminary salient region detector, or even the final detector, especially in the applications, in which the computational speed is of utmost importance or for resource-constrained applications, e.g. simple, battery powered smart cameras and mesh-like camera networks [11,14]. Random forest, being an ensemble of decision trees, has achieved an accuracy that is on par with the accuracy of SVM-based solutions. Please note, however, that the training and running time for the random forest approach are significantly shorter (up to two orders of magnitude in our experiments). Moreover, the tuning of parameters is generally considered easier than it is the case with the SVM-based approaches.

6   Conclusions

An evaluation of the selected classifiers working in conjunction with the HoG feature descriptor for the task of human silhouette detection was presented in the paper. The results indicate, that the default choice of classifier for HoG – the linear SVM – is indeed a favorable solution in terms of accuracy. The use of the more complex kernel SVM does not result in any performance improvement and has an additional computational cost. The simple decision tree classifier is the fastest, but performs noticeably worse in terms of accuracy. Interestingly, the random forest classifier, which is an uncommon choice in this context, was proven to be comparable in terms of accuracy, yet considerably faster and easier to train. We believe the results may provide useful guidelines for classifier choice to be paired with HoG.

References

1. Breiman, L.: Random forests. Mach. Learn. **45**(1), 5–32 (2001). http://dx.doi.org/10.1023/A:1010933404324
2. Bristow, H., Lucey, S.: Why do linear svms trained on HOG features perform so well? arXiv preprint abs/1406.2419 (2014). http://arxiv.org/abs/1406.2419
3. Cortes, C., Vapnik, V.: Support-vector networks. Mach. Learn. **20**(3), 273–297 (1995). http://dx.doi.org/10.1007/BF00994018
4. Cristianini, N., Shawe-Taylor, J.: An Introduction to Support Vector Machines and Other Kernel-Based Learning Methods. Cambridge University Press, New York (2000)
5. Dalal, N., Triggs, B.: Histograms of oriented gradients for human detection. In: IEEE Computer Society Conference on Computer Vision and Pattern Recognition (CVPR 2005), vol. 1, pp. 886–893 (2005)
6. Felzenszwalb, P.F., Girshick, R.B., McAllester, D., Ramanan, D.: Object detection with discriminatively trained part-based models. IEEE Trans. Pattern Anal. Mach. Intell. **32**(9), 1627–1645 (2010)
7. Fernández-Delgado, M., Cernadas, E., Barro, S., Amorim, D.: Do we need hundreds of classifiers to solve real world classification problems? J. Mach. Learn. Res. **15**(1), 3133–3181 (2014). http://dl.acm.org/citation.cfm?id=2627435.2697065
8. Greenhalgh, J., Mirmehdi, M.: Real-time detection and recognition of road traffic signs. IEEE Trans. Intell. Transp. Syst. **13**(4), 1498–1506 (2012)
9. Hsu, C.W., Chang, C.C., Lin, C.J., et al.: A practical guide to support vector classification (2003)
10. Li, M., Zhang, Z., Huang, K., Tan, T.: Estimating the number of people in crowded scenes by mid based foreground segmentation and head-shoulder detection. In: 19th International Conference on Pattern Recognition, pp. 1–4, December 2008

11. Magno, M., Tombari, F., Brunelli, D., Stefano, L.D., Benini, L.: Multimodal video analysis on self-powered resource-limited wireless smart camera. IEEE J. Emerg. Sel. Topics Circ. Syst. **3**(2), 223–235 (2013)

12. Malisiewicz, T., Gupta, A., Efros, A.A.: Ensemble of Exemplar-SVMs for object detection and beyond. In: 2011 International Conference on Computer Vision, pp. 89–96, November 2011

13. Marin, J., Vázquez, D., Gerónimo, D., López, A.M.: Learning appearance in virtual scenarios for pedestrian detection. In: IEEE Conference on Computer Vision and Pattern Recognition (CVPR), pp. 137–144. IEEE (2010)

14. Miller, L., Abas, K., Obraczka, K.: SCmesh: solar-powered wireless smart camera mesh network. In: 24th International Conference on Computer Communication and Networks (ICCCN), pp. 1–8, August 2015

15. Pedregosa, F., Varoquaux, G., Gramfort, A., Michel, V., Thirion, B., Grisel, O., Blondel, M., Prettenhofer, P., Weiss, R., Dubourg, V., Vanderplas, J., Passos, A., Cournapeau, D., Brucher, M., Perrot, M., Duchesnay, E.: Scikit-learn: machine learning in python. J. Mach. Learn. Res. **12**, 2825–2830 (2011)

16. Quinlan, J.: Induction of decision trees. Mach. Learn. **1**(1), 81–106 (1986). http://dx.doi.org/10.1023/A:1022643204877

17. Vazquez, D., Lopez, A.M., Marin, J., Ponsa, D., Geronimo, D.: Virtual and real world adaptation for pedestrian detection. IEEE Trans. Pattern Anal. Mach. Intell. **36**(4), 797–809 (2014)

18. Zhang, C., Ma, Y.: Ensemble Machine Learning: Methods and Applications. Springer, New York (2012)

Virtual Cameras and Stereoscopic Imaging for the Supervision of Industrial Processes

Paweł Rotter[(✉)]

Department of Automatics and Biomedical Engineering,
AGH-University of Science and Technology, Kraków, Poland
rotter@agh.edu.pl

Abstract. In this article we present a concept of visualisation of some industrial processes, such as glass melting. The process is observed from a single camera at fixed location but when the scene geometry is known, like for example in case of glass furnaces, the image can be transformed to a perspective view from a virtual camera at arbitrary location and to stereoscopic views based on two virtual cameras. We applied automatic analysis of the input image to present to the supervisor of the process a synthetic image, which only contains features that are important for the process control. We have developed a prototype visualization system, tested in cooperation with the glass industry.

Keywords: Stereoscopy · Orthoimage · Image transformation · Image segmentation · Industrial process visualisation

1 Introduction

In many industrial processes image is an important source of information about the process. For example, in glass melting the operator controls the process based on distribution of batch (raw material) floating on the surface of molten glass. The image is captured by a camera located inside the furnace, in the upper part of the furnace chamber. There is usually one camera in the furnace, and because of high temperature inside the furnace and requirements for cooling system, installation of additional cameras would be expensive. On the other hand, the possibility of looking into the furnace from different viewpoints is very desirable. Especially relevant views are:

- Orthoimage, i.e. view from above with a uniform scale. Orthoimage presents real proportions between different features of the image, independently on the location of the physical camera.
- View from the batch hoppers side, which is natural to glass melting technologists. In glass technology, terms "left side" and "right side" of the furnace are related to the view from batch hoppers (beginning of the furnace) towards the place where melted glass flows out (the end of the furnace). The physical camera is always located at the end of the furnace because of technological reasons, see Fig. 1.

© Springer International Publishing AG 2017
L. Rutkowski et al. (Eds.): ICAISC 2017, Part I, LNAI 10245, pp. 561–569, 2017.
DOI: 10.1007/978-3-319-59063-9_50

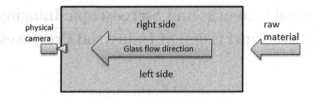

Fig. 1. Scheme of a glass furnace, view from above

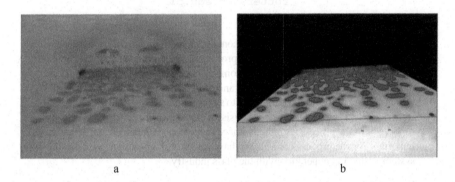

a   b

Fig. 2. An image from the furnace camera (a) and edges of the segmented image superimposed on the original (b)

- Side view, where the virtual camera is located on the side of the furnace, above the line where batch should disappear. Observation of this part of the furnace is especially important.

In this article we present the concept and prototype software with the following functionalities:

- Segmentation of the image, which leads to a synthetic image that only contains relevant elements of the real image (batch, molten glass and areas behind the field of view).
- Transforming the camera image or synthetic image to the orthoimage.
- Transforming the camera image or synthetic image to a view from any location defined by the user.
- Generation of a stereoscopic image: the image from a single physical camera is transformed to two images from virtual cameras corresponding to the left and the right eye.

The algorithms presented in this paper have been developed as an extension of the system for automatic analysis of the symmetry of the glass melting process and extraction of the process parameters, proposed in [1, 2], which we develop in cooperation with the glass industry. Nevertheless, imaging methods presented here can also be used in imaging of other industrial processes, where geometry of the scene allows to extract parameters of transformation between the camera image and real-word 3D coordinates.

2 Segmentation

The operator who controls temperature in the furnace mostly takes into account distribution of batch. If raw material appears at the end of the furnace, it may cause deterioration of the quality of production and even make it unusable. Hence, segmentation of batch is the first step to produce a synthetic image that presents only important information without unnecessary details. Before segmentation, we apply a linear lowpass filter, and then medial filter, to remove noise. Parameters of filtration depend on quality of the image: when it comes from older models of analog cameras, which are still in use in some glassworks, strong filtration is needed, while if a high quality camera is used, excessive filtration would deteriorate the image quality.

Segmentation of the glass surface is carried out based on its brightness. Calibration involves indicating by a user several sample points in batch area and several point in clear glass area, in various parts of the image. Since there are no variations in lighting conditions, and all the light in the furnace is emitted by a molten glass because of its high temperature, calibration is performed once for the specific furnace and camera.

There are many methods for automatic selection of the threshold, for overview see [3]. On our specific problem there are two classes, each corresponding to a distinct local maximum of the image histogram, therefore one of adequate methods is to select the threshold at minimum of the histogram between two peaks. Precisely, we select as the threshold grayscale level $i_g \in \left(i_{Bmax} + 1, i_{Gmin} - 1 \right)$ that fulfils:

$$\forall j \in \left(i_{Bmax} + 1, i_{Gmin} - 1 \right) : h\left(i_g - 1\right) + h(i_g) + h\left(i_g + 1\right) \le h(j-1) + h(j) + h(j+1), \tag{1}$$

where $h(i)$ is the number of pixels with grayscale i in the polygon corresponding to the glass surface area, i_{Bmax} is the maximum grayscale of the set of points marked by the user as batch and i_{Gmin} is the minimum grayscale of points marked by the user as molten glass. The user may correct the threshold manually in the interactive mode, observing segmentation changes superimposed on the original image (Fig. 4b).

3 Transformation to Orthoimage

The camera is located in the upper part of the furnace and the image from it is a perspective projection of the glass surface. After identification of the parameters of this 2D homography, the camera image can be transformed back to the coordinate system related to the glass surface [4]. In such transformed image the user can observe batch distribution in the image with true proportions, not distorted by the perspective projection and independent on the camera location.

In order to extract the parameters of transformation, the user indicates in the image four points that define edges of the glass surface (two corners and any point from each of two side edges) and two points that define the bubbling line – see Fig. 3.

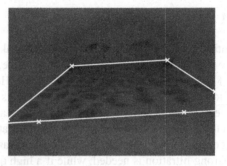

Fig. 3. Edges of the glass surface and the bubbling line indicated by the user allow calculation of the parameters of perspective transformation

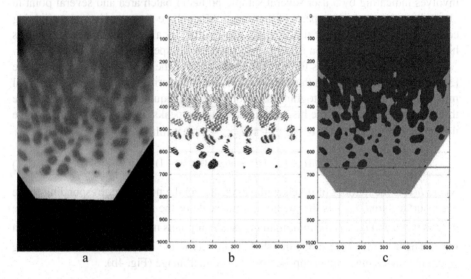

Fig. 4. Orthoimages of the glass surface: transformations of the real image (a), sparse image (b) and segmented image (c)

The corners of the rectangle defined by three edges of the glass surface and the bubbling line allow calculation of the perspective transformation parameters. In Fig. 4a we present a grayscale camera image transformed to the orthoimage. An important property of the orthoimage is that areas closer to the real camera are represented with higher accuracy. This feature is especially visible in sparse image presented in Fig. 4b, where each pixel of the original image is transformed into a single pixel of the target image and all other pixels of the target image are left white.

Inhomogeneous accuracy of representation also applies to images from any virtual camera, which will be presented in the next section. It is however desirable property because the real camera is located at the end of the furnace, where exact information about the distribution of batch is more important than for areas at the beginning of the furnace. In Fig. 4c we present the orthoimage after segmentation, where dark grey denotes batch, light grey – molten glass. Areas behind the field of view of the physical camera are left white.

Orthoimage is very important for automatic calculation of parameters of batch distribution [1] because areas and distances in the orthoimage are proportional to corresponding values in the real world. Nevertheless, a human operator may prefer to look at the glass surface from natural perspective. Perspective views from virtual cameras located at any point and generation of stereoscopic views are the subject of the next chapter.

4 Synthetic View from Arbitrary Location and Stereoscopic Imaging

In the related literature much research is reported on automatic generation of stereoscopic images or virtual views from arbitrary location. Some methods are based on 3D information obtained from several cameras [5], other use advanced but often ambiguous methods of analysis of 2D image content, like in [6], where image segmentation and analysis of perspective is used to calculate the map of depth. Finally, stereopairs or virtual views can be generated from full 3D models of the scene [7]. In our specific problem, the input to our algorithms is a single monoscopic image of the scene but based on camera location and geometry of the scene we are able to calculate 3D coordinates of any point of the glass surface or furnace walls, therefore it is possible to render an image seen from arbitrary viewpoint. In case of glass melting process this is particularly useful because technologists are used to define the left and the right side of the furnace according to the direction of glass flow – from batch hoppers, which define the beginning of the furnace, towards the end of the furnace. In Fig. 5 we present a synthetic view obtained from the real image presented in Fig. 2. Grid density is 10 cm, bold grid lines are every meter. The real walls of the furnace have been replaced with synthetic image, first because the detailed image of walls is irrelevant to the process, and second because

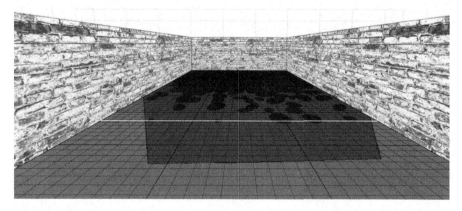

Fig. 5. A synthetic view from a viewpoint located at the end of the furnace, close to the real camera

only a part of walls is observed by the camera. Light grey colour denotes areas beyond the field of view of the physical camera. The white line is the bubbling line, which approximately corresponds to 2/3 of the furnace length.

In Fig. 6 we present the image transformed to the view from a virtual camera located at the beginning of the furnace, above the batch hoppers. Let us underline that the accuracy of this image corresponds to the original image. This means that batch distribution is more precise for areas closer to the physical camera (so farther from the virtual camera). This meets requirements of the operators, because precise calculation of batch distribution is more important at the end of the furnace, where even small pieces of raw material should not appear.

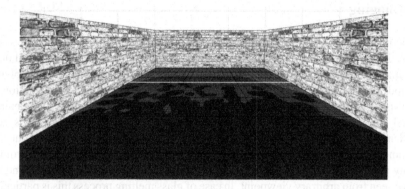

Fig. 6. A synthetic view from a viewpoint located at the beginning of the furnace, above batch hoppers

Another viewpoint important for technologists is the side view from virtual camera located exactly above the bubbling line (see Fig. 7), which is an important marker for operators: only a small fractions of batch can cross this line, like in Fig. 7, otherwise temperature in the furnace should be increased.

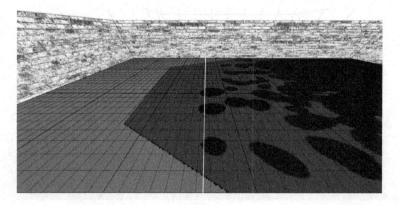

Fig. 7. A synthetic side view. Viewpoint is located above the bubbling line

An image of the furnace can be synthetized from any viewpoint, therefore two viewpoints can be used to create a stereo pair and enable looking at the process in three dimensions. 3D visualisation provides more information than a monoscopic image and it exploits natural mechanics of perception of the human vision system. Many application have been developed in robotics [8] and medical imaging [9], while in related literature there is no evidence of using stereoscopy for visualisation of industrial processes.

In Fig. 8 we present the left and the right image of a synthetic stereopair. Images are generated for a pair of virtual cameras located at the beginning of the furnace, above the batch hoppers. Based on experiments we chose parameters of the stereoscopic system which ensures optimal image.[1] Because of specific type of scene, which is composed of a set of planes, we do not need to consider common problems that usually corrupt stereoscopic imaging, like cardboard effect [12] or puppet theatre effect [13]. We decided to use a stereoscopic system with toed-in camera configuration (converging optical axes). This configuration leads to nonlinear perception of depth [13, 14] but the impression of depth it is more natural because geometry of the camera set is closer to that of the human visual system. Based on experiments we set the point of convergence at 75% of the furnace length. In most stereoscopic systems the distance between two cameras is constant, in the human visual system the average distance between eyes is 6.2 cm [15]. However, the optimal base of stereoscopic systems depends on the range of distance to objects of interest. For example when the viewpoint is located near hoppers, the closest relevant details of the image are at the distance of about 3–4 meters and the base 8 cm yields high quality stereogram. For viewpoints close to the real camera the base should be shorter, around 5 cm because area with high diversification of batch structure is closer to the virtual camera, while for a side view good results are achieved for base of 3 cm.

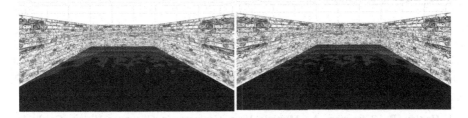

Fig. 8. A stereopair of images from two virtual cameras located at the beginning of the furnace. The distance between the virtual cameras is 8 cm

3D image can be observed by process supervisors and technologists using any kind of stereoscopic display, like a passive or active 3D monitor, 3D TV set or virtual reality glasses, see [8, 15, 16] for more information and comparison of different 3D devices. We recommend a console based on a simple passive 3D TV set, mostly because passive glasses are unfailing, what is relevant in industrial environment [17]. Examples of

---

[1] Viewing experience and optimal parameters of the stereoscopic system depend on equipment [10, 11]. We used 3D passive 24″ 16:10 monitor Hyundai S243A. For other equipment optimal parameters may be different, especially if size of the screen is substantially different.

stereoscopic images of the furnace generated by our software are available in Multi Picture Object (mpo) format at: http://home.agh.edu.pl/~rotter/furnace3D.

5 Conclusions

The goal of this article was to demonstrate that in industrial processes, in which location of the observed surface is known, a single camera at fixed position can be used to synthetize monoscopic and stereoscopic images of the process from any viewpoint. We synthetized artificial images, which only contain features of the image that are important for the assessment of the process course.

The system was implemented and tested on the glass melting process, where a camera installed inside the furnace is the main source of information about the process for the operator who controls it. In stereoscopic systems it is difficult to define an objective quantitative measure of the algorithm performance, which is usually estimated by users. Such visual assessment of images generated by our algorithm confirms that the quality of the synthetized images, including 3D views, is sufficient for practical applications in the process control.

Acknowledgments. The author would like to thank the president Andrzej Skowiniak and employees of Techglass Ltd for their help in the acquisition of test images and to the board of Huta Szkla Orzesze glassworks for allowing us access to its installations. We also thank to Dr. Maciej Klemiato who is currently working on integration of the software presented in this article with the system for furnace control at Huta Szkla Orzesze glassworks.

References

1. Rotter, P., Skowiniak, A.: Image-based analysis of the symmetry of the glass melting process. Glass Technol. Eur. J. Glass Sci. Technol. Part A **54**, 119–131 (2013)
2. Rotter, P.: Extraction of relevant glass melting parameters based on the pairwise comparisons of sample images from a furnace. Glass Technol. Eur. J. Glass Sci. Technol. Part A **55**, 55–62 (2014)
3. Russ, J.C.: The Image Processing Handbook. CRC Press, Boca Raton (2011)
4. Hartley, R., Zisserman, A.: Multiple View Geometry in Computer Vision. Cambridge University Press, New York (2003)
5. Lara, J.T.: 3D visualization using virtual view generation for stereoscopic hardware. M.Sc. Escola Tècnica Superior d'Enginyeria de Telecomunicació de Barcelona (2010)
6. Battiato, S., Capra, A., Curti, S., Cascia, M.: 3D stereoscopic image pairs by depth map generation. In: 2nd International Symposium on 3D Data Processing, Visualization and Transmission (2004)
7. Rojas, G.M., Gálvez, M., Potler, N.V., Craddock, R.C., Margulies, D.S., Castellanos, F.X., Milham, M.P.: Stereoscopic three-dimensional visualization applied to multimodal brain images: clinical applications and a functional connectivity atlas. Front. Neurosci. **8**, 1–14 (2014)
8. Ferre, M., Aracil, R., Sanchez-Uran, M.A.: Stereoscopic human interfaces. IEEE Robot. Autom. Magaz. **15**, 50–57 (2008)

9. Trzupek, M., Ogiela, M.R., Tadeusiewicz, R.: Intelligent image content semantic description for cardiac 3D visualisations. Eng. Appl. Artif. Intell. **24**, 1410–1418 (2011)
10. Shibata, T., Kim, J., Hoffman, D.M., Banks, M.S.: The zone of comfort: predicting visual discomfort with stereo displays. J. Vis. **11**, 1–29 (2011)
11. Banks, M.S., Read, J.C.A., Allison, R.S., Watt, S.J.: Stereoscopy and the human visual system. SMPTE Motion Imaging J. **121**, 24–43 (2012)
12. Yamanoue, H., Okui, M., Yuyama, I.: A study on the relationship between shooting conditions and cardboard effect of stereoscopic images. IEEE Trans. Circ. Syst. Video Technol. **10**, 411–416 (2000)
13. Yamanoue, H., Okui, M., Okano, F.: Geometrical analysis of puppet-theater and cardboard effects in stereoscopic HDTV images. IEEE Trans. Circ. Syst. Video Technol. **16**, 744–752 (2006)
14. Emoto, M., Yamanoue, H.: Working towards developing human harmonic stereoscopic systems. In: Javidi, B., Okano, F., Son, J.Y. (eds.) Three-Dimensional Imaging, Visualization, and Display. Springer, New York (2009)
15. Held, R.T., Banks, M.S.: Misperceptions in stereoscopic displays: a vision science perspective. In: 5th Symposium on Applied Perception in Graphics and Visualization (APGV 2008), pp. 23–31 (2008)
16. Kim, E.S.: Three-dimensional projection display system. In: Poon, T.C. (ed.) Digital Holography and Three-Dimensional Display, pp. 293–332. Springer, Heidelberg (2006)
17. Rotter, P.: Why did the 3D revolution fail? The present and future of stereoscopy. IEEE Technol. Soc. Magaz. **36**, 81–85 (2017)

Object Detection with Few Training Data: Detection of Subsiding Troughs in SAR Interferograms

Paweł Rotter[1(✉)], Jacek Strzelczyk[2], Stanisława Porzycka-Strzelczyk[2], and Claudio Feijoo[3,4]

[1] Department of Automatics and Biomedical Engineering,
AGH-University of Science and Technology, Kraków, Poland
rotter@agh.edu.pl
[2] Department of Geoinformatics and Applied Computer Science,
AGH-University of Science and Technology, Kraków, Poland
{strzelcz,porzycka}@agh.edu.pl
[3] Telecommunications College, Tongji University, Shanghai, China
cfeijoo@cedint.upm.es
[4] Department of Signals, Systems and Radiocommunications,
Technical University of Madrid, Madrid, Spain

Abstract. Subsiding troughs that are the result of mining activities can be detected in SAR interferograms as approximately elliptic shapes against the noisy background. Despite large areas being covered by interferogram, the number of positive samples, which can be used for automatic learning, is limited. In this paper we propose two alternative methods for the detection of subsiding troughs: the first one is designed to detect any circular shapes and does not require any learning set and the second is based on automatic learning but requires a reduced number of positive samples. The two proposed methods can support manual inspection of large areas in SAR interferograms.

Keywords: Subsiding troughs · SAR interferograms · Gabor filter

1 Introduction

In this work we propose and test two methods for the detection of subsidence troughs in interferograms. The interferograms are obtained as a result of satellite-based SAR (Synthetic Aperture Radar) images processing. SAR is an active system, usually placed on satellite or aircraft boards that allows to monitor Earth surface regardless of time of day or weather conditions with high temporal and spatial resolutions [1]. For each pixel of radar image the information about the amplitude and phase of the backscattered signal is saved [2]. Based on those parameters ground subsidence on the order of even some few centimetres can be detected. Those results can be achieved using DInSAR (Differential Interferometry SAR) method. It consists of exploiting two SAR images that cover the same area but were acquired at different times [3]. Based on them, and on information about topography of the region under study, a differential interferogram is generated. The interferometric fringes that appear in this interferogram represent the wrapped values of ground deformations in satellite LOS (Line of Sight) direction. For mining areas, these fringes,

© Springer International Publishing AG 2017
L. Rutkowski et al. (Eds.): ICAISC 2017, Part I, LNAI 10245, pp. 570–579, 2017.
DOI: 10.1007/978-3-319-59063-9_51

with approximately circular or elliptical shapes, correspond to subsidence troughs that are the result of underground exploitation [4]. Automatic learning was used in mining sciences for a number of applications, mostly related to the modelling of mining processes [5]. In this article we will apply it for the detection of ground deformations, which are the result of mining activities. The automatic detection of fringes that represent subsidence troughs plays a crucial role in the acceleration of the warning process, improving the security of people endangered by the ground deformation process. However, the automation of the detection procedure is challenging, especially when a very small set of learning data is available. Additionally, the high level of noise and the irregular shape of troughs greatly hinder the detection [6]. In this work we used two SAR interferograms (42.6 MPix and 4.8 MPix), each containing several positive samples only. They were generated based on radar images acquired from European radar imaging satellite Sentinel-1A. Exploited interferograms cover the region of Upper Silesian Coal Basin for which extensive coal exploitation is characteristic. The DInSAR processing of SAR data was performed using SNAP - Sentinel Application Platform software.

Because of such a small set of data, there is a need for algorithms that may work either on a small learning set or do not require any learning at all. In the paper we propose two alternative methods:

- Detection of troughs' centres without learning, presented in Sect. 2. This method is based on image convolution with circular wavelets, where a complex convolution masks are derived from Gabor wavelets. The goal is to detect circular features in the input image.
- Detection of troughs' areas with learning on a small set of samples, presented in Sect. 3. The algorithm is based on Gabor features calculated directly from the image in Cartesian coordinates. The goal is to detect the whole area of troughs and not only the centre, as it was the case in the first method.

In Table 1 we list the images with SAR interferograms that we have used. The shortage of interferograms of troughs that result from underground exploitation imposed limitations not only on learning algorithms but also on the assessment of the results. The amount of data was insufficient for application of such measures as precision-recall charts, so we judged the usability of methods based on comparing the results with formations visible in the input images.

Table 1. SAR interferograms used for tests

Image name	Number of pixels	Description
Q	11 215 × 3 984	Relatively good quality (distinctive troughs, areas with noise of high-amplitude and low frequency are separated from areas of troughs)
Q_{zoom1}	2 102 × 1 187	Sub-image of Q – area of occurrence of troughs
Q_{zoom2}	1 324 × 623	Sub-image of Q_{zoom1} with particular density of troughs and without low-frequency and high-amplitude noise
P	2 501 × 2 001	Example of bad quality image: areas of troughs are mixed with high-amplitude noise. Areas of noise are irregular and sometimes their shape is similar to the shape of troughs

2 Detection of Troughs' Centres Without Learning Samples

In this section we propose a method for the detection of circular shapes in noisy images. An important feature of this method is that it is not based on automatic learning, therefore there is no need for a training set. It was noticed as early as in 1960s [7] that the human visual system uses multi-channel filtering and that the retinal image is decomposed into a number of images corresponding to narrow ranges of frequency and orientation. Based on this observation, multi-channel filtering approach using Gabor filters was developed [8, 9]. In the proposed method we convolve the input image with a circular kernel derived from the Gabor impulse response. Let us recall that the 2D Gabor kernel used for extraction of local frequency features along x axis (see Fig. 1a) is given by:

$$g(x, y) = \frac{1}{2\pi\sigma_x\sigma_y} \exp\left[-\frac{1}{2}\left(\frac{x^2}{\sigma_x^2} + \frac{y^2}{\sigma_y^2}\right) + 2\pi j f_{centr} x\right], \tag{1}$$

and its 1D intersection along x axis is:

$$g(x) = \frac{1}{2\pi\sigma} \exp\left[-\frac{x^2}{2\sigma^2} + 2\pi j f_{centr} x\right], \tag{2}$$

where f is the central frequency of the filter and σ is the standard deviation. Based on (2) we propose a circular kernel (see Fig. 1b):

$$g(r) = \frac{1}{2\pi\sigma} \exp\left[-\frac{r^2}{2\sigma^2} + 2\pi j f_{centr} r\right], \tag{3}$$

where r is the distance from the centre (x_0, y_0) of the mask:

$$r = \sqrt{(x - x_0)^2 + (y - y_0)^2}, \tag{4}$$

see Fig. 1. The idea is that convolution of the kernel (3) with the input image should yield magnitude peaks in centres of circular shapes. Based on initial experiments, we set spatial frequency bandwidth to 0.3. Note that decreasing this parameter results in increasing filter kernel. For example in experiments with spatial frequency bandwidth = 0.3 the size of kernel for the longest wave we used ($T = 60.63$) was 766×766.

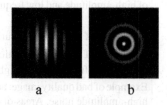

a   b

Fig. 1. Real part of Gabor impulse response (a) and the corresponding circular sine wave (b)

This implies the necessity to convolve the input image in Fourier domain to keep reasonable computation time.

Centres of troughs are detected at points where the module of filter output exceeds the threshold. Initially we set a threshold for each frequency based on quantiles of the magnitude of the filter response but experiments proved that a defined percentage of the maximum value of the module yields better results. In Fig. 2 we present an interferogram Q_{zoom2} that we used in the first test, with several clearly visible troughs denoted with letters A–I.

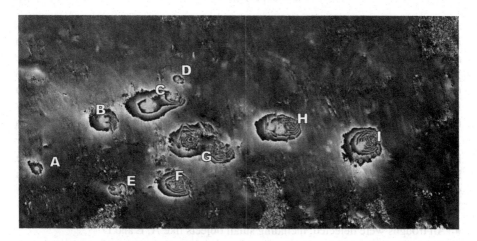

Fig. 2. Image Q_{zoom2} and detection of trough's centres (Color figure online)

The magnitude of response for two different wavelengths is presented in Fig. 3. Depending on the size of troughs and frequencies of grayscale changes in its interferograms, different troughs yield maxima of response for different wavelengths. For example, there is clear maximum of response corresponding to the centre of trough F for wavelength $T = 34.2$ and there is no corresponding maximum for $T = 51.2$, while for trough I situation is exactly opposite (compare Fig. 3a and b). This example shows that there is no common wavelength that could allow detecting all troughs and it is necessary to apply a series of filters.

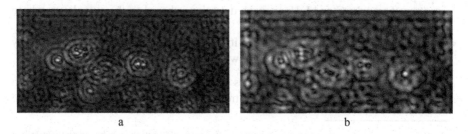

a   b

Fig. 3. Examples of magnitude of response for wavelengths: $T = 34.2$ and $T = 51.2$ for spatial frequency bandwidth set to 0.5

The results of trough detection are superimposed on interferogram in Fig. 2. Green points are spots where for any frequency the magnitude response is above the threshold. We use a series of wavelengths:

$$T_i = T_0 k^i \tag{5}$$

where:

$$k = \exp(\ln 2/S), \tag{6}$$

and S is the number of scales per octave. Based on measurements of the size of objects in interferograms, we decide that the wavelength vector should cover spectrum from around 10 to 60 pixels. Therefore assuming $S = 5$, $k = 1.149$, the wavelength vector is: $T = [10.00\ 11.49\ 13.20\ 15.16\ 17.41\ 20.00\ 22.97\ 26.39\ 30.31\ 34.82\ 40.00\ 45.95\ 52.78\ 60.63]$. We selected parameters of the Gabor filter bank used to generate convolution filters based on a series of experiments: spatial aspect ratio $= 1$ and spatial frequency bandwidth $= 0.3$.[1] A pixel is classified as the central area of a trough if for any wavelength T_i the response magnitude exceeds $\theta_i \max(|g|)$, where θ_i is the threshold for i-th wavelength and max(lgl) is the maximum magnitude of i-th filter response over the whole image. Based on experiments we resigned from adjusting individual threshold for each wavelength and for simplicity we set all thresholds $\theta_i = 0.8$, $i = 1, \ldots, 14$.

The result can be regarded successful, since:

- All troughs were detected, including small objects like A, E and D
- There are only two false detections: one between troughs B and C, and one below trough I. Both false detections are caused by proximity of a trough's edge
- There are no false detections in the noisy areas of the image (e.g. top right corner)

There are however several issues, which call into question usefulness of this method in case of particularly noisy images:

- Our filter is designed for detecting circular shapes but not for ellipses. Basins in our image are close to circles. In our test image only trough C is elongated but its left part is circular and this is sufficient to detect the trough. However, if all layers of the trough are elongated, the trough may remain undetected.
- In Fig. 2 we can see that the maxima of magnitude response do not lay precisely in centres of troughs G and I. Here again, detection is based on the outer layer, whose response is much stronger.
- The trough G is correctly detected but it composed of two smaller troughs. If they were separated, the trough on the right would not be detected.

When we tested the method on images that contain a large amount of noise with diversified spectral features, where some troughs are difficult to distinguish from the

---

[1] We did not use optimisation procedure, because: (i) any goal function (e.g. based on the number of undetected troughs and false detections) defined for such small amount of data would be constant in large areas of the parameter's space, (ii) it was not necessary because of small sensitivity of the method to parameters' changes.

background, around a third part of troughs remained undetected and the noise generates a large number of false alarms. Therefore the applicability of this method depends on the quality of the input interferograms.

3 Gabor Features Calculated Directly from the Image in Cartesian Coordinates

In this section we present a method for the detection of troughs based on their texture, described by Gabor features calculated directly from the image in Cartesian coordinates. We used Support Vector Machine (SVM) to classify pixels of image into two categories: areas of troughs and the background.

We used Gabor filter bank with wavelengths: $T_i = T_0 k^i$, where T_0 is the shortest wavelength. In input images the distance between consecutive ridges of troughs varies between 4 and 45 pixels, so we set the shortest wavelength at 3 pixels (to ensure a certain margin). We set frequency spacing at $k = 1.5$, so the vector of wavelengths is:

$$T = [3\ 4.5\ 6.8\ 10\ 15\ 22.8\ 34.2\ 51.2]. \tag{7}$$

Note that frequency spacing is sparser than in the previous method and the filter bank covers a different part of the spectrum. In the previous section wavelengths had precise interpretation as the distance between consecutive maxima of the trough's image, while here we use Gabor descriptors as features for texture-based classification. In general purpose systems for texture analysis wavelength spacing is even sparser, compare [10]. We used four orientations:

$$\Phi = [0\ 45\ 90\ 135], \tag{8}$$

so there are 32 filters in the bank in total.

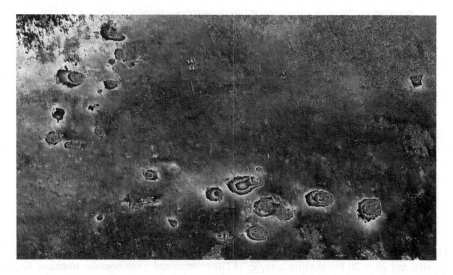

Fig. 4. Original image Q_{zoom1} used in the first test.

After filtration, each pixel of the input interferogram is described with a set of 32 features, where each feature is the magnitude of the corresponding Gabor filter output.

In the first series of tests we used image Q_{zoom1}, presented in Fig. 4.

For training SVM we indicated manually areas of troughs – see the mask of area used as positive samples in Fig. 5a. The whole learning set consists of 8 troughs only. In Fig. 5b we show the mask for areas that were excluded from training, in order to be used in tests. The rest of image was used as negative examples.

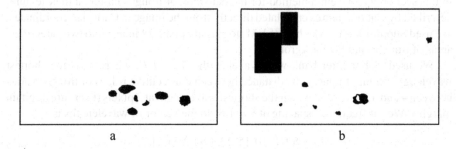

a   b

Fig. 5. Mask for positive examples (a) and for the part of the image excluded from training (b). The part of the image not covered by any of these two masks was used for training as negative examples.

In order to make the algorithm more effective, we did not use for learning each pixel but only pixels lying on $N \times N$ grid, where $N = 2$ or $N = 3$ for positive samples and N is between 10 and 20 for negative samples. In case of negative samples the grid can be sparser because the background covers a larger area of image than troughs. In Fig. 6 we present the results of classification. When comparing the results with similar experiment

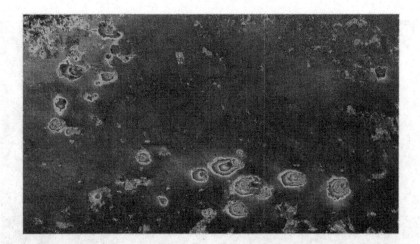

Fig. 6. Classification of interferogram Q with linear SVM classifier, learned on grid 3×3 for positive and 20×20 for negative samples (5230 positive and 5726 negative samples). Green colour denotes areas classified as troughs. (Color figure online)

for $N = 2$ for positive samples and $N = 10$ for negative samples we can deduce that increasing the density of learning samples improved the classification of training data but in areas excluded from training hardly any improvement can be noticed.

In the next series of experiments we test the method on noisy data. Images with graphical presentation of the results are included in the supplementary material available at:

http://home.agh.edu.pl/~rotter/sar/SAR_supplementary.pdf.

We performed six experiments:

1. The classifier was learned on image Q_{zoom1} and tested on image P. Linear SVM classifier was used, learned on grid 2×2 for positive samples and on grid 10×10 for negative samples (total 11 760 positive and 22 606 negative samples. Results: large areas of noise were classified as trough areas and some troughs were classified as meaningless background. Quality of classification is insufficient for practical applications.
2. The classifier was learned on image P and tested on the same image. Linear SVM classifier was used, learned on grid 2×2 for positive samples and on grid 15×15 for negative samples (total 18 243 positive and 22 056 negative samples). Results: Quality of classification is similar to the previous experiment. We can conclude that linear SVM classifier is not able to discriminate between areas of troughs and the background in the space of Gabor filter descriptors.
3. The classifier was learned on image P and tested on the same image. SVM classifier with RBF kernel was used, learned on grid 3×3 for positive samples and on grid 20×20 for negative samples (total 8 099 positive and 12 535 negative samples). Then we did two experiments when the same classifier was trained on a part of image P (approximately half of the image) and tested on the remaining part. Results: Classification was 100% correct for areas used for training because of adjustment of the classifier to the training data, while for the other part of the image all area was classified as background.
4. Similar experiments to the previous one were made but with increasing value of *KernelScale* parameter to avoid overfitting. Results: The results were unsatisfactory and we can conclude that in case of a noisy image the SVM classifier based on RBF kernel is not able to achieve good results. This is because for so noisy input data the vector of outputs of the Gabor filter bank is insufficient as a feature vector for classification.
5. The linear SVM classifier was trained on image P (8 099 positive samples and 12 535 negative samples) and tested on image Q. Results: The results are, as expected, worse than when learning data were taken from Q but they can be regarded as almost satisfactory and after some processing they could be used in next stages. We can conclude that the main reason for incorrect classification for image P is that in some areas the background has similar spectral properties to trough areas.
6. In the last experiment we checked the relation between the parameters of the Gabor filter bank and the quality of classification. We increased the number of frequencies to 5 per octave, recommended for some popular frequency-based descriptors, e.g. in SIFT method [11]. The shortest and the longest wavelengths were set to 4 and 50 pixels. Based on Eqs. (5) and (6) there are 19 wavelengths, from 4 to 48.50. The

number of orientations remains 4, so there are 76 filters in the bank in total. Results: The quality of classification is very similar to the result from the filter bank with 8 frequencies (and 32 features) so we can conclude that increasing the number of frequencies is not relevant for discriminative properties of the feature set. Moreover, for wavelength vector: $T = [6.8\ 10\ 15\ 22.8\ 34.2\ 51.2]$ the results remained unchanged, so the highest frequencies in our filter bank that correspond to wavelet lengths 3 and 4.5 do not influence the discriminative hyperplane parameters.

4 Conclusions

In the article we presented two alternative methods for the detection of subsiding troughs in SAR interferograms. This first method detects centres of circular objects in the input image. It is based on the convolution of the image with circular wavelets and does not require any learning set. The second classifies the area of image into trough areas and the background based on Gabor features and it needs at least several learning samples.

The performance, and therefore applicability, of our methods strongly depends on the quality of SAR interferograms. The first method performed very well when tested on an image with distinctive troughs and small amount of noise: all troughs were detected and the percentage of false detection was kept at 20%. These results are sufficient for practical applications, where all troughs found automatically are manually verified, so some false detections are acceptable. However, in tests on low-quality image, with high amount of noise and faint troughs, only 9 out of 13 troughs were detected and the number of false alarms in areas of the image with low-frequency and high-amplitude noise is high.

Similarly, the second method when applied to a high quality image, i.e. image with distinctive troughs and without low-frequency noise, yields acceptable results but in case of low quality input some areas of noise are misclassified as troughs. Experiments with different SVM classifiers (linear and RBF, with different sets of parameters) demonstrate that areas of troughs and some types of noisy background are not separable in the space of Gabor features. In future research the results may be improved by using modified Gabor features that are invariant to rotation, like methods proposed in [12, 13] or [14]. Also, in the future interactive learning based on relevance feedback can be applied. Such methods, which can be used for image retrieval when a small set of learning samples is available were proposed in [15–17].

As a general conclusion, since the quality of SAR interferograms is rather variable, the performance of the proposed methods is insufficient to be used for fully automatic detection of subsiding troughs. Nevertheless, both methods can support manual inspection of large areas in interferograms.

References

1. Weng, Q.: Remote Sensing of Impervious Surfaces. CRC Press Taylor and Francis Group, Boca Raton (2008)
2. Gupta, R.P.: Remote Sensing Geology. Springer, Heidelberg (2003)

3. Przyłucka, M., Herrera, G., Graniczny, M., Colombo, D., Bejar-Pizzaro, M.: Combination of conventional and advanced DInSAR to monitor very fast mining subsidence with TerraSAR-X Data: Bytom City (Poland). Remote Sens. **7**, 5300–5328 (2015)
4. Leśniak, A., Porzycka, S.: Comprehensive interpretation of satellite and surface measurements for hazard assessment on mining and post mining areas. Mineral Resour. Manage. **24**, 147–159 (2008)
5. Tadeusiewicz, R.: Neural networks in mining sciences – general overview and some representative examples. Arch. Min. Sci. **60**, 971–984 (2015)
6. Bała, J., Porzycka-Strzelczyk, S., Strzelczyk, J.: Subsidence troughs detection for SAR images - preliminary results. In: SGEM 2015: Informatics, Geoinformatics and Remote Sensing: 15th International Multidisciplinary Scientific Geoconference, pp. 829–836, Bulgaria (2015)
7. Campbell, F.W., Robson, J.G.: Application of fourier analysis to the visibility of gratings. J. Physiol. **197**, 551–566 (1968)
8. Jain, A.K., Farrokhnia, F.: Unsupervised texture segmentation using Gabor filters. Pattern Recognit. **24**, 1167–1186 (1991)
9. Vyas, V.S., Rege, P.: Automated texture analysis with gabor filters. GVIP J. **1**, 35–41 (2006)
10. Manjunath, B.S., Ma, W.Y.: Texture features for browsing and retrieval of image data. IEEE Trans. Pattern Anal. Mach. Intell. **18**, 837–842 (1996)
11. Lowe, D.G.: Distinctive image features from scale-invariant keypoints. Int. J. Comput. Vis. **60**, 91–110 (2004)
12. Sastry, C.S., Ravindranath, M., Pujari, A.K., Deekshatulu, B.L.: A modified Gabor function for content based image retrieval. Pattern Recognit. Lett. **28**, 293–300 (2007)
13. Rahman, M.H., Pickering, M.R., Frater, M.R.: Scale and rotation invariant Gabor features for texture retrieval. In: International Conference on Digital Image Computing Techniques and Applications (DICTA), pp. 602–607 (2011)
14. Manthalkar, R., Biswas, P.K., Chatterji, B.N.: Rotation invariant texture classification using even symmetric Gabor filters. Pattern Recognit. Lett. **24**, 2061–2068 (2003)
15. Rotter, P.: Relevance feedback based on n-tuplewise comparison and the ELECTRE methodology and an application in content-based image retrieval. Multimed. Tools Appl. **72**, 667–685 (2014)
16. Rotter, P.: Multimedia information retrieval based on pairwise comparison and its application to visual search. Multimed. Tools Appl. **60**, 573–587 (2012)
17. Rotter, P., Skulimowski, Andrzej M.J.: A new approach to interactive visual search with rbf networks based on preference modelling. In: Rutkowski, L., Tadeusiewicz, R., Zadeh, Lotfi A., Zurada, Jacek M. (eds.) ICAISC 2008. LNCS, vol. 5097, pp. 861–873. Springer, Heidelberg (2008). doi:10.1007/978-3-540-69731-2_82

FPGA-Based System for Fast Image Segmentation Inspired by the Network of Synchronized Oscillators

Michal Strzelecki[1]([✉]), Przemyslaw Brylski[1], and H. Kim[2]

[1] Institute of Electronics, Lodz University of Technology,
Wolczanska 211/215, 90-924 Lodz, Poland
michal.strzelecki@p.lodz.pl
[2] Division of Electronics and Information Engineering,
Chonbuk National University, 561-756 Jeonju, Korea
hskim@jbnu.ac.kr

Abstract. This paper presents an FPGA-based system for fast and parallel image segmentation. Implemented segmentation method is inspired by operation of the network of synchronized oscillators - a robust tool for image processing and analysis. The architecture of parallel digital image processor was presented and discussed. It was optimized to enable fully synchronized parallel processing along with reduction of FPGA resources. The developed system is able to analyze both binary and monochrome images with size of 64×64 pixels. It was demonstrated that it can perform region growing image segmentation, edge detection, labelling of binary objects, and basic morphological operations. Sample analysis results were also presented and discussed.

Keywords: Image segmentation · Synchronized oscillator network · FPGA implementation

1 Introduction

This paper describes hardware implantation of the image segmentation technique inspired by oscillator neural network. It is an important part of image processing flow, with a key influence on quantitative results of its further analysis. An important area of image segmentation application is medicine. Medical imaging techniques allow to create visual representations of internal human organs while the segmentation is aimed at detection of lesions or the presence of pathological structures (e.g. tumours) in acquired images. The information obtained by the analysis of detected pathologies makes medical diagnostics more objective; it is also very useful in treatment monitoring and for evaluation of rehabilitation. Due to often necessity of analyzing a large number of images in a limited time (e.g. during screening), it's a need to focus on fast segmentation approaches, like GPU based or implemented in hardware. A segmentation method that is suitable for hardware implementation is a network of synchronized oscillators (SON).

© Springer International Publishing AG 2017
L. Rutkowski et al. (Eds.): ICAISC 2017, Part I, LNAI 10245, pp. 580–590, 2017.
DOI: 10.1007/978-3-319-59063-9_52

Its operation is based on "temporary correlation" theory, which attempts to explain scene recognition as performed by a human brain [1]. This theory assumes that different groups of neural cells code different properties of homogeneous image regions (e.g. texture). Monitoring of temporal activity of cell groups allows for scene segmentation. To implement this theory an oscillator model to emulate brain neural cell and network of connected oscillators for image segmentation was proposed [2,3]. Computer simulations of this network demonstrated that it is a reliable tool for texture segmentation of a wide class of images (including biomedical) as well as for other image processing algorithms (e.g. morphological operation or edge detection) [4]. Improved oscillator model, described in [5] is very suitable for hardware realization to speed up the image segmentation process.

Such a network was already designed as an ASIC CMOS VLSI chip. It contains a matrix of an active image processing elements with 32×32 size. It was demonstrated that image analysis system that implemented such chip was able to segment sample biomedical images [6]. However, the analogue realization suffers of many drawbacks due to differences between network oscillators caused by the scattering of transistors channel length. This leads to the significant changes of oscillator's output characteristics. Another problem is a complex design process and small matrix size which limits the practical applications of the network circuit [6]. Thus a digital implementation of such a network is considered. Compared to analogue form, a FPGA based version of the network would have following advantages: no need for calibration, scalable size of processing matrix, less complicated design process, simpler control mechanism and additionally available information about the number of pixels in each identified object. First attempt of digital SON implementation was presented in [7]. Resulted FPGA circuit allowed segmentation of monochrome images with size of 16×16 pixels. This work presents more mature FPGA based image processor that besides of image segmentation enables object labelling, edge detection along with basic morphological operations. Thanks to optimization of the FPGA resources the network size was increased providing analysis of 64×64 images.

2   Network Architecture

Network of synchronized oscillators was an inspiration to develop an architecture of the parallel digital image processor. Such processor enables image segmentation (based on region growing and edge detection), labelling and selected morphological operations. Main segmentation algorithm is region growing which is a basic operation performed by oscillator network [8]. Figure 1 presents a block diagram of the system for parallel image processing [9]. It contains microcontroller, control unit and a matrix of NxN nodes, where N defines the analyzed image size.

Microcontroller loads the image from the host computer, evaluates node weights and neighborhood mask values, controls the segmentation process and finally delivers processed image to the host computer. Central unit downloads

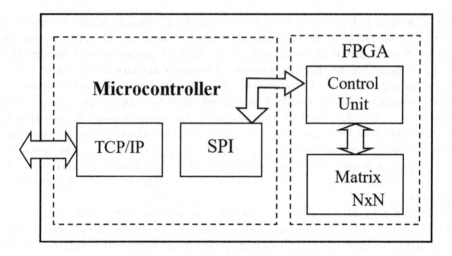

Fig. 1. Block diagram of image processing system

the preprocessed image data from the microcontroller into processing matrix, ensures appropriate clock and control signals and delivers analyzed image back to the microcontroller.

Nodes, which correspond to the image pixels, are combined into a matrix of active elements. These nodes are connected by weights that depend on brightness similarity between adjacent pixels. In the proposed digital network, nodes labeling replaces oscillations. Nodes connected to homogeneous image region possess the same unique label that is assigned to them during iterative network operation. During each propagation cycle, the label is assigned to all these neighbors of given active node (which already has been labelled), for which connection weight is sufficiently high.

Weights are transformed and stored in each particular node in a form of neighborhood mask shown in Fig. 2. The N_ADDR signal selects the direction of label propagation for each particular propagation cycle.

Neighborhood mask is evaluated for every node by the microcontroller based on weights values. It selects possible neighbors to receive its label. Logic '0' value on the position corresponding to local address of node's neighbor excludes it from the process of label propagation. It is due to performing a logical AND operation on the acquired label with mask value for a selected neighbor.

Neighborhood mask allows to store node's local neighborhood data more efficiently. It reduces (to one for each neighbor) the number of necessary registers (bits) for the information storage related to node neighborhood and saves usage of FPGA resources. This solution enables for a very effective implementation in distributed RAM shift registers (an alternative functionality of LUTs).

The node block diagram is shown in Fig. 3. It consists of two parts. The first part (at the top of the figure), is responsible for addressing the local neighborhood and contains the mask register along with integrated multiplexer.

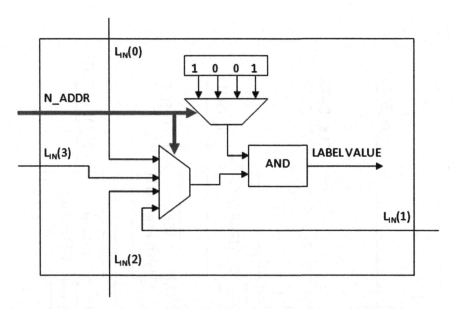

Fig. 2. The idea of the mask register for neighborhood size N = 4

The second part (at the bottom) enables label exchange. It consists of a series of multiplexers that determine the label propagation path. The role of comparator is to ensure that the new label value (if acquired) will be greater than the previously stored in the node. This requires that propagation of label starts from the least significant bit.

As shown in Fig. 1 the microcontroller is connected to the FPGA device via Serial Peripheral Interface (SPI). This interface is also used inside FPGA device to deliver data to each node. One of the advantages of the applied solution is the ability of transferring neighborhood masks to all nodes in the same time. It is possible thanks to cascade connection of nodes. Data are transferred to (i,k) node by shifting via all nodes in the chain. This allows output the segmentation result from the matrix during the transfer of neighborhood masks for the next image. The SPI bus can be effectively implemented in FPGA with usage of distributed RAM shift registers.

A digital image processor performs the following segmentation algorithm [9,10]:

Step 1:
1. Image information is downloaded to the external processor
2. Weights between each node neighbors are computed by the external processor using the formula:

$$W_{ik,n} = \frac{255}{|I_{ik} - I_n| + 1}, \quad n = 1,..,NoN \tag{1}$$

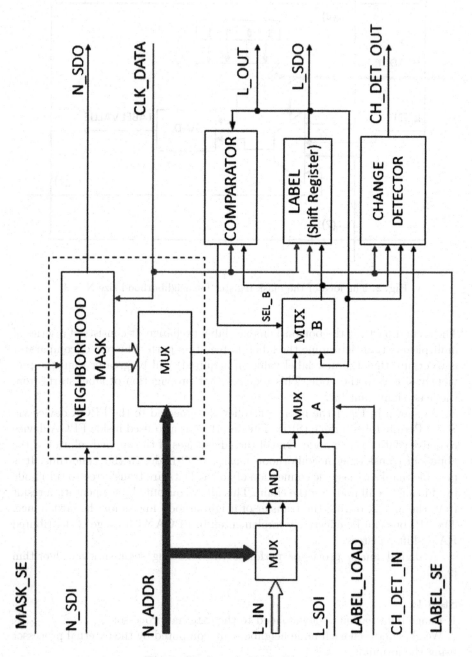

Fig. 3. Single node block diagram

for region growing-based segmentation, and

$$W_{ik,n} = |I_{ik} - I_n|, \quad n = 1, .., NoN \tag{2}$$

for edge detection. In (1) and (2) i,k are node coordinates, NoN is the number of all neighbors connected to each node, I_{ik} represents pixel value in image point (i,k), In is the intensity of the n-th neighbor pixel of node (i,k).
3. Evaluation of neighborhood mask value for each node using the formula

$$M_{ik,n}(W_{ik,n}) = \begin{cases} \text{``1''} & \text{if } W_{ik,n} > T \\ \text{``0''} & \text{if } W_{ik,n} <= T \end{cases} \quad n = 1, .., NoN \tag{3}$$

where i,k are node coordinates, $W_{ik,n}$ is weight value computed for the n-th neighbor pixel of node (i,k), NoN is the number of all neighbors connected to each node, T is predefined threshold.

Step 2:
4. Node initialization performed by the CU according to the formula:

$$L_{ik} = \begin{cases} L+1 & \text{if } M_{ik} \ != 0 \\ 0, & otherwise \end{cases} \tag{4}$$

where i,k are node coordinates, L is the initial label value, M_{ik} is mask value computed for the n-th neighbor pixel of node (i,k).

Node becomes a leader when is connected to at least one neighbor by weight value that satisfies the similarity criterion. It has assigned a unique value of the initial label.
5. Segmentation takes place until there is no label change in any node.

Step 3:
6. Labels L_{ik} of all nodes are transferred to CU.
7. The labeled image is uploaded to the microcontroller using the output bus.

Morphological operations are performed together with segmentation by appropriate defining of neighboring mask M. For example, in the case of erosion, the following equation is applied to check if given pixel (that belongs to binary object) should remain in the image as a results of this operation, assuming that structuring element B is applied.

$$\sum_s I_{ik+s} B_s = N \tag{5}$$

where I_{ik} means image gray level value of pixel (i,k), N is a number of B elements and s defines neighborhood defined by size and shape of B. If (5) is not satisfied, it means that pixel I_{ik} should be not considered in further region growing since it is eroded by the morphological filter. In this case corresponding neighborhood mask elements $M_{ik,s}$ should be equal to zero, otherwise this mask remains unchanged. Analysis of (5) corresponds to defining mask M using the following formula

$$M = \begin{cases} M, & \text{when B = B AND M} \\ 0, & otherwise \end{cases} \tag{6}$$

where B is coded an N-bit word in microcontroller memory.

FPGA-based image processor was implemented XUPV5 board equipped with XC5VLX110T FPGA chip that belongs to Virtex-5 Xilinx family. Processor implements 64×64 matrix, analysis is performed for 8 pixel neighbors, with 12 bit labels. Table 1 presents FPGA resources of developed implementation.

Table 1. Resource allocation for 64×64 image processor based on XC5VLX110T FPGA.

Type of resources	Resources allocated	Resources available	Allocation [%]
SLICES	16343	17280	94,5
6-input LUT	37371	69120	54
As logic circuits	29174	69120	42,2
As shift registers	8194	17920	45,7
SLICE FF	20898	69120	30,2
BRAM (kb)	72	5328	1,3
DCM	2	12	16,6
BUFG	5	32	15,6
IOB	10	640	1,5

Where: SLICES – number of FPGA configuration blocks, 6-input LUT – number of logic function generators/RAM blocks, SLICE FF – number FPA flipflops, BRAM – amount of RAM, DCM – number of clock circuits, BUFG – number of clock buffers, IOB – number of I/O ports.

3   System Validation - Segmentation Results

FPGA-implemented segmentation method was tested on sample images, artificial and biomedical. Segmentation results are shown in Fig. 4.

Images from Fig. 4(a, d) represent optical microscopic skin cell samples used in dermatology for psoriasis assessment. Image from Fig. 4(g) is a numerical phantom used for testing of vessel segmentation techniques on MR angiography [11], since it contains nonhomogeneous intensity artifact along with Gaussian noise (both are typical distortions for ToF MRA images used for visualization of e.g. brain vascular trees). Sample results of morphological operations are shown

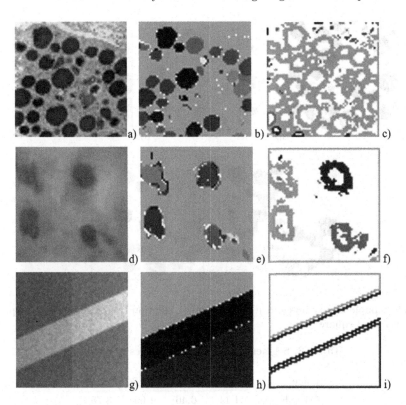

Fig. 4. Segmentation results od sample images (a, d, g): region based (b, e, h), edge detection (c, f, i).

Table 2. Number of resulting objects (labels) and segmentation times for images from Fig. 4.

Image	Fig. 4b	Fig. 4c	Fig. 4e	Fig. 4f	Fig. 4h	Fig. 4i
Number of labels	63	48	77	18	3	3
TMASK [ms]	3.54	1.97	3.37	1.92	2.72	1.92
TP [μs]	63.23	73.53	182.23	50.8	102.33	44.87

in Fig. 5. Image from Fig. 5(a) is a binarized version of skin cells, while Fig. 5(b-e) present results of erosion, dilation, closing and opening, respectively. As a structuring element, the 3×3 square was used.

Numbers of detected objects (in terms of applied labels) along with analysis times of the FPGA chip are presented in Tables 2 and 3. T_{MASK} is the estimation time of neighborhood mask while T_p means segmentation time.

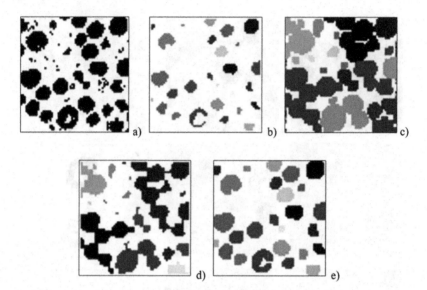

Fig. 5. Sample binary image (a), its erosion (b), dilation (c), closing (d), and opening (e) with 3×3 square.

Table 3. Segmentation times for images from Fig. 5.

Image	Fig. 5b	Fig. 5c	Fig. 5d	Fig. 5e
TMASK [ms]	1.11	6.46	1.69	3.75
TP [µs]	21.07	44.53	39.43	18.52

4 Discussion and Conclusion

Obtained segmentation results, shown in Figs. 4 and 5 confirm correct operation of implemented image analysis algorithms. This regards image segmentation by region growing (Figs. 4(b, e, h)) and by edge detection (Figs. 4(c, f, i)) as well as morphological filtering (erosion, dilation, closing, and opening, as shown in Fig. 5(b-e)). These results depend on threshold value T that is used as a similarity measure between given pixel and their neighbors, as defined in (3). T value (that is arbitrarily selected) influences number of objects detected during segmentation thus also affects analysis time T_p. The number of objects is increasing with rising T. The threshold is increased until maximum number of detected objects is reached maintaining on the same time not fragmented, uniform background. Thus the T value is a comprise between number of correctly detected objects and number of background artifacts (expressed as fragmented background elements that create false objects). Since image objects are analyzed in parallel, segmentation time does not depend on number of detected objects but on object size. Thus the maximum T_pP is defined by the size of largest object in the image since number of region growing iterations increases with object size.

It is also worth to mention that execution of morphological operations does not require any additional time. Morphological filtering is performed together with labelling of objects in the binary image based on appropriately defined neighboring masks, as described in Sect. 2.

Total image analysis time of designed system consists of the following elements: estimation of neighboring mask (T_{MASK}), initialization of leaders and FPGA matrix (T_{LOAD}) and readout of the analyzed image from the FPGA to microcontroller. Times T_{LOAD} and T_{READ} have the same values for any image analysis type. Their value is equal to 2.64 ms and depends on clock of SPI bus (connecting microcontroller and the FPGA) and the size of transferred data. Thus T_{LOAD} and T_{READ} contribute significantly to the total analysis time and are much larger than segmentation time T_p. However, when the sequence of images is analyzed, FPGA initialization for the next analyzed image in the sequence is performed in parallel with transferring of previously processed image to the microcontroller. Then TREAD can be omitted in estimation of total analysis time of the image. T_{MASK} depends on analysis type (on computational complexity of weight estimation algorithm and further weight transformation to neighboring masks M). T_{MASK} achieves the highest values for morphological operations while the lowest for region growing.

To summarize the properties of developed FPGA-based system for image segmentation, one can state that it can perform segmentation based on region growing and edge detection along with selected morphological operations. The system supports binary and monochrome images with size is 64×64 pixels. Minimum size of analyzed object equals 2 pixels while maximum number detected object is 255. Analysis time is defined by the size and shape of the largest image object.

The following topics will be considered in the future improvement of the proposed system:

- increasing image size by application of more advanced FPGA devices;
- enabling color image analysis by updating weight estimation formulas to consider color components;
- improving leader selection procedure by optimizing initial leader label distribution to minimize number of region growing iterations.

References

1. von der Malsburg, C., Buhmann, J.: Sensory segmentation with coupled neural oscillators. Biol. Cybern. **67**(3), 233–242 (1992)
2. Wang, G., Terman, D.: Image segmentation based on oscillatory correlation. Neural Comput. **9**, 805–836 (1997)
3. Cesmeli, E., Wang, D.: Texture segmentation using Gaussian-Markov random fields and neural oscillator networks. IEEE Trans. Neural Networks **12**(2), 394–404 (2001)
4. Strzelecki, M.: Texture boundary detection using network of synchronised oscillators. Electron. Lett. **40**(8), 466–467 (2004)

5. Strzelecki, M., Kowalski, J., Kim, H., Ko, S.: A new CNN Oscillator Model for parallel image segmentation. Int. J. Bifurcat. Chaos **18**(7), 1999–2015 (2008)
6. Kowalski, J., Strzelecki, M., Kim, H.: Implementation of synchronized oscillator circuit for fast sensing and labeling of image objects. Sensors-Basel **11**(4), 3401–3417 (2011)
7. Girau, B., Torres-Huitzi, C.: FPGA implementation of an integrate and fire LEGION model for image segmentation In: European Symposium on Artificial Neural Networks Bruges Belgium, pp. 173-178 (2006)
8. Shareef, N., Wang, D., Yagel, R.: Segmentation of medical images using LEGION. IEEE Trans. Med. Imaging **18**(1), 74–91 (1999)
9. Brylski, P., Strzelecki, M.: FPGA implementation of parallel digital image processor. In: Proceeding of IEEE SPA, 23–25 September 2010, Poznan, Poland, pp. 25–28 (2010)
10. Brylski, P., Strzelecki, M.: Optimisation of the FPGA parallel digital image processor. In: Proceeding of IEEE SPA/NTAV 2012, 27-29 September 2012, Lodz, Poland, pp. 183-188 (2012)
11. Kociski, M., Klepaczko, A., Materka, A., Chekenya, M., Lundervold, A.: 3D image texture analysis of simulated and real-world vascular trees. Comput. Methods Programs Biomed. **107**(2), 140–154 (2012)

From Pattern Recognition to Image Understanding

Piotr S. Szczepaniak and Arkadiusz Tomczyk[(⊠)]

Institute of Information Technology, Lodz University of Technology,
Wolczanska 215, 90-924 Lodz, Poland
{piotr.szczepaniak,arkadiusz.tomczyk}@p.lodz.pl

Abstract. This paper shows the trend in the transformation of the classic image recognition via the interpretation of the image content towards automatic shape and image understanding. The approach presented combines the mechanism proposed by Tadeusiewicz in [1] with the theory of granular computing introduced by Pedrycz in [2]. Its name, active partitions, is related to active contour techniques, from which it originates. It provides the ability to transfer the well-known concepts of object localization from the pixel level to image representations with meaningful image granules. Thus, the approach offers a great potential for the development of human-like image content interpretation.

Keywords: Pattern recognition · Automatic image understanding · Granular computing · Justifiable granularity · Active partitions

1 Introduction

Pattern recognition can be considered as a process which involves the assignment of interpretations to recognized objects. In this work, we will focus on the classification task, in which the interpretation means the assignment of labels from a finite set of labels.

The recognized object may either be perceived as a whole or analyzed with respect to its inner structure. In the latter case, meaningful components of the object must be distinguished. Typical examples of objects with an inner structure include: texts, images, sounds, etc., where the spatial or temporal distribution of the constituent elements is of great importance. The present paper concentrates on images (Fig. 1). However, many of the presented concepts may be applied to other object types.

The analysis of an object structure typically involves substructure localization. The goal of this task, which is closely related to recognition, is to select a subset of structure elements that can be assigned a specific interpretation within a considered domain. Its relationship with the recognition task can be discussed at several levels. Firstly, the interpretation itself, regardless of whether the given substructure represents a semantically important item or not, constitutes a recognition task. It may be aimed at identifying, for example, the image

L. Rutkowski et al. (Eds.): ICAISC 2017, Part I, LNAI 10245, pp. 591–602, 2017.
DOI: 10.1007/978-3-319-59063-9_53

region that represents the interior of the heart ventricle (Fig. 2a). In this case, structure recognition usually translates directly into the interpretation of the whole object. Secondly, the localization task can be formulated as a decision of whether an element of a structure is an element of the substructure that is looked for. For example, the goal of this task might be to determine which pixels represent the interior of the heart ventricle (Fig. 2b).

Recognition requires knowledge which, in the context of the present discussion, can be understood in two different ways. The first type of knowledge is used to select the optimal parameters for the classifier model. It corresponds to the human experience collected while completing previous tasks. In supervised classification, it usually has a form of a training set. Less often it is available as a set of formulas and rules translatable into a machine language (e.g. linguistic description). In unsupervised classification, this type of knowledge is usually reduced to the similarity measure between classified objects. The second type of knowledge refers to additional information about the recognized object. This information, on the one hand, constitutes a context which influences the functioning of the recognition system, and on the other hand, may be interpreted as an element of the recognized structure.

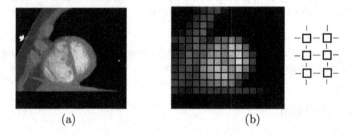

(a)   (b)

Fig. 1. Sample CT heart image: (a) - image, (b) - inner structure of the image: pixels (squares) and neighborhood relationship (line segments). Pixels were enlarged for better legibility.

In this paper, the classic approaches to the above-mentioned problems are discussed. A new group of methods, called active partitions, is then presented. The application of active partitions provides the ability to create systems of automatic image understanding. This approach may be further generalized and interpreted as a possible practical implementation of data analysis based on information granules. The paper is organized as follows: Sect. 2 discusses the basic approaches to pattern recognition; Sect. 3 focuses on the image understanding paradigm which is founded on the concept of cognitive resonance; in Sect. 4 granular computing is described and image information granules are proposed and, finally, in Sect. 5 active partitions are presented. The paper concludes with a short summary.

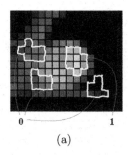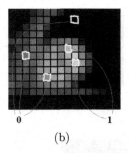

(a)	(b)

Fig. 2. Relationship between localization of a substructure and pattern recognition task ($C = \{0, 1\}$): (a) - substructure classification (label is assigned to image regions), (b) - classification of structure elements (every structure element is labeled separately).

2 Machine Pattern Recognition and Localization

Depending on the way we look upon the object, two approaches to pattern recognition may be distinguished, namely global and structural. In both cases, regardless of whether the inner structure of the object is taken into consideration or not, the recognition process [3–5] can be defined as a transformation which consists of three stages, as presented in Fig. 3.

Thus, the aim is to develop algorithms for implementing transformations T_1 and T_2. Although the process of feature extraction T_1 is also crucial for proper recognition results, this section focuses on the mapping $T_2 : X \rightarrow C$ (decision algorithm, decision rule or recognition mechanism) from feature space X into the decision space C. Function T_2 assigns to object described with features $x \in X$ decision $c \in C$, i.e. $T_2(x) = c$, where $c \in \{0, 1, 2, \ldots, L\}$ is a decision label. It provides an answer to the practical question: *what is (are) the object (objects) presented on the image?*. Interpretation of the numbers depends on the considered task.

The difference between global and structural approach lies in the character of feature space elements. In the case of global recognition, the object is usually described by means of a finite and constant number of attributes (Fig. 3a). It does not matter here whether the inner structure of the object is considered or not. There are many classification techniques that can be applied in this case, e.g., bayesian classifiers, neural networks, support vector machines, etc. [4]. Structural approach may be applied if the elements composing the object are known (Fig. 1b). In this case, feature extraction T_1 usually does not concern the object as a whole, but rather its integral elements and relations between them. Those relations are defined on the basis of element features. However, those features need not be specified explicitly if the corresponding relation is known (e.g. there is no need to know the exact coordinates of the objects position if the knowledge about their relative location is satisfactory). Typical techniques used for such object representation are the approaches based on template matching. These can be single templates or their groups. An example of description of such

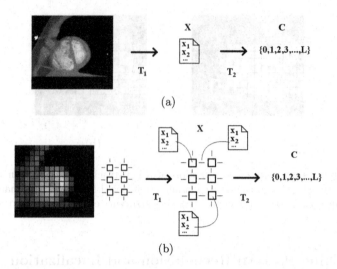

(a)

(b)

Fig. 3. Basic machine pattern recognition scheme with feature extraction T_1 and classification T_2: (a) - global approach, (b) - structural approach.

template groups are formal languages [1] where template matching is performed by parsers corresponding to a given language grammar. In the case of images, the convolutional neural networks are also worthy of noticing [10]. On the one hand, they resemble global techniques in that they treat the input image as a whole. On the other hand, however, in hidden layers the spatial structure of features is revealed.

As mentioned above, the localization task is closely related to the recognition task. Two approaches can be distinguished. The first one requires searching through a set of structure elements to find those subsets that represent semantically important image regions. The decision of whether a given region fulfills the expectations may be made with one of the above-mentioned techniques. However, the analysis of all the subsets, with regard to their number, must take into account additional constraints, reducing the number of possible solutions. At a pixel level, the typical methods used for that purpose are: the classic Hough transform [11] and the active contours. The latter will be described in more detail below. The second approach aims to recognize every element of a structure separately and determine if it is a part of the localized object or the background. Of course, it is necessary to consider a context (other elements, relations) in which those elements are classified. A sample group of methods that can solve this problem is called structured prediction [12]. It includes, among others, the methods that employ probabilistic graphical models [13] and collective classification [14].

3   Image Understanding

So far, image understanding has been assumed to be a human-specific ability to explain and interpret image contents on the basis of image perception,

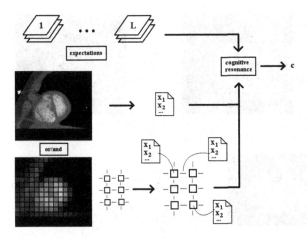

Fig. 4. The main paradigm of image understanding. To properly understand image content a cognitive resonance between image and expectation streams is required.

object recognition, knowledge, experience, reasoning, culture, association, context analysis, etc. Thus, it seems to be a complicated and technically demanding process for which in [15] the term visual understanding is used.

The problem of automatic intelligent analysis and interpretation of medical image contents was addressed by Tadeusiewicz et al. [1] and Ogiela et al. [6–9]. The general idea advocated by those authors is as follows: the input (image) data stream is compared with the stream of demands available from the dedicated source of knowledge (Fig. 4). This is the most obvious difference between this approach and the classic pattern recognition process (Fig. 3). The demands or expectations come from the expert knowledge in a particular field, for example, medical imaging. They are a kind of postulates for the desired values of some selected image features. The semantic interpretation of the image content is validated as possibly true when selected parameters of the examined image have desired values (determined exactly or approximately). When the parameters of the input image are different then this is interpreted as a partial falsification of one of possible hypotheses about the meaning of the image content. During the comparison of the features calculated for the input image and the knowledge-based demands, one can observe an amplification of some hypotheses about the meaning of the image content, while other hypotheses lose their importance. This mechanism is called cognitive resonance.

In structural recognition the hypotheses can be expressed using formal languages which provide the ability to perform linguistic description of image structures. A number of languages, dedicated to diverse medical problems, was already proposed in such domains as: spinal cord diagnosis [6], radiological palm diagnostics [7], coronary artery analysis [8], and bone fractures analysis [9]. This approach, however, requires the design of a grammars with a corresponding, effective parser. This is performed mainly by human experts after a deep analysis

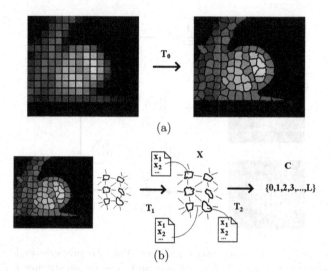

(a)

(b)

Fig. 5. Pattern recognition scheme using information granules. In fact, this scheme does not reflect any new concept. It just considers possibly more informative space X of information granules: (a) - detection of image granules (in this case these are superpixels), (b) - structural pattern recognition scheme at higher level of granulation.

of the problem domain. Automatic creation of such grammars (inference) based on a training set have also been considered but the existing solutions can cope only with relatively simple tasks. Moreover, parsers are prepared for specific types of objects. It means that the analysis does not usually include the whole image but focuses only on the previously identified substructure. The substructure identification process is a separate computation task, if it is to be performed automatically. A natural solution to this kind of problem are the active partitions. Thanks to this method, cognitive resonance is possible not only in the recognition but also in the localization task.

4   Granular Computing

A basic structure describing image content is imposed by a spatial grid of pixels (Fig. 1). However, sometimes this structure is difficult to process using structural recognition methods. This results from the fact that the image is described by a large number of primitive elements (pixels carrying information about their color) which are connected by simple and local neighborhood relations.

In recent years, a considerable amount of research has been conducted into the subject of forming, collecting and machine processing of information granules. The term granule is understood as a group, cluster, or a class of clusters in a universe of discourse. Information granulation of images can be performed at diverse levels. Assuming that pixels create the zero level and that one aims to group the pixels according to their gray level, we can say that the vector composed of pixel coordinates and the value of its gray level is the zero level information. By applying a low-level method to the pixel image, one may detect a set of elements of higher granularity - additional T_0 mapping. Consequently, the search for first level information can be performed. For example, superpixels (small patches) can be detected (Fig. 5a). Other examples will be presented in the next section. Generalization of the granulation for 3D is also possible.

In other words, image granules can be patches, or, more generally, spatial patches [24–26]. A spatial patch can be literally anything from single pixels, through model-based objects, like line segments or circles, to any set of pixels. Moreover, patches may or may not have a direct translation to image pixel space. Nevertheless, recognized objects are considered to be describable by sets of patches. The recognition and localization can be performed then in the set of patches and not in the set of pixels.

Within the granular computing, the theory, methodologies and techniques which deal with the processing of information granules are conceptually located [2, 17]. Moreover, there are two contradictory requirements for an information granule to be considered meaningful [18]:

(a) Experimental evidence. The numeric evidence accumulated within the bounds of the granule has to be as high as possible. In order to make the data set legitimate, the evidence should reflect as big amount of data as possible.
(b) Well-articulated semantics. At the same time, the granule should be as specific as possible. It should carry a well-defined semantics (meaning). The agreement with our perception of knowledge about the problem is desired.

Note that the method of finding a compromise between (a) and (b) is of practical importance [19]. Semantically meaningful granules are very informative. In practice, to find meaningful image granules, one can consider a parametrized operation T_0 and use a training set with segmented images to select appropriate parameters to enable the localized structures to be described as precisely as possible. The formal pattern recognition scheme, involving the identification of image information granules, is shown in Fig. 5b.

The recognition (classification or localization) can be performed using any method which is able to operate on information granules. A powerful approach to this issue is the use of generalization of active partitions because they can operate on elements of higher granularity, while also allowing the implementation of human requirements or expert knowledge into the objective function.

5   Active Partitions

One of the classic object localization techniques at a pixel level of image description are active contours. This group of methods assumes that the space of contours can be defined where a contour uniquely identifies regions in the image. Next, the energy function E is defined. This function is able to evaluate the

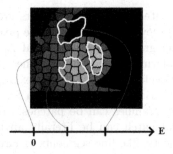

Fig. 6. In active contour as well as in active partition the problem of substructure localization is defined as an optimization problem where energy function E expressing expectations about sought substructure plays are role of objective function.

contour and, consequently, expresses the expectations about the searched structures. With the search space and the objective function defined, the problem of object localization can be formulated. It consists in searching for the optimal argument of the energy function (Fig. 6). Of course, this requires a proper choice of the optimization algorithm - contour evolution.

There are many approaches to practical realization of active contours. In snakes [20], the contour is defined as a parametric curve. Consequently, energy is a functional, in which the optimum can be found using the calculus of variations. This leads to partial derivative equations that, after discretization, can be solved iteratively. In geometric active contours, the contour is defined as a level set of 3D surface [21]. Contour evolution is controlled by forces perpendicular to the local contour orientation. Active shape models [22] define a contour as a set of landmarks and relative distribution of them is trained using examples of correctly segmented images. The contour evolution aims to find optimal landmark positions such that do not violate the trained constraints. Finally, in potential active contours, the contour is located where there is an equilibrium of potential fields generated by sources of two types [23]. To find the optimal parameters describing the position and strength of those sources simulated annealing was used. This short presentation shows the variety of the existing methods.

The knowledge about the object can be encoded in an active contour algorithm in three ways. Firstly, it can be used to select a proper contour model and, consequently, to reduce the considered contour space (e.g. a potential contour can be useful if rounded shapes are expected). Secondly, directly in the energy function, it may provide the ability to express the expectations about the objects (e.g. smoothing components in snakes). Finally, additional soft and hard constraints can be introduced into the optimization process (e.g. model constraints in active shape models). What is important is that knowledge does not need to be connected with image content. In practical applications, in particular in medical ones, it is often the case that the image information is not sufficient for its proper interpretation (e.g. the expected shape and its smoothness must

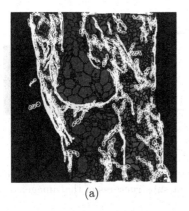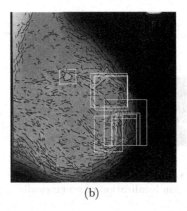

(a)   (b)

Fig. 7. Sample applications of active partitions: (a) - localization of thin, elongated structures in MR knee images (in this case ellipses found with CEC algorithm are image granules), (b) - localization of circular structures in mammograms (here line segments detected with modified LSD algorithm constitute image granules).

be taken into account). An additional knowledge connected with the analyzed image can also be easily utilized to set the parameters of all the above elements.

So far, active contours have usually been defined in cases where the pixel granularity of the image was considered. Then, it was relatively natural to define different contour models on the image plane. Lately, however, a novel active partition approach has been proposed [24–26] which transfers the concepts known from active contours to image content representation at a higher granularity level. The term partition comes from the observation that the contour divides the pixel structure of an image into two partitions - one representing the object of interest and one representing the background. Those partitions evolve in line with contour evolution. The term contour is hardly applicable in the context of image representations based on elements of higher granularity, since the neighborhood relation between granules does not have to be as regular as it is between pixels. Thus, the term *active partition* was proposed instead, in order to emphasize a more general nature of the new approach.

The application of active partitions to different types of image granules has already been discussed in the literature:

- In [24] granularity using superpixels was proposed (Fig. 5). The SLIC algorithm was used [27] to generate this representation.
- In [25] the MR knee images were represented with ellipses generated using the CEC algorithm [28] (Fig. 7a).
- In [26] the content of mammograms was described with line segments generated using a modified LSD algorithm [29] (Fig. 7b).

Those examples prove that active partitions can be successfully applied to different image granules, providing the ability to identify interesting substructures within an image.

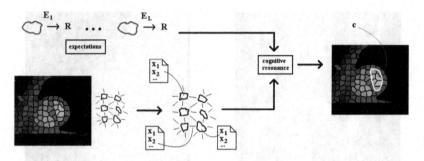

Fig. 8. Cognitive resonance with active partitions allows to understand image content basing on localization of semantically important substructures of that image.

The active partition approach can be used to extend the image understanding scheme presented in Sect. 3. So far, the expectations considered as an information stream required for cognitive resonance have expressed only the expectations about the recognized objects. If those expectations are replaced with energy E function one can speak of cognitive resonance in the localization problem (Fig. 8) and, consequently, active partitions can be considered as an image understanding algorithm.

6 Summary

This paper has presented the active partitions approach which connects the image understanding mechanism proposed in [1] with the theory of granular computing described in [2].

The main postulate of image understanding is the cognitive resonance of two information streams: the first one coming from the image itself (visual information) and the second one representing recognition expectations (domain knowledge and experience of the expert). In [1], a practical realization of this postulate with the use of formal languages able to describe diverse image structures was proposed. Respective parsers were applied to check if a given structure corresponds with the expectations or not. However, this required a prior identification of the recognized structures. In active partitions, the processes of identification (localization) and decision (recognition) are performed simultaneously. The expectations are encoded in the partition model, energy function and evolution scheme. After proper parametrization, those elements need not be designed manually but can be trained automatically on the basis of previously segmented images.

In [2], the term granule is understood as a group of elements in a universe of discourse. Active partition can operate with different representations of image content. Elements of that representation (superpixels, ellipses, line segments, etc.) can be considered as image information granules. Thus, active partition may be seen as a practical realization of the reasoning mechanism within granular computing theory. What is more, since the results of active partition are groups of granules with some semantic interpretations assigned, they can also be regarded as granules. This leads to the concept of hierarchical granular computing, implemented in cognitive hierarchical active partitions [30].

Acknowledgement. This project has been partly funded with support from the National Science Centre, Republic of Poland, decision number DEC-2012/05/D/ST6/03091. The authors would like to also express their gratitude to the Department of Radiology of the Barlicki University Hospital in Lodz for making medical images available.

References

1. Tadeusiewicz, R., Ogiela, M.R.: Medical Image Understanding Technology. Studies in Fuzziness and Soft Computing, vol. 156. Springer-Verlag, Berlin (2004)
2. Pedrycz, W.: Granular Computing in Data Mining. In: Last, M., Kandel, A. (eds.) Data Mining and Computational Intelligence. Springer Verlag, Singapore (2001)
3. Pal, S.K., Mitra, P.: Pattern Recognition Algorithms for Data Mining. Chapman & Hall/CRC, Boca Raton, London, New York, Washington, D.C. (2004)
4. Bishop, C.: Pattern Recognition and Machine Intelligence. Springer, Heidelberg (2006)
5. Maji, P., Pal, S.K.: Rough-Fuzzy Pattern Recognition. Applications in Bioinformatics and Medical Imaging. Wiley, IEEE Press, Hoboken (2012)
6. Ogiela, L., Tadeusiewicz, R., Ogiela, M.R.: Cognitive techniques in medical information systems. Comput. Biol. Med. **38**(4), 501–507 (2008)
7. Ogiela, M.R., Tadeusiewicz, R., Ogiela, L.: Image languages in intelligent radiological palm diagnostics. Pattern Recogn. **39**(11), 2157–2165 (2006)
8. Ogiela, M.R., Tadeusiewicz, R.: Syntactic reasoning and pattern recognition for analysis of coronary artery images. Int. J. Artifi. Intell. Med. (Elsevier) **26**(1–2), 145–159 (2002)
9. Tadeusiewicz, R., Ogiela, M.R.: Medical pattern understanding based on cognitive linguistic formalisms and computational intelligence methods. In: Wang, J. (ed.) 2008 IEEE World Congress on Computational Intelligence WCCI, pp. 1729–1733. IEEE Piscataway (2008)
10. LeCun, Y., Bengio, Y.: Convolutional networks for images, speech, and time-series. In: Arbib, M.A. (ed.) The Handbook of Brain Theory and Neural Networks. MIT Press, Cambridge (1995)
11. Hough, P.V.C.: Method and means for recognizing complex patterns, U.S. Patent 3,069,654 (1962)
12. Nowozin, S., Gehler, P.V., Jancsary, J., Lampert, C.: Advanced Structured Prediction. The MIT Press, Cambridge (2014)
13. Koller, D., Friedman, N.: Probabilistic Graphical Models. Principles and Techniques. The MIT Press, Cambridge (2009)
14. Sen, P., Namata, G., Bilgic, M., Getoor, L., Galligher, B., Eliassi-Rad, T.: Collective Classification in Network Data. AI Mag. **29**(3), 93–106 (2008)
15. Les, Z., Les, M.: Shape Understanding System. SCI, vol. 588. Springer, Cham (2015)
16. Tadeusiewicz, R., Szczepaniak, P.S.: Basic concepts of knowledge-based image understanding. In: Nguyen, N.T., Jo, G.S., Howlett, R.J., Jain, L.C. (eds.) KES-AMSTA 2008. LNCS, vol. 4953, pp. 42–52. Springer, Heidelberg (2008). doi:10.1007/978-3-540-78582-8_5

17. Lin, T.Y., Yao, Y.Y., Zadeh, L.A. (eds.): Data mining, rough sets and granular computing. Physica-Verlag, Berlin (2002)
18. Pedrycz, W., Al-Hamouz, R., Morfeq, A., Balamash, A.: The design of free structure granular mappings: the use of the principle of justifiable granularity. IEEE Trans. Cybern. (2013)
19. Szczepaniak, P.S.: Interpretation of image segmentation in terms of justifiable granularity. In: Rutkowski, L., Korytkowski, M., Scherer, R., Tadeusiewicz, R., Zadeh, L.A., Zurada, J.M. (eds.) ICAISC 2015. LNCS, vol. 9119, pp. 638–648. Springer, Cham (2015). doi:10.1007/978-3-319-19324-3_57
20. Kass, M., Witkin, W., Terzopoulos, S.: Snakes: active contour models. Int. J. Comput. Vis. 1(4), 321–333 (1988)
21. Caselles, V., Kimmel, R., Sapiro, G.: Geodesic active contours. Int. J. Comput. Vis. 22(1), 61–79 (2000)
22. Cootes, T., Taylor, C., Cooper, D., Graham, J.: Active shape models - their training and application. CVGIP Image Underst. 61(1), 8–59 (1994)
23. Tomczyk, A., Szczepaniak, P.S.: Adaptive potential active contours. Pattern Anal. Appl. 14, 425–440 (2011)
24. Tomczyk, A., Szczepaniak, P.S.: Knowledge based active partition approach for heart ventricle recognition. In: 10th International Conference on Computer Recognition Systems, CORES (2017, in press)
25. Tomczyk, A., Spurek, P., Podgórski, M., Misztal, K., Tabor, J.: Detection of elongated structures with hierarchical active partitions and CEC-based image representation. In: Burduk, R., Jackowski, K., Kurzyński, M., Woźniak, M., Żołnierek, A. (eds.) Proceedings of the 9th International Conference on Computer Recognition Systems CORES 2015. AISC, vol. 403, pp. 159–168. Springer, Cham (2016). doi:10.1007/978-3-319-26227-7_15
26. Jadczyk, M., Tomczyk, A.: Object localization using active partitions and structural description. In: Rutkowski, L., Korytkowski, M., Scherer, R., Tadeusiewicz, R., Zadeh, L.A., Zurada, J.M. (eds.) ICAISC 2015. LNCS (LNAI), vol. 9119, pp. 727–736. Springer, Cham (2015). doi:10.1007/978-3-319-19324-3_65
27. Achanta, R., Shaji, A., Smith, K., Lucchi, A., Fua, P., Susstrunk, S.: SLIC superpixels compared to state-of-the-art superpixel methods. IEEE Trans. Pattern Anal. Mach. Intell. 34(11), 2274–2281 (2012)
28. Tabor, J., Spurek, P.: Cross-entropy clustering. Pattern Recogn. 47(9), 3046–3059 (2014)
29. von Gioi, R.G., Jakubowicz, J., Morel, J.-M., Randall, G.: LSD: a line segment detector. Image Process. Line 2, 35–55 (2012)
30. Tomczyk, A., Szczepaniak, P.S., Pryczek, M.: Cognitive hierarchical active partitions in distributed analysis of medical images. J. Ambient Intell. Humanized Comput. 4(3), 357–367 (2012). open access, Springer

Linguistic Description of Color Images Generated by a Granular Recognition System

Krzysztof Wiaderek[1]([⊠]), Danuta Rutkowska[1,2],
and Elisabeth Rakus-Andersson[3]

[1] Institute of Computer and Information Sciences,
Czestochowa University of Technology, 42-201 Czestochowa, Poland
{krzysztof.wiaderek,danuta.rutkowska}@icis.pcz.pl
[2] Information Technology Institute,
University of Social Sciences, 90-113 Lodz, Poland
[3] Department of Mathematics and Natural Sciences, Blekinge Institute
of Technology, 37179 Karlskrona, Sweden
elisabeth.andersson@bth.se

Abstract. The paper proposes a new method employed in an intelligent
pattern recognition system that generates linguistic description of color
digital images. The linguistic description is produced based on fuzzy rules
and information granules concerning colors as most important among
image attributes. With regard to the color, the CIE chromaticity color
model is applied, with the concept of fuzzy color areas. The linguistic
description uses information about location of color granules in input
images.

Keywords: Image recognition · Information granulation · Linguistic
description · Fuzzy sets · Knowledge-based system · CIE chromaticity
color model

1 Introduction

Linguistic description of a color image – that can be produced by an intelligent
system – may be very useful in many applications. Therefore, such a system with
a new method of image analysis and inference is proposed in this paper. The
idea of the system refers to the granular pattern recognition system (GPRS),
introduced in [22] and developed in [23], as well as our previous articles [20,21]
where the granulation approach is presented.

The GPRS system uses information granules that can be created by means
of fuzzy sets [24] or rough sets [7]. For details, see [22] where color, location, size,
and shape granules are considered within the concept of the object information
granule (OIG). The color attribute is treated as most important and others
(location, size, shape) may exist along with the color granule in an OIG.

Color is a very important attribute of digital images. It carries significant
information that helps to distinguish, recognize, compare, and classify different

© Springer International Publishing AG 2017
L. Rutkowski et al. (Eds.): ICAISC 2017, Part I, LNAI 10245, pp. 603–615, 2017.
DOI: 10.1007/978-3-319-59063-9_54

pictures or objects presented on various images. As a matter of fact, color should be considered as a triplet, i.e. hue (pure color), saturation, and lightness. In spite of the fact that the word "color" is commonly used as a synonym of "hue" it is worth emphasizing that "hue" means the pure color. Thus, hue is one of the main properties of a color. Saturation (also called chroma) and lightness (also called brightness, value, or tone) are two additional properties of a color. Hue is the term for the pure spectrum of colors that appear in the rainbow as well as in the visible spectrum of white light separated by a prism. Usually, colors with the same hue are distinguished with descriptive adjectives such as "light blue", "pastel blue", "vivid blue", "dark blue", which refer to their lightness and/or chroma (saturation). Exceptions include "brown" which is a dark "orange", and "pink" that is a light red with reduced chroma.

Theoretically all hues can be mixed from three basic hues, known as primaries. There are different definitions of the primary colors, i.e. painters primaries, printers primaries, and light primaries; for details, see e.g. [1]. The well known RGB (red, green, blue) refers to the application in computer screens where colored light is mixed. If all three light primaries are mixed the theoretical result is white light. The RGB is an additive color model, combining red, green, and blue light. In computers the RGB color model is used in numerical color specifications.

It has been observed that the RGB colors have some limitations. The RGB is hardware-oriented and non-intuitive which means that people can easily learn how to use the RGB but they rather think of hue, saturation and lightness, and how to translate them to the RGB.

Two most common representations of points in the RGB color model, based on hue, saturation and lightness, are HSL and HSV. The former stands for "hue", "saturation", and "lightness", while the latter for: "hue", "saturation", and "value". The HSV is also called HSB (where B stands for "brightness"). A third model, common in computer vision applications, is HSI, for "hue", "saturation", and "intensity". The HSV, HSB, HSL color models are slight variations on the HSI theme.

In a color space, colors can be identified numerically by their coordinates. There are precise rules for converting between the HSL and HSV spaces, defined as mappings of the RGB. The conversion between them should remain the same color; however it is not always true with regard to different color spaces (e.g. RGB to CMYK that is a subtractive color model, used in color printing). Since RGB and CMYK are both device-dependent spaces, there is no simple or general conversion formula that converts between them. The CMYK color model is based on the printers primaries, i.e. cyan, magenta, and yellow. In addition, the key (black) component is used. Color printing typically employ ink of the four colors (including black). Mixing the three printers primaries theoretically results in black, but imperfect ink formulations do not give true black, which is why the additional key component is needed. It is worth noticing that secondary mixtures of the CMY primaries (cyan, magenta, yellow) results in red, green, blue.

It should be emphasized that the RGB model is usually employed for production of colors while the HSI for description of colors. Conversion between RGB and HSI is also possible; see e.g. [3,4].

2 Color Granules in the CIE Chromaticity Diagram

In this paper, we focus our attention on the color areas of the CIE chromaticity triangle (diagram), viewed as fuzzy regions (fuzzy sets) characterized by membership functions [20,21]. The CIE color model was developed to be completely independent of any device or other means of emission or reproduction and is based as closely as possible on how humans perceive color. This model was introduced in 1931 by the CIE that stands for Commission Internationale de l'Eclairage (International Commission on Illumination). The original CIE 1931 color space was updated in 1960 and 1976 for practical reasons.

For our consideration and applications, the CIE 1931 chromaticity triangle (Fig. 1) is suitable. In addition, the color areas are depicted in the figure, and denoted by numbers 1,2, ... ,23 where two of them are divided for smaller regions (1a, 1b, 1c, and 7a, 7b) for computational convenience. Thus, in the next sections, especially in the example described in Sect. 5, we always use 26 color areas (granules).

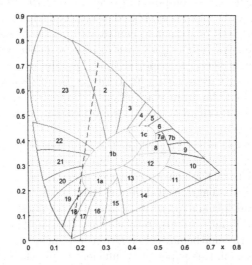

Fig. 1. The CIE chromaticity diagram (Color figure online)

The CIE chromaticity diagram represents the mapping of human color perception in terms of two CIE parameters x and y, called the chromaticity coordinates, which map a color with respect to hue and saturation. Color names have been assigned to different regions of the CIE color space (chromaticity triangle) by various researchers; see e.g. [2,3]. These are approximate colors that represent vague categories, and not to be taken as precise statements of color. Therefore, we can treat them as fuzzy sets, and boundaries between the regions may be viewed as not crisp but belonging to the distinct areas with a certain membership value.

With regard to the color areas, we apply the original CIE chromaticity triangle, with the labelled regions presented in [2,3]. Thus, we employ 23 fuzzy regions of the CIE color space, associated with the following colors (hues): white, yellowish green, yellow green, greenish yellow, yellow, yellowish orange, orange, orange pink, reddish orange, red, purplish red, pink, purplish pink, red purple, reddish purple, purple, bluish purple, purplish blue, blue, greenish blue, bluegreen, bluish green, green (see also [20,21]). As mentioned earlier, for computational convenience, we use 26 color regions $C_1, C_2, ..., C_{26}$ that are fuzzy sets.

The CIE chromaticity diagram shows the range of perceivable hues for the normal human eye. We can say that the chromaticity diagram plots the entire gamut of human-perceivable colors by their x, y coordinates. The inverted-U shaped locus boundary (that is the upper part of the horseshoe shaped boundary) represents spectral colors (wavelengths in nm). The lower-bound of the locus is known as the "line of purples" and represents non-spectral colors obtained by mixing light of red and blue wavelengths. Colors on the periphery of the locus are saturated, and become progressively desaturated in the direction towards white somewhere in the middle of the plot.

Color digital images are composed of pixels. Using the RGB color model in computers, the color of a pixel is expressed as an RGB triplet (r, g, b) where each of the components (RGB coordinates) can vary from zero to a defined maximum value (e.g. 1 or 255). An RGB triplet (r, g, b) represents the 3-dimensional coordinate of the point of the given color within the cube created by 3 axes (red, blue, and green) with values within $[0, 1]$ range. In this model, every point in the cube denotes the color from black $(0, 0, 0)$ to white $(1, 1, 1)$. The triplets (r, g, b) are viewed as ordinary Cartesian coordinates in a euclidean space.

The (r, g, b) coordinates can be transformed into the CIE chromaticity triangle, i.e. to the color areas located on the 2-dimensional space (of the CIE diagram) with (x, y) coordinates. Detailed information concerning the transformation from RGB to XYZ space and vice versa as well as color gamut representation in the CIE diagram of different RGB color spaces is presented e.g. in [6]. Mathematical formulas describing the transformation from XYZ space to xyY and can be found in many publications, e.g. [3]. The transformation is also explained and the mathematical equations are included in [19].

For considerations in this paper, it is sufficient to use the following equations

$$x = f_1(r, g, b), \quad y = f_2(r, g, b) \tag{1}$$

which in this general form describe the transformation from the RGB color space (3-dimensional) to the 2-dimensional xy space of the CIE chromaticity diagram. As a matter of fact, the chromaticity coordinates x and y are used in the xyY space, in the CIE color model, where the brightness parameter Y is a measure of luminance [3]. Of course, for the calculations we employ the precisely defined functions (1), presented in the publications cited above.

It is worth emphasizing that chromaticity is an objective specification of the quality of a color regardless of its luminance. This means that the CIE diagram removes all intensity information, and uses its two dimensions to describe hue and saturation.

Knowing the functions (1), we can transform each triplet (r, g, b) associated with particular pixels of a digital color image to the CIE chromaticity triangle (the gamut). In this way, we can assign a proper color area (Fig. 1) of the CIE diagram to every pixel of the image.

It should be emphasized that the main advantage of using the CIE color model is the fuzzy granulation of the color space, so we can employ the granular recognition system introduced in [22] and developed in [23]. It is also important that the CIE color model is suitable from artificial intelligence point of view because the intelligent recognition system should imitate the way of human perception of colors.

3   Image Recognition by the Granular System

By the granular recognition system we understand the system considered in [22, 23] based on the granulation approach introduced in [20, 21]. However, in this paper we use the system in a different way. Instead of recognizing a picture from a collection of color digital images, now we want to employ the system in order to generate linguistic description of particular pictures as well as concerning the collection of images.

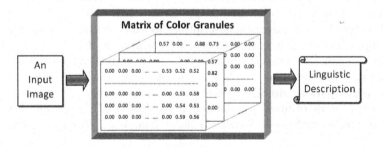

Fig. 2. Illustration of the process of generating linguistic description of an input image

The granular system – for generating linguistic description of color images – recognizes pixels that belong to the color granules of the CIE chromaticity triangle. The color granules are viewed as fuzzy sets defined by their membership functions (see [20, 21]). Thus, for every pixel of an input image, the system determines membership values that is the degree of membership to the color granules of the CIE chromaticity triangle. In this way, a matrix with elements representing the membership values is produced by the system (see Fig. 3 in the next section).

4   Location of the Color Granules in Input Images

Based on the membership matrix (Fig. 3) the granular system (see Sect. 3) can recognize location of the color granules in the input image. According to [20,21], the location is considered with regard to fuzzy approach. This means that fuzzy regions of an image are defined by use of fuzzy sets represented by fuzzy macropixels introduced in [20].

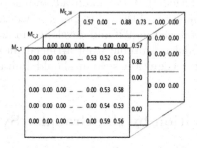

Fig. 3. Matrix of membership values of color granules in an input image

Let us assume that the matrix portrayed in Fig. 3 concerns the input image presented in Fig. 5(a). Each matrix $M_{C\_1}, M_{C\_2}, ..., M_{C\_26}$ includes elements equal to membership values of the pixels to fuzzy color granules $C_1, C_2, ..., C_{26}$, respectively, of the CIE chromaticity triangle. Let us notice that values 0.00 denote zero membership to the color granule C_k, for $k = 1, 2, ...26$.

With regard to the linguistic description, we can calculate participation rate of the input image pixels in the particular color granule $(C_k; k = 1, 2, ...26)$. In addition, we can also count the pixels of the color granules in smaller regions of the image, e.g. in the areas of $1/9$ of the whole image as shown in Fig. 5(b).

The participation rate of the pixels concerning the particular color granules is expressed by use of fuzzy sets defined as illustrated in Fig. 4 where VS, S, M, B, VB denote *Very Small, Small, Middle, Big, Very Big*, respectively.

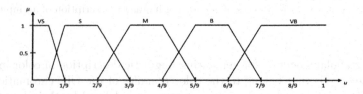

Fig. 4. Membership functions of fuzzy sets VS, S, M, B, VB

The same or similar membership functions may be employed for fuzzy location of macropixels in the image [21]. The areas of $1/9$ of the whole image may be viewed as a special case of large macropixels. Each of them can be associated with the following labels: *Left Upper*, *Middle Upper*, *Right Upper*, *Left Central*, *Middle Central*, *Right Central*, *Left Down*, *Middle Down*, *Right Down* [20], according to the upper, middle (central), down, left, and right parts of the image shown in Fig. 5(b).

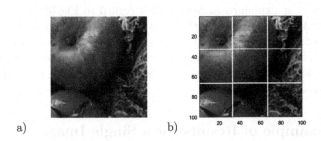

a)   b)

Fig. 5. (a) An input image (b) Distinct location regions in the input image

5 Linguistic Description Produced by the Granular System

The knowledge concerning location of the color granules in input images (see Sect. 4) can be presented by the system in the form of linguistic description. The process of generating the description is performed based on the matrix shown in Fig. 3 as illustrated in Fig. 2.

Each matrix $M_{C\_1}, M_{C\_2}, ..., M_{C\_26}$ is visualized as an image of pixels of particular color granules $C_1, C_2, ..., C_{26}$. This is portrayed in Fig. 6 for the example of input image presented in Fig. 5(a). As we see in the figure, not every color granule is visible; this means that some of them correspond to zero matrix M_{C_k}, for specific $k = 1, 2, ...26$. In this case, the visualizations from number 13 to 22 are empty; thus there are not pixels of the colors, e.g. *Purplish pink* $- C_{13}$, *Blue* $- C_{19}$, in the input image.

Apart from the information concerning empty visualizations (zero matrix M_{C_k}), the system can generate description about participation rate of the input image pixels in the particular color granule, for the case of non empty matrix; see Sect. 4. Moreover, linguistic description with regard to the location and size of the color groups of pixels can be produced by the system.

The participation rate of pixels in color granules may be expressed by linguistic values associated with membership functions of fuzzy sets as illustrated in Fig. 4. According to Fig. 2, the linguistic description is produced based on the information included in each matrix $M_{C\_1}, M_{C\_2}, ..., M_{C\_26}$. With regard to the participation rate, we take into account only the membership values equal or greater then 0.5 as elements of the matrix M_{C_k}, for $k = 1, 2, ...26$. We ignore the pixels with membership values less then 0.5 as located beyond the boundaries of the color region C_k in the CIE chromaticity triangle (Fig. 1). As a matter of fact, we consider the fuzzy color granules with the α-cut level equal to 0.5; see e.g. [14]. The number of elements of M_{C_k} equal or greater then 0.5 determines the participation rate as crisp value that is matched to the proper fuzzy set from Fig. 4. In this way, the linguistic label of the fuzzy set expresses the description of the participation rate of the pixels in the color granule.

Concerning the location and size of the image color granules, and the linguistic description, as explained in Sect. 4 with regard to Fig. 5(b), we can employ the linguistic labels corresponding to the upper, middle (central), down, left, and right parts of the image. We may consider the image areas composed of more then one of the nine parts of the picture. In this way, we can describe bigger regions, e.g. left part of the image. This is illustrated in the example presented in the next section.

6   An Example of Results for a Single Image

For the example of the input image shown in Fig. 5(a), the visualizations of the M_{C_k}, for $k = 1, 2, ...26$, are portrayed in Fig. 6. As we see, the color "Yellowish green" is located in the right part of the image. This means that this color appears in the *Right Upper, Right Central, Right Down* regions.

Similarly, we can say that the color "Yellow green" is located in the right part of the image, and partially in the *Middle Down*. Moreover, the system may infer the conclusion that both colors "Yellowish green" and "Yellow green", viewed as one color granule, approximately cover the right part of the picture.

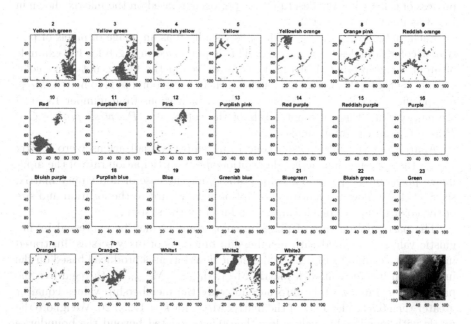

Fig. 6. Color granules in the input image (Color figure online)

This inference is realized by use of the operation of union of fuzzy sets (fuzzy color granules). The inferred region corresponds to the area where we see the cabbage in Fig. 5.

Analogously, the system generates the description concerning color "Red" in the *Left Down* corner and partially in the *Middle Upper* region. However, in this case, pixels of the red color are located in separate clusters of different sizes. The smaller one can be combined in one color granule ("Red" and "Pink") located in the *Middle Upper* area.

The participation rate of the pixels in a color granule gives information about concentration of the color clusters in the visualizations shown in Fig. 6. It is obvious that the bigger participation rate in a smaller region the more concentrated is the color cluster. According to this rule, the system generates linguistic description of the color granules in the picture.

Of course, the empty visualizations produce information: No colors "Purplish pink", "Red purple", "Reddish purple", "Purple", "Bluish purple", "Purplish blue", "Blue", "Greenish blue", "Bluegreen", "Bluish green" in the image. This is already mentioned in Sect. 5.

The linguistic description generated by the system (Fig. 2), for the input image presented in Fig. 5(a), is of the following form

LD I – The picture contains:

1. A medium size granule of color "Yellowish green" and "Yellow green", located in the right part.
2. A medium size granule of color "Red" in the *Left Down* corner.
3. A small granule of color "Red" and "Pink" in the *Middle Upper* part.
4. A low concentrated granule of colors "Yellow", "Yellowish orange", "Orange pink", "Reddish orange", "Orange", in the left middle part.
5. A medium concentrated granule of color "White", in the upper part, and low concentrated in the right and down parts.

A linguistic summarization inferred from the above description can be formulated as follows

LD II – The picture includes two main ares of color granules:

1. "Yellowish green" and "Yellow green", in the right part.
2. "Yellow", "Yellowish orange", "Orange pink", "Reddish orange", "Orange", "Red", and "Pink", in the left middle part.

The linguistic description LD I.1 corresponds to LD II.1. The conclusion LD II.2 is inferred from LD I.2–4, by use of the union operation of fuzzy granules. Ignoring the small granule of color "Greenish yellow", as well as very small granules of color "Purplish red", and the "White" color, we can say that LD II describes main features of the example image shown in Fig. 5(a). As we see, the system presented in Fig. 2 realizes a clustering algorithm discovering groups of clusters of similar colors in an input image. LD II.1 refers to the area of colors close to green (rather cold color), LD II.2 – warm colors.

The information in the form of LD II (summarization) is sufficient for many tasks of image recognition, and recovering a picture from a collection of images [20–23].

7 Directions of Further Research

7.1 Inference and Linguistic Description for a Collection of Images

The linguistic descriptions presented in Sects. 5 and 6 concern a single input image. The system illustrated in Fig. 2 can also generate similar descriptions for a collection of images. In this case, for each picture from the collection the matrix of color granules (Fig. 3) is produced and visualized like in Fig. 6. This may be realized in parallel. Then, the system makes two-step inference: First – for every single image, in the way explained in Sects. 5 and 6; Second – aggregation of the results obtained from the first step.

The inferred conclusion may include e.g. the following linguistic description: There are *many* pictures with *big* granules of color "Blue" in upper or down parts. Of course, the words *many* and *big* are linguistic values of fuzzy sets. The inference is realized according to fuzzy IF-THEN rules.

7.2 Inference Concerning Shape Granules

In order to recognize shape of color granules in an input image, macropixels of different sizes [21] must be considered. With regard to the example shown in Fig. 5, each of the nine parts of the picture 5(b) can be divided for nine smaller macropixels, and so on until the smallest size, depending on the image resolution.

7.3 Image Understanding

The linguistic description generated by the system presented in this paper is a first step to study the problem of understanding an image – described by words and sentences of natural language. The theory of computing with words (and perceptions), introduced by Zadeh [25], should be employed along with the conception of fuzzy and rough granulation [5, 8–11, 15, 26].

Interesting publications concerning image understanding [16, 17] refer to medical images (that are not color pictures) as well as to semantic content. Another aspect of image understanding and semantics is discussed in [18].

With regard to the color digital images, and the information granules considered in [22], as well as in [9], relations between the granules should be studied and developed within the framework of fuzzy and rough sets theory. Specific membership functions and inference rules must be defined; methods presented in [12, 13] may be applied.

7.4 Luminance

Further research may also concern the third dimension of the CIE chromaticity triangle, that is the luminance. As explained in Sect. 2, the CIE diagram represents the mapping of human color perception in terms of two chromaticity coordinates x and y that correspond to hue and saturation.

The CIE color model is a color space that separates the three dimensions of color into one luminance dimension and a pair of chromaticity dimension. In this paper, we apply the CIE diagram with two chromaticity coordinates x and y, and this is sufficient for the problem under consideration. However, it is also important to employ the intelligent granular recognition system in order to solve tasks where the CIE color space with the luminance dimension is applied. In [21] fuzzy granulation of the three-dimensional xyY space is introduced. In this case, instead of 23 fuzzy regions (color granules) of the CIE color space presented in Sect. 2, we have e.g. 23 times 3 granules with distinct fuzzy values of the luminance (low, medium, high).

8 Conclusions and Final Remarks

The linguistic description of a color image, presented in this paper, is produced by the granular recognition system that is a knowledge based system. The inference is realized by use of fuzzy IF-THEN rules that represent knowledge concerning color granules in the CIE chromaticity color model.

The system does not recognize specific objects in images but generates a description about color granules located in a picture. The description uses natural language with words representing linguistic values of fuzzy sets that define color granules and they location in the image.

As mentioned in Sect. 7.1, the use of the CIE chromaticity color model allows to solve the problems presented in this paper, and [20–23], in the way of parallel processing. This is very important from the computational point of view, and can be employed in various levels of the image processing.

References

1. Briggs, D.: The Dimensions of Colour (2012). http://www.huevaluechroma.com
2. Fortner, B.: Number by color, Part 5. SciTech J. **6**, 30–33 (1996)
3. Fortner, B., Meyer, T.E.: A Guide to Using Color to Undersdand Technical Data. Springer, New York (1997)
4. Moeslund, T.B.: Building Real Systems and Applications. Springer, London (2012)
5. Pal, S.K., Meher, S.K., Dutta, S.: Class-dependent rough-fuzzy granular space, dispersion index and classification. Pattern Recogn. **45**, 2690–2707 (2012)
6. Pascale, D.: RGB Coordinates of the Macbeth ColorChecker. The BabelColor Company (2006). http://www.babelcolor.com
7. Pawlak, Z.: Theoretical Aspects of Reasoning About Data. Kluwer Academic Publishers, Dordrecht (1991)

8. Pawlak, Z.: Granularity of knowledge, indiscernibility and rough sets. In: Fuzzy Systems Proceedings, IEEE World Congress on Computational Intelligence, vol. 1, pp. 106–110 (1998)
9. Pedrycz, W., Vukovich, G.: Granular computing in pattern recognition. In: Bunke, H., Kandel, A. (eds.) Neuro-Fuzzy Pattern Recofnition, pp. 125–143. World Scientific, Singapore (2000)
10. Pedrycz, W., Park, B.J., Oh, S.K.: The design of granular classifiers: a study in the synergy of interval calculus and fuzzy sets in pattern recognition. Pattern Recogn. **41**, 3720–3735 (2008)
11. Peters, J.F., Skowron, A., Synak, P., Ramanna, S.: Rough sets and information granulation. In: Bilgiç, T., Baets, B., Kaynak, O. (eds.) IFSA 2003. LNCS, vol. 2715, pp. 370–377. Springer, Heidelberg (2003). doi:10.1007/3-540-44967-1_44
12. Rakus-Andersson, E.: Fuzzy and Rough Techniques in Medical Diagnosis and Medication. Springer, Heidelberg (2007)
13. Rakus-Andersson, E.: Approximation and rough classification of letter-like Polygon shapes. In: Skowron, A., Suraj, Z. (eds.) Rough Sets and Intelligent Systems, pp. 455–474. Springer, Heidelberg (2013)
14. Rutkowska, D.: Neuro-Fuzzy Architectures and Hybrid Learning. Springer, Heidelberg (2002)
15. Skowron, A., Stepaniuk, J.: Information granules: towards foundations of granular computing. Int. J. Intell. Syst. **16**(1), 57–85 (2001)
16. Tadeusiewicz, R., Ogiela, M.R.: Why automatic understanding? In: Beliczynski, B., Dzielinski, A., Iwanowski, M., Ribeiro, B. (eds.) ICANNGA 2007. LNCS, vol. 4432, pp. 477–491. Springer, Heidelberg (2007). doi:10.1007/978-3-540-71629-7_54
17. Tadeusiewicz, R., Ogiela, M.R.: Semantic content of the images. In: Image Processing and Communications Challenges, pp. 15–29. Academic Publishing House EXIT, Warsaw (2009)
18. Wei, H.: A bio-inspired integration method for object semantic representation. J. Artif. Intell. Soft Comput. Res. **6**(3), 137–154 (2016)
19. Wiaderek, K.: Fuzzy sets in colour image processing based on the CIE chromaticity triangle. In: Rutkowska D., Cader A., Przybyszewski K. (eds.) Selected Topics in Computer Science Applications, pp. 3–26. Academic Publishing House EXIT, Warsaw, Poland (2011)
20. Wiaderek, K., Rutkowska, D.: Fuzzy granulation approach to color digital picture recognition. In: Artificial Intelligence and Soft Computing, Part I. LNAI, vol. 7894, pp. 412–425. Springer, Heidelberg (2013)
21. Wiaderek, K., Rutkowska, D., Rakus-Andersson, E.: Color digital picture recognition based on fuzzy granulation approach. In: Rutkowski, L., Korytkowski, M., Scherer, R., Tadeusiewicz, R., Zadeh, L.A., Zurada, J.M. (eds.) ICAISC 2014. LNCS, vol. 8467, pp. 319–332. Springer, Cham (2014). doi:10.1007/978-3-319-07173-2_28
22. Wiaderek, K., Rutkowska, D., Rakus-Andersson, E.: Information granules in application to image recognition. In: Rutkowski, L., Korytkowski, M., Scherer, R., Tadeusiewicz, R., Zadeh, L.A., Zurada, J.M. (eds.) ICAISC 2015. LNCS, vol. 9119, pp. 649–659. Springer, Cham (2015). doi:10.1007/978-3-319-19324-3_58
23. Wiaderek, K., Rutkowska, D., Rakus-Andersson, E.: New algorithms for a granular image recognition system. In: Rutkowski, L., Korytkowski, M., Scherer, R., Tadeusiewicz, R., Zadeh, L.A., Zurada, J.M. (eds.) ICAISC 2016. LNCS, vol. 9693, pp. 755–766. Springer, Cham (2016). doi:10.1007/978-3-319-39384-1_67
24. Zadeh, L.A.: Fuzzy sets. Inf. Control **8**, 338–353 (1965)

25. Zadeh, L.A.: Fuzzy logic = computing with words. IEEE Trans. Fuzzy Syst. **4**, 103–111 (1996)
26. Zadeh, L.A.: Toward a theory of fuzzy information granulation and its centrality in human reasoning and fuzzy logic. Fuzzy Sets Syst. **90**, 111–127 (1997)

Bioinformatics, Biometrics and Medical Applications

Classification of Physiological
Data for Emotion Recognition

Philip Gouverneur[1], Joanna Jaworek-Korjakowska[2]([✉]), Lukas Köping[1],
Kimiaki Shirahama[1], Pawel Kleczek[2], and Marcin Grzegorzek[1,3]

[1] Research Group for Pattern Recognition, University of Siegen, Siegen, Germany
[2] Department of Automatics and Biomedical Engineering,
AGH University of Science and Technology, Krakow, Poland
jaworek@agh.edu.pl
[3] Faculty of Informatics and Communication,
University of Economics in Katowice, Katowice, Poland

Abstract. Emotion recognition is seen to be important not only for
computer science or sport activity but also for old and sick people to
live independently in their own homes as long as possible. In this paper
Empatica E4 wristband is used to collect the date and assess the stress
level of the user. We describe an algorithm for the classification of physiological data for emotion recognition. The algorithm has been divided into
the following steps: data acquisition, signal preprocessing, feature extraction, and classification. The data acquired during various daily activities
consist of more than 3 h of wristband signal. Through various stress tests
we achieve a maximum accuracy of 71% for a stressed/relaxed classification. These results lead to the conclusion that Empatica E4 wristband
can be used as a device for emotion recognition.

1   Introduction

For many years psychologist, researchers and therapists have debated the nature
of emotions, understanding exactly what emotions are, and tried to identify and
classify the different types of emotions. In 1972, a psychologist Paul Eckman suggested that there are six basic emotions universal throughout human cultures:
fear, disgust, anger, surprise, happiness, and sadness [6]. In 1999, he expanded
this list to include a number of other basic emotions which can be then be combined in a variety of ways. A very interesting definition of emotions has been
stated in 2007 by Hockenbury and Hockenbury [7]:*"An emotion is a complex
psychological state that involves three distinct components: a subjective experience, a physiological response, and a behavioural or expressive response"*. In this
paper we focus on the evaluation and classification of emotions of a user based
on his/her physiological responses.

To collect data, Empatica E4 wristband has been used. This wearable wireless
device is designed for comfortable, continuous, real-time data acquisition in daily
life [1] and contains four sensors: photoplethysmography (PPG), electrodermal
activity (EDA), 3-axis accelerometer and infrared thermopile (Fig. 1).

© Springer International Publishing AG 2017
L. Rutkowski et al. (Eds.): ICAISC 2017, Part I, LNAI 10245, pp. 619–627, 2017.
DOI: 10.1007/978-3-319-59063-9_55

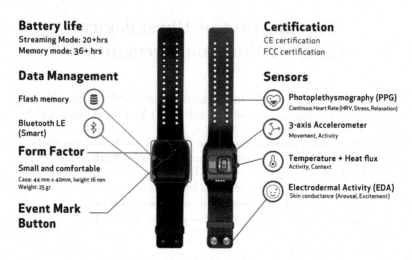

Fig. 1. Empatica E4 specifications [1].

1.1 Motivation

Interest in the field of emotion recognition has been motivated by the unbiased nature of signals, which are generated autonomously from the central nervous system. In general, these signals can be collected from the activity of subject's cardiovascular system, respiratory system, electrodermal activities, muscular system and brain activities as well as from body language and behavioral response. Emotion recognition is an important part of the artificial intelligence and is being applied in such fields as healthcare or automatics as well as for the detection of critical situations while driving a car. The analysis of emotions in speech has been proposed in 1988 by Tadeusiewicz in [13]. He suggested that the recognition of speech and emotions can be used to identify and analyse situations during an attack, space flight or interrogations of criminals. Since then, different applications of emotion recognition have been proposed. In 'Cognitive Village' emotion recognition can be used to identify the general sensation and the health state. During everyday life the analysis of stress level and behavior may permit to detect depression and other illnesses. Sensors can be employed to recognize stress situations, to inform and avoid an accident. These are just a few of many examples of potential applications of small sensors which can help us detect stressful situations, analyse our mood and recognize emotions.

1.2 Related Works

Recently much research has been undertaken in assessment of stress based on the analysis of physiological signals [14,15]. The published methods allow for stress classification with 80–90% accuracy [5,8–12,16]. These methods adopt a similar framework where signals measured during relaxation and in stressful situations caused by some stressors are then used to train a classifier. But, these methods differ in the selection of physiological signals, stressors and classifiers.

Methods described in literature are based mostly on the analysis of electrodermal activity (EDA) [9–12,16], temperature of skin and the environment [10,12,16], blood volume pressure [9,16] and heart rate [5,10]. Other classification models like also, researchers analyse electroencephalograms (EEGs) [8], pupil diameter [16] and respiration [10]. In order to trigger stress the subject is usually asked to calculate in head [5,10–12], to keep hands in icy water [10,12], to recite in front of the audience [10,12] or to carry out the Stroop test [5,16]. In some studies a set of photos [8] or Trier Social Stress Test [9] have been used as a stressor. The considerable number of methods predicts stress using support vector machines (SVMs) [8–12,16]. Other classification models like decision trees [10,16], naive Bayes [16] and linear discriminant analysis [11] have also been applied. The highest "stress/rest" classification accuracy (in laboratory conditions) have been achieved by [10,16], which scored 90.1% (for a SVM) and 90%, respectively. Systems in which predictions were based solely on the analysis of a single signal (e.g., heart rate [5], EDA [11] or EEG [8]) exhibited lower accuracy – about 83%.

In the majority of the above-mentioned experiments individual signals have been registered using separate devices. As such a solution lowers subject's comfort and is not handy. In our study, we collected data using only an armband. To detect stress most of the quoted methods used only SVMs [5,9,12]. In our study we compare results obtained using a variety of classifiers: naive Bayes, decision trees, SVMs, and random forests.

This paper is organized in 4 sections as follows: The next Section (Materials and Methods) specifies the implemented system including signal preprocessing, feature selection and classification steps. In Sect. 3 (Results) the data acquisition process is described as well as conducted tests are presented. Section 4 (Conclusion and future works) closes the paper and highlights future directions.

2   Materials and Methods

The proposed methodology of stress level assessment is shown in Fig. 2. The automated system is divided into three main stages, preprocessing (signal enhancement), feature selection, and classification, which are described below.

2.1   Signal Preprocessing

Empatica E4 uses a pair of electrodes to measure skin conductance between them. Consequently, the results of its measurements are highly dependent on the movement of its wearer. Motion artefacts can occur when the wristband is not worn tightly or when the subject does rapid or fast movements. Especially, errors can arise when the wristband shifts or hits against an object. As such artifacts may arise even if the user wears the device in an appropriate way, we implemented a method to remove those motion artifacts. Figure 3 presents a recorded

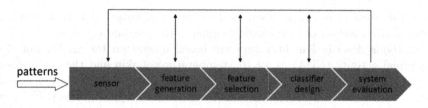

Fig. 2. An overview of our stress level classification method based on the data collected with Empatica E4.

session with different measuring failures (high blue spikes). To detect the start of an artifact, our method sequentially analyses the rate signal change at each data point. If a change is greater then a predefined threshold and proportional to the standard derivation of the signal, the corresponding data point is considered a beginning of an artifact. From there until the nearest point under the following twenty points, the signal is deleted. The signal with artifacts removed is shown in Fig. 3. The green and red points represent the start point and the end point of the detected artefact, respectively.

Moreover, an **electrodermal activity** (hereafter referred to as **EDA**) signal consists of two components [2]: skin conductance level and the tonic signal. The skin conductance level represents the basic level of the signal and the rapid differences in the signal called phasic changes. Each peak in this signal is also referred to as **skin conductance response** (hereafter referred to as **SCR**), which appears naturally in the rate of 1–3 per minute. Any additional SCR occurrence may be the result of stress. This conclusion is shown in Fig. 4: the original signal (red) is separated into its two components, the tonic level (yellow, for better visualisation shifted down) and phasic changes (purple). The tonic signal is calculated as follows: for every data point in the original EDA signal the mean value of values in a window is calculated. The window size is set to 8 s (i.e., 32 data points). The phasic component is obtained by subtracting the tonic signal from the original signal.

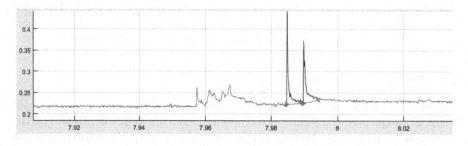

Fig. 3. EDA signal (orange) with removed motion artefacts (blue) (Color figure online)

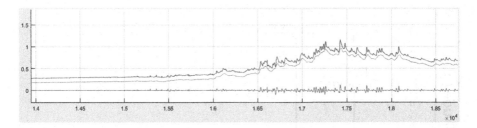

Fig. 4. EDA signal with its different components (original signal: red, tonic signal: yellow, phasic changes: purple). (Color figure online)

2.2 Feature Selection

Features are calculated for EDA and heart rate (hereafter referred to as HR) signals. The training data consist of samples of those signals sliced into 60 second-long intervals. Then statistical features such as mean, standard derivation, variance and maximum value are calculated for different windows for both EDA and HR. Since the number of SCRs and the characteristics of the peaks can give various information about the stress level of the person, the number of peaks, the mean amplitude and the mean rise time are calculated as well. These features and their computation are explained in the following paragraph.

For each SCR the starting point (onset), the top (peak) and ending point (offset) are computed. These variables are used to compute the rise time, the time from the onset to the peak and the peak amplitude (which is the difference in signal between the peak and onset). To find onsets, the phasic signal is passed through sequentially. An onset is detected if the signal is non-negative and the rate of change exceeds a predefined threshold. That threshold does not have only one correct value, for instance [3] uses a value of $0.01\,\mu S$, whereas [4] uses a value of $0.05\,\mu S$. In this study the threshold is set to $0.05\,\mu S$.

An offset is identified, when the phasic signal is negative for the first time again. The local maximum between the onset and offset represents the peak. This relation is illustrated in Fig. 5. Figure 6 shows an EDA signal with onsets (green), peaks (yellow) and offsets (red) marked.

Moreover, a Fourier transformation is computed for the EDA signal and its first 100 coefficients are also used as features. These features are used independently from the others to train an additional classifier.

2.3 Classification

Training data for classifiers consist of pairs of EDA- and HR-related feature vectors (computed according to the aforementioned procedure), each pair labelled as either "Stress" or "NoStress". The following classifiers are trained: Naive Bayes, a support vector machine (SVM), a decision tree, and a random forest. Additionally, random forest is trained on vectors of Fourier coefficients. Table 1 summarizes the features used for training and classification process.

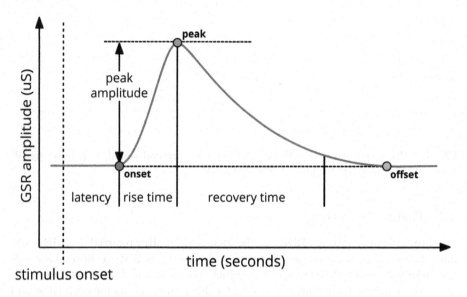

Fig. 5. An illustration of skin conductance response with the different characteristics [2]

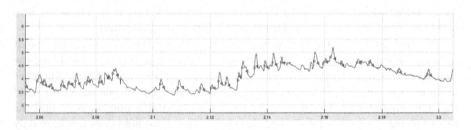

Fig. 6. An example of EDA signal with computed onsets (green), peaks (yellow) and offsets (red) (Color figure online)

Each of trained classifiers can be used to predict stress (i.e., to classify as either "Stress" or "NoStress") in any additional session data from Empatica E4. To accomplish this, the recorded signal is again cut into 60 seconds-long samples, the features described in Table 1 are computed and fed into classifiers in order to obtain the predicted class.

Table 1. Features used for stress detection

EDA	Heart rate
Mean, SD, Variance, Max, # Peaks, Mean amplitude, Mean rise time	Mean, SD, Variance, Max
Fourier feature	–

3   Results

In this section, we firstly present how data are acquired using Empatica E4 wristband. Then, emotion recognition results on these data are presented.

3.1   Data Acquisition Process

The E4 operates in either a streaming mode (for real-time data viewing on a mobile device using Bluetooth low energy) or in a recording mode (using its internal memory). The data from each sensor are stored in CSV files, making it convenient to separate and evaluate them. The dataset contains more than one hour of sensor signals collected during various activities. The period for each activity has been chosen according to its difficulty and incidence rate (15 min – no movement, 35 min – normal working, 15 min – walking, 10 min – running).

3.2   Data Analysis Results

To evaluate our stress level classification system, physiological signals are captured during various stress tests. Then for each classifier the predictions for stress phases were compared with the resting and stressing phases of those test. Figure 7 presents such a comparison performed on prediction results by random forest. According to the comparisons, the random forest model outperforms other classifiers achieving a maximum accuracy of 71%. The accuracies of other models are as follows: support vector machine – 67%, decision tree – 63%, Naive Bayes – 62%. The testes are repeated and in each test the performance of Naive Bayes model is inferior to the performance of other models. Although the classifier based on Fourier transformation coefficients occasionally give good results, it exhibits a significant variability in its performance across tests.

Since the random forest model leads to the best performance, an additional analysis is carried out to determine the number of individual trees needed to achieve the highest accuracy. The analysis consists in computing the out-of-bag classification error for forests grown from a different number of trees (from 1 up to 200 trees). As the random forest classifier always uses only a part of training data for training, the remaining part can be used to evaluate the performance of the model.

Using this data, a prediction by the classifier is compared to the known label. The failure rate is called out-of-bag classification (OOB) error. Based on the examination of OOB error for 1 to 200 trees the following two conclusions are drawn: (1) the mean error rate is reached with 7 trees, and (2) the minimum rate is reached with 27 trees. Therefore growing the forest from 200 trees yields no benefit and a number of 30 trees should be sufficient to achieve a near optimal performance.

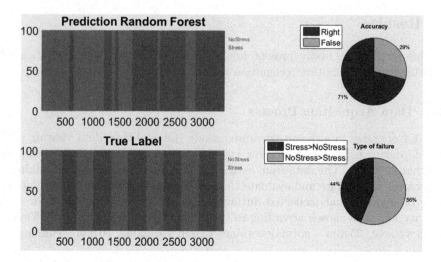

Fig. 7. Result of a classification of a stress test: predictions of the random forest (upper left), true labels of the stress test (bottom left), classification accuracy (upper right), and type of errors (bottom right).

4   Conclusion and Future Works

The main research aim of this study was to assess the usefulness of physiological signals captured by Empatica E4 wristband for emotion recognition, especially for stress recognition. The results of our study suggest that Empatica E4 is indeed a valuable tool for stress recognition, as the maximum classification accuracy of 71% was achieved for a classifier (a random forest) which predicted stress using solely features obtained by the analysis of signal from Empatica E4.

Nevertheless, this study also shows some important issues in stress categorization. Firstly, the sensors are highly dependent on the movement level of a subject. For this reason, a better method for artefact detection and removal need to be implemented. Secondly, there are significant changes in data for different subjects. Additional variations in data are expected to be caused by such factors as whether the wristband is worn on the left or right arm. Finally, session data captured during daily activities performed in real-life conditions vary from the data collected in laboratory conditions. All these problems can be tackled by collecting a larger and more diverse training dataset (e.g., recordings of different subjects) to improve the classification process.

Another task would be to develop a real-time application, based on the proposed method, which could be used in daily life and give an instant feedback to the user.

Furthermore, the usefulness of the proposed method could be increased, by differentiating not only between "stress" and "no stress", but also between different kinds of stress (e.g., positive stress versus negative stress and stressed caused by sport versus mental stress).

Acknowledgement. Research and development activities leading to this article have been supported by the German Federal Ministry of Education and Research within the project "Cognitive Village: Adaptively Learning Technical Support System for Elderly" (Grant Number: 16SV7223K).

References

1. E4 wristband, June 2016. https://www.empatica.com/e4-wristband. Accessed 29 Sept 2016
2. Imotions pocket guide GSR, June 2016. https://imotions.com/blog/galvanic-skin-response/. Accessed 29 Sept 2016
3. Benedek, M., Kaernbach, C.: A continuous measure of phasic electrodermal activity. J. Neurosci. Methods **190**(1), 80–91 (2010)
4. Braithwaite, J.J., Watson, D.G., Jones, R., Rowe, M.: A guide for analysing electrodermal activity (EDA) and skin conductance responses (SCRs) for psychological experiments. Psychophysiology **49**, 1017–1034 (2013)
5. Choi, J., Gutierrez-Osuna, R.: Using heart rate monitors to detect mental stress. In: Proceedings of the 2009 Sixth International Workshop on Wearable and Implantable Body Sensor Networks, pp. 219–223 (2009)
6. Ekman, P.: Basic Emotions. In: Dalgleish, T., Power, M. (eds.) Handbook of Cognition and Emotion, pp. 45–60. Wiley, New York (1999)
7. Hockenbury, D., Hockenbury, S.: Discovering Psychology. Worth Publishers, New York (2007)
8. Hosseini, S.A., Naghibi-Sistani, M.B.: Classification of emotional stress using brain activity. In: Gargiulo, G.D., McEwan, A. (eds.) Applied Biomedical Engineering, chap. 14, pp. 313–336. InTech (2011)
9. Mozos, O.M., Sandulescu, V., Andrews, S., Ellis, D., Bellotto, N., Dobrescu, R., Ferrandez, J.M.: Stress detection using wearable physiological and sociometric sensors. Int. J. Neural Syst. (2016)
10. Plarre, K., Raij, A., Hossain, S.M., Ali, A.A., Nakajima, M., Al'absi, M., Ertin, E., Kamarck, T., Kumar, S., Scott, M., Siewiorek, D., Smailagic, A., Wittmers, L.E.: Continuous inference of psychological stress from sensory measurements collected in the natural environment. In: Proceedings of the 10th ACM/IEEE International Conference on Information Processing in Sensor Networks, pp. 97–108. IEEE (2011)
11. Setz, C., Arnrich, B., Schumm, J., La Marca, R., Trster, G., Ehlert, U.: Discriminating stress from cognitive load using a wearable EDA device. IEEE Trans. Inf. Tech. Biomed. **14**, 410–417 (2010)
12. Shi, Y., Nguyen, M.H., Blitz, P., French, B., Fisk, S., De La Torre, F., Smailagic, A., Siewiorek, D.P., Al Absi, M., Ertin, E., Kamarck, T., Kumar, S.: Personalized stress detection from physiological measurements. In Proceedings of the International Symposium on Quality of Life Technology, pp. 28–29 (2010)
13. Tadeusiewicz, R.: Sygnal mowy. Wydaw, Komunikacji i cznoci, Warszawa (1988)
14. Tadeusiewicz, R.: New trends in neurocybernetics, vol. 10, pp. 1–7. Akademia Gorniczo-Hutnicza (2010)
15. Tadeusiewicz, R., Augustyniak, P.: Ubiquitous Cardiology - Emerging Wireless Telemedical Application. IGI Global, Hershey (2009)
16. Zhai, J., Barreto, A.: Stress detection in computer users based on digital signal processing of noninvasive physiological variables. In: Proceedings of the IEEE Engineering in Medicine and Biology Society, pp. 1355–1358 (2006)

Biomimetic Decision Making in a Multisensor Assisted Living Environment

Piotr Augustyniak$^{(\boxtimes)}$ and Magdalenia Smoleń

AGH University of Science and Technology,
30 Mickiewicza Ave., 30-059 Kraków, Poland
{august,msmolen}@agh.edu.pl

Abstract. Various sensors were found adequate to monitor human behavior in natural habitat. Besides metrological factors they were adopted to specific conditions of unobtrusive acquisition in human (e.g. an elder or child). An initial approach focused on use of a single sensor to detect a particular event evolved to behavioral studies based on complex recordings in multisensor environments. In such environment sensors based on different physical principles play complementary role and the resultant detection outperforms any of single component-based if combines the detail information correctly. We follow the rules of neural modulation in human sensory system to propose a biomimetic decision making in multisensor assisted living environment. In our approach each sensor contributes to the final detection accordingly to its presumed reliability in particular human behavior. Moreover, the system learns human habits, predicts most probable future actions from the history and anticipates accordingly the importance of particular sensors.

1 Introduction

Multisensor environments are widely used in technical measurement and surveillance. Designers of such measurement platforms take into account wide range of possible actions to predict the reliability of results. In video surveillance a human operator is often in charge of selecting best sensors (e.g. cameras), some advanced automatic systems detect motion or particular patterns in images to select and record the most informative one. The adaptation is always implemented as a binary one-out-of-many selection of raw data source and not as a modulated contribution of several processed data streams.

Studying the rules of information propagation in living neural systems, we focused on two different types of chemical synapses:

- ionotropic, with quick and short synaptic response, specialized in fast sensory or executory, excitatory or inhibitory pulse messaging, and
- metabotropic, with delayed and long standing response, which role is modulation of the pulse conduction.

The dual mechanism of synaptic junctions explains how animals select the sense they actually use to perceive the surroundings. It also discloses possibilities to

© Springer International Publishing AG 2017
L. Rutkowski et al. (Eds.): ICAISC 2017, Part I, LNAI 10245, pp. 628–637, 2017.
DOI: 10.1007/978-3-319-59063-9_56

remove pain by local mechanical (i.e. acupuncture) or electrical stimulation. The rules of nerve sensitivity modulation are proposed in this paper as a background of sensor selection in technical infrastructure for assisted living.

The proposed surveillance system consists of several complementary sensors based on different physical principles. Consequently they not only differ by size, cost and signal processing requirement, but also show different detection performance for particular human actions. Considering an optimal performance of the system in wide range of possible human actions, we propose a dynamic modulation of sensors' contribution, which is the main novelty of this paper.

The remaining part of this paper is organized as follows: Sect. 2 presents most relevant related work, Sect. 3 describes the existing multisensor assisted living system, Sect. 4 summarizes most important results on performance of isolated sensors, Sect. 5 presents rules of adaptive modulation of sensors' contribution, Sect. 6 presents the use case and Sect. 7 provides discussion and conclusions.

2   Related Work

Physiological measurements in real living conditions are performed with different aims, among of which telemedical services [19,27], safety prevention [10,29] and behavioral studies [9,16] seem the most relevant. Although several systems still focus on a single monitoring aspect and use the most appropriate sensor [2,8], the others follow the example from technical measurements [12] and create a multisensor environment employing different physical principles [7]. Such approach requires an appropriate information selection. In most such systems sensors are selectively activated depending on the supervised action, however some propose a linear combination of data contribution [15,19], and some others - recognition of behavioral primitives [3] followed by a graph-based calculation of their contribution [21].

The fundamental rule of metrology to not influence the observed phenomena by the measurement equipment has to be particularly observed in a multisensor environment. To this point sensors of various types are embedded into objects of everyday use, like garment [11] or kitchen appliances [5] and all remaining infrastructure is designed accordingly to a maintenance-free paradigm [13]. An example of such contactless monitoring system using pulse detector, bed-embedded accelerometers, camera and microphone was reported in [23]. Along with unobtrusive operation, the environment gathering sensitive behavior or health-related information should also comply with all digital safety requirements to guarantee privacy of the monitored person and his or her domicile [14].

Personalized interpretation of behavioral data from home-embedded and wearable sensor network is also a topic currently focused by scientists worldwide. Some systems use adaptive learning for continuous screening of human behavior and report irregularities as possible markers of health setback. First studied example by Vu et al. [28] uses predictive model to exploit the regularity of human motion found in the real traces, places visited and contacts made in daily activities. Another example by Ros et al. [20] employs learning automata and fuzzy temporal windows approach for the behavior recognition problem. Both systems learn the normal behaviors of supervised human, and use that knowledge to recognize normal and abnormal activities in real time. In addition they are self-adaptable to environmental variations, changes in human habits, and various time intervals when the behavior should be performed.

All existing systems studied as far attribute time-invariant importance to data streams originating from particular sensors. Even self-adaptive systems have separate learning and supervising phases, beneficial in initial adaptation of the monitoring to the person and his or her environment, but making the contribution from particular sensors unrelated to the actual actions. In fact the use of senses in animals and the human subjects seamless modulation depending on many factors, the present action among others. Consequently, we propose to implement the modulation mechanism and rules based on our previous research in the prototype surveillance environment.

3   Multisensory Surveillance System for Elderly

Multimodal surveillance home care system requires use of varied sensors with adaptive functionalities and application of dedicated algorithms [4,6].

Raw signals data acquired simultaneously from the sensors need to be wirelessly transmitted to the station, properly filtered, analyzed and assigned to different detection groups by means of automatic classification methods.

Multisensor environment could be considered with the following appliance aspects:

- automatic detection of falls, atypical behavior or situations [18],
- automatic detection of health risk factors in the human body,
- automatic interpretation of state and amount of activity of the human body [25,26] and mind,
- recognition of daily life activities of the supervised person [17],
- remote nightlong sleep analysis [22–24],
- biofeedback for rehabilitation process - both psychological and physical,
- HCI (Human-Computer Interaction) and interaction between the human and his or her environment [1],
- remote interaction between supervised person and his or her caregivers, therapists and doctors.

Sensors used in assisted living environment could be wearable or mounted in home care area [1]. They should enable to measure and detect many different biomedical, behavior, activity or human presence signals:

- electrophysiological: electroencephalographic (EEG), electromyographic (EMG), electrocardiographic (ECG), electrooculographic (EOG),
- motor: accelerometric, feet pressure, electromyographic (EMG), video, Kinect sensor, monochrome camera with IR light sources,
- eyetracking: video, IR reflection sensors, electrooculographic (EOG),
- sound: single microphones or matrices of microphones.

4 Reliability of Pose and Activity Detection

Our previous studies concerned automatic recognition of 12 selected human motor activities were performed with simultaneous acquisition by means of 4 different sensors:

- wireless (WLAN) biopotentials amplifier - ME6000 (Mega Electronics),
- wireless feet pressure measurement system - ParoLogg with Hydrocell technology based on the piezoresistive pressure sensors placed in the silicone cells,
- digital video camera - Sony HDR-FX7E (720×576 pixels, 25 frames per second),
- 3-axis accelerometer built in Revitus system.

Electromyographic (EMG) signals were recorded from 4 selected surface muscles of both lower limbs: quadriceps (vastus lateralis), biceps femoris, tibialis anterior and gastrocnemius (medial head). EMG data were sampled with the frequency of 2 kHz.

Spatio-temporal signals of feet pressure were registered with 32 independent pressure sensors (from each foot) with sampling frequency of 100 Hz.

Online sternum triaxial acceleration (ACC) signal was measured with the sampling frequency of 100 Hz.

Video signals with human silhouette were registered with a camera placed from the left side of the supervised volunteer.

In the experiments 20 healthy volunteers participated. Each of them was asked to perform about 30 repetitions of 12 different motor activities:

- squatting (**1a**) and getting up from a squat (**1b**),
- sitting on (**2a**) and getting up from a chair (**2b**),
- reaching the upper limb forward in the sagittal plane (**3a**) and return from reaching (**3b**),
- reaching the upper limb upwards in the sagittal plane (**4a**) and return from reaching (**4b**),
- bending the trunk forward in the sagittal plane (**5a**) and straightening the trunk from bend forward (**5b**),
- single step for the right (**6a**) and left lower limb (**6b**).

The reliability of each activity detection counted for all volunteers together are presented below in Table 1 and Fig. 1. On the basis of results presented below it can be concluded that the best classifier for most of the activities is EMG sensor (with recognition rate of **98.4%**) and then Video (**96.7%**), ACC (**95.5%**) and Pressure (**93.1%**) sensors respectively.

For each of the activity we can distinguish the sensor by means of which the detection is most accurate. Table 2 presents the order of the best individual sensors for recognition of the specific types of activities.

Regarding to the obtained results of recognition rate, EMG sensor is the best choice for detection of the following activities:

Table 1. Reliability of recognition (in %) of activities 1a ÷ 6b for all volunteers for each sensor.

	1a	1b	2a	2b	3a	3b	4a	4b	5a	5b	6a	6b	ALL
EMG	96.9	100.0	99.5	98.5	99.0	99.3	99.1	98.6	97.9	98.2	96.0	97.6	**98.4**
Pressure	90.8	91.8	95.7	96.7	94.9	92.4	95.3	93.6	87.0	88.2	94.9	97.1	**93.1**
Video	95.2	97.2	95.5	94.2	96.6	95.1	98.4	97.6	97.9	99.3	96.5	96.0	**96.7**
ACC	99.7	99.5	95.5	95.5	99.3	97.6	96.0	79.8	99.3	99.3	91.7	92.3	**95.5**

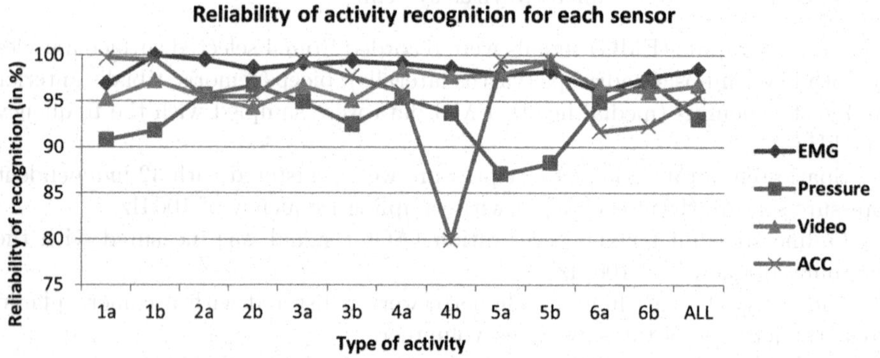

Fig. 1. Reliability of recognition (in %) of activities 1a ÷ 6b for all volunteers for each sensor.

Table 2. Order of sensors (from the best to the worst) for specific activities recognition.

Activity	Best	Good	Moderate	Worst
1a	ACC	EMG	Video	Pressure
1b	EMG	ACC	Video	Pressure
2a	EMG	Pressure	ACC	Video
2b	EMG	Pressure	ACC	Video
3a	ACC	EMG	Video	Pressure
3b	EMG	ACC	Video	Pressure
4a	EMG	Video	ACC	Pressure
4b	EMG	Video	Pressure	ACC
5a	ACC	EMG	Video	Pressure
5b	ACC	Video	EMG	Pressure
6a	Video	EMG	Pressure	ACC
6b	EMG	Pressure	Video	ACC

- getting up from a squat (**1b**),
- sitting on (**2a**) and getting up from a chair (**2b**),
- return from reaching the upper limb forward in the sagittal plane (**3b**),
- reaching the upper limb upwards in the sagittal plane (**4a**) and return from reaching (**4b**),
- single step for the left lower limb (**6b**).

The ACC sensor is most appropriate for recognition of movements such as:

- squatting (**1a**),
- reaching the upper limb forward in the sagittal plane (**3a**) and return from reaching (**3b**),
- bending the trunk forward in the sagittal plane (**5a**) and straightening the trunk from bend forward (**5b**).

A sensor based on video recordings is most suitable for (**6a**) activity - a single step for the left lower limb (**6b**).

Considering the above results and implementing adequate biomimetic decision making algorithms enables the selection of the most appropriate sensor for each of the activity and thus the most accurate recognition.

5   Reliability-Driven Rough Decision Making

5.1   General Assumptions and System Design

General architecture of multisensor environment for assisted living consists of sensors, dedicated feature extraction methods and modality selector. The proposed innovation replaces the selector by a modulator using weight coefficients W_k (Fig. 2) to prefer the most pertinent features while dicrimination the others.

Accordingly to the currently detected subject's action, the system automatically adapts the sensor set (accordingly to Table 2) to optimally detect the present action. The modification closes the information loop and, like all kinds of feedback, raises the stability issue.

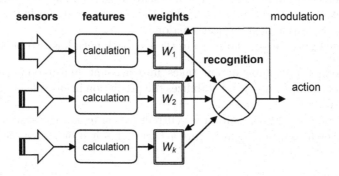

Fig. 2. Diagram flow of an elderly surveillance system using energy usage sensor to extract behavioral data.

5.2 Stability Condition for Modulated Sensor Set

The stability issue in a sensor set with modulated contribution can be solved by limitation of weight modulation range. Let f be a function $t = f(S_k, W_k)$ assigning a unique subject's action t to specific sensor outputs S_k modulated by W_k. Let m be a function $W_k = m(t)$ modulating the contributions from sensors S_k to maximize the reliability of the recognition of t accordingly to Table 2. Therefore, the modulator is stable if:

$$\forall f : f(S_k, W_k) = f(S_k, m(t)) \tag{1}$$

which means the modulation does not influence the current recognition result.

Since we cannot expect the recognition result be a linear function of the modulation depth, we propose an iterative try and fail algorithm finding the modulation limits. To find the value of W_k, between the original W_{k1} and the desired target W_{k2} the algorithm repeatedly bisects an interval and then selects a subinterval in which both ends yield different actions for further processing.

$$\begin{cases} W_k = W_{k1} \text{ when } f(S_k, W_k) = f(S_k, W_{k1}) \\ W_k = W_{k2} \text{ when } f(S_k, W_k) = f(S_k, W_{k2}) \end{cases} \tag{2}$$

All necessary steps of the modulation algorithm are performed within the subject state sampling interval. New data gathered from the sensors are processed with optimized sensors' contribution and confirm the detected subject's action.

5.3 Predictive Modulation of Sensors' Contribution

One may question the purpose of improvement of recognition already made. Fortunately in most assisted living environments, prevention of dangerous events is stressed as a primary goal, their architecture usually includes an artificial intelligence-based system for learning of subject's habits and detecting unusual behavior as potential sign of danger. We propose to use the information from the habits database to predict the subject's upcoming action and adjust the sensor's contribution accordingly (Fig. 3). The modulation is still made according to the stability requirements (see Sect. 5.2), but the sensor's contribution now adapts to the most probable next subject's action.

Introducing the habits database in the feedback path has two benefits:

- prediction of upcoming action takes into account multimodal time series instead of single points, what stabilizes the prediction in case of singular recognition error,
- focusing on optimal recognition for current action makes the system conservative, whereas optimizing for future action makes it progressive (i.e. awaiting changes of the status).

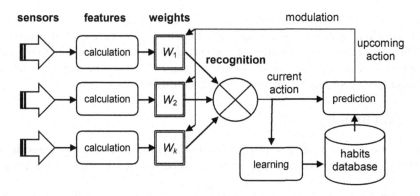

Fig. 3. Diagram flow of an elderly surveillance system using energy usage sensor to extract behavioral data.

6 Case Study

The proposed sensor's contribution modulation technique was analyzed in a previously proposed multisensor environment for assisted living [4]. We also used previously recorded data from 20 volunteers (8 women and 12 men, age between 22 ÷ 61 years), acting accordingly to predefined realistic scenarios. Table 3. presents an example compound action of searching a book on a wall-mounted shell, consisting of elementary poses (defined in Sect. 4): squatting (1a, 1b), reaching forward (3a, 3b), reaching upward (4a, 4b) and bending (5a, 5b).

Table 3. Contribution from sensors (in %) to the compound action recognition, 'searching on the shell'.

Time [s]	Pose	EMG	ACC	Video	Pressure
0	4a	60.0	7.5	25.0	7.5
1.4	4b	42.5	35.0	17.5	7.5
1.9	3a	42.5	42.5	10.0	5.0
3.2	3b	42.5	42.5	10.0	5.0
3.7	5a	17.5	60.0	17.5	5.0
4.9	5b	17.5	60.0	17.5	5.0
5.1	1a	42.5	42.5	10.0	5.0
6.7	1b	60.0	25.0	10.0	5.0

Multiple repetitions of patterns in the habits learning phase and opposed direction of elementary poses labeled with **a** and **b** facilitate correct prediction of subsequent poses and respective adaptation of sensors' contribution. In the studied case no abrupt corrections in sensor set were necessary, consequently changes of weighting coefficients were linear and not restricted by stability limits. Smoothing influence of prediction on sensors' modulation is also revealed in Table 3. Nevertheless, studies of correct work of the system for unexpected activities and possible errors of stabilizing algorithm need recording of human performance according to purposely designed misbehavior.

7   Conclusion

Biomimetic modulation of sensor's contribution in a multisensor assisted living environment puts forward their advantages accordingly to subject's behavior.

Acknowledgement. This scientific work is supported by the AGH University of Science and Technology in years 2016–2017 as a research project No. 11.11.120.612.

References

1. Augustyniak, P., Smoleń, M., Broniec, A., Chodak, J.: Data integration in multimodal home care surveillance and communication system. In: Pietka, E., Kawa, J. (eds.) Information Technologies in Biomedicine. Advances in Intelligent and Soft Computing, vol. 69, pp. 391–402. Springer, Heidelberg (2010). doi:10.1007/978-3-642-13105-9_39
2. Augustyniak, P.: Wearable wireless heart rate monitor for continuous long-term variability studies. J. Electrocardiol. **44**(2), 195–200 (2011)
3. Augustyniak, P.: Description of human activity using behavioral primitives. In: Burduk, R., et al. (eds.) Proceedings of the 8th International Conference on Computer Recognition Systems CORES 2013, pp. 661–670. Springer, Heidelberg (2013). doi:10.1007/978-3-319-00969-8_65
4. Augustyniak, P., Smoleń, M., Mikrut, Z., Kańtoch, E.: Seamless tracing of human behavior using complementary wearable and house-embedded sensors. Sensors **14**(5), 7831–7856 (2014)
5. Augustyniak, P., Kańtoch, E.: Turning domestic appliances into a sensor network for monitoring of activities of daily living. J. Med. Imaging Health Inf. **5**, 1662–1667 (2015)
6. Augustyniak, P., Barczewska, K., Broniec, A., Izworski, A., Kańtoch, E., Orzechowski, T., Przybyło, J., Smoleń, M., Tadeusiewicz, R.: Technical Systems Forming Intelligent Environment for a Disabled Person. EXIT, Warsaw (2015). (In Polish)
7. Augustyniak, P.: Remotely programmable architecture of a multi-purpose physiological recorder. Microprocess. Microsyst. **46**, 55–66 (2016)
8. Badura, P.: Accelerometric signals in automatic balance assessment. Comput. Med. Imaging Graph. **46**, 169–177 (2015)
9. Brdiczka, O., Crowley, J.L., Reignier, P.: Learning situation models in a smart home. IEEE Trans. Syst. Man Cybern. Part B Cybern. **39**(1), 56–63 (2009)
10. Bujnowski, A., Skalski, L., Wtorek, J.: Monitoring of a bathing person. J. Med. Imag. Health Inform. **2**, 27–34 (2012)
11. Das, A., Beatty, P., Dutta, R.: Estimation of physiological body parameters from smart garment data. In: Proceedings of IEEE International Instrumentation and Measurement Technology Conference (I2MTC) (2014). doi:10.1109/I2MTC.2014.6860706

12. Głowacz, A.: Recognition of acoustic signals of induction motors with the use of MSAF10 and Bayes classifier. Arch. Metall. Mater. **61**(1), 153–158 (2016)

13. Kańtoch, E.: Telemedical human activity monitoring system based on wearable sensors network. Comput. Cardiol. **41**, 469–472 (2014)

14. Kańtoch, E.: BAN-based health telemonitoring system for in-home care. Comput. Cardiol. **42**, 113–116 (2015)

15. Koshmak, G., Loutfi, A., Linden, M.: Challenges and issues in multisensor fusion approach for fall detection: review paper. J. Sens., p. 12 (2016). doi:10.1155/2016/6931789

16. Luhr, S., West, G., Venkatesh, S.: Recognition of emergent human behavior in a smart home: a data mining approach. Pervasive Mob. Comput. **3**, 95–116 (2007)

17. Mikrut, Z., Smoleń, M.: A neural network approach to recognition of the selected human motion patterns. Automatics **1**(3), 535–543 (2011)

18. Mikrut, Z., Pleciak, P., Smoleń, M.: Combining pattern matching and optical flow methods in home care vision system. In: Pietka, E., Kawa, J. (eds.) ITIB 2012. LNCS, vol. 7339, pp. 537–548. Springer, Heidelberg (2012). doi:10.1007/978-3-642-31196-3_54

19. Robben, S., Krse, B.: Longitudinal residential ambient monitoring: correlating sensor data to functional health status. In: Proceedings of 7th International Conference on Pervasive Computing Technologies for Healthcare (Pervasive Health), pp. 244–247 (2013)

20. Ros, M., Cullar, M.P., Delgado, M., Vila, A.: Online recognition of human activities and adaptation to habit changes by means of learning automata and fuzzy temporal windows. Inf. Sci. **220**, 86–101 (2013)

21. Ślusarczyk, G., Augustyniak, P.: A graph representation of subject's time-state space. In: Pietka, E., Kawa, J. (eds.) Information Technologies in Biomedicine. Advances in Intelligent and Soft Computing, vol. 69, pp. 379–390. Springer, Heidelberg (2010). doi:10.1007/978-3-642-13105-9_38

22. Smoleń, M.: Analysis of EEG activity during sleep - brain hemisphere symmetry of two classes of sleep spindles. Pol. J. Med. Phys. Eng. **15**(2), 65–75 (2009)

23. Smoleń, M., Czopek, K., Augustyniak, P.: Sleep evaluation device for home-care. In: Pietka, E., Kawa, J. (eds.) Information Technologies in Biomedicine. Advances in Intelligent and Soft Computing, vol. 69, pp. 367–378. Springer, Heidelberg (2010). doi:10.1007/978-3-642-13105-9_37

24. Smoleń, M., Czopek, K., Augustyniak, P.: Non-invasive sensors based human state in nightlong sleep analysis for home-care. Comput. Cardiol. **37**, 45–48 (2010)

25. Smoleń, M., Kańtoch, E., Augustyniak, P., Kowalski, P.: Wearable patient home monitoring based on ECG and ACC sensors. IFMBE Proc. **37**, 941–944 (2011)

26. Smoleń, M., Kańtoch, E., Augustyniak, P.: Wireless body area network system based on ECG and accelerometer pattern. Comput. Cardiol. **38**, 245–248 (2011)

27. Tamura, Y.: Home geriatric physiological measurements. Physiol. Meas. **33**(10), R47–R65 (2012)

28. Vu, L., Do, Q., Nahrstedt, K.: Jyotish: constructive approach for context predictions of people movement from joint Wifi/Bluetooth trace. Pervasive Mob. Comput. **7**, 690–704 (2011)

29. Wójtowicz, B., Dobrowolski, A., Tomczykiewicz, K.: Fall detector using discrete wavelet decomposition and SVM classifier. Metrol. Meas. Syst. **22**(2), 303–314 (2015)

Classification of Splice-Junction DNA Sequences Using Multi-objective Genetic-Fuzzy Optimization Techniques

Marian B. Gorzałczany[(✉)] and Filip Rudziński

Department of Electrical and Computer Engineering, Kielce University
of Technology, Al. 1000-lecia P.P. 7, 25-314 Kielce, Poland
{m.b.gorzalczany,f.rudzinski}@tu.kielce.pl

Abstract. The main goal of this paper is the application of our fuzzy
rule-based classification technique with genetically optimized accuracy-
interpretability trade-off to the classification of the splice-junction DNA
sequences coming from the *Molecular Biology (Splice-junction Gene
Sequences)* benchmark data set (available from the UCI repository).
Two multi-objective evolutionary optimization algorithms are employed
and compared in the framework of our technique, i.e., the well-known
Strength Pareto Evolutionary Algorithm 2 (SPEA2) and our SPEA2's
generalization (referred to as SPEA3) characterized by a higher spread
and a better-balanced distribution of solutions. A comparative analysis
with 15 alternative approaches is also performed.

Keywords: Splice-junction DNA sequence classification · Fuzzy rule-
based classifier · Multi-objective evolutionary optimization · Accuracy-
interpretability trade-off optimization

1   Introduction

The development of high-throughput sequencing technologies has resulted in a
huge amount of biological data such as DNA (deoxyribonucleic acid) sequences,
gene expression patterns, chemical structures, etc. The meaningful interpretation
and knowledge discovery in such a large (and still growing) volume of biological
data belong presently to essential and difficult tasks in bioinformatics [1]. In
particular, classification of sequence data can provide a valuable insight into
the function of genes and proteins as well as their relations and similarities. A
separate task is the classification of splice-junction sequences considered in this
paper. It is an important problem because the splice junctions in DNA sequence
indicate which parts of such a sequence carry protein-coding information (see
the next section of the paper for brief characteristics of this problem). Automatic
classification systems for the interpretation and knowledge discovery in biological
data should be evaluated in terms of their accuracy and interpretability [2].

   The main goal of this paper is the application of our fuzzy rule-based classifiers (FRBCs) with genetically optimized accuracy-interpretability trade-off

© Springer International Publishing AG 2017
L. Rutkowski et al. (Eds.): ICAISC 2017, Part I, LNAI 10245, pp. 638–648, 2017.
DOI: 10.1007/978-3-319-59063-9_57

(see, e.g., [3–6]) to knowledge discovery in the splice-junction DNA sequences data. We employ multi-objective evolutionary optimization algorithms (MOEOAs) as the FRBC's structure- and parameter-optimization tools yielding the FRBC's accuracy-interpretability trade-off optimization (see [7] for the related-work review and [8,9] for a single-objective optimization approach). First, the splice-junction classification problem is briefly presented. Then, main components of our FRBC and its genetic learning and MOEOA-based optimization are outlined. Two MOEOAs are considered and compared: our generalization (referred to as SPEA3 [10–12]) of the well-known Strength Pareto Evolutionary Algorithm 2 (SPEA2) [13] and SPEA2 itself. Finally, the application of our approach to classification of the splice-junction DNA sequences coming from the *Molecular Biology (Splice-junction Gene Sequences)* benchmark data set (available from the UCI repository: http://archive.ics.uci.edu/ml) is presented. A comparative analysis with 15 alternative approaches is also performed.

2   Splice-Junction Classification Problem

Splice-junction classification is an important stage in the genetic information processing within a biological system (to be more specific - in an eukaryotic cell, i.e., the cell that contains a nucleus). Figure 1 illustrates the flow of such an information (also referred to as a gene expression); it is based on the so-called central dogma of molecular biology [14], which is also described as [15]: "DNA makes RNA and RNA makes protein" (RNA stands for ribonucleic acid). In its first stage (i.e., transcription), in eukaryotic cells the information contained in a section of DNA is replicated in the form of a piece of pre-mRNA (precursor-messenger RNA). In pre-mRNA, regions that code for proteins (called exons) are interrupted by the non-coding regions (called introns). In the second stage (i.e., splicing), all introns are removed from pre-mRNA and the remaining exons are joined together to make one continuous mRNA strand. Finally, the spliced mRNA is exported out of the cell nucleus for translation (the third stage) to different kinds of proteins. Splice junctions or splice sites are the boundary points where the splicing takes place, i.e., they are the meeting points of exons and introns.

Concluding, the splice-junction classification problem consists in recognition of splice sites in DNA sequences (represented by the corresponding pre-mRNA sequences). The classification aims at recognizing the exon-intron (EI) junction, the intron-exon (IE) junction, or none of the junction sites.

3   Brief Characteristics of Main Components of the Proposed Fuzzy Rule-Based Classifiers (FRBCs) and Their Multi-objective Genetic Optimization

In this section, we outline the following basic components of the proposed approach: FRBC's knowledge base and approximate inference engine, learning data

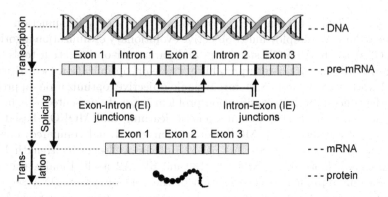

Fig. 1. Illustration of the flow of genetic information within a biological system (a gene expression)

set, optimization objectives, original genetic operators and the selection of a multi-objective evolutionary optimization algorithm (MOEOA). More details on our approach can be found in [3–6].

FRBC's knowledge base is a set of linguistic fuzzy classification rules discovered and optimized during the genetic learning process from the learning data set (a FRBC with n input attributes $x_1, x_2, \ldots, x_n$ and an output, which has the form of a fuzzy set over the set $Y = \{y_1, y_2, \ldots, y_c\}$ of c class labels is considered). The r-th rule, $r = 1, 2, \ldots, R$ (R changes during the learning process), has the following form:

$$\mathbf{IF}[x_1 \text{ is } [\text{not}]_{(sw_1^{(r)}<0)} \ A_{1,|sw_1^{(r)}|}]_{(sw_1^{(r)}\neq 0)} \ \mathbf{AND...AND}$$
$$[x_n \text{ is } [\text{not}]_{(sw_n^{(r)}<0)} \ A_{n,|sw_n^{(r)}|}]_{(sw_n^{(r)}\neq 0)} \tag{1}$$
$$\mathbf{THEN} \ y \text{ is } B_{(singl.)j^{(r)}},$$

where: (a) $[expression]_{(condition)}$ in (1) denotes conditional inclusion of $[expression]$ into a given rule iff $(condition)$ is fulfilled, (b) $|\cdot|$ returns the absolute value, and (c) $sw_i^{(r)}$ is a switch which controls the presence/absence of the i-th input attribute in the r-th rule, $i = 1, 2, \ldots, n$. $sw_i^{(r)} \in \{0, \pm 1, \pm 2, \ldots, \pm a_i\}$, where a_i is the number of fuzzy sets (linguistic terms) defined for the i-th attribute. $sw_i^{(r)} = 0$ excludes the i-th attribute from the r-th rule, whereas $sw_i^{(r)} > 0$ includes the component $[x_i$ is $A_{ik_i}]$ ($k_i = |sw_i^{(r)}|$) into the r-th rule and $sw_i^{(r)} < 0$ includes the component $[x_i$ is not $A_{ik_i}]$ into that rule (not $A_{ik_i} = \bar{A}_{ik_i}$ and $\mu_{\bar{A}_{ik_i}}(x_i) = 1 - \mu_{A_{ik_i}}(x_i)$; $\mu_{A_{ik_i}}(x_i)$ and $\mu_{\bar{A}_{ik_i}}(x_i)$ represent membership functions of fuzzy sets A_{ik_i} and $\bar{A}_{ik_i}$, respectively).

In the splice-junction DNA sequences considered in this paper, only categorical attributes occur (they represent particular nucleotides in the DNA sequence - see the next section of the paper). For this reason - despite the fact that our approach can process both categorical and numerical input attributes - we briefly characterize here only the categorical-attribute representation. Let $F(X_i)$, $i = 1, 2, \ldots, n$, and $F(Y)$ denote families of all fuzzy sets defined in the universes X_i and Y, respectively. Each categorical attribute x_i ($x_i \in X_i = \{x_{i1}, x_{i2}, \ldots, x_{ia_i}\}$) is characterized by a_i fuzzy singletons $A_{(singl.)ik_i} \in F(X_i)$, $k_i = 1, 2, \ldots, a_i$ defined for particular "values" x_{ik_i} of x_i as follows: $\mu_{A_{(singl.)ik_i}}(x_i) = 1$ for $x_i = x_{ik_i}$ and 0 elsewhere. Similarly, class labels $y_{j^{(r)}}$ in particular rules (1) ($j^{(r)} \in \{1, 2, \ldots, c\}$) are represented by appropriate fuzzy singletons $B_{(singl.)j^{(r)}} \in F(Y)$.

The FRBC's knowledge base consists of a linguistic rule base (RB) and a data base (DB). We propose direct and computationally efficient RB's representation as follows (original, dedicated crossover and mutation operators have been defined for its processing; see comments later in this section):

$$RB = \{sw_1^{(r)}, sw_2^{(r)}, \ldots, sw_n^{(r)}, j^{(r)}\}_{r=1}^{R}. \tag{2}$$

DB, in general, contains: (a) parameters of fuzzy sets for numerical attributes (not occurring in the DNA-sequence data), (b) domains of categorical attributes $X_i = (x_{i1}, x_{i2}, \ldots, x_{ia_i})$, $i \in \{1, 2, \ldots, n\}$, and c) the set of class labels $Y = \{y_1, y_2, \ldots, y_c\}$ (parameters of (b) and (c), obviously, are not being tuned during the learning of FRBC from data).

FRBC's approximate inference engine generates a FRBC's response to input data. It is used both in the learning phase (to evaluate particular individuals - i.e., fuzzy knowledge bases - competing with each other) and in the phase of FRBC's use (when decisions for new-coming data must be generated). Applying the most widely used Mamdani's model (although the alternative approaches - see, e.g., [3] - can also be used), we obtain - for the input data $x' = (x_1', x_2', \ldots, x_n')$ - a FRBC's fuzzy-set response B' characterized by its membership function $\mu_{B'}(y)$, $y \in Y = \{y_1, y_2, \ldots, y_c\}$:

$$\mu_{B'}(y) = \max_{r=1,2,\ldots,R} \mu_{B'^{(r)}}(y) = \max_{r=1,2,\ldots,R} \min[\alpha^{(r)}, \mu_{B_{(singl.)j^{(r)}}}(y)], \tag{3}$$

where

$$\alpha^{(r)} = \min_{\substack{i=1,2,\ldots,n, \\ sw_i^{(r)} \neq 0}} \alpha_i^{(r)} = \begin{cases} \min_{i=1,2,\ldots,n} \mu_{A_{i,sw_i^{(r)}}}(x_i'), & \text{for } sw_i^{(r)} > 0, \\ \min_{i=1,2,\ldots,n} \mu_{\bar{A}_{i,|sw_i^{(r)}|}}(x_i'), & \text{for } sw_i^{(r)} < 0. \end{cases} \tag{4}$$

Usually, a non-fuzzy FRBC's response y' is required; it is calculated in the following way:

$$y' = \arg\max_{y \in Y} \mu_{B'}(y). \tag{5}$$

Learning data set contains K input-output learning samples:

$$L = \{x_k^{(lrn)}, y_k^{(lrn)}\}_{k=1}^{K}, \tag{6}$$

where $x_k^{(lrn)} = (x_{1k}^{(lrn)}, x_{2k}^{(lrn)}, \ldots, x_{nk}^{(lrn)}) \in X = X_1 \times X_2 \times \cdots \times X_n$ ($\times$ stands for Cartesian product of ordinary sets) is the set of input attributes and $y_k^{(lrn)}$ is the corresponding class label ($y_k^{(lrn)} \in Y$) for the k-th data sample.

Optimization objectives - FRBC's accuracy: The accuracy measure (the objective function subject to maximization) has the following form:

$$Q_{ACC}^{(lrn)} = \frac{1}{K} \sum_{k=1}^{K} \delta(y_k', y_k^{(lrn)}) \text{ and } \delta(y_k', y_k^{(lrn)}) = \begin{cases} 1, \text{ for } y_k' = y_k^{(lrn)}, \\ 0, \text{ elsewhere,} \end{cases} \quad (7)$$

where y_k' is the class label which is the system's non-fuzzy response (5) for the learning data sample $\boldsymbol{x}_k^{(lrn)}$ and $y_k^{(lrn)}$ is the desired class label from that sample taken from $\boldsymbol{L}$ (6); $Q_{ACC}^{(lrn)} \in [0,1]$. An analogous measure $Q_{ACC}^{(tst)}$ for the test data can also be defined.

Optimization objectives - FRBC's interpretability: In general, FRBC's complexity-related and semantics-related interpretabilities should be considered. However, the second one is related to linguistic terms describing numerical attributes which do not occur in the DNA data. Hence, only the complexity-related interpretability will be considered. Its measure Q_{INT} (the objective function subject to maximization) is defined as follows:

$$Q_{INT} = 1 - Q_{CPLX}, \quad \text{where} \quad Q_{CPLX} = \frac{Q_{RINP} + Q_{INP} + Q_{FS}}{3} \quad (8)$$

and

$$Q_{RINP} = \frac{1}{R} \sum_{r=1}^{R} \frac{n_{INP}^{(r)} - 1}{n - 1}, \ Q_{INP} = \frac{n_{INP} - 1}{n - 1}, \ Q_{FS} = \frac{n_{FS} - 1}{\sum_{i=1}^{n} a_i - 1}, \ n > 1. \quad (9)$$

Q_{CPLX} in (8) is the FRBC's complexity measure ($Q_{CPLX} \in [0,1]$; 0 and 1 represent minimal and maximal complexity levels, respectively). It is an average of three sub-indices that measure an average complexity of particular rules Q_{RINP} as well as the complexity of the whole system in terms of its active inputs Q_{INP} and active fuzzy sets Q_{FS} (9). $n_{INP}^{(r)}$ in (9) is the number of input attributes included in the r-th rule. n_{INP} and n_{FS} in (9) are the numbers of inputs and fuzzy singletons, respectively, in the whole system.

Original genetic operators - some comments: Original crossover and mutation operators for the processing of RBs (2) have been developed by us. We also adopted some specialized crossover and mutation operators for the DB processing. However, as already mentioned, in the DNA-sequence data processing, DB contains only parameters that are not tuned. Thus, in the considered case, DB is not subject to processing. Additionally, the genetic operations on RBs are followed by (a) removal of possible empty rules (i.e., the rules without antecedents), (b) removal of rule duplicates, and (c) adding - for each class label that is not represented in RB - one rule with that class label - see [3–6] for details.

Selection of a multi-objective evolutionary optimization algorithm (MOEOA): The performance of MOEOAs is usually evaluated in terms of the following aspects [16]: (a) the accuracy of generated solutions (their closeness to Pareto optimal or reference solutions), (b) the spread of the solution set (the distance between extreme solutions), and (c) the balance of the distribution of solutions. For obvious reasons, the sets of solutions which are more accurate, of higher spread and better-balanced distribution are preferred. For comparison purposes, two MOEOAs are used in our experiments presented in the paper. As already mentioned in the Introduction of the paper, they include one of the most advanced and well-known methods, i.e., SPEA2 as well as our generalization of SPEA2, referred to as SPEA3. Our SPEA3 generates sets of non-dominated solutions characterized by higher spread and better-balanced distribution than the SPEA2's solutions. The essence of our generalization of SPEA2 consists in replacing its environmental selection mechanism by our original procedure improving the spread and distribution balance of generated solutions - see [10] for a detailed presentation and also [11] for a discussion.

4   Application to DNA Splice-Junction Classification Based on *Molecular Biology* (*Splice-Junction Gene Sequences*) data

The *Molecular Biology* (*Splice-junction Gene Sequences*) data set - available from http://archive.ics.uci.edu/ml - and taken from GenBank 64.1 (ftp site: genbank.bio.net) is one of the benchmark sets in splice-junction recognition and classification problems. This data set contains 3190 DNA samples. Each sample is a sequence of 60 nucleotides, which we denote by x_1 through x_{60}, and the splice junction or splice site (if any) is located in the middle of the sequence, i.e., between x_{30} and x_{31}. 25% of the overall number of samples contain exon-intron (EI) sites, another 25% - intron-exon (IE) sites, and the remaining 50% - no sites at all (it will be referred to as 'NO'). The nucleotides are commonly represented by symbols A, T, G, and C.

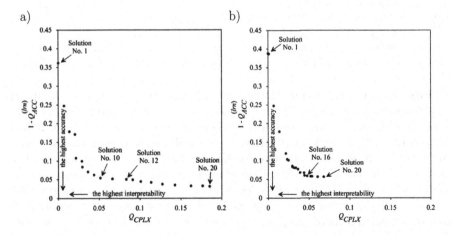

Fig. 2. The best Pareto-front approximations generated by our SPEA3 (a) and SPEA2 (b) for the considered splice-junction data set

Table 1. Interpretability and accuracy measures of solutions from Fig. 2a (our SPEA3) and Fig. 2b (SPEA2)

No.	Objective functions complements		Interpretability measures				Accuracy measures	
	$1 - Q_{INT} = \dfrac{Q_{INT}}{Q_{CPLX}}$	$1 - Q_{ACC}^{(lrn)}$	R	n_{INP}	n_{FS}	$n_{INP/R}$	$ACC^{(lrn)}$	$ACC^{(tst)}$
SPEA3:								
1.	0	0.36247	3	1	1	1	63.7%	59.7%
2.	0.00681	0.24751	3	2	2	1	75.2%	71.5%
3.	0.01363	0.17862	3	3	3	1	82.1%	80.0%
4.	0.02045	0.17102	4	4	4	1	82.9%	81.1%
5.	0.02158	0.10784	5	4	4	1.2	89.2%	89.9%
6.	0.02889	0.09928	7	5	5	1.3	90.1%	91.1%
7.	0.02916	0.08361	6	5	5	1.3	91.6%	90.4%
8.	0.03621	0.07125	8	6	6	1.4	92.9%	91.9%
9.	0.04342	0.06270	9	7	7	1.4	93.7%	93.4%
10.	0.05150	0.05415	9	8	8	1.7	94.6%	94.0%
11.	0.06599	0.05178	11	10	10	1.8	94.8%	93.7%
12.	0.08300	0.05035	11	12	14	2	95.0%	94.1%
13.	0.09099	0.04940	8	12	16	3	95.1%	93.3%
14.	0.10110	0.04513	8	13	18	3.4	95.5%	93.4%
15.	0.11375	0.04180	10	15	20	3.2	95.8%	93.3%
16.	0.12728	0.03705	11	17	22	3.1	96.3%	93.3%
17.	0.14267	0.03515	13	19	26	3.0	96.5%	93.9%
18.	0.16234	0.03325	16	22	29	2.9	96.7%	93.1%
19.	0.17668	0.03277	16	24	31	3.1	96.7%	93.1%
20.	0.18498	0.03135	17	25	33	3.1	96.9%	93.1%
SPEA2:								
1.	0	0.38765	3	1	1	1	61.2%	65.0%
2.	0.00117	0.38717	3	1	2	1	61.3%	65.0%
3.	0.00681	0.24751	3	2	2	1	75.2%	71.5%
4.	0.01363	0.17862	3	3	3	1	82.1%	80.0%
5.	0.02158	0.11971	5	4	4	1.2	88.0%	86.0%
6.	0.02324	0.10451	7	4	5	1.3	89.5%	87.7%
7.	0.02468	0.10166	9	4	6	1.3	89.8%	88.4%
8.	0.02939	0.08646	8	5	5	1.4	91.3%	91.0%
9.	0.02978	0.08361	9	5	5	1.4	91.6%	90.4%
10.	0.03158	0.08218	9	5	6	1.5	91.8%	90.3%
11.	0.03401	0.08171	9	5	7	1.8	91.8%	90.3%
12.	0.03660	0.07791	9	6	6	1.4	92.2%	92.2%
13.	0.03886	0.06888	11	6	7	1.6	93.1%	92.6%
14.	0.04399	0.06698	11	7	7	1.5	93.3%	93.3%
15.	0.04451	0.06033	11	7	7	1.6	94.0%	93.7%
16.	0.04749	0.05890	12	7	9	1.7	94.1%	94.0%
17.	0.05184	0.05795	11	8	8	1.7	94.2%	94.0%
18.	0.05314	0.05748	12	8	9	1.7	94.2%	94.0%
19.	0.06043	0.05700	12	9	10	1.8	94.3%	93.9%
20.	0.06772	0.05653	12	10	11	1.9	94.3%	94.0%

First, in order to illustrate the operation of our approach in some detail, the genetic learning experiment for a single learning-test data split (with 9:1 ratio) of the considered data set is performed. Figure 2a presents a collection of 20 non-dominated solutions (optimized FRBCs) obtained in a final generation of a single run of the FRBC's design technique implemented in the framework of our SPEA3 method. They define the best SPEA3-based approximation of Pareto-optimal solutions characterized by various levels of optimized accuracy-interpretability trade-off. The user can select a single solution (a specific FRBC) characterized by a desired level of compromise between its accuracy and interpretability. The interpretability and accuracy measures for the solutions from Fig. 2a are

Table 2. Fuzzy rule base of solution (FRBC) No. 10 from Fig. 2a and Table 1 (SPEA3 part)

No	Fuzzy classification rules
1.	**IF** x_{31} *is* G **AND** x_{32} *is* T **AND** x_{35} *is* G **THEN** Class "EI"
2.	**IF** x_{30} *is not* G **THEN** Class "NO"
3.	**IF** x_{28} *is not* G **AND** x_{29} is A **AND** x_{32} *is not* T **THEN** Class "IE"
4.	**IF** x_{32} *is not* T **THEN** Class "NO"
5.	**IF** x_{31} *is* G **AND** x_{33} is A **THEN** Class "EI"
6.	**IF** x_{29} *is not* A **THEN** Class "NO"
7.	**IF** x_{23} *is not* A **AND** x_{28} *is not* G **THEN** Class "IE"
8.	**IF** x_{31} *is not* G **THEN** Class "NO"
9	**IF** x_{35} *is* G **THEN** Class "NO"

presented in Table 1 (SPEA3 part), where $n_{INP/R}$ is the number of input attributes per rule, $ACC^{(lrn)} = 100 \cdot Q_{ACC}^{(lrn)}[\%]$ and $ACC^{(tst)} = 100 \cdot Q_{ACC}^{(tst)}[\%]$ are the percentages of correct decisions in the learning and test sets, respectively; the remaining parameters were defined earlier in the paper. Figure 2b presents an analogous set of optimized solutions generated by means of SPEA2. Their numerical details are collected in the SPEA2 part of Table 1. Figures 2a and b clearly demonstrate that our SPEA3-based approach generates the set of solutions (FRBCs) characterized by much higher spread and much-better-balanced distribution in the objective space than SPEA2-based solutions.

A single solution of particular interest might be the solution which is, first, the most accurate for the test data (i.e., the data not seen during the learning process) and, second, the most transparent and interpretable (i.e., the simplest in terms of its rule-base complexity). In SPEA3 case, it is the solution No. 12 of Fig. 2a and Table 1 (underlined - using a solid line - in Table 1). It is the most accurate one for the test data (94.1%) and requires 11 simple rules (with only 2 input attributes per rule on average). It is worth remembering that the overall number of input attributes is equal to 60. Slightly worse solution No. 10 (94.0% on test data) requires only 9 rules with 1.7 input attribute per rule on average (see Table 2 for its fuzzy rule base). In SPEA2 case, the same accuracy requires 12 rules (see solution No. 16 in Fig. 2b and the SPEA2-part of Table 1).

The results of our main multiple-cross-validation-based experiment and the results of 15 alternative approaches applied to the considered data set are collected in Table 3 for the purpose of comparative analysis. The following alternative methods are used: (i) KBANN (Knowledge-Based Artificial Neural Networks), GDS (Gradient Descent Symbolic rule generation), Backprop (Backpropagation-based feed-forward neural network), and the well-known ID3 - see [17], (ii) UCS (sUpervised Classifier System), GAssist (Genetic Algorithm based claSSIfier sySTem), and cAnt-Miner (an algorithm inspired by behavior of real ant colonies) - see [18], (iii) MLP (Multi-Layer Perceptrons) with

Table 3. Results of our approach and comparison with alternative approaches

Source	Method	Learn-to-test ratio	Number of single k-fcv experiments	Average accuracy measures for learning and test data		Average interpretability measures			
				$ACC^{(lrn)}$	$ACC^{(tst)}$	R	n_{ATR}	n_{FS}	$n_{ATR/R}$
[17] (1994) and UCI repository	KBANN	9:1	1	N/a	93.68%	N/a	N/a	N/a	N/a
	GDS	9:1	1	N/a	93.25%	N/a	N/a	N/a	N/a
	Backprop	9:1	1	N/a	93.23%	-	-	-	-
	ID3	9:1	1	N/a	89.44%	N/a	N/a	N/a	N/a
[18] (2009)	UCS	9:1	1	N/a	57.30%	N/a	N/a	N/a	N/a
	GAssist	9:1	1	N/a	92.45%	N/a	N/a	N/a	N/a
	cAnt-Miner	9:1	1	N/a	83.80%	N/a	N/a	N/a	N/a
[19] (2009)	MLP with regular error function (MSE)	9:1	10	N/a	91.66%	628.6	N/a	N/a	N/a
	MLP with new error function	9:1	10	N/a	89.85%	298.9	N/a	N/a	N/a
[20] (2010)	C4.5	2:1	1	N/a	93.6%	N/a	N/a	N/a	N/a
	Naive Bayes	2:1	1	N/a	96.6%	–	–	–	–
	SVM	2:1	1	N/a	92.0%	–	–	–	–
[21] (2013)	BEXA	9:1	1	N/a	92.7%	N/a	N/a	N/a	N/a
	JRip	9:1	1	N/a	94.14%	N/a	N/a	N/a	N/a
	PART	9:1	1	N/a	92.51%	N/a	N/a	N/a	N/a
This paper	Our approach based on SPEA3	2:1	10	93.4%	93.6%	5.0	6.3	7.0	2.2
		9:1	10	93.4%	94.3%	5.4	7.2	8.0	2.2

N/a stands for not available;
– stands for not applicable, e.g., the number of rules for non-rule-based systems.

regular and new error functions - see [19], (iv) the well-known C4.5, Naive Bayes, and SVM methods - see [20], and (v) BEXA (Basic EXclusion Algorithm), JRip (Java implementation of RIPPER (Repeated Incremental Pruning to Produce Error Reduction) algorithm), and PART (PARTial C4.5 method) - see [21].

The k-fold cross-validation (k-fcv) with two values of k is considered: $k = 10$ (i.e., learn-to-test ratio 9:1) and $k = 3$ (ratio 2:1) in order to compare our results with the results generated by all aforementioned approaches. Each learning experiment generates a Pareto-front approximation. A single solution characterized, first, by the highest accuracy in the test data set and, second, by the highest interpretability is selected from that front approximation. Then, the average results from all partial experiments are computed. They are presented in Table 3 (a "single k-fcv experiment" in Table 3 represents k partial experiments for particular folds; 10 single k-fcv experiments correspond to 10 different divisions of the original data set).

For the 9:1 learn-to-test ratio, our approach with 94.3% average accuracy for the test data and the interpretability measures as in Table 3 outperforms all alternative approaches. Their interpretability measures are either non-available or take very high values (see, e.g., the average number of rules for "MLP with new error function" equal to 298.9 comparing to 5.4 for our approach). Obviously, the black-box approaches are not interpretable at all. Similar regularity holds for the 2:1 ratio. In this case, only the black-box Naive-Bayes approach gives higher accuracy than our method.

Concluding, in general, our approach generates - for decision-support purposes - fuzzy classification systems characterized by comparable or better accuracy and significantly better interpretability than various alternative approaches.

5 Concluding Remarks

The main goal of this paper is the application of our MOEOA-based FRBCs to the classification of the splice-junction DNA sequences data coming from the *Molecular Biology (Splice-junction Gene Sequences)* benchmark data set (available from the UCI repository: http://archive.ics.uci.edu/ml). The splice-junction classification problem and the main components of our FRBC and its genetic learning and MOEOA-based optimization are outlined. Two MOEOAs are employed and compared, i.e., the well-known SPEA2 and our SPEA2's generalization (referred to as SPEA3) characterized by a higher spread and a better-balanced distribution of solutions. A comparative analysis with 15 alternative approaches is also performed.

References

1. Maji, P., Paul, S.: Scalable Pattern Recognition Algorithms: Applications in Computational Biology and Bioinformatics. Springer, Cham (2014)
2. Towell, G., Shavlik, J.W.: Interpretation of artificial neural networks: mapping knowledge-based neural networks into rules. In: Proceedings of the 4th International Conference on Neural Information Processing Systems (NIPS 1991), pp. 977–984. Morgan Kaufmann Publishers Inc., Denver (1991)
3. Rudziński, F.: A multi-objective genetic optimization of interpretability-oriented fuzzy rule-based classifiers. Appl. Soft Comput. **38**, 118–133 (2016)
4. Gorzałczany, M.B., Rudziński, F.: A multi-objective genetic optimization for fast, fuzzy rule-based credit classification with balanced accuracy and interpretability. Appl. Soft Comput. **40**, 206–220 (2016)
5. Gorzałczany, M.B., Rudziński, F.: Interpretable and accurate medical data classification - a multi-objective genetic-fuzzy optimization approach. Expert Syst. Appl. **71**, 26–39 (2017)
6. Gorzałczany, M.B., Rudziński, F.: Handling fuzzy systems' accuracy-interpretability trade-off by means of multi-objective evolutionary optimization methods - selected problems. Bull. Pol. Acad. Sci. Tech. Sci. **63**(3), 791–798 (2015)
7. Fazzolari, M., Alcala, R., Nojima, Y., Ishibuchi, H., Herrera, F.: A review of the application of multiobjective evolutionary fuzzy systems: current status and further directions. IEEE Trans. Fuzzy Syst. **21**(1), 45–65 (2013)

8. Gorzałczany, M.B., Rudziński, F.: Accuracy vs. interpretability of fuzzy rule-based classifiers: an evolutionary approach. In: Rutkowski, L., Korytkowski, M., Scherer, R., Tadeusiewicz, R., Zadeh, L.A., Zurada, J.M. (eds.) EC/SIDE -2012. LNCS, vol. 7269, pp. 222–230. Springer, Heidelberg (2012). doi:10.1007/978-3-642-29353-5_26

9. Gorzałczany, M.B., Rudziński, F.: A modified pittsburg approach to design a genetic fuzzy rule-based classifier from data. In: Rutkowski, L., Scherer, R., Tadeusiewicz, R., Zadeh, L.A., Zurada, J.M. (eds.) ICAISC 2010. LNCS (LNAI), vol. 6113, pp. 88–96. Springer, Heidelberg (2010). doi:10.1007/978-3-642-13208-7_12

10. Rudziński, F.: Finding sets of non-dominated solutions with high spread and well-balanced distribution using generalized strength Pareto evolutionary algorithm. In: Proceedings of the 16th World Congress of the International Fuzzy Systems Association (IFSA) and the 9th Conference of the European Society for Fuzzy Logic and Technology (EUSFLAT), IFSA-EUSFLAT 2015. Advances in Intelligent System Research, vol. 89, pp. 178–185. Atlantis Press, June 2015

11. Gorzałczany, M.B., Rudziński, F.: An improved multi-objective evolutionary optimization of data-mining-based fuzzy decision support systems. In: Proceedings of 2016 IEEE International Conference on Fuzzy Systems (FUZZ-IEEE), Vancouver, Canada, pp. 2227–2234, 25–29 July 2016

12. Gorzałczany, M.B., Rudziński, F.: A multi-objective-genetic-optimization-based data-driven fuzzy classifier for technical applications. In: Proceedings of 2016 IEEE 25th International Symposium on Industrial Electronics (ISIE), Santa Clara, CA, USA, pp. 78–83, 8–10 June 2016

13. Zitzler, E., Laumanns, M., Thiele, L.: SPEA2: improving the strength Pareto evolutionary algorithm for multiobjective optimization. In: Proceeding of the Evolutionary Methods for Design, Optimisation, and Control, CIMNE, Barcelona, Spain, pp. 95–100 (2002)

14. Crick, F.: Central dogma of molecular biology. Nature **227**, 561–563 (1970)

15. Leavitt, S.A.: Deciphering the Genetic Code: Marshall Nirenberg. Office of NIH History (2010). https://history.nih.gov/exhibits/nirenberg/glossary.htm

16. Okabe, T., Jin, Y., Sendhoff, B.: A critical survey of performance indices for multi-objective optimisation. In: Proceedings of 2003 Congress on Evolutionary Computation, pp. 878–885. IEEE Press (2003)

17. Blasig, R.: GDS: gradient descent generation of symbolic classification rules. In: Cowan, J.D., Tesauro, G., Alspector, J. (eds.) Advances in Neural Information Processing Systems, vol. 6, pp. 1093–1100. Morgan-Kaufmann, San Mateo (1994)

18. Tanwani, A.K., Farooq, M.: Performance evaluation of evolutionary algorithms in classification of biomedical datasets. In: Proceedings of the 11th Annual Conference Companion on Genetic and Evolutionary Computation Conference (GECCO 2009), New York, NY, USA, pp. 2617–2624 (2009)

19. Huynh, T.Q., Reggia, J.A.: Improving rule extraction from neural networks by modifying hidden layer representations. In: Proceedings of 2009 International Joint Conference on Neural Networks, pp. 1316–1321 (2009)

20. Kerdprasop, N., Kerdprasop, K.: A high recall DNA splice site prediction based on association analysis. In: Proceedings of the 10th WSEAS International Conference on Applied Computer Science (ACS 2010), Stevens Point, Wisconsin, USA, pp. 484–489 (2010)

21. Gasparovica, M., Aleksejeva, L., Gersons, V.: The use of BEXA family algorithms in bioinformatics data classification. Inf. Technol. Manage. Sci. **15**(1), 120–126 (2013)

Automatic Detection of Blue-Whitish Veil as the Primary Dermoscopic Feature

Joanna Jaworek-Korjakowska[1]([⊠]), Paweł Kłeczek[1], Marcin Grzegorzek[2,3], and Kimiaki Shirahama[2]

[1] Department of Automatics and Biomedical Engineering,
AGH University of Science and Technology in Krakow, Kraków, Poland
jaworek@agh.edu.pl
[2] Research Group for Pattern Recognition, University Siegen, Siegen, Germany
[3] Faculty of Informatics and Communication,
University of Economics in Katowice, Katowice, Poland

Abstract. Dermoscopy is one of the major imaging modalities used in the diagnosis of melanoma and other pigmented skin lesions. Owing to the difficulty and subjectivity of human interpretation, dermoscopy image analysis has become an important research area. One of the most important local structure that is likely to appear in malignant melanoma is the blue-whitish veil. In this article, we present an unsupervised approach to the blue-whitish veil detection in dermoscopy images of pigmented skin lesions based on the analysis of HSV color space. The method is tested on a set of 179 dermoscopy images and the detection error rate is lower than 15%. The results demonstrate that the presented method achieves both fast and accurate blue structure segmentation in dermoscopy images.

1 Introduction

Malignant melanoma, the most deadly form of skin cancer, is nowadays one of the most rapidly increasing cancers in the world among many white-skinned populations. Since investigations have shown that the curability rate of melanomas in their early stage of development is nearly 100% [9], its accurate early diagnosis is a very important issue.

Dermoscopy is a non-invasive skin imaging technique which allows to observe features of pigmented melanocytic lesions which are not discernible during an examination with the naked eye (Fig. 1). When practiced by experienced observers, this imaging technique improves accuracy in diagnosing pigmented skin lesions from 10% to 27% compared to the clinical diagnosis with the naked eye [11].

In this paper we present an approach towards the issue of blue-whitish veil detection in dermoscopic images of melanomas with the help of supervised image processing and analysis techniques.

This paper is organized in 4 sections as follows. Section 1 (Introduction) presents the issue of emotion recognition, motivation to undertake the work

© Springer International Publishing AG 2017
L. Rutkowski et al. (Eds.): ICAISC 2017, Part I, LNAI 10245, pp. 649–657, 2017.
DOI: 10.1007/978-3-319-59063-9_58

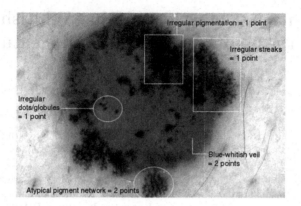

Fig. 1. Different local structures present within a pigmented lesion visible during a dermoscopy examination [1].

and presents state-of-the art. Section 2 (Materials and Methods) specifies the implemented system including signal preprocessing, feature selection and classification steps. In Sect. 3 (Results) the data acquisition process is described as well as conducted tests are presented. Section 4 (Conclusion and future works) closes the paper and highlights future directions.

1.1   Motivation

Malignant melanoma is likely to become one of the most common malignant tumors in the future, with even a ten times higher incidence rate. Since the early 1970s, malignant melanoma incidence has increased significantly; for example, in the USA it grows approximately by 4% every year. Due to the high skin cancer incidence, dermatological oncology has become a quickly developing branch of medicine. Nowadays the progress is visible both in primary research concerning pathogenesis of tumors (the role of genes or viruses in tumor development) and in the development of new, more efficient methods of computer-aided diagnosis. The development of computer-aided diagnosis (CAD) systems for automated diagnosis of melanoma is particularily important as young inexperienced dermatologists and family physicians have huge difficulties in the correct visual assessment of skin lesions. The CAD systems for dermatology help to increase the specificity and sensitivity and make the assessment of the skin mole more simple [15]. Many different algorithms have been proposed by dermatologists and researchers and are nowadays used in common medical practice. The main ones being the ABCD rule, the Menzies method, the 7-point checklist and global pattern analysis [1]. Dermatoscopy permits to identify a number of morphological patterns not visible with the naked eye which may be used for the diagnosis of skin melanocytic lesions. In particular, three diagnostic models have become more widely accepted by clinicians: (i) Pattern analysis, which is based on the expert qualitative assessment of numerous individual ELM criteria; (ii) the ABCD-rule of dermatoscopy which is based on a semi-quantitative analysis of the following groups of criteria: asymmetry (A), border (B), color (C) and different dermatoscopic (D) structures; (iii) the ELM 7-point checklist scoring diagnosis analysis, proposed by Argenziano et al. [1], defining only seven standard ELM criteria. In particular, the ELM 7-point checklist provides a simplification of standard pattern analysis and, if compared to ABCD, allows less experienced observers to achieve higher diagnostic accuracy values.

1.2 Clinical Significance

One of the most important indicators of invasive malignant melanoma is the blue-white veil (irregular, structureless areas of confluent blue pigmentation with an overlying white "ground-glass" film [1]) (Fig. 2). The blue-whitish veil is almost exclusively found in malignant melanomas and Spitz/Reed nevi. No differentiation between the veil in melanomas and Spitz/Reed nevi has been observed. Blue veil is a local structure that helps to distinguishing melanoma from Clark nevus [1]. With a sensitivity of 51% and a specificity of 97% [12]. The blue-whitish veil is one of the major criteria of the 7-point checklist with the highest odds ratio >11.

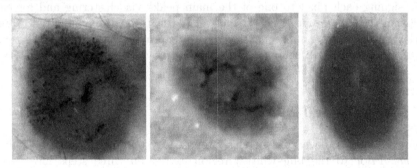

Fig. 2. Examples of blue-whitish veil [1]. (Color figure online)

1.3 Related Works

In recent years a few approaches for a computer-aided detection of blue-white veil in dermoscopy images has been proposed. Most of them use supervised machine learning techniques, mainly decision tree classifiers, to detect the relevant dermoscopic pattern.

The method proposed by Celebi et al. [3] is based on classifying each image pixel as either belonging to the blue veil or not depending on values of color and texture features computed for its neighborhood. Out of fifteen absolute and relative color features and three texture features (based on the gray level co-occurrence matrix) only two features, both belonging to color features, were finally selected for the classification model. To classify the data, a decision tree was trained using C4.5 algorithm. For manually selected test pixels, the classifier achieved sensitivity of 84.33% and specificity of 96.19%.

Ogorzałek et al. [13] used thresholding in RGB space to find grey-bluish areas. Those areas were identified as pixels which satisfied the following constraint: $R > 60 \wedge R - 46 < G < R + 15 \vee G > B - 30 \wedge R - 30 < B < R + 45$ (the thresholds were determined manually, based on past experience and joint work with clinical dermatologists). However, the paper does not provide any performance evaluation data of the proposed technique.

Arroyo et al. [2] proposed an approach similar to [3], they first compute a number of color features for individual pixels and then apply C4.5 algorithm to generate the decision tree used for pixel classification to state whether a given pixel may constitute part of the blue veil or not. For the obtained region masks another set of twelve shape-based features is extracted and used by a classifier to determine whether a given region is part of the blue veil. The technique yield sensitivity of 80.50% and a specificity of 90.93% for the classification of lesion images as either cases of melanoma with blue veil or other cases.

In their recent work, Di Leo et al. [10] took a different approach. Firstly, the lesion was subdivided into color regions via principal component analysis (PCA). A two dimensional histogram is computed with the two first principal components and the most significant peaks in the histogram were chosen as representatives of color regions in the input image. A lesion map was created by assigning each pixel to one of the main peaks via clustering and for each region in the lesion map a set of color-related features were computed. Values of those features were subsequently used to determine the presence or absence of the blue-whitish veil using a logistic model tree, which achieved sensitivity of 87% and specificity of 85%.

2 Materials and Methods

Figure 3 shows an overview of this proposed system to detect blue-whitish regions. Prior to the blue-whitish region extraction three steps have to be performed. In first stage the dermoscopy image has to be enhanced and pre-processed. In the second step the skin lesion is extracted from the surrounding. In the third step the basic features for pixels are obtained. Steps, namely the image enhancement and the determination of the background color, were performed on the images. In the last, fourth step, the selected features are calculated for each pixel belonging to the segmented skin lesion and based on the outcome the blue-whitish region is extracted. Those steps are typical for tasks of medical image recognition [7,8]. The main goal of this research is to present the blue-whitish veil area feature extraction and detection steps, so the preprocessing and segmentation stages will be described shortly. A detailed description of these steps can be found in our works [5,6].

2.1 Dermoscopy Image Preprocessing

Dermoscopic images are inhomogeneous and complex and furthermore they contain extraneous artifacts, such as skin lines, air bubbles, and hairs which appear

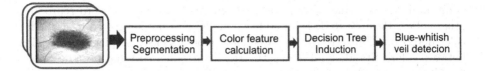

Fig. 3. An overview of our blue-whitish veil detection method. (Color figure online)

in almost every image. The black frame is detected on the basis of the lightness component value in the HSL color space. The removing of thick black hair is very important because the remaining thick black lines can influence the segmentation step. The detection of dark hairs is performed while using the top-hat transform. The air bubbles and remaining light-thin hairs are removed with the Gaussian filter.

2.2 Skin Mole Segmentation

In dermoscopy images we can observe two regions. The healthy skin which is s mostly homogeneous and the skin mole that contains a variety of colors and different local structures. Based on this assumption that the skin surrounding the mole is alike in each part of the image the region-growing algorithm can be applied to detect the desired area. The main advantage of the region-growing algorithm is that it is capable of correctly segmenting regions that have the same properties and are spatially separated. The implemented algorithm has been described in [6]. Figure 4 presents the results of the segmentation step.

Fig. 4. Examples of the segmentation step results.

2.3 Feature Calculation

The detection of specific colors can be performed while analyzing the values of channels in a certain color space. A color space is the type and number of colors which originate from the combinations of color components of a color model. The most known, and used color model is RGB and it defines a color space in terms of three components (Red, Green, and Blue). The RGB color model is an additive one which means that the channels are combined to reproduce other colors. Because RGB is an additive model it is not suggested to use it for color detection. Unlike RGB, the HSV color model seeks to depict relationships between colors. Standing for hue, saturation, and value, HSV has been described by Alvy Ray Smith in 1978. Figure 5 presents the Hue-Saturation-Value color model.

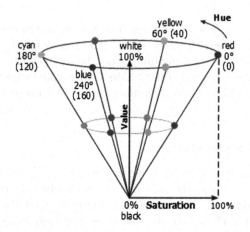

Fig. 5. The hue-saturation-value (HSV) color model [4]. (Color figure online)

In order to segment the blue-whitish veil regions a number of features is calculated to obtain the best classification result. Based on the three dimensional representation of HSV color model we analyze the following parameters: P_hue, P_saturation and P_value. Hue is defined as an angle in the range $[0, 2\pi]$ where the blue color lies between 240–300°. Saturation is the depth or purity of the color and is measured as a radial distance from the central axis with value between 0 at the center to 1 at the outer surface. The blue color can be identified for P_saturation higher than 0.5. Additionally, a 7×7 neighborhood of a pixel has been analyzed. Texture descriptors based on the Gray Level Co-occurrence Matrix (GLCM) [14] were calculated including entropy, contrast, and correlation. The GLCM parameters have been computed for each of the 4 directions 0, 45, 90, 135 and the statistics calculated from these matrices were averaged.

2.4   Detection of Blue-Whitish Veil

For the detection of the blue-whitish veil regions the most suitable features have been selected and described in the previous section. In this work we propose to use the decision tree as a predictive model to map the observations (pixel rules). The training data contained over 700 pixels corresponding to blue-whitish veil and 700 pixels chosen from other structures in skin lesion. The C4.5 algorithm is used to generate the decision tree. Decision Trees are often fast to train and generate easy to understand rules. Thus, Logistic Model Tree has been proposed as solution for the classification of the blue regions. The results of the classification process are presented in Fig. 6.

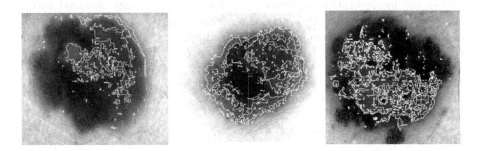

Fig. 6. Automatic detection of the blue-whitish veil regions. (Color figure online)

3   Results

The image set used in this study consists of 179 digital dermoscopy images obtained from the Interactive Atlas of Dermoscopy [1]. Images for this atlas have been provided by two university hospitals (University of Naples, Italy, and University of Graz, Austria) and stored on a CD-ROM in a JPEG format. These were true-color images with a typical resolution of 768×512 pixels. The diagnosis distribution of the cases was as follows: 132 with blue-whitish veil and 47 without searched area. The lesions were biopsied and diagnosed histopathologically in cases where significant risk for melanoma was present; otherwise they were diagnosed by follow-up examination All of the images have been assessed manually by a dermoscopic expert with extensive clinical experience.

The XOR measure was firstly used by Hance et al. and quantifies the percentage area detection error as:

$$Error = \frac{Area(Automatic\_BV \oplus Manual\_BV)}{Area(Manual\_BV)} \tag{1}$$

where Automatic_BV and Manual_BV are the binary images obtained by filling the automatic and manual areas of blue-whitish veil, respectively, $\oplus$ is the XOR operation that gives the pixels for which Automatic_BV and Manual_BV disagree, and Area (Manual_BV) denotes the number of pixels in the binary image Manual_BV. The area error rate for the detection of blue-whitish veil area obtained 15%.

4   Conclusion and Future Work

In this research we propose a new approach to the detection of blue-whitish veil in pigmented skin lesion. Due to the difficulty and subjectivity of human interpretation, the computerized image analysis techniques have become an important tool in the interpretation of dermoscopic images. The results obtained within this study indicate that the proposed algorithm can be used for the detection of the most important local structure in the diagnostic algorithm called 7-point checklist.

Acknowledgments. This work was supported by AGH University of Science and Technology statutory funds (No. 11.11.120.612).

References

1. Argenziano, G., Soyer, P.H., Giorgio, V.D., Piccolo, D., Carli, P., Delfino, M., Ferrari, A., Hofmann-Wellenhof, R., Massi, D., Mazzocchetti, G., Scalvenzi, M., Wolf, I.H.: Interactive Atlas of Dermoscopy. Book and CD/Web Resource, Edra Medical Publishing and New Media, Milan (2000)
2. Arroyo, J.L.G., Zapirain, B.G., Zorrilla, A.M.: Blue-white veil and dark-red patch of pigment pattern recognition in dermoscopic images using machine-learning techniques. In: 2011 IEEE International Symposium on Signal Processing and Information Technology (ISSPIT), pp. 196–201 (2011)
3. Celebi, M.E., Iyatomi, H., Stoecker, W.V., Moss, R.H., Rabinovitz, H.S., Argenziano, G., Soyer, H.P.: Automatic detection of blue-white veil and related structures in dermoscopy images. Comput. Vis. Pattern Recogn. **32**(8), 670–677 (2008)
4. Environmental Systems Research Institute Inc: ArcMap. 10.3 color model conversion function (2016). http://desktop.arcgis.com
5. Jaworek-Korjakowska, J.: Computer-aided diagnosis of micro-malignant melanoma lesions applying support vector machines. BioMed Res. Int. **2016**, 1–8 (2016)
6. Jaworek-Korjakowska, J.: Novel method for border irregularity assessment in dermoscopic color images. Comput. Math. Methods Med. **2015**, 1–11 (2016)
7. Kowal, M., Filipczuk, P.: Nuclei segmentation for computer-aided diagnosis of breast cancer. Int. J. Appl. Math. Comput. Sci. **24**(1), 19–31 (2014)
8. Kowal, M., Filipczuk, P., Obuchowicz, A., Korbicz, J., Monczak, R.: Computer-aided diagnosis of breast cancer based on fine needle biopsy microscopic images. Comput. Biol. Med. **43**, 1563–1572 (2013)
9. Leiter, U., Buettner, P.G., Eigentler, T.K., Garbe, C.: Prognostic factors of thin cutaneous melanoma: an analysis of the central malignant melanoma registry of the german dermatological society. J. Clin. Oncol. **22**(18), 3660–3667 (2004). PMID: 15302905, http://ascopubs.org/doi/abs/10.1200/JCO.2004.03.074
10. Leo, G.D., Fabbrocini, G., Paolillo, A., Rescigno, O., Sommella, P.: Toward an automatic diagnosis system for skin lesions: estimation of bluewhitish veil and regression structures. In: 6th International Multi-Conference on Systems, Signals and Devices (2009)
11. Mayer, J.: Systematic review of the diagnostic accuracy of dermatoscopy in detecting malignant melanoma. Med. J. Aust. **167**(4), 206–210 (1997)

12. Menzies, S.W., Crotty, K.A., Ingwar, C., McCarthy, W.H.: An Atlas of Surface Microscopy of Pigmented Skin Lesions: Dermoscopy, 2nd edn. McGraw-Hill, Sydney (2002)

13. Ogorzałek, M., Surówka, G., Nowak, L., Merkwirth, C.: Computational intelligence and image processing methods for applications in skin cancer diagnosis. In: Fred, A., Filipe, J., Gamboa, H. (eds.) Communications in Computer and Information Science. CCIS, vol. 52, pp. 3–20. Springer, Berlin, Heidelberg (2010)

14. Śmietański, J., Tadeusiewicz, R., Łuczyńska, E.: Texture analysis in perfusion images of prostate cancer a case study. Int. J. Appl. Math. Comput. Sci. **20**(1), 149–156 (2010)

15. Tadeusiewicz, R.: Place and role of intelligent systems in computer science. Comput. Methods Mater. Sci. **10**(4), 193–206 (2010)

Bio-inspired Topology of Wearable Sensor Fusion for Telemedical Application

Eliasz Kantoch$^{(\boxtimes)}$, Dominik Grochala, and Marcin Kajor

AGH Univeristy of Science and Technology, Kraków, Poland
kantoch@agh.edu.pl

Abstract. Application of wearable sensors is a promising approach in building novel telemedical services. In this paper, we propose the biologically inspired method for monitoring human activity in living conditions. The solution is based on the set of sensors integrated in the single wearable device and imitates the natural arrangement of human perception system. The designed wearable device enables to acquire physiological and environmental parameters. With the use of proposed appliance it is possible to collect body and ambient temperature, barometric pressure, light intensity and acceleration. In the experimental part, the signals were recorded during selected activities of daily living (ADL). The sitting activity classification was implemented using perceptron.

1 Introduction

During the last decade we witnessed significant improvement in the development of wearable sensors in the scope of power consumption, miniaturization and resolution. The world population is ageing rapidly. The fraction of people aged 65 and more was 7% in 2000 and will probably increase to 16% in 2050 as the global population grows [1]. This requires development of the efficient technologies supporting the healthcare. One approach is to transform selected healthcare services into the telemedicine solution based on wearable sensors. Appling wearable devices and telemedicine drives the development of more and more sophisticated assisted living systems supporting the elderly and chronically ill or disabled patients. The most commonly used architecture of such system contains the sensor measuring one of the patient's physiological parameters communicating with the control unit which is capable of acquiring the signal and sending data to the personal computer or remote server. Further processed data may be then analyzed with dedicated algorithms to extract some characteristic features and obtain the heart rate, temperature or blood oxygen saturation. The idea of the application of wearable sensors for activity monitoring, was raised by many researchers. More than 10 years ago Kara E. Bliley et al. stated justification for the use of triaxial accelerometers in this type of systems [2]. By using MEMS technology, this solution allowed significant miniaturization and reduction of costs in contrast to traditional uniaxial accelerometers. Motoi et al. noticed that measurement of patient activity is widely used in the process of assessing the progress of rehabilitation. It can also be helpful in planning

L. Rutkowski et al. (Eds.): ICAISC 2017, Part I, LNAI 10245, pp. 658–667, 2017.
DOI: 10.1007/978-3-319-59063-9_59

the long term treatment [3]. It is stated that data recorded from different areas and in different conditions carry the most relevant information. Studies found the usefulness of the application of multiple sensors that measure various parameters. Ince et al. presented a solution based on wearable and stationary sensors distributed in the system for monitoring activity at home [4]. Simultaneous acquisition of signals from the accelerometer, magnetometer, pressure sensor, in conjunction with the data from the camera and acoustic sensors, create a complete information allowing for the identification of certain activities. The researchers also confirmed that such system may be not very expensive, which increases its availability. Maeneka et al. noted that monitoring of daily activity is useful in maintaining health [5]. Researchers have developed a prototype device that integrates multiple sensors and measures acceleration, pressure, humidity and ambient temperature. The use of sensors based on MEMS technology has allowed the system miniaturization. To increase usability, the extension of additional communication standards was proposed [6]. Mukhopadhyay noted that non-hospital care, where telemedical systems plays a key role, may significantly limit the funds spent on healthcare. The development of technology allows to build smaller and more efficient devices. The author describes the basic configurations and system architectures in terms of advantages and disadvantages of their use. Further development of these technologies as well as their importance is anticipated in the future. Etemadi et al. presented a dedicated solution of wearable monitoring for patients with cardiovascular diseases [7]. Apart from the typical accelerometer and pressure sensor, the logger is equipped with a chip allowing electrodiagnostic signals measurement. Data carrier is microSD card, and the power source is a lithium coin battery. Simultaneous acquisition of ECG signals and the SCG (seismocardiogram) is a source of information about the work of the heart while the barometric pressure sensor provides data for calculation of the patient's altitude. Previous studies have focused on human monitoring using various wearable sensors including heart rate monitors, body temperature sensors or accelerometers [4,6,9,10]. It was also demonstrated that house embedded sensors can be successfully used to monitor human behavior [4,5,8,11]. Therefore, in this study we investigate the possibility of integrating both wearable and environmental sensors into single device for telemedical application. The nature constitutes the perfect source of inspiration in the design of technical systems [12]. The number of bioinspired sensors systems were described in [13]. Sometimes the researchers are racking their brains to find the solution for the complex problem before realizing that it actually have already been solved by the natural process of evolution throughout the thousands of years. Such approach may lead to optimization of the design process and strongly contribute to reducing the time and research costs. Undoubtedly, the challenge is to wisely adapt the biologically inspired concepts into technical field. The biocybernetic modelling is certainly vital for current development and may be the great tool for considering and solving mentioned problems. Our approach was based on the direct observation of the anatomy and physiology of chordate neural system, in particular the human one. According to this, we proposed the topology of the system consisting of the sensors set and the processing unit. In order to optimize the patient monitoring process, we performed the fusion of sensors providing the information similar to the one receiving by the receptors and human neural system. The proposed device was equipped with accelerometers and barometric pressure sensor, which imitate the patient motion and position sensing. The state of the environment in which the person is being monitored is supervised by the light intensity sensor and the set of digital thermometers. Supplying the system with the touch sensor in the form of polymeric piezofilm gives the possibility of mechanic pressure perception inspired by the skin mechanoreceptors [12]. The main processing unit is in charge of combining provided signals and enables data acquisition on the memory carrier. Such approach would be meaningless and incomplete from the biological point of view if it had not been for the remote communication aspect. The realization of this functionality was implemented by the wireless protocol, Bluetooth Low Energy. In order enable the development and tests execution of described concept, the compact measurement device was designed and prototyped Fig. 1.

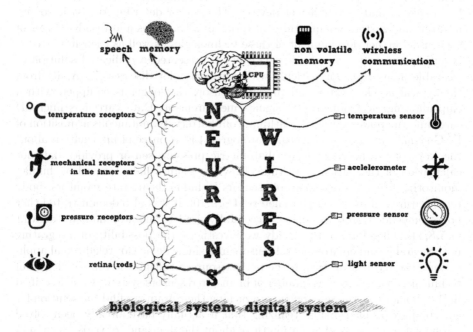

Fig. 1. System concept.

The remaining part of the paper is organized as follows. Section 2 presents hardware setup. Section 3 presents and discusses the results. Section 4 concludes the paper.

2   Hardware Setup

2.1   Architecture of the Developed Prototype

The device consists of MEMS and analog sensors connected to the microcontroller and the circuits supporting the power supply from Li-Poly battery which can be easily charged with the DC jack or USB port. The prototype is embedded inside the plastic casing with the microSD slot, control microswitch and RGB diode serving as the simple indicator. The architecture of the developed prototype is shown in the Fig. 2.

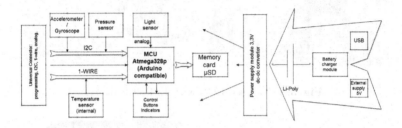

Fig. 2. Architecture of the developed prototype.

2.2   Power Supply and Interfaces

The logger is equipped with the universal IDC connector supporting the ISP programming and external sensors such as DS18b20 temperature sensor. The USB charger design is based on the galvanic separator preventing the user from the direct voltage impact during the recharging procedure. The power supply circuits involve DC/DC converters enabling whole device to be supplied by the 3.3 V, which contributes to the lower power consumption. The average battery life reaches approximately 24 h, which is the main factor determining the possibility of long-lasting monitoring. In order to prevent the user from the eventual battery damage effects, the accumulator is wrapped in the flame resistant composite based on the PTFE and glass fiber. Developed PCB with Li-Poly battery is shown in the Fig. 3.

The measurement data is recorded on the microSD card in the .txt file format which can be easily imported into Excel or MATLAB and further processed or analyzed. The significant advantage of proposed system is its versatility Fig. 4. The universal connector enables to apply different sensors for further development of the device. Moreover, the possibility to equip the prototype with additional RS-232 module seems to be the convenient way to establish the direct communication with the computer of other extension boards.

Fig. 3. Developed PCB with Li-Poly battery.

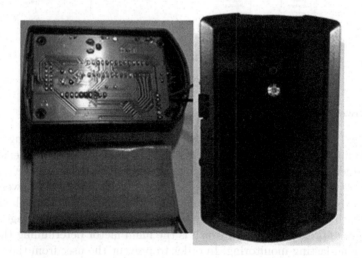

Fig. 4. The developed device embedded inside the plastic casing with dimensions: $90 \times 58 \times 18$ mm.

2.3 Sensors

The prototype takes advantage of the advanced MEMS sensors dedicated to the specific measurement of environmental parameters. The MPU-6050, used for motion detection, is the integrated 6-axis MotionTracking device combining a 3-axis gyroscope, 3-axis accelerometer and Digital Motion Processor. It features high resolution and can be used to perform advanced motion analysis with the use of motion processor. The BMP-085 sensor performs the barometric pressure measurement and in conjunction with altitude calculations may be used in the systems monitoring the height above sea level. DS18b20 digital thermometer is utilized for ambient temperature measurements. This device features programmable resolution of the measurements and uses 1-Wire communication protocol which minimalizes the wire connections on the PCB. The sample rate of the signals acquired from digital sensors is limited by the hardware conversion time of single measurement. The Table 1 exhibits the most important technical parameters of the sensors used in the project. For light intensity monitoring the basic analog photoresistor was used in the prototype. It is configured as the voltage divider and provides the analog signal to the analog to digital converter of the microcontroller in 10-bit resolution setup.

Table 1. The characteristics of the used sensors.

Parameter	Sensors		
	MPU-6050	BMP-085	DS18b20
Range	± 16 g	700–1100 hPa	−10–85°C
Accuracy	-	± 2.5 hPa	± 0.5°C
ADC resolution	16 bits	16–19 bits	9–12 bits
Communication protocol	I2C	I2C	1-Wire
Max. sampling rate	1 kHz	200 Hz	1–10 Hz

3 Methods, Results and Discussion

In the experimental part, we recorded signals during selected activities of daily living. We selected the following: sitting, walking, going up the stairs. Sensors were placed on the chest and fastened using elastic belt. A set of signals was acquired from healthy volunteers in laboratory conditions. Table 2 gives an overview of obtained measurement data record form the device.

We choose sitting as an example of daily activity which can be easily detected with designed prototype and simple perceptron neural network. We investigated a set of different methods for extracting features from acceleration sensor.

Table 2. The piece of data record form the device

Acc X [g]	Acc Y [g]	Acc Z [g]	Ext temp [C]	Int temp [C]	Press. [Pa]	Light [%]
−0.27	−0.22	1.13	26.75	25.11	98042	99.12
−0.36	−0.32	0.90	26.79	25.13	98042	99.12
−0.50	−0.21	0.80	26.86	25.18	98042	99.12
−0.46	−0.20	0.86	26.87	25.26	98042	95.80

Finally, the output vector for the perceptron was calculated from the x-axis of
the acceleration signal as the index corresponding to the frequency with maximum energy. We used Fast Fourier Transform for spectrum energy calculation.
Figure 5 gives an overview of obtained selected acceleration data record from the
device and Fig. 6 depicts the plot the of selected activity (going upstairs). The
output vector of the network was ones for sitting activity and zeros for others.
The neural network was trained on 75% of collected samples and tested on the
25% of data. The perceptron was capable of separating an input space with
a straight line into two categories: one sitting and zero other activity (Fig. 7).
This classification may have clinical motivation as it can be used for exclusion of
pathological movement or falling accident, especially in case of disabled patients.

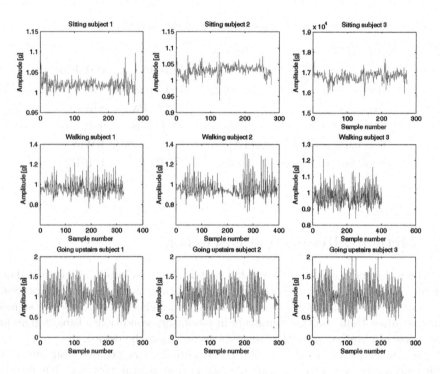

Fig. 5. The plot of acceleration data record from the device.

However, the device can provide additional parameters as an input of neural
network which can be easily extended with further neurons to gain the ability
of classification of more complex activities. Calculation of altitude, based on the
barometric pressure, can be the fundamental for detection of going up or descending the stairs without complex computational methods. On the other hand, the
motivation for usage of mentioned sensors, apart from their high accuracy and
overall performance, was mainly the possibility of integrating the physiological
(body temperature), movement (acceleration) and environmental measurements

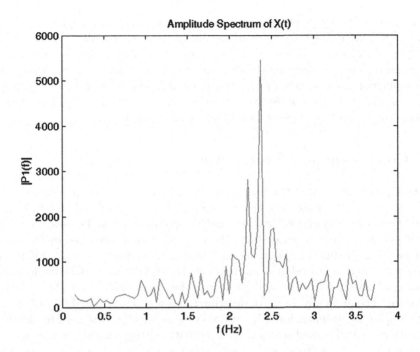

Fig. 6. Amplitude spectrum for going upstairs.

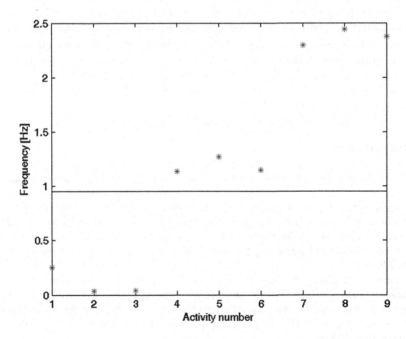

Fig. 7. The plot of separated input space into two categories (1–3 - sitting, 4–9 - other).

(barometric pressure, light and external temperature) into one single data record. This approach may be interesting especially for data synchronization but also for analysis of different activities. Another application of described system can be the simultaneous acquisition of body temperature and 3-axis acceleration which in conjunction may be used for detection of epilepsy attacks or fever occurrence. We can also measure the eventual weather changes which may in turn exert strong impact on the elderlies mood and physical condition.

4   Conclusion and Future Works

In this paper we showed the feasibility of building a system for remote multi-parameter patient monitoring in home conditions. The result of this study indicated that selected physiological and ambient parameters can be measured using the single device and integrated into data record. Signals provided by the device can be used to support diagnosis and treatment of patients who suffer from numerous diseases. Developed method can be utilized as a platform for monitoring selected parameters and provides a lot of valuable data about patient behavior. The main advantage of described sensor fusion is possibility of providing complex multidimensional feature vectors as an input for machine learning algorithms. We classified selected ADL activity (sitting) using perceptron. The feature extraction method and classification model can be used for real-time activity recognition on resource-constrained devices. The designed prototype features small size, intuitive operation and it can be easily reprogram. This gives the system usability attributes and makes it user friendly. Future work will focus on carrying out more experiments on the larger elderly group in home environment and integrating developed device within more complex telemedical system.

Acknowledgment. This project was funded by the AGH University of Science and Technology project no. 11.11.120.612.

References

1. World Health Organisation: WHO | Ageing. http://www.who.int/topics/ageing/en/
2. Bliley, K.E., Holmes, D.R., Kane, P.H., Foster, R.C., Levine, J.A., Daniel, E.S., Gilbert, B.K.: A miniaturized low power personal motion analysis logger utilizing mems accelerometers and low power microcontroller. In: 2005 3rd IEEE/EMBS Special Topic Conference on Microtechnology in Medicine and Biology, pp. 92–93. IEEE (2005)
3. Motoi, K., Higashi, Y., Kuwae, Y., Yuji, T., Tanaka, S., Yamakoshi, K.: Development of a wearable device capable of monitoring human activity for use in rehabilitation and certification of eligibility for long-term care. In: Conference on Proceedings of the IEEE Engineering in Medicine and Biology Society, vol. 1, pp. 1004–1007 (2005)

4. Ince, N.F., Min, C.H., Tewfik, A.H.: Integration of wearable wireless sensors and non-intrusive wireless in-home monitoring system to collect and label the data from activities of daily living. In: Proceedings of the 3rd IEEE-EMBS International Summer School and Symposium on Medical Devices and Biosensors, ISSS-MDBS 2006, pp. 28–31. IEEE (2006)
5. Maenaka, K., Masaki, K., Fujita, T.: Application of multi-environmental sensing system in MEMS technology - monitoring of human activity. In: 4th International Conference on Networked Sensing Systems, INSS, pp. 47–52. IEEE (2007)
6. Mukhopadhyay, S.C.: Wearable sensors for human activity monitoring: a review (2015). http://ieeexplore.ieee.org/document/6974987/
7. Etemadi, M., Inan, O.T., Heller, J.A., Hersek, S., Klein, L., Roy, S.: A wearable patch to enable long-term monitoring of environmental, activity and hemodynamics variables. IEEE Trans. Biomed. Circuits Syst. **10**, 280–288 (2016)
8. Augustyniak, P., Smolen, M., Mikrut, Z., Kantoch, E.: Seamless tracing of human behavior using complementary wearable and house-embedded sensors. Sensors (Switzerland) **14**, 7831–7856 (2014)
9. Kantoch, E.: Technical verification of applying wearable physiological sensors in ubiquitous health monitoring. In: Computing in Cardiology Conference (CinC) 2013 (2013)
10. Kantoch, E., Augustyniak, P., Markiewicz, M., Prusak, D.: Monitoring activities of daily living based on wearable wireless body sensor network. In: Annual International Conference of the IEEE Engineering in Medicine and Biology Society 2014, pp. 586–589 (2014)
11. Augustyniak, P., Kantoch, E.: Turning domestic appliances into a sensor network for monitoring of activities of daily living. J. Med. Imaging Heal. Informatics. **5**, 1662–1667 (2015)
12. Barbarossa, S., Scutari, G.: Bio-inspired sensor network design. IEEE Signal Process. Mag. **24**, 26–35 (2007)
13. Valle, M.: Bioinspired sensor systems (2011). http://www.mdpi.com/1424-8220/11/11/10180/

An Evaluation of Fuzzy Measure for Face Recognition

Paweł Karczmarek[1]([✉]), Adam Kiersztyn[1], and Witold Pedrycz[2,3,4]

[1] Institute of Mathematics and Computer Science, The John Paul II Catholic
University of Lublin, Ul. Konstantynów 1H, 20-708 Lublin, Poland
{pawelk,adam.kiersztyn}@kul.pl
[2] Department of Electrical and Computer Engineering, University of Alberta,
Edmonton, AB T6R 2V4, Canada
[3] Department of Electrical and Computer Engineering, Faculty of Engineering,
King Abdulaziz University, Jeddah 21589, Saudi Arabia
[4] Systems Research Institute, Polish Academy of Sciences, Warsaw, Poland
wpedrycz@ualberta.ca

Abstract. In this paper, we analyze the properties and performance of
the Choquet integral and fuzzy measure, particularly λ–fuzzy measure
in the context of an aggregation of classifiers based on various facial
areas. The fuzzy measure and Choquet integral have been shown to be an
efficient aggregation techniques. However, in practice reported so far, the
choice of the initial values of the measure corresponding to the saliency
of facial features has been dependent upon the expert decision. Here, we
propose an algorithmic way of finding these values. For this purpose a
Particle Swarm Optimization (PSO) method is considered. The reported
experimental results show that the method is more effective than the
expert – centered approach.

Keywords: Aggregation · Facial features · Face recognition · Choquet
integral · Fuzzy measure · Particle Swarm Optimization

1 Introduction

Face recognition has been one of the most considered problems in the area of
image recognition. This is because of its wide applicability in many areas of life
such as access control, driver's license and passport verification, border control,
or identification of mobile devices owners. The plethora of approaches is very
rich. One can here refer to [13], Eigenfaces [31], Fisherfaces [4], local descriptors
[1,5,15], elastic bunch graph matching [33], Granular Computing [20], sparse
representation [34], deep learning [11,30], information fusion and aggregation
[17,21], etc. The latter lets us save the amount of computational memory needed
to execute the classification process and improve the result of classification.
However, the main question arises here: How to treat the particular facial areas
contributing to the process of classification?

© Springer International Publishing AG 2017
L. Rutkowski et al. (Eds.): ICAISC 2017, Part I, LNAI 10245, pp. 668–676, 2017.
DOI: 10.1007/978-3-319-59063-9_60

In the literature, there are many approaches related to the estimation of the importance of particular facial parts. In general, they are two main groups. The first approach is based on psychological experiments and observations of ways people recognize others. The main methods here are comparing the facial parts by subjects taking parts in the experiments, recognition of faces with covered or changed parts of face, and similar, cf. [7,10,12,16,22,26,27,32]. From the computational point of view, the methods are based on checking the recognition rate with an application of a method used to the cropped region of the face [6,9, 35] or applying intuitive observation made by an expert in the field (supported by trial–and–error test), cf. [17,21]. Usually, all the described methods show some regularity in the produced results confirming that the most salient region of the face is eye and eyebrows area. However, in the context of a particular aggregation method the weights associated with all the subregions on which the classifiers are built should be obtained in a different way depending on the structure of the aggregation operator.

The main objective of this study is to present a novel approach to determine the parameters of the fuzzy measure which are related to particular facial parts or regions without the necessity to conduct psychological experiments as well as with relatively low computational cost. We are interested in obtaining these values by applying the well–known Particle Swarm Optimization method. More-over, our aim is to carry out a comprehensive comparison of our proposal with other methods.

The paper is organized as follows. The role of the Particle Swarm Optimization and the general processing scheme are covered in Sect. 2. Section 3 presents the experimental results while Sect. 4 covers the conclusions and the future work directions.

2   The Role of Choquet Integral and PSO in the Process of Aggregation of Classifiers

Let us assume that in the process of classification we consider n facial features such as eyes, eyebrows, nose, mouth, and others depending on our preferences. Formally, the overall face area can be represented as $X = \{x_1, \ldots, x_n\}$ with x_i, $i = 1, \ldots, n$ being the corresponding facial regions. If the dataset consists of N subjects and each person k has N_k images in the dataset then we can assume that a new unknown face can likely belong to the class k for which the following Choquet integral attains in maximal value (cf. [23]):

$$\int g \circ h = \sum_{i=1}^{n} (h(x_{ik}) - h(x_{i+1,k})) g(A_i), \quad h(x_{n+1,k}) = 0, \tag{1}$$

where the function $g : P(X) \to [0, 1]$ is a fuzzy measure satisfying the following conditions

$$g(\emptyset) = 0, \tag{2}$$

$$g(X) = 1, \tag{3}$$

$$g(A) \leq g(B) \quad \text{for} \quad A \subset B, \quad \text{where} \quad A, B \in P(X). \tag{4}$$

In the Sugeno parametric version of the fuzzy measure [29] one has

$$g(A \cup B) = g(A) + g(B) + \lambda g(A) g(B), \quad \lambda > -1. \tag{5}$$

Then

$$g(A_1) = g(x_1) \tag{6}$$

and

$$g(A_i) = g(x_i) + g(A_{i-1}) + \lambda g(x_i) g(A_{i-1}) \quad \text{for} \quad i = 2, \ldots, n. \tag{7}$$

$g(x_i), i = 1, \ldots, n$ are the values of λ–fuzzy measure to be found and representing the atomic cues saliency. The parameter λ can be easily obtained from the following equality [29]

$$1 + \lambda = \prod_{i=1}^{n} (1 + \lambda g(x_i)). \tag{8}$$

Finally, the values $h(x_{ik})$ of the kth class with maximum membership grade for the ith feature (i.e., in the ith classifier) are obtained as

$$h(x_{ik}) = 1/N_k \sum_{\mu_{ij} \in C_k} \mu_{ij}, \tag{9}$$

where

$$\mu_{ij} = \frac{1}{1 + \frac{d_{ij}}{\overline{d}_i}}. \tag{10}$$

Here, d_{ij} denotes the distance between a test image and the vector representing ith feature for a jth training image, $\overline{d}_i$ is an average distance between the images in the ith classifier. C_k denotes the kth class. Moreover, the values $h(\cdot)$ appearing in the formula (1) are sorted in decreasing order. Hence, according to the classification rule that returns a class for which the Choquet integral returns the maximal value, a way of finding the initial values of function $g(\cdot)$ invokes an optimization technique maximizing the values between for the same class images and minimizing the inter–class values.

Here, a sound alternative is a well–known Particle Swarm Optimization method [19] which is a socially–inspired approach where the parameters (values) to be optimized are gathered to form a particle. The group of particles is involved in the optimization process. The process is initialized by random setting of particles locations and velocities. The following rules are used to update the positions of particles $\mathbf{y_i}, i = 1, \ldots, n$, (in our case, a particle is a set of the parameters of the fuzzy measure corresponding to the facial parts) in the next generations:

$$\overline{\mathbf{v}}_\mathbf{i} = \overline{\mathbf{v}}_\mathbf{i} + 2\mathbf{r_1} \otimes (\mathbf{p_i} - \mathbf{y_i}) + 2\mathbf{r_2} \otimes (\mathbf{p_g} - \mathbf{y_i}), \tag{11}$$

$$\mathbf{y_i} = \mathbf{y_i} + \overline{\mathbf{v}}_\mathbf{i}, \tag{12}$$

where $\overline{\mathbf{v}}_i$ denote velocities, $\mathbf{r}_1$ and $\mathbf{r}_2$ constitute random vectors with values from $[0,1]$, $\mathbf{p}_i$ is the ith particle personal best position while $\mathbf{p_g}$ is global best position of the whole swarm. A point-wise vector multiplication is represented by $\otimes$ symbol. The cognitive and social part are $2\mathbf{r}_1 \otimes (\mathbf{p_i} - \mathbf{y_i})$ and $2\mathbf{r}_2 \otimes (\mathbf{p_g} - \mathbf{y_i})$, respectively. They are used to change the positions of the particles [18,19]. In our case we use the concept of inertia weight which helps to better control exploration and exploitation by the PSO [3,28]. We modify the equation (11) using so called random inertia weight [8]

$$\overline{\mathbf{v}}_i = (0.5 + r/2)\,\overline{\mathbf{v}}_i + 2\mathbf{r}_1 \otimes (\mathbf{p_i} - \mathbf{y_i}) + 2\mathbf{r}_2 \otimes (\mathbf{p_g} - \mathbf{y_i}), \qquad (13)$$

where r is a random number from $[0,1]$. This way of algorithm modification lets us to speed up the execution time by obtaining the optimized result by reaching the satisfying level of convergence after a few generations. The numbers r, r_1, and r_2 are randomly obtained during each generation for each particle.

As mentioned, our main goal is to find the fuzzy measure initial values. However, in our initial part of experiments we found that different parts of face are responsible for *positive* and for *negative* classification, i.e., for the conclusion that a given face belongs or does not belong to a given class, respectively. Therefore, the optimization has to be conducted separately for the case where the Choquet integral values are maximized, and separately for the case when they are minimized. The final result of this optimization process is yielded as the weighted average result of *positive* and *negative* optimization, respectively. An overall scheme is presented in Fig. 1.

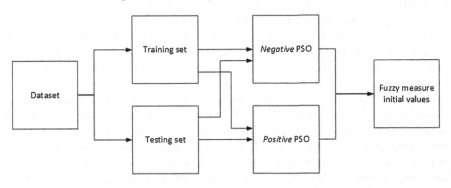

Fig. 1. An overall flow of processing

3   Experimental Results

For the purpose of our experimental analysis we use the AT&T dataset [2] which is one of the most commonly utilized face datasets. It consists of 400 images of 40 people (10 per subject). Each of the images are initially preprocessed (scaled, cropped, and the histogram was equalized). Next, we extract 6 parts of the face, namely eyebrows, eyes, nose, mouth, left and right cheek area, respectively

Fig. 2. Chosen facial parts: eyebrows, eyes, nose, mouth, left cheek, and right cheek (from the observer point of view)

Table 1. Recognition rates for particular facial parts using LBP

Eyebrows	Eyes	Nose	Mouth	Left cheek	Right cheek
67.30	57.94	70.79	58.43	60.17	56.57

(see Fig. 2). The method of obtaining the distances to be used in the optimization processes is a classic Local Binary Pattern algorithm (see [1]) with no partition of the images onto the subregions. It means, that the resulting histogram lengths are 256. The recognition rates obtained for particular facial parts separately are given in Table 1. To obtain the parameters of the fuzzy measure we divide randomly the dataset of images on the gallery (5 images of each person) and the testing set (5 images per face). Next, we set the number of particles to be 50. The number of generations is 10 only. We assign the weight 0.2 for the *positive* and 0.8 for the *negative* PSO result. The experiment is repeated 100 times. The final parameters of the fuzzy measure are collected in Table 2.

To compare our proposal with other methods of finding the initial weights assigned to the parts of face we conducted a few additional experiments with the presence of experts which are the members of our lab, friends, and people experienced in estimating the face party using well–known Analytic Hierarchy Process (see [24,25]). The first group of 18 people estimated the percentage importance of particular facial regions in the process of face recognition. Two experts were asked to conduct the AHP processes based on pairwise comparisons of the 6 abstract facial features. Finally, the results obtained by these experts were optimized as described in [14]. All the weights gathered from the experts or obtained in other experiments (after normalization) are listed in Table 2. An interesting result is listen in the first two rows of the table. They show that (in case of LBP algorithm) the eyebrows, right, and left cheek areas are responsible for a statement that an image belongs to a specific class while the eyes (without eyebrows), nose, and mouth areas are responsible for a statement that an image does not belong to a class it is compared to.

Table 3 consists the average results of 100 iterations of experiments based on aggregation of LBP classifiers applied to the particular facial parts when, again, 5 images of each person are taken to the training set and 5 images are taken to the testing set, and with initial fuzzy measure weights given in Table 2, respectively. Each iteration is completed separately by dividing the dataset randomly onto the equal sets, namely gallery and testing set, to proceed the simulation

Table 2. Parameters of the fuzzy measure obtained in different computational and psychological experiments

Method	Eyebrows	Eyes	Nose	Mouth	Left cheek	Right cheek
Negative PSO	0.02	0.32	0.34	0.26	0.03	0.02
Positive PSO	0.59	0.03	0.05	0.03	0.20	0.10
Av. of *neg.* PSO & *pos.* PSO	0.31	0.17	0.20	0.15	0.11	0.06
Our proposal	0.14	0.26	0.28	0.22	0.06	0.04
Experts votes	0.13	0.32	0.20	0.21	0.07	0.07
LBP for particular face parts	0.18	0.16	0.19	0.16	0.16	0.15
AHP by expert 1 [16]	0.32	0.32	0.10	0.16	0.05	0.05
AHP by expert 2 [16]	0.02	0.54	0.10	0.26	0.04	0.04
Average AHP	0.17	0.43	0.10	0.21	0.04	0.05
AHP by expert 1 + opt. [14]	0.32	0.32	0.10	0.16	0.05	0.05
AHP by expert 2 + opt. [14]	0.03	0.47	0.11	0.29	0.04	0.06
Average optimized AHP	0.18	0.40	0.10	0.22	0.05	0.05
Eigenfaces results [17]	0.20	0.21	0.19	0.16	0.12	0.13
Fisherfaces results [17]	0.19	0.19	0.16	0.14	0.16	0.16

Table 3. Results with an application of different weights (see Table 2)

Weight set	Rank 1	Rank 5	Rank 10
Negative PSO	85.65	96.97	99.00
Positive PSO	74.35	91.18	95.79
Average of *negative* PSO & *positive* PSO	86.79	96.42	98.71
Our proposal	**88.92**	**97.65**	**99.12**
Experts votes	88.63	97.54	99.03
LBP for particular face parts	86.94	96.71	98.71
AHP by expert 1 [16]	85.08	95.78	98.48
AHP by expert 2 [16]	81.82	95.57	98.48
Average AHP	86.48	96.67	98.72
AHP by expert 1 + optimization [14]	85.11	95.80	98.49
AHP by expert 2 + optimization [14]	83.79	96.23	98.71
Average optimized AHP	87.09	96.84	98.77
Eigenfaces results [17]	87.90	97.01	98.87
Fisherfaces results [17]	86.79	96.54	98.62

(see The results show the effectiveness of our proposal which is slightly better than the weights obtained from 18 experts for rank 1, rank 5, and rank 10 recognition rates. Good results are produced with the weights obtained by AHP process followed by optimization with PSO of the result to preserve the consistency condition [24,25]. This method was described in details in [14]. The weights being the outcomes of the experiments with Eigenfaces [31] for the AT&T dataset can be useful as well here.

4   Conclusions and Future Studies

In this study, we have discussed a novel approach to obtaining fuzzy measure coefficients required in aggregation techniques. This method offers an efficient vehicle to substitute a way of weight determination realized by a human expert. To illustrate the effectiveness of our proposal we have conducted a series of experiments and the results clearly demonstrated the usefulness of the algorithmic approach. The potential future work directions include an extension of the method to the multimodal biometrics, an application of other algorithms of recognition as well as optimization methods, or an application to other datasets and domains. The studies on the facial features saliency in the context of *positive* and *negative* classification are definitely worth pursuing.

Acknowledgments. The authors are supported by National Science Centre, Poland (grant no. 2014/13/D/ST6/03244). Support from the Canada Research Chair (CRC) program and Natural Sciences and Engineering Research Council is gratefully acknowledged (W. Pedrycz).

References

1. Ahonen, T., Hadid, A., Pietikäinen, M.: Face recognition with local binary patterns. In: Pajdla, T., Matas, J. (eds.) ECCV 2004. LNCS, vol. 3021, pp. 469–481. Springer, Heidelberg (2004). doi:10.1007/978-3-540-24670-1_36
2. AT&T Laboratories Cambridge: AT&T Database of Faces. http://www.cl.cam.ac.uk/research/dtg/attarchive/facedatabase.html
3. Bansal, J.C., Singh, P.K., Saraswat, M., Verma, A., Jadon, S.S., Abraham, A.: Inertia weight strategies in particle swarm optimization. In: 2011 Third World Congress on Nature and Biologically Inspired Computing. Salamanca, pp. 633–640 (2011)
4. Belhumeur, P.N., Hespanha, J.P., Kriegman, D.J.: Eigenfaces vs. Fisherfaces: recognition using class specific linear projection. IEEE Trans. Pattern Anal. Mach. Intell. **19**, 711–720 (1997)
5. Bereta, M., Pedrycz, W., Reformat, M.: Local descriptors and similarity measures for frontal face recognition: a comparative analysis. J. Vis. Commun. Image Represent. **24**, 1213–1231 (2013)
6. Brunelli, R., Poggio, T.: Face recognition: features versus templates. IEEE Trans. Pattern Anal. Mach. Intell. **15**, 1042–1052 (1993)
7. Davies, G., Ellis, H., Shepherd, J.: Cue saliency in faces as assessed by the Photofit technique. Perception **6**, 263–269 (1977)
8. Eberhart, R.C., Shi, Y.: Tracking and optimizing dynamic systems with particle swarms. In: Proceedings of the 2001 IEEE Congress on Evolutionary Computation, vol. 1, pp. 94–100 (2001)

9. Ekenel, H.K., Stiefelhagen, R.: Generic versus salient region-based partitioning for local appearance face recognition. In: Tistarelli, M., Nixon, M.S. (eds.) ICB 2009. LNCS, vol. 5558, pp. 367–375. Springer, Heidelberg (2009). doi:10.1007/978-3-642-01793-3_38

10. Haig, N.D.: Exploring recognition with interchanged facial features. Perception **15**, 235–247 (1986)

11. Huang, G.B., Lee, H., Learned-Miller, E.: Learning hierarchical representations for face verification with convolutional deep belief networks. In: IEEE Conference on Computer Vision and Pattern Recognition (CVPR), pp. 2518–2525 (2012)

12. Johnston, R.A., Edmonds, A.J.: Familiar and unfamiliar face recognition: a review. Mem. **17**, 577–596 (2009)

13. Kanade, T.: Computer Recognition of Human Faces. Birkhauser Verlag, Basel (1977)

14. Karczmarek, P., Kiersztyn, A., Pedrycz, W., Dolecki, M., Linguistic descriptors in face recognition. Appl. Soft Comput. (in press)

15. Karczmarek, P., Pedrycz, W., Kiersztyn, A., Dolecki, M.: An application of chain code-based local descriptor and its extension to face recognition. Pattern Recognit. **65**, 26–34 (2017)

16. Karczmarek, P., Pedrycz, W., Kiersztyn, A., Rutka, P.: A study in facial features saliency in face recognition: an analytic hierarchy process approach. Soft Comput. (in press)

17. Karczmarek, P., Pedrycz, W., Reformat, M., Akhoundi, E.: A study in facial regions saliency: a fuzzy measure approach. Soft Comput. **18**, 379–391 (2014)

18. Kacprzyk, J., Pedrycz, W.: Springer Handbook of Computational Intelligence. Springer, Heidelberg (2015)

19. Kennedy, J.F., Eberhart, R.C., Shi, Y.: Swarm Intelligence. Academic Press, San Diego (2001)

20. Kurach, D., Rutkowska, D., Rakus-Andersson, E.: Face classification based on linguistic description of facial features. In: Rutkowski, L., Korytkowski, M., Scherer, R., Tadeusiewicz, R., Zadeh, L.A., Zurada, J.M. (eds.) ICAISC 2014. LNCS, vol. 8468, pp. 155–166. Springer, Cham (2014). doi:10.1007/978-3-319-07176-3_14

21. Kwak, K.-C., Pedrycz, W.: Face recognition: a study in information fusion using fuzzy integral. Pattern Recognit. Lett. **26**, 719–733 (2005)

22. Matthews, M.L.: Discrimination of Identikit constructions of faces: evidence for a dual processing strategy. Percept. Psychophys. **23**, 153–161 (1978)

23. Pedrycz, W., Gomide, F.: An Introduction to Fuzzy Sets: Analysis and Design. The MIT Press, Cambridge (1998)

24. Saaty, T.L.: The Analytic Hierarchy Process. McGraw-Hill, New York (1980)

25. Saaty, T.L., Vargas, L.G.: Models, Methods, Concepts & Applications of the Analytic Hierarchy Process. Springer, New York (2012)

26. Sadr, J., Jarudi, I., Sinha, P.: The role of eyebrows in face recognition. Perception **32**, 285–293 (2003)

27. Shepherd, J., Davies, G., Ellis, H.: Studies of cue saliency. In: Davies, G., Ellis, H.D., Shepherd, J.W. (eds.) Perceiving and Remembering Faces, pp. 105–131. Academic Press, New York (1981)

28. Shi, Y., Eberhart, R.: A modified particle swarm optimizer. In: The 1998 IEEE International Conference on Evolutionary Computation Proceedings, IEEE World Congress on Computational Intelligence, pp. 69–73 (2002)

29. Sugeno, M.: Theory of fuzzy integral and its applications. Dissertation, Tokyo Institute of Technology (1974)

30. Sun, Y., Wang, X., Tang, X.: Deep learning face representation from predicting 10.000 classes. In: The IEEE Conference on Computer Vision and Pattern Recognition (CVPR), pp. 1891–1898 (2014)
31. Turk, M., Pentland, A.: Eigenfaces for recognition. J. Cogn. Neurosci. **3**, 71–86 (1991)
32. Vignolo, L.D., Milone, D.H., Scharcanski, J.: Feature selection for face recognition based on multi-objective evolutionary wrappers. Expert Syst. Appl. **40**, 5077–5084 (2013)
33. Wiskott, L., Fellous, J.-M., Krüger, N., von der Malsburg, C.: Face recognition by elastic bunch graph matching. IEEE Trans. Pattern Anal. Mach. Intell. **19**, 775–779 (1997)
34. Wright, J., Yang, A.Y., Ganesh, A., Sastry, S.S., Ma, Y.: Robust face recognition via sparse representation. IEEE Trans. Pattern Anal. Mach. Intell. **31**, 210–227 (2009)
35. Yan, Y., Osadciw, L.A.: Intra-difference based segmentation and face identification. In: Jain, A.K., Ratha, N.K. (eds.) Biometric Technology for Human Identification, Proceedings of SPIE 5404, pp. 502–510 (2004)

Analysis of Dermatoses Using Segmentation and Color Hue in Reference to Skin Lesions

Lukasz Was[1], Piotr Milczarski[2(✉)], Zofia Stawska[2], Marcin Wyczechowski[1],
Marek Kot[3], Slawomir Wiak[1], Anna Wozniacka[3], and Lukasz Pietrzak[1]

[1] Institute of Mechatronics and Information Systems, Technical University of Lodz,
Stefanowski str. 18/22, 90-924 Lodz, Poland
{lukasz.was,marcin.wyczechowski,slawomir.wiak,lukasz.pietrzak}@p.lodz.pl
[2] Faculty of Physics and Applied Informatics, University of Lodz,
Pomorska str. 149/153, 90-236 Lodz, Poland
{piotr.milczarski,zofia.stawska}@uni.lodz.pl
[3] Department of Dermatology and Venereology, Medical University of Lodz,
Haller Square 1, 90-647 Lodz, Poland
{marek.kot,anna.wozniacka}@umed.lodz.pl
http://www.imsi.pl
http://www.wfis.uni.lodz.pl
http://umed.pl

Abstract. In dermatology there are several well known algorithms of melanocytic lesions recognition but there are not automated algorithms of skin lesion identification and classification. The main aim of this paper is to examine skin changes based on the skin analysis in the chosen model color spaces. With the help of that analysis, the authors show how to extract information from the skin images that will be useful for a future dermatology expert system. In the paper, the authors introduce a novel clinical feature extraction and segmentation method based on modified dermatologists' approach to diagnose skin lesions. We have also prepared a database (DermDB) of dermoscopic images with the reference data prepared and validated by expert dermatologists.

Keywords: Skin diseases · Melanocytic lesions · Color spaces · Principal Component Analysis · Contour analysis · Dermoscopic database DermDB · Lesions Clinical Feature Segmentation Method (LesionCFSM)

1 Introduction

In dermatology we distinguish many melanocytic and non-melanocytic skin lesions. Recognition of such skin changes can be difficult and the final diagnosis may have a great clinical value. One of melanocytic lesions is melanoma. There are well known algorithms of melanocytic changes recognition. But there are not automated algorithms of skin lesions recognition and classification or at least a set of them.

In the paper, we present a new approach to the automated skin lesions processing based on the skin analysis in chosen model color spaces. With the

© Springer International Publishing AG 2017
L. Rutkowski et al. (Eds.): ICAISC 2017, Part I, LNAI 10245, pp. 677–689, 2017.
DOI: 10.1007/978-3-319-59063-9_61

help of that analysis, we show how to segment clinical features from the skin images that will be useful for a future expert system. That is why we introduce and describe *Lesions Clinical Feature Segmentation Method (LesionCFSM)*.

The paper shows the research that is an introduction and an integral part of the task of creating a classifier and an automated support tool for the diagnosis of skin dermatological changes.

The paper is organized into six sections. In Sect. 2 the revision of melanocytic and non-melanocytic lesions are described. In Sect. 3 databases of dermoscopic images are described and DermDB database is presented. In Sect. 4 a description of Lesions Clinical Feature Segmentation Method (LesionCFSM) as a procedure of image analysis with the help of Principal Component Analysis and modified pattern analysis are presented. In the next Section the research methodology is given. Conclusions are drawn in Sect. 6.

2   Revision of Melanocytic Lesions

Skin is the largest organ in the human body and most adults' skin weighs in at 10 kg. Therefore, it is not surprising that in many internal diseases skin changes may occur and their correct recognition can help to make proper final diagnosis.

Melanocytic and non-melanocytic skin lesions are often very difficult to diagnose because there are many factors that can lead to misdiagnosis which could cause a very long and expensive clinical treatment. Skin cancers may derive from the individual skin layers, the skin appendages, neurogenic blood and lymph derivatives [4]. One of the most difficult skin lesion to recognize is papilla which can vary in size, shape and appearance [3].

The proper dermatological diagnosis is based on clinical symptoms, experience of doctors and correct differentiation of skin lesions. Often histological examinations are needed to make a proper diagnosis. To provide correct diagnosis may be very difficult because in many cases skin eruptions are polymorphic and may change over time. In the clinical course of the disease, various skin lesions could also appear for a short time. Hence, the time, in which an appropriate assessment of such a change has been performed, is very important factor [1].

Semiology and recognition of the pattern and colors of skin lesions are usually the first step toward diagnosis [3]. Pattern recognition is particularly important in the melanocytic lesions to properly perform the differentiation with melanoma. In the first stage of diagnosing very specific features like a single dot or a single papilla are not the most important. If necessary, you can then incorporate these detailed information for further analysis.

Chaos is defined as the asymmetry of color, shape and structure of the lesions [13]. However, in nature there is no perfect symmetry which makes difficult to create a proper algorithm. If chaos is not present you need to go to the next assessment. Hence, a correct segmentation of the lesion is very important. We do not take into account only the atypical melanocytic lesions.

Below there is a brief summary of melanocytic and non-melanocytic lesions [14]:

Melanocytic lesions: Melanoma, Hamartoma, Lentigo, Congenital and acquired skin moles, Typical and atypical naevus.

Non-melanocytic lesions: Vascular lesions (angiomas), Seborrheic papilla, Actinic keratosis, BCC, SCC, Bowen's disease, Fibroma, Urticaria pigmentosa (mastocytosis), Inflammatory skin diseases: lichen planus, contact dermatatis, psoriasis, tinea etc.

As a result of our work, we expect to create an expert system examining specific photos/images with lesions of skin diseases by introducing proprietary LesionCFSM method of segmentation and feature extraction using a modified method of factor analysis.

3 Dermoscopic Databases

Computer-aided diagnosis (CAD) systems are usually based on three stages [6] (see Fig. 1): image segmentation, feature extraction/selection and lesion classification. Each of these stages has its own challenges and therefore, they need to have proper evaluation and validation, which require reliable reference data. The reference data has to be prepared and validated by expert dermatologists.

The availability of manually segmented skin lesions, performed by expert dermatologists, is of crucial importance since they give essential information for the evaluation of the segmentation step of a CAD system. The manual segmentation of each image of the database is available in a binary format, more specifically as a binary mask with the same size as the original image.

3.1 Examples of Dermoscopic Databases Validated by Dermatologists

Below we summarize briefly the two databases with dermoscopic images and their reference data.

PH^2 [11] 200 dermoscopic 8-bit RBG color images with a resolution of 768×560 pixels along with the corresponding medical annotations, comprising 80 common nevi, 80 atypical nevi, and 40 malignant melanomas acquired using a magnification of 20x under unchanged conditions.

DB of Warsaw Memorial Cancer Center (Poland) [9] 176 dermoscopic 8-bit JPEG color images with a resolution from 465×599 to 1077×1899 pixels along with the corresponding medical annotations, comprising 92 non-melanoma and 84 melanoma images acquired using a dermatoscope of the magnification of 20x.

The presented databases are effects of a joint collaboration between hospitals, dermatology departments and IT institutions.

3.2 DermDB

As a joint collaboration between Department of Dermatology and Venereology of Medical University of Lodz, Institute of Mechatronics and Information Systems of Technical University of Lodz and Faculty of Physics and Applied Informatics of University of Lodz, we have prepared a database DermDB. DermDB database contains images with the reference data prepared and validated by expert dermatologists from Department of Dermatology and Venereology of Medical University of Lodz.

For each image in the DermDB database, the manual segmentation and the clinical diagnosis of the skin lesion as well as the identification of other important dermoscopic criteria are available. These dermoscopic criteria are based on *3-point checklist of dermoscopy* (3PCLD) that includes:

1. Assessment of asymmetry of shape and structures.
2. Identification of colors and several differential structures, such as pigment network, dots, globules, streaks, regression areas.
3. Blue-white structures (veil).

Apart from 3PCLD, in dermatology Glasgow 7-point checklist (G7PCL) is widely used. G7PCL includes change in size, irregular shape, irregular color, diameter > 7 mm, inflammation, bleeding, erosion, change in sensation (itch, lack of feeling). Occuerrnce of three or more points suggests the diagnosis of atypical/malignant lesion.

For enhancing our research methodology we extended above checklists by adding a mesh generated for the whole lesion and received *extended 3-point checklist* (x3PCLD) and *extended Glasgow 7-point checklist* (xG7PCL).

The database currently consists of 56 dermoscopic images along with the corresponding medical annotations, comprising 22 common nevi, 22 atypical nevi, and 16 malignant melanomas. The DermDB database is being developed. The dermoscopic images were carefully acquired using a magnification of 20x under unchanged conditions. They are 8-bit RBG color images with a resolution from 800×600 to 1280×960 pixels. The set of images available in the DermDB database was selected with some constraints, regarding their quality, resolution, and dermoscopic features, so that they are suitable enough to be used as a dermoscopic reference database.

4   Research Methodology

The general approach to the skin lesions processing and classification is presented in Fig. 1. The raw images and dermographic data of the patient can be kept in a single or distributed database.

The process is divided into 3 stages starting from the raw RGB dermoscopic image:

Stage 1. Image processing to achieve lesions segmentation with their borders. The methods examples are described in [5,17,18]. The methods and algorithms that can be used in filtering skin are presented in e.g. [12,16].

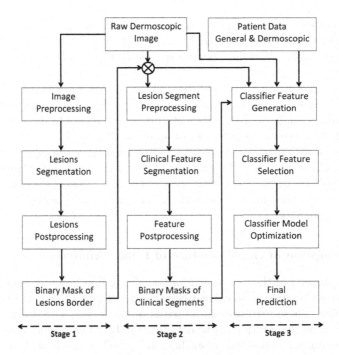

Fig. 1. The general approach to the skin lesions processing, segmentation and classification

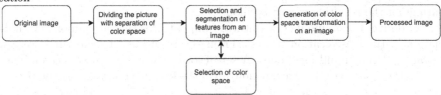

Fig. 2. The diagram of Lesions Clinical Feature Segmentation Method (LesionCFSM)

Stage 2. Clinical feature segmentation with resulting binary masks of clinical feature segments basing on the **Stage 1**. The methods examples are presented in e.g. [7,15].

Stage 3. Final classification based on the feature selection and classifier model optimization taking into account **Stage 1–2** and patient dermographic data. The methods examples are presented in e.g. [10,18].

In our research we use PH^2 and DermDB as reference datasets. In the paper, we describe the part of the research that is finished at the **Stage2** and one of the results is Lesions Clinical Feature Segmentation Method (LesionCFSM).

The diagram presented in Fig. 2 summarizes the approach that is described in the paper that is a part of a general approach to the skin processing and segmentation described as LesionCFSM.

The proposed algorithm of information transfer between the working color space within a single pixel or/and its surroundings depends on the initial conditions given to the algorithm, returning each R, G and B component from the RGB pixel.

The second approach involves mapping from the RGB color space into a chosen color space i.e. YCbCr, and subsequent transfer of its components to RGB representations using original coefficients, and not to return to the original image. As a result of these operations a color space is displayed for each component. It is a kind of a heat map, which significantly facilitates the preparation of the image for further analysis. The example of this is the edge detection algorithms using different Sobel operator [8] as well as the further image segmentation. In the further step, the authors make all the heat maps and pigment distribution with all components.

4.1   Description of the Procedure of Image Analysis

The main purpose of the mechanism of transformation lesion images into some kind of diverse heatmap is to extract certain important features, from the expert point of view. These features can be paths in the lesion. The other aim of the extration can be preparation of the data/material for further feature extraction.

Manipulation of individual colors/hues also affects contour analysis of specific fragments of lesion. These contours in conjunction with the modified algorithm of contours approximation can give complement effects and can introduce a broader outline for the resulting expert opinion.

We derive certain number of features from patterns that are important information in the classification process. From the resulting set of features we can extract the global and local ones. A set of local features represents color (hue) and texture (pattern). For each region/area, the features representing colors consist of the mean and variance of each channel of the RGB and HSV color spaces. Some results of the transformations are presented in Figs. 3, 5, 6, 7 and 8 are described in Sect. 5.

4.2   The Process of Image Segmentation

The main objective of the research is to develop original algorithms for studying the variability using local features with the help e.g. methods of factor analysis in the processing and segmentation of dermatological images. Basing of the reasearch assumptions, an expert decision support system (EDSS) will be developed for specialists dermatologists.

The other main objective is to study these skin lesions with the help of that expert system EDSS and factor analysis. In the application examining specific images with lesion changes we use proprietary algorithms of feature segmentation and a modified method of factor analysis. The method of factor analysis and additional elements of numerical methods will also be a subject of the research of finding the edges and contours of possible lesions.

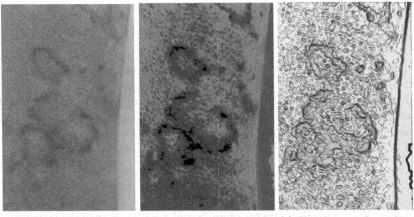

(a) Original picture   (b) Color transformations (c) Skin lesion after edging

Fig. 3. Skin lesion after color transformation and edging algorithm of #1 (Color figure online)

Skin lesions have typical pattern (image pattern) and are described by the five basic elements: lines, pseudopodia, circles, dots and papulaes. They will be discussed further in that Section.

Figure 3b shows the lesion in the revised set of colors. It is a result of conversion of each component, starting from the R component of the RGB, then components G and B for RGB subsequently.

Figure 3c shows how an algorithm of finding shapes according to a defined model.

4.3   Lesions Clinical Feature Segmentation Method Using PCA

Principal Component Analysis (PCA) is one of the methods of factor analysis. In the Lesions Clinical Feature Segmentation Method we use a weighted PCA method and the procedure is described below. We only present a part of the equations in PCA method because it is well-known method [2].

Step 1. Build feature vectors $x_1 \ldots x_n$ from the segmented lesions using x3PCLD and xG7PCL described in Sect. 3.2 where mesh helps to separate "normal" skin. Place the vectors into a matrix X. In each of n vectors we have p variables.

Step 2. Normalize your data vectors finding their mean and deviation. As a result, we receive vectors u and the matrix U and B as $B = X - U^T$.

Step 3. Calculate correlation matrix C and covariation matrix (for non-normalised data) as an outer product:

$$C = E[B \otimes B] = [B \cdot B^*] = \frac{1}{N} B \cdot B^*, \tag{1}$$

where E is the expected value in Eq. 1.

Step 4. Derive eigen values and eigen vectors as the matrix V:

$$V^{-1}CV = D, \tag{2}$$

where D is the diagonal matrix of eigen values of C [2].

Step 5. Sort in descending order eigen values.

Step 6. Choose first L eigen vectors corresponding to first L eigen values (sorted). As a result we achieve a subset of the eigenvectors as basis vectors.

Step 7. Convert the source data to z-scores

$$Z = \frac{B}{h \cdot s^T}, \tag{3}$$

where h is an $n1$ column vector of all 1s and s p-dimentional deviation vector defined as

$$s = \{\sqrt{C_{jj}}\}, \tag{4}$$

Step 8. Project the z-scores of the data onto the new basis

$$T = Z \cdot W, \tag{5}$$

As a result, the rows of matrix T are segmented features in the lesion. These features are a new data set for further classification process. That is why the properly evaluated new features are extremely important for the classification process.

4.4   Modified Pattern Analysis

Cutaneous lesions have typical pattern (image pattern) that can be described by the five basic elements: lines, pseudopodia, circles, dots and papulaes. Each of these elements can be either part of the whole pattern or the only one in the lesion. The proposed modified analysis can be described shortly as:

Model + Color (Hue) + Pattern = Diagnosis

The proposed pattern analysis is a part of Lesions Clinical Feature Segmentation Method.

The following list presents basic patterns that mightbe included in the skin lesion:

Line patterns - reticularis, branched, parallel, radial of the curved portion. Two-dimensional structure in which the length substantially exceeds the width.

Pseudopodium - linear structures not clearly combined with pigment network/mesh lines.

Wheel pattern -   a   curved   line   equidistant   from   a   central   point.

Papulae - each structure clearly demarcated, dense object which can take any shape and has diameter less than 1 cm.

Dot - black or brown oval structures, irregularly distributed within the lesion, diameter is less than 1 cm.

Area without structure - a compact area that lacks the dominant basic element.

The number of colors and the occurence of specific colorsin the lesions are of great importance in dermoscopy. Efflorescence pigment may consist of one or several colors, see Fig. 4. As in the case of patterns, colors may be arranged symmetrically or asymmetrically. With the exception of a brown one, different shades of the same color should not be interpreted as a separate color. Brown shades should be interpreted as separate ones. The distinction between a light and dark brown color is important for diagnosis, but only in the clear situations. Although, many eruptions are slightly brighter on the perimeter or have clearly colored areas around the hair follicle. Then, it should be still classified as a rash.

The color of "normal" skin varies both in the individuals and depends on the part of the body. Although, natural skin color is not considered a separate color, it serves as a reference when you want to define the color "white". The white structures are clearly brighter than the surrounding normal skin.

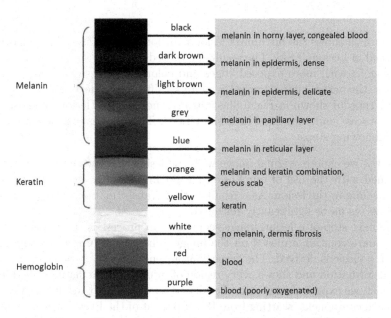

Fig. 4. The combination of colors (Color figure online)

5   Results of LesionCFSM Method

Figures 5a-b show the birthmark lesion which has been processed using our algorithms contained in LesionCFSM method.

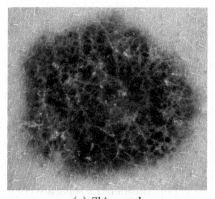
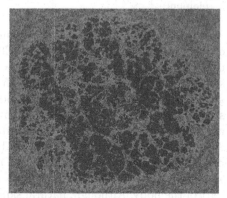

(a) Skin mark   (b) Astructural map of the lesion

Fig. 5. Skin lesion and its map (Color figure online)

Figure 5a shows the birthmark lesion transformed with use of our algorithms contained in LesionCFSM. The results presented in Fig. 5b show the skin change from Fig. 5a with more exposed edges and colors. As a result of that transformations, the actual shape (range) is more significantly reflected.

The contour shown in Fig. 6 illustrates the next step of lesion image segmentation process and contained in LesionCFSM. The process can be summarized in the following steps:

1. Firstly, we choose original image with the lesion mark which has been transformed with the use of our LesionsCFSM in which we exposed more edges and colors in the lesion. As a result of that transformations, the actual shape (range) is more significantly reflected.
2. The next step of a lesion image segmentation process is the implementation of square contour imposed on the image. Then, by a process of approximation, a circle is derived. The circle contour enhances the process of the lesion transformation and shows every region of interest in the distinguished lesion.
3. At last, we expose the lesion in the revised set of colors. It results in conversion of each component, starting from R component of the RGB, then components G and B for RGB subsequently.

As we mentioned in Sect. 4.1, the aim of that mechanism of transformation lesion images into a diverse heatmap is to extract important features from the expert point of view.

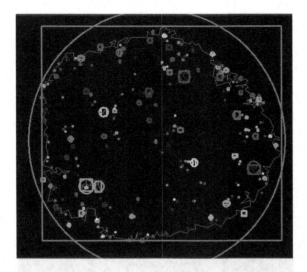

Fig. 6. The contour changes - segment image for further analysis

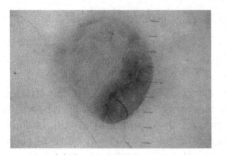
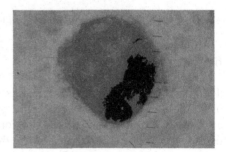

(a) Original skin lesion   (b) Skin lesion afterx G7CL

Fig. 7. Skin lesion and its tranformations #2

For each image we perform the segmentation to establish the clinical diagnosis of the skin lesion as the identification of criteria we use x3PCLD and xG7PCL by adding mesh generated for the whole lesion (see Fig. 5b). Three or more points in xG7PCL suggest the diagnosis of atypical/malignant lesion.

For each image we perform the segmentation to establish the clinical diagnosis of the skin lesion as the identification of criteria we use x3PCLD and xG7PCL by adding mesh generated for the whole lesion (see Fig. 5b). Three or more points in xG7PCL suggest the diagnosis of atypical/malignant lesion.

The results of described operations are also shown in Fig. 7 and the resulting contour in Fig. 8. In Fig. 7b asymmetry can be seen easily.

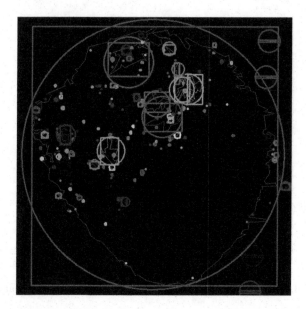

Fig. 8. The contour changes - segment image for further analysis #2

6   Summary

The basis of dermatological diagnosis is a correct assessment and a differentiation of skin lesion. It is often very difficult, because of the original skin lesions, which are usually a direct result of the development of the disease in the skin. In some cases, even in the early stages of the disease, it can not be said much of these lesions, as they may occur for short periods. Hence, a very important factor is the time in which an appropriate assessment of such a change is found.

Presented DermDB database and LesionCFSM method will be able to help to assess more accurately whether the respondent's clipping skin should immediately be referred for further testing, eg. histopathology or it can be initially assessed as a harmless or not posing a threat to her/his health and life.

Our further aim of the research is to develop the DermDB database with more cases and build classifiers correctly (Stage 3 in Sect. 4.2). On that structures we will be able to implement a dermotological expert system.

References

1. Argenziano, G., Fabbrocini, G., Carli, P., De Giorgi, V., Sammarco, E., Delfino, M.: Epiluminescence microscopy for the diagnosis of doubtful melanocytic skin lesions. Comparison of the ABCD rule of dermatoscopy and a new 7-point checklist based on pattern analysis. Arch. Dermatol. **134**, 15631570 (1998)
2. Bharati, M.H., MacGregor, J.F.: Texture analysis of images using Principal Component Analysis. In: Proceedings of the SPIE - Process Imaging for Automatic Control, vol. 4188, pp. 27–33 (2001)

3. Cardili, R.N., Roselino, A.M.: Elementary lesions in dermatological semiology: literature review. An Bras Dermatol. **91**(5), 629–633 (2016)
4. Chummun, S., McLean, N.R.: The management of malignant skin cancers. Surgery **29**(10), 529–533 (2011)
5. Celebi, M.E., Wen, Q., Iyatomi, H., Shimizu, K., Zhou, H., Schaefer, G.: A state-of-the-art survey on lesion border detection in dermoscopy images. In: Celebi, M.E., Mendonca, T., Marques, J.S. (eds.) Dermoscopy Image Analysis, pp. 97–129. CRC Press, Boca Raton (2015)
6. Deserno, T.M.: Biomedical Image Processing. Springer, Heidelberg (2011)
7. Gómez, D.D., Butakoff, C., Ersbll, B.K., Stoecker, W.: Independent histogram pursuit for segmentation of skin lesions. IEEE Trans. Biomed. Imaging **55**(1), 157–161 (2008)
8. Jain, A.K.: Fundamental of Digital Image Processing. Prentice Hall of India (2002)
9. Kruk, M., Świderski, B., Osowski, S., Kurek, J., Słowińska, M., Walecka, I.: Melanoma recognition using extended set of descriptors and classifiers. J. Image Video Proc. **2015**(1), Article 43 (2015)
10. Liu, J.J., Bharati, M.H., Dunn, K.G., MacGregor, J.F.: Automatic masking in multivariate image analysis using support vector machines. Chemometr. Intell. Lab. Syst. **79**, 42–54 (2005)
11. Mendonça, T., Ferreira, P.M., Marques, J.S., Marcal, A.R.S., Rozeira, J.: PH^2 - a dermoscopic image database for research and benchmarking. In: 35th Annual International Conference of the IEEE Engineering in Medicine and Biology Society (EMBC), Osaka, pp. 5437–5440 (2013)
12. Milczarski, P., Stawska, Z.: Complex colour detection methods used in skin detection systems. ISIM **3**(1), 40–52 (2014)
13. Rosendahl, C., Cameron, A., McColl, I., Wilkinson, D.: Dermatoscopy in routine practice "Chaos and Clues". Aust. Fam. Physician **41**(7), 482–487 (2012)
14. Rosendahl, C., Tschandl, P., Cameron, A., Kittler, H.: Diagnostic accuracy of dermatoscopy for melanocytic and nonmelanocytic pigmented lesions. J. Am. Acad. Dermatol. **64**(6), 10681073 (2011)
15. Schmid, P.: Segmentation of digitized dermatoscopic images by two-dimensional color clustering. IEEE Trans. Med. Imaging **18**(2), 164–171 (1999)
16. Stawska, Z., Milczarski, P.: Algorithms and methods used in skin and face detection suitable for mobile applications. ISIM **2**(3), 227–238 (2013)
17. Wighton, P., Lee, T.K., Lui, H., McLean, D.I., Atkins, M.S.: Generalizing common tasks in automated skin lesion diagnosis. IEEE Trans. Inf Technol. Biomed. **15**, 622–629 (2011)
18. Xie, F., Bovik, A.C.: Automatic segmentation of dermoscopy images using self-generating neural networks seeded by genetic algorithm. Pattern Recognit. **46**, 1012–1019 (2013)

Improving Data Locality of RNA Secondary Structure Prediction Code

Marek Palkowski$^{(\boxtimes)}$, Wlodzimierz Bielecki, and Piotr Skotnicki

Faculty of Computer Science and Information Systems, West Pomeranian University
of Technology in Szczecin, Zolnierska 49, 71210 Szczecin, Poland
{mpalkowski,wbielecki,pskotnicki}@wi.zut.edu.pl
http://www.wi.zut.edu.pl

Abstract. An approach allowing us to improve the locality of RNA Secondary Structure Prediction code is proposed. We discuss the application of this technique to automatic loop nest tiling for the Nussinov algorithm. The approach requires an exact representation of dependences in the form of tuple relations and calculating the transitive closure of dependence graphs. First, to improve code locality, 3-d rectangular tiles are formed within the 3-d iteration space of the Nussinov loop nest. Then tiles are corrected to establish code validity by means of applying the exact transitive closure of a dependence graph. The approach has been implemented as a part of the polyhedral TRACO compiler. The experimental results presents the speed-up factor of optimized code. Related work and future tasks are outlined.

Keywords: Computational biology · The Nussinov algorithm · Automatic loop tiling · RNA folding · Transitive closure

1 Introduction

Optimization of molecular biology programs is still a challenging task for developers and researchers. The cost of moving data from main memory is orders of magnitude higher than the cost of computation for modern architectures. Fortunately, many of bioinformatics algorithms, such as the dynamic programming core of the Nussinov algorithm for prediction of the secondary structure of RNA, involve mathematical operations over affine control loops whose iteration space can be represented by polyhedral models [9].

In this paper, we focus on the automatic code locality improvement of RNA folding realized with the Nussinov algorithm. This algorithm searches an optimal structure with a minimal free energy and is an example of "nonserial polyadic dynamic programming" (NPDP). The term "nonserial polyadic" stands for family of dynamic programming codes exposing non-uniform data dependences, which is more difficult to be optimized [8].

The presented approach applies tiling to the Nussinov loop nest. Tiling is a very important iteration reordering transformation for both improving data

© Springer International Publishing AG 2017
L. Rutkowski et al. (Eds.): ICAISC 2017, Part I, LNAI 10245, pp. 690–699, 2017.
DOI: 10.1007/978-3-319-59063-9_62

locality and extracting loop nest parallelism. Tiling for improving locality groups loop nest statement instances in a loop nest iteration space into smaller blocks (tiles) allowing reuse when the block fits in local memory.

To our best knowledge, well-known tiling techniques are based on linear or affine transformations [4,5,16]. In paper [3], we presented a novel algorithm to tile affine loop nests which is based on the transitive closure of program dependence graphs. First, rectangular tiles are formed, then they are corrected to establish tiling validity. The algorithm is able to tile non-fully permutable loops with non-uniform data dependences which are exposed for NPDP code.

For the purpose of this paper, we adopt that technique to tile each loop of the Nussinov code whereas other classical affine transformations leave un-tiled the innermost loop nest. For large problem sizes, tiling only outer loops may not remain data in cache between different iterations, resulting in poor locality [15].

The rest of the paper is organized as follows. Section 2 introduces background. Section 3 presents the structure of the Nussinov code exposing non-uniform data dependences. Tiling is explained in Sect. 4. Section 5 presents results of experiments and demonstrates that tiled (optimized) code is dramatically faster than original one for modern processors. This section discusses the following three factors: time of code execution, cache-misses, and instruction numbers per cycle. Section 6 explores related work. Section 7 considers future work to make the approach useful for multi-core processors.

2   Background

Dependence analysis produces execution-order constraints between statements/ statement instances and is required to guarantee the validity of loop nest transformations. Two statement instances I and J are *dependent* if both access the same memory location and if at least one access is a write. The presented approach requires an exact representation of loop-carried dependences and consequently an exact dependence analysis which detects a dependence if and only if it actually exists. The dependence analysis proposed by Pugh and Wonnacott [11] was chosen, where dependences are represented with dependence relations.

A dependence relation is a tuple relation of the form $[input\ list] \rightarrow [output\ list]$: *formula*, where *input list* and *output list* are the lists of variables and/or expressions used to describe input and output tuples and *formula* describes the constraints imposed upon *input list* and *output list* and it is a Presburger formula built of constraints represented with algebraic expressions and using logical and existential operators [11].

Standard operations on relations and sets are used, such as intersection ($\cap$), union ($\cup$), difference ($-$), domain (dom R), range (ran R), relation application ($S' = R(S)$: $e' \in S'$ iff exists e s.t. $e \rightarrow e' \in R, e \in S$). The detailed description of these operations is presented in [7,11].

The positive transitive closure of a given relation R, R^+, is defined as follows [7]:

$$R^+ = \{e \rightarrow e' :\ e \rightarrow e' \in R \vee \exists e''s.t.\ e \rightarrow e'' \in R \wedge e'' \rightarrow e' \in R^+\}. \quad (1)$$

It describes which vertices e' in a dependence graph (represented by relation R) are connected directly or transitively with vertex e.

Transitive closure, R^*, is defined as below:

$$R^* = R^+ \cup I, \tag{2}$$

where I is the identity relation. It describes the same connections in a dependence graph (represented by R) that R^+ does plus connections of each vertex with itself.

In sequential loop nests, the iteration i executes before j if i is *lexicographically less* than j, denoted as $i \prec j$, i.e., $i_1 < j_1 \lor \exists k \geq 1 : i_k < j_k \land i_t = j_t$, *for* $t < k$.

3   The Nussinov Algorithm

One of the first attempts at predicting RNA secondary structure in a computationally efficient way is the base pair maximization approach, developed by Nussinov in 1978 [15]. Given an RNA sequence $x_1, x_2, ..., x_n$, the Nussinov algorithm solves the problem of RNA non-crossing secondary structure prediction by means of computing the maximum number of base pairs for subsequences $x_i, ..., x_j$, starting with subsequences of length 1 and building upwards, storing the result of each subsequence in a dynamic programming array.

Let N be a $n \times n$ Nussinov matrix and $\sigma(i, j)$ be a function which returns 1 if (x_i, x_j) match and $i < j - 1$, or 0 otherwise, then the following recursion $N(i, j)$ (the maximum number of base-pair matches of $x_i, ..., x_j$) is defined over the region $1 \leq i \leq j \leq n$ as

$$N(i, j) = max(N(i+1, j-1) + \sigma(i, j), \max_{1 \leq j \leq n} (N(i, k) + N(k+1, j))) \tag{3}$$

and zero elsewhere [15].

The equation leads directly to triple-nested loops to compute N with the k loop innermost. The following C/C++ code represents the RNA folding computations [9]:

Listing 1. Nussinov loop nest.

```
for (i = N-1; i >= 0; i--) {
 for (j = i+1; j < N; j++) {
  for (k = 0; k < j-i; k++) {
   S[i][j] = MAX(S[i][k+i] + S[k+i+1][j], S[i][j]); //s0
  }
  S[i][j] = MAX(S[i][j], S[i+1][j-1] + can_pair(RNA,i,j));//s1
 }
}
```

Dependence relations calculated with the Petit analyser [11] for the Nussinov loop nest are presented below and represent non-uniform dependences.

$$
R = \begin{cases}
s0 \to s0 : \{[i,j,k] \,-\!\!>[i,j',j-i] : j < j' < N \cap 0 \leq k \cap i+k < j \\
\quad \cap\, 0 \leq i\} \cup \{[i,j,k] \,-\!\!>[i',j,i-i'-1] : 0 \leq i' < i \cap j < N \\
\quad \cap\, 0 \leq k \cap i+k < j\} \cup \{[i,j,k] \,-\!\!>[i,j,k'] : 0 \leq k < k' \\
\quad \cap\, j < N \cap 0 \leq i \cap i+k' < j\} \\
s0 \to s1 : \{[i,j,k] \,-\!\!>[i-1,j+1] : j \leq N-2 \cap 0 \leq k \cap i+k < j \cap \\
\quad 1 \leq i\} \cup \{[i,j,k] \,-\!\!>[i,j] : j < N \cap 0 \leq k \cap i+k < j \cap 0 \leq i\} \\
s1 \to s0 : \{[i,j] \,-\!\!>[i,j',j-i] : 0 \leq i < j < j' < N\} \cup \\
\quad \{[i,j] \,-\!\!>[i',j,i-i'-1] : 0 \leq i' < i < j < N\} \\
s1 \to s1 : \{[i,j] \,-\!\!>[i-1,j+1] : 1 \leq i < j \leq N-2\}
\end{cases}
$$

4   Tiling of the Nussinov Loop Nest

To generate valid tiled code for the Nussinov loop nest, we adopt the approach presented in paper [3], which is based on the transitive closure of dependence graphs. We slightly modified the algorithm to be applied for NPDP codes where loops are imperfectly nested with incrementing or decrementing index values.

Let vector $I = [i,j,k]$ represent indices of the Nussinov loop nest, matrix $B = [b_1, b_2, b_3]$ define tile sizes, vectors $II = [ii, jj, kk]$ and $II' = [iip, jjp, kkp]$ specify tile identifiers. Each tile identifier is represented with a non-negative integer vector, i.e., the following constraint $II \geq 0$ has to be satisfied.

First, we form a parametric set $TILE(I, B)$ including statement instances belonging to a rectangular parametric tile as follows

$$
TILE = \begin{cases}
i : N-1 - b_1 * ii \geq i \geq max(-b_1 * (ii+1), N-1) \\
j : b_2 * jj + i + 1 \leq j \leq min(b_2 * (jj+1) + i, N-1) \\
k : \begin{cases} s0 : b_3 * kk \leq k \leq min(b_3 * (kk+1) - 1, j-i-1) \\ s1 : k = 0 \end{cases}
\end{cases} \cap II \geq 0 \quad (4)
$$

Let us note that for index i, iterations constraints are defined inversely because the value of i is decremented.

Below, we present the lexicographical relation $II \prec II'$ on vectors, defining tile identifiers, as follows.

$$
II' \prec II = \begin{cases}
s0 : \begin{cases} s0 : ii > iip \cup (ii = iip \cap jj > jjp) \cup \\ \quad (ii = iip \cap jj = jjp \cap kk > kkp)) \\ s1 : ii > iip \cup (ii = iip \cap jj > jjp) \end{cases} \\
s1 : \begin{cases} s0 : ii > iip \cup (ii = iip \cap jj > jjp) \cup (ii = iip \cap jj = jjp)) \\ s1 : ii > iip \cup (ii = iip \cap jj > jjp) \end{cases}
\end{cases}
$$

$$(5)$$

Then, we define constraints $CONSTR(II, B)$ for tile identifiers which have to be satisfied for given values $b1$, $b2$, $b3$, defining a tile size, and parameter N specifying the upper loop index bound.

$$CONSTR = \begin{cases} ii, b_1 : N - 1 - b1 * ii >= 0 \\ jj, b_2 : (i+1) + b2 * jj <= N - 1 \\ kk, b_3 : b3 * kk + 0 <= j - i - 1 \end{cases} \quad (6)$$

Next, we build sets $TILE\_LT$ and $TILE\_GT$ that are the unions of all the tiles whose identifiers are lexicographically less and greater than that of $TILE(II, B)$, respectively:

$$TILE\_LT = \{[I] |\ \exists\ II'\ s.t.\ II' \prec II \cap II \geq 0 \cap CONSTR(II, B) \cap \quad (7)$$
$$II' \geq 0 \cap CONSTR(II', B) \cap I \in TILE(II', B)\},$$

$$TILE\_GT = \{[I] |\ \exists\ II'\ s.t.\ II' \succ II \cap II \geq 0 \cap CONSTR(II, B) \cap \quad (8)$$
$$II' \geq 0 \cap CONSTR(II', B) ANDI \in TILE(II', B)\}.$$

Then, the transitive closure of the union of all dependence relations, R^+, is computed. Techniques aimed at calculating the transitive closure of a dependence graph, which in general is parametric, are presented in papers [7,14] and they are out of the scope of this paper. We would like only to note that those commonly known algorithms return only an over-approximation of the union of all dependence relations for the Nussinov loop nest, i.e., transitive closure describes both all existing and false (non-existing) dependences. Such representations can be used in the presented algorithm but the tiled code will be not optimal.

To obtain exact transitive closure, we applied the iterative method presented in paper [2]. The approach uses basis dependence distance vectors in the modified Floyd-Warshall algorithm and is able to compute R^+ without non-existing dependences.

Listing 2. Tiled Nussinov loop nest for the original tile sizes [16,16,16].

```
for( c1 = 0; c1 <= floord(N - 1, 16); c1 += 1)
  for( c3 = 0; c3 <= min(c1, floord(N - 2, 16)); c3 += 1){
// tiles   with the instances of statement s0 not including any dependence ta-
// rget whose source  is within the tiles including instances of statement s1
    for( c5 = 0; c5 <= c3; c5 += 1)
      for( c7 = max(-N + 16 * c1 + 1, -N + 15 * c1 + c3 + 2);
                         c7 <= min(0, -N + 16 * c1 + 16); c7 += 1) {
        if (N + c7 >= 16 * c1 + 2) {
          for( c11 = 16 * c5; c11 <= min(15*c3 + c5, 16*c5 + 15); c11 += 1)
            S[-c7][16*c3-c7+1] = max(S[-c7][c11+-c7]
                      + S[c11+-c7+1][16*c3-c7+1], S[-c7][16*c3-c7+1]);
        } else
          for( c9 = N-16*c1 + 16*c3; c9 <= N - 16*c1 + 16*c3 + 15; c9 += 1)
            for( c11 = 16 * c5; c11 <= min(15*c3 + c5, 16*c5+15); c11 += 1)
              S[N-16*c1-1][c9] = max(S[N-16*c1-1][c11+N-16*c1-1]
                        + S[c11+N-16*c1-1+1][c9], S[N-16*c1-1][c9]);
      }
// tiles  with instances of  statement s1 and the rest of the instances of
```

```
// statement s0 which include the dependence targets whose sources are within
// tiles  including instances of statement s1
   for( c7 = max(-N + 16 * c1 + 1, -N + 16 * c3 + 2);
                                 c7 <= min(0, -N + 16 * c1 + 16); c7 += 1)
     for( c9 = 16 * c3 - c7 + 1; c9 <= min(N - 1, 16 * c3 - c7 + 16); c9 += 1)
       for( c10 = max(0, 16 * c3 - c7 - c9 + 2); c10 <= 1; c10 += 1) {
         if (c10 == 1) {
           S[-c7][c9] = max(S[-c7][c9],S[-c7+1][c9-1] + can_pair(RNA,-c7,c9));
         } else {
           if (N + c7 >= 16 * c1 + 2)
             for( c11 = 0; c11 <= 16 * c3; c11 += 1)
               S[-c7][c9] = max(S[-c7][c11-c7] + S[c11-c7+1][c9], S[-c7][c9]);
           for( c11 = 16 * c3 + 1; c11 < c7 + c9; c11 += 1)
             S[-c7][c9] = max(S[-c7][c11-c7] + S[c11-c7+1][c9], S[-c7][c9]);
         }
       }
   }
```

Using the exact form of R^+, we calculate set

$$TILE\_ITR = TILE - R^+(TILE\_GT), \tag{9}$$

which does not include any invalid dependence target, i.e., it does not include any dependence target whose source is within set $TILE\_GT$.

The following set

$$TVLD\_LT = (R^+(TILE\_ITR) \cap TILE\_LT) - R^+(TILE\_GT) \tag{10}$$

includes all the iterations that (i) belong to the tiles whose identifiers are lexicographically less than that of set $TILE\_ITR$, (ii) are the targets of the dependences whose sources are contained in set $TILE\_ITR$, and (iii) are not any target of a dependence whose source belong to set $TILE\_GT$. Target tiles are defined by the following set

$$TILE\_VLD = TILE\_ITR \cup TVLD\_LT. \tag{11}$$

Finally, we form set $TILE\_VLD\_EXT$ by means of inserting (i) into the first positions of the tuple of set $TILE\_VLD$ elements of vector II: $ii_1, ii_2, ..., ii_d$; (ii) into the constraints of set $TILE\_VLD$ the constraints defining tile identifiers $II \geq 0$ and $CONSTR(II, B)$. Target code is generated by means of applying any code generator allowing for scanning elements of set $TILE\_VLD\_EXT$ in the lexicographic order, for example, isl AST [14]. The proof of the validity of such an approach is presented in paper [3]. Listing 2 presents tiled code. After scrutinizing this code, we may conclude that it represents both 3-d and 2-d tiles. However, they are not rectangular and some of them are parametric (their size depends on parameter N). The amount of 3-d tiles depends on the value of parameter N, for example, for N = 500 and N = 1000, 3-d tiles constitute 91% and 95% of all tiles, respectively.

5   Experiments

This section presents results of an experimental study on speed-up of tiled code presented in Listing 2. To carry out experiments, we have used a machine with a

Table 1. Execution time of original and tiled codes (in seconds).

N	Original	Tiled-ij	Tiled-ijk
500	0.237	0.240	0.208
1000	1.755	1.737	1.514
1500	5.341	5.381	4.604
2000	18.918	16.339	14.187
2500	43.268	29.789	26.781
3000	93.184	63.940	57.466
3500	161.355	101.015	90.849
4000	249.710	186.658	170.162
4500	376.430	237.803	216.889
5000	487.625	369.907	338.191

processor Intel Xeon E5-2695 v2 (2.4 Ghz, 24 cores, 30 MB Cache) and 128 GB RAM. A code generation script and target codes of the examined programs are available as a part of the publicly available TRACO compiler[1].

All programs were compiled with the −O3 flag of optimization. The performance of the generated code was studied to compare it with that of the original one compiled by means of the Intel C++ Compiler (*icc* 15.0.2). Experiments were carried out for ten sizes of the problem defined with parameter N.

The execution time of the tiled code presented in Listing 2 is always shorter than that of the original one. We examined also 2-d tiled code (without k loop tiling). The results presented in Table 1 show that tiling k loop of the Nussinov loop nest accelerates executions for all examined problem sizes.

We carried out additional performance measurements by means of the *perf* profiler tool [12] for Linux 2.6+ based systems. It supports an interface to performance counters of hardware events stored in CPU registers and exported by recent versions of the Linux kernel. Table 2 presents cache-misses for tiled and original codes. For larger sizes of the problem, the cache-misses factor for the original codes can exceed ten percent, while for the tiled ones exceeds one percent only for $N = 2500$. Table 3 shows the numbers of instructions per cycle for the studied codes. We can observe that for the tiled codes this factor is always higher than that for the original ones. For larger sizes of the problem, one cycle is sufficient to execute at least one instruction in the tiled codes.

Summing up, we conclude that the presented approach successfully reduces time of execution and cache-misses as well as increases the number of instructions per cycle for the Nussinov loop nest.

---

[1] traco.sourceforge.net.

Table 2. Cache-misses for original and tiled codes (in %)

N	Original	Tiled-ij	Tiled-ijk
500	0.13	0.56	0.92
1000	0.22	0.39	0.35
1500	0.44	0.56	0.59
2000	4.38	0.47	0.43
2500	13.44	1.08	1.11
3000	9.34	0.69	0.72
3500	10.61	0.76	0.76
4000	6.36	0.52	0.52
4500	8.77	0.61	0.62
5000	6.46	0.47	0.47

Table 3. Instructions per cycle for original and tiled codes

N	Original	Tiled-ij	Tiled-ijk
500	2,02	2,46	2,42
1000	1,75	2,13	2,09
1500	1,92	2,23	2,20
2000	1,25	1,68	1,74
2500	1,07	1,78	1,82
3000	0,85	1,43	1,39
3500	0,78	1,39	1,48
4000	0,75	1,16	1,20
4500	0,71	1,22	1,31
5000	0,75	1,09	1,17

6   Discussion and Related Work

A number of authors have developed theoretical approaches to tiling NPDP codes [1,6,13]. A commonly known and very efficient source-to-source compiler, Pluto [4] tiles and parallelizes the Nussinov loop nest (since version 0.11.4). However, Pluto is able to produce only 2-d tiles for this code without tiling k loop.

In paper [10], run-time scheduling of the RNA secondary structure prediction for parallel machines is discussed. Although the algorithm does not require the transitive closure calculation, it is limited only to program loop nests with known parameters during compilation time.

Mullapudi and Bondhugula presented dynamic tiling for the Zukers optimal RNA secondary structure prediction [9]. 3-d iterative tiling for dynamic scheduling is calculated by means of reduction chains. Operations along each chain can be reordered in order to eliminate cycles in an inter-tile dependence graph.

However, this approach allows for only manually produced tiled code, i.e., tiling is not automated.

Wonnacott et al. introduced 3-d tiling of "mostly-tileable" loop nests of RNA secondary-structure prediction codes in paper [15]. For mostly-tileable loop nests, the number of "problematic" iterations grows with the tile size, but not with the problem size, whereas the number of non-problematic iterations grows with the problem size, i.e., the loop nest iteration space is dominated by non-problematic iterations. The approach is readily implemented with polyhedral tools, however the authors presented only the script for the RNA folding and did not explore the possibility of implementing the algorithm more fully into any general tiling tool, e.g. AlphaZ [15].

So, we may conclude that the presented approach is the first one allowing for automatic tiling of the Nussinov loop nest with 3-d tiles by means of the general-purpose tiling framework.

7   Conclusion

In this paper, we presented automatic loop nest tiling for the RNA secondary structure prediction. We demonstrated that more efficient codes are generated by means of the transitive closure of dependence graphs. Generated tiled code exposes reduced: execution time, numbers of instructions per cycle, and cache-misses. The approach can be aimed at other NPDP codes with imperfectly nested loops.

In future, we are going to study the following problems: (i) how to parallelize the tiled code discussed in this paper; (ii) exploring the performance of the Nussinov parallel tiled code executed on multi-core machines.

We strongly believe that transitive closure or its approximation can be useful to optimize other bioinformatics algorithms. These challenges deserve more attention from the research community accelerating applications of computational biology.

Acknowledgments. Thanks to the Miclab Team (miclab.pl) from the Technical University of Czestochowa (Poland) that provided access to high performance multi-core machines for the experimental study presented in this paper.

References

1. Almeida, F., Andonov, R., Gonzalez, D., Moreno, L.M., Poirriez, V., Rodriguez, C.: Optimal tiling for the RNA base pairing problem. In: Proceedings of the Fourteenth Annual ACM Symposium on Parallel Algorithms and Architectures, SPAA 2002, pp. 173–182. ACM, New York (2002)
2. Bielecki, W., Kraska, K., Klimek, T.: Using basis dependence distance vectors in the modified Floyd-Warshall algorithm. J. Comb. Optim. **30**(2), 253–275 (2015)
3. Bielecki, W., Palkowski, M.: Tiling arbitrarily nested loops by means of the transitive closure of dependence graphs. Appl. Math. Comput. Sci. **26**(4), 919–939 (2016)

4. Bondhugula, U., Hartono, A., Ramanujam, J., Sadayappan, P.: A practical automatic polyhedral parallelizer and locality optimizer. SIGPLAN Not. **43**(6), 101–113 (2008)
5. Griebl, M.: Automatic parallelization of loop programs for distributed memory architectures (2004)
6. Jacob, A.C., Buhler, J.D., Chamberlain, R.D.: Rapid RNA folding: analysis and acceleration of the Zuker recurrence. In: 2010 18th IEEE Annual International Symposium on Field-Programmable Custom Computing Machines (FCCM), pp. 87–94 (2010)
7. Kelly, W., Maslov, V., Pugh, W., Rosser, E., Shpeisman, T., Wonnacott, D.: The omega library interface guide. Technical report, College Park, MD, USA (1995)
8. Liu, L., Wang, M., Jiang, J., Li, R., Yang, G.: Efficient nonserial polyadic dynamic programming on the cell processor. In: 25th IEEE International Symposium on Parallel and Distributed Processing, IPDPS 2011, Workshop Proceedings, Anchorage, Alaska, USA, 16–20 May 2011, pp. 460–471 (2011)
9. Mullapudi, R.T., Bondhugula, U.: Tiling for dynamic scheduling. In: Rajopadhye, S., Verdoolaege, S. (eds.) Proceedings of the 4th International Workshop on Polyhedral Compilation Techniques, Austria, Vienna, January 2014
10. Palkowski, M.: Finding free schedules for RNA secondary structure prediction. In: Rutkowski, L., Korytkowski, M., Scherer, R., Tadeusiewicz, R., Zadeh, L.A., Zurada, J.M. (eds.) ICAISC 2016. LNCS (LNAI), vol. 9693, pp. 179–188. Springer, Cham (2016). doi:10.1007/978-3-319-39384-1_16
11. Pugh, W., Wonnacott, D.: An exact method for analysis of value-based array data dependences. In: Banerjee, U., Gelernter, D., Nicolau, A., Padua, D. (eds.) LCPC 1993. LNCS, vol. 768, pp. 546–566. Springer, Heidelberg (1994). doi:10.1007/3-540-57659-2_31
12. de Melo, A.C.: The new linux perf tools. Linux Kongress, Georg Simon Ohm University Nuremberg/Germany. Technical report (2010)
13. Tan, G., Feng, S., Sun, N.: Locality and parallelism optimization for dynamic programming algorithm in bioinformatics. In: 2006 Proceedings of the ACM/IEEE Conference on SC, p. 41 (2006)
14. Verdoolaege, S.: Integer set library - manual. Technical report (2011). www.kotnet.org/skimo//isl/manual.pdf
15. Wonnacott, D., Jin, T., Lake, A.: Automatic tiling of "mostly-tileable" loop nests. In: 5th International Workshop on Polyhedral Compilation Techniques, IMPACT 2015, Amsterdam, The Netherlands (2015)
16. Xue, J.: Loop Tiling for Parallelism. Kluwer Academic Publishers, Norwell (2000)

Robust Detection of Systolic Peaks in Arterial Blood Pressure Signal

Tomasz Pander[✉], Robert Czabański, Tomasz Przybyła, Stanisław Pietraszek, and Michał Jeżewski

Faculty of Automatic Control, Electronics and Computer Science,
Institute of Electronics, Silesian University of Technology,
Akademicka Str. 16, 44-100 Gliwice, Poland
{tpander,rczabanski,tprzybyla,spietraszek,mjezewski}@polsl.pl

Abstract. The heart rate signal is one of the most important physiological signals characterizing the human heart. The heart bits are usually determined on the basis of the electrocardiographic (ECG) signal. However, they can be also detected by monitoring systolic peaks in a arterial blood pressure (ABP) signal. The pressure signal, as other physiological signals, may be disturbed with noise. In this work we propose the method of precise location of the systolic peaks in ABP signal in the presence of noise, by applying the detection function waveform and fuzzy clustering. The new method is tested using real signals from the MIT-BIH Polysomnographic Database. The results obtained during experiments show the high effectiveness of the proposed method in relation to reference methods.

Keywords: ABP signal · Systolic peak detection · Fuzzy clustering

1 Introduction

Today's technology allows for simultaneous registration of multiple physiological signals over a long period of time. Hence, the assumption of quasi-stationarity can not always be accepted. The automatic detection of heart beats is essential for the analysis of multiple biomedical signals and patient monitoring. The literature describes many heart beats detection algorithms based on an analysis of electrocardiographic (ECG) signal. However, not only electrical, but also mechanical activity of the heart can be analyzed.

There are three basic features of arterial blood pressure (ABP), which have their representation in the ABP signal waveform i.e. the systolic peak, dichrotic notch and dichrotic peak [1]. An example of the pressure wave without noise is shown in Fig. 1(a), whereas Fig. 1(b) presents the noisy ABP signal. The ABP signal is used for: (i) estimate cardiac output, (ii) verify and eliminate ECG-based false alarms, (iii) monitor an average blood pressure, (iv) determine the resistance of the peripheral circulatory system, (v) assess the flexibility of the large arteries, such as the thoracic aorta (i.e. arterial compliance), and finally

© Springer International Publishing AG 2017
L. Rutkowski et al. (Eds.): ICAISC 2017, Part I, LNAI 10245, pp. 700–709, 2017.
DOI: 10.1007/978-3-319-59063-9_63

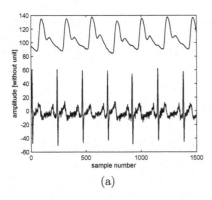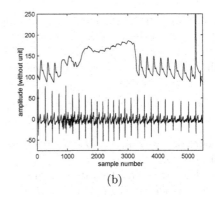

(a)   (b)

Fig. 1. An example of the undisturbed ABP signal (a) and the noisy ABP signal (b), accompanied by the electrocardiographic signal. For the purpose of presentation clarity, the ECG signal was amplified 50 times.

(vi) to determine a heart rate [1]. The accurate recognition of the waveform peaks is an essential step of the ABP signal processing and many algorithms of the APB signal analysis have been described in the literature [1,7,9,12–14] so far.

This paper presents a new method of detecting systolic peaks of the ABP signal. Firstly, the detection function, which local maxima correspond to ABP waveform peaks, is determined. Only the maxima whose amplitude are greater than a threshold value are considered as corresponding to systole. In order to improve the detection accuracy, the procedure for adaptive selection of the amplitude threshold using fuzzy clustering was proposed. Moreover, to increase the detection accuracy, the ABP quality signal QS was formed. It allows for eliminating those parts of the ABP signal, wherein the signal is degraded due to interferences of various kinds. The proposed and reference methods were tested using real ABP signals from the MIT-BIH Polysomnographic Database [2,3].

2   Algorithm Description

The proposed method of systolic peaks detection in the ABP signal consists of three main stages: creation the detection function waveform corresponding to the ABP signal [10], estimation an appropriate value of the amplitude threshold and finally, qualification whether the designated peak locations come from the undisturbed ABP signal. The detailed description of the algorithm is presented in the next subsections.

2.1   Signal Pre-processing

The initial filtration is performed using a derivative Gaussian filter whose coefficients are equal to the first derivative of the Gaussian function [4,13]. It allows for eliminating of the trend line and high frequency noise components. The high amplitude of the resulting signal corresponds to the systolic peaks of blood pressure. The length and width of the filter are adjusted empirically [13] and the filter coefficients $w(m)$ are determined as

$$w(m) = \exp\left(-\frac{\left(m - \frac{M}{2}\right)^2}{2\sigma^2}\right),$$ (1)

where $m = 1, 2, 3, \ldots, M$, M denotes the length of Gaussian kernel and σ is the dispersion.

The Gaussian derivative kernel can be computed as $w_d(m) = w(m + 1) - w(m)$, where $m = 1, 2, 3, \ldots, M - 1$. The signal processing is defined as the convolution of the filter coefficients $w_d(m)$ and the raw ABP signal $x_{raw}(n)$

$$x_f(n) = \sum_{k=-\infty}^{\infty} x_{raw}(k) \cdot w_d(n - k).$$ (2)

We assumed $M = 101$ and $\sigma = 12.5$ [13].

The next stage of the pre-processing procedure includes the non-linear operation (the square function) and smoothing of the resulting signal

$$x(n) = \frac{1}{2N_{ma} + 1} \sum_{i=-N_{ma}}^{N_{ma}} (x_f(n + 1))^2,$$ (3)

where $x_f(n)$ denotes signal after Gaussian filtering and $2N_{ma} + 1$ is the length of the moving average filter. We assumed $N_{ma} = 30$ which ensures that the detection function waveform has only one peak at the location of ABP systolic peak.

The detection function waveform which corresponds to a proper (without any distortion) ABP signal has characteristic peaks, which locations correspond to the locations of the systolic peaks in the pressure signal. In contrast, if the ABP signal does not contain deterministic components (a part of signal without a characteristic beats), then the detection function waveform is approximately equal to zero.

However, there are situations when the pre-processing fails. High peaks of the ABP signal (corresponding to, e.g. inflating the cuff) may also result in significant increase of values of the detection function waveform. This is not a typical thick error (an outlier) which could be removed with the robust filter. It is due to enlargement to a large width of the resulting detection function (approx. 100 samples). Therefore, amplitude of samples of the detection function waveform needs to be verified. In the proposed method we removed the samples of detection function satisfying the condition $\underset{1 \leq n \leq N}{\forall} x(n) > 4 \cdot \sigma_{x(n)}$, where $\sigma_{x(n)}$ is a standard deviation of all samples of the detection function $x(n)$. The resulting detection function $(x^*(n))$ is used only for estimating the amplitude threshold A_{th}. The systolic peaks locations are always carried out using samples of the unmodified function $x(n)$.

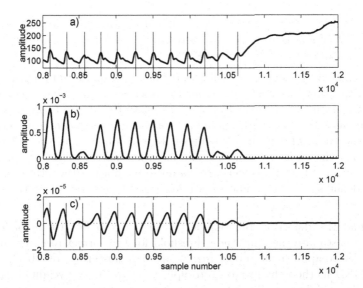

Fig. 2. Illustration of the various stages of the pressure signal processing: (a) the ABP signal (vertical lines denote detected peaks), (b) the detection function waveform (the dashed line denotes the value of amplitude threshold A_{th}), (c) the first derivative of the detection function waveform after smoothing (vertical lines denote the zero-crossings).

The process of localization of peaks of the detection function waveform is performed using its first derivative. Additionally, the first derivative signal is smoothed by the moving average filter of the length 51 [10] and the events of zero-crossings are detected. If the amplitude of the selected sample is greater than the threshold A_{th} then the systolic peak in the ABP signal is identified. An example of the ABP signal, the detection function waveform and its first derivative is shown in Fig. 2(a), (b) and (c).

Despite the advanced noise suppression techniques, the interferences may also lead to local increases of the detection function. Consequently, only the peaks which amplitude exceed the assumed threshold level A_{th} should be considered. The calculation of the appropriate threshold level is hence crucial to the correct systolic peaks identification. Careful analysis of the shape of the detection function waveform enabled to identify several different levels of samples amplitudes with the lowest corresponding to distortions. The amplitudes levels vary depending on the ABP waveform and as a result the application of a fixed amplitude threshold A_{th} does not provide sufficient accuracy of the automatic recognition. Hence, the identification of groups of signal's samples can be done by using fuzzy clustering [10].

2.2 Estimation of the Amplitude Threshold

Data clustering consists in a partition of a set of N elements (objects) into c groups (subsets) of similar objects. Our task is to find subsets of detection function samples of similar values. The similarity criterion is usually defined on the basis of comparison of the selected object properties, represented by the so-called feature vector $\underset{1 \leq n \leq N}{\forall} \mathbf{x}(n) \in \mathbb{R}^p$.

A class of the clustering methods is defined as a minimization problem of a criterion function (a scalar index) that represents the quality of the partition. Among them, the subclass of algorithms based on the idea of fuzzy sets [17] can be distinguished. Fuzzy clustering allows for partial membership of objects into groups and the degree of membership of the n-th object into the i-th group is described as the element $u_{in} \in [0, 1]$ of a partition matrix $\mathbf{U} \in \mathbb{R}^{c \times N}$. Zero value of u_{in} indicates that the object $\mathbf{x}(n)$ is not a member of the i-th group, while $u_{in} = 1$ denotes the full membership. Each group is represented by the so-called prototype $\underset{1 \leq i \leq c}{\forall} \mathbf{v}(i) \in \mathbb{R}^p$.

To find the detection function level that corresponds to distortions and thereby to estimate the amplitude threshold for the correct recognition of systolic peaks we applied the robust fuzzy c-median clustering (FCMED) [5,6]. In the FCMED method the group prototypes $\mathbf{v}(i)$ are fuzzy (weighted) medians. The median weights are defined as the r-th power of membership values $(u_{in})^r$, where $r \in (1, +\infty)$. Fuzzy medians ensure the robustness to outliers, which may be the results of the ABP signal noise. The classic approach to calculate the fuzzy median requires sorting the entire set of objects. Hence, to reduce the computational time we applied the bisection method to estimate the fuzzy medians [5].

The partition is determined as a result of alternating calculations of prototypes and the partition matrix (Picard algorithm). The degree of membership of the object into the particular group is a function of the distance of the feature vector from the group prototype. The closer object $\mathbf{x}(n)$ is to the prototype $\mathbf{v}(i)$, the higher is its membership u_{in} to the i-th group. In the proposed approach, the feature vectors are directly the values of detection function samples, i.e. $p = 1$, hence

$$\underset{1 \leq i \leq c}{\forall} \underset{1 \leq n \leq N}{\forall} u_{in} = \frac{|v(i) - x(n)|^{\frac{1}{1-r}}}{\sum_{j=1}^{c} |v(j) - x(n)|^{\frac{1}{1-r}}}. \tag{4}$$

The algorithm starts with a random partition matrix $\mathbf{U}^{(0)}$. On the basis of $\mathbf{U}^{(0)}$, the group prototypes are calculated $\mathbf{V}^{(0)} = \left[v_1^{(0)}, v_2^{(0)}, \cdots, v_c^{(0)} \right]$ as fuzzy medians. The new location of prototypes provides the new degrees of membership $\mathbf{U}^{(1)}$, calculated on the basis of (4). The process is repeated until the maximum number of iterations ($t_{\max}$) is reached, or if the change of the scalar index

$$J = \sum_{n=1}^{n} \sum_{i=1}^{c} u_{in}^r |x(n) - v(i)|, \tag{5}$$

in the subsequent iterations is less than a pre-set value ε, i.e. $|J(t+1) - J_r(t)| < \varepsilon$, where t is the iteration index. In the proposed solution $\varepsilon = 10^{-5}$ was assumed.

Since the systolic peak of the ABP wave must correspond to maximum of the detection function waveform, only the $M < N$ local maxima of the detection function are clustered

$$x(n-1) \leq x_l(m) = x(n) \leq x(n+1), \tag{6}$$

where $x_l(m)$ is the m-th local maximum of the detection function. Such an approach reduces the computational time of the amplitude threshold estimation.

The thorough analysis of the detection function waveform allowed us to distinguish five various levels of samples amplitudes. Hence, the number of groups c for the FCMED was set to 5. The values of the detection function related to noise are the smallest thus, the prototype representing a group of samples from noise is determined as

$$v(\eta) = \min(v(1), v(2), \ldots, v(5)). \tag{7}$$

The amplitude threshold was set to exceed the maximum value of the detection function of ABP samples, that were identified as originating from disturbances. However, the sample is considered as representing the noise component only if its membership degree to the η-th group is higher than δ (in the numerical experiments $\delta = 0.75$ was assumed). Consequently, the amplitude threshold of the detection function is defined as

$$A_{th} = \max_{1 \leq n \leq M} \left(x_l(n) |_{u_{\eta n} > \delta} \right). \tag{8}$$

If there are no samples of the detection function that are characterized by the high membership degree to the group η, i.e. $\forall_{1 \leq n \leq M} u_{\eta n} < \delta$, then A_{th} is calculated using scaled $\mathbf{U}_\eta = [s_{\eta 1}, s_{\eta 2}, \cdots, s_{\eta M}]$, where

$$\mathop{\forall}_{1 \leq n \leq M} s_{\eta n} = \frac{u_{\eta n}}{\max\limits_{1 \leq n \leq M} (u_{\eta n})}. \tag{9}$$

2.3 Determination of ABP Quality Signal

The ABP signals are subjected to disturbances [8] hence, various methods were proposed [8,15,18] to determine ABP signal quality. The outcome of the quality check is usually presented as so-called "quality signal" (QS) or the signal abnormality index (SAI). The logical conditions to be met by ABP signal, which allow for determining its SAI were discussed in [15]. In this paper, we present a method of calculation of the quality signal based on the ABP signal only. Samples of the QS take the logical zero values if poor quality of the ABP signal is recognized i.e. the signal is disturbed, or it contains a transducer calibration beats, etc. The QS is equal to logical one for the ABP samples of good quality. The proposed algorithm is as follows:

1. The raw ABP signal $x_{raw}(n)$ is filtered by the robust simplified Cauchy-based p-norm filter to remove an impulsive noise (of very short duration) [11].

The weights of the filter equal to values of Hamming function, the linear parameter $K = 1$, the filter width equals 25, and $p = 1.7$. As a result the signal $x_r(n)$ is calculated.

2. The power-line interference (50 Hz) is removed from $x_r(n)$ with IIR notch filter forming the $x_{50\,Hz}(n)$ signal.

3. The baseline $x_{iso}(n)$ of the ABP signal is estimated with the myriad filter. The filter width is 250, filter weights are $w_i = \frac{1}{250}$ where $i = 1, \ldots, 250$, and the linear parameter $K = 1$ [16].

4. The de-trended signal $x_f(n) = x_{50\,Hz}(n) - x_{iso}(n)$ is calculated.

5. The envelope signal $x_{env}(n)$ is formed using the Hilbert transformation of $x_f(n)$ signal. Then $x_{env}(n)$ is smoothed with the moving average filter of the length equals 350.

6. The resulting QS signal is determined by assuming the threshold Δ and checking the following condition

$$\mathop{\forall}_{1 \leq n \leq N} \quad \text{if} \quad \Delta \leq x_{env}(n) \leq 4 \cdot \Delta, \text{ then } QS(n) = 1, \text{ else } QS(n) = 0.$$

Our experiments showed, that the best results are achieved if $\Delta = 10$.

Additionally, we included the condition to prevent rapid changes of QS signal. Thus, if the duration of the series of 0 or 1 is too short (less than 3 s), then the QS is corrected (QS') by setting its value as equal to the quality assessment preceding the episode of too short QS. An example of the raw ABP signal and the corresponding quality signal $QS(n)$ is shown in Fig. 3.

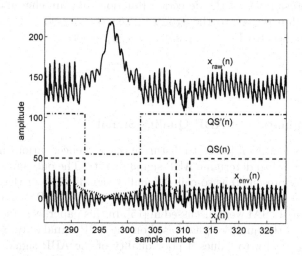

Fig. 3. An example of ABP signal ($x_{raw}(n)$ - solid line), the envelope signal ($x_{env}(n)$ - dotted line), the quality signal ($QS(n)$ - dashed line) and the quality signal after correction (QS'(n) - dashed-dot line).

3   Numerical Experiment and Results

To investigate the performance of the proposed method we used real signals from
the MIT-BIH Polysomnographic Database [2,3]. The ABP signals are accompanied by an annotation of each heart evolution in the ECG signal. The ECG
signals were used to establish a reference for assessing the efficacy of the systolic peak detection. Three records from the database were selected for testing
providing five ABP signals of various lengths and number of systolic peaks. The
detailed characteristics of testing signals are presented in Table 1. The sampling
rate is $f_s = 250\,\mathrm{Hz}$.

Table 1. The short characteristics of the testing ABP signals

No.	The record name	Duration [sec]	Number of ABP peaks
1	slp02a	153	237
2	slp02b (1st file)	354	257
3	slp02b (2nd file)	947	1162
4	slp41 (1st file)	225	210
5	slp41 (2nd file)	3306	3752
Summary		4985	5618

The result of systolic peaks detection is positive if the systolic peak is found
and negative otherwise. Therefore, the detection accuracy can be assessed using
numbers of: true-positive detections (TP – that specifies the number of correctly
identified peaks), false positive detections (FP – that specifies the number of
incorrectly detected peaks) and false negative detections (FN – that specifies
the number of undetected peaks). Consequently, the detection efficacy can be
evaluated on the basis of the following performance measures:

1. sensitivity SEN = TP/(TP+FN),
2. positive predictivity P^+ = TP/(TP+FP),
3. detection error rate DER = (FP+FN)/TP,
4. accuracy ACC = TP/(TP+FP+FN),
5. F-measure F = $(2 \cdot \mathrm{SEN} \cdot P^+)/(\mathrm{SEN} + P^+)$.

During our experiments the QS was used to indicate the parts of ABP signal
with a valid ECG reference (i.e. parts of the ABP signal where the systolic peaks
were possible to be detected correctly). We used two reference methods, which
were described in details in [13] (*RefMethod1*) and [7] (*RefMethod2*). Because
unambiguous evaluation of each algorithm using several performance measures
simultaneously is difficult, our conclusions focused primarily on the F-measure
analysis. The F-measure combines the information about the detection sensitivity and positive predictivity, being the basic measure of the detection efficacy.

708   T. Pander et al.

Table 2. Summary of the detection efficacy for all tested signals

Method	TP	FP	FN	SEN	P$^+$	DER	ACC	F
NewMethod	5386	324	232	**0.9587**	0.9433	**0.1032**	**0.9064**	**0.9509**
RefMethod1	4939	113	679	0.8791	**0.9776**	0.1604	0.8618	0.9258
RefMethod2	5358	393	260	0.9537	0.9317	0.1219	0.8914	0.9426

Table 3. Average values of the performance measures calculated for all signals

Method	SEN	P$^+$	DER	ACC	F
NewMethod	**0.9811**	0.9536	**0.0702**	**0.9373**	**0.9669**
RefMethod1	0.9103	**0.9788**	0.1399	0.8912	0.9390
RefMethod2	0.9734	0.9404	0.0943	0.9184	0.9563

Table 2 summarizes the results. It shows that the proposed method achieved the best detection quality for the considered set of signals. However, the highest positive predictivity was obtained for the first reference method [13]. Nevertheless, with the increase of P$^+$ the decrease of sensitivity and, consequently, the decrease of F-measure was noticed. The second [7] reference procedure provided slightly worse results than the proposed algorithm. At the same time the *RefMethod2* was characterized with higher SEN, ACC and F as well as smaller DER when comparing to *RefMethod1*. The averaged values of the performance measures of all tested methods are presented in Table 3.

4   Conclusions

The paper describes a new method for the precise location of systolic peaks in the arterial blood pressure (ABP) signal. The first step of the proposed procedure involves pre-processing of ABP signal providing the detection function waveform. The detection function is obtained with the derivative Gaussian filter, which removes a trend line and high-frequency components. In the next stage, the signal is squared as well as smoothed with the moving average filter. The obtained smoothed peaks of the detection function waveform correspond to the systolic peaks of ABP signal. Their position is determined based on the amplitude thresholding. In order to improve the accuracy of the detection we apply the fuzzy median clustering, wherein the samples of the detection function were clustered into five classes. We proposed also the procedure of the ABP signal quality evaluation based on the analysis of the envelope of the ABP signal. The new method was tested on real signals from the MIT-BIH Polysomnographic Database. The results showed the improvement of the efficacy of the systolic peaks detection in relation to the reference methods.

Acknowledgments. This work was partially supported by the Ministry of Science and Higher Education funding for statutory activities (decision no. 8686/E-367/S/2015 of 19 February 2015) and the Ministry of Science and Higher Education funding for statutory activities of young researchers (BKM-508/RAu-3/2016).

References

1. Aboy, M., McNames, J., Thong, T., Tsunami, D., Ellenby, M., Goldstein, B.: An automatic beat detection algorithm for pressure signals. IEEE Trans. Biomed. Eng. **52**, 1662–1670 (2005)
2. Goldberger, A., Amaral, L., Glass, L., Hausdorff, J., Ivanov, P., Mark, R., Mietus, J., Moody, G., Peng, C.K., Stanley, H.: Physiobank, physiotoolkit, and physionet: components of a new research resource for complex physiologic signals. Circulation **101**(23), e215–e220 (2000)
3. Ichimaru, Y., Moody, G.: Development of the polysomnographic database on CD-ROM. Psychiatr. Clin. Neuros. **53**, 175–177 (1999)
4. Kathirvel, P., Manikandan, M., Prasanna, S., Soman, K.: An efficient r-peak detection based on new nonlinear transformation and first-order gaussian differentiator. Cardiovasc. Eng. Technol. **2**, 408–425 (2011)
5. Kersten, P.: Implementation issues in the fuzzy c-medians clustering algorithm. In: Proceedings of the Sixth IEEE International Conference on Fuzzy Systems, vol. 2, pp. 957–962 (1997)
6. Kersten, P.: Fuzzy order statistics and their application to fuzzy clustering. IEEE Trans. Fuzzy Syst. **7**(6), 708–712 (1999)
7. Li, B., Dong, M., Vai, M.: On an automatic delineator for arterial blood pressure waveforms. Biomed. Signal Process. Contr. **5**, 76–81 (2010)
8. Li, Q., Mark, R., Clifford, G.: Artificial arterial blood pressure artifact models and an evaluation of a robust blood pressure and heart rate estimator. Biomed. Eng. Online **89**, 1–13 (2009)
9. Pachauri, A., Bhuyan, M.: ABP peak detection using energy analysis technique. In: International Conference on Multimedia, Signal Processing and Communication Technologies, 17–19 December 2011 (2011)
10. Pander, T., Czabański, R., Przybyła, T., Pojda-Wilczek, D.: An automatic saccadic eye movement detection in an optokinetic nystagmus signal. Biomed. Tech. (Berl) **59**(6), 529–543 (2014)
11. Pander, T., Przybyła, T.: Impulsive noise cancelation with simplified cauchy-based p-norm filter. Signal Process. **92**, 2187–2198 (2012)
12. Perfetto, J., Ruiz, A., Sirne, R., D'Attellis Klapuri, A.: Pressure-detection algorithms. IEEE Eng. Med. Biol. **28**(5), 35–40 (2009)
13. Raju, D., Manikandan, M., Ramkumar, B.: An automated method for detecting systolic peaks from arterial blood pressure signals. In: Proceeding of the 2014 IEEE Student's Technology Symposium, pp. 41–46 (2014)
14. Scholkmann, F., Boss, J., Wolf, M.: An efficient algorithm for automatic peak detection in noisy periodic and quasi-periodic signals. Algorithms **5**, 588–603 (2012)
15. Sun, J., Reisner, A., Mark, R.: A signal abnormality index for arterial blood pressure waveforms. Comput. Cardiol. **101**(33), 13–16 (2006)
16. Wróbel, J., Horoba, K., Pander, T., Jeżewski, J., Czabański, R.: Improving fetal heart rate signal interpretation by application of myriad filtering. Biocybern. Biomed. Eng. **33**(4), 211–221 (2013)
17. Zadeh, L.: Fuzzy sets. Inform. Contr. **8**(4), 338–353 (1965)
18. Zong, W., Moody, G., Mark, R.: Reduction of false arterial blood pressure alarms using signal quality assessment and relationships between the electrocardiogram and arterial blood pressure. Med. Biol. Eng. Comput. **42**(5), 698–706 (2004)

Fuzzy System as an Assessment Tool for Analysis of the Health-Related Quality of Life for the People After Stroke

Piotr Prokopowicz[1(✉)], Dariusz Mikołajewski[1], Emilia Mikołajewska[2], and Piotr Kotlarz[1]

[1] Institute of Mechanics and Applied Computer Science, Kazimierz Wielki University, ul. Kopernika 1, 85-074 Bydgoszcz, Poland
{piotrekp,dmikolaj,piotrk}@ukw.edu.pl
[2] Ludwik Rydygier Collegium Medicum, Department of Physiotherapy, Nicolaus Copernicus University, ul. Jagiellonska 13-15, 85-094 Bydgoszcz, Poland
emiliam@cm.umk.pl

Abstract. Stroke remains one of the leading causes of long-term disability in both developed and developing countries. Prevalence and impact of the stroke-related disability on Health-Related Quality of Life (HRQoL) as a recognized and important outcome after stroke is huge. Quick, valid and reliable assessment of the HRQoL in people after stroke constitutes a significant worldwide problem for scientists and clinicians - there are many tools, but no one fulfills all requirements or has prevailing advantages. This paper presents proposition of an evaluation of HRQoL based on the two-level hierarchical fuzzy system. It uses five clinical scores and scales as the inputs and gives in result value from the interval [0; 1]. It may constitute a useful semi-automated tool for supplementary initial assessment of patient functioning and further cyclic re-assessment for rehabilitation process and patient-centered goals of rehabilitation shaping purposes.

Keywords: Fuzzy system · Hierarchical fuzzy system · Linguistic modeling · Clinimetrics · Stroke rehabilitation · Health-Related Quality of Life

1 Introduction

Quality and quantity of survival in patients with severe illnesses are common in the evaluation of treatment. Conceptual and methodological issues of patients quality of life (QoL) measurement are not easy and are still under debate. Many developed tools such as Ferrans and Powers QOL Index-Stroke Version, Niemi QOL scale, SA-SIP30, Sickness Impact Profile, etc. have advantages and disadvantages (or even some unresolved issues), thus selection for the proper tool need particular caution, and taking into consideration goals, context, and limitations of the particular application. Measurement of the patient-centered outcomes such as functional status and health-related quality of life (HRQoL) are very important within current health care, especially in rehabilitation after severe diseases, injuries and cerebrovascular accidents such as stroke [16].

© Springer International Publishing AG 2017
L. Rutkowski et al. (Eds.): ICAISC 2017, Part I, LNAI 10245, pp. 710–721, 2017.
DOI: 10.1007/978-3-319-59063-9_64

The HRQoL is a very imprecise element. There are so many features which are difficult to calculate. Therefore, it is worth to consider the methods which let us analyze imprecise data. If the precise model is out of reach, we can use the tools for the imprecise information processing - fuzzy systems. Their main advantage is the flexibility, intuitiveness, and clarity of rules that are easy to describe linguistically (e.g. see [2,15]).

The purpose of this paper is a presentation of the tool under workname Multicriteria Fuzzy Evaluator of Health-Related Quality of Life (abbr. MuFE-HRQoL). It is an algorithm for evaluation of a general quality of life of people after stroke. As the main evaluating mechanism, an idea of the hierarchical fuzzy systems [11,22,23] is used. The fuzzy systems are a well-known and popular tool which allow modeling knowledge represented linguistically. By using them in the evaluation process, we gain a possibility to propose a model basing on experience and intuitions of medical scientists who deal with given problems - in our case with the rehabilitation of patients after stroke.

The order of the paper is as follows: presentation of medical basics of the model of evaluation; short analysis of a problem of an evaluation using a fuzzy system; presentation of the evaluator with description of the structure of used hierarchical fuzzy system and details about the linguistic model; presentation of the practical results of evaluation then its analysis and conclusions.

2   Clinical Scores to Evaluate

Bobath Scale (to assess hand functions), Barthel Index (to assess activities of daily living - ADL), and normalized values of the gait parameters (normalized gait velocity, normalized cadence, and normalized stride length) were applied to assess functional status and independence of the subjects. Aforementioned measures are often used in everyday clinical practice, assessed as valid and reliable. Measurements were performed in every post-stroke patient (i.e. belonging to the study group) twice:

- before the therapy (before the first session of the therapy),
- after the therapy (after the last session of the therapy) - to compare results and assess rehabilitation effects.

Ten sessions of the NDT-Bobath therapy were provided during the course of 2 weeks (10 days of the therapy rehabilitation was performed every day for 5 days a week). Each session lasted 30 min.

Bobath Concept (neurodevelopmental treatment) constitutes the most popular treatment approach applied in stroke rehabilitation, despite the superiority of the one particular approach has not been established yet due to methodological limitations and few compartmental studies. Current evidence syntheses are weak, pose too many methodological shortcomings and lack of the high-quality trials. Lack of detailed clinical guidelines in the area of post-stroke physiotherapy cause that preferences and experience of the therapist constitute the framework of the most effective treatment (so-called mixed/eclectic approach) [6,8,12,13].

Patients were treated according to the rules of the method by experienced (>15 years to experience) therapists of NDT-Bobath method for adults with international certificates:

– IBITA recognized Basic Course "Assessment and Treatment of Adults with Hemiplegia The Bobath Concept",
– IBITA recognized Advanced Course "Assessment and Treatment of Adults with Neurological Conditions The Bobath Concept",
– additionally, EBTA recognized NDT-Bobath Basic Course and EBTA recognized NDT-Bobath Baby Course.

All measurements were performed for every member of the reference group (healthy people) once. The study was accepted by the appropriate Bioethical Committee. The subjects gave written informed consent before entering the study, in accordance with the recommendations of the Bioethical Committee, acting on the rules of Good Clinical Practice and the Helsinki Declaration.

3 Basics of the Fuzzy Evaluation Model

Fuzzy logic allows for transferring a linguistic description into a computer algorithm. As there is a lack of a mathematical model of the evaluation, the fuzzy systems seems to be a good direction. A linguistic model of rules which describes expected evaluations depending on input values comes from the experience health scientists who work with the post-stroke patients.

An evaluation is a very general concept. We can understand it as a procedure which should grant us a single value which is representing a group of some elements. A general purpose of evaluation is comparing the results generated for the same classes of elements. Thus the real numbers interval as a set of possible results is a good choice.

Generally, it can be defined as a function which assigns one value to the n-input data. We could say the result is a singular value representing all inputs. In such view, an evaluation is close to the idea of an aggregation function (see [1,5]). However, there are key differences:

1. for evaluation process input values can be from various domains,
2. the evaluation must not meet a basic requirement for aggregations nondecreasing [1,5].

A first difference is easy to overcome by normalization of all input domains into [0; 1] interval.

The second case - nondecreasing - means that increase in any input can not cause a decrease of the result. It cannot be fulfilled by many situations where evaluation is needed. In many aspects of life, we have to deal with the evaluation which nature is non-monotonic. It happens that we estimate poorly for both low-value, as well as the large as e.g. ideal outdoor temperature for relaxing. We feel non-comfortable both if it is too low or too high.

A similar situation occurs with the gait parameters, which are components of HRQoL assessment presented in further part of this paper. Therefore for the construction, we can not base directly on aggregation functions.

4   Fuzzy Evaluator of HRQoL

There is lack of comparable simple, quick and cost-effective tools ready to use in everyday clinical practice including home rehabilitation. Current assessment of the QoL rely on questionnaires and is not objective.

In this section, the tool under workname Multicriteria Fuzzy Evaluator of Health-Related Quality of Life (abbr. MuFE-HRQoL) is presented. This tool evolved from Multicriteria Fuzzy Evaluator proposed in [21] which purpose was an evaluation of Multicast Routing Algorithms.

The evaluator was implemented using the Mathlab FuzzyLogic Toolbox. It is a two level hierarchical fuzzy system with five inputs and one output value. As the input values are used the medical scores for Bobath Test, Barthel Index and three descriptors of gait: velocity, cadence and stride length. The output - general quality range is [0; 1] interval, which can be easily transformed to any other.

A potential for the linguistic description is in this research very important. Thus MuFE-HQRoL uses fuzzy systems of the Mamdani-type. Ideas presented in this paper are strictly connected with [14, 19], where we use to evaluate gait quality special model the Ordered Fuzzy Numbers [7, 9, 10, 17] alternatively called also Kosinski's Fuzzy Numbers [18, 20].

4.1   Hierarchical Model of Evaluation

The fuzzy system used in the MuFE-HRQoL works in two steps. First one uses a group of small fuzzy systems to evaluate separately every singular feature describing HRQoL. Next step uses outputs of the first one as inputs, where the final result - the general quality of HRQoL is calculated. The structure is presented on the Fig. 1.

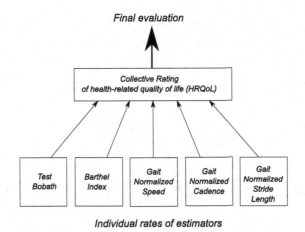

Fig. 1. Hierarchy of fuzzy model of evaluation.

The proposed structure of evaluation is a kind of hierarchical fuzzy system [11,22,23]. Generally, the hierarchical organization of fuzzy systems is used to decrease the total number of rules. However additionally, in our proposition, the hierarchy let us to separate context of the medical properties what makes easier to formulate the model of evaluation linguistically.

4.2 Evaluator - Fuzzy Sets for First Level Fuzzy Systems

The fuzzy sets forming linguistic values in the first level of the hierarchical structure of the evaluator came from the specificity of chosen five medical scores. The Bobath Test and Barthel Index are strictly defined tests. Their quality results represent a typical monotonic tendency, 'the more the better', where maximum means a normal and healthy condition, and minimum very bad condition. Therefore these two input linguistic values are divided into two triangular fuzzy sets each. They represent 'bad/low' and 'good/high' opinion on values - see Fig. 2 left side. Right side shows the output linguistic variable of all fuzzy systems at every level of hierarchy presented on the Fig. 1.

The remaining three linguistic parameters describing gait and are based on technical measures of human body behavior. The model of their evaluation is also more complicated. Each of gait parameters is represented by a separate linguistic variable. The fuzzy sets are defined basing on the measures of reference group of healthy (non post-stroke) people.

The remaining three linguistic parameters describing gait are based on technical measures of human body behavior. The model of their evaluation is also more complicated. Each of gait parameters is represented by a separate linguistic variable. The fuzzy sets are defined basing on the measures of a reference group of healthy (non post-stroke) people.

The fuzzy set representing the ideal value for each gait parameter is constructed from the data given for the reference group - people without stroke. Each of them is a triangular fuzzy set (see LR fuzzy sets notation in [4]) and is determined by all the available data. All three describing gait linguistic variables are modeled in the same pattern. For example, lets look at set *Proper Gait Velocity* - ($prop^{GV}$):

$$prop^{GV} = \Lambda(x; x_{mean} - 2 \cdot \Delta_L, x_{mean}, x_{mean} + 2 \cdot \Delta_R) \tag{1}$$

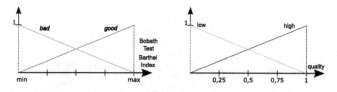

Fig. 2. Left - two input linguistic variables 'Bobath Test' and 'Barthel Index', right - output linguistic variables of all fuzzy systems in the first level of MuFE-HRQoL

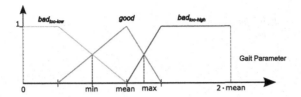

Fig. 3. A pattern of fuzzy sets for the gait describing linguistic variables.

where $\Delta_L = x_{mean} - x_{min}$, $\Delta_R = x_{max} - x_{mean}$, $x_{min}/x_{max}/x_{mean}$ – the minimum/maximum/mean value of the Gait Velocity parameters for the available data about healthy (non post-stroke) people.

The range of linguistic variable is defined as interval $[0; 2 \cdot x_{mean}]$. It is enough to cover the all available data for healthy and post-stroke people. There are two another fuzzy sets which represent 'bad' quality of gait. The '$bad^{GV}_{too-low}$' and '$bad^{GV}_{too-high}$' are trapezoidal fuzzy sets (see fuzzy intervals [3,4]) defined by four parameters as follows:

$$bad^{GV}_{too-low} = \Pi(x; x_0, x_0, x_{mean} - 2 \cdot \Delta_L, x_{mean}),$$
$$bad^{GV}_{too-high} = \Pi(x; x_{mean}, x_{mean} + 2 \cdot \Delta_R, 2 \cdot x_{mean}, 2 \cdot x_{mean}), \tag{2}$$

where $x_0 = MIN(0, x_{mean} - 2 \cdot \Delta_L)$.

Above method of constructing fuzzy sets guarantee that the gait of any healthy people from the reference group will be evaluated at least as 0,5 true value.

The fuzzy output sets were presented on the Fig. 2. The result will be a number from the continuous interval [0, 1] where the higher value means better quality. Such interval allows for the representation of results in a convenient and intuitive percentage scale. However, it should be stressed that the upper bound stands for the ideal gait parameters, but in real life, there are natural individual differences between healthy people. Thus, the model is formulated to point evaluation each of this persons at least as 50% result value. On the Fig. 3 is presented a general pattern for the all linguistic variables describing gait. The fuzzy set 'good' is a key element and is defined as presented in the formula 1 for 'gait velocity' variable.

4.3 Evaluator - Fuzzy Sets for Output Fuzzy System

The purpose of the final fuzzy system - second level - in the hierarchy is an aggregation of first level outputs. The input variables are simplified to two triangular values 'low' and 'high' defined on the interval [0; 1]. However the output 'Quality' is divided into six fuzzy values (see Fig. 4). They represent terms from an 'extremal low' to a 'normal'.

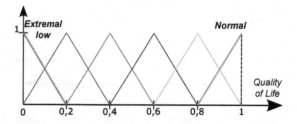

Fig. 4. Output fuzzy sets of the final fuzzy system

4.4 Rules of First Level of Fuzzy Evaluator

The rules for the Bobath Test and Barthel Index are simply an illustration of the idea 'the more, the better'.

As was mentioned at the end of Sect. 3, measuring the gait quality have certain specificity. This is related to not monotonic dependence. The bad quality is either a parameter is too low or too high. Therefore each of variable that represents gait parameter consists of three linguistic values: 'too low', 'proper' and 'too high'. The output variable - quality - consists of two values: 'good' and 'bad'. So the rules follow the patterns:

1. IF the parameter is 'too low' THEN quality is 'bad',
2. IF the parameter is 'proper' THEN quality is 'good',
3. IF the parameter is 'too high' THEN quality is 'bad'.

4.5 Rules of Second Level of Fuzzy Evaluator

The fuzzy system of the second level of evaluator aggregates outputs from the first level. There are five input variables with two fuzzy sets ('low' and 'high') each. Therefore we have $2^5 = 32$ rules in this system. To shorten the description, we will use the abbreviations of qualities: qV - quality of gait velocity, qC - of gait cadence, qS - of gait stride length, qBT - of Bobath Test, qBI - of Barthel Index.

The output variable is represented by six values (see Fig. 4). We will notate them as Out_i where $i = 0..5$ and Out_0 represent the worst condition of the patient and Out_5 means the best condition, understand as the 'normal quality of life'.

In general, the rules follow the pattern:

$$\text{IF } qV \text{ and } qC \text{ and } qS \text{ and } qBT \text{ and } qBI \text{ THEN } Out_s \tag{3}$$

where s is a number of times the term 'high' was used in the premise part of the rule.

Such set of rules could be described as a kind of 'accumulative character', where the output is higher if more inputs are in 'high' state. For example, the highest Out_5 is used only in one rule:

$$\text{IF } qV = h \text{ and } qC = h \text{ and } qS = h \text{ and } qBT = h \text{ and } qBI = h \text{ THEN } Out = Out_5 \tag{4}$$

the letter 'h' stands for 'high'. The output is Out_5 because the 'high' was used five times in premise of the rule. Adequately the Out_0 also is used only in one rule ('l' stands for 'low'):

$$\text{IF } qV = l \text{ and } qC = l \text{ and } qS = l \text{ and } qBT = l \text{ and } qBI = l \text{ THEN } Out = Out_0 \tag{5}$$

It is worth to notice that the general evaluation problem could be intuitively related to the classification process where the fuzzy output sets represent a linguistically expressed classes and those classes are discrete levels of evaluation. In such view, the MuFEG extremal low and extremal high fuzzy output would be elitist classes. Thus the only one rule point each of them. It is understandable as evaluating HRQoL we concentrate on modeling the nuances and dependencies on the area between lowest and highest results. Therefore in our case, the evaluation can not be directly the same as the classification. We do not group the results in the classes because we especially want a mechanism which will represent the output values even a small differences in inputs.

4.6   The Parameters of Fuzzy System

The particular parameters of the fuzzy systems used in evaluator are as follows:

- fuzzification of all inputs of all fuzzy systems - the singleton type,
- implication operator for all fuzzy systems MIN,
- aggregation of fuzzy outputs (accumulation) for all fuzzy systems MAX,
- aggregation of premise parts in second level fuzzy system MIN,
- defuzzification of outputs for the first level (in the hierarchy) fuzzy systems - the Middle of Maxima (MOM) method,
- defuzzification of outputs for the second level fuzzy system - the Center of Area COA type.

We may notice that fuzzy systems for the first level do not use an aggregation of premise parts due to the fact that there is only one input variable. It should also be stressed that the COA type defuzzification gives more flexibility than MOM, but it also narrows a range of the results. Therefore the defuzzified outputs of the final fuzzy system are normalized into $[0, 1]$ interval.

5   Practical Results

The tool MuFE-HRQoL was used to evaluate a condition of 40 post-stroke patients. We evaluated their results before starting a cycle of Bobath-based rehabilitation, and also after it was finished. The Table 1 presents the results. For compartment/validation purposes we also evaluated a reference group (20 healthy people), which was the source of a pattern of the quality of gait.

Table 1. Practical results of evaluation with the MuFE-HRQoL

s.n	Before rehab. quality	After rehab. quality	s.n.	Reference group quality
1	0,325573264	0,431471438	1	0,754498216
2	0,367753549	0,713399741	2	0,799898955
3	0,382564125	0,739715052	3	0,870636771
4	0,403269273	0,589430218	4	0,887701817
5	0,423703744	0,557905203	5	0,794935902
6	0,540633331	0,428769306	6	0,75076171
7	0,195263771	0,347582707	7	0,883172977
8	0,374546372	0,606537676	8	0,81821456
9	0,428856491	0,565574129	9	0,764061584
10	0,54295519	0,586609385	10	0,787773336
11	0,161698596	0,389457669	11	0,792370447
12	0,574538221	0,542789823	12	0,798028811
13	0,415103592	0,385372554	13	0,71592842
14	0,528958294	0,607686962	14	0,832546303
15	0,103012635	0,173181876	15	0,799953625
16	0,375523905	0,563692038	16	0,83445063
17	0,402974854	0,443984289	17	0,801948629
18	0,575180327	0,576361835	18	0,835638691
19	0,1515976	0,198442339	19	0,845103201
20	0,153944726	0,215072763	20	0,758346443
21	0,541128775	0,401489236	avg.	0,806298551
22	0,529374365	0,449475527		
23	0,40355343	0,555672344		
24	0,390093754	0,399943642		
25	0,412535381	0,437140092		
26	0,44145176	0,560678825		
27	0,372291169	0,606179309		
28	0,333076183	0,359804512		
29	0,584276451	0,608122899		
30	0,608210916	0,652253653		
31	0,385372554	0,402974854		
32	0,250248977	0,385310566		
33	0,392543337	0,622175941		
34	0,516757048	0,444998125		
35	0,405280398	0,55675543		
36	0,361757857	0,518881459		
37	0,250258254	0,479208805		
38	0,553733149	0,503002778		
39	0,390093754	0,623150371		
40	0,51000365	0,424720362		
avg.	0,401492326	0,491374393		

The results show that rehabilitation gives an improvement in the quality of life, but not in all cases. It is consistent with the practical situations where rehabilitation sometimes gives short-term negative results (often followed by more favorable long-term results). However, if we compare the averages, we see that the rehabilitation, in general, improves the patient's life quality. As we can see from the reference group, their results are much higher than patients. However, values are noticeably less than maximal. It is a consequence of the triangular fuzzy sets used in the gait evaluation. We assume there that the norm for gait is at least 0,5 result, but in the next stage of evaluation, we define the norm as a consequence of all full memberships (values equal one) on the inputs. It needs improvement in future.

6   Conclusions and Future Actions

In this paper, we presented the tool MuFE-HRQoL for evaluation of HQRoL for the people after stroke. The practical results show its usefulness. However, that tool has also a potential to compare the different rehabilitation methods, therefore, in fact, it could also be a tool for evaluation of their short- and long-term outcomes. The results show that rehabilitation gives an improvement in the quality of life, but not in all cases. It is consistent with the practical situations where rehabilitation sometimes gives short-term negative results (often followed by more favorable long-term results). Presently by using triangular fuzzy sets, we interpret as a norm also the values with 0,5 what have consequences in the final result. New sets should have a trapezoidal shape. It allows for modeling the range of input results as representations of the norm with full membership.

There is also a place for improvement in the structure of evaluator's hierarchical fuzzy system. As gait is only one aspect of life, it could be modeled by the independent level of the hierarchy which as a result gives one value representing the gait quality. Then, it could be used as one independent value among other features that affects the final result. From the practical point of view clinical application of the new tool needs for deeper research on its:

- reliability, i.e. measurement error associated with an instrument,
- validity, i.e. extent to which an instrument measures HQRoL itself in the absence of the "gold standard" of HQRoL measurement tool (including compartmental studies with other existing clinical scores and scales measuring HQRoL),
- responsiveness, i.e. the minimal degree of change that is thought to be clinically significant as far as "floor" and "ceiling" effects,
- acceptability of new tool to patients and assessors.

Aforementioned studies may strengthen results taking into consideration evidence-based medicine (EBM) paradigm. Finally, an improvement of evaluator would be using more medical scores in the evaluation of the quality of life. However, it is restricted by the availability of the another test results, which should be performed on the same group of patients.

References

1. Beliakov, G., Pradera, A., Calvo, T.: Aggregation Functions: A Guide for Practitioners. Studies in Fuzziness and Soft Computing, vol. 221. Springer, Heidelberg (2007). http://dx.doi.org/10.1007/978-3-540-73721-6
2. Buckley, J.J., Eslami, E.: Advances in Soft Computing: An Introduction to Fuzzy Logic and Fuzzy Sets. Physica-Verlag GmbH, Heidelberg (2002)
3. Dubois, D., Kerre, E., Mesiar, R., Prade, H.: Fuzzy interval analysis. In: Dubois, D., Prade, H. (eds.) Fundamentals of Fuzzy Sets. The Handbooks of Fuzzy Sets Series, vol. 7, pp. 483–581. Springer, Heidelberg (2000). http://dx.doi.org/10.1007/978-1-4615-4429-6_11
4. Dubois, D.: Fuzzy Sets and Systems: Theory and Applications. Mathematics in Science and Engineering. Elsevier Science, Amsterdam (1980)
5. Grabisch, M., Marichal, J.L., Mesiar, R., Pap, E.: Aggregation functions: means. Inf. Sci. 181(1), 1–22 (2011). http://www.sciencedirect.com/science/article/pii/S002002551000424X
6. Klimkiewicz, P., Kubsik, A., Woldańska-Okońska, M.: NDT-bobath method used in the rehabilitation of patients with a history of ischemic stroke. Wiad. Lek. 65(2), 102–107 (2012)
7. Koleśnik, R., Prokopowicz, P., Kosiński, W.: Fuzzy calculator – useful tool for programming with fuzzy algebra. In: Rutkowski, L., Siekmann, J.H., Tadeusiewicz, R., Zadeh, L.A. (eds.) ICAISC 2004. LNCS, vol. 3070, pp. 320–325. Springer, Heidelberg (2004). doi:10.1007/978-3-540-24844-6_45
8. Kollen, B.J., Lennon, S., Lyons, B., Wheatley-Smith, L., Scheper, M., Buurke, J.H., Halfens, J., Geurts, A.C., Kwakkel, G.: The effectiveness of the Bobath concept in stroke rehabilitation: what is the evidence? Stroke 40(4), 89–97 (2009)
9. Kosinski, W., Prokopowicz, P.: Fuzziness - representation of dynamic changes? In: Stepnicka, M., Novak, V., Bodenhofer, U. (eds.) New Dimensions in Fuzzy Logic and Related Technologies, Proceedings, 5th Conference of the European-Society-for-Fuzzy-Logic-and-Technology, Ostrava, Czech Republic, vol. 1, pp. 449–456. European Society for Fuzzy Logic & Technology, Univ. Ostrava, Ostravska Univ. & Ostrave, Dvorakova 7, Ostrava 1, 701 03, Czech Republic, 11–14 September 2007 (2007)
10. Kosiński, W., Prokopowicz, P., Ślęzak, D.: On algebraic operations on fuzzy numbers. In: Kłopotek, M.A., Wierzchoń, S.T., Trojanowski, K. (eds) Intelligent Information Processing and Web Mining. Advances in Soft Computing, vol 22. Springer, Heidelberg (2003). http://dx.doi.org/10.1007/978-3-540-36562-4_37
11. Lee, M.L., Chung, H.Y., Yu, F.M.: Modeling of hierarchical fuzzy systems. Fuzzy Sets Syst. 138(2), 343–361 (2003). http://www.sciencedirect.com/science/article/pii/S0165011402005171
12. Mikołajewska, E.: NDT-Bobath method in normalization of muscle tone in post-stroke patients. Adv. Clin. Exp. Med. 21(4), 513–517 (2012)
13. Mikołajewska, E.: Associations between results of post-stroke NDT-Bobath rehabilitation in gait parameters, ADL and hand functions. Adv. Clin. Exp. Med. 22(5), 731–738 (2013)
14. Mikołajewska, E., Prokopowicz, P., Mikolajewski, D.: Computational gait analysis using fuzzy logic for everyday clinical purposes – preliminary findings. Bioalg. Medsyst. 13(1), 37–42 (2017). https://doi.org/10.1515%2Fbams-2016-0023
15. Pedrycz, W., Gomide, F.: Fuzzy Systems Engineering: Toward Human-Centric Computing. Wiley-IEEE Press, New York (2007)

16. Pickard, A.S., Johnson, J.A., Feeny, D.H.: Responsiveness of generic health-related quality of life measures in stroke. Qual. Life Res. **14**(1), 207–219 (2005)
17. Prokopowicz, P.: Flexible and simple methods of calculations on fuzzy numbers with the ordered fuzzy numbers model. In: Rutkowski, L., Korytkowski, M., Scherer, R., Tadeusiewicz, R., Zadeh, L.A., Zurada, J.M. (eds.) ICAISC 2013. LNCS, vol. 7894, pp. 365–375. Springer, Heidelberg (2013). doi:10.1007/978-3-642-38658-9_33
18. Prokopowicz, P.: Analysis of the changes in processes using the Kosinski's Fuzzy Numbers. In: Ganzha, M., Maciaszek, L., Paprzycki, M. (eds.) Proceedings of the 2016 Federated Conference on Computer Science and Information Systems, Annals of Computer Science and Information Systems, vol. 8, pp. 121–128. IEEE (2016). http://dx.doi.org/10.15439/2016F140
19. Prokopowicz, P., Mikolajewska, E., Mikolajewski, D., Kotlarz, P.: Traditional vs OFN-based analysis of temporo-spatial gait parameters. In: Prokopowicz, P., Czerniak, J., Mikolajewski, D., Apiecionek, L., Slezak, D. (eds.) Theory and Applications of Ordered Fuzzy Numbers - A Tribute to Professor Witold Kosinski. Studies in Fuzziness and Soft Computing, vol. 356. Springer, Heidelberg (2017, in press)
20. Prokopowicz, P., Pedrycz, W.: The directed compatibility between ordered fuzzy numbers - a base tool for a direction sensitive fuzzy information processing. In: Rutkowski, L., Korytkowski, M., Scherer, R., Tadeusiewicz, R., Zadeh, L.A., Zurada, J.M. (eds.) ICAISC 2015. LNCS, vol. 9119, pp. 249–259. Springer, Cham (2015). doi:10.1007/978-3-319-19324-3_23
21. Prokopowicz, P., Piechowiak, M., Kotlarz, P.: The linguistic modeling of fuzzy system as multicriteria evaluator for the multicast routing algorithms. In: Rutkowski, L., Korytkowski, M., Scherer, R., Tadeusiewicz, R., Zadeh, L.A., Zurada, J.M. (eds.) ICAISC 2014. LNCS (LNAI), vol. 8468, pp. 665–675. Springer, Cham (2014). doi:10.1007/978-3-319-07176-3_58
22. Raju, G.V.S., Zhou, J., Kisner, R.A.: Hierarchical fuzzy control. Int. J. Contr. **54**(5), 1201–1216 (1991). http://dx.doi.org/10.1080/00207179108934205
23. Torra, V.: A review of the construction of hierarchical fuzzy systems. Int. J. Intell. Syst. **17**(5), 531–543 (2002). http://dx.doi.org/10.1002/int.10036

Exploratory Analysis of Quality Assessment of Putative Intrinsic Disorder in Proteins

Zhonghua Wu[1], Gang Hu[1], Kui Wang[1], and Lukasz Kurgan[2(✉)]

[1] School of Mathematical Sciences and LPMC, Nankai University,
Tianjin, People's Republic of China
{wuzhh,huggs,wangkui}@nankai.edu.cn
[2] Department of Computer Science, Virginia Commonwealth University,
Richmond, VA, USA
lkurgan@vcu.edu

Abstract. Intrinsically disorder proteins are abundant in nature and can be accurately identified from sequences using computational predictors. While predictions of disorder are relatively easy to obtain there are no tools to assess their quality for a particular amino acid or protein. Quality assessment (QA) scores that quantify correctness of the predictions are not available. We define QA for the prediction of intrinsic disorder and use a large dataset of over 25 thousand proteins and ten modern predictors of disorder to empirically assess the first approach to quantify QA scores. We formulate the QA scores based on the readily available propensities of the intrinsic disorder generated by the ten methods. Our evaluation reveals that these QA scores offer good predictive performance for native structured residues (AUC > 0.74) and poor predictive performance for native disordered residues (AUC < 0.67). Specifically, we show that most of the native disordered residues that are incorrectly predicted as structured have high QA values that inaccurately suggest that these predictions are correct. Consequently, more research is needed to develop high-quality QA scores. We also outline three possible future research directions.

1 Introduction

Intrinsically disordered proteins lack stable tertiary structure under physiological conditions along their entire amino acid chain or in specific region(s) [1, 2]. They are abundant in nature, with recent estimates showing that about 19% of amino acids in eukaryotic proteins are disordered [3], and up to 50% of eukaryotic proteins have at least one long (≥30 consecutive amino acids) intrinsically disordered region [4, 5]. Intrinsically disordered proteins are crucial for a diverse range of cellular functions including transcription, translation, signaling, protein-protein, protein-nucleic acids and virus-host interactions, to name just a few [2, 3, 6, 7]. A large number of computational methods that predict intrinsic disorder in protein sequences was developed. A study from 2012 estimates this number to be at about 60 [8]. The predictions that these methods generate are utilized to support and plan experimental studies and to quantify prevalence and analyze functions of disorder on a large, genomic scale [3, 9–13]. They are also used in other research

© Springer International Publishing AG 2017
L. Rutkowski et al. (Eds.): ICAISC 2017, Part I, LNAI 10245, pp. 722–732, 2017.
DOI: 10.1007/978-3-319-59063-9_65

areas including structural genomics [14]. In recent years two large databases that offer access to putative annotations of intrinsic disorder for millions of proteins were developed: MobiDB [15, 16] and D^2P^2 [17].

In spite of the popularity and wide-spread use of these predictors and the fact that their predictive performance was evaluated in a number of studies [18–23], there are no studies that investigate quality assessment of these predictions. While the users nowadays can easily collect predictions of disorder, there are no methods that quantify quality of these predictions for a particular amino acid or protein. In other words, quality assessment methods that assign a numeric score to each prediction that quantifies whether it is correct are lacking. This is in stark contrast to the prediction of the tertiary protein structure where many tools for the quality assessment were developed in recent years [24–28]. To this end, we define the quality assessment in the context of the prediction of intrinsic disorder. Using a large dataset of proteins, we also empirically assess whether the propensities of the intrinsic disorder generated by the modern predictors of disorder can be used as a proxy for the quality assessment scores.

2   Materials and Methods

2.1   Dataset

The dataset was originally developed in ref. [22] and can be downloaded from http:// mobidb.bio.unipd.it/lsd. Proteins from the UniProt resource [29] were mapped into the MobiDB database [15] to obtain their annotations of native intrinsic disorder. All proteins for which the annotations were found are included and a majority vote was used to assign disorder in cases when multiple annotations are found in MobiDB, i.e., a given residue is assumed to be disordered if most of the annotations for this residues indicate that it is disordered. This approach arguably allows to filter out conflicts due to variations in experimental conditions [22]. Similar sequences were removed at 90% pairwise sequence identity using the CD-HIT program [30] resulting in a set of 25,833 annotated proteins. Each residue in these proteins is annotated as disordered, structured or unknown, the latter in the case when MobiDB does not provide an annotation. Our analysis is based on the residues that are annotated as either disordered or structured. We exclude the residues with the unknown annotations. Moreover, the dataset is further reduced to 25,717 proteins for which we were able to secure putative intrinsic disorder with all considered predictors of disorder. In total, our dataset includes 7,049,517 annotated residues with 6,700,101 and 349,416 that are structured and disordered, respectively. This corresponds to the overall disorder content (defined as a fraction of disordered residues among all residues) of 5%.

2.2   Putative Annotations of Intrinsic Disorder

Putative annotations of intrinsic disorder were generated with ten methods: three version of the ESpritz method [31] that are designed to predict disorder annotated using X-ray crystallography (Espritz-X-ray), NMR (Espritz-NMR), and DisProt database [32] (Espritz-Disprot); two versions of IUPred [33] that are optimized to predict short

(IUPred-short) and long (IUPred-long) disordered regions; two versions of the DisEMBL method [34] that predict disordered regions defined as hot loops (DisEMBL-HL) and based on the remark 465 from Protein Data Bank (PDB) [35] (DisEMBL-465), RONN [36],VSL2b [37], and GlobPlot that predicts globular regions [38]. These methods represent a comprehensive selection of modern predictors that cover various flavors of disorder and that are sufficiently runtime-efficient to provide results at the scale of our large dataset; the runtime of these methods is under 1 min for an average size protein sequence. The predictors of disorder typically generate two outputs for each residue in the input protein sequence: a real-valued propensity score and a binary prediction. The score is a putative likelihood that a given residue is in a disordered conformation. The binary value is usually derived from the propensity based on a method-specific threshold and it categorizes the residue as either disordered or structured. Residues with propensities > threshold are classified as disordered and the remaining residues are classified as structured. The ranges of values of the propensities for the ten predictors together with the native and putative disorder content, the latter estimated from the putative binary values, are summarized in Table 1. Interestingly, the Pearson correlation coefficients (PCCs) between predicted propensities generated by different predictors range between 0.07 and 0.81, with average of 0.46. This demonstrates that these methods in fact offer substantially different predictions.

Table 1. The native and predicted amount of intrinsic disorder for the benchmark set of 25717 proteins. We also list the minimal and maximal values of propensity and the threshold value used to convert these propensities into binary scores for the 10 predictors of intrinsic disorder.

Native annotations and predictors	Putative propensity of disorder			Disorder content
	Min	Max	Threshold	
Native annotations	NA	NA	NA	5.0%
DisEMBL-465	0.000	0.968	0.500	6.4%
DisEMBL-HL	0.000	0.585	0.086	28.9%
Espritz-Disprot	0.004	0.978	0.507	2.6%
Espritz-NMR	0.002	0.997	0.309	9.1%
Espritz-X-ray	0.003	0.997	0.143	16.5%
GlobPlot	−0.329	0.513	0.000	13.5%
IUPred-long	0.000	0.995	0.500	6.0%
IUPred-short	0.000	1.000	0.500	6.7%
RONN	0.070	1.000	0.500	16.2%
VSL2b	0.002	1.000	0.500	21.0%

2.3 Definition of Quality Assessment for Putative Intrinsic Disorder

The putative annotations of intrinsic disorder are typically derived based on the binary values where residue are categorized as either structured or disordered. The putative propensities can be used to quantify confidence that accompanies the binary predictions. The putative disordered residues predicted with high propensity scores should be more accurately predicted compared to the residues that are associated with propensities that are just slightly higher than the threshold. The same is true for the structured residues where the putative structured residues that have low propensities should be more accurately predicted than the structured residues with propensities just below the threshold. However, while predictive performance of the disorder predictors was evaluated extensively [18–23], the use of the propensities as a proxy to quantify quality of these predictions was not yet researched.

The quality assessment (QA) boils down to computation of a score that quantifies correctness of a given prediction. More specifically, in the QA scenario each prediction, whether it suggests that a given residue is disordered or structured, is associated with a propensity score that is high when the prediction is correct and low when it is incorrect. In other words, native disordered residues predicted as disordered and native structured residues predicted as structured should have high QA scores, while residues that are incorrectly predicted (native disordered as structured or native structured as disordered) should have low QA scores. One immediately available option to generate these QA scores is to use the predicted propensities for disorder to generate QA scores for the binary disorder predictions:

$$\text{IF } D_{prop} > \text{THR THEN } QA_{score} = \left\{ (D_{prop} - \text{THR}) / (\max(D_{prop}) - \text{THR}) \right\}$$

$$\text{IF } D_{prop} \leq \text{THR THEN } QA_{score} = \left\{ (D_{prop} - \text{THR}) / (\min(D_{prop}) - \text{THR}) \right\}$$

where D_{prop} is the putative propensity for disorder and THR is the threshold used to convert D_{prop} into the binary disorder prediction. This definition ensures that high and low values of the putative propensity for disorder (that denote likely correct predictions of disordered and structured residues, respectively) correspond to high QA scores, while QA scores for values of the predicted propensity for disorder that are close to the threshold are low. The relation between values of D_{prop} and QA_{score} is visualized in Fig. 1.

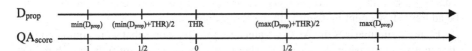

Fig. 1. The relation between the values of the putative propensity for intrinsic disorder (D_{prop}) and the values of the quality assessment score (QA_{score}).

2.4 Evaluation Measures

Quality of the predicted propensity for the intrinsic disorder is typically evaluated using ROC curves and the corresponding AUC values [18–23]. More specifically, the propensities are used to compute a curve defined by FPR = FP/(FP + TN) and TPR = TP/(TP + FN) values where TP is the number of correctly predicted disordered residues, FP is the number of structured residues predicted as disordered, TN is the number of correctly predicted structured residues, and FN is the number of disordered residues predicted as structured; multiple values of FPR and TPR are generated by using different thresholds on the value of the propensity. AUC is the area under the ROC curve and its values range between 0.5 for a random-like prediction and 1 for the perfect prediction. We denote this measure as AUCd (AUC for the prediction of disordered residues).

We similarly utilize the AUC values to assess the predictive quality of the QA scores. In this case TP is the number of correctly predicted correct predictions (of both disordered and structured residues) based on the QA scores, FP is the number incorrect predictions that are predicted as correct using QA scores, TN is the number of correctly predicted incorrect predictions (of both disordered and structured residues) and FN is the number of correct predictions that are predicted as incorrect utilizing the QA scores. We denote this measure as AUCqa (AUC for the quality assessment). We also compute the AUCqa values specifically for the native disordered residues (AUCqa_d) and for the native structured residues (AUCqa_s). The latter two values quantify how well the quality assessment scores work when they are used for the native disordered and native structured residues.

3   Results

The AUCd values of the considered ten predictors of the intrinsic disorder are in agreement with the results in [22]. The values are shown in Fig. 2 and they range between 0.63 and 0.81 with average of 0.75. These results are also similar to the findings in [21] where the AUCd values of 19 predictors are shown to be between 0.70 and 0.82. Collectively, these studies conclude that the predictors of intrinsic disorder offer relatively strong predictive performance. However, the binary predictions of some of these methods disagree with the native annotations of disorder. Table 1 reveals that the native disorder content in our large dataset is at 5% while the putative disorder content generated by the ten predictors varies between 2.6% and 28.9%, with an average of 12.7%. This suggests that the putative binary annotation require improvements, and this could be addressed by coupling them with the QA scores.

Figure 2 summarizes the AUCqa values for the QA scores that were computed from the putative propensities for disorder. These AUCqa values quantify how well the QA scores predict correctness of the binary predictions of disorder. The size of the circles represent relative values of the AUCqa and the absolute values are shown next to the circles. The AUCqa values range between 0.74 for VSL2b and 0.90 for Espritz-X-ray, with an average value of 0.81. Figure 2 also shows the AUCqa_d and AUCqa_s values (the AUC for the QA scores for the native disordered and structured residues, respectively) as the y- and x-axis coordinates, respectively. The bubbles located below the 0.5 value on the y-axis correspond to seven methods that perform very poorly for the disordered residues: IUPred-short, IUPred-long, Espritz-NMR, DisEMBL-HL, GlobPlot, Espritz-X-ray, and Espritz-Disprot. While their overall AUCqa values are relatively high (between 0.77 and 0.90), they provide high quality QA scores only for the structured residues. In other words, the QA scores for these seven methods successfully identify correctly vs. incorrectly predicted structured residues, while they largely fail to identify correctly predicted disordered residues. Their AUCqa values are high in spite of the low

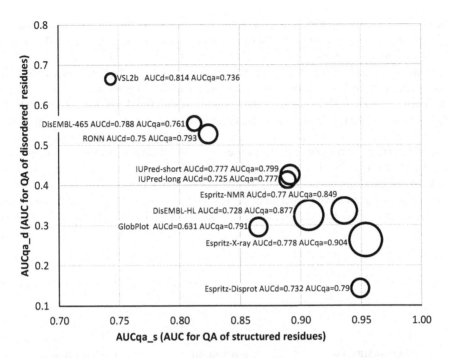

Fig. 2. Relation between AUC for the quality assessment of the disordered residues (AUCqa_d on the y-axis) and structured residues (AUCqa_s on the x-axis). Each predictor is represented by a circle; sizes of the circles represent relative values of AUC of the quality assessment of all residues (AUCqa). The names of the predictors together with the numeric values of AUCqa and AUCd (AUC for the prediction of disordered residues) are shown next to the circles.

values of AUCqa_d because a significant majority of the residues in the dataset is structured. Two methods, DisEMBL-465 and RONN, achieve modest values of AUCqa < 0.8 coupled with relatively high AUCqa_s at about 0.82 and slightly above-random values of AUCqa_d at about 0.55. The only method for which the QA scores are reasonably balanced between the structured and disordered residues is VSL2b. It secures AUCqa = 0.74, AUCqa_d = 0.67 and AUCqa_s = 0.74. However, these are rather modest values of predictive performance, particularly the AUCqa_d for the QA scores for the intrinsically disordered residues.

Figure 3 that gives the ROC curves for the QA scores offers further insights. The brown and red lines that denote the ROC curves for the quality assessment of all and native structured residues, respectively, reveal a favorable trade-off between the TPR and FPR values, i.e., TPR values are substantially higher than the corresponding FPR values. The three exceptions include DisEMBL-465, RONN and VSL2b methods (Figs. 3A, I and J) for which the two curves are relatively flat for the low FPR values, resulting in low AUCqa and AUCqa_s values. More importantly, the blue ROC curves for the quality assessment of native disordered residues, which for most predictors are located below the diagonal line, demonstrate that the corresponding FPR values are higher than the TPR values. Consequently, the QA scores produce more incorrect predictions than the number of correct predictions. More specifically, the ratios of incorrect predictions of disordered residues that are predicted as correct using the QA scores among all incorrect predictions (FPR values) are higher than the ratios of correct predictions of disordered residues that are predicted as correct using the QA scores among all correct predictions (TPR values). In other words, the high FPR means that many native disordered residues that are incorrectly predicted as structured are associated with high QA values. Such high QA values inaccurately suggest that the associated with them predictions are correct. In turn, the high QA values result from the fact that the corresponding putative propensities for disorder are low for these disordered residues. We observe that virtually all of the considered methods, except for VSL2b, generate low putative propensities for a majority of the disordered residues. This is a significant drawback of the putative propensities for disorder generated by the considered representative set of disorder predictors. It effectively renders the corresponding QA scores

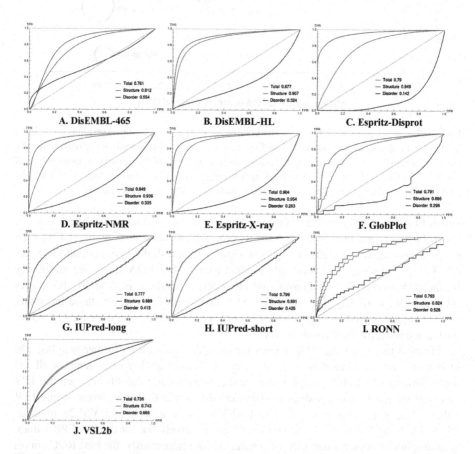

Fig. 3. The ROC curves for the quality assessment of the considered ten predictors. The brown, red and blue lines correspond to ROC curves for the quality assessment of all, structured and disordered residues, respectively. The legends show the corresponding three AUC values. (Color figure online)

useless when applied to the native disordered residues. Finally, although the three ROC curves for the QA scores for VLS2b are all above the diagonal line (Fig. 3J), the curves are flat suggesting that the predictive performance of these scores is rather low.

4   Conclusions

Our analysis that examines ten modern predictors of intrinsic disorder on a large set of close to 26 thousand proteins reveals that although the overall predictive performance of these methods is relatively high, the putative annotations that they generate would benefit from the inclusion of QA scores. These scores could be used to indicate which predictions could be trusted more than others and to identify correct vs. incorrect predictions. We are the first to attempt to define the QA scores based on the readily available putative propensities for disorder generated by the ten predictors. We empirically evaluate whether these propensities can be used to derive accurate QA values for the assessment of the corresponding binary predictions of disorder. Our analysis demonstrates that the QA scores that we define provide accurate results for the native structured residues for majority of the considered methods. However, the QA scores for the native disordered residues are inaccurate. For 9 out of 10 methods most of the native disordered residues that are incorrectly predicted as structured have high QA values (and low putative propensities for disorder) which falsely indicate that the corresponding predictions are correct. The only method for which QA scores perform reasonably well for both structured and disordered residues is VSL2b. However, the predictive quality of its QA scores is relatively modest, with the AUC values equal 0.74 and 0.67 for the structured and disordered residues, respectively.

Our results suggest that the QA scores generated based on the propensity for intrinsic disorder generated by modern, high-throughput predictors do not offer desirable levels of predictive performance. Further research to develop high-quality QA scores for the putative intrinsic disorder is needed. This is particularly urgent in the context of the recent emergence of large databases, such as MobiDB and D^2P^2, which offer easy access to predicted disorder for dozens of millions of proteins. Three possible directions could be pursued. The first option is to build one methodology that will provide QA scores for any disorder predictor using predictions from a single method. This would be very challenging given the relatively high degree of differences between the predictions from different methods for the same protein sequence. The second alternative is to develop one methodology that will provide QA scores for any disorder predictor using predictions from multiple methods. In other words, predictions from several disorder predictors would be used to derive a generic QA score which could be used to assess predictions of any of the input methods. While this should be easier than the first alternative, it will also require availability of multiple disorder predictions. This is feasible when employing the MobiDB and D^2P^2 databases that provides access to multiple predictions for each protein. The third option is to build QA methodologies that are coupled with specific disorder predictors. This option would be perhaps the easiest to develop but it would also require designing multiple QA schemes, as many as the number of the corresponding disorder predictors.

Acknowledgments. We thank Dr. Silvio Tosatto and his research group from University of Padova for sharing their dataset and predictions of disorder, which they published in ref. [22]. This research was supported in part by the National Science Foundation grant 1617369 and by the Qimonda Endowed Chair from Virginia Commonwealth University to L.K.

References

1. Dunker, A.K., Babu, M.M., Barbar, E., Blackledge, M., Bondos, S.E., Dosztányi, Z., Dyson, H.J., Forman-Kay, J., Fuxreiter, M., Gsponer, J., Han, K.-H., Jones, D.T., Longhi, S., Metallo, S.J., Nishikawa, K., Nussinov, R., Obradovic, Z., Pappu, R.V., Rost, B., Selenko, P., Subramaniam, V., Sussman, J.L., Tompa, P., Uversky, V.N.: What's in a name? Why these proteins are intrinsically disordered. Intrinsically Disord. Proteins **1**, e24157 (2013)
2. van der Lee, R., Buljan, M., Lang, B., Weatheritt, R.J., Daughdrill, G.W., Dunker, A.K., Fuxreiter, M., Gough, J., Gsponer, J., Jones, D.T., Kim, P.M., Kriwacki, R.W., Oldfield, C.J., Pappu, R.V., Tompa, P., Uversky, V.N., Wright, P.E., Babu, M.M.: Classification of intrinsically disordered regions and proteins. Chem. Rev. **114**, 6589–6631 (2014)
3. Peng, Z., Yan, J., Fan, X., Mizianty, M.J., Xue, B., Wang, K., Hu, G., Uversky, V.N., Kurgan, L.: Exceptionally abundant exceptions: comprehensive characterization of intrinsic disorder in all domains of life. Cell. Mol. Life Sci. **72**, 137–151 (2015)
4. Xue, B., Dunker, A.K., Uversky, V.N.: Orderly order in protein intrinsic disorder distribution: disorder in 3500 proteomes from viruses and the three domains of life. J. Biomol. Struct. Dyn. **30**, 137–149 (2012)
5. Ward, J.J., Sodhi, J.S., McGuffin, L.J., Buxton, B.F., Jones, D.T.: Prediction and functional analysis of native disorder in proteins from the three kingdoms of life. J. Mol. Biol. **337**, 635–645 (2004)
6. Fuxreiter, M., Toth-Petroczy, A., Kraut, D.A., Matouschek, A., Lim, R.Y., Xue, B., Kurgan, L., Uversky, V.N.: Disordered proteinaceous machines. Chem. Rev. **114**, 6806–6843 (2014)
7. Xue, B., Blocquel, D., Habchi, J., Uversky, A.V., Kurgan, L., Uversky, V.N., Longhi, S.: Structural disorder in viral proteins. Chem. Rev. **114**, 6880–6911 (2014)
8. Kozlowski, L.P., Bujnicki, J.M.: MetaDisorder: a meta-server for the prediction of intrinsic disorder in proteins. BMC Bioinform. **13**, 1–11 (2012)
9. Peng, Z., Oldfield, C.J., Xue, B., Mizianty, M.J., Dunker, A.K., Kurgan, L., Uversky, V.N.: A creature with a hundred waggly tails: intrinsically disordered proteins in the ribosome. Cell. Mol. Life Sci. **71**, 1477–1504 (2014)
10. Xue, B., Mizianty, M.J., Kurgan, L., Uversky, V.N.: Protein intrinsic disorder as a flexible armor and a weapon of HIV-1. Cell. Mol. Life Sci. **69**, 1211–1259 (2012)
11. Pentony, M.M., Jones, D.T.: Modularity of intrinsic disorder in the human proteome. Proteins **78**, 212–221 (2010)
12. Wang, C., Uversky, V.N., Kurgan, L.: Disordered nucleiome: abundance of intrinsic disorder in the DNA- and RNA-binding proteins in 1121 species from Eukaryota. Bacteria and Archaea. Proteomics **16**, 1486–1498 (2016)
13. Peng, Z., Xue, B., Kurgan, L., Uversky, V.N.: Resilience of death: intrinsic disorder in proteins involved in the programmed cell death. Cell Death Differ. **20**, 1257–1267 (2013)
14. Oldfield, C.J., Xue, B., Van, Y.Y., Ulrich, E.L., Markley, J.L., Dunker, A.K., Uversky, V.N.: Utilization of protein intrinsic disorder knowledge in structural proteomics. Biochim. Biophys. Acta **1834**, 487–498 (2013)
15. Potenza, E., Domenico, T.D., Walsh, I., Tosatto, S.C.E.: MobiDB 2.0: an improved database of intrinsically disordered and mobile proteins. Nucleic Acids Res. **43**, D315–D320 (2015)

16. Di Domenico, T., Walsh, I., Martin, A.J.M., Tosatto, S.C.E.: MobiDB: a comprehensive database of intrinsic protein disorder annotations. Bioinformatics **28**, 2080–2081 (2012)

17. Oates, M.E., Romero, P., Ishida, T., Ghalwash, M., Mizianty, M.J., Xue, B., Dosztányi, Z., Uversky, V.N., Obradovic, Z., Kurgan, L., Dunker, A.K., Gough, J.: D2P2: database of disordered protein predictions. Nucleic Acids Res. **41**, D508–D516 (2013)

18. Deng, X., Eickholt, J., Cheng, J.: A comprehensive overview of computational protein disorder prediction methods. Mol. BioSyst. **8**, 114–121 (2012)

19. Monastyrskyy, B., Fidelis, K., Moult, J., Tramontano, A., Kryshtafovych, A.: Evaluation of disorder predictions in CASP9. Proteins **79**(Suppl 10), 107–118 (2011)

20. Monastyrskyy, B., Kryshtafovych, A., Moult, J., Tramontano, A., Fidelis, K.: Assessment of protein disorder region predictions in CASP10. Proteins **82**(Suppl 2), 127–137 (2014)

21. Peng, Z.L., Kurgan, L.: Comprehensive comparative assessment of in-silico predictors of disordered regions. Curr. Protein Pept. Sci. **13**, 6–18 (2012)

22. Walsh, I., Giollo, M., Di Domenico, T., Ferrari, C., Zimmermann, O., Tosatto, S.C.: Comprehensive large-scale assessment of intrinsic protein disorder. Bioinformatics **31**, 201–208 (2015)

23. Noivirt-Brik, O., Prilusky, J., Sussman, J.L.: Assessment of disorder predictions in CASP8. Proteins **77**(Suppl 9), 210–216 (2009)

24. Kihara, D., Chen, H., Yang, Y.D.: Quality assessment of protein structure models. Curr. Protein Pept. Sci. **10**, 216–228 (2009)

25. Skwark, M.J., Elofsson, A.: PconsD: ultra rapid, accurate model quality assessment for protein structure prediction. Bioinformatics **29**, 1817–1818 (2013)

26. McGuffin, L.J., Buenavista, M.T., Roche, D.B.: The ModFOLD4 server for the quality assessment of 3D protein models. Nucleic Acids Res. **41**, W368–W372 (2013)

27. Cao, R., Bhattacharya, D., Adhikari, B., Li, J., Cheng, J.: Massive integration of diverse protein quality assessment methods to improve template based modeling in CASP11. Proteins **84**(Suppl 1), 247–259 (2016)

28. Kryshtafovych, A., Fidelis, K.: Protein structure prediction and model quality assessment. Drug Discov. Today **14**, 386–393 (2009)

29. UniProt Consortium: UniProt: a hub for protein information. Nucleic Acids Res. **43**, D204–D212 (2015)

30. Fu, L., Niu, B., Zhu, Z., Wu, S., Li, W.: CD-HIT: accelerated for clustering the next-generation sequencing data. Bioinformatics **28**, 3150–3152 (2012)

31. Walsh, I., Martin, A.J., Di Domenico, T., Tosatto, S.C.: ESpritz: accurate and fast prediction of protein disorder. Bioinformatics **28**, 503–509 (2012)

32. Sickmeier, M., Hamilton, J.A., LeGall, T., Vacic, V., Cortese, M.S., Tantos, A., Szabo, B., Tompa, P., Chen, J., Uversky, V.N., Obradovic, Z., Dunker, A.K.: DisProt: the database of disordered proteins. Nucleic Acids Res. **35**, D786–D793 (2007)

33. Dosztanyi, Z., Csizmok, V., Tompa, P., Simon, I.: IUPred: web server for the prediction of intrinsically unstructured regions of proteins based on estimated energy content. Bioinformatics **21**, 3433–3434 (2005)

34. Linding, R., Jensen, L.J., Diella, F., Bork, P., Gibson, T.J., Russell, R.B.: Protein disorder prediction: implications for structural proteomics. Structure **11**, 1453–1459 (2003)

35. Berman, H.M., Westbrook, J., Feng, Z., Gilliland, G., Bhat, T.N., Weissig, H., Shindyalov, I.N., Bourne, P.E.: The protein data bank. Nucleic Acids Res. **28**, 235–242 (2000)

36. Yang, Z.R., Thomson, R., McNeil, P., Esnouf, R.M.: RONN: the bio-basis function neural network technique applied to the detection of natively disordered regions in proteins. Bioinformatics **21**, 3369–3376 (2005)

37. Peng, K., Radivojac, P., Vucetic, S., Dunker, A.K., Obradovic, Z.: Length-dependent prediction of protein intrinsic disorder. BMC Bioinform. **7**, 208 (2006)
38. Linding, R., Russell, R.B., Neduva, V., Gibson, T.J.: GlobPlot: exploring protein sequences for globularity and disorder. Nucleic Acids Res. **31**, 3701–3708 (2003)

Stability Evaluation of the Dynamic Signature Partitions Over Time

Marcin Zalasiński[1(✉)], Krzysztof Cpałka[1], and Meng Joo Er[2]

[1] Institute of Computational Intelligence, Częstochowa University of Technology,
Częstochowa, Poland
{marcin.zalasinski,krzysztof.cpalka}@iisi.pcz.pl
[2] Nanyang Technological University, Singapore, Singapore
emjer@ntu.edu.sg

Abstract. Analysis of biometric attributes' changes is an important
issue of behavioral biometrics. It seems to be very important in the case
of identity verification. In this paper the analysis of features describing
the dynamic signature was performed. The dynamic signature is represented by a set of nonlinear waveforms describing dynamics of signing
process. The proposed analysis is based on a set of coefficients defined
in the context of the dynamic signature partitioning. The partitioning is
performed in order to facilitate analysis of the signature. It consists in
division of the signature into parts which can be related to e.g. high and
low velocity of pen in the initial and final phase of signing. The proposed
method was tested using ATVS-SLT DB dynamic signature database.

Keywords: Dynamic signature verification · Signature partitioning ·
Evaluation of signature stability

1 Introduction

Dynamic signature is a biometric feature described by signals changing over time
(see e.g. [17, 18, 21]). Its acquisition is performed using digital input device, e.g.
graphic tablet. Identity verification on the basis of the dynamic signature is more
effective than verification using static signature (see e.g. [6, 32, 45, 64–68]), which
is only an image of the signature.

Partitions of the dynamic signature are areas containing fragments of signals
describing the signature, e.g. horizontal and vertical shape trajectories (i.e. x
and y). The partitions can be determined on the basis of the time step value of
signing process, value of the pen velocity, value of the pen pressure and on the
basis of all these approaches (it is so-called hybrid approach).

Solutions based on partitioning presented in the literature are mainly focused
on methods used for partition generation and identity verification (see e.g.
[14, 15, 27]). Identity verification may also be performed using computational
intelligence methods (see e.g. [7, 8, 11, 24, 25, 29–31, 33–38, 40, 43, 46, 51–55, 58–
62]), including systems based on fuzzy logic (see e.g. [1, 12, 13, 22, 26, 39, 44, 48–
50, 57]), evolutionary algorithms (see e.g. [9, 41, 42, 56]), genetic programming

© Springer International Publishing AG 2017
L. Rutkowski et al. (Eds.): ICAISC 2017, Part I, LNAI 10245, pp. 733–746, 2017.
DOI: 10.1007/978-3-319-59063-9_66

(see e.g. [2–5]), neural-networks (see e.g. [10,19,20,28,47]), etc. However, an important issue concerning the dynamic signature verification is also an analysis of the variability of biometric attributes changes over time. During analysis we assume that we have a reference signature of each signer and test signatures used in the verification process are compared to this signature. We do not consider a complete change of the signature.

In the literature an issue of the dynamic signature stability has been considered in the context of so-called global features (see e.g. [23]), but it has not been considered in the context of the signature partitioning. In this paper we perform an analysis of the dynamic signature partitions' stability over time.

Structure of the paper is as follows: Sect. 2 contains an introduction from the field of the dynamic signature partitioning, Sect. 3 presents description of adopted criteria for evaluating the dynamic signature variability over time, Sect. 4 shows simulation results, conclusions are drawn in Sect. 5.

2   Introduction to the Partitioning of the Dynamic Signature

Identity verification method used in this paper is based on the selection of the most characteristic templates of the user signature. The templates are a part of the partitions created on the basis of the characteristic values of pen velocity and pressure signals and characteristic time moments of signing process. General assumptions of this method can be characterized as follows:

- **Creation of the signature partitions.** The partitions are created individually for each user on the basis of his/her reference signatures. In this process values of pen velocity and pressure signals and values of time moments of signing process are used.
- **Determination of templates in the partitions.** Each partition contains signals of trajectories x and y, for which templates $\mathbf{tc}_{i,p,r}^{\{s,a\}}$ are created, where i is the user index, $\{p,r\}$ are indices indicating the partition (p is the vertical section index, r is the horizontal section index), s is the type of signal used to create partition (velocity v or pressure z) and a is the type of trajectory used to create the template (x or y). The templates are average values of trajectory signals a of the reference signatures of the user i in the partition denoted by indices $\{p,r\}$.
- **Creation of the classifier.** For each considered user a flexible fuzzy one-class classifier is created. Parameters of the classifier are determined on the basis of the values of reference signatures' signals in partitions.
- **Identity verification.** This process is performed by the classifier using distance values $d_{i,j,p,r}^{\{s,a\}}$ between trajectory signals of the signature j and the template, determined for each partition.

A more detailed description of the partitioning procedure can be found in our previous works (see e.g. [16,69]). The remainder of this article presents a way of variability analysis of the templates in the partitions (Sect. 3). The analysis has not previously been presented in the literature in the context of partitioning signals (in particular, used for the biometrics).

3   Description of the Adopted Criteria for Evaluation of the Dynamic Signature Partitions Variability Over Time

Analysis of the variability of the dynamic signature partitions over time is based on the defined criteria. Values of the criteria are determined for each acquisition session nS. They are presented in Table 1. Remarks on the way of their definition can be summarized as follows:

- Coefficient $\bar{d}_{i,p,r,nS}^{\{s,a\}}$ from Table 1 is interpreted as the average distance between template created on the basis of reference signatures from acquisition session number 1 and trajectory signal of the signatures acquired in session nS. The distance refers to the template determined for the trajectory a in the partition denoted by indices $\{p,r\}$ and created on the basis of the signal s for the user i. It is used to determine variability level of the dynamic signature features in subsequent acquisition sessions in relation to the reference feature (the template). This coefficient is defined as follows:

$$\bar{d}_{i,p,r,nS}^{\{s,a\}} = \frac{1}{J} \sum_{j=1}^{J} d_{i,j,p,r,nS}^{\{s,a\}}. \tag{1}$$

- Coefficient $\sigma_{i,p,r,nS}^{\{s,a\}}$ from Table 1 is interpreted as the standard deviation of distances between template created on the basis of reference signatures from acquisition session number 1 and trajectory signal of the signatures acquired in session nS. The standard deviation refers to the template determined for the trajectory a in the partition denoted by indices $\{p,r\}$ and created on the basis of the signal s for the user i. It is used to determine dispersion level of the dynamic signature features in subsequent acquisition sessions. This coefficient is defined as follows:

$$\sigma_{i,p,r,nS}^{\{s,a\}} = \sqrt{\frac{1}{J} \sum_{j=1}^{J} \left(\bar{d}_{i,p,r,nS}^{\{s,a\}} - d_{i,j,q,nS}^{\{s,a\}} \right)^2}. \tag{2}$$

- Coefficient $VC_{i,p,r,nS}^{\{s,a\}}$ from Table 1 is interpreted as the product of the average and variance relative variation of the mentioned distances between two acquisition sessions. It has been proposed in the paper [23]. It refers to the template determined for the trajectory a in the partition denoted by indices $\{p,r\}$ and created on the basis of the signal s for the user i. It is used to determine the most stable features if the signature. This coefficient is defined as follows:

$$VC_{i,p,r,nS}^{\{s,a\}} = \left| \bar{d}_{i,p,r,nS}^{\{s,a\}} - \bar{d}_{i,p,r,nS-1}^{\{s,a\}} \right| \cdot \left| \frac{\sigma_{i,p,r,nS}^{\{s,a\}}}{\bar{d}_{i,p,r,nS}^{\{s,a\}}} - \frac{\sigma_{i,p,r,nS-1}^{\{s,a\}}}{\bar{d}_{i,p,r,nS-1}^{\{s,a\}}} \right|. \tag{3}$$

Table 1. A way of definition of the criteria used for evaluation of the dynamic signature partitions' variability over time.

No	Name	Notation
1	Average	$\bar{d}_{i,p,r,nS}^{\{s,a\}}$
2	Standard deviation	$\sigma_{i,p,r,nS}^{\{s,a\}}$
3	VC coefficient [23]	$VC_{i,p,r,nS}^{\{s,a\}}$

Next Sections of the paper present simulations scenario, simulation results and conclusions.

4   Simulation Results

Simulations were performed using ATVS-SLT database [23] which contains signatures of 27. users, created in 6. sessions. First 4 sessions contain 4 signatures of each user and 2. last sessions contain 15 signatures of each user.

Training phase of the considered system for identity verification on the basis of the dynamic signature (described in Sect. 2) contains: creation of partitions, determination of templates in the partitions and determination of the classifier. It was performed individually for each user, taking into account signatures from the session number 1. We assumed that each signature was partitioned into 2. vertical sections and 2. horizontal sections ($p = 2$, $r = 2$). Next, criteria for evaluation of the dynamic signature partitions' variability over time were determined for remaining sessions, taking into account all signatures of individuals. They are presented in Tables 2, 3 and 4. Moreover, pie charts shown in Figs. 2, 3 and 4 contain percentage values of all determined criteria, averaged in the context of all users.

During simulations we also performed identity verification for each user and sessions number 2–6, taking into account genuine signatures of other users as forged signatures (so-called random forgeries). Results of verification process averaged in the context of all users are presented in Table 5 in the form of FAR, FRR and EER coefficients [63]. Moreover, Fig. 1 shows the trend of changes of verification error in comparison to the changes of the criteria evaluating the dynamic signature partitions' variability over time. Values shown in this figure were normalized to the unit range.

It should be noted that the simulations were carried out five times and the results were averaged.

Conclusions from the simulations can be summarized as follows:

- Value of all defined criteria evaluating the dynamic signature partitions' variability (presented in Table 1) increases over time.
- The lowest variability level (associated with the value of coefficient $\bar{d}_{i,p,r,nS}^{\{s,a\}}$) is related to the trajectory x from partition denoted by indices $\{p = 0, r = 1\}$, created on the basis of the signal v (see Table 2).

- Variability level of features in the partitions determined on the basis of the signal v is lower than in the partitions determined on the basis of the signal z (see Fig. 2).
- The lowest dispersion level (associated with the value of coefficient $\sigma_{i,p,r,nS}^{\{s,a\}}$) is related to the trajectory x from partition denoted by indices $\{p = 0, r = 1\}$, created on the basis of the signal v (see Table 3).
- Dispersion level of features in the partitions determined on the basis of the signal v is lower than in the partitions determined on the basis of the signal z (see Fig. 3).
- The most stable feature (determined on the basis of the value of coefficient $VC_{i,p,r,nS}^{\{s,a\}}$) is the trajectory y from partition denoted by indices $\{p = 0, r = 1\}$, created on the basis of the signal v (see Table 4).
- Stability of features in the partitions determined on the basis of the signal v is lower than in the partitions determined on the basis of the signal z (see Fig. 4).
- The system used for identity verification tends to decrease the verification accuracy over time (see Table 5). Trend of increasing verification error over time is consistent with the trend of increasing values of the coefficients used for evaluation of the dynamic signature partitions' variability over time (see Fig. 1).

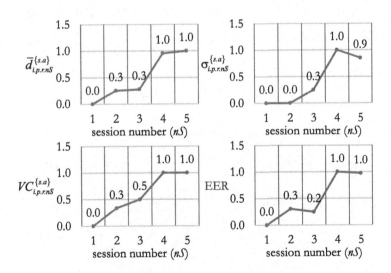

Fig. 1. Trend of verification accuracy changing in relation to the averaged values of the coefficients used for evaluation of the dynamic signature partitions' variability over time.

Table 2. Values of the coefficient $\bar{d}_{i,p,r,nS}^{\{s,a\}}$ averaged in the context of all users from the database ATVS-SLT DB.

Template (s,a,p,r)	$nS = 2$	$nS = 3$	$nS = 4$	$nS = 5$	$nS = 6$	Average
$v,x,0,0$	0.79	0.87	0.74	1.22	1.19	0.96
$v,x,0,1$	0.72	0.78	0.73	1.03	1.06	0.87
$v,x,1,0$	0.89	0.89	0.87	1.37	1.41	1.09
$v,x,1,1$	0.82	0.87	0.84	1.27	1.23	1.01
$v,y,0,0$	1.14	1.24	1.21	1.56	1.60	1.35
$v,y,0,1$	0.96	1.05	1.11	1.33	1.34	1.16
$v,y,1,0$	1.26	1.45	1.40	1.77	1.86	1.55
$v,y,1,1$	0.98	1.10	1.10	1.34	1.35	1.17
$z,x,0,0$	1.10	1.14	1.31	1.47	1.62	1.33
$z,x,0,1$	1.52	1.57	1.76	1.84	1.98	1.73
$z,x,1,0$	1.34	1.49	1.44	1.84	1.87	1.60
$z,x,1,1$	1.77	1.91	1.83	2.12	2.04	1.93
$z,y,0,0$	1.57	1.76	1.70	1.92	1.95	1.78
$z,y,0,1$	1.97	2.14	2.21	2.38	2.26	2.19
$z,y,1,0$	1.51	1.68	1.81	2.11	2.10	1.85
$z,y,1,1$	1.90	2.18	2.19	2.32	2.35	2.19
Average	1.27	1.38	1.39	1.68	1.70	–

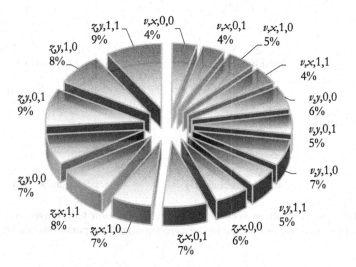

Fig. 2. Percentage average values of variability criterion $\bar{d}_{i,p,r,nS}^{\{s,a\}}$ for each template. The templates are described using the following parameters: s,a,p,r.

Table 3. Values of the coefficient $\sigma_{i,p,r,nS}^{\{s,a\}}$ averaged in the context of all users from the database ATVS-SLT DB.

Template (s,a,p,r)	$nS = 2$	$nS = 3$	$nS = 4$	$nS = 5$	$nS = 6$	Average
$v,x,0,0$	0.30	0.27	0.25	0.52	0.30	0.33
$v,x,0,1$	0.23	0.16	0.20	0.34	0.23	0.23
$v,x,1,0$	0.27	0.21	0.30	0.53	0.42	0.35
$v,x,1,1$	0.21	0.19	0.26	0.38	0.37	0.28
$v,y,0,0$	0.38	0.32	0.36	0.51	0.41	0.40
$v,y,0,1$	0.29	0.23	0.25	0.40	0.36	0.31
$v,y,1,0$	0.33	0.39	0.52	0.59	0.52	0.47
$v,y,1,1$	0.24	0.27	0.34	0.46	0.34	0.33
$z,x,0,0$	0.35	0.29	0.37	0.57	0.68	0.45
$z,x,0,1$	0.40	0.40	0.51	0.64	0.73	0.54
$z,x,1,0$	0.44	0.39	0.47	0.61	0.68	0.52
$z,x,1,1$	0.48	0.58	0.51	0.73	0.74	0.61
$z,y,0,0$	0.36	0.43	0.45	0.53	0.54	0.46
$z,y,0,1$	0.40	0.45	0.54	0.72	0.62	0.55
$z,y,1,0$	0.43	0.44	0.51	0.61	0.64	0.53
$z,y,1,1$	0.42	0.55	0.56	0.69	0.70	0.58
Average	0.35	0.35	0.40	0.55	0.52	–

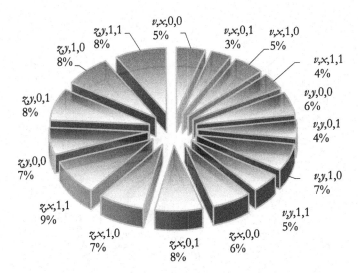

Fig. 3. Percentage average values of variability criterion $\sigma_{i,p,r,nS}^{\{s,a\}}$ averaged in the context of all users from the database ATVS-SLT DB.

Table 4. Values of the coefficient $VC_{i,p,r,nS}^{\{s,a\}}$ averaged in the context of all users from the database ATVS-SLT DB.

Template (s,a,p,r)	$nS = 2$	$nS = 3$	$nS = 4$	$nS = 5$	$nS = 6$	Average
$v,x,0,0$	0.04	0.13	0.07	0.16	0.16	0.11
$v,x,0,1$	0.06	0.05	0.07	0.14	0.08	0.08
$v,x,1,0$	0.08	0.06	0.10	0.18	0.13	0.11
$v,x,1,1$	0.05	0.07	0.07	0.14	0.12	0.09
$v,y,0,0$	0.11	0.08	0.08	0.12	0.11	0.10
$v,y,0,1$	0.06	0.06	0.06	0.10	0.10	0.07
$v,y,1,0$	0.10	0.15	0.21	0.16	0.18	0.16
$v,y,1,1$	0.06	0.08	0.09	0.15	0.10	0.09
$z,x,0,0$	0.08	0.07	0.10	0.14	0.19	0.11
$z,x,0,1$	0.07	0.09	0.12	0.12	0.14	0.11
$z,x,1,0$	0.06	0.12	0.12	0.14	0.14	0.11
$z,x,1,1$	0.09	0.14	0.08	0.11	0.16	0.12
$z,y,0,0$	0.05	0.10	0.09	0.09	0.12	0.09
$z,y,0,1$	0.06	0.08	0.09	0.14	0.11	0.10
$z,y,1,0$	0.11	0.14	0.16	0.16	0.18	0.15
$z,y,1,1$	0.08	0.10	0.13	0.13	0.16	0.12
Average	0.07	0.09	0.10	0.13	0.13	–

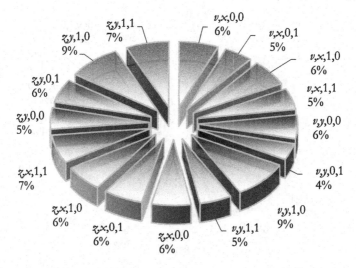

Fig. 4. Percentage average values of variability criterion $VC_{i,p,r,nS}^{\{s,a\}}$ averaged in the context of all users from the database ATVS-SLT DB.

Table 5. Identity verification errors averaged in the context of all users from the database ATVS-SLT DB.

nS	FAR	FRR	EER
2	4.11%	2.85%	3.48%
3	4.75%	5.37%	5.06%
4	6.35%	3.19%	4.77%
5	4.42%	13.00%	8.71%
6	5.40%	11.72%	8.56%

5   Conclusions

In this paper we analyzed criteria evaluating the dynamic signature variability over time in the context of the signature partitions. This analysis can be used wherever the interval between sessions of signatures' acquisition is sufficiently long. It was assumed that the basic shape of the signature of each user is not radically changing but it only evolves. The described approach can have informative meaning or be a component of the method for identity verification based on the dynamic signature and partitioning.

In further research in the field of stability of the dynamic signature partitions we are planning to: (a) develop a procedure for updating the templates, taking into account the trend of changes in signing, (b) develop a method determining importance of the partitions on the basis of their changing over time, (c) develop a method for the dynamic signature verification taking into account the trends of changes in the partitions.

Acknowledgments. The project was financed by the National Science Centre (Poland) on the basis of the decision number DEC-2012/05/B/ST7/02138.

References

1. Almohammadi, K., Hagras, H., Alghazzawi, D., Aldabbagh, G.: Users-centric adaptive learning system based on interval type-2 fuzzy logic for massively crowded E-learning platforms. J. Artif. Intell. Soft Comput. Res. **6**(2), 81–101 (2016)
2. Bartczuk, Ł., Przybył, A., Koprinkova-Hristova, P.: New method for non-linear correction modelling of dynamic objects with genetic programming. In: Rutkowski, L., Korytkowski, M., Scherer, R., Tadeusiewicz, R., Zadeh, L.A., Zurada, J.M. (eds.) ICAISC 2015. LNCS, vol. 9120, pp. 318–329. Springer, Cham (2015). doi:10.1007/978-3-319-19369-4_29
3. Bartczuk, Ł.: Gene expression programming in correction modelling of nonlinear dynamic objects. In: Borzemski, L., Grzech, A., Świątek, J., Wilimowska, Z. (eds.) Information Systems Architecture and Technology: Proceedings of 36th International Conference on Information Systems Architecture and Technology – ISAT 2015 – Part I. AISC, vol. 429, pp. 125–134. Springer, Cham (2016). doi:10.1007/978-3-319-28555-9_11

4. Bartczuk, Ł., Łapa, K., Koprinkova-Hristova, P.: A new method for generating of fuzzy rules for the nonlinear modelling based on semantic genetic programming. In: Rutkowski, L., Korytkowski, M., Scherer, R., Tadeusiewicz, R., Zadeh, L.A., Zurada, J.M. (eds.) ICAISC 2016. LNCS, vol. 9693, pp. 262–278. Springer, Cham (2016). doi:10.1007/978-3-319-39384-1_23

5. Bartczuk, Ł., Galushkin, A.I.: A new method for generating nonlinear correction models of dynamic objects based on semantic genetic programming. In: Rutkowski, L., Korytkowski, M., Scherer, R., Tadeusiewicz, R., Zadeh, L.A., Zurada, J.M. (eds.) ICAISC 2016. LNCS, vol. 9693, pp. 249–261. Springer, Cham (2016). doi:10.1007/978-3-319-39384-1_22

6. Batista, L., Granger, E., Sabourin, R.: Dynamic selection of generative discriminative ensembles for off-line signature verification. Pattern Recognit. **45**, 1326–1340 (2012)

7. Bilski, J., Galushkin, A.I.: A new proposition of the activation function for significant improvement of neural networks performance. In: Rutkowski, L., Korytkowski, M., Scherer, R., Tadeusiewicz, R., Zadeh, L.A., Zurada, J.M. (eds.) ICAISC 2016. LNCS, vol. 9692, pp. 35–45. Springer, Cham (2016). doi:10.1007/978-3-319-39378-0_4

8. Bilski, J., Wilamowski, B.M.: Parallel learning of feedforward neural networks without error backpropagation. In: Rutkowski, L., Korytkowski, M., Scherer, R., Tadeusiewicz, R., Zadeh, L.A., Zurada, J.M. (eds.) ICAISC 2016. LNCS, vol. 9692, pp. 57–69. Springer, Cham (2016). doi:10.1007/978-3-319-39378-0_6

9. Brasileiro, Í., Santos, I., Soares, A., Rabelo, R., Mazullo, F.: Ant colony optimization applied to the problem of choosing the best combination among M combinations of shortest paths in transparent optical networks. J. Artif. Intell. Soft Comput. Res. **6**(4), 231–242 (2016)

10. Cierniak, R., Rutkowski, L.: On image compression by competitive neural networks and optimal linear predictors. Signal Process. Image Commun. **156**, 559–565 (2000)

11. Cpałka, K.: Design of Interpretable Fuzzy Systems. Springer, Cham (2017)

12. Cpałka, K., Rebrova, O., Nowicki, R., Rutkowski, L.: On design of flexible neuro-fuzzy systems for nonlinear modelling. Int. J. Gen. Syst. **42**(6), 706–720 (2013)

13. Cpałka, K., Rutkowski, L.: Flexible Takagi-Sugeno. fuzzy systems. In: Proceedings of the 2005 IEEE International Joint Conference on Neural Networks (IJCNN 2005), vol. 3, pp. 1764–1769 (2005)

14. Cpałka, K., Zalasiński, M.: On-line signature verification using vertical signature partitioning. Expert Syst. Appl. **41**, 4170–4180 (2014)

15. Cpałka, K., Zalasiński, M., Rutkowski, L.: New method for the on-line signature verification based on horizontal partitioning, Pattern Recognition. vol. **47**, 2652–2661 (2014)

16. Cpałka, K., Zalasiński, M., Rutkowski, L.: A new algorithm for identity verification based on the analysis of a handwritten dynamic signature. Appl. Soft Comput. **43**, 47–56 (2016)

17. Diaz, M., Fischer, A., Ferrer, M.A., Plamondon, R.: Dynamic signature verification system based on one real signature. IEEE Trans. Cyber. **PP**, 1–12 (2016)

18. Doroz, R., Porwik, P., Orczyk, T.: Dynamic signature verification method based on association of features with similarity measures. Neurocomputing **171**, 921–931 (2016)

19. Duda, P., Hayashi, Y., Jaworski, M.: On the strong convergence of the orthogonal series-type kernel regression neural networks in a non-stationary environment. In: Rutkowski, L., Korytkowski, M., Scherer, R., Tadeusiewicz, R., Zadeh, L.A., Zurada, J.M. (eds.) ICAISC 2012. LNCS, vol. 7267, pp. 47–54. Springer, Heidelberg (2012). doi:10.1007/978-3-642-29347-4_6

20. Er, M.J., Duda, P.: On the weak convergence of the orthogonal series-type kernel regresion neural networks in a non-stationary environment. In: Wyrzykowski, R., Dongarra, J., Karczewski, K., Waśniewski, J. (eds.) PPAM 2011. LNCS, vol. 7203, pp. 443–450. Springer, Heidelberg (2012). doi:10.1007/978-3-642-31464-3_45

21. Fierrez, J., Ortega-Garcia, J., Ramos, D., Gonzalez-Rodriguez, J.: HMM-based on-line signature verification: feature extraction and signature modeling. Pattern Recognit. Lett. 28, 2325–2334 (2007)

22. Gabryel, M., Cpałka, K., Rutkowski, L.: Evolutionary strategies for learning of neuro-fuzzy systems. In: Proceedings of the I Workshop on Genetic Fuzzy Systems, Granada, pp. 119–123 (2005)

23. Galbally, J., Martinez-Diaz, M., Fierez, J.: Aging in biometrics: an experimental analysis on on-line signature. PLoS ONE 8(7), e69897 (2013)

24. Galkowski, T., Pawlak, M.: Nonparametric extension of regression functions outside domain. In: Rutkowski, L., Korytkowski, M., Scherer, R., Tadeusiewicz, R., Zadeh, L.A., Zurada, J.M. (eds.) ICAISC 2014. LNCS (LNAI), vol. 8467, pp. 518–530. Springer, Cham (2014). doi:10.1007/978-3-319-07173-2_44

25. Galkowski, T., Pawlak, M.: Nonparametric function fitting in the presence of non-stationary noise. In: Rutkowski, L., Korytkowski, M., Scherer, R., Tadeusiewicz, R., Zadeh, L.A., Zurada, J.M. (eds.) ICAISC 2014. LNCS (LNAI), vol. 8467, pp. 531–538. Springer, Cham (2014). doi:10.1007/978-3-319-07173-2_45

26. Harmati, I.Á., Bukovics, Á., Kóczy, L.T.: Minkowski's inequality based sensitivity analysis of fuzzy signatures. J. Artif. Intell. Soft Comput. Res. 6(4), 219–229 (2016)

27. Ibrahim, M.T., Khan, M.A., Alimgeer, K.S., Khan, M.K., Taj, I.A., Guan, L.: Velocity and pressure-based partitions of horizontal and vertical trajectories for on-line signature verification. Pattern Recognit. 43, 2817–2832 (2010)

28. Jaworski, M., Er, M.J., Pietruczuk, L.: On the application of the parzen-type kernel regression neural network and order statistics for learning in a non-stationary environment. In: Rutkowski, L., Korytkowski, M., Scherer, R., Tadeusiewicz, R., Zadeh, L.A., Zurada, J.M. (eds.) ICAISC 2012. LNCS, vol. 7267, pp. 90–98. Springer, Heidelberg (2012). doi:10.1007/978-3-642-29347-4_11

29. Kapustianyk, V., Shchur, Y., Kityk, I., Rudyk, V., Lach, G., Laskowski, Ł., Tkaczyk, S., Swiatek, J., Davydov, V.: Resonance dielectric dispersion of TEA-CoCl2Br 2 nanocrystals incorporated into the PMMA matrix. J. Phys. Condens. Matter 20(36), 365215–365223 (2008). IOP Publishing

30. Kitajima, R., Kamimura, R.: Accumulative information enhancement in the self-organizing maps and its application to the analysis of mission statements. J. Artif. Intell. Soft Comput. Res. 5(3), 161–176 (2015)

31. Korytkowski, M., Rutkowski, L., Scherer, R.: Fast image classification, by boosting fuzzy classifiers. Inf. Sci. 327, 175–182 (2016)

32. Kumar, R., Sharma, J.D., Chanda, B.: Writer-independent off-line signature verification using surroundedness feature. Pattern Recognit. Lett. 33, 301–308 (2012)

33. Laskowska, M., Laskowski, Ł., Jelonkiewicz, J.: SBA-15 mesoporous silica activated by metal ions-verification of molecular structure on the basis of Raman spectroscopy supported by numerical simulations. J. Mol. Struct. 1100, 21–26 (2015). Elsevier

34. Laskowski, Ł.: A novel hybrid-maximum neural network in stereo-matching process. Neural Comput. Appl. **23**(7–8), 2435–2450 (2013). Springer
35. Laskowski, Ł., Laskowska, M., Jelonkiewicz, J., Boullanger, A.: Spin-glass implementation of a hopfield neural structure. In: Rutkowski, L., Korytkowski, M., Scherer, R., Tadeusiewicz, R., Zadeh, L.A., Zurada, J.M. (eds.) ICAISC 2014. LNCS, vol. 8467, pp. 89–96. Springer, Cham (2014). doi:10.1007/978-3-319-07173-2_9
36. Laskowski, Ł., Laskowska, M., Jelonkiewicz, J., Boullanger, A.: Molecular approach to hopfield neural network. In: Rutkowski, L., Korytkowski, M., Scherer, R., Tadeusiewicz, R., Zadeh, L.A., Zurada, J.M. (eds.) ICAISC 2015. LNCS, vol. 9119, pp. 72–78. Springer, Cham (2015). doi:10.1007/978-3-319-19324-3_7
37. Laskowski, Ł., Laskowska, M., Jelonkiewicz, J., Dulski, M., Wojtyniak, M., Fitta, M., Balanda, M.: SBA-15 mesoporous silica free-standing thin films containing copper ions bounded via propyl phosphonate units-preparation and characterization. J. Solid State Chem. **241**, 143–151 (2016). Elsevier
38. Laskowski, Ł., Laskowska, M., Jelonkiewicz, J., Gałkowski, T., Pawlik, P., Piech, H., Doskocz, M.: Iron doped SBA-15 mesoporous silica studied by Mössbauer spectroscopy. J. Nanomater. **2016**, 1–6 (2016). Hindawi Publishing Corporation
39. Li, X., Er, M.J., Lim, B.S., Zhou, J.H., Gan, O.P., Rutkowski, L.: Fuzzy regression modeling for tool performance prediction and degradation detection. Int. J. Neural Syst. **2005**, 405–419 (2010)
40. Lin, C., Dong, F., Hirota, K.: Common driving notification protocol based on classified driving behavior for cooperation intelligent autonomous vehicle using vehicular ad-hoc network technology. J. Artif. Intell. Soft Comput. Res. **5**(1), 5–21 (2015)
41. Łapa, K., Cpałka, K., Wang, L.: New method for design of fuzzy systems for nonlinear modelling using different criteria of interpretability. In: Rutkowski, L., Korytkowski, M., Scherer, R., Tadeusiewicz, R., Zadeh, L.A., Zurada, J.M. (eds.) ICAISC 2014. LNCS, vol. 8467, pp. 217–232. Springer, Cham (2014). doi:10.1007/978-3-319-07173-2_20
42. Łapa, K., Szczypta, J., Venkatesan, R.: Aspects of structure and parameters selection of control systems using selected multi-population algorithms. In: Rutkowski, L., Korytkowski, M., Scherer, R., Tadeusiewicz, R., Zadeh, L.A., Zurada, J.M. (eds.) ICAISC 2015. LNCS, vol. 9120, pp. 247–260. Springer, Cham (2015). doi:10.1007/978-3-319-19369-4_23
43. Nobukawa, S., Nishimura, H., Yamanishi, T., Liu, J.: Chaotic states induced by resetting process in Izhikevich neuron model. J. Artif. Intell. Soft Comput. Res. **5**(2), 109–119 (2015)
44. Prasad, M., Liu, Y., Li, D., Lin, C., Shah, R.R., Kaiwartya, O.P.: A new mechanism for data visualization with Tsk-type preprocessed collaborative fuzzy rule based system. J. Artif. Intell. Soft Comput. Res. **7**(1), 33–46 (2017)
45. Radhika, K.R., Venkatesha, M.K., Sekhar, G.N.: Signature authentication based on subpattern analysis. Appl. Soft Comput. **11**, 3218–3228 (2011)
46. Rutkowski, L.: Identification of MISO nonlinear regressions in the presence of a wide class of disturbances. IEEE Trans. Inf. Theor. **37**(1), 214–216 (2002)
47. Rutkowski, L.: Adaptive probabilistic neural networks for pattern classification in time-varying environment. IEEE Trans. Neural Netw. **15**(4), 811–827 (2004)
48. Rutkowski, L., Cpałka, K.: A neuro-fuzzy controller with a compromise fuzzy reasoning. Control Cybern. **31**(2), 297–308 (2002)

49. Rutkowski, L., Cpałka, K.: Compromise approach to neuro-fuzzy systems. In: Proceedings of the 2nd Euro-International Symposium on Computation Intelligence. Frontiers in Artificial Intelligence and Applications, vol. 76, pp. 85–90 (2002)

50. Rutkowski L., Cpałka K.: Neuro-fuzzy systems derived from quasi-triangular norms. In: Proceedings of the IEEE International Conference on Fuzzy Systems, Budapest, 26–29 July, vol. 2, pp. 1031–1036 (2004)

51. Rutkowski, L., Jaworski, M., Pietruczuk, L., Duda, P.: Decision trees for mining data streams based on the Gaussian approximation. IEEE Trans. Knowl. Data Eng. **26**(1), 108–119 (2014)

52. Rutkowski, L., Jaworski, M., Pietruczuk, L., Duda, P.: The CART decision tree for mining data streams. Inf. Sci. **266**, 1–15 (2014)

53. Rutkowski, L., Jaworski, M., Pietruczuk, L., Duda, P.: A new method for data stream mining based on the misclassification error. IEEE Trans. Neural Netw. Learn. Syst. **26**, 1048–1059 (2015)

54. Rutkowski, L., Przybył, A., Cpałka, K.: Novel online speed profile generation for industrial machine tool based on flexible neuro-fuzzy approximation. IEEE Trans. Ind. Electron. **59**(2), 1238–1247 (2012)

55. Saitoh, D., Hara, K.: Mutual learning using nonlinear perceptron. J. Artif. Intell. Soft Comput. Res. **5**(1), 71–77 (2015)

56. Szczypta, J., Łapa, K., Shao, Z.: Aspects of the selection of the structure and parameters of controllers using selected population based algorithms. In: Rutkowski, L., Korytkowski, M., Scherer, R., Tadeusiewicz, R., Zadeh, L.A., Zurada, J.M. (eds.) ICAISC 2014. LNCS, vol. 8467, pp. 440–454. Springer, Cham (2014). doi:10.1007/978-3-319-07173-2_38

57. Starczewski, J., Rutkowski, L.: Connectionist Structures of Type 2 Fuzzy Inference Systems. In: Wyrzykowski, R., Dongarra, J., Paprzycki, M., Waśniewski, J. (eds.) PPAM 2001. LNCS, vol. 2328, pp. 634–642. Springer, Heidelberg (2002). doi:10.1007/3-540-48086-2_70

58. Sugiyama, H.: Pulsed power network based on decentralized intelligence for reliable and lowloss electrical power distribution. J. Artif. Intell. Soft Comput. Res. **5**(2), 97–108 (2015)

59. Tabellout, M., Kassiba, A., Tkaczyk, S., Laskowski, Ł., Świątek, J.: Dielectric and EPR investigations of stoichiometry and interface effects in silicon carbide nanoparticles. J. Phys. Condens. Matter **18**(4), 11–43 (2006). IOP Publishing

60. Tennyson, M.F., Kuester, D.A., Casteel, J., Nikolopoulos, C.: Accessible robots for improving social skills of individuals with autism. J. Artif. Intell. Soft Comput. Res. **6**(4), 267–277 (2016)

61. Wang, G., Zhang, S.: ABM with behavioral bias and applications in simulating China stock market. J. Artif. Intell. Soft Comput. Res. **5**(4), 257–270 (2015)

62. Weerakoon, T., Ishii, K., Nassiraei, A.A.F.: An artificial potential field based mobile robot navigation method to prevent from deadlock. J. Artif. Intell. Soft Comput. Res. **5**(3), 189–203 (2015)

63. Yeung, D.-Y., Chang, H., Xiong, Y., George, S., Kashi, R., Matsumoto, T., Rigoll, G.: SVC2004: first international signature verification competition. In: Zhang, D., Jain, A.K. (eds.) ICBA 2004. LNCS, vol. 3072, pp. 16–22. Springer, Heidelberg (2004). doi:10.1007/978-3-540-25948-0_3

64. Zalasiński M., Cpałka K.: A new method of on-line signature verification using a flexible fuzzy one-class classifier, pp. 38–53. Academic Publishing House EXIT (2011)

65. Zalasiński, M., Cpałka, K.: Novel algorithm for the on-line signature verification using selected discretization points groups. In: Rutkowski, L., Korytkowski, M., Scherer, R., Tadeusiewicz, R., Zadeh, L.A., Zurada, J.M. (eds.) ICAISC 2013. LNCS, vol. 7894, pp. 493–502. Springer, Heidelberg (2013). doi:10.1007/978-3-642-38658-9_44

66. Zalasiński, M., Cpałka, K., Hayashi, Y.: New method for dynamic signature verification based on global features. In: Rutkowski, L., Korytkowski, M., Scherer, R., Tadeusiewicz, R., Zadeh, L.A., Zurada, J.M. (eds.) ICAISC 2014. LNCS, vol. 8468, pp. 231–245. Springer, Cham (2014). doi:10.1007/978-3-319-07176-3_21

67. Zalasiński, M., Cpałka, K., Er, M.J.: New method for dynamic signature verification using hybrid partitioning. In: Rutkowski, L., Korytkowski, M., Scherer, R., Tadeusiewicz, R., Zadeh, L.A., Zurada, J.M. (eds.) ICAISC 2014. LNCS, vol. 8468, pp. 216–230. Springer, Cham (2014). doi:10.1007/978-3-319-07176-3_20

68. Zalasiński, M., Łapa, K., Cpałka, K.: New algorithm for evolutionary selection of the dynamic signature global features. In: Rutkowski, L., Korytkowski, M., Scherer, R., Tadeusiewicz, R., Zadeh, L.A., Zurada, J.M. (eds.) ICAISC 2013. LNCS, vol. 7895, pp. 113–121. Springer, Heidelberg (2013). doi:10.1007/978-3-642-38610-7_11

69. Zalasiński, M., Cpałka, K., Rakus-Andersson, E.: An idea of the dynamic signature verification based on a hybrid approach. In: Rutkowski, L., Korytkowski, M., Scherer, R., Tadeusiewicz, R., Zadeh, L.A., Zurada, J.M. (eds.) ICAISC 2016. LNCS, vol. 9693, pp. 232–246. Springer, Cham (2016). doi:10.1007/978-3-319-39384-1_21

A Method for Genetic Selection of the Most Characteristic Descriptors of the Dynamic Signature

Marcin Zalasiński[1]([✉]), Krzysztof Cpałka[1], and Yoichi Hayashi[2]

[1] Institute of Computational Intelligence, Częstochowa University
of Technology, Częstochowa, Poland
{marcin.zalasinski,krzysztof.cpalka}@iisi.pcz.pl
[2] Department of Computer Science, Meiji University, Tokyo, Japan
hayashiy@cs.meiji.ac.jp

Abstract. Dynamic signature verification is an important area of biometrics. In this area methods from the field of computational intelligence can be used. In this paper we propose a new method for genetic selection of the most characteristic descriptors of the dynamic signature. The descriptors are global features of the signature and components created within its partitions. Selection of the descriptors is realized individually for each user of the biometric system. Its purpose is to increase the precision of the biometric system by eliminating the descriptors which do not increase efficiency of verification procedure. Number of descriptors (their combination) can be high, so the use of genetic algorithm to reduce their number seems to be justied. Moreover, reduction of descriptors increases interpretability of fuzzy mechanism for evaluation of signatures' similarity. Proposed method was tested using known dynamic signatures database-MCYT-100.

Keywords: Biometric system · Dynamic signature verification · Descriptors selection · Genetic algorithm

1 Introduction

Dynamic signature is behavioral biometric feature. This kind of features bases on a characteristic, learned behavior of every individual. This feature, instead of information about shape of the signature, also contains important data about a way of signing. Acquisition of the dynamic signature can be performed using digital input device, e.g. graphic tablet or any touch screen device.

The dynamic signature is described by different signals changing over time. These signals are, among others, trajectories related to horizontal and vertical pen movement, pen pressure, pen velocity. Most of the described signals can be read directly at the stage of the signature acquisition using capabilities of the input device. In order to verify the identity on the basis of the dynamic signature, some characteristic information describing the signature should be

© Springer International Publishing AG 2017
L. Rutkowski et al. (Eds.): ICAISC 2017, Part I, LNAI 10245, pp. 747–760, 2017.
DOI: 10.1007/978-3-319-59063-9_67

extracted from the signals. They are descriptors of the signature, which can be e.g. so-called global features or components of so-called partitions of the signature.

In the literature we can find many solutions aimed at generating descriptors of the dynamic signature in the form of: global features (see e.g. [42–44, 80, 83]) and components of partitions (see e.g. [15–17, 76, 77, 79]). The number of generated descriptors is usually high, so it can: (1) decrease interpretability of fuzzy system (see e.g. [11, 16, 26, 38, 39]) used for the signature verification, (2) decrease efficiency of verification procedure due to using the descriptors which are not enough characteristic for individuals (they do not have good discriminatory properties).

Therefore, an interesting solution would be to propose a method for selecting the most characteristic descriptors of the user signature. Action by trial and error might be difficult due to the large number of combinations of descriptors, so using algorithms from the field of artificial intelligence (see e.g. [1, 7, 9, 18–20, 27, 28, 30, 31, 54, 55, 64–66]), especially genetic algorithms, seems to be better solution. Artificial intelligence includes such areas as fuzzy logic (see e.g. [2–5, 13, 14, 22]), evolutionary computation (see e.g. [12, 21, 33, 34]), neural networks (see e.g. [6, 8, 10, 35, 57]), robotics (see e.g. [32, 36, 69, 70]), etc. Genetic algorithms are based on biological evolution and search the area of the problem in order to find solutions close to the optimal in the acceptable time.

In this paper we made an attempt to select the most characteristic (in the context of individual users) descriptors of the dynamic signature in the form of global features and components of the partitions using the genetic algorithm. Efficiency of the selection process was tested using MCYT-100 dynamic signature database (see [45]).

The structure of the paper is as follows: in Sect. 2 an introduction to determination of the dynamic signature descriptors was presented, in Sect. 3 a description of the proposed method for genetic selection of the descriptors was presented, in Sect. 4 simulation results are presented and in Sect. 5 conclusions are drawn.

2   Introduction to Determination of the Dynamic Signature Descriptors

In this section types of used descriptors and a way of identity verification on the basis of the selected (the most characteristic) descriptors were described.

2.1   Descriptors of the Dynamic Signature Expressed in the Form of Global Features

Global features of the dynamic signature are important group of descriptors (see e.g. [25, 37, 75]). These features are e.g. total time of signing or number of pen-ups. The features determined for all J reference signatures of the user i can be stored in the matrix $\mathbf{G}_i$, which has the following structure:

$$\mathbf{G}_i = \begin{bmatrix} g_{i,1,1} & \cdots & g_{i,N,1} \\ \vdots & & \vdots \\ g_{i,1,J} & \cdots & g_{i,N,J} \end{bmatrix}, \tag{1}$$

where I is the number of the users, J is the number of the signatures created by the user in the acquisition phase, N is the number of used global features, and $g_{i,n,j}$ is the value of the global feature n $(n = 1, \ldots, N)$ determined for the signature j $(j = 1, \ldots, J)$ created by the user i $(i = 1, \ldots, I)$. The method of determining the values of global features used in the simulations was described in detail in [25] and it will not be considered in this paper.

To simplify the discussion, we can average the values of the corresponding global features describing reference signatures of the user i. Averaged values of the features can be stored in the vector $\bar{\mathbf{g}}_i = [\bar{g}_{i,1}, \ldots, \bar{g}_{i,N}]$, where $\bar{g}_{i,n}$ is the average value of n-th global feature for all J reference signatures of the user i:

$$\bar{g}_{i,n} = \frac{1}{J} \sum_{j=1}^{J} g_{i,n,j}. \tag{2}$$

2.2 Descriptors of the Dynamic Signature Expressed in the Form of Partitions' Components

An important group of descriptors are also components of a partition. The partitions are areas of the signature determined e.g. on the basis of values of pen velocity or pressure signals and values of time step of signing process. The method of determining partitions used in simulations was described in detail in [17] and it will not be considered in this paper. For each signature j in the partition q created for the user i on the basis of the signal s (velocity v or pressure z) we can determine descriptor $c_{i,q,j}^{\{s,a\}}$ which is a component of the partition associated with the trajectory signal a (x or y):

$$c_{i,q,j}^{\{s,a\}} = \frac{1}{K_{i,q}^{\{s\}}} \sum_{k=1}^{K_{i,q}^{\{s\}}} a_{i,q,j,k}^{\{s\}}, \tag{3}$$

where $K_{i,q}^{\{s\}}$ is the number of discretization points of the signal in the partition q determined for the user i on the basis of the signal s, $a_{i,q,j,k}^{\{s\}}$ is the value of the trajectory signal (x or y) at discretization point k of the trajectory signal of the signature in the partition q determined for the user i on the basis of the signal s.

Components of the partitions determined for all J reference signatures of the user i can be stored in the matrix $\mathbf{C}_i$, which has the following structure:

$$\mathbf{C}_i = \begin{bmatrix} c_{i,1,1}^{\{v,x\}} & \cdots & c_{i,Q,1}^{\{z,y\}} \\ \vdots & & \vdots \\ c_{i,1,J}^{\{v,x\}} & \cdots & c_{i,Q,J}^{\{z,y\}} \end{bmatrix}. \tag{4}$$

750 M. Zalasiński et al.

To simplify the discussion, we can average the values of the corresponding components of the partitions describing reference signatures of the user i. Averaged values of the components can be stored in the vector $\bar{c}_i = \left[\bar{c}_{i,1}^{\{v,x\}}, \ldots, \bar{c}_{i,Q}^{\{z,y\}}\right]$, where $\bar{c}_{i,q}^{\{s,a\}}$ is an average value of the component of the trajectory a from the partition q determined on the basis of the signal s for all J reference signatures of the user i:

$$\bar{c}_{i,q}^{\{s,a\}} = \frac{1}{J} \sum_{j=1}^{J} c_{i,q,j}^{\{s,a\}}. \tag{5}$$

2.3 Dynamic Signature Verification Using Descriptors

Dynamic signature verification using descriptors proceeds according to the following steps:

- **Step 1.** Determination of the classifier parameters. In this step weights of importance of selected descriptors and parameters of fuzzy one-class classifier are determined. These parameters are computed on the basis of the reference signatures' descriptors of the user.
- **Step 2.** Determination of input values of the classifier. These values are distances between average values of the reference signatures' descriptors of the user and values of the test signature descriptors.
- **Step 3.** Identity verification. This process works using fuzzy system (see e.g. [23,29,34,40,46–48,53,59,61,62]) which is one-class classifier. Verification is performed on the basis of the output value of the system.

A detailed description of how to use descriptors of the dynamic signature in the verification process can be found in our previous works (see e.g. [78,81,82]).

3 Genetic Selection of the Descriptors

In this section we describe assumptions of genetic selection of the most characteristic descriptors of the dynamic signature. Remarks on this process can be summarized as follows:

- The most characteristic descriptors are selected from the set of global features and components of partitions describing the dynamic signature.
- Selection is performed individually for each user of the system.
- Selection is performed using classic genetic algorithm with binary encoding, which uses crossover and mutation operators.

3.1 Encoding of Solutions

Each individual from the population encodes a full set of descriptors of the dynamic signature of the user i. It consists of two parts: $\mathbf{X}_{i,ch}^{\mathrm{gfeat}}$ and $\mathbf{X}_{i,ch}^{\mathrm{part}}$

$\left(\mathbf{X}_{i,ch} = \left\{ \mathbf{X}_{i,ch}^{\text{gfeat}}, \mathbf{X}_{i,ch}^{\text{part}} \right\} \right)$. Part $\mathbf{X}_{i,ch}^{\text{gfeat}}$ encodes the set of descriptors of the dynamic signature of the user i in the form of global features and it is expressed as follows:

$$\mathbf{X}_{i,ch}^{\text{gfeat}} = \{\bar{g}_{i,1}, \ldots, \bar{g}_{i,N}\} = \left\{ X_{i,ch,1}^{\text{gfeat}}, \ldots, X_{i,ch,L^{\text{gfeat}}}^{\text{gfeat}} \right\}, \tag{6}$$

where each gene of the individual $\mathbf{X}_{i,ch}^{\text{gfeat}}$ encodes information whether global feature related to this gene would be included to the subset of the most characteristic descriptors of the dynamic signature of the user i (value of each gene is equal to 0 or 1), L^{gfeat} is the number of considered global features.

Part $\mathbf{X}_{i,ch}^{\text{part}}$ encodes the set of descriptors of the dynamic signature of the user i in the form of partitions' components and it is expressed as follows:

$$\mathbf{X}_{i,ch}^{\text{part}} = \left\{ \bar{c}_{i,1}^{\{v,x\}}, \ldots, \bar{c}_{i,Q}^{\{z,y\}} \right\} = \left\{ X_{i,ch,1}^{\text{part}}, \ldots, X_{i,ch,L^{\text{part}}}^{\text{part}} \right\}, \tag{7}$$

where each gene of the individual $\mathbf{X}_{i,ch}^{\text{part}}$ encodes information whether partition component related to this gene would be included to the subset of the most characteristic descriptors of the dynamic signature of the user i (value of each gene is equal to 0 or 1), L^{part} is the number of considered partitions' components.

3.2   Processing of Solutions

A purpose of the genetic algorithm (see e.g. [24,41,72]) is selection of a subset of the descriptors which values determined for reference signatures of the user i are similar. Considered method works according to the algorithm shown in Fig. 1. First, random initialization of individuals $\mathbf{X}_{i,ch}$ is performed. They are interpreted as chromosomes of the population and encode subsets of the descriptors. Next, chromosomes are evaluated by determination of the value of their fitness function (see Sect. 3.3). Having values of fitness function, stopping criterion is checked. The criterion takes into account the threshold value of the fitness function or execution of the specified number of steps (generations) by the algorithm. If the stopping criterion is satisfied, then the procedure of evolutionary features selection quits and returns the information about the best chromosome from the population. It is rarely directly after the initialization of the population, so the population must be processed in the process of evolution (see e.g. [49–52,56,67,68,71,74]). First step of evolution is drawing of individuals in order to apply genetic operators. Individuals that have a better value of fitness function have also higher chance to be drawn. Next, pairs of chromosomes exchange genes (crossover is performed) in randomly selected points and finally a mutation of randomly selected genes is performed (value of gene changes from 0 to 1 or vice versa). During use of genetic operators the probabilities of crossover and mutation are taken into account. They are parameters of the algorithm selected from the unit range. In this way, the descendant population is created from the parent population. This new population is evaluated again and the whole process is repeated. A detailed description of the algorithm can be found, among others, in [58,60,63].

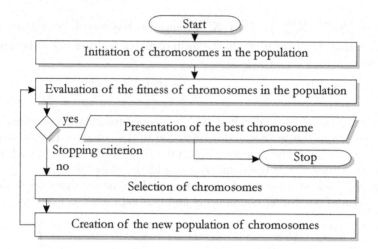

Fig. 1. Scheme of the genetic algorithm.

3.3 Evaluation of Solutions

To determine value of fitness function we use L^* descriptors chosen during genetic selection process. They are descriptors for which values of genes in the chromosome $\mathbf{X}_{i,ch}$ are equal 1.

In order to simplify description of determination of the fitness function values, the following variables will be used: (a) $\mathbf{d}_{i,j} = [d_{i,j,1}, \ldots, d_{i,j,L^*}]$-the vector containing the values of the selected L^* descriptors of the signature j of the user i, (b) $\bar{\mathbf{d}}_i = [\bar{d}_{i,1}, \ldots, \bar{d}_{i,L^*}]$-the vector containing averaged in the context of the reference signatures of the user i values of selected L^* descriptors, (c) $\mathbf{D}_i = [\mathbf{d}_{i,1}, \ldots, \mathbf{d}_{i,J}]$-the matrix containing values of selected L^* descriptors of J reference signatures of the user i.

Determination of fitness function value is based on the Mahalanobis distance between values of selected descriptors of all reference signatures of the user i and their average values. It is realized as follows:

$$\text{ff} \left(\mathbf{X}_{i,ch} \right) = \frac{1}{J} \sum_{j=1}^{J} \sqrt{\left(\mathbf{d}_{i,j} - \bar{\mathbf{d}}_i \right) \left(\text{cov} \left(\mathbf{D}_i \right) \right)^{-1} \left(\mathbf{d}_{i,j} - \bar{\mathbf{d}}_i \right)^{\text{T}}}. \tag{8}$$

Use of Mahalanobis distance allows us to take into account mutual correlation and individual variance (expressed by the arithmetic mean of the squared deviations from the arithmetic mean) of descriptors. Lower value of the fitness function $\text{ff} \left(\mathbf{X}_{i,ch} \right)$ means that the chromosome $\mathbf{X}_{i,ch}$ is "better" (subset of global features encoded in the chromosome $\mathbf{X}_{i,ch}$ is the most characteristic for the user i).

4   Simulation Results

Simulations were performed using public MCYT-100 database which contains
signatures of 100 users. During training phase we used 5 randomly selected
genuine signatures of each signer. During test phase we used 15 genuine signatures and 15 skilled forgeries of each signer. The process was performed five
times, and the results were averaged. The test was performed using the authorial testing environment implemented in C# language. During the simulations
the following assumptions have been adopted: (a) population contains 100 chromosomes, (b) algorithm stops after the lapse of a determined number of 1000
generations, (c) during selection of chromosomes tournament selection method
is used, (d) crossover is performed with probability equal to 0.8 at three points,
(e) mutation is performed for each gene with probability equal to 0.02.

Results obtained by the presented method and other methods using global
features are presented in Fig. 2 and in Table 1. The proposed method for the
considered MCYT-100 database enables a significant reduction in the descriptors
necessary for the successful signature verification (see Fig. 2). It allows us to
eliminate descriptors which negatively affect the accuracy (see Table 1). Due to
this, the method works with high accuracy in comparison with the methods
presented in the Table 1. The comparison criterion was the value of the error
EER (Equal Error Rate), which is commonly used to evaluate the accuracy of
biometric methods (see e.g. [73]).

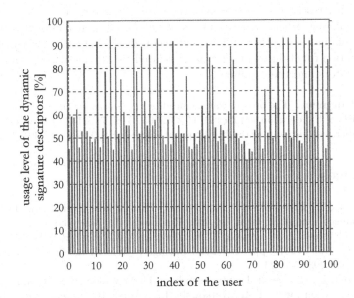

Fig. 2. The percentage comparison of the number of selected descriptors of the dynamic
signature for users of the MCYT-100 database.

Table 1. Comparison of the results for the dynamic signature verification methods taking into account methods based on global features using maximum 5 reference signatures and tested using so-called skilled forgeries.

Method	Average FAR	Average FRR	Average error
Nanni and Lumini (b) [43]	–	–	8.40%
Nanni and Lumini (a) [44]	–	–	7.60%
Fierrez-Aguilar et al. [25]	–	–	5.61%
Nanni [42]	–	–	5.20%
Cpałka and Zalasiński [15]	5.28%	4.87%	5.20%
Lumini and Nanni [37]	–	–	4.50%
Zalasiński et al. [81]	3.85%	1.99%	2.92%
Our method	**3.07%**	**3.13%**	**2.60%**

5   Conclusions

In this paper a new method for genetic selection of the most characteristic descriptors of the dynamic signature was presented. Simulation results confirm that the reduction of the number of descriptors used in the verification phase has a positive effect on the biometric system accuracy and interpretability of fuzzy mechanism evaluating similarity of signatures.

In further research on selection of the dynamic signature descriptors we will consider, among others: (a) the use of other population based algorithms for selection of the descriptors and (b) extension of the set of criteria included in the evaluation function by other criteria related to, among others, readability of the descriptors and their number.

Acknowledgment. The project was financed by the National Science Centre (Poland) on the basis of the decision number DEC-2012/05/B/ST7/02138. The work presented in this paper was also supported by the grant number BS/MN 1-109-301/16/P.

References

1. Bartczuk, Ł., Dziwiński, P., Starczewski, J.T.: A new method for dealing with unbalanced linguistic term set. In: Rutkowski, L., Korytkowski, M., Scherer, R., Tadeusiewicz, R., Zadeh, L.A., Zurada, J.M. (eds.) ICAISC 2012. LNCS (LNAI), vol. 7267, pp. 207–212. Springer, Heidelberg (2012). doi:10.1007/978-3-642-29347-4_24
2. Bartczuk, Ł., Dziwiński, P., Starczewski, J.T.: New method for generation type-2 fuzzy partition for FDT. In: Rutkowski, L., Scherer, R., Tadeusiewicz, R., Zadeh, L.A., Zurada, J.M. (eds.) ICAISC 2010. LNCS (LNAI), vol. 6113, pp. 275–280. Springer, Heidelberg (2010). doi:10.1007/978-3-642-13208-7_35

3. Bartczuk, Ł., Galushkin, A.I.: A new method for generating nonlinear correction models of dynamic objects based on semantic genetic programming. In: Rutkowski, L., Korytkowski, M., Scherer, R., Tadeusiewicz, R., Zadeh, L.A., Zurada, J.M. (eds.) ICAISC 2016. LNCS, vol. 9693, pp. 249–261. Springer, Cham (2016). doi:10. 1007/978-3-319-39384-1_22

4. Bartczuk, Ł., Przybył, A., Cpałka, K.: A new approach to nonlinear modelling of dynamic systems based on fuzzy rules. Int. J. Appl. Math. Comput. Sci. 3, 603–621 (2016)

5. Bartczuk, Ł., Przybył, A., Koprinkova-Hristova, P.: New method for non-linear correction modelling of dynamic objects with genetic programming. In: Rutkowski, L., Korytkowski, M., Scherer, R., Tadeusiewicz, R., Zadeh, L.A., Zurada, J.M. (eds.) ICAISC 2015. LNCS, vol. 9120, pp. 318–329. Springer, Cham (2015). doi:10. 1007/978-3-319-19369-4_29

6. Bas, E.: The training of multiplicative neuron model based artificial neural networks with differential evolution algorithm for forecasting. J. Artif. Intell. Soft Comput. Res. 6(1), 5–11 (2016)

7. Bello, O., Holzmann, J., Yaqoob, T., Teodoriu, C.: Application of artificial intelligence methods in drilling system design and operations: a review of the state of the art. J. Artif. Intell. Soft Comput. Res. 5(2), 121–139 (2015)

8. Bertini, J.J.R., Nicoletti, M.D.C.: Enhancing constructive neural network performance using functionally expanded input data. J. Artif. Intell. Soft Comput. Res. 6(2), 119–131 (2016)

9. Bilski, J., Kowalczyk, B., Żurada, J.M.: Application of the givens rotations in the neural network learning algorithm. In: Rutkowski, L., Korytkowski, M., Scherer, R., Tadeusiewicz, R., Zadeh, L.A., Zurada, J.M. (eds.) ICAISC 2016. LNCS, vol. 9692, pp. 46–56. Springer, Cham (2016). doi:10.1007/978-3-319-39378-0_5

10. Cierniak, R., Rutkowski, L.: On image compression by competitive neural networks and optimal linear predictors. Signal Process. Image Commun. 156, 559–565 (2000)

11. Cpałka, K.: Design of Interpretable Fuzzy Systems. Springer, Cham (2017)

12. Cpałka, K., Łapa, K., Przybył, A.: A new approach to design of control systems using genetic programming. Inf. Technol. Control 44(4), 433–442 (2015)

13. Cpałka, K., Rebrova, O., Nowicki, R., Rutkowski, L.: On design of flexible neurofuzzy systems for nonlinear modelling. Int. J. Gener. Syst. 42(6), 706–720 (2013)

14. Cpałka, K., Rutkowski, L.: Flexible Takagi-Sugeno. fuzzy systems. In: Proceedings of the 2005 IEEE International Joint Conference on Neural Networks, IJCNN 2005, vol. 3, pp. 1764–1769 (2005)

15. Cpałka, K., Zalasiński, M.: On-line signature verification using vertical signature partitioning. Expert Syst. Appl. 41, 4170–4180 (2014)

16. Cpałka, K., Zalasiński, M., Rutkowski, L.: New method for the on-line signature verification based on horizontal partitioning. Pattern Recognit. 47, 2652–2661 (2014)

17. Cpałka, K., Zalasiński, M., Rutkowski, L.: A new algorithm for identity verification based on the analysis of a handwritten dynamic signature. Appl. Soft Comput. 43, 47–56 (2016)

18. Colchester, K., Hagras, H., Alghazzawi, D.: A survey of artificial intelligence techniques employed for adaptive educational systems within E-learning platforms. J. Artif. Intell. Soft Comput. Res. 7(1), 47–64 (2017)

19. Duda, P., Hayashi, Y., Jaworski, M.: On the strong convergence of the orthogonal series-type kernel regression neural networks in a non-stationary environment. In: Rutkowski, L., Korytkowski, M., Scherer, R., Tadeusiewicz, R., Zadeh, L.A., Zurada, J.M. (eds.) ICAISC 2012. LNCS, vol. 7267, pp. 47–54. Springer, Heidelberg (2012). doi:10.1007/978-3-642-29347-4_6

20. Duda, P., Jaworski, M., Pietruczuk, L.: On pre-processing algorithms for data stream. In: Rutkowski, L., Korytkowski, M., Scherer, R., Tadeusiewicz, R., Zadeh, L.A., Zurada, J.M. (eds.) ICAISC 2012. LNCS, vol. 7268, pp. 56–63. Springer, Heidelberg (2012). doi:10.1007/978-3-642-29350-4_7

21. Dziwiński, P., Avedyan, E.D.: A new approach to nonlinear modeling based on significant operating points detection. In: Rutkowski, L., Korytkowski, M., Scherer, R., Tadeusiewicz, R., Zadeh, L.A., Zurada, J.M. (eds.) ICAISC 2015. LNCS (LNAI), vol. 9120, pp. 364–378. Springer, Cham (2015). doi:10.1007/978-3-319-19369-4_33

22. Dziwiński, P., Avedyan, E.D.: A new approach for using the fuzzy decision trees for the detection of the significant operating points in the nonlinear modeling. In: Rutkowski, L., Korytkowski, M., Scherer, R., Tadeusiewicz, R., Zadeh, L.A., Zurada, J.M. (eds.) ICAISC 2016. LNCS (LNAI), vol. 9693, pp. 279–292. Springer, Cham (2016). doi:10.1007/978-3-319-39384-1_24

23. Dziwiński, P., Avedyan, E.D.: A new method of the intelligent modeling of the nonlinear dynamic objects with fuzzy detection of the operating points. In: Rutkowski, L., Korytkowski, M., Scherer, R., Tadeusiewicz, R., Zadeh, L.A., Zurada, J.M. (eds.) ICAISC 2016. LNCS (LNAI), vol. 9693, pp. 293–305. Springer, Cham (2016). doi:10.1007/978-3-319-39384-1_25

24. El-Samak, A.F., Ashour, W.: Optimization of traveling salesman problem using affinity propagation clustering and genetic algorithm. J. Artif. Intell. Soft Comput. Res. **5**, 239–245 (2015)

25. Fierrez-Aguilar, J., Nanni, L., Lopez-Peñalba, J., Ortega-Garcia, J., Maltoni, D.: An on-line signature verification system based on fusion of local and global information. In: Kanade, T., Jain, A., Ratha, N.K. (eds.) AVBPA 2005. LNCS, vol. 3546, pp. 523–532. Springer, Heidelberg (2005). doi:10.1007/11527923_54

26. Gabryel, M., Cpałka, K., Rutkowski, L.: Evolutionary strategies for learning of neuro-fuzzy systems. In: Proceedings of the I Workshop on Genetic Fuzzy Systems, Granada, pp. 119–123 (2005)

27. Gałkowski, T., Rutkowski, L.: Nonparametric fitting of multivariate functions. IEEE Trans. Autom. Control. **31**(8), 785–787 (1986)

28. Greblicki, W., Rutkowski, L.: Density-free Bayes risk consistency of nonparametric pattern recognition procedures. Proc. IEEE **69**(4), 482–483 (1981)

29. Harmati, I.Á., Bukovics, Á., Kóczy, L.T.: Minkowski's inequality based sensitivity analysis of fuzzy signatures. J. Artif. Intell. Soft Comput. Res. **6**(4), 219–229 (2016)

30. Jaworski, M., Er, M.J., Pietruczuk, L.: On the application of the Parzen-type kernel regression neural network and order statistics for learning in a non-stationary environment. In: Rutkowski, L., Korytkowski, M., Scherer, R., Tadeusiewicz, R., Zadeh, L.A., Zurada, J.M. (eds.) ICAISC 2012. LNCS, vol. 7267, pp. 90–98. Springer, Heidelberg (2012). doi:10.1007/978-3-642-29347-4_11

31. Jaworski, M., Pietruczuk, L., Duda, P.: On resources optimization in fuzzy clustering of data streams. In: Rutkowski, L., Korytkowski, M., Scherer, R., Tadeusiewicz, R., Zadeh, L.A., Zurada, J.M. (eds.) ICAISC 2012. LNCS, vol. 7268, pp. 92–99. Springer, Heidelberg (2012). doi:10.1007/978-3-642-29350-4_11

32. Jimenez, F., Yoshikawa, T., Furuhashi, T., Kanoh, M.: An emotional expression model for educational-support robots. J. Artif. Intell. Soft Comput. Res. **5**(1), 51–57 (2015)

33. Kitajima, R., Kamimura, R.: Accumulative information enhancement in the self-organizing maps and its application to the analysis of mission statements. J. Artif. Intell. Soft Comput. Res. **5**(3), 161–176 (2015)

34. Korytkowski, M., Rutkowski, L., Scherer, R.: Fast image classification, by boosting fuzzy classifiers. Inf. Sci. **327**, 175–182 (2016)

35. Korytkowski, M., Scherer, R., Rutkowski, L.: On combining backpropagation with boosting. In: Proceedings of the 2006 International Joint Conference on Neural Networks, IEEE World Congress on Computational Intelligence, pp. 1274–1277 (2006)

36. Lan, K., Sekiyama, K.: Autonomous viewpoint selection of robot based on aesthetic evaluation of a scene. J. Artif. Intell. Soft Comput. Res. **6**(4), 255–265 (2016)

37. Lumini, A., Nanni, L.: Ensemble of on-line signature matchers based on overcomplete feature generation. Expert Syst. Appl. **36**, 5291–5296 (2009)

38. Łapa, K., Cpałka, K., Galushkin, A.I.: A new interpretability criteria for neuro-fuzzy systems for nonlinear classification. In: Rutkowski, L., Korytkowski, M., Scherer, R., Tadeusiewicz, R., Zadeh, L.A., Zurada, J.M. (eds.) ICAISC 2015. LNCS, vol. 9119, pp. 448–468. Springer, Cham (2015). doi:10.1007/978-3-319-19324-3_41

39. Łapa, K., Cpałka, K., Wang, L.: New method for design of fuzzy systems for nonlinear modelling using different criteria of interpretability. In: Rutkowski, L., Korytkowski, M., Scherer, R., Tadeusiewicz, R., Zadeh, L.A., Zurada, J.M. (eds.) ICAISC 2014. LNCS, vol. 8467, pp. 217–232. Springer, Cham (2014). doi:10.1007/978-3-319-07173-2_20

40. Łapa, K., Przybył, A., Cpałka, K.: A new approach to designing interpretable models of dynamic systems. In: Rutkowski, L., Korytkowski, M., Scherer, R., Tadeusiewicz, R., Zadeh, L.A., Zurada, J.M. (eds.) ICAISC 2013. LNCS, vol. 7895, pp. 523–534. Springer, Heidelberg (2013). doi:10.1007/978-3-642-38610-7_48

41. Łapa, K., Szczypta, J., Venkatesan, R.: Aspects of structure and parameters selection of control systems using selected multi-population algorithms. In: Rutkowski, L., Korytkowski, M., Scherer, R., Tadeusiewicz, R., Zadeh, L.A., Zurada, J.M. (eds.) ICAISC 2015. LNCS, vol. 9120, pp. 247–260. Springer, Cham (2015). doi:10.1007/978-3-319-19369-4_23

42. Nanni, L.: An advanced multi-matcher method for on-line signature verification featuring global features and tokenised random numbers. Neurocomputing **69**, 2402–2406 (2006)

43. Nanni, L., Lumini, A.: Ensemble of Parzen window classifiers for on-line signature verification. Neurocomputing **68**, 217–224 (2005)

44. Nanni, L., Lumini, A.: Advanced methods for two-class problem formulation for on-line signature verification. Neurocomputing **69**, 854–857 (2006)

45. Ortega-Garcia, J., Fierrez-Aguilar, J., Simon, D., Gonzalez, J., Faundez-Zanuy, M., Espinosa, V., Satue, A., Hernaez, I., Igarza, J.J., Vivaracho, C., Escudero, D., Moro, Q.I.: MCYT baseline corpus: a bimodal biometric database. IEE Proc. Vis. Image Signal Process **150**, 395–401 (2003)

46. Patgiri, C., Sarma, M., Sarma, K.K.: A class of neuro-computational methods for assamese fricative classification. J. Artif. Intell. Soft Comput. Res. **5**(1), 59–70 (2015)

47. Prasad, M., Liu, Y., Li, D., Lin, C., Shah, R.R., Kaiwartya, O.P.: A new mechanism for data visualization with Tsk-type preprocessed collaborative fuzzy rule based system. J. Artif. Intell. Soft Comput. Res. **7**(1), 33–46 (2017)
48. Przybył, A., Er, M.J.: The idea for the integration of neuro-fuzzy hardware emulators with real-time network. In: Rutkowski, L., Korytkowski, M., Scherer, R., Tadeusiewicz, R., Zadeh, L.A., Zurada, J.M. (eds.) ICAISC 2014. LNCS (LNAI), vol. 8467, pp. 279–294. Springer, Cham (2014). doi:10.1007/978-3-319-07173-2_25
49. Przybył, A., Er, M.J.: A new approach to designing of intelligent emulators working in a distributed environment. In: Rutkowski, L., Korytkowski, M., Scherer, R., Tadeusiewicz, R., Zadeh, L.A., Zurada, J.M. (eds.) ICAISC 2016. LNCS, vol. 9693, pp. 546–558. Springer, Cham (2016). doi:10.1007/978-3-319-39384-1_48
50. Przybył, A., Er, M.J.: The method of hardware implementation of fuzzy systems on FPGA. In: Rutkowski, L., Korytkowski, M., Scherer, R., Tadeusiewicz, R., Zadeh, L.A., Zurada, J.M. (eds.) ICAISC 2016. LNCS, vol. 9692, pp. 284–298. Springer, Cham (2016). doi:10.1007/978-3-319-39378-0_25
51. Przybył, A., Jelonkiewicz, J.: Genetic algorithm for observer parameters tuning in sensorless induction motor drive. In: Rutkowski, L., Kacprzyk, J. (eds.) Neural Networks and Soft Computing. Advances in Soft Computing, vol. 19, pp. 376–381. Springer, Heidelberg (2003)
52. Przybył, A., Smoląg, J., Kimla, P.: Distributed control system based on real time ethernet for computer numerical controlled machine tool. Przeglad Elektrotechniczny **86**(2), 342–346 (2010). (in Polish)
53. Rutkowska, A.: Influence of membership functions shape on portfolio optimization results. J. Artif. Intell. Soft Comput. Res. **6**(1), 45–54 (2016)
54. Rutkowski, L.: Sequential estimates of probability densities by orthogonal series and their application in pattern classification. IEEE Trans. Syst. Man Cybern. **10**(12), 918–920 (1980)
55. Rutkowski, L.: Sequential pattern-recognition procedures derived from multiple Fourier-series. Pattern Recognit. Lett. **8**(4), 213–216 (1988)
56. Rutkowski, L.: Non-parametric learning algorithms in time-varying environments. Signal Process. **182**, 129–137 (1989)
57. Rutkowski, L.: Adaptive probabilistic neural networks for pattern classification in time-varying environment. IEEE Trans. Neural Netw. **15**(4), 811–827 (2004)
58. Rutkowski, L.: Computational Intelligence. Springer, Heidelberg (2008)
59. Rutkowski L., Cpałka K.: A general approach to neuro-fuzzy systems. In: The 10th IEEE International Conference on Fuzzy Systems, Melbourne, pp. 1428–1431 (2001)
60. Rutkowski, L., Cpałka, K.: Flexible weighted neuro-fuzzy systems. In: Proceedings of the 9th International Conference on Neural Information Processing (ICONIP 2002), Orchid Country Club, Singapore, 18–22 November 2002
61. Rutkowski, L., Cpałka, K.: A neuro-fuzzy controller with a compromise fuzzy reasoning. Control Cybern. **31**(2), 297–308 (2002)
62. Rutkowski, L., Cpałka, K.: Compromise approach to neuro-fuzzy systems. In: Proceedings of the 2nd Euro-International Symposium on Computation Intelligence. Frontiers in Artificial Intelligence and Applications, vol. 76, pp. 85–90 (2002)
63. Rutkowski, L., Cpałka, K.: Neuro-fuzzy systems derived from quasi-triangular norms. In: Proceedings of the IEEE International Conference on Fuzzy Systems, Budapest, 26–29 July, vol. 2, pp. 1031–1036 (2004)
64. Rutkowski, L., Jaworski, M., Pietruczuk, L., Duda, P.: A new method for data stream mining based on the misclassification error. IEEE Trans. Neural Netw. Learn. Syst. **26**, 1048–1059 (2015)

65. Rutkowski, L., Pietruczuk, L., Duda, P., Jaworski, M.: Decision trees for mining data streams based on the McDiarmid's bound. IEEE Trans. Knowl. Data Eng. **25**(6), 1272–1279 (2013)

66. Rutkowski, L., Przybył, A., Cpałka, K.: Novel online speed profile generation for industrial machine tool based on flexible neuro-fuzzy approximation. IEEE Trans. Ind. Electron. **59**(2), 1238–1247 (2012)

67. Stanovov, V., Semenkin, E., Semenkina, O.: Self-configuring hybrid evolutionary algorithm for fuzzy imbalanced classification with adaptive instance selection. J. Artif. Intell. Soft Comput. Res. **6**(3), 173–188 (2016)

68. Szczypta, J., Przybył, A., Cpałka, K.: Some aspects of evolutionary designing optimal controllers. In: Rutkowski, L., Korytkowski, M., Scherer, R., Tadeusiewicz, R., Zadeh, L.A., Zurada, J.M. (eds.) ICAISC 2013. LNCS, vol. 7895, pp. 91–100. Springer, Heidelberg (2013). doi:10.1007/978-3-642-38610-7_9

69. Tennyson, M.F., Kuester, D.A., Casteel, J., Nikolopoulos, C.: Accessible robots for improving social skills of individuals with autism. J. Artif. Intell. Soft Comput. Res. **6**(4), 267–277 (2016)

70. Weerakoon, T., Ishii, K., Nassiraei, A.A.F.: An artificial potential field based mobile robot navigation method to prevent from deadlock. J. Artif. Intell. Soft Comput. Res. **5**(3), 189–203 (2015)

71. Wei, H.: A bio-inspired integration method for object semantic representation. J. Artif. Intell. Soft Comput. Res. **6**(3), 137–154 (2016)

72. Yang, C.H., Moi, S.H., Lin, Y.D., Chuang, L.Y.: Genetic algorithm combined with a local search method for identifying susceptibility genes. J. Artif. Intell. Soft Comput. Res. **6**, 203–212 (2016)

73. Yeung, D.-Y., Chang, H., Xiong, Y., George, S., Kashi, R., Matsumoto, T., Rigoll, G.: SVC2004: first international signature verification competition. In: Zhang, D., Jain, A.K. (eds.) ICBA 2004. LNCS, vol. 3072, pp. 16–22. Springer, Heidelberg (2004). doi:10.1007/978-3-540-25948-0_3

74. Yin, Z., O'Sullivan, C., Brabazon, A.: An analysis of the performance of genetic programming for realised volatility forecasting. J. Artif. Intell. Soft Comput. Res. **6**(3), 155–172 (2016)

75. Zalasiński, M.: New algorithm for on-line signature verification using characteristic global features. In: Wilimowska, Z., Borzemski, L., Grzech, A., Świątek, J. (eds.) ISAT 2015. AISC, vol. 432, pp. 137–146. Springer, Cham (2016). doi:10.1007/978-3-319-28567-2_12

76. Zalasiński, M., Cpałka, K.: A new method of on-line signature verification using a flexible fuzzy one-class classifier, pp. 38–53. Academic Publishing House EXIT (2011)

77. Zalasiński, M., Cpałka, K.: Novel algorithm for the on-line signature verification using selected discretization points groups. In: Rutkowski, L., Korytkowski, M., Scherer, R., Tadeusiewicz, R., Zadeh, L.A., Zurada, J.M. (eds.) ICAISC 2013. LNCS, vol. 7894, pp. 493–502. Springer, Heidelberg (2013). doi:10.1007/978-3-642-38658-9_44

78. Zalasiński, M., Cpałka, K.: New algorithm for on-line signature verification using characteristic hybrid partitions. In: Wilimowska, Z., Borzemski, L., Grzech, A., Świątek, J. (eds.) ISAT 2015. AISC, vol. 432, pp. 147–157. Springer, Cham (2016). doi:10.1007/978-3-319-28567-2_13

79. Zalasiński, M., Cpałka, K., Er, M.J.: New method for dynamic signature verification using hybrid partitioning. In: Artificial Intelligence and Soft Computing. Lecture Notes in Computer Science, vol. 8467, pp. 236–250. Springer (2014)

80. Zalasiński, M., Cpałka, K., Hayashi, Y.: New method for dynamic signature verification based on global features. In: Artificial Intelligence and Soft Computing. Lecture Notes in Computer Science, vol. 8467, pp. 251–265. Springer (2014)
81. Zalasiński, M., Cpałka, K., Hayashi, Y.: A new approach to the dynamic signature verification aimed at minimizing the number of global features. In: Rutkowski, L., Korytkowski, M., Scherer, R., Tadeusiewicz, R., Zadeh, L.A., Zurada, J.M. (eds.) ICAISC 2016. LNCS, vol. 9693, pp. 218–231. Springer, Cham (2016). doi:10.1007/978-3-319-39384-1_20
82. Zalasiński, M., Cpałka, K., Rakus-Andersson, E.: An idea of the dynamic signature verification based on a hybrid approach. In: Rutkowski, L., Korytkowski, M., Scherer, R., Tadeusiewicz, R., Zadeh, L.A., Zurada, J.M. (eds.) ICAISC 2016. LNCS, vol. 9693, pp. 232–246. Springer, Cham (2016). doi:10.1007/978-3-319-39384-1_21
83. Zalasiński, M., Łapa, K., Cpałka, K.: New algorithm for evolutionary selection of the dynamic signature global features. In: Rutkowski, L., Korytkowski, M., Scherer, R., Tadeusiewicz, R., Zadeh, L.A., Zurada, J.M. (eds.) ICAISC 2013. LNCS, vol. 7895, pp. 113–121. Springer, Heidelberg (2013). doi:10.1007/978-3-642-38610-7_11

A Method for Changes Prediction of the Dynamic Signature Global Features over Time

Marcin Zalasiński[1(✉)], Krystian Łapa[1], Krzysztof Cpałka[1], and Takamichi Saito[2]

[1] Institute of Computational Intelligence,
Częstochowa University of Technology, Częstochowa, Poland
{marcin.zalasinski,krystian.lapa,krzysztof.cpalka}@iisi.pcz.pl
[2] Department of Computer Science, Meiji University, Tokyo, Japan
saito@cs.meiji.ac.jp

Abstract. Dynamic signature can be represented by a set of global features. These features are interpreted as e.g. number of pen ups, time of signing process, etc. Values of global features can be determined on the basis of non-linear waveforms defining dynamics of the signature. They are acquired using graphic tablet or a device with a touch screen. In this paper we present a method for prediction values of the dynamic signature global features changing over time. The purpose of the prediction is, among others, improving the efficiency of the dynamic signature verification process when the interval between acquisition sessions is large. This interval causes a slight change in the way of signing, which can affect change in the value of global features. In this case the effectiveness of the signature verification also changes (decreases). The possibility of predicting the values of global features can result in a partial elimination of the described problem. Tests of the proposed method were performed using ATVS-SLT DB database of the dynamic signatures.

Keywords: Dynamic signature verification · Global features · Prediction of global features' values

1 Introduction

The dynamic signature is behavioral biometric characteristic, used to identity verification. It is described by signals changing over time, which are acquired using digital input device, e.g. graphic tablet or the device with a touch screen. The signals contain information about, among others, shape of the signature, pen pressure or pen angle during signing process.

Identity verification process using the dynamic signature is the most often based on some characteristic features extracted from the signals describing the signature. These features are, among others, global features, which can specify e.g. time of signing, number of pen-ups, etc. In the literature we can find many

© Springer International Publishing AG 2017
L. Rutkowski et al. (Eds.): ICAISC 2017, Part I, LNAI 10245, pp. 761–772, 2017.
DOI: 10.1007/978-3-319-59063-9_68

methods using global features for the signature verification (see e.g. [19,40–42,77,82,83,85]).

In this paper we propose a method for changes prediction of the dynamic signature global features over time. The purpose of the prediction is, among others, improvement of the verification process efficiency when the interval between sessions in which signature was acquired is large. This interval causes a slight change in the way of signing, which can also change the values of global features. In this case effectiveness of the verification process usually decreases. The possibility of predicting the global features can result in a partial elimination of the described problem. In the prediction process possibilities of fuzzy systems are used. However, other methods used for prediction can be also used (see. e.g. [43,69,74]). Fuzzy systems (see e.g. [4,14,18,24,37,70,78–81,84]) belong to computational intelligence methods (see e.g. [1,3,26,27,29,48,52,65,68]). These methods include both the data structures used to solve problems in the field of control [12,15,45,50,57,72,73], modelling [6,8–10,23,32,53,54] or classification [30,55,67] (e.g. neural networks [5,11,13,20,21,28,71], decision trees [22,47,62–64], support vector machines [75], etc.) and optimization algorithms (e.g. gradient algorithms [31], evolutionary algorithms [2,33,39,44,49,76], etc.). Fuzzy systems are characterized by a clear representation of knowledge in the form of fuzzy rules. Their parameters and structure (also form of fuzzy rules) can be selected during gradient and evolutionary learning.

Structure of the paper is as follows: Sect. 2 describes the proposed method for prediction values of the dynamic signature global features, Sect. 3 characterizes obtained simulation results and Sect. 4 contains conclusions.

2   Method for Prediction Values of the Dynamic Signature Global Features

Remarks on the proposed method for prediction values of the dynamic signature global features can be summarized as follows:

- It allows us for prediction values of the dynamic signature global features of the individual user. It takes into account values of the features determined during the previous training acquisition sessions of the signature (or sessions during which the signature was verified positively) and stored in the database.
- It can work for any number of the dynamic signature global features. The system used for prediction can process signals associated with any number of previous training sessions. However, in this paper we assume that values of the features from one session are given to the system input (Fig. 1).
- It uses possibilities of fuzzy system (see e.g. [7,16,17,34,46,51,58,66]) learned by the algorithm based on population [35,36]. In this paper learning process was performed using evolutionary strategy $(\mu + \lambda)$.

2.1   Preparation of Learning and Testing Data

The proposed method is based on the set of global features determined for signatures created in subsequent training sessions which took place at certain time intervals. Values of these features are denoted as $g_{i,n,j,s}$, where i is the index of the user ($i = 1, ..., I$), I is the number of the users, n is the index of the feature ($n = 1, ..., N$), N is the number of considered features, j is the index of the signature ($j = 1, ..., J_s$), J_s is the number of signatures of the user created in the session s ($s = 1, ..., S$), S is the number of sessions.

The range of the global features' values within each session is usually different (even though it is considered for each user independently). Due to this, values of features should be normalized:

$$g'_{i,n,j,s} = \frac{\max_{\substack{s=1,...,S \\ j=1,...,J_s}} \{g_{i,n,j,s}\} - g_{i,n,j,s}}{\max_{\substack{s=1,...,S \\ j=1,...,J_s}} \{g_{i,n,j,s}\} - \min_{\substack{s=1,...,S \\ j=1,...,J_s}} \{g_{i,n,j,s}\}}. \tag{1}$$

Next, values of features should be averaged:

$$\bar{g}'_{i,n,s} = \frac{1}{J_s} \sum_{j=1}^{J_s} g'_{i,n,j,s}. \tag{2}$$

Averaged values of global features are part of learning ant testing sequence, used in training and testing phase of the system used for prediction (fuzzy system). It is created for each user independently taking into account the assumption that prediction is based on the values of features only from previous session (in order to increase the accuracy, more sessions can be used during prediction). Therefore, learning sequence has the following form:

$$\{\mathbf{x}_{i,s=1}, \mathbf{d}_{i,s=1}\}, \{\mathbf{x}_{i,s=2}, \mathbf{d}_{i,s=2}\}, ..., \{\mathbf{x}_{i,s=S-2}, \mathbf{d}_{i,s=S-2}\}, \tag{3}$$

where $\mathbf{x}_{i,s} = [\bar{g}'_{i,n=1,s}, \bar{g}'_{i,n=2,s}, ..., \bar{g}'_{i,n=N,s}]$ represent input vectors, $\mathbf{d}_{i,s} = \mathbf{x}_{i,s+1} = [\bar{g}'_{i,n=1,s+1}, \bar{g}'_{i,n=2,s+1}, ..., \bar{g}'_{i,n=N,s+1}]$ represent reference vectors. Data from the last session can be used in testing phase, so the test set has the following form:

$$\{\mathbf{x}_{i,s=S-1}, \mathbf{d}_{i,s=S-1}\}. \tag{4}$$

In case of the dynamic signature verification issue the number of acquisition sessions is small, which forced the proposed approach to the division of data. In practice, the fuzzy system used for prediction can be systematically trained using the data of signatures classified as genuine.

2.2 Training and Testing

Prediction can be implemented using e.g. Mamdani-type neuro-fuzzy system [59–61] (with many inputs and many outputs). Its operation can be expressed in a symbolic way as follows:

$$\mathbf{y}_{i,s} = \mathbf{f}_i(\mathbf{x}_{i,s}), \tag{5}$$

where $f_i(\cdot)$ is a function representing system for the user i and $\mathbf{y}_{i,s} = [y_{i,n=1,s}, y_{i,n=2,s}, ..., y_{i,n=N,s}]$ represents a vector of real answers of the fuzzy system for input vector $\mathbf{x}_{i,s}$.

Fuzzy system has to be learned to work properly (perform correct prediction of the user i signature global features' values). Learning can be achieved using gradient or evolutionary algorithm, which minimizes differences between reference vectors $(\mathbf{d}_{i,s})$ and output vectors $\mathbf{y}_{i,s}$ of the system (5).

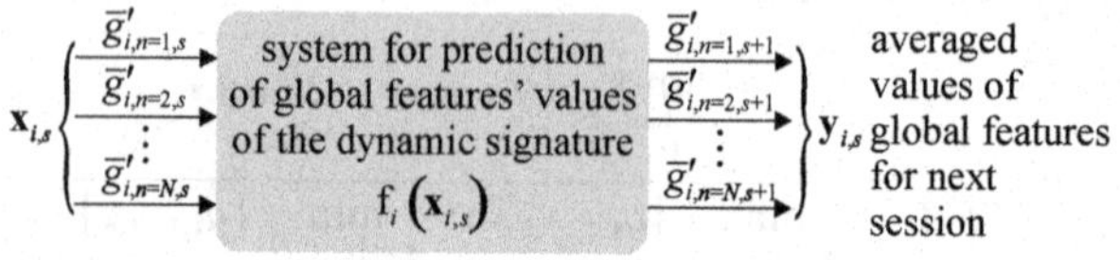

Fig. 1. Idea of prediction of the dynamic signature global features' values.

Evaluation of the fuzzy system operation in the learning phase can be realized using standard RMSE error. Use of it makes sense due to the normalization of features performed earlier. The error is expressed as follows:

$$RMSE_i = \frac{1}{S-2} \cdot \sum_{n=1}^{N} \sqrt{\sum_{s=1}^{S-2} (d_{i,n,s} - y_{i,n,s})^2} \qquad (6)$$

Error of the form (6) is used in evolutionary learning phase in order to evaluate individuals encoding parameters of the fuzzy system used for prediction. The purpose of the learning algorithm is minimization of the error. In order to better show the accuracy of the system, we can also use percentage measure of accuracy defined as follows:

$$ACC_i = \left(1 - \frac{1}{N \cdot (S-2)} \cdot \sum_{n=1}^{N} \sum_{s=1}^{S-2} |d_{i,n,s} - y_{i,n,s}|\right) \cdot 100\%. \qquad (7)$$

Formula (7) takes into account normalization of features according to the formula (2). Formulas (6) and (7) are related to the learning phase (taking into account learning sequence of the form (3)), but analogous formulas can be created for testing phase (taking into account testing sequence of the form (4)).

In this paper details related to the fuzzy system structure, aspects of learning (e.g. selection of parameters) and interpretability of fuzzy rules are omitted. They can be found in our previous papers (see e.g. [37,38]).

3   Simulations

Details of the simulations can be summarized as follows:

- They were performed in authorial testing environment implemented in C#.
- They were performed using ATVS-SLT DB (see [25]) dynamic signature database which has the following structure: $I = 27$, $S = 6$, $J_1 = 4$, $J_2 = 4$, $J_3 = 4$, $J_4 = 4$, $J_5 = 15$, $J_6 = 15$.

– Prediction was performed for 10. the best global features ($N = 10$) pointed out in [25]. Indices of these features are as follows: 3, 7, 17, 38, 45, 58, 59, 72, 93, 97.
– Prediction was performed using Mamdani-type fuzzy system characterized by the following parameters: number of rules: 3, number of inputs: 10, number of outputs: 10, fuzzy sets type: Gaussian, triangular norms type: algebraic (see e.g. [56]).
– For each user from the database learning of the fuzzy system is performed independently. The following settings of used evolutionary strategy $(\mu + \lambda)$ were adopted: number of individuals in the population (primary and temporary): 100, number of steps (generations) of the algorithm: 1000, crossing probability: 0.9, mutation probability: 0.3, mutation range: 0.15, method of parents selection: roulette wheel method, number of repetitions of the simulation for each user: 25 (results were averaged).

Simulation results are presented in Table 1 and in Figs. 2 and 3. Conclusions can be summarized as follows:

Table 1. Averaged simulation results for individual users.

i	Training sequence		Testing sequence		Prediction errors for individual features (%)									
	RMSE	ACC (%)	RMSE	ACC (%)	3	7	17	38	45	58	59	72	93	97
1	0.0989	97.8	0.7298	80.4	4.9	28.8	15.6	36.3	5.5	39.9	23.2	22.8	4.1	14.6
2	0.0964	97.6	0.5847	84.2	13.7	8.2	5.2	28.0	6.0	27.4	24.0	27.8	14.0	4.0
3	0.0932	97.9	0.5779	82.5	13.6	17.0	25.1	16.0	10.8	20.4	18.9	21.4	7.7	24.0
4	0.1405	96.9	0.5022	84.9	11.0	22.0	22.2	15.0	11.0	14.0	14.2	18.9	17.1	5.4
5	0.1205	97.2	0.8487	76.5	30.7	51.4	16.2	16.8	17.6	20.5	42.0	16.2	15.5	8.5
6	0.1168	97.4	0.7381	79.3	24.7	16.6	16.1	24.2	26.2	22.9	46.7	13.0	11.8	5.3
7	0.1419	96.6	0.5978	82.0	12.8	15.5	10.6	26.8	22.6	20.9	24.3	22.9	11.8	12.1
8	0.1153	97.5	0.3546	89.6	14.0	16.6	2.5	14.4	13.1	10.6	8.3	10.7	5.7	8.6
9	0.1723	95.9	0.8614	75.4	46.1	11.8	14.5	31.2	32.2	47.9	28.6	7.8	7.2	6.4
10	0.1483	96.7	0.6967	81.7	12.3	9.0	10.8	24.4	12.2	12.2	13.4	27.9	7.2	23.2
11	0.1030	97.6	0.4839	86.6	4.3	21.2	11.9	13.8	13.4	16.4	1.9	22.2	5.5	23.8
12	0.0891	98.0	0.5034	84.5	17.4	23.0	20.0	16.1	12.3	15.5	10.3	15.7	12.1	12.4
13	0.1412	96.7	0.6007	82.8	26.4	8.5	15.3	14.9	11.5	35.7	12.0	19.8	9.8	18.5
14	0.1218	97.1	0.5992	81.8	23.9	13.3	13.9	10.9	12.0	21.2	28.5	20.0	18.1	19.9
15	0.1239	97.1	0.7418	78.6	19.6	23.1	2.5	18.1	7.8	26.7	25.8	31.5	35.8	23.2
16	0.0868	97.9	0.8366	76.9	30.9	29.2	3.5	33.7	38.6	10.0	17.2	39.4	3.8	24.2
17	0.1248	97.1	0.4591	86.4	6.8	6.9	14.8	17.2	18.6	17.0	22.2	8.9	14.0	9.7
18	0.1295	97.1	0.6515	82.6	19.2	30.0	2.6	38.1	13.2	23.9	20.3	3.5	5.3	17.8
19	0.1144	97.6	0.7621	78.4	29.8	25.9	15.5	26.2	7.2	36.1	38.0	6.5	14.6	16.3
20	0.0854	97.9	0.3614	90.0	14.8	18.5	5.0	16.4	3.9	14.9	2.9	11.5	5.0	7.0
21	0.1283	97.0	0.6870	80.8	11.6	38.7	22.4	36.7	9.5	17.3	7.1	13.7	16.3	18.5
22	0.1617	96.3	0.5025	85.2	7.4	13.7	14.8	19.6	17.5	18.1	9.8	22.2	4.5	20.8
23	0.0756	98.2	0.3588	89.4	8.3	18.0	16.0	8.2	15.8	9.6	7.3	8.7	7.7	6.6
24	0.0878	98.0	0.5130	86.8	10.0	19.1	2.2	19.8	4.8	33.9	4.1	18.4	4.3	15.0
25	0.1065	97.6	0.5895	82.4	31.3	13.7	16.0	14.8	10.6	19.7	11.2	25.2	19.0	14.5
26	0.1361	96.7	0.6702	80.3	17.7	35.5	20.7	25.7	18.7	22.3	3.9	17.0	22.4	13.4
27	0.1220	97.2	0.4821	86.0	10.9	8.7	19.6	27.1	14.7	7.7	16.1	10.0	17.7	7.9
avg	0.1178	97.3	0.6035	82.8	17.6	20.1	13.2	21.9	14.3	**22.3**	18.5	17.4	**12.5**	14.1

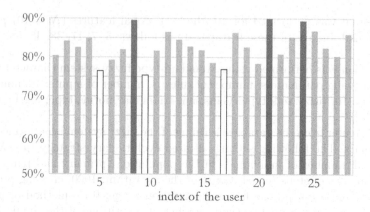

Fig. 2. Global features' values prediction accuracy for individual users (black color means users for which achieved accuracy is high).

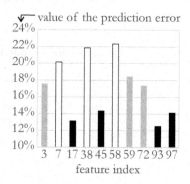

Fig. 3. Prediction errors for considered global features of the dynamic signature averaged for all users (black color means features for which achieved accuracy is high).

- For each user a set of features can be predicted with a good accuracy, however, changes of features' values are individual for each user. Average prediction accuracy is equal about 83%, so prediction error is quite low (about 17%) (Table 1).
- Accuracy of prediction for individual users (Fig. 2) is in the range from 75% to 90%. We should remember that values associated with prediction do not concern effectiveness of verification process. However, the effectiveness can be related with prediction accuracy.
- Prediction error for individual features is in the range (Fig. 3) from 12.5% to 22.3%. Results of this analysis can be a base for selection of features' subset, which changes can be easily predicted. However, we should remember that features which values can be easily predicted has not be the ones which characterize dynamic signature of individual users in the best way possible.

4   Conclusions

In this paper we proposed a method for prediction changes of biometric features' values over time. It may be particularly important for applications in which the interval between the adjacent dynamic signature acquisition sessions is long. It results in changes of biometric features' values which we can try to predict. This situation takes place in case of the dynamic signature. Our simulations performed using the database containing real signatures show that accuracy of the prediction of considered dynamic signature global features is adequate.

In further research on global features' values prediction we would like to: (a) develop new approach to the dynamic signature verification using global features prediction mechanism, (b) develop mechanism predicting different descriptors describing dynamics of signing process.

Acknowledgments. The project was financed by the National Science Centre (Poland) on the basis of the decision number DEC-2012/05/B/ST7/02138.

References

1. Abbas, J.: The bipolar choquet integrals based on ternary-element sets. J. Artif. Intell. Soft Comput. Res. **6**(1), 13–21 (2016)
2. Aghdam, M.H., Heidari, S.: Feature selection using particle swarm optimization in text categorization. J. Artif. Intell. Soft Comput. Res. **5**(4), 231–238 (2015)
3. Akimoto, T., Ogata, T.: Experimental development of a focalization mechanism in an integrated narrative generation system. J. Artif. Intell. Soft Comput. Res. **5**(3), 177–188 (2015)
4. Almohammadi, K., Hagras, H., Alghazzawi, D., Aldabbagh, G.: Users-centric adaptive learning system based on interval type-2 fuzzy logic for massively crowded e-learning platforms. J. Artif. Intell. Soft Comput. Res. **6**, 81–101 (2016)
5. Bas, E.: The training of multiplicative neuron model based artificial neural networks with differential evolution algorithm for forecasting. J. Artif. Intell. Soft Comput. Res. **6**(1), 5–11 (2016)
6. Bartczuk, Ł.: Gene expression programming in correction modelling of nonlinear dynamic objects. In: Borzemski, L., Grzech, A., Świątek, J., Wilimowska, Z. (eds.) Information Systems Architecture and Technology: Proceedings of 36th International Conference on Information Systems Architecture and Technology – ISAT 2015 – Part I. AISC, vol. 429, pp. 125–134. Springer, Cham (2016). doi:10.1007/978-3-319-28555-9_11
7. Bartczuk, Ł., Dziwiński, P., Starczewski, J.T.: New method for generation type-2 fuzzy partition for FDT. In: Rutkowski, L., Scherer, R., Tadeusiewicz, R., Zadeh, L.A., Zurada, J.M. (eds.) ICAISC 2010. LNCS (LNAI), vol. 6113, pp. 275–280. Springer, Heidelberg (2010). doi:10.1007/978-3-642-13208-7_35
8. Bartczuk, Ł., Przybył, A., Cpałka, K.: A new approach to nonlinear modelling of dynamic systems based on fuzzy rules. Int. J. Appl. Math. Comput. Sci. **3**, 603–621 (2016)
9. Bartczuk, Ł., Przybył, A., Koprinkova-Hristova, P.: New method for non-linear correction modelling of dynamic objects with genetic programming. In: Rutkowski, L., Korytkowski, M., Scherer, R., Tadeusiewicz, R., Zadeh, L.A., Zurada, J.M. (eds.) ICAISC 2015. LNCS, vol. 9120, pp. 318–329. Springer, Cham (2015). doi:10.1007/978-3-319-19369-4_29

10. Bartczuk, Ł., Łapa, K., Koprinkova-Hristova, P.: A new method for generating of fuzzy rules for the nonlinear modelling based on semantic genetic programming. In: Rutkowski, L., Korytkowski, M., Scherer, R., Tadeusiewicz, R., Zadeh, L.A., Zurada, J.M. (eds.) ICAISC 2016. LNCS, vol. 9693, pp. 262–278. Springer, Cham (2016). doi:10.1007/978-3-319-39384-1_23

11. Bertini, J.J.R., Nicoletti, M.D.C.: Enhancing constructive neural network performance using functionally expanded input data. J. Artif. Intell. Soft Comput. Res. **6**(2), 119–131 (2016)

12. Chen, Q., Abercrombie, R.K., Sheldon, F.T.: Risk assessment for industrial control systems quantifying availability using Mean Failure Cost (MFC). J. Artif. Intell. Soft Comput. Res. **5**(3), 205–220 (2015)

13. Cierniak, R., Rutkowski, L.: On image compression by competitive neural networks and optimal linear predictors. Signal Process. Image Commun. **156**, 559–565 (2000)

14. Cpałka, K.: Design of Interpretable Fuzzy Systems. Springer, Cham (2017)

15. Cpałka, K., Łapa, K., Przybył, A.: A new approach to design of control systems using genetic programming. Inf. Technol. Control **44**(4), 433–442 (2015)

16. Cpałka, K., Rebrova, O., Nowicki, R., Rutkowski, L.: On design of flexible neuro-fuzzy systems for nonlinear modelling. Int. J. Gener. Syst. **42**(6), 706–720 (2013)

17. Cpałka, K., Rutkowski, L.: Flexible Takagi-Sugeno fuzzy systems. In: Proceedings of the 2005 IEEE International Joint Conference on Neural Networks (IJCNN 2005), vol. 3, pp. 1764–1769 (2005)

18. Cpałka, K., Zalasiński, M., Rutkowski, L.: A new algorithm for identity verification based on the analysis of a handwritten dynamic signature. Appl. Soft Comput. **43**, 47–56 (2016)

19. Fierrez-Aguilar, J., Nanni, L., Lopez-Peñalba, J., Ortega-Garcia, J., Maltoni, D.: An on-line signature verification system based on fusion of local and global information. In: Kanade, T., Jain, A., Ratha, N.K. (eds.) AVBPA 2005. LNCS, vol. 3546, pp. 523–532. Springer, Heidelberg (2005). doi:10.1007/11527923_54

20. Duda, P., Hayashi, Y., Jaworski, M.: On the strong convergence of the orthogonal series-type kernel regression neural networks in a non-stationary environment. In: Rutkowski, L., Korytkowski, M., Scherer, R., Tadeusiewicz, R., Zadeh, L.A., Zurada, J.M. (eds.) ICAISC 2012. LNCS, vol. 7267, pp. 47–54. Springer, Heidelberg (2012). doi:10.1007/978-3-642-29347-4_6

21. Er, M.J., Duda, P.: On the weak convergence of the orthogonal series-type kernel regresion neural networks in a non-stationary environment. In: Wyrzykowski, R., Dongarra, J., Karczewski, K., Waśniewski, J. (eds.) PPAM 2011. LNCS, vol. 7203, pp. 443–450. Springer, Heidelberg (2012). doi:10.1007/978-3-642-31464-3_45

22. Dziwiński, P., Avedyan, E.D.: A new approach for using the fuzzy decision trees for the detection of the significant operating points in the nonlinear modeling. In: Rutkowski, L., Korytkowski, M., Scherer, R., Tadeusiewicz, R., Zadeh, L.A., Zurada, J.M. (eds.) ICAISC 2016. LNCS (LNAI), vol. 9693, pp. 279–292. Springer, Cham (2016). doi:10.1007/978-3-319-39384-1_24

23. Dziwiński, P., Avedyan, E.D.: A New method of the intelligent modeling of the nonlinear dynamic objects with fuzzy detection of the operating points. In: Rutkowski, L., Korytkowski, M., Scherer, R., Tadeusiewicz, R., Zadeh, L.A., Zurada, J.M. (eds.) ICAISC 2016. LNCS (LNAI), vol. 9693, pp. 293–305. Springer, Cham (2016). doi:10.1007/978-3-319-39384-1_25

24. Gabryel, M., Cpałka, K., Rutkowski, L.: Evolutionary strategies for learning of neuro-fuzzy systems. In: Proceedings of the I Workshop on Genetic Fuzzy Systems, Granada, pp. 119–123 (2005)

25. Galbally, J., Martinez-Diaz, M., Fierez, J.: Aging in biometrics: an experimental analysis on on-line signature. PLoS ONE **8**(7), e69897 (2013)

26. Gałkowski, T., Rutkowski, L.: Nonparametric fitting of multivariate functions. IEEE Trans. Autom. Control **31**(8), 785–787 (1986)

27. Held, P., Dockhorn, A., Kruse, R.: On merging and dividing social graphs. J. Artif. Intell. Soft Comput. Res. **5**(1), 23–49 (2015)

28. Jaworski, M., Er, M.J., Pietruczuk, L.: On the application of the parzen-type kernel regression neural network and order statistics for learning in a non-stationary environment. In: Rutkowski, L., Korytkowski, M., Scherer, R., Tadeusiewicz, R., Zadeh, L.A., Zurada, J.M. (eds.) ICAISC 2012. LNCS, vol. 7267, pp. 90–98. Springer, Heidelberg (2012). doi:10.1007/978-3-642-29347-4_11

29. Kasthurirathna, D., Piraveenan, M., Uddin, S.: Evolutionary stable strategies in networked games: the influence of topology. J. Artif. Intell. Soft Comput. Res. **5**(2), 83–95 (2015)

30. Korytkowski, M., Rutkowski, L., Scherer, R.: Fast image classification, by boosting fuzzy classifiers. Inf. Sci. **327**, 175–182 (2016)

31. Korytkowski, M., Scherer, R., Rutkowski, L.: On combining backpropagation with boosting. In: Proceedings of the 2006 International Joint Conference on Neural Networks, IEEE World Congress on Computational Intelligence, pp. 1274–1277 (2006)

32. Li, X., Er, M.J., Lim, B.S., Zhou, J.H., Gan, O.P., Rutkowski, L.: Fuzzy regression modeling for tool performance prediction and degradation detection. Int. J. Neural Syst. **2005**, 405–419 (2010)

33. Leon, M., Xiong, N.: Adapting differential evolution algorithms for continuous optimization via greedy adjustment of control parameters. J. Artif. Intell. Soft Comput. Res. **6**(2), 103–118 (2016)

34. Łapa, K., Przybył, A., Cpałka, K.: A new approach to designing interpretable models of dynamic systems. In: Rutkowski, L., Korytkowski, M., Scherer, R., Tadeusiewicz, R., Zadeh, L.A., Zurada, J.M. (eds.) ICAISC 2013. LNCS, vol. 7895, pp. 523–534. Springer, Heidelberg (2013). doi:10.1007/978-3-642-38610-7_48

35. Łapa, K., Szczypta, J., Venkatesan, R.: Aspects of structure and parameters selection of control systems using selected multi-population algorithms. In: Rutkowski, L., Korytkowski, M., Scherer, R., Tadeusiewicz, R., Zadeh, L.A., Zurada, J.M. (eds.) ICAISC 2015. LNCS, vol. 9120, pp. 247–260. Springer, Cham (2015). doi:10.1007/978-3-319-19369-4_23

36. Łapa, K., Szczypta, J., Saito, T.: Aspects of evolutionary construction of new flexible PID-fuzzy controller. In: Rutkowski, L., Korytkowski, M., Scherer, R., Tadeusiewicz, R., Zadeh, L.A., Zurada, J.M. (eds.) ICAISC 2016. LNCS, vol. 9692, pp. 450–464. Springer, Cham (2016). doi:10.1007/978-3-319-39378-0_39

37. Łapa, K., Cpałka, K., Galushkin, A.I.: A new interpretability criteria for neuro-fuzzy systems for nonlinear classification. In: Rutkowski, L., Korytkowski, M., Scherer, R., Tadeusiewicz, R., Zadeh, L.A., Zurada, J.M. (eds.) ICAISC 2015. LNCS, vol. 9119, pp. 448–468. Springer, Cham (2015). doi:10.1007/978-3-319-19324-3_41

38. Łapa, K., Cpałka, K., Wang, L.: New approach for interpretability of neuro-fuzzy systems with parametrized triangular norms. In: Rutkowski, L., Korytkowski, M., Scherer, R., Tadeusiewicz, R., Zadeh, L.A., Zurada, J.M. (eds.) ICAISC 2016. LNCS, vol. 9692, pp. 248–265. Springer, Cham (2016). doi:10.1007/978-3-319-39378-0_22

39. Miyajima, H., Shigei, N., Miyajima, H.: Performance comparison of hybrid electromagnetism-like mechanism algorithms with descent method. J. Artif. Intell. Soft Comput. Res. **5**(4), 271–282 (2015)
40. Nanni, L.: An advanced multi-matcher method for on-line signature verification featuring global features and tokenised random numbers. Neurocomputing **69**, 2402–2406 (2006)
41. Nanni, L., Lumini, A.: Ensemble of Parzen window classifiers for on-line signature verification. Neurocomputing **68**, 217–224 (2005)
42. Nanni, L., Lumini, A.: Advanced methods for two-class problem formulation for on-line signature verification. Neurocomputing **69**, 854–857 (2006)
43. Nikulin, V.: Prediction of the shoppers loyalty with aggregated data streams. J. Artif. Intell. Soft Comput. Res. **6**, 69–79 (2016)
44. Nguyen, K.P., Fujita, G., Dieu, V.N.: Cuckoo search algorithm for optimal placement and sizing of static var compensator in large-scale power systems. J. Artif. Intell. Soft Comput. Res. **6**(2), 59–68 (2016)
45. Nonaka, S., Tsujimura, T., Izumi, K.: Gain design of quasi-continuous exponential stabilizing controller for a nonholonomic mobile robot. J. Artif. Intell. Soft Comput. Res. **6**(3), 189–201 (2016)
46. Nowicki, R., Scherer, R., Rutkowski, L.: A Method for Learning of Hierarchical Fuzzy Systems. Intelligent Technologies - Theory and Applications. IOS Press, Amsterdam (2002)
47. Pietruczuk, L., Duda, P., Jaworski, M.: Adaptation of decision trees for handling concept drift. In: Rutkowski, L., Korytkowski, M., Scherer, R., Tadeusiewicz, R., Zadeh, L.A., Zurada, J.M. (eds.) ICAISC 2013. LNCS, vol. 7894, pp. 459–473. Springer, Heidelberg (2013). doi:10.1007/978-3-642-38658-9_41
48. Pietruczuk, L., Rutkowski, L., Jaworski, M., Duda, P.: How to adjust an ensemble size in stream data mining? Inf. Sci. **381**, 46–54 (2017)
49. Przybył, A., Jelonkiewicz, J.: Genetic algorithm for observer parameters tuning in sensorless induction motor drive. In: Rutkowski, L., Kacprzyk, J. (eds.) Neural Networks and Soft Computing. Advances in Soft Computing, vol. 19, pp. 376–381. Springer, Heidelberg (2003)
50. Przybył, A., Smoląg, J., Kimla, P.: Distributed control system based on real time ethernet for computer numerical controlled machine tool. Przeglad Elektrotechniczny **86**(2), 342–346 (2010). (in Polish)
51. Rutkowska, A.: Influence of membership functions shape on portfolio optimization results. J. Artif. Intell. Soft Comput. Res. **6**(1), 45–54 (2016)
52. Rutkowski, L.: Sequential pattern-recognition procedures derived from multiple Fourier-series. Pattern Recognit. Lett. **8**(4), 213–216 (1988)
53. Rutkowski, L.: Multiple Fourier series procedures for extraction of nonlinear regressions from noisy data. IEEE Trans. Signal Process. **41**(10), 3062–3065 (1993)
54. Rutkowski, L.: Identification of MISO nonlinear regressions in the presence of a wide class of disturbances. IEEE Trans. Inf. Theor. **37**(1), 214–216 (2002)
55. Rutkowski, L.: Adaptive probabilistic neural networks for pattern classification in time-varying environment. IEEE Trans. Neural Netw. **15**(4), 811–827 (2004)
56. Rutkowski, L.: Computational Intelligence. Springer, Heidelberg (2008)
57. Rutkowski, L., Cpałka, K.: A neuro-fuzzy controller with a compromise fuzzy reasoning. Control Cybern. **31**(2), 297–308 (2002)
58. Rutkowski, L., Cpałka, K.: Compromise approach to neuro-fuzzy systems. In: Proceedings of the 2nd Euro-International Symposium on Computation Intelligence. Frontiers in Artificial Intelligence and Applications, vol. 76, pp. 85–90 (2002)

59. Rutkowski, L., Cpałka, K.: Flexible neuro-fuzzy systems. IEEE Trans. Neural Netw. **14**, 554–574 (2003)
60. Rutkowski, L., Cpałka, K.: Flexible weighted neuro-fuzzy systems. In: Proceedings of the 9th International Conference on Neural Information Processing (ICONIP 2002), Orchid Country Club, Singapore, 18–22 November 2002
61. Rutkowski, L., Cpałka, K.: Neuro-fuzzy systems derived from quasi-triangular norms. In: Proceedings of the IEEE International Conference on Fuzzy Systems, Budapest, 26–29 July, vol. 2, pp. 1031–1036 (2004)
62. Rutkowski, L., Jaworski, M., Pietruczuk, L., Duda, P.: Decision trees for mining data streams based on the Gaussian approximation. IEEE Trans. Knowl. Data Eng. **26**(1), 108–119 (2014)
63. Rutkowski, L., Jaworski, M., Pietruczuk, L., Duda, P.: The CART decision tree for mining data streams. Inf. Sci. **266**, 1–15 (2014)
64. Rutkowski, L., Pietruczuk, L., Duda, P., Jaworski, M.: Decision trees for mining data streams based on the McDiarmid's bound. IEEE Trans. Knowl. Data Eng. **25**(6), 1272–1279 (2013)
65. Sakurai, S., Nishizawa, M., Soft, C.R.: A new approach for discovering top-k sequential patterns based on the variety of items. J. Artif. Intell. Soft Comput. Res. **5**(2), 141–153 (2015)
66. Scherer, R.: Designing boosting ensemble of relational fuzzy systems. Int. J. Neural Syst. **20**, 381–388 (2010)
67. Scherer, R.: Multiple Fuzzy Classification Systems. Springer, Heidelberg (2012)
68. Serdah, A.M., Ashour, W.M., Soft, C.R.: Clustering large-scale data based on modified affinity propagation algorithm. J. Artif. Intell. Soft Comput. Res. **6**(1), 23–33 (2016)
69. Song, J., Romero, C.E., Yao, Z.: A globally enhanced general regression neural network for on-line multiple emissions prediction of utility boiler. Knowl. Based Syst. **118**, 4–14 (2017)
70. Stanovov, V., Semenkin, E., Semenkina, O.: Self-configuring hybrid evolutionary algorithm for fuzzy imbalanced classification with adaptive instance selection. J. Artif. Intell. Soft Comput. Res. **6**, 173–188 (2016)
71. Szarek, A., Korytkowski, M., Rutkowski, L., Scherer, R., Szyprowski, J.: Application of neural networks in assessing changes around implant after total hip arthroplasty. In: Rutkowski, L., Korytkowski, M., Scherer, R., Tadeusiewicz, R., Zadeh, L.A., Zurada, J.M. (eds.) ICAISC 2012. LNCS, vol. 7268, pp. 335–340. Springer, Heidelberg (2012). doi:10.1007/978-3-642-29350-4_40
72. Szczypta, J., Łapa, K., Shao, Z.: Aspects of the selection of the structure and parameters of controllers using selected population based algorithms. In: Rutkowski, L., Korytkowski, M., Scherer, R., Tadeusiewicz, R., Zadeh, L.A., Zurada, J.M. (eds.) ICAISC 2014. LNCS, vol. 8467, pp. 440–454. Springer, Cham (2014). doi:10.1007/978-3-319-07173-2_38
73. Szczypta, J., Przybył, A., Cpałka, K.: Some aspects of evolutionary designing optimal controllers. In: Rutkowski, L., Korytkowski, M., Scherer, R., Tadeusiewicz, R., Zadeh, L.A., Zurada, J.M. (eds.) ICAISC 2013. LNCS, vol. 7895, pp. 91–100. Springer, Heidelberg (2013). doi:10.1007/978-3-642-38610-7_9
74. Xiao, Q.: Time series prediction using dynamic Bayesian network. Opt. Int. J. Light Electron Opt. **135**, 98–103 (2017)
75. Villmann, T., Bohnsack, A., Kaden, M.: Can learning vector quantization be an alternative to SVM and deep learning? - Recent trends and advanced variants of learning vector quantization for classification learning. J. Artif. Intell. Soft Comput. Res. **7**(1), 65–81 (2017)

76. Wei, H.: A bio-inspired integration method for object semantic representation. J. Artif. Intell. Soft Comput. Res. **6**(3), 137–154 (2016)
77. Zalasiński, M.: New algorithm for on-line signature verification using characteristic global features. In: Wilimowska, Z., Borzemski, L., Grzech, A., Świątek, J. (eds.) Information Systems Architecture and Technology: Proceedings of 36th International Conference on Information Systems Architecture and Technology – ISAT 2015 – Part IV. AISC, vol. 432, pp. 137–146. Springer, Cham (2016). doi:10.1007/978-3-319-28567-2_12
78. Zalasiński, M., Cpałka, K.: A new method of on-line signature verification using a flexible fuzzy one-class classifier, pp. 38–53. Academic Publishing House EXIT (2011)
79. Zalasiński, M., Cpałka, K.: Novel algorithm for the on-line signature verification using selected discretization points groups. In: Rutkowski, L., Korytkowski, M., Scherer, R., Tadeusiewicz, R., Zadeh, L.A., Zurada, J.M. (eds.) ICAISC 2013. LNCS, vol. 7894, pp. 493–502. Springer, Heidelberg (2013). doi:10.1007/978-3-642-38658-9_44
80. Zalasiński, M., Cpałka, K.: New algorithm for on-line signature verification using characteristic hybrid partitions. In: Wilimowska, Z., Borzemski, L., Grzech, A., Świątek, J. (eds.) Information Systems Architecture and Technology: Proceedings of 36th International Conference on Information Systems Architecture and Technology – ISAT 2015 – Part IV. AISC, vol. 432, pp. 147–157. Springer, Cham (2016). doi:10.1007/978-3-319-28567-2_13
81. Zalasiński, M., Cpałka, K., Er, M.J.: New method for dynamic signature verification using hybrid partitioning. In: Rutkowski, L., Korytkowski, M., Scherer, R., Tadeusiewicz, R., Zadeh, L.A., Zurada, J.M. (eds.) ICAISC 2014. LNCS, vol. 8468, pp. 216–230. Springer, Cham (2014). doi:10.1007/978-3-319-07176-3_20
82. Zalasiński, M., Cpałka, K., Hayashi, Y.: New method for dynamic signature verification based on global features. In: Rutkowski, L., Korytkowski, M., Scherer, R., Tadeusiewicz, R., Zadeh, L.A., Zurada, J.M. (eds.) ICAISC 2014. LNCS, vol. 8468, pp. 231–245. Springer, Cham (2014). doi:10.1007/978-3-319-07176-3_21
83. Zalasiński, M., Cpałka, K., Hayashi, Y.: A new approach to the dynamic signature verification aimed at minimizing the number of global features. In: Rutkowski, L., Korytkowski, M., Scherer, R., Tadeusiewicz, R., Zadeh, L.A., Zurada, J.M. (eds.) ICAISC 2016. LNCS, vol. 9693, pp. 218–231. Springer, Cham (2016). doi:10.1007/978-3-319-39384-1_20
84. Zalasiński, M., Cpałka, K., Rakus-Andersson, E.: An idea of the dynamic signature verification based on a hybrid approach. In: Rutkowski, L., Korytkowski, M., Scherer, R., Tadeusiewicz, R., Zadeh, L.A., Zurada, J.M. (eds.) ICAISC 2016. LNCS, vol. 9693, pp. 232–246. Springer, Cham (2016). doi:10.1007/978-3-319-39384-1_21
85. Zalasiński, M., Łapa, K., Cpałka, K.: New algorithm for evolutionary selection of the dynamic signature global features. In: Rutkowski, L., Korytkowski, M., Scherer, R., Tadeusiewicz, R., Zadeh, L.A., Zurada, J.M. (eds.) ICAISC 2013. LNCS, vol. 7895, pp. 113–121. Springer, Heidelberg (2013). doi:10.1007/978-3-642-38610-7_11

Author Index

Printed in the United States
By Bookmasters

Printed in the United States
By Bookmasters